CURRENT ISSUES IN PLANT MOLECULAR AND CELLULAR BIOLOGY

Current Plant Science and Biotechnology in Agriculture

VOLUME 22

Aims and Scope
The book series is intended for readers ranging from advanced students to senior research scientists and corporate directors interested in acquiring in-depth, state-of-the-art knowledge about research findings and techniques related to plant science and biotechnology. While the subject matter will relate more particularly to agricultural applications, timely topics in basic science and biotechnology will also be explored. Some volumes will report progress in rapidly advancing disciplines through proceedings of symposia and workshops while others will detail fundamental information of an enduring nature that will be referenced repeatedly.

The titles published in this series are listed at the end of this volume.

Current Issues in Plant Molecular and Cellular Biology

Proceedings of the VIIIth International Congress on Plant Tissue and Cell Culture, Florence, Italy, 12-17 June, 1994

Edited by

M. TERZI
Department of Biology,
University of Padova,
Italy

R. CELLA
Department of Genetics and Microbiology,
University of Pavia,
Italy

and

A. FALAVIGNA
Research Institute of Vegetable Crops,
Montanaso Lombardo (Milano),
Italy

KLUWER ACADEMIC PUBLISHERS
DORDRECHT / BOSTON / LONDON

A C.I.P. Catalogue record for this book is available from the Library of Congress.

ISBN 0-7923-3322-5

Published by Kluwer Academic Publishers,
P.O. Box 17, 3300 AA Dordrecht, The Netherlands.

Kluwer Academic Publishers incorporates
the publishing programmes of
D. Reidel, Martinus Nijhoff, Dr W. Junk and MTP Press.

Sold and distributed in the U.S.A. and Canada
by Kluwer Academic Publishers,
101 Philip Drive, Norwell, MA 02061, U.S.A.

In all other countries, sold and distributed
by Kluwer Academic Publishers Group,
P.O. Box 322, 3300 AH Dordrecht, The Netherlands.

Printed on acid-free paper

Printed in the Netherlands

TABLE OF CONTENTS

3.10 Agronomic Traits

3.11 Somatic Embryogenesis

3.12 Meristems

3.15 Reception and Transduction of signals

3.16 Gene Expression under extreme conditions

3.17 Primary metabolism

3.20 Large Scale Production

Preface

From June 12 to June 17, 1994, the 8th International Congress of Plant Tissue and Cell Culture was held in Florence.
About 1300 scientists, coming from more than 50 Countries attended the plenary lectures, keynote lectures, oral presentations to symposia and workshops. All the invited lectures are collected here in the sameway as they appeared in the general program.
We are sure that this collection will be of much use as a reference for many state-of-the art presentations. Moreover for those that could not partecipate to the Congress, it will be useful reading.
For everybody, a comparison of this with the proceedings of the previous meetings will allow a measurement of the impressive speed of discoveries of facts and techniques characterizing the dynamic field of plant tissue and cell culture.

M. Terzi
R. Cella
A. Falavigna

INTERNATIONAL PROGRAMMES OF FAO AND IAEA IN MUTATION BREEDING AND IN VITRO TECHNOLOGY

B. Sigurbjornsson & M. Maluszynski
Joint FAO/IAEA Division of Nuclear Techniques in Food and Agriculture, IAEA, Vienna, Austria

Exactely 30 years ago, in June, 1964, a benchmark conference in plant science was held here in Italy. The symposium on induced mutations in plant breeding, sponsored by FAO and IAEA was held in Rome. It was attended by nearly all leading radiobiologists, radiogeneticists, plant breeders and others who were interested in the new techniques of artificial creation of variability in plants. The main topics dealt with effects of radiation dose and dose rates and chemical mutagens on plant cells and tissues.

Some partecipants reported on useful mutants which had been discovered - almost as a biproduct of their studies. The number of marketable varieties of induced mutant origin at that time were under 30.

It was clearly brought out at the conference that our knowledge of how to apply induced mutagenesis was indeed limited and there was not even agreement on the use of terminology, e.g. if one should use $X_{1,2,3}$ or $M_{1,2,3}$ for generations following mutagenic treatment.

At that time the senior author was working as a plant geneticist at the IAEA in Vienna and was co-scientific secretary of the symposium. He called together a few of the participants, including Prof. D'Amato, and they decided to form an international advisory group on mutation breeding in order to stimulate research to fill the gaps in our knowledge and to standardize terminology. Later that year FAO and IAEA combined their efforts and founded the Joint FAO/IAEA Division of Nuclear (then "atomic energy") Techniques in agriculture, and the Advisory Group was established as an official international group of the two agencies.

The Group met every year for the next five years to review progress and to attempt to fill in more gaps. The result was that, 5 years later, the first edition of the Manual on Mutation Breeding was published and a second edition a few years later.

In the meantime the Joint Division had established several international groups of scientists, dealing with the use of induced mutations in different crops, rice, barley, wheat, legumes, oil crops and others. One of the first international undertaking was to test induced mutants of durum wheat, developed by Prof. Scarascia-Mugnozza and his team in 24 countries of the Middle East and North Africa. Some of these mutants later became leading durum varieties here in Italy, where a large part of the durum acreage is still under varieties, descendants of these induced mutants.

The results of all these international activities, coordinated research groups, extensive training workshops and other meetings, were rather dramatic. The number of mutant varieties rose steadily until now we have records of nearly 1900 released cultivars which have induced mutants in their backgrounds. These mutant varieties have been obtained in 154 plant species and by plant breeders in more than 50 countries, the most active being in China, India, the former USSR, Holland, Japan and the USA. Of the released mutants nearly 1300 are in agricultural crops, with over 500 in ornamental crops. For many ornamental crops this method of breeding has become a method of choice. In many agricultural crops, such as rice, barley, cotton and durum wheat, induced mutants have come to occupy major parts of the crop area with enormous economic returns to growers and society as a whole.

M. Terzi et al. (eds.), Current Issues in Plant Molecular and Cellular Biology, 1-3.

If can be safely said now that the use of induced mutants in plant breeding has become a standard method along with other classical methods of plant breeding.

The development of new technologies in plant research and particularly cell and tissue culture is now profoundly affecting breeding techniques and has become a major part of the activities of the Plant Breeding and Genetics Section of the Joint Division. These activities which are carried out world-wide through coordinated research programmes with participating scientists from both developing and developed countries are supported by our own research and training laboratory located at Seibersdorf, near Vienna. A great deal of the early developmental work on mutagen dose and conditions of seed and plant parts undergoing treatment was carried out by this laboratory. In 1984 we recruited Dr. Franticek Novak to establish an *in vitro* tissue culture facility at the laboratory. Under his leadership this quicky developed into a major international centre, pioneering in the combination of mutagenesis and cell and tissue culture. Each year we operate 6-weeks training courses, one of which with plant scientists from 23 developing countries will be concluded next week. A number of fellowship holders from developing countries undergo long-term training in our laboratory. Dr. Novak's research on the use of radiation treatment in tissue culture in bananas resulted in the development of several superior mutants, one of which, named Novaria, is being released in Malaysia this year.

It was a great tragedy when Dr. Novak was killed in a car accident a year ago at the age of 51. We are finding it very difficult to replace a scientist of his knowledge, skill and experience to put our programme on track again.

A variety of *in vitro* techniques and molecular approaches are used in combination with mutagenesis in our programme. These include anther culture, shoot organogenesis, somatic embryogenesis and protoplast fusion which are employed with mutation techniques in both seed and vegetatively propagated crops. This combination can speed up breeding programmes by more efficient generation of variability, more effective selection and more rapid multiplication of mutant genotypes. The expression of induced mutations as a pure homozygote obtained through microspore, anther or ovary culture can enhance the rapid recovery of the desired traits. In vegetatively propagated species, mutation in combination with *in vitro* culture techniques may be the only method of improving an existing cultivar.

Another use of this combination is the use of induced mutation and *in vitro* techniques in model species, such as *Arabidopsis* for construction and saturation of genetic maps, understanding of developmental genetics and elucidation of biochemical pathways.

In seed propagated plants the application of induced mutations in doubled haploid systems has been found to be of great promise.

Our activities thus increasingly rely on the use of induced mutant technology in *in vitro* systems of cell and plant tissue culture and also the use of radioactive isotope labelling of DNA in a variety of molecular genetic studies involving DNA based markers, and fingerprinting.

At this 8th International Congress on Plant Tissue and Cell Culture a number of the most advanced scientists in this field will be reporting on their wide variety of studies, involving cell and tissue culture. These techniques provide the plant scientist with extraordinarily efficient tools for elucidating the many pathways and the nature of the plants genetics, biochemistry and physiology. On the surface the presented papers appear to be of a foundamental nature, only to satisfy our curiosity about what makes the Plant Kingdom function. But underneath, as revealed in so many of the titles, there is an underlying desire, common to all plant scientists to help create plants that are better able to serve man, keep him better fed and clothed, and -to ease the pains of hunger experienced by far too many of our species on this earth.

FAO is at present time looking into the future and what challenges lie ahead. A book on Agriculture towards 2010 is being prepared. In short the outlook is as follows:

The present situation is that **800 million people go hungry every day, 150 million children are undernourished** to the extent that their future physical and mental health may be jeopardized.

This is spite of the fact that there has been enormous progress in agricultural production world-wide, achieved by massive use of fertilizers and pesticides and the exploitation of marginal lands. The result is that each year 22 billion tons of soil is lost. The half-life of the worlds soils has been calculated to be 100 years! The availability of arable land per person has decreased by 32% in the last 30 years.

This is the present day situation. The world population is increasing by one million every 4 days. Every decade another billion is added. All this means that the future challenge is blow to double food production in the next 30 years. With hardly any more land visitable for extra food production, this means that it is up to plant and soil scientists to prive crops and soil conditions capable of sustaining such enormous yields without harming our environment. This must be accomplished on a sustainable basis.

Our message to you as plant scientists working at the edge of modern research is this:

Keep this future challenge in mind. Make it your overall, long-term goal to help plant breeders design crops capable of meeting this enormous challenge. It has been calculated that if population keeps growing at the present rate and no more arable land will become available, average grain yields will have to reach 14 tons per hectare by the year 2050! Clearly, only with good plant science and good plant scientists working on well funded research programmes can we hope to meet this challenge.

We are proud that the Joint FAO/IAEA Division is one of the sponsors of this Congress.

CELLULAR AND MOLECULAR GENETIC IMPROVEMENT OF CEREALS

INDRA K. VASIL
Laboratory of Plant Cell and Molecular Biology
1143 Fifield Hall
University of Florida
Gainesville, FL 32611-0690, USA

ABSTRACT. Efficient methods for the regeneration of plants from cultured cells of gramineous species were developed during the 1980's, based on the culture of tissue/organ explants comprised largely of undifferentiated cells on media containing high concentrations of 2,4-dichlorophenoxyacetic acid (2,4-D). These advances, combined with the later development of methods for the direct delivery of DNA, led to the production of the first transgenic cereals in 1988. Since then, in the relatively short period of six years, all major cereal crops have been transformed, often with agronomically useful genes that confer resistance to non-selective herbicides, or viruses and insects. Such improved crops can reduce or even eliminate the huge losses in crop productivity caused by weeds, pathogens and pests. Further molecular improvement of cereals will depend on the availability of genes that determine the quality and productivity of cereal crops, and protect them from biological and environmental stresses.

INTRODUCTION

Plants account for 93% of the human diet, and even the remaining 7% is indirectly contributed by plants through animal products (Borlaug 1983, Vasil 1990). Of the more than 250,000 species of angiosperms, a mere 29 provide most of our calories and proteins. The most important of these are the eight cereal grains (wheat, rice, maize, barley, oats, rye, sorghum, pearl millet) of the family Gramineae (Poaceae), which together constitute the centerpiece of world agriculture by providing 52% of the total food calories. The primitive forms of many cereals were amongst the handful of plants that were first domesticated during the Neolithic age nearly 10,000 years ago, when agriculture gradually replaced hunting and gathering. In retrospect, the choices made by the Neolithic man have not only proven to be wise and remarkably durable, but have also helped to sustain the development of the human civilization by providing a rich and dependable source of nutrition for the ever increasing world population.

The dramatic increases in the yields and production of cereal grains achieved by breeding and selection during 1950-1984, kept pace with the rapidly growing world population and prevented widespread hunger and famine. Such increases can not be realistically sustained indefinitely, and indeed during the past ten years grain yields and production have begun to decline. Yet by various estimates world population will double to nearly 11 billion some time between 2030 to 2050, of which more than 90% will reside in the already overpopulated and undernourished countries of Africa, Asia and Latin America. This huge increase in population will cause even more rapid degradation of the already fragile environment, and will further decrease per capita supplies of crop land, grain and water,

M. Terzi et al. (eds.), Current Issues in Plant Molecular and Cellular Biology, 5-18.

all at a time when food supplies should be actually more than doubled just to meet the basic needs of human nutrition.

Plant improvement by breeding is severely restricted by the availability of a rather limited gene pool owing to natural incompatibilities, even between related species, and by the time scale of most breeding programs. Therefore, much attention has been directed recently to the newly emerging and novel technologies of plant cell and molecular biology (biotechnology), which provide a powerful means to supplement and complement the traditional methods of plant improvement, by permitting access to an unlimited gene pool through the transfer of desirable genes between any two species of interest, irrespective of their evolutionary or taxonomic relationships.

Advances in plant cell culture research, especially of major crop species, have played an increasingly critical role in the development of modern plant biotechnology (Vasil 1990). Widely used protocols for the regeneration of plants from cultured cells of dicotyledonous species (dicots) were already available in the early 1980's, when *Agrobacterium*-based methods for DNA delivery and integration were developed, leading to the production of the first transgenic plants in 1983 (Fraley et al. 1983, Zambryski et al. 1983). The difficulties faced with the adaptation of these procedures for the transformation of cereals led to the development of alternative strategies and technologies for the improvement of cereals. In this brief review, I describe the key components of these strategies, and highlight the significant advances in the molecular improvement of cereal crops.

PLANT REGENERATION FROM CULTURED CELLS

Prior to 1980, there were only scattered reports of plant regeneration from tissue cultures of gramineous species. In most instances regeneration was sporadic and transient, limited to obscure genotypes of a few species, and was based either on the *de novo* formation of shoot meristems in callus cultures or the 'microtillering' of preexisting meristems (Vasil 1987). Four significant discoveries were made in 1980-1981 that form the basis of much of the modern work on tissue culture of cereals: (a) That the culture of immature embryos, and segments of young inflorescences and bases of immature leaves at defined stages of development, (b) on simple nutrient media containing high concentrations of strong auxins like 2,4-D, (c) gives rise to long-term embryogenic callus cultures in which plant regeneration takes place via the formation of somatic embryos, and that (d) embryogenic cell suspension cultures derived from the embryogenic calli yield totipotent protoplasts (Dale 1980, Vasil and Vasil 1980, 1981, Wernicke and Brettell 1980, Haydu and Vasil 1981, Lu and Vasil 1981). During the past 15 years, these strategies have been successfully used to develop reliable and efficient methods for the regeneration of plants from all of the important species of cereals as well as grasses (Vasil and Vasil 1994).

The importance of the physiological condition as well as the developmental state of the explant has been found to be related to the endogenous levels of plant growth regulators (Rajasekaran et al. 1987a,b, Carnes and Wright 1988, Wenck et al. 1988). Although regeneration *in vitro* is a function of the genotype of the donor plant, regeneration has been obtained by a judicious choice of explant and nutrient media from a wide variety of germplasm in most species, including those which were once considered to be recalcitrant.

Somatic embryogenesis is the predominant mode of regeneration in cereal species (Vasil and Vasil 1986, 1994). This is significant, because the somatic embryos, like their zygotic counterparts, are derived directly or indirectly from single cells, are non-chimeric in nature and possess a root-shoot axis which is critical for their transfer to, and survival in, soil. Another useful characteristic of plants derived from somatic embryos is their uniformity and genetic fidelity. These are attributed to a stringent selection during somatic embryogenesis, and the inability of somatic embryos derived from genetically aberrant cells to develop into mature viable embryos (Swedlund and Vasil 1985, Hanna et al. 1984, Rajasekaran et al. 1986, Cavallini et al. 1987, Kobayashi 1987, Breiman et al. 1989, Cavallini and Natali 1989, Morrish et al. 1990, Gmitter et al. 1991, Shimron-Abarbanell and Breiman 1991, Shenoy and Vasil 1992, Chowdhury and Vasil 1993, Isabel et al. 1993, Valles et al. 1993, Chowdhury et al. 1994). This unique characteristic of plants derived from somatic embryos makes them ideally suited not only for clonal propagation but also for genetic transformation.

Until 1987, the only possibility to obtain transgenic cereals was by the direct delivery of DNA into protoplasts. Yet, exhaustive attempts made during 1975-1980 to induce sustained divisions in protoplasts isolated from leaves of cereal species proved to be entirely unsuccessful (Potrykus 1980, Vasil 1983, Vasil and Vasil 1992). Even today, no reliable and reproducible method is available for the recovery of plants from mesophyll protoplasts of any grass or cereal species. Therefore, based on the induction of sustained divisions in protoplasts isolated from non-morphogenic cell suspension

Table 1. Regeneration of plants from protoplasts isolated from embryogenic cell suspension cultures.

Species	Reference
Pennisetum americanum	Vasil and Vasil 1980
Panicum maximum	Lu et al. 1981
Pennisetum purpureum	Vasil et al. 1983, Wan and Vasil 1994
Saccharum sp.	Srinivasan and Vasil 1986, Chen et al. 1988
Oryza sativa	Yamada et al. 1986, Kyozuka et al. 1987, Datta et al. 1990a
Polypogon fugax	Chen and Xia 1987
Dactylis glomerata	Horn et al. 1988a
Festuca arundinacea	Dalton 1988
Lolium perenne	Dalton 1988, Creemers-Molenaar et al. 1989
Lolium multiflorum	Dalton 1988, Wang et al. 1993a
Zea mays	Rhodes et al. 1988a, Prioli and Sondahl 1989, Shillito et al. 1989, Morocz et al. 1990
Triticum aestivum	Vasil et al. 1990, Ahmed and Sagi 1993
Setaria italica	Dong and Xia 1990
Agrostis alba	Asano and Sugiura 1990
Sorghum vulgare	Wei and Xu 1990
Hordeum vulgare	Jahne et al. 1991, [a]Holm et al. 1994
Oryza rufipogon	Baset et al. 1991
Agrostis palustris	Terakawa et al. 1992
Paspalum dilatatum	Akashi and Adachi 1992
Festuca pratensis	Wang et al. 1993b
Oryza granulata	Baset et al. 1993
Poa pratensis	Nielsen et al. 1993
Triticum durum	Yang et al. 1993
Hordeum murinum	Wang and Lorz 1994
Festuca rubra	Spangenberg et al. 1994

[a]From mechanically isolated fertilized egg cell protoplasts.

cultures, attempts were made to establish embryogenic suspension cultures as a source of totipotent protoplasts (Vasil and Vasil 1980, 1981). This strategy proved to be most useful, as plants have been regenerated from protoplasts isolated from embryogenic suspension cultures of a large number of gramineous species, including all the major cereal crops (Table 1). It should be noted, however, that the establishment of regenerable embryogenic suspension cultures is still a difficult and tedious procedure. Furthermore, in some species (eg. maize and wheat), such cultures can be established only from a special type of callus. Nevertheless, once established, such cultures can be cryopreserved to provide a permanent and reliable source of totipotent protoplasts (Shillito et al. 1989, Gnanapragasam and Vasil 1990, Meijer et al. 1991, Fretz et al. 1992, Lu and Sun 1992, Wan and Vasil 1994).

For almost twenty years, somatic hybridization has been billed as a potentially powerful tool for creating new and useful hybrids that can not otherwise be obtained by breeding. Unfortunately, with a few exceptions, this has generally not proven to be true. In cereals, few somatic hybrids have been obtained, as of *Oryza sativa* (+) *Echinochloa oryzicola* (Terada et al. 1987), or *Festuca arundinacea* (+) *Lolium multiflorum* (Takamizo et al. 1991). Perhaps the best use of this procedure is in the transfer of cytoplasmic male sterility, as was demonstrated in rice (Kyozuka et al. 1989, Yang et al. 1989). Also, *in vitro* fusion of egg and sperm cell protoplasts of maize, leading to the generation of fertile plants (Kranz and Lorz 1993), provides a powerful new tool for the study of cell to cell interaction and the process of fertilization.

TRANSFORMATION OF CEREALS

Genetic transformation of dicot species has been greatly aided by the natural transformation system provided by *Agrobacterium tumefaciens.* It has become a preferred and popular method because it requires minimal exposure of cells to tissue culture conditions, and provides high frequencies of transformation with single copy insertions which appear to be less prone to methylation and silencing. It is not surprising, therefore, that extensive attempts have been made to transform cereals with *Agrobacterium.* However, in spite of many attempts and a few reports (Hess et al. 1990, Gould et al. 1991, Chan et al. 1993), no unambiguous and clear evidence for *Agrobacterium*-mediated transformation of cereals has ever been presented. Indeed, it has been suggested that the reported claims of transformation may be actually artifacts caused by the transformation of an endophytic organism associated with the host plants (Langridge et al. 1992, Chen et al. 1994). It is known, nevertheless, that *Agrobacterium* can deliver viral genomic sequences to cereal cells resulting in systemic viral infection (Grimsley et al. 1987, Dale et al. 1989, Shen et al. 1993). This process, termed "agroinfection", does not result in the integration of the viral genes into the plant genome. It is thus likely that the difficulties faced with *Agrobacterium*-mediated transformation of cereals lie in the integration, rather than the delivery of genes, and may be eventually overcome with further understanding of the process of transformation.

In the mean time, other useful methods, based on the direct delivery of DNA into protoplasts as well as intact cells and tissues, have been developed for the transformation of cereals. All of the early successes, including the first transgenic cereal (maize, Rhodes et al. 1988b) and grass (Orchard grass, Horn et al. 1988b) plants, were based on DNA delivery into protoplasts by osmotic (polyethylene glycol treatment) or electric (electroporation) shock. Although this procedure has proven to be very useful for rice, maize and some grass species (Table 2), its use is limited because of dependence on totipotent protoplasts isolated from embryogenic suspension cultures, both of which are difficult to obtain and maintain in culture.

The development of the biolistics procedure, based on the high velocity bombardment of DNA-coated microprojectiles into intact cells and tissues (Sanford et al. 1987, 1991, Klein et al. 1992), has provided an important alternative to the use of protoplasts for cereal transformation. Within a short period of four years, from 1990 to 1994, it was used to produce transgenic plants of maize, rice, wheat, oats, barley, sorghum, rye, and important grass species, like sugarcane (Table 2). The delivery of DNA directly into regenerable cells or tissues substantially reduces the time that the cells must be maintained in culture. The immature embryo, which is the most widely used explant for plant regeneration in cereals, has proven to be particularly useful for the biolistic delivery of DNA and the production of transgenic plants.

More recently, transgenic maize and rice plants have been also obtained by electroporating DNA into enzymatically or mechanically wounded suspension culture cells, callus or embryos (D'Halluin et al. 1992, Laursen et al. 1994, Xu and Li 1994).

Table 2 lists all transgenic cereal and grass species produced to date. It includes many examples where agronomically useful genes for resistance to non-selective herbicides, viruses and

Table 2. Transgenic cereals and grasses obtained by the direct delivery of DNA into protoplasts (P), by high velocity microprojectile bombardment (B), and by tissue electroporation (E).

Species	Method of Transformation	Reference
Dactylis glomerata	P	Horn et al. 1988b
Zea mays	P	Rhodes et al. 1988b, Golovkin et al. 1993, Omirulleh et al. 1993, Sukhapinda et al. 1993
	B	[a]Fromm et al. 1990, [a]Gordon-Kamm et al. 1990, [b]Koziel et al. 1993, [c]Murry et al. 1993
	E	D'Halluin et al. 1992, [a]Laursen et al. 1994
Oryza sativa	P	Shimamoto et al. 1989, Datta et al. 1990b, [a]1992, [c]Hayakawa et al. 1992, [a,d]Uchimiya et al. 1993, [b]Fujimoto et al. 1993
	B	[a]Christou et al. 1991, Li et al. 1993
	E	Xu and Li 1994
Avena sativa	B	[a]Somers et al. 1992
Festuca arundinacea	P	[a]Wang et al. 1992, Ha et al. 1992
Saccharum sp. hybrid	B	Bower and Birch 1992
Triticum aestivum	B	[a]Vasil et al. 1992, 1993, [a]Weeks et al. 1993, [a]Becker et al. 1994, [a]Nehra et al. 1994
Sorghum bicolor	B	[a]Cassas et al. 1993
Agrostis palustris	B	Zhong et al. 1993
Agrostis alba	P	Asano and Ugaki 1994
Hordeum vulgare	B	Ritala et al. 1994, [a]Wan and Lemaux 1994
Secale cereale	B	[a]Castillo et al. 1994
Festuca rubra	P	[a]Spangenberg et al. 1994
Tritordeum	B	Barcelo et al. 1994

[a]herbicide resistant, [b]insect resistant, [c]virus resistant, [d]fungus resistant

insects, have been introduced. In several instances field evaluations of transgenic cereals have been performed, but most of this information is proprietary and has not been published, except for insect resistant maize which performed well under field conditions (Koziel et al. 1993)

A number of other strategies have been used for the direct delivery of DNA into cereal cells (Vasil 1994). Some of these have resulted in the production of transgenic callus tissues (Kaeppler et al. 1992, Rasmussen et al. 1994). Attempts to transform germ line cells by microtargetting (Sautter et al. 1991, Iglesias et al. 1994) or microinjection (Simmonds et al. 1992) have so far not been successful. Claims of transgenic plants produced by macroinjection of plasmid DNA into flowering tillers of rye (de la Pena et al. 1987) and barley (Rogers and Rogers 1992), or of transgenic plants obtained by the pollen tube pathway (Luo and Wu 1988, Hess et al. 1990), did not provide evidence for the integration of the transgene into germ line cells and its Mendelian segregation. Some of these results could be attributed to the transformation of an endophyte residing in the host plant (Langridge et al. 1992, Chen et al. 1994).

The success achieved in the transformation of cereals is due not only to the development of novel means of DNA delivery, but also to the use of efficient reporter and selectable marker genes, and powerful promoters and introns, to ensure high levels of gene expression (Vasil 1994). The most widely used reporter gene is the β-glucuronidase (*gus*) gene of *Escherichia coli*, that can be readily evaluated by both histochemical and fluorometric assays (Jefferson et al. 1987). The neomycin phosphotransferase II (*nptII*) and hygromycin phosphotransferase (*hph*) genes of *E. coli*, which confer resistance to kanamycin and hygromycin, respectively, have also proven to be useful for cereal transformation (Shimamoto et al. 1989, Datta et al. 1990b, Battraw and Hall 1992, d'Halluin et al. 1992, Walters et al. 1992, Murry et al. 1993, Sukhapinda et al. 1993, Nehra et al. 1994). The *dhfr* gene conferring resistance to methotrexate has been used to select maize transformants (Golovkin et al. 1993).

The high levels of natural resistance to antibiotics exhibited by cereal species (Hauptmann et al. 1988), and continuing concerns regarding the presence of antibiotic resistance genes in food crops, have prompted the use of other selectable marker genes. Amongst these, the *bar* gene of *Streptomyces hygroscopicus* has been used most successfully and widely in cereal transformation (Fromm et al. 1990, Gordon-Kamm et al. 1990, Christou et al. 1991, Somers et al. 1992, Vasil et al. 1992, 1993, Cassas et al. 1993, Weeks et al. 1993, Castillo et al. 1994, Wan and Lemaux 1994). It encodes the enzyme phosphinothricin acetyltransferase (PAT), providing resistance to the non-selective broad-spectrum herbicide basta (glufosinate, bialaphos). The use of herbicide resistance genes as selectable markers has the advantage that it helps to produce agronomically useful transgenic plants without the presence of any unnecessary DNA sequences, such as those for antibiotic resistance. However, in order to eliminate the possibility of weeds becoming resistant to herbicides, the use of such genes is not recommended in crops like oats (Somers et al. 1992) and sorghum (Cassas et al. 1993), which can interbreed with weeds like wild oat and Johnson grass, respectively.

In order to be useful, the introduced genes must be expressed at high levels during the selection process (selectable marker genes) as well as *in planta* (agronomically useful genes). This depends largely on the promoters used to drive the genes (Vasil 1994). The constitutively expressed 35S promoter of cauliflower mosaic virus (CaMV) is the most widely used promoter in plant transformation. However, the levels of gene expression obtained in cereals by this promoter are upto 100-fold less than in dicots (Fromm et al. 1985, Hauptmann et al. 1987). Duplication of the CaMV 35S promoter sequences, known to enhance gene expression (Kay et al. 1987), has been used to obtain transgenic maize plants (Omirulleh et al. 1993). However, the highest level of gene expression in cereals has been obtained by the use of introns, such as intron 1 of the *Adh 1* gene (Callis et al. 1987) or intron 1 of the *Sh 1* gene (Vasil et al. 1989), of maize. In addition, three monocot promoters have been shown to increase gene expression: pEMU (Last et al. 1991), Actin 1 (*Act 1*) gene of rice (Zhang et al. 1991), and the Ubiquitin 1 (*Ubi 1*) gene of maize (Christensen et al. 1992). Amongst these, *Ubi 1* has been shown to provide the highest levels of gene expression (Cornejo et al. 1993, Taylor et al. 1993), and has been used to obtain transgenic plants of wheat, rice and barley (Toki et al. 1992, Vasil et al. 1993, Weeks et al. 1993, Wan and Lemaux 1994). Nevertheless, the efficiency of transformation obtained in cereals is still low compared to the results in dicots with *Agrobacterium*. In most instances, it is in the range of 0.1-1.0%, although higher frequencies have been reported in a few cases (Li et al. 1993, Wan and Lemaux 1994). Transformation efficiency must be substantially increased in order to produce the large numbers of independently transformed fertile lines which are needed for field evaluations. This will require improvements in all aspects of transformation technology, including gene integration and expression, selection of transformed lines and regeneration of fertile transgenic plants.

Conclusions

Efficient regeneration of fertile plants from cultured cells, combined with novel methods of DNA delivery and selection of transformed cells, has resulted in the production of transgenic plants in all of the major cereal crops. In most instances agronomically useful genes - which confer resistance to broad-spectrum, environmentally desirable, and non-selective herbicides, or to viral pathogens and insect pests - have been introduced. This is particularly noteworthy, keeping in mind that nearly a third of the productivity of most cereal crops is lost to weeds, pests and pathogens. Therefore, the introduction of such engineered crops in agricultural production should help to increase food productivity. However, further molecular improvement of cereals in the near future will be limited most by the lack of our knowledge about, and access to, important and useful genes. Therefore, high priority should be given to the development of DNA-based maps of cereals (Phillips and Vasil 1994) and the identification and cloning of agronomically important genes (eg. those controlling multigenic traits like yield, and resistances to biotic and abiotic stresses). In this respect, the conservation of gene order along chromosomes among various cereal species, as well as similarity of gene composition and map linearity (Bennetzen and Freeling 1993, Kurata et al. 1994), should be of considerable advantage. Improvements in transformation technology (higher efficiency of transformation and generation of large numbers of independently transformed fertile lines) must also take place simultaneously, in order to allow the introduction of several genes regulating a multigenic trait. This will require more precise control of copy number and sites of integration to avoid deleterious position effects. Although we can take pride in the rapid progress made in the molecular improvement of cereals, we should remember that the future benefits of this novel and powerful technology for mankind will depend on at least three factors: (a) Demonstrated safety and superiority of the engineered crops, (b) Their contribution to the development of a sustainable agricultural system, and (c) Public acceptance of the engineered plants and their products.

Acknowledgements. I wish to acknowledge and pay tribute to the more than fifty graduate students and research associates who made significant and often pioneering contributions to the research on cell culture and genetic transformation of cereals in my laboratory. It has been my pleasure, privilege and good fortune to be associated with such a dedicated group of scientists.

References

Ahmed KZ, F Sagi (1993) Culture of and fertile plant regeneration from regenerable embryogenic cell-derived protoplasts of wheat (*Triticum aestivum* L.). Plant Cell Rep 12:175-179.

Akashi R, T Adachi (1992) Plant regeneration from suspension culture-derived protoplasts of apomictic dallisgrass (*Paspalum dilatatum* Poir.). Plant Sci 82:219-225.

Asano Y, K Sugiura (1990) Plant regeneration from suspension culture-derived protoplasts of *Agrostis alba* L. (Redtop). Plant Sci 72:267-273.

Asano Y, M Ugaki (1994) Transgenic plants of *Agrostis alba* obtained by electroporation-mediated direct gene transfer into protoplasts. Plant Cell Rep 13:243-246.

Barcelo P, C Hagel, D Becker, A Martin, H Lorz (1994) Transgenic cereal (tritordeum) plants obtained at high efficiency by microprojectile bombardment of inflorescence tissue. The Plant J 5:583-592.

Baset A, EC Cocking, RP Finch (1993) Regeneration of fertile plants from protoplasts of the wild rice species *Oryza granulata*. J Plant Physiol 141:245-247.

Baset A, RP Finch, EC Cocking (1991) Plant regeneration from protoplasts of wild rice (*Oryza rufipogon* Griff.). Plant Cell Rep 10:200-203.

Battraw M, TC Hall (1992) Expression of a chimeric neomycin phosphotransferase II gene in first and second generation transgenic rice plants. Plant Sci 86:191-202.

Becker D, R Brettschneider, H Lorz (1994) Fertile transgenic wheat from microprojectile bombardment of scutellar tissue. Plant J 5:299-307.

Bennetzen JL, M Freeling (1993) Grasses as a single genetic system: genome composition, collinearity and compatibility. Trends Genet 9:259-261.

Borlaug NE (1983) Contributions of conventional plant breeding to food production. Science 219:689-693.

Bower R, RG Birch (1992) Transgenic sugarcane plants via mircoprojectile bombardment. Plant J 2:409-416.

Breiman A, T Felsenburg, E Galun (1989) Is *Nor* region variability in wheat invariably caused by tissue culture? Theor Appl Genet 77:809-814.

Callis J, ME Fromm, V Walbot (1987) Introns increase gene expression in cultured maize cells. Genes Dev 1:1183-1200.

Carnes MG, MS Wright (1988) Endogenous hormone levels of immature corn kernels of A188, Missouri-17, and DeKalb XL-12. Plant Sci 57:195-203.

Cassas AM, AK Kononowicz, UB Zehr, DT Tomes, JD Axtell, LG Butler, RA Bressan, PM Hasegawa (1993) Transgenic sorghum plants via microprojectile bombardment. Proc Nat Acad Sci USA 90:11212-11216.

Castillo AM, V Vasil, IK Vasil (1994) Rapid production of fertile transgenic plants of rye (*Secale cereale* L.). In prepraration.

Cavallini A, MC Lupi, R Cremonini, A Benici (1987) *In vitro* culture of *Bellevalia romana* (L.) Rchb. III. Cytological study of somatic embryos. Protoplasma 139:66-70.

Cavallini A, L Natali (1989) Cytological analyses of *in vitro* somatic embryogenesis in *Brimura amethystina* Salisb. (Liliaceae). Plant Sci 62:255-261.

Chan M, H Chang, S Ho, W Tong, S Yu (1993) *Agrobacterium*-mediated production of transgenic rice plants expressing a chimeric α-amylase promoter/β-glucuronidase gene. Plant Mol Biol 22:491-506.

Chen D, Z Xia (1987) Mature plant regeneration from cultured protoplasts of *Polypogon fugax* Nees ex Steud. Sci Sinica B 30:698-704.

Chen DF, PJ Dale, JS Heslop-Harrison, JW Snape, W Harwood, S Bean, P Mullineaux (1994) Stability of transgenes and presence of N^6 methyladenine DNA in transformed wheat cells. Plant J 5:429-436.

Chen WH, MR Davey, JB Power, EC Cocking (1988) Sugarcane protoplasts: factors affecting division and plant regeneration. Plant Cell Rep 7:344-347.

Chowdhury MKU, IK Vasil (1993) Molecular analysis of plants regenerated from embryogenic cultures of hybrid sugarcane cultivars (*Saccharum* sp.) Theor Appl Genet 86:181-188.

Chowdhury MKU, V Vasil, IK Vasil (1994) Molecular analysis of plants regenerated from embryogenic cultures of wheat (*Triticum aestivum* L.). Theor Appl Genet 87:821-828.

Christensen AH, RA Sharrock, PH Quail (1992) Maize polyubiquitin genes: thermal perturbation of expression and transcript splicing, and promoter activity following transfer to protoplasts by electroporation. Plant Mol Biol 18:675-689.

Christou P, TL Ford, M Kofron (1991) Production of transgenic rice (*Oryza sativa* L.) plants from agronomically important Indica and Japonica varieties via electric discharge particle acceleration of exogenous DNA into immature zygotic embryos. Bio/Technology 9:957-962.

Cornejo M, D Luth, KM Blankenship, OD Anderson, AE Blechl (1993) Activity of a maize ubiquitin promoter in transgenic rice. Plant Mol Biol 23:567-581.

Creemers-Molenaar J, P van der Valk, JPM Loeffen, MACM Zaal (1989) Plant regeneration from suspension cultures and protoplasts of *Lolium perenne* L. Plant Sci 63:167-176.

Dale PJ (1980) Embryoids from cultured immature embryos of *Lolium multiflorum*. Z. Pflanzenphysiol 100:73-77.

Dale PJ, MS Marks, MM Brown, CJ Woolston, HV Gunn, PM Mullineaux, DM Lewis, JM Kempt, DF Chen, DM Gilmour, RB Flavell (1989) Agroinfection of wheat: inoculation of in vitro grown seedlings and embryos. Plant Sci 63:237-245.

Dalton SJ (1988) Plant regeneration from cell suspension protoplasts of *Festuca arundinacea* Schreb. (Tall fescue) and *Lolium perenne* L. (perennial ryegrass). J Pl Physiol 132:170-175.

Datta SK, K Datta, N Soltanifar, G Donn, I Potrykus (1992) Herbicide-resistant Indica rice plants from IRRI breeding line IR72 after PEG-mediated transformation of protoplasts. Plant Mol Biol 20:619-629.

Datta SK, K Datta, I Potrykus (1990a) Fertile Indica rice plants regenerated from protoplasts isolated from microspore derived cell suspensions. Plant Cell Rep 9:253-256.

Datta SK, A Peterhans, K Datta, I Potrykus (1990b) Genetically engineered fertile Indica rice recovered from protoplasts. Bio/Technology 8:736-740.

De la Pena A, H Lorz, J Schell (1987) Transgenic rye plants obtained by injecting DNA into young floral tillers. Nature 235:274-276.

D'Halluin K, E Bonne, M Bossut, M De Beuckeleer, J Leemans (1992) Transgenic maize plants by tissue electroporation. Plant Cell 4:1495-1505.

Dong J, Z Xia (1990) Protoplast culture and plant regeneration of foxtail millet. Chinese Sci Bull 35:1560-1564.

Fraley RT, SG Rogers, RB Horsch, PR Sanders, JS Flick (1983) Expression of bacterial genes in plant cells. Proc Nat Acad Sci USA 80:4803-4807.

Fretz A, A Jahne, H Lorz (1992) Cryopreservation of embryogenic suspension cultures of barley (*Hordeum vulgare* L.). Bot Acta 105:140-145.

Fromm ME, LP Taylor, V Walbot (1985) Expression of genes transferred into monocot and dicot plant cells by electroporation. Proc Nat Acad Sci USA 882:5824-5828.

Fromm ME, F Morrish, C Armstrong, R Williams, J Thomas, TM Klein (1990) Inheritance and expression of chimeric genes in the progeny of transgenic maize plants. Bio/Technology 8:833-839.

Fujimoto H, K Itoh, M Yamamoto, J Kyozuka, K Shimamoto (1993) Insect resistant rice generated by introduction of a modified δ-endotoxin gene of *Bacillus thuringiensis*. Bio/Technology 11:1151-1155.

Gmitter Jr FG, X Ling, C Cai, JW Grosser (1991) Colchicine induced polyploidy in *Citrus* embryogenic cultures, somatic embryos and regenerated plantlets. Plant Sci 74:135-141.

Gnanapragasam S, IK Vasil (1990) Plant regeneration from a cryopreserved embryogenic cell suspension of a commercial sugarcane hybrid (*Saccharum* spp.) Plant Cell Rep 9:419-423.

Golovkin MV, M Abraham, S Morocz, S Bottka, A Feher, D Dudits (1993) Production of transgenic maize plants by direct DNA uptake into embryogenic protoplasts. Plant Sci 90:41-52.

Gordon-Kamm WJ, TM Spencer, ML Mangano, TR Adams, RJ Daines, WG Start, JV O'Brien, SA Chambers, WR Adams, NG Willets, TB Rice, CJ Mackey, RW Krueger, AP Kausch, PG Lemaux (1990) Transformation of maize cells and regeneration of fertile transgenic plants. Plant Cell 2:603-618.

Gould J, M Devey, O Hasegawa, EC Ulian, G Peterson, RH Smith (1991) Transformation of *Zea mays* using *Agrobacterium tumefaciencs* and the shoot apex. Plant Physiol 95:426-434.

Grimsley N, T Hohn, JW Davies, B Hohn (1987) *Agrobacterium*-mediated delivery of infectious maize streak virus into maize plants. Nature 325:177-179.

Ha S, F Wu, TK Thorne (1992) Transgenic turf-type tall fescue (*Festuca arundinacea* Schreb.) plants regenerated from protoplasts. Plant Cell Rep 11:601-604.

Hanna WW, C Lu, IK Vasil (1984) Uniformity of plants regenerated from somatic embryos of *Panicum maximum* Jacq. (Guinea grass). Theor Appl Genet 67:155-159.

Hauptmann, RM, P Ozias-Akins, V Vasil, Z Tabaeizadeh, SG Rogers, RB Horsch, IK Vasil, RT Fraley (1987) Transient expression of electroporated DNA in monocotyledonous and dicotyledonous species. Plant Cell Rep 6:265-270.

Hauptmann RM, V Vasil, P Ozias-Akins, Z Tabaeizadeh, SG Rogers, RT Fraley, RB Horsch, IK Vasil (1988) Evaluation of selectable markers for obtaining stable transformants in the Gramineae. Plant Physiol 86:602-606.

Hayakawa T, Y Zhu, K Itoh, Y Kimura, T Izawa, K Shimamoto, S Toriyama (1992) Genetically engineered rice resistant to rice stripe virus, an insect-transmitted virus. Proc Nat Acad Sci USA 89:9865-9869.

Haydu Z, IK Vasil (1981) Somatic embryogenesis and plant regeneration from leaf tissues and anthers of *Pennisetum purpureum*. Theor Appl Genet 59:269-273.

Hess D, K Dressler, R Nimmrichter (1990) Transformation experiments by pipetting *Agrobacterium* into the spikelets of wheat (*Triticum aestivum* L.). Plant Sci 72:233-244.

Holm PB, S Knudsen, P Mouritzen, D Negri, FL Olsen, C Roue (1994) Regeneration of fertile barley plants from mechanically isolated protoplasts of the fertilized egg cell. Plant Cell 6:531-543.

Horn ME, BV Conger, CT Harms (1988a) Plant regeneration from protoplasts of embryogenic suspension cultures of orchardgrass (*Dactylis glomerata* L.). Plant Cell Rep 7:371-374.

Horn ME, RD Shillito, BV Conger, CT Harms (1988b) Transgenic plants of orchardgrass (*Dactylis glomerata* L.) from protoplasts. Plant Cell Rep 7:469-472.

Iglesias VA, A Gisel, R Bilang, N Leduc, I Potrykus, C Sautter (1994) Transient expression of visible marker genes in meristem cells of wheat embryos after ballistic micro-targetting. Planta 192:84-91.

Isabel N, L Tremblay, M Michaud, FM Tremblay, J Bousquet (1993) RAPDs as an aid to evaluate the genetic integrity of somatic embryogenesis-derived populations of *Picea mariana* (Mill.) B.S.P. Theor Appl Genet 86:81-87.

Jähne A, PA Lazzeri, H Lorz (1991) Regeneration of fertile plants from protoplasts derived from embryogenic cell suspensions of barley (*Hordeum vulgare* L.). Plant Cell Rep 10:1-6.

Jefferson RA, TA Kavanagh, MW Bevan (1987) GUS fusions: β-glucuronidase as a sensitive and versatile gene fusion marker in plants. EMBO J 6:3901-3907.

Kaeppler HF, DA Somers, HW Rines, AF Cockburn (1992) Silicon carbide fiber-mediated stable transformation of plant cells. Theor Appl Genet 84:560-566.

Kay R, A Chan, M Daly, J McPherson (1987) Duplication of CaMV 35S promoter sequences creates a strong enhancer for plant genes. Science 230:1299-1302.

Klein TM, R Arentzen, PA Lewis, S Fitzpatrick-McElligott (1992) Transformation of microbes, plants and animals by particle bombardment. Bio/Technology 10:286-291.

Kobayashi S (1987) Uniformity of plants regenerated from orange (*Citrus sinensis* Osb.) protoplasts. Theor Appl Genet 74:10-14.

Koziel MG, GL Beland, C Bowman, NB Carozzi, R Crenshaw, L Crossland, J Dawson, N Desai, M Hill, S Kadwell, K Launis, K Lewis, D Maddox, K McPherson, MR Meghji, E Merlin, R Rhodes, GW Warren, M Wright, SV Evola (1993) Field performance of elite transgenic maize plants expressing an insecticidal protein derived from *Bacillus thuringiensis*. Bio/Technology 11:194-200.

Kranz E, H Lorz (1993) In vitro fertilization with isolated, single gametes results in zygotic embryogenesis and fertile maize plants. Plant Cell 5:739-746.

Kurata N, G Moore, Y Nagamura, T Foote, M Yano, Y Minobe, M Gale (1994) Conservation of genome structure between rice and wheat. Bio/Technology 12:276-278 (1994).

Kyozuka J, Y Hayashi, K Shimamoto (1987) High efficiency plant regeneration from rice protoplasts by novel nurse culture methods. Mol Gen Genet 206:408-413.

Kyozuka J, K Taneda, K Shimamoto (1989) Production of cytoplasmic male sterile rice (*Oryza sativa* L.) by cell fusion. Bio/Technology 7:1171-1174.

Langridge P, R Brettschneider, P Lazzeri, H Lorz (1992) Transformation of cereals via *Agrobacterium* and the pollen pathway: a critical assessment. Plant J 2:631-638.

Last DI, RIS Brettell, DA Chamberlain, AM Chaudhury, PJ Larkin, EL Marsh, WJ Peacock, ES Dennis (1991) pEmu: an improved promoter for gene expression in cereal cells. Theor Appl Genet 81:581-588.

Laursen CM, RA Krzyzek, CE Flick, PC Anderson, TM Spencer (1994) Production of fertile transgenic maize by electroporation of suspension culture cells. Plant Mol Biol 24:51-61.

Li L, R Qu, A de Kochko, C Fauquet, RN Beachy (1993) An improved rice transformation system using the biolistic method. Plant Cell Rep 12:250-255.

Lu C, IK Vasil (1981) Somatic embryogenesis and plant regeneration from leaf tissues of *Panicum maximum* Jacq. Theor Appl Genet 59:275-280.

Lu C, V Vasil, IK Vasil (1981) Isolation and culture of protoplasts of *Panicum maximum* Jacq. (Guinea grass): somatic embryogenesis and plantlet formation. Z. Pflanzenphysiol. 104:311-318.

Lu TG, CS Sun (1992) Cryopreservation of millet (*Setaria italica* L.). J Plant Physiol 139:295-298.

Luo Z, R Wu (1988) A simple method for the transformation of rice via the pollen tube pathway. Plant Mol Biol Rep 6:165-174.

Meijer EGM, F van Iren, E Schrijnemakers, LAM Hensgens, M van Zijderveld, RA Schilperoort (1991) Retention of capacity to produce plants from protoplasts in cyropreserved cell lines of rice (*Oryza sativa* L.). Plant Cell Rep.. 10:171-174.

Morocz S, G Donn, J Nemeth, D Dudits (1990) An improved system to obtain fertile regenerants via maize protoplasts isolated from a highly embryogenic suspension culture. Theor Appl Genet 80:721-726.

Morrish FM, WW Hanna, IK Vasil (1990) The expression and perpetuation of inherent somatic variation in regenerants from emrbyogenic cultures of *Pennisetum glaucum* (L.) R.Br. (pearl millet). Theor Appl Genet 80:409-416.

Murry LE, LG Elliott, SA Capitant, JA West, KK Hanson, L Scarafia, S Johnston, C DeLuca-Flaherty, S Nichols, D Cunanan, PS Dietrich, IJ Mettler, S Dewald, DA Warnick, C Rhodes, RM Sinibaldi, KJ Brunke (1993) Transgenic corn plants expressing MDMV strain B coat protein are resistant to mixed infections of maize dwarf mosaic virus and maize chlorotic mottle virus. Bio/Technology 11: 559-564.

Nehra NS, RN Chibbar, N Leung, K Caswell, C Mallard, L Steinhauer, M Baga, KK Kartha (1994) Self-fertile transgenic wheat plants regenerated from isolated scuellar tissues following microprojectile bombardment with two distinct gene constructs. Plant J 5:285-297.

Nielsen KA, E Larsen, E Knudsen (1993) Regeneration of protoplast-derived green plants of Kentucky blue grass (*Poa pratensis* L.). Plant Cell Rep 12:537-540.

Omirulleh S, M Abraham, M Golovkin, I Stefanov, MK Karabaev, L Mustardy, S Morocz, D Dudits (1993) Activity of a chimeric promoter with the doubled CaMV 35S enhancer element in protoplast-derived cells and transgenic plants of maize. Plant Mol Biol 21:415-428.

Phillips RL, IK Vasil (eds) (1994) DNA-Based Markers in Plants. Kluwer Academic Publishers, Dordrecht.

Potrykus I (1980) The old problem of protoplast culture: cereals. In: Ferenczy L, GL Farkas, G Lazar (eds), Advances in Protoplast Research, pp 243-254. Akademiai Kiado, Budapest.

Prioli LM, MR Sondahl (1989) Plant regeneration and recovery of fertile plants from protoplasts of maize (*Zea mays* L.). Bio/Technology 7:589-594.

Rajasekaran K, MB Hein, GC Davis, MG Carnes, IK Vasil (1987a) Endogenous plant growth regulators in leaves and tissue cultures of Napier grass (*Pennisetum purpureum* Schum.). J Plant Physiol 130:13-25.

Rajasekaran K, MB Hein, IK Vasil (1987b) Endogenous abscisic acid and indole-3-acetic acid and somatic embryogenesis in cultured leaf explants of *Pennisetum purpureum* Schum.: effects *in vivo* and *in vitro* of glyphosate, fluridone and paclobutrazol. Plant Physiol 84:47-51.

Rajasekaran K, SC Schank, IK Vasil (1986) Characterization of biomass production and phenotypes of plants regenerated from embryogenic callus cultures of *Pennisetum americanum* x *P. purpureum* (hybrid Napier grass). Theor Appl Genet 73:4-10.

Rasmussen JL, JR Kikkert, MK Roy, JC Sanford (1994) Biolistic transformation of tobacco and maize suspension cells using bacterial cells as microprojectiles. Plant Cell Rep 13:212-217.

Rhodes CA, KS Lowe, KL Ruby (1988a) Plant regeneration from protoplasts isolated from emrbyonic maize cell cutlures. Bio/Technology 6:56-60.

Rhodes CA, DA Pierce, IJ Mettler, D Mascarenhas, JJ Detmer (1988b) Genetically transformed maize plants from protoplasts. Science 240:204-207.

Ritala A, K Aspergen, U Kurten, M Salmenkallio-Marttila, L Mannone, R Hannus, V Kauppinen, TH Teeri, T-M Enari (1994) Fertile transgenic barley by particle bombardment of immature embryos. Plant Mol Biol 24:317-325.

Rogers SW, JC Rogers (1992) The importance of DNA methylation for stability of foreign DNA in barley. Plant Mol Biol 18:945-961.

Sanford JC, TM Klein, ED Wolf, N Allen (1987) Delivery of substances into cells and tissues using a particle bombardment process. J Part Sci Technol 5:27-37.

Sanford JC, MJ DeVit, JA Russell, FD Smith, PR Harpending, MK Roy, SA Johnson (1991) An improved, helium-driven biolistic device. Technique 3:3-16.

Sautter C, H Waldner, G Neuhaus-Url, A Galli, G Neuhaus, I Potrykus (1991) Micro-targetting: high efficiency gene transfer using a novel approach for the acceleration of microprojectiles. Bio/Technology 9:1080-1085.

Shen W, J Escudero, M Schlappi, C Ramos, B Hohn, Z Koukolikova-Nicola (1993) T-DNA transfer to maize cells: histochemical investigation of β-glucuronidase activity in maize tissues. Proc Nat Acad Sci USA 90:1488-1492.

Shenoy VB, IK Vasil (1992) Biochemical and molecular analysis of plants derived from embryogenic tissue cultures of Napier grass (*Pennisetum purpureum* Schum.). Theor Appl Genet 83:947-955.

Shimamoto K, R Terada, T Izawa, H Fujimoto (1989) Fertile transgenic rice plants regenerated from transformed protoplasts. Nature 338:274-276.

Shimron-Abarbanell D, A Breiman (1991) Comprehensive molecular characterization of tissue culture-derived *Hordeum marinum* plants. Theor Appl Genet 83:71-80.

Shillito RD, GK Carswell, CM Johnson, JJ DiMaio, CT Harms (1989) Regeneration of fertile plants from protoplasts of elite inbred maize. Bio/Technology 7:581-587.

Simmonds J, P Steward, D Simmonds (1992) Regeneration of *Triticum aestivum* apical explants after microinjection of germ line progenitor cells with DNA. Physiol Plant 85:197-206.

Somers DA, HW Rines, W Gu, HF Kaeppler, WR Bushnell (1992) Fertile, transgenic oat plants. Bio/ Technology 10:1589-1594.

Spangenberg G, ZY Wang, J Nagel, I Potrykus (1994) Protoplast culture and generation of transgenic plants in red fescue (*Festuca rubra* L.). Plant Sci 97:83-94.

Srinivasan C, IK Vasil (1986) Plant regeneration from protoplasts of sugarcane (*Saccharum officinarum* L.). J Plant Physiol 126:41-48.

Sukhapinda K, ME Kozuch, B Rubin-Wilson, WM Ainley, DJ Merlo (1993) Transformation of maize (*Zea mays* L.) protoplasts and regeneration of haploid transgenic plants. Plant Cell Rep 13:63-68.

Swedlund B, IK Vasil (1985) Cytogenetic characterization of embryogenic callus and regenerated plants of *Pennisetum americanum* (L.) K. Schum. Theor Appl Genet 69:575-581.

Taylor MG, V Vasil, IK Vasil (1993) Enhanced GUS gene expression in cereal/grass cell suspensions and immature embryos using the maize ubiquitin-based plasmid pAHC25. Plant Cell Rep 12:491-495.

Takamizo T, G Spangenberg, K Suginobu, I Potrykus (1991) Intergeneric somatic hybridization in Gramineae: somatic hybrid plants between tall fescue (*Festuca arundinacea* Schreb.) and Italian ryegrass (*Lolium multiflorum* Lam.). Mol Gen Genet 231:1-6.

Terada R, J Kyozuka, S Nishibayashi, K Shimamoto (1987) Plantlet regeneration from somatic hybrids of rice (*Oryza sativa* L.) and barnyard grass (*Echinochloa oryzicola* Vasing.). Mol Gen Genet 210:39-43.

Terakawa T, T Sato, M Koike (1992) Plant regeneration from protoplasts isolated from embryogenic suspension cultures of creeping bentgrass (*Agrostis palustris* Huds.). Plant Cell Rep 11:457-461.

Toki S, S Takamatsu, C Nojiri, S Ooba, H Anzai, M Iwata, AH Christensen, PH Quail, H Uchimiya (1992) Expression of a maize ubiquitin gene promoter-*bar* chimeric gene in transgenic rice plants. Plant Physiol 100:1503-1507.

Uchimiya H, M Iwata, C Nojiri, PK Samarajeewa, S Takamatsu, S Óoba, H Anzai, AH Christensen, PH Quail, S Toki (1993) Bialaphos treatment of transgenic rice plants expressing a *bar* gene prevents infection by the sheath blight pathogen (*Rhizoctonia solani*). Bio/Technology 11:835-836.

Valles MP, ZY Wang, P Montavon, I Potrykus, G Spangenberg (1993) Analysis of genetic stability of plants regenerated from suspension cultures and protoplasts of meadow fescue (*Festuca pratensis* Huds.). Plant Cell Rep 12:101-106.

Vasil IK (1983) Isolation and culture of protoplasts of grasses. Int Rev Cytol Supp 16:79-88.

Vasil IK (1987) Developing cell and tissue culture systems for the improvement of cereal and grass crops. J Plant Physiol 128:193-218.

Vasil IK (1990) The realities and challenges of plant biotechnology. Bio/Technology 8:296-301.

Vasil IK (1994) Molecular improvement of cereals. Plant Mol Biol (In Press).

Vasil IK, V Vasil (1986) Regeneration in cereal and other grass species. In: Vasil IK (ed), Cell Culture and Somatic Cell Genetics of Plants, Vol 3, Plant Regeneration and Genetic Variability, pp 121-150. Academic Press, Orlando.

Vasil IK, V Vasil (1992) Advances in cereal protoplast research. Physiol Plant 85:279-283.

Vasil IK, V Vasil (1994) In vitro culture of cereals and grasses. In: Vasil IK, TA Thorpe (eds), Plant Cell and Tissue Culture, pp 293-212. Kluwer Academic Publishers, Dordrecht.

Vasil V, IK Vasil (1980) Isolation and culture of cereal protoplasts. II. Embryogenesis and plantlet formation from protoplasts of *Pennisetum americanum*. Theor Appl Genet 56:97-99.

Vasil V, IK Vasil (1981) Somatic embryogenesis and plant regeneration from suspension cultures of pearl millet (*Pennisetum americanum*). Ann Bot 47:669-678.

Vasil V, AM Castillo, ME Fromm, IK Vasil (1992) Herbicide resistant fertile transgenic wheat plants obtained by microprojectile bombardment of regenerable embryogenic callus. Bio/Technology 10:667-674.

Vasil V, M Clancy, RJ Ferl, IK Vasil, LC Hannah (1989) Increased gene expression by the first intron of maize *Shrunken-1* locus in grass species. Plant Physiol 91:1575-1579.

Vasil V, FA Redway, IK Vasil (1990) Regeneration of plants from embryogenic suspension culture protoplasts of wheat (*Triticum aestivum* L.). Bio/Technology 8:429-433.

Vasil V, V Srivastava, AM Castillo, ME Fromm, IK Vasil (1993) Rapid production of transgenic wheat plants by direct bombardment of cultured immature embryos. Bio/Technology 11:1553-1558.

Vasil V, D Wang, IK Vasil (1983) Plant regeneration from protoplasts of *Pennisetum purpureum* Schum. (Napier grass). Z Pflanzenphysiol 111:319-325.

Walters DA, CS Vetsch, DE Potts, RC Lundquist (1992) Transformation and inheritance of a hygromycin phosphotransferase gene in maize plants. Plant Mol Biol 18:189-200.

Wan C, IK Vasil (1994) Plant regeneration from protoplasts, emrbyogenic suspensions and cryopreserved cells of Napier grass (*Pennisetum purpureum* Schum.). In Preparation.

Wan Y, PG Lemaux (1994) Generation of large numbers of independently transformed fertile barley plants. Plant Physiol 104:37-48.

Wang X-H, H Lorz (1994) Plant regeneration from protoplasts of wild barley (*Hordeum murinum* L.). Plant Cell Rep 13:139-144.

Wang ZY, J Nagel, I Potrykus, G Spangenberg (1993a) Plants from cell suspension-derived protoplasts in *Lolium* species. Plant Sci 94:179-193.

Wang Z, T Takamizo, VA Iglesias, M Osusky, J Nagel, I Potrykus, G Spangenberg (1992) Transgenic plants of tall fescue (*Festuca arundinacea* Schreb.) obtained by direct gene transfer to protoplasts. Bio/Technology 10: 691-696.

Wang Z, MP Valles, P Montavon, I Potrykus, G Spangenberg (1993b) Fertile plant regeneration from protoplasts of meadow fescue (*Festuca pratensis* Huds.). Plant Cell Rep 12:95-100.

Weeks IT, OD Anderson, AE Blechl (1993) Rapid production of multiple independent lines of fertile transgenic wheat (*Triticum aestivum*). Plant Physiol 102:1077-1084.

Wei Z, Z Xu (1990) Regeneration of fertile plants from emrbyogenic suspension culture protoplasts of *Sorghum vulgare*. Plant Cell Rep 9:51-53.

Wenck AR, BV Conger, RN Trigiano, CE Sams (1988) Inhibition of somatic embryogenesis in orchard grass by endogenous cytokinins. Plant Physiol 88:990-992.

Wernicke W, R Brettell (1980) Somatic embryogenesis from *Sorghum bicolor* leaves. Nature 287:138-139.

Xu X, B Li (1994) Fertile transgenic Indica rice plants obtained by electroporation of the seed embryo cells. Plant Cell Rep 13:237-242.

Yamada Y, ZQ Yang, DT Tang (1986) Plant regeneration from protoplast derived callus of rice (*Oryza sativa* L). Plant Cell Rep 4:85-88.

Yang YM, DD He, KJ Scott (1993) Plant regeneration from protoplasts of durum wheat (*Triticum durum* Desf. cv. D6962). Plant Cell Rep 12:320-323.

Yang Z, T Shikanai, K Mori, Y Yamada (1989) Plant regeneration from cytoplasmic hyrids of rice (*Oryza sativa* L.). Theor Appl Genet 77:305-310.

Zambryski P, H Joss, C Genetello, J Leemans, M Van Montagu, J Schell (1983) Ti plasmid vector for the introduction of DNA into plant cells without alteration of their normal regeneration capacity. EMBO J 2:2143-2150.

Zhang W, D McElroy, R Wu (1991) Analysis of rice *Act1* 5' region activity in transgenic rice plants. Plant Cell 3:1155-1165.

Zhong H, MG Bolyard, C Srinivasan, MB Sticklen (1993) Transgenic plants of turfgrass (*Agrostis palustris* Huds.) from microprojectile bombardment of embryogenic callus. Plant Cell Rep 13:1-6.

PLASMA MEMBRANE-CELL-WALL ADHESION AND ITS ROLE IN RESPONSE TO BIOTIC AND ABIOTIC STRESSES

SARAH E. WYATT, ALICE L. DOLPH, ALEX A. AVERY, AND N. C. CARPITA
Department of Botany and Plant Pathology
Purdue University
West Lafayette, IN 47907
U. S. A.

ABSTRACT. The concept that the cytoskeleton and cell wall form an interactive scaffolding for perception and transduction of positional information is relatively new. Researchers now realize that a cytoskeleton-exocellular matrix continuum may be a feature common to all eukaryotic cells--plant, fungal and animal--and that many of the molecules participating in this continuum may be homologous. Cells in liquid culture have proven to be suitable models to study the structural and functional aspects of this continuum. Tobacco cells adapted to grow in high concentrations of NaCl develop tight zones of adhesion between the plasma membrane and cell wall. This adhesion is revealed by concave plasmolysis in osmotic solutions. *Zinnia* mesophyll cells develop into tracheary elements *in vitro*, and these cells also exhibit an asymmetric adhesion during development. Cells and protoplasts exposed to fungal mycelia form localized sites of adhesion to the cell wall. Proteins immunologically related to human vitronectin (Vn) and fibronectin (Fn) are found in cell walls of tobacco cells. Onion epidermal cells also contain the Fn- and Vn-like proteins and an additional protein that binds specifically to human fibronectin.

1. Introduction

Polarity, cell shape, recognition, and movement or migration of animal cells is established by delicate sensing mechanisms involving an interaction of the cytoskeleton and its exocellular matrix (ECM) via transmembrane linking proteins (Edelman 1983). The concept that the cytoskeleton and ECM form an interactive scaffolding for perception and transduction of positional information is relatively new, but there are numerous examples where the contact of transmembrane receptors to exocellular glycoproteins or glycosaminoglycans elicits specific developmental responses (Bissell et al. 1982). All ECMs are complex networks of polysaccharides, proteoglycans, and proteins, yet are quite diverse. Plant cells are enveloped by a fibrillar matrix of cellulose interwoven with complex polysaccharides and structural proteins. The plant ECM, the cell wall, is depicted as three independent but dynamic and interacting networks of pectin, cellulose-cross linking glycan, and protein (McCann and Roberts 1991; Carpita and Gibeaut 1993).

Because the cell walls of plant are fused, cell migration is not involved in morphogenesis, and differentiation of cells into functional types occurs through coordinated cell expansion. In animal cells, several complex glycans and proteoglycans form extensive substrata with matrix proteins, such as fibronectin (Fn), vitronectin (Vn), fibrinogen, type I collagen and laminin, to which transmembrane linking integrin proteins form selective cell-substratum adhesions (Hynes 1987; Ruoslahti 1988; Ruoslahti and Pierschbacher 1986). An Arg-Gly-

M. Terzi et al. (eds.), Current Issues in Plant Molecular and Cellular Biology, 19-29.

Asp (RGD) tripeptide is common to the cell-binding domain of these matrix proteins. Contact of some cells with these proteins can elicit developmental pattern formation, and small peptides containing the RGD sequence can strongly interfere with this developmental patterning (Ruoslahti 1988; Ruoslahti and Pierschbacher 1986). Actin filaments terminate at these adhesion sites and are linked to integrins via several small proteins (Burridge et al. 1988). Actin filaments are also connected to the plasma membrane in numerous other ways, and many of these interacting molecules also culminate in attachment of the membrane to the ECM (Stossel 1993).

Many of the molecules that participate in this continuum may be homologous among plants, fungi, and animals, but there is also growing evidence that plants possess unique proteins that may be of specialized importance in adaptation to stress or in normal development of specialized cells and tissues (Wyatt and Carpita 1993). Cells in liquid culture are good models to illustrate specific membrane-wall and wall-substrate adhesion in plants. These adhesions may function directly in environmental sensing of biotic and abiotic stress, polarity and tropic movement, differentiation, and even special cases of cell migration.

2. The cytoskeleton-cell wall continuum

2.1 PLASMOLYTIC BEHAVIOR AND ADHESION

A tight interaction of the cytoskeleton and cell wall is not readily apparent in plant cells because the plasma membrane is forced tightly against the cell wall by turgor. When most mature plant cells are plasmolyzed, the membrane pulls away from the cell wall, demonstrating a lack of physical attachment between the plasma membrane and the wall. However, localized adhesion of the membrane to the wall is found in some cells after differentiation. The adhesions are revealed by "concave" plasmolysis, a pulling in of the membrane in hypertonic solutions which leaves sites of attachment between the cell wall and membrane. "Concave" plasmolysis has been documented numerous times in reports that date to the mid 19th century (see Stadelmann 1956). Cells within the apical meristem are extremely difficult to plasmolyze, and the walls often collapse around the shrinking protoplast. A localized adhesion is seen during differentiation of *Zinnia* tracheary elements *in vitro*. Mesophyll cells isolated aseptically from young leaves of *Zinnia* and incubated in liquid culture develop secondary wall thickenings, and the cells take on the appearance of true water-conducting cells of xylem (Fukuda and Kobayashi 1989). Plasmolysis of the *Zinnia* cells during this development reveals tight adhesion of the membrane to the primary wall specifically in these gap regions between thickenings (Roberts and Haigler 1989). Gomez-Lepe et al. (1979) noted that epidermal cells of *Allium cepa* show "concave" plasmolysis and that this interaction between the protoplast surface and the cell wall is disrupted by octylguanidine (OG). They suggest that the adhesion is the result of loose binding between cell wall Ca^{2+} and the membrane surface and that the protonized form of OG replaces Ca^{2+} ions in the wall, thus loosening the attachment sites. We have observed "concave" plasmolysis in onion epidermal tissue (Wyatt and Carpita 1993) and in tobacco suspension culture cells adapted to grow in high NaCl (Zhu et al. 1993). In both cases, the plasma membrane to cell wall adhesion is unaffected by 50 mM EGTA, indicating the interaction is Ca^{2+}-independent. However, as with Gomez-Lepe's work (1979), the protoplasts are immediately rounded in the presence of OG, and all sites of adhesion are lost (Fig. 1).

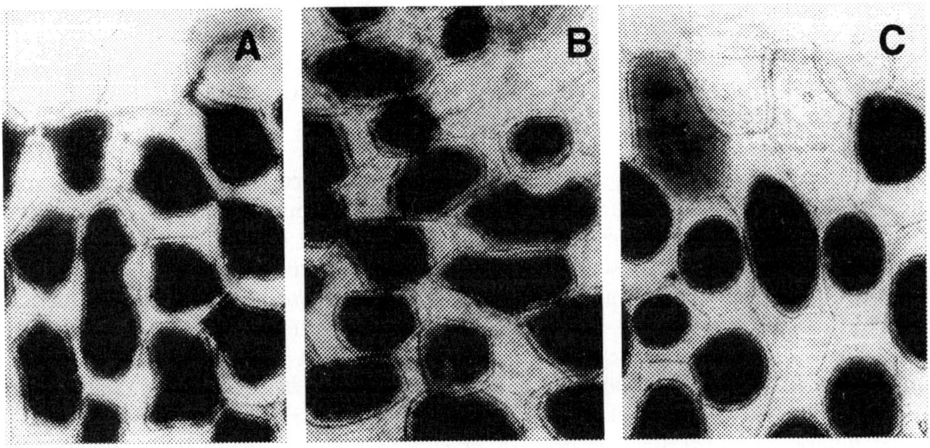

Figure 1. Plamolytic behavior of onion epidermal tissue. A) Tissue plasmolyzed in 1 M NaCl showing several sites of adhesion between the plasma membrane and the cell wall. B) Tissue plasmolyzed in 0.5 M NaCl in the presence of 50 mM EGTA. The adhesive character of the protoplast is maintained. C) Tissue plasmolyzed in 0.5 M NaCl in the presence of 90 µM octylguanadine. The protoplasts round immediately as sites of adhesion are lost.

2.2 MEMBRANE-WALL ADHESION IS INDUCED BY BIOTIC AND ABIOTIC STRESS

The plasma membrane also develops zones of tight adhesion to the cell wall in cells adapted to grow under conditions of environmental stress, and, in some cases, a physiological advantage may come with such a tight adhesion. Cells acclimated to deep freezing tolerance develop adhesive membranes (Johnson-Flanagan and Singh 1986). Guy (1990) points out that stiffened cell walls and firm attachment of the membrane to the cell wall are needed to withstand the enormous tensions that develop as a result of the low water potential of the extracellular ice. By this adhesive mechanism, cell volume is maintained and water remains as a liquid up to a spontaneous nucleation temperature of below -40°C.

Cells invaded by pathogenic fungi also develop adhesive membranes (Lee-Stadelmann et al. 1984). Hohl and his colleagues (Hohl and Balsiger 1986; Odermatt et al. 1988) developed a co-culture system where fungal hyphae overgrow and contact protoplasts of soybean or barley embedded in agar. This system permits observation of the cellular responses of both host and pathogen. Interestingly, only rarely do fungal hyphae recognize and attempt penetration of a plant protoplast without its cell wall, and, in most instances of contact, the hyphae traverse a portion of the plasma membrane and continue on (Odermatt et al. 1988). In contrast, the protoplast recognizes the fungal hypha, sticks tightly to it, and develops tracks of callose and other material specifically at the sites of contact and adhesion.

We discovered an exceptionally strong adhesion between the plasma membrane and cell wall in tobacco cells in liquid culture adapted to grow under NaCl stress (Zhu et al. 1993). As with the onion epidermal cells, the adhesion of the membrane to the cell wall is not disrupted by treatment with EGTA, and, thus, is not Ca^{2+}-dependent but is sensitive to treatment with OG.

2.3 DIRECT INTERACTIONS BETWEEN CYTOSKELETAL COMPONENTS AND THE CELL WALL OCCUR DURING PLANT DEVELOPMENT

There is considerable cytological evidence for a dynamic interaction between the cytoskeleton and the cell wall during development. Except under rare experimental conditions, cell division does not occur in the absence of the cell wall (Meyer and Herth 1982), and wounding or inhibition of cellulose biosynthesis prevents normal functioning of the actin filaments in positioning of the nucleus--indirect evidence for a transmembrane interaction of cytoskeleton and wall (Goodbody and Lloyd 1990; Katsuta and Shibaoka 1988). Cortical microtubules are implicated in biosynthesis of the cell wall. A "pre-prophase" thick band of microtubules tightly associated with the plasma membrane forms precisely at the eventual plane of cell division (Goddard et al. 1994), and the distribution of cortical microtubules parallels the pattern of cellulose deposition in both primary and secondary walls (Goddard et al. 1994; Roberts and Haigler 1989; Schneider and Herth 1986). Actin filaments are also associated with the plasma membrane (Roberts 1990; Sonesson and Widell 1993). In *Zinnia* cells undergoing tracheary element formation *in vitro*, actin filaments are seen attached to the membranes underlying the gaps of the developing secondary walls (Fukuda and Kobayashi 1989). Actin filaments also form what appear to be focal adhesions in pollen tubes developing *in vivo* (Pierson et al. 1986). Physical connections of actin and tubulin to the plasma membrane can be seen with an electron microscope (Roberts 1990). Unfortunately, the molecules responsible for the interaction have been elusive.

3. Asymmetric adhesion of the plasma membrane to the wall occurs during normal development

3.1 POLLEN TUBE GROWTH IS A SPECIAL CASE OF PLANT CELL "MIGRATION"

One possible example of cell "migration" in plants is the growth of the pollen tube, a gametophytic cell that often extends several centimeters along the transmitting tract of the style of a flower to deliver the sperm nuclei to the ovule (Lord and Sanders 1992). Sanders and Lord (1989) discovered that latex beads move directionally in the stylar matrix, much like Fn- and Vn-mediated transport of latex particles *in vitro* and in the chicken neural crest (Bronner-Fraser 1985; Underwood and Bennett 1989). They have since uncovered evidence for RGD-dependent interactions between the pollen tube and style matrix (Sanders et al. 1991). Pollen grains germinate *in vitro* but do not grow as rapidly they do *in vivo*. Lord and Sanders (1992) propose a model of pollen tube extension where actin bundles organize into a focal adhesion behind the tip forming a continuum between cell wall and the generative nucleus. This provides the scaffolding needed to maintain the germ unit in its proximal position and forms a firm bond between the tube and the transmitting tract behind the growing tip (Lord and Sanders 1992).

3.2 EVIDENCE FOR CYTOSKELETON-CELL WALL INTERACTIONS TO ESTABLISH POLARITY AND GRAVISENSING

A good example of induced polarity in plants is seen in the site of rhizoid emergence in fertilized eggs of the brown algae *Pelvetia* or *Fucus*. The emergence is polarized by several environmental cues and is "fixed" several hours before protrusion of the rhizoid is detected (Jaffe 1968). The cell wall is necessary for this polar-axis fixation to occur, and transmembrane bridges that connect the cytoskeleton to the cell wall are implicated to stabilize membrane components of this axis fixation (Kropf et al. 1989). Interestingly, a localized adhesion of the membrane to the cell wall at the rhizoid tip is observed after plasmolysis of the embryo. A similar phenomenon is seen in the early development of the fern *Adiantum capillus-veneris* (Kagawa et al. 1992). Adhesion of the cell wall and plasma membrane is seen as a ring-like band in the growing tip cells. This band coincides with a circular array of both microtubules and microfilaments.

The green alga, *Chara*, exhibits an unusual unilateral constriction of the cytoplasm in the elongated internodal cells in response to changes in gravity. This constriction results in approximately a 10% increase in the downward flow rate of organelles along their actin cables (Wayne et al. 1992). As in the polar-fixed axis of *Fucus*, local adhesion of the membrane to the end walls can be seen after plasmolysis, indicating attachment of the cytoskeleton to the wall via transmembrane linkages.

4. Proteins of the cytoskeleton-cell wall continuum

4.1 SOME POSSIBLE MATRIX PROTEINS INVOLVED IN CELL WALL-CYTOSKELETON INTERACTIONS

In each example of a developmentally regulated or stress-induced interaction of the cytoskeleton with the cell wall, the interaction is associated with the appearance of animal-like matrix proteins. Affinity-purified polyclonal antisera to human Vn strongly recognize a *Fucus* protein of 65 kDa. The *Fucus* Vn also binds to heparin and fucoidin sulfate, a native *Fucus* cell wall polymer similar to heparin. The *Fucus* Vn extends beyond the cell wall of the rhizoid tip where it may function in adhesion of the rhizoid to the substratum (Wagner et al. 1992). The proteins involved in the gravitropic sensing and the adhesion of the plasma membrane to the end walls of *Chara* have not been elucidated, but an RGD-containing tetrapeptide interferes with gravity perception (Wayne et al. 1992). Involvement of a cytoskeleton-cell wall interaction in gravisensing by higher plants has not been examined in this respect.

Using an *in vitro* adhesion test, Ding et al. (1994) found that antibodies against the mammalian matrix adhesion molecules Vn, Fn, collagen, and laminin, as well as antisera against IgG, inhibit the attachment and subsequent colonization of soybean cells by the fungal pathogen *Phytophthora*. They identified a 65 kDa glycoprotein, common to both the plant and fungus, that is recognized by purified rabbit IgG, and the IgG alone is able to block adhesion and infection of plant cells by fungal cells. Wagner and Matthysse (1992) have found that both human Vn and anti-Vn antibodies inhibit the binding of

Agrobacterium tumefaciens to carrot cells. They suggest that *A. tumefaciens* recognizes a Vn-like protein on the plant cell surface as the receptor for its initial attachment to host cells.

Sanders et al. (1991) showed that affinity-purified rabbit antisera against human Vn recognize a 55 kDa polypeptide in all tissues of several higher plant species, and this protein is particularly enriched on the surfaces of the stylar transmitting tract. More recently, a 41 kDa protein is recognized by a different antibody preparation against human Vn (Wang et al. 1994). Both the 55- and 41-kDa protein have been purified by 2-dimensional electrophoresis, and the amino acid composition of both proteins is similar to that of human Vn. Attempts to sequence either protein or to obtain a cDNA by a variety of methods have so far been unsuccessful. The only plant protein known to have any significant homology with Vn is a pea seed albumin with its four hemopexin-like repeats (Jenne 1991).

Protoplasts isolated from these adapted cells remain adhesive and adhere to one another, whereas those isolated from unadapted cells do not. The most effective inhibitor of protoplast adhesion is an RGD peptide (Zhu et al. 1993). A 55 kDa polypeptide recognized by antisera against human Vn is enriched in cell walls and membranes of both adapted and unadapted cells, whereas a 59 kDa Fn-like protein, localized by antisera against human Fn, is found only in the walls of the adapted cells (Zhu et al. 1993). However, screening of an expression library with antibodies prepared against the Vn-like protein resulted in the selection of clones homologous to EF-1α (Zhu et al. 1994), a cytosolic protein unlikely to be involved in membrane-wall adhesion. Wang et al. (1994) have demonstrated that the amino acid compositions of the two Vn-like proteins they purified vary widely from that of EF-1α. The tobacco EF-1α gene selected by the putative Vn-specific antisera undoubtedly encodes a cytosolic protein, whereas the antisera also recognize an epitope in the cell wall (Zhu et al. 1994), and that protein may be the true Vn-like protein. The identification of a higher plant Vn protein or gene has been unsuccessful.

4.2 A FIBRONECTIN-BINDING PROTEIN IN ONION IS A CELL-WALL PROTEIN

In addition to proteins that cross react with antisera to human Fn and Vn, a 47 kD, and possibly a 40 kD protein, that specifically bind human Fn *in vitro* have been found in onion bulb tissue (S. Wyatt et al., unpublished results). The proteins were identified by SDS-PAGE and Western analysis by incubating the nitrocellulose membrane with human Fn, then antisera to Fn. Cell fractionation has shown the proteins to be localized in the cell wall and solubilized only by boiling in SDS or by extraction with $CaCl_2$.

4.3 TRANSMEMBRANE PROTEINS INVOLVED IN ATTACHMENT OF ACTIN TO THE PLASMA MEMBRANE

Antisera against the β-1 subunit of integrin and chicken vinculin faintly recognize *Fucus* 92 kDa and 116 kDa polypeptides, respectively, indicating that the alga may possess an actin binding protein and an associated transmembrane receptor, two necessary components of an animal-like focal adhesion (Quatrano et al. 1991). While there are now several confirmations of proteins antigenically related to Vn, there is only one report of an integrin-like protein in plants, a 72 kDa polypeptide purified by an RGD-affinity matrix and detected by antisera against human β-1 integrin (Schindler et al. 1989). Attempts to prepare antisera against the plant integrin and to clone its gene by heterologous probes have been unsuccessful. Meanwhile, M. Schindler (personal communication) has uncovered another possible integrin candidate, a 95 kDa protein deduced from an *Arabidopsis* cDNA with significant homology to animal β-1 integrin subunit genes. Using anti-integrin antisera, we were unable to identify integrin-like subunits from purified plasma membrane of the adapted tobacco cells.

In the absence of a *bona fide* integrin-like protein in plants, other transmembrane linking proteins unique to plants must be considered. One potential candidate was deduced from a cDNA clone of wheat, called *wir1*. The *wir1* transcript appears after inoculation of wheat with the non-pathogen *Erysiphe graminis* f. sp. *hordei* and is thought to encode one of several proteins involved in a localized induced resistance to *E. grammis f. sp. tritici* (Dudler et al. 1991). The *wir1* product is predicted to have a membrane-spanning domain of 5 kD and an exocellular domain of about 6 kDa (Dudler et al. 1991). The hydrophilic exocellular domain has several Pro residues, which would interrupt globule-forming α-helices, and several Arg residues that may bind to negatively charged molecules in the wall. Antisera directed against a synthetic peptide corresponding to a portion of the exocellular domain recognize an 11 kDa polypeptide that appears after inoculation with spores of the non-pathogenic fungus and is extracted from cell wall with 1 M NaCl.

4.4 OTHER POSSIBLE COMPONENTS OF THE ADHESION COMPLEX IN HIGHER PLANTS

Several proteins are unique to the plant cell wall. Diverse families of hydroxyproline- rich glycoproteins (HRGP), glycine-rich proteins (GRP), and proline-rich glycoproteins have been described, but a precise function has been ascribed to only a few of them (Showalter 1993). Pont-Lezica et al. (1993) proposed these well-established wall glycoproteins to be candidates for the wall-membrane linkers. Polyclonal antibodies prepared against wall HRGP bind not only to onion walls but also bind to the plasma membrane after the wall is enzymatically digested. The distribution of the HRGP is punctate. The points tend to be scattered more or less uniformly on the plasma membrane but can cluster at termini of large streaming strands. These streaming strands appear to exert tension on the membrane.

Some GRPs may also have a role in adhesion. The French bean GRP 1.8 has been localized in the cell walls of developing primary and secondary xylem vessels (Keller *et al.* 1989; Ye and Varner 1991), but the location of the other GRPs is uncertain. Using models that predict membrane protein topology (von Heijne and Gavel 1988; Hartmann et al. 1989), we found that the sequences flanking the putative signal peptides of some GRPs (deOliviera et al. 1990; Lei and Wu 1991) predict an interior, rather than an exterior, location at the plasma membrane (Wyatt and Carpita 1993).

5. Identification of new proteins of the cytoskeleton-cell wall continuum

There is good visual evidence in plant cells for cytoskeletal connections to the plasma membrane, for focal adhesions, and for localized membrane-cell wall contacts, but molecules homologous to animal cells have been elusive. "Antigenically related" reflects neither gene homology nor protein structure and function. Not a single matrix protein, transmembrane linker, or cytoskeletal attachment protein homologous to the animal cytoskeleton-ECM continuum has been demonstrated in higher plants. Stossel (1993) has illustrated several actin-plasma membrane interactions that may function in cell migration. Evidence for a few of these proteins has been found in plants. There is good evidence for annexin proteins which function in actin-membrane attachment (Blackbourn et al. 1992; Smallwood et al.

1990). Connexin proteins, which function in animal gap junctions, have been localized in association with plant plasmodesmata (Meiners and Schindler 1989; Meiners et al. 1991). Spectrin-like membrane skeletal proteins have been detected with antisera against the animal protein (deRuijter and Emons 1993; Faraday and Spanswick 1992). The challenge for plant biologists is to purify these proteins and to clone their genes. Experiments can then be devised to localize these proteins and demonstrate their functions. The plant cytoskeleton, plasma membrane, and cell wall also have unique structural elements. Only by devising methodologies to preserve and purify these elements will scientists be able to fully characterize the cell wall - cytoskeleton continuum in plants.

6. Acknowledgements

Our work is supported by BARD, the Bi-National Israel-U.S. Agriculture Research and Development Administration, the U.S. Department of Agriculture, and McKnight Foundation fellowships to S.E.W and A.A.A. Journal paper No. 14,259 of the Purdue University Agriculture Experiment Station.

7. References

Bissell, M. J., Hall, H. G., and Parry, G. (1982) 'How does the extracellular matrix direct gene expression?', J. Theor. Biol. 99, 31-68.

Blackbourn, H.D. Barker, P.J. Huskisson, N.S., and Battey, N.H. (1992) 'Properties and partial protein sequence of plant annexins', Plant Physiol. 99, 864-871.

Bronner-Fraser, M. (1985) 'Effects of different fragments of the fibronectin molecule on latex bead translocation along neural crest migratory pathways', Devel. Biol. 108, 131-145.

Burridge, K., Fath, K., Kelly, T., Nuckolls, G., and Turner, C. (1988) 'Focal adhesions: transmembrane junctions between the extracellular matrix and the cytoskeleton', Ann. Rev. Cell Biol. 4, 487-525.

Carpita , N.C., and Gibeaut, D.M. (1993) 'Structural models of primary cell walls in flowering plants: consistency of molecular structure with the physical properties of the walls during growth', Plant J. 3, 1-30.

deOliveira, D. E., Seurinck, J., Inzé, D., Van Montague, M., and Botterman, J. (1990) 'Differential expression of five *Arabidopsis* genes encoding glycine-rich proteins', Plant Cell 2, 427-436.

deRuijiter, N.C.A., and Emmons, A.M.C. (1993) 'Immunodetection of spectrin antigens in plant cells', Cell Biol. Internat. 17, 169-182.

Ding, H., Balsiger, S., Guggenbühl, and Hohl, H. R. (1994) A putative IgG-binding 65 kDa adhesin involved in adhesion and infection of soybeans by *Phytophthera megasperma* f. sp. *glycinea,* Physiol. Molec. Plant Pathol., in press

Dudler, R., Hertig, C., Rebmann, G., Bull, J., and Mauch, F. (1991) 'Sequence and expression of a wheat gene that encodes a novel protein associated with pathogen defense', Mol. Plant Microbe Int. 5, 516-519.

Edelman, G, M. (1983) 'Cell adhesion molecules', Science 219, 450-457.

Faraday, C.D., and Spanswick, R.M. (1993) 'Evidence for a membrane skeleton in higher plants: A spectrin-like polypeptide co-isolates with rice root plasma membranes', FEBS Letters 318, 313-316.

Fukuda, H., and Kobayashi, H. (1989) 'Dynamic organization of the cytoskeleton during tracheary-element differentiation', Develop. Growth Differ. 31, 9-16.

Goddard, R.H., Wick, S.M., and Silflow, C.D. (1994) 'Microtubule components of the plant cell cytoskeleton', Plant Physiol. 104, 1-6.

Gomez-Lepe, B.E. Lee-Stadelmann. O.Y., Palta, J.P., and Stadelmann, E.J. (1979) 'Effects of octylguanidine on cell permeability and other protoplasmic properties of *Allium cepa* epidermal cells', Plant Physiol. 64, 131-138.

Goodbody, D.C. and Lloyd, C.W. (1990) 'Actin filaments line up across *Tradescantia* epidermal cells, anticipating wound-induced division planes', Protoplasma 157, 92-101.

Guy, C. L. (1990) 'Cold acclimation and freezing tolerance: role of protein metabolism', Annu. Rev. Plant Physiol. Plant Molec. Biol. 41, 187-223.

Hartmann, E., Rapoport, T. A., and Lodish, H. F. (1989) 'Predicting the orientation of eukaryotic membrane-spanning proteins', Proc. Natl. Acad. Sci. USA 86, 5786-5790.

Hohl, H.R. and Balsiger, S. (1986) 'Probing the surfaces of soybean protoplasts and of germ tubes fo the soybean pathogen *Phytophthora megasperma* f. sp. *glycinea* with lectins', Botanica Helvetica 96, 289-297.

Hynes, R. O. (1987) 'Integrins: a family of cell surface receptors', Cell 48, 549-554.

Jaffe, L. (1968) 'Localization in the developing *Fucus* egg and the general role of localizing currents", Adv. Morphogen. 7, 295-328.

Jenne, D. (1991) 'Homology of placental protein 11 and pea seed albumin 2 with vitronectin', Biochem. Biophys. Res. Comm. 176, 1000-1006.

Johnson-Flanagan, A.M., and Singh, J. (1986) 'Membrane deletion during plasmolysis in hardened and non-hardened plant cells', *Plant Cell and Environment* 9, 199-305.

Kagawa, T., Kadota, A., and Wada, M. (1992) 'The junction between the plasma membrane and the cell wall in fern protonemal cells, as visualized after plasmolysis, and its dependence on arrays of cortical microtubules', Protoplasma 170, 186-190.

Katsuta, J., and Shibaoka, H. (1988) 'The roles of the cytoskeleton and the cell wall in nuclear positioning in tobacco BT-2 cells', Plant Cell Physiol. 29, 403-413.

Keller, B., Templeton, M. D., and Lamb, C. J. (1989) 'Specific localization of a plant cell wall glycine-rich protein in protoxylem cells of the vascular system', Proc. Natl. Acad. Sci. USA 86, 1529-1533.

Kropf, D. L., Berge, S. K., and Quatrano, R. S. (1989) 'Actin localization during *Fucus* embryogenesis', Plant Cell 1,191-200.

Lee-Stadelmann, O. Y., Bushnell, W. R., and Stadelmann, E. J. (1984) 'Changes of plasmolysis form in epidermal cells of *Hordeum vulgare* infected by *Erysiphe graminis'*, Can. J. Bot. 62, 1714-1723.

Lei, M., and Wu, R. (1991) 'A novel glycine-rich cell wall protein gene in rice', Plant Mol. Biol., 16, 187-198.

Lord, E.M., and Sanders, L.C. (1992) 'Roles for the extracellular matrix in plant development and pollination: a special case of cell movement in plants', Devel. Biol. 153, 16-28.

McCann, M.C., and Roberts, K. (1991) 'Architecture of the primary cell wall', in C. W. Lloyd (ed.), Cytoskeletal Basis of Plant Growth and Form, Academic Press, London, pp. 109-129.

Meiners, S., and Schindler, M. (1989) 'Characterization of a connexin homologue in cultured soybean cells and diverse plant organs', Planta 179, 148-155.

Meiners, S., Xu, A., and Schindler, M. (1991) 'Gap junction protein homologue from *Arabidopsis thaliana*: evidence for connexins in plants', Proc. Natl. Acad. Sci. USA 88, 4119-4122.

Meyer, Y., and Herth, W. (1982) 'Interaction of cell-wall formation and cell division in higher plant cells', in R.M. Jr. Brown (ed.), Cellulose and Other Natural Polymer Systems, Plenum Press, New York, pp. 149-165.

Odermatt, M., Rothlisberger, A., Werner, C., and Hohl, H. R. (1988) 'Interaction between agarose-embedded plant protoplasts and germ tubes of *Phytophthora*', Physiol. Mol. Plant Pathol. 33, 209-220.

Pierson, E.S. Derksen, J., and Traas, J.A. (1986) 'Organization of microfilaments and microtubules in pollen tubes grown *in vitro* or *in vivo* in various angiosperms', Eur. J. Cell Biol. 41, 14-18.

Pont-Lezica, R.F., McNally, J.G., and Pickard, B.G. (1993) 'Wall-to-membrane linkers in onion epidermis: some hypotheses', Plant Cell Environ. 16, 111-123.

Quatrano, R. S., Brian, L., Aldridge, J., and Schultz, T. (1991) 'Polar axis fixation is *Fucus* zygotes: components of the cytoskeleton and extracellular matrix', Development, Suppl. 1, 11-16.

Roberts, K. (1990) 'Structures at the plant cell surface', Curr. Opinion Cell Biol. 2, 920-928.

Roberts, A.W., and Haigler, C.H. (1989) 'Rise in chlorotetracycline fluorescence accompanies tracheary element differentiation in suspension cultures of *Zinnia*', Protoplasma 152, 37-45.

Ruoslahti, E. (1988) 'Fibronectin and its receptors', Annu. Rev. Biochem. 57, 375-413.

Ruoslahti, E., and Pierschbacher, M. D. (1986) 'Arg-gly-asp: a versatile cell recognition signal', Cell 44, 517-518.

Sanders, L. C., and Lord, E. M. (1989) 'Directed movement of latex particles in the gynoecia of three species of flowering plants', Science 243, 1606-1608.

Sanders, L. C., Wang, C-S., Walling, L. L., and Lord, E. M. (1991) 'A homolog of the substrate adhesion molecule vitronectin occurs in four species of flowering plants', Plant Cell 3, 629-635.

Schneider, B. and Herth, W. (1986) 'Distribution of plasma membrane rosettes and kinetics of cellulose formation in xylem development of higher plants', Protoplasma 131, 142-152.

Schindler, M., Meiners, S., and Cheresh, D. A. (1989) 'RGD-dependent linkage between plant cell wall and plasma membrane: consequences for growth', J. Cell Biol. 108, 1955-1965.

Showalter, A.M. (1993) 'Structure and function of plant cell wall proteins', Plant Cell 5, 9-23.

Smallwood, M., Keen J.M., and Bowles, D.J. (1990) 'Purification and partial sequence analysis of plant annexins', Biochem. J. 270, 157-161.

Sonesson, A. and Widell, S. (1993) 'Cytoskeleton components of inside-out and right-side-out plasma membrane vesicles from plants', Protoplasma 177, 45-52.

Stadelmann, E. (1956) 'Plasmolyse und deplasmolyse', in W. Ruhland (ed.) Handbuch der Pflanzenphysiologie, Springer-Verlag, Berlin, pp. 71-115.
Stossel, T.P. (1993) 'On the crawling of animal cells', Science 260, 1086-1093.
Underwood, P. A., and Bennett, F. A. (1989) 'A comparison of the biological activities of the cell-adhesive proteins vitronectin and fibronectin', J. Cell Sci. 93, 641-649.
von Heijne, G., and Gavel, Y. (1988) 'Topogenic signals in integral membrane proteins', Eur. J. Biochem. 174, 671-678.
Wagner, V., Brian, L., and Quatrano, R. (1992) 'Role of vitronectin-like molecule in embryo adhesion of the brown alga *Fucus*', Proc. Natl. Acad. Sci. USA 90, 3644-3648.
Wagner, V.T., and Matthysse, A.G. (1992) 'Involvement of a vitronectin-like protein in attachment of *Agrobacterium tumefaciens* to carrot suspension culture cells', J. Bacteriol. 174, 5999-6003.
Wang, C.-S., Walling, L.L., Gu, Y.Q., Ware, C.F., and Lord, E.M. (1994) 'Two classes of proteins and mRNAs in *Lilium longiflorum* L. identified by human vitronectin probes', Plant Physiol. 104, 711-717.
Wayne, R., Staves, M. P., and Leopold, A. C. (1992) 'The contribution of the extracellular matrix to gravisensing in characean cells', J. Cell Sci. 101, 611-623.
Wyatt, S. E., and Carpita, N. C. (1993) 'The plant cytoskeleton-cell wall continuum', Trends Cell Biol. 3, 413-417.
Ye, Z-H., and Varner, J. E. (1991) 'Tissue-specific expression of cell wall proteins in developing soybean tissues', Plant Cell 3, 23-37.
Zhu, J.-K., Damsz, B., Kononowicz, A.K., Bressan, R.A., and Hasegawa, P.M. (1994) 'A higher plant extracellular vitronectin-like adhesion protein is related to the translational elongation factor-1α', Plant Cell 6, 393-404.
Zhu, J-K., Shi, J., Singh, U., Wyatt, S.E., Bressan, R. A., Hasegawa, P. M., and Carpita, N. C. (1993) 'Enrichment of vitronectin- and fibronectin-like proteins in NaCl-adapted plant cells and evidence for their involvement in plasma membrane-cell wall adhesion', Plant J. 3, 637-646.

MOLECULAR AND CELLULAR EVENTS IN THE FORMATION OF NEW MERISTEMS

I. M. SUSSEX, J. A. GODOY, N. M. KERK, M. J. LASKOWSKI, H. C. NUSBAUM, J. A. WELSCH and M. E. WILLIAMS
Department of Plant Biology
University of California
Berkeley, CA 94720, USA

ABSTRACT. We have developed an efficient system in which to analyse cellular and molecular events during the initiation and organization of lateral root meristems as a way of obtaining markers for regeneration from cell and tissue cultures. From subtracted cDNA libraries we isolated approximately 40 genes whose expression patterns are altered during these processes. We determined spatial and temporal patterns of expression by Northern, *in situ* and GUS assays, and we have identified 2 mutants in lateral root initiation.

1. Introduction

A conventional approach to obtaining regeneration *in vitro* by organogenesis or embryogenesis has been to vary components of the culture system, either medium or environment, until regeneration is observed. The basis of this approach was the discovery by Skoog and Miller (1957) that the auxin to cytokinin ratio in the culture medium was crucial for shoot or root regeneration in tobacco tissue explants. Despite the fact that regeneration of many species has been achieved using this approach few other generalizations concerning the requirements for regeneration have emerged from these studies. Thus, successful regeneration of one species has not generally predicted the conditions required for regeneration of other related species, or even of other cultivars of the same species. Furthermore, there still remain a large number of so-called "recalcitrant species" that have not yielded to this or any other approach.

An alternative strategy is to identify molecular markers of meristem or embryo initiation and early development, and to monitor the expression of these in the tissue culture as the system is manipulated to achieve regeneration. We have adopted this approach by identifying genes that are expressed preferentially in newly organizing root meristems. Not only should this approach provide molecular markers for regeneration, but it should also contribute to an understanding of the molecular organization of plant meristems in general.

2. The Experimental System

Our requirements for an optimal experimental system on which to carry out this study were that a large number of meristems be available for analysis, that the development of these meristems be relatively synchronous, and that initiation and organization of meristems be rapid. After examining several regeneration systems, including the tobacco thin cell layer system (Tran Thanh Van et al., 1974) which did not meet any of the above criteria, we focused attention on indoleacetic acid (IAA)-induced lateral root formation in radish (Blakely et al., 1982). Since radish is taxonomically related to *Arabidopsis* information should be readily transferable between them.

M. Terzi et al. (eds.), Current Issues in Plant Molecular and Cellular Biology, 31-35.

Radish seed (*Raphanus sativus* cv Scarlet Globe) was sterilized in Clorox, water washed, and planted on germination paper separated by pleated aluminum sheets for aeration. Sheets were held vertical in sterile plastic dishpans and were moistened with MS medium or water. After 3 days in the dark, primary roots were approximately 6 cm long. Culture medium was replaced with MS + IAA at various concentrations and roots were collected at specified times for analysis. Alternatively, root segments 1 cm long were cut 0.5 cm behind the tip and incubated in MS + IAA.

3. Development of Lateral Roots

Lateral roots (LR) in radish are initiated in the pericycle, the outermost cell layer of the vascular system of the root. Pericycle cells are vacuolated, axially elongated postmitotic cells arrested in the G2 phase of the cell cycle (Blakely and Evans, 1979). 5-10 hours after the start of IAA treatment transverse divisions occur in pericycle cells that lie on the two protoxylem radii of the vascular cylinder. Transverse divisions soon occur in pericycle cells on either side of the protoxylem radius, subdividing them into isodiametric cells. In each vertical file of pericycle cells divisions occur in 1-3 cells so that for the initiation of each LR there are approximately 20 founder cells (as defined by Poethig, 1987). By 15 hours after the start of IAA treatment lateral expansion has commenced in the founder cell derivatives, which then undergo the first of a series of periclinal divisions. By 24 hours the LR primordium consists of a bulge of approximately isodiametric, cytologically uniform cells, but by 72 hours the LR has an organized meristem in which a quiescent center, a proximal meristem and a one cell layer thick root cap can be identified. In untreated control roots LRs are initiated in acropetal sequence behind the primary root tip. In IAA treated roots and root segments LRs are initiated and develop synchronously from numerous sites along the root. LR spacing is sensitive to auxin concentration and duration of application (Kerk, 1990).

4. Identification and Temporal Expression of Genes in Organizing LR Meristems

Three subtracted cDNA libraries of radish were constructed to enrich for genes that were expressed at specific stages (times) of LR meristem development. These were: a) a 4 hr library, from mRNA collected from roots 4 hours after the start of auxin treatment, and enriched in genes expressed before the first division in pericycle founder cells by subtraction against untreated mature root cells; b) a 24 hr library, from mRNA collected from roots 24 hours after the start of auxin treatment, and enriched in genes expressed before regional differentiation in the LR tip is evident by subtraction against roots in treatment a) and untreated mature root tissue; c) a primary root tip library, from mRNA collected from fully developed root tips and subtracted against mature root tissue.

From these libraries we have cloned and sequenced cDNAs that represent 51 genes. cDNAs were used as probes to investigate temporal patterns of expression by Northern blot analysis, and spatial expression patterns by *in situ* hybridization.

The most abundant class of cDNAs that we have identified represent ribosomal protein (RP) genes. To date we have identified 14 different RP genes. Protein products of 10 of these are associated with the large ribosomal subunit and 4 with the small subunit. Since there are 60-80 RPs in cytosolic ribosomes (Mager, 1988), at a first approximation this suggests that we have identified approximately 20% of the sequences present in the subtracted library if we assume that at least one member c each RP gene family is transcriptionally active in IAA treated roots. Other cDNAs thε have been identified correspond to cellular functions required for continued cell growt (ribosomal RNA synthesis, mRNA translation, and tRNA synthesis), as well as cDNAs that correspond to other metabolic enzymes. About 10 sequences have no similarity to sequences in any data bank that we have searched.

mRNA from roots exposed to auxin for 0, 4, 24, 48, and 96 hours, and from primary root tips was slot blotted and probed with each of the cDNAs. All were expressed in primary root tips. Expression was low in the 0 hr roots, increased in expression to maxima at 24 or 48 hr and generally declined after that time. However, one clone continued to be expressed at a high level at 96 hr. *In situ* hybridization was carried out on primary root tip longitudinal sections using clones representative of the various classes of genes that had been identified. Four major expression patterns were identified: a) mRNA was expressed in all cells of the root tip including root cap, all cells of the terminal meristem, and differentiating cells of the primary root; b) mRNA was expressed uniformly in the root tip but was absent from the root cap; c) mRNA was expressed uniformly in the root tip but was absent from the root cap and the quiescent center; d) mRNA was expressed only in distal cells of the meristem and superficial cells of differentiating root tissue.

Recently we have used differential display (Liang and Pardee, 1992) to identify a gene in *Arabidopsis* roots that is upregulated as soon as 10-20 minutes after the start of auxin treatment. However, we have not yet determined whether this gene is expressed in pericycle cells.

5. Spatial Expression Patterns of mRNAs During LR Development in Radish

We examined the pattern of expression of 2 mRNAs during LR initiation and early development by *in situ* hybridization. A protein translation factor gene was expressed in pericycle LR founder cells prior to their entry into the cell cycle following auxin treatment, and at all subsequent stages examined. In early stages of LR formation when histological differentiation was not observable in the LR all cells of the LR were uniformly reactive to the probe. However, by 72 hours there was no reaction in cells of the LR nearest the vascular system of the primary root, and in the tip of the LR a quiescent center with diminished reaction was observed together with a root cap one cell thick that did not react. The expression pattern of a ribosomal protein gene was also examined by *in situ* hybridization. This mRNA was expressed in LRs maximally at 24 hours after the start of auxin treatment and subsequently declined so that by 48 hours expression level was very low.

6. Isolation and Expression of a Ribosomal Protein Gene from *Arabidopsis*

A fragment of the radish ribosomal protein cDNA described in the previous section was used to isolate corresponding cDNAs from an *Arabidopsis* library and one of these was used to isolate corresponding genomic sequences. We identified 2 members of a gene family. Each gene contained 3 introns whose positions were precisely conserved. The predicted protein sequences of the 2 genes differed by 3 amino acids. These genes showed considerable protein sequence similarity to comparable RP genes of *Dictyostelium* and yeast, and the *Arabidopsis* and radish sequences were highly conserved.

To study spatial and temporal expression of these genes during development of *Arabidopsis*, sequences 5' to the coding region containing the presumed promoter of each gene were subcloned into a plasmid containing the GUS coding sequence and transformed into *Arabidopsis* according to Clarke et al., 1992. T1 and T2 progeny were assayed for GUS activity at various stages of the life history. One of the genes of this family is expressed at a high level in meristematic tissues of the shoot and the root including the terminal meristems, leaf primordia, expanding cotyledons, flowers, and vascular tissue. Auxin treatment of seedlings resulted in increased GUS expression in xylem radius pericycle cells and in cells of the developing LR. The second gene in this family has a much more restricted expression pattern. In 3 day old seedlings it is expressed in only a few xylem radius pericycle cells just behind the root tip. Auxin treatment of seedlings results in GUS activity in dividing pericycle cells, cells of the forming LR, and adjacent cells in the vascular tissue of the primary root. Expression of this gene in the shoot was limited to flowers starting at stage 8 (Smyth et al., 1990). Expression occurred first in cells of the tapetum and later in pollen. No other cells in the flower reacted positively.

7. *Arabidopsis* Mutants in LR Initiation

To identify components of a genetic pathway for LR development we chemically mutagenized seed and screened M2 seedling families for LR mutant phenotypes. We have identified two recessive mutations that appear to be in this class. We have tentatively mapped them to chromosomal location as a first step in cloning the genes.

One of the mutations results in alteration of LR patterning in both the longitudinal and radial dimension. The mutant produces a larger number of LRs per unit primary root length than does the wild type in response to IAA treatment, and LRs appear not to be restricted to the xylem radius of the primary root. The second mutation results in failure to initiate LR primordia, and the terminal meristem of the primary root is not maintained as a functional and structurally distinct zone beyond 3-4 days after germination.

8. Discussion and Conclusions

We have developed a system in which large numbers of genes whose expression patterns are altered during early stages of development of the terminal meristem for a lateral root can be identified and isolated for further study. So far none of the genes whose expression patterns we have identified by Northern, *in situ*, or GUS analysis can be described as "meristem specific". That is, all show some level of expression in parts of the plant outside of the terminal meristems. This fact seems to be an emerging theme in studies of gene expression because other genes that have been described as meristem specific in their expression are also expressed in other parts of the plant (Medford, 1992, Fleming et al., 1993, Jackson et al., 1994). Nonetheless, the present results indicate that there are distinctive patterns of gene expression that characterize early stages of meristem formation. Further analysis of these patterns could form the basis for designing a rational strategy for regeneration in plant tissue and cell culture studies.

9. References

Blakely, L. M.,Durham, M., Evans, T. A., and Blakely, R. (1982) Experimental studies on lateral root formation in radish seedling roots. I. General methods, developmental stages, and spontaneous formation of laterals. Bot. Gaz. 143, 341-352.

Blakely, L. M. and Evans, T. A. (1979) Cell dynamics studies on the pericycle of radish seedling roots. Plant Sci. Lett. 14, 79-83.

Clarke, M. C., Wei, W. and Lindsey, K. (1992) High-frequency transformation of *Arabidopsis thaliana* by *Agrobacterium tumefaciens*. Plant Mol. Biol. Rep. 10, 178-189.

Fleming, A. J., Mandel, T., Roth, I. and Kuhlemeier, C. (1993) The patterns of gene expression in tomato shoot apical meristem. Plant Cell 5, 297-309.

Jackson, D., Veit, B. and Hake, S. (1994) Expression of maize *KNOTTED1* related homeobox genes in the shoot apical meristem predicts patterns of morphogenesis in the vegetative shoot. Devel. 120, 405-413.

Kerk, N. M. (1990) Gene expression during meristem initiation and organization. Ph. D. thesis, Yale University, New Haven, CT, USA.

Mager, W. H. (1988) Control of ribosomal protein gene expression. Biochem. Biophys. Acta 949, 1-15.

Medford, J. I. (1992) Vegetative apical meristems. Plant Cell 4, 1029-1039.

Liang, P. and Pardee, A. B. (1992) Differential display of eukaryotic messenger RNA by means of the polymerase chain reaction. Science 257, 967-971.

Poethig, R. S. (1987) Clonal analysis of cell lineage patterns in plant development. Amer. J. Bot. 74, 581-594.

Smyth, D. R. Bowman, J. L. and Meyerowitz, E. M. (1990) Early flower development in *Arabidopsis*. Plant Cell 2, 755-767.

Tran Thanh Van, K. M., Nguyen, T. D. and Chlyah, A. (1974) Regulation of organogenesis in small explants of superficial tissues of *Nicotiana tabacum* L. Planta 119, 149-159.

Skoog, F. and Miller, C. O. (1957) Chemical regulation of growth and organ formation in plant tissues cultured *in vitro*. Symp. Soc. Exper. Biol. 11, 118-131.

OLIGOSACCHARIDE ELICITORS AND ELICITOR RECEPTORS

MICHAEL G. HAHN
The University of Georgia
Complex Carbohydrate Research Center
and Department of Botany
220 Riverbend Road
Athens, GA 30602-4712
USA

ABSTRACT. Oligosaccharide elicitors capable of inducing one or more plant defense responses have been prepared from plant (homogalacturonan) and fungal (β-glucan, chitin, chitosan) cell wall polysaccharides, and fungal glycoproteins. An overview of the structures and activities of these elicitors will be presented. In addition, recent biochemical investigations of the cellular signaling pathway(s) triggered by these signal molecules will be reviewed. The latter will focus on our studies of the signaling pathway leading to the biosynthesis and accumulation of phytoalexins in soybean tissues. This pathway is triggered by nanomolar concentrations of a branched hepta-β-glucoside elicitor originating from mycelial walls. Current research is focused on the first step in the signaling pathway, the recognition of the elicitor by a specific receptor. A radio-labeled derivative of the elicitor has been prepared and used to demonstrate the presence of specific, high-affinity binding protein(s) (EBPs) (putative receptors) for the elicitor in soybean root plasma membranes. The EBPs are solubilized from the soybean root membranes using non-ionic detergents with retention of their high affinity and specificity for the hepta-β-glucoside elicitor. Purification of the EBPs using affinity chromatography is in progress, and current results suggest that the EBPs exist as a multimeric protein complex. The EBPs recognize the same structural elements of the hepta-β-glucoside elicitor that are essential for its phytoalexin-inducing activity, suggesting that the EBPs are physiological elicitor receptors.

1. Introduction

Plants utilize a multi-faceted array of defense responses when confronted by invasive microorganisms [18,67]. These defense responses include the synthesis and accumulation of anti-microbial phytoalexins [49,52], the production of glycosylhydrolases capable of attacking surface polymers of pathogens [22], the synthesis of proteins that inhibit degradative enzymes produced by pathogens [67,117], and the modification of plant cell walls by deposition of callose [1,75], hydroxyproline-rich glycoproteins [125] and/or lignin [112].

Biochemical analysis of the induction of plant defense responses has been facilitated by the recognition that cell-free extracts of microbial and plant origin are capable of inducing defense responses when applied to plant tissues. The active components in the extracts are commonly referred to as "elicitors." The term "elicitor" was originally

M. Terzi et al. (eds.), Current Issues in Plant Molecular and Cellular Biology, 37-58.

used to refer to molecules and other stimuli that induce the synthesis and accumulation of antimicrobial compounds (phytoalexins) in plant cells [77], but is now commonly used for molecules that stimulate any plant defense mechanism [48,51,52,64]. A number of different biotic elicitors, including oligosaccharides isolated from fungal and plant cell walls, which are the focus of this overview, (glyco)proteins [5,103,111], and lipids [23], as well as abiotic elicitors (such as heavy metal salts or UV light) are known. Abiotic elicitors are thought to result in the release of biotic elicitors from the cell walls of the plant [10,69].

This article will provide a brief overview of the structures and activities of five oligosaccharide elicitors whose structures are shown in Figure 1. These elicitors are derived from either plant or fungal cell wall polysaccharides, or from fungal glycoproteins. The reader is referred to several recent reviews [40,53,118] for a more detailed discussion of oligosaccharide elicitors and their activities than space permits here. In addition, we will provide a brief overview of recent progress toward identifying and characterizing specific binding proteins (putative receptors) for each of these elicitors, highlighting our studies on hepta-β-glucoside elicitor-binding protein(s) in soybean plasma membranes.

2. Oligogalacturonide Elicitors

The presence of elicitor-active molecules in plants was first noted in experiments in which the placement of frozen-thawed bean stem segments in contact with healthy stem segments resulted in the accumulation of phytoalexins in the healthy stem segments [71]. Subsequently, elicitor-active material was observed in boiling water extracts of bean [72] or soybean tissue [65] and isolated soybean cell walls [69]. Elicitor-active molecules present in these extracts have been identified as linear oligomers of (1→4)-linked α-galactosyluronic acid residues [69,100] (structure shown in Figure 1).

2.1 PURIFICATION OF OLIGOGALACTURONIDE ELICITORS

Oligogalacturonide elicitors can be released from homogalacturonans present in plant primary cell walls either by partial acid hydrolysis of the walls [69,100] or by treatment of the walls with pectic-degrading enzymes such as endopolygalacturonase [74,114] or endopectate lyase [42,43]. A more readily available starting material for the preparation of oligogalacturonide elicitors is polygalacturonic acid (methyl de-esterified pectin) obtainable from commercial sources. The active oligogalacturonides released from either starting material by either chemical or enzymatic means appear to be similar if not identical [43,126].

The mixture of oligogalacturonides released from either plant cell walls or poly-galacturonic acid is very heterogeneous, containing oligomers ranging in size between a degree of polymerization (DP) of 2 and a DP of 20. The released oligogalacturonides are routinely separated by low-pressure anion-exchange chromatography [69,74,100, 114] into fractions enriched in a specific size of oligogalacturonide. Such size-fractionated oligomers have been used in many of the bioactivity studies published to date. However, recent evidence indicates that the oligomers, particularly those of DP > 10,

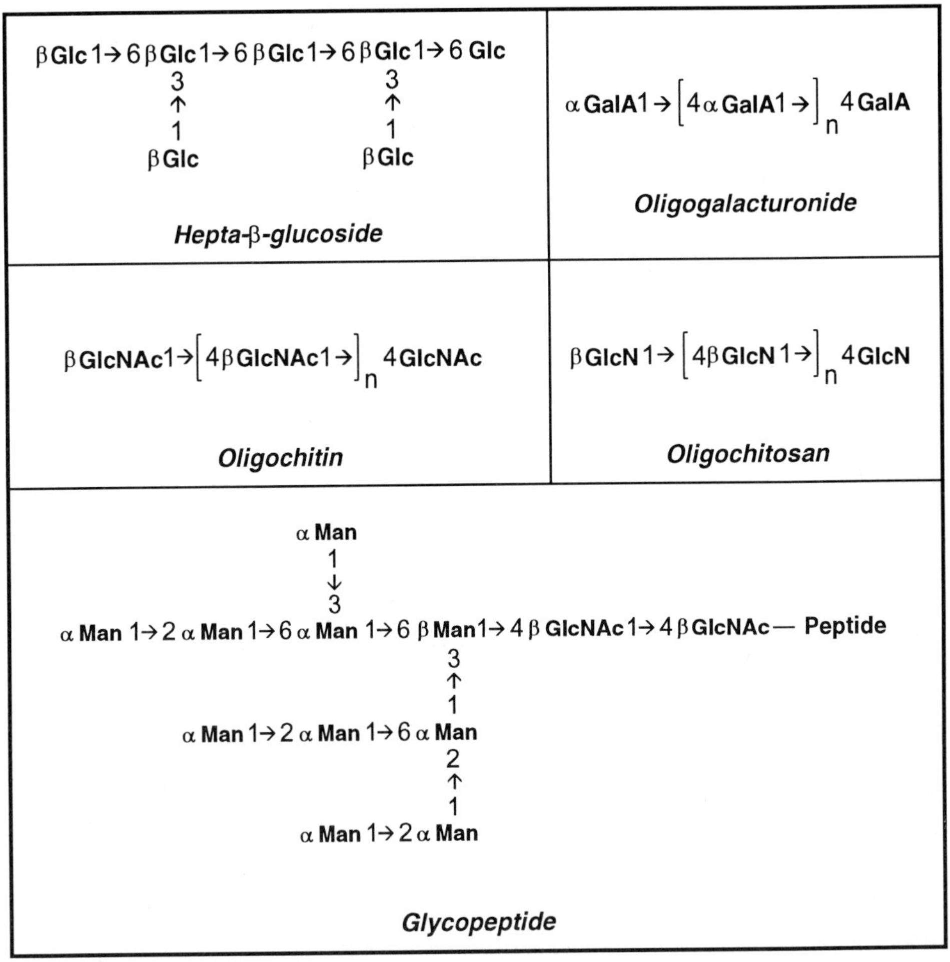

Figure 1. Structures of oligosaccharide elicitors. The oligogalacturonide elicitor is a fragment of a plant cell wall polysaccharide, the hepta-β-glucoside, oligochitin, and oligochitosan elicitors are fragments of mycelial wall polysaccharides, and the glycopeptide elicitor is derived from yeast invertase. The most active oligogalacturonides have a degree of polymerization between 11 and 14 ($n = 9$-12). A minimum size of at least a tetramer ($n = 2$) or a heptamer ($n = 5$) is required, respectively, for activity of the oligochitin and oligochitosan elicitors.

contain significant amounts (~30%) of modified oligogalacturonides [126]. These modified oligogalacturonides have been separated from the predominant homo-oligogalacturonide by high-performance anion exchange chromatography in combination with pulsed amperometric detection (HPAEC-PAD) [126]. Several of the modified oligogalacturonides have been shown to contain galactaric acid (the C-1 oxidized derivative of galacturonic acid); none of the galactaric acid-containing oligogalacturonides elicit phytoalexin accumulation in soybean tissue (A. Koller, J.-J. Cheong and M. O'Neill, unpublished results). The HPAEC-purified, apparently homogeneous tridecagalacturonide is an active elicitor [126]. These results emphasize the importance of obtaining homogeneous oligogalacturonides in order to avoid uncertainty in the attribution of biological activity to homo-oligogalacturonides.

2.2 ELICITATION OF PLANT DEFENSE RESPONSES BY OLIGOGALACTURONIDES

The nature of the plant defense response(s) induced by oligogalacturonides depends on the plant being studied. Thus, oligogalacturonides have been shown to induce the accumulation of phytoalexins in soybean [42,43,69,100], castor bean [74], bean [50,129], pea [135] and parsley [44]. Oligogalacturonides also induce the accumulation of anti-microbial shikonins in suspension-cultured *Lithospermum erythrorhizon* cells [128]. In soybean and parsley, oligogalacturonide and oligo-β-glucoside elicitors (see below) act synergistically, i.e. the concentrations required to elicit phytoalexins when both elicitors are present are less than the concentration required for each elicitor to elicit phytoalexins individually [41,44]. This synergistic effect has been observed with a synthetic dodecagalacturonide, but not with the decagalacturonide [68]. Other oligo-galacturonide-induced defense responses include the induction of glycosylhydrolases (β-glucanase, chitinase, lysozyme) in parsley [44] and tobacco [25], the increased deposition of lignin in cucumber [113] and castor bean [26], and the accumulation of proteinase inhibitors in tomato [21,55].

The size range of oligogalacturonides that activate defense responses is usually quite narrow. For example, oligogalacturonides with DP between 10 and 15 are generally required to elicit most of the plant defense responses listed above. Although the reason for the frequently observed size-dependence of the response to oligogalact-uronides is not known, this requirement suggests that oligogalacturonides need ten or more galactosyluronic acid residues in order to assume a solution conformation or oligomeric structure that is biologically active. Evidence that oligogalacturonides undergo a conformational transition at DP > 10 has been obtained from binding studies with a monoclonal antibody that preferentially binds oligogalacturonides of DP ≥ 10 [93]. However, di- and trigalacturonides induce proteinase inhibitor production in tomato [99], suggesting that in some plants there may be a different size requirement for biological activity.

It is thought that the oligogalacturonide elicitors are released during plant-pathogen interactions by the action of microbial pectic-degrading enzymes, in particular, endo-polygalacturonase and endopectate lyase. However, the action of these enzymes *in vitro* results in the rapid and total fragmentation of polygalacturonic acid to short, biologically inactive oligomers. Therefore, regulatory mechanisms must exist *in planta* in order for the action of pectic-degrading enzymes to result in the production of longer, biologically active oligogalacturonides. *In vitro* experiments using purified pectic-degrading enzymes have suggested two possible regulatory mechanisms; whether or not these mechanisms actually function at the host-pathogen interface awaits further experimentation. One proposed regulatory mechanism invokes the physiological pH of the plant cell wall (pH 5-6) as a means to regulate the activity of endopectate lyases, many of which are maximally active between pH 8 and 9 [16]. Thus, evidence has been presented that, at pH 5.75, the activity of a bacterial endopectate lyase is reduced to such an extent that the formation of biologically active oligogalacturonides of DP > 10 is favored [45].

The involvement of a polygalacturonase-inhibiting protein (PGIP), which has broad specificity against fungal endopolygalacturonases [2,28], has been proposed as another mechanism to favor the production of biologically active oligogalacturonides. PGIP can reduce the activity of fungal endopolygalacturonases *in vitro* by ~99.7% [27]; bacterial endopectate lyases and plant endopolygalacturonases are not affected by PGIP [28]. The reduced activity of the endopolygalacturonase results in the accumulation of biologically active oligogalacturonides with DP > 10 that have half-lives of hours rather than minutes [29]. The gene encoding bean PGIP has been cloned [131] and its expression has been studied [19]. Interestingly, PGIP mRNA levels increase after treatment with elicitors such as oligogalacturonides, oligoglucosides and salicylic acid, and after wounding. Western blot analysis using an antibody specific to PGIP, confirmed the increased expression in salicylic acid-treated and wounded tissues. Moreover, an immunocytological study using the same antibody in bean infected with *Colletotrichum lindemuthianum* has shown that PGIP accumulates preferentially in cells adjacent to the infection site [19].

2.3 OLIGOGALACTURONIDE ELICITOR PERCEPTION

The results of several studies suggest that oligogalacturonide perception occurs at the plasma membrane of responsive cells. For example, oligogalacturonide mixtures rapidly (within 5 min) depolarize the membranes of tomato leaf mesophyll cells, albeit at relatively high concentrations (1 mg/mL) [130]. Lower concentrations of size-specific oligogalacturonides (DP 12 to 15) induce, within 5 min, a transient stimulation of K^+ efflux, alkalinization of the extracellular medium, depolarization of the plasma membrane, and a decrease in the external Ca^{2+} concentration in suspension-cultured tobacco cells [97]. These effects are specific for oligogalacturonides; size-heterogeneous mixtures of oligomannuronides and oligoguluronides are approximately 400-fold less effective. Lastly, treatment of suspension-cultured soybean cells with heterogeneous oligogalacturonide fractions results in the rapid production of hydrogen peroxide (oxidative burst) [6,90].

The rapid responses at the cell surface described above have not yet been causally linked to any of the known plant responses induced by oligogalacturonides. One critical gap in our knowledge is the absence of information about the molecules responsible for initial recognition of the active oligogalacturonides. Derivatives of biologically active oligogalacturonides have been prepared that could be radiolabeled to yield ligands for use in binding studies to identify and characterize specific oligogalacturonide-binding sites in plants. The synthesis of the derivatized oligogalacturonides is complicated by the occurrence of side reactions involving the carboxylic acid groups in the oligogalacturonides. Homogeneous tyraminylated and biotinylated derivatives of the tridecagalacturonide have been prepared (M. Spiro and B. Ridley, personal communication). The preparation of a tyrosine hydrazone of the dodecagalacturonide has also been described [73], although the purity and structure of the derivative have not been documented. Saturable binding of the radio-iodinated form of the latter derivative to intact soybean cells has been reported [95] and evidence suggesting uptake of the bound oligogalacturonide via receptor-mediated endocytosis has been presented [73]. Most recently, a fluorescein derivative of a tetradeca-galacturonide-enriched fraction was prepared and binding of this derivative to the surface of soybean protoplasts was visualized using silver-enhanced immunogold epi-polarization microscopy [47]. However, in no case has the specificity of the observed binding of the labeled oligogalacturonides been demonstrated, raising questions about the significance of the findings with respect to oligogalacturonide elicitor signal transduction. These preliminary studies clearly need to be followed up with more rigorous experiments to identify and characterize oligogalacturonide elicitor-binding proteins.

3. Oligochitin Elicitors

Chitin, a linear polysaccharide composed of (1→4)-linked β-*N*-acetylglucosaminyl residues, and oligosaccharides derived from this polysaccharide have been shown to induce several plant defense responses. Chitin is an important structural component of the mycelial walls of many fungi, including a number of phytopathogens [12,105].

3.1 PURIFICATION OF OLIGOCHITIN ELICITORS

Oligomers of chitin (structure shown in Figure 1) are generated by partial acid hydrolysis of the polysaccharide, and the products fractionated on high resolution size-exclusion columns. Pure chitin oligomers are separated from partially *N*-acetylated fragments by cation-exchange chromatography. The chitin fragments are then further purified by HPLC on amine-modified silica columns [57]. Alternatively, *N*-acetyl-chitooligosaccharides can be prepared by re-*N*-acetylation of the corresponding chitosanoligosaccharides [137]. Chitin-derived oligosaccharides generated from fungal spores and mycelial walls have also been purified by liquid chromatography [79]. Chitin fragments of DP 3 to 6 are commercially available [17,88].

3.2 INDUCTION OF PLANT DEFENSE RESPONSES BY OLIGOCHITINS

Major plant defense responses that can be induced by chitin include lignification in wheat leaves [104] and suspension-cultured cells of slash pine (*Pinus elliottii*) [91], induction of chitinase and glucanase activity [98], and increased activity of an enzyme in the putative biosynthetic pathway of phytoalexins in rice [110]. More recently, oligosaccharide fragments of chitin have also been shown to induce defense responses in plant cells, including lignification [11], and accumulation of chitinases [84,115] and phytoalexins [137].

Structure-activity studies using purified chitin oligomers indicate that oligomers having a degree of polymerization (DP) of less than four are not active, while oligomers with a DP of six or more are active [11,84,110,115,137]. Furthermore, the de-*N*-acetylated (chitosan) oligomers are not active in those plants where the chitin oligomers are active.

3.3 OLIGOCHITIN ELICITOR PERCEPTION

Several studies have reported results that suggest that perception of chitin fragments occurs at the cell surface of responsive cells. Thus, the elicitation of phytoalexin production in suspension-cultured rice cells by chitin oligomers of DP $\geq$ 6 is accompanied by a rapid (within 1 min.) and transient membrane depolarization [88]. Chitin oligomers of DP $\geq$ 4 also stimulate alkalinization of the medium of suspension-cultured tomato cells [57]; this alkalinization is accompanied by changes in the pattern of protein phosphorylation. The reducing end of the chitin fragments could be modified without any loss of activity as long as four *N*-acetylglucosaminyl residues in their pyranose form remained intact [17]. Biologically inactive, shorter chitin oligomers, as well as chitosan oligomers, do not induce these rapid cellular responses.

The activity of oligochitins on tomato cells is only transient, and it is followed by a refractory state where the cells are no longer responsive to subsequent treatment with the same signal [57]. Moreover, oligochitins and a lipochitooligosaccharide nodulation factor from *Rhizobium leguminosarum* desensitize tomato cells to each other's action [127]. These results suggest the operation of mechanism(s) to remove the signal from the cell surface, for example via endocytosis or via degradation by induced extra-cellular chitinases.

The rapid responses of plant cells to oligochitins have yet to be directly linked to a signal transduction pathway leading to activation of a plant defense response. A significant step toward establishing such a link would be the isolation and characterization of specific oligochitin-binding proteins. Studies demonstrating the presence of such binding proteins in the plasma membrane of responsive plant cells were reported recently. Radiolabeled chitin oligomers have been prepared by either reductive amination with tyramine followed by radio-iodination [123] or by derivatization with *t*-butoxycarbonyl-L-[^{35}S]methionine [17]. Both methods result in the incorporation of the radiolabel at the reducing end of the oligosaccharides. These radio-labeled ligands were used to demonstrate the presence of a single class of high-affinity binding sites for biologically active chitin oligomers in membranes prepared from suspension-cultured cells of rice (apparent $K_d = 5$ nM) [123] and of tomato (apparent $K_d = 32$ nM) [17]. High-affinity binding (apparent $K_d = 1.7$ nM) of the radiolabeled pentamer to intact tomato cells was also observed [17], suggesting that the binding site is localized to the plasma membrane. The observed binding to tomato membranes and cells is reversible [17]. Binding of radiolabelled chitin octamer to rice membranes is abolished by pre-treatment of the membranes with protease [123], suggesting that the binding site is proteinaceous.

The specificity of these membrane-localized oligochitin-binding sites has been investigated using oligochitins of various degrees of polymerization and oligomers of chitosan, the de-*N*-acetylated form of chitin. Oligochitins of DP 4 and 5 are effective competitors of the radiolabeled pentamer, while smaller chitin oligomers are significantly less effective [17]. Chitosan oligomers are unable to compete for the oligochitin-binding sites, even at concentrations as high as 100 μM [17,123]. Thus, there is a direct correlation between the ability of chitin oligosaccharides to induce biological responses in treated plant cells and the ability of those oligosaccharides to bind to the membrane-localized binding sites, providing evidence that the oligochitin-binding proteins are physiological receptors. It will be of interest to determine the identity and nature of the oligochitin-binding protein(s) and how they connect with the remainder of the signal transduction pathway to induce the physiological defense response(s) observed in affected plant cells.

4. Oligochitosan Elicitors

Chitosan is the de-*N*-acetylated form of chitin, and is a linear polysaccharide composed of (1→4)-linked β-glucosaminyl residues. This polysaccharide is a component of the mycelial walls of many fungi, including various phytopathogens [12]. Chitosan and oligomers derived from this polysaccharide (structure shown in Figure 1) can induce various plant defense responses.

4.1 INDUCTION OF PLANT DEFENSE RESPONSES BY OLIGOCHITOSANS

Plant defense responses induced by chitosan include the accumulation of phytoalexins in pea pods [62,134] and in suspension-cultured soybean [87] and parsley [34] cells, the induction of chitinases and glucanases [54,96,98], and the accumulation of defense-related proteinase inhibitors in tomato and potato leaves [106,134-136]. Chitosan has also been reported to induce, in mechanically wounded leaves, protection from subsequent infection with viruses [108,109], though the mechanisms underlying this induced resistance are unknown. Chitosan elicits free radical formation that is thought to lead to cell death and cell wall thickening in suspension-cultured rice cells [96]; such responses have been associated with the hypersensitive response leading to disease resistance.

Heterogeneous preparations of chitosan-derived oligosaccharide fragments of chitosan have been shown to induce defense-related lignification of the walls of suspension-cultured slash pine (*Pinus elliottii*) cells [91,92]. Other plant defenses induced by chitosan-derived oligosaccharides include and the synthesis of callose, a β-1,3-glucan, in suspension-cultured parsley [34], tomato [60] and *Catharanthus roseus* [76] cells. The deposition of both callose and lignin are thought to function in plant defense by strengthening the plant cell wall.

The structural requirements for bioactive chitosan-derived oligosaccharides have not, until recently, been investigated in detail. The oligosaccharides generally must have a DP > 4 to induce a biological response, but beyond that requirement, it is not possible to generalize about structural features essential for functionality. For example, chitosan-derived with DP > 6 elicit pisatin accumulation in pea pods, while the *N*-acetylated (chitin) oligosaccharides are not active in this system [62,79,80]. Partial *N*-acetylation or chemical fragmentation also reduces the ability of chitosan to elicit callose formation in *C. roseus* cells [76]. In contrast, de-*N*-acetylated and partially *N*-acetylated oligoglucosamines with DP > 4 are both effective elicitors of proteinase inhibitors in tomato and potato leaves [134,135].

Recently, the stereo-controlled chemical synthesis of chitosan oligomers [89] has allowed a more precise examination of the effect of DP on the ability of these oligosaccharides to induce phytoalexin accumulation in pea pods [63]. Oligochitosans of DP ≤ 6 are not active. An unexpected result of these studies was the observation that the nature of the substituent at the reducing end of the oligochitosans has a major impact on the activity of the oligosaccharides. Oligochitosans of DP = 8 carrying an *O*-methyl group at the reducing end are active, while those carrying an *O*-methoxyphenyl group are inactive. These results stand in notable contrast to experiments with chitin- and glucan-oligomers, where the nature of the substituent at the reducing end of the oligosaccharide has no significant impact on the abilities of the oligosaccharides to induce plant defense responses.

4.2 OLIGOCHITOSAN ELICITOR PERCEPTION

The cellular signaling pathway triggered by oligochitosan elicitors is not known. Treatment of suspension-cultured soybean cells with chitosan induces electrolyte fluxes across the plasma membrane [133,142,143] that were proposed to result in the activation of the Ca^{2+}-dependent callose synthase [86]. In contrast, chitosan-induced defense responses in pea pods are not correlated with membrane leakage and Ca^{2+} flux [81,82]. Evidence has also been presented suggesting that protein phosphorylation is not involved in the chitosan-induced synthesis of callose in tomato cells [60]. Based on studies of the correlations between the DP of the chitosan oligomers, the degree of *N*-acetylation of these oligosaccharides, and the abilities of oligochitosans to elicit callose synthesis, it has been suggested that the oligochitosan elicitors interact primarily with regularly spaced negative charges on the plant plasma membrane rather than with a discrete receptor [76].

5. Glycopeptide elicitors

A number of glycoproteins have been identified that induce defense responses in plants (reviewed in [5]). In some cases, the intact native glycoprotein is required (e.g., for enzyme activity), and for others the activity resides in a non-glycosylated peptide (e.g., the peptide elicitor from *Phytophthora sojae* [101,103]). Recent studies of some of these glycoproteins have demonstrated that their carbohydrate portion is essential for the bioactivity, and resulted in the identification of glycopeptides as a new class of oligosaccharide elicitors.

5.1 INDUCTION OF PLANT DEFENSE RESPONSES BY GLYCOPEPTIDE ELICITORS

A partially purified extract from yeast was shown to induce ethylene biosynthesis and phenylalanine-ammonia-lyase (PAL) activity, but not callose deposition, in suspension-cultured tomato cells [56,60]. The activity of the yeast extract differs from that of chitosan oligomers, which induce all of those responses [60]. The molecules in the yeast extract that are responsible for these activities were subsequently shown to be glycopeptides [14]. Similar glycopeptide fragments could be generated from yeast invertase by chymotrypsin treatment, and were shown to be elicitors of ethylene biosynthesis [13]. Structure-activity studies using different glycan side chains and peptide sequences have shown that the elicitor activity depends more on the glycan than on the peptide structure [13,14]. The most active glycopeptide elicitors, at concentration of 5-10 nM, induce ethylene biosynthesis and PAL activity . The peptides are glycosylated with $Man_{10}GlcNAc_2$ and $Man_{11}GlcNAc_2$ glycan side chains (see Figure 1) [13]. Release of the oligosaccharides from the glycopeptides by treatment with endo-β-*N*-acetylglucosaminidase H (endo-H) results in the loss of elicitor activity. However, the free *N*-glycans suppress the activity of the glycopeptides, suggesting that the oligosaccharides interact with the same binding site as the glycopeptide [13,14]. Such high mannose oligosaccharides have not been found in plant glycoproteins.

Two biologically active glycopeptides have been purified from the spore germination fluid of the pea pathogen, *Mycosphaerella pinodes*. These two molecules, suppressins A (GalNAc-O-Ser-Ser-Gly) and B (Gal-GalNAc-O-Ser-Ser-Gly-Asp-Glu-Thr), suppress the elicitation of the accumulation of the pea phytoalexin, pisatin, by crude fungal elicitor preparations [124,141], although the concentrations required for suppression are quite high (≥ 220 and 80 μM respectively). In addition to its suppressor activity, suppressin B at a concentration of 320 μM inhibits plasma membrane ATPase activity by 80%. The relative importance of the peptide and the carbohydrate portions of the molecules have not been established, and no structure-activity studies have been performed.

5.2 GLYCOPEPTIDE ELICITOR PERCEPTION

Biochemical analysis of the signal transduction pathway(s) stimulated by glycopeptide elicitors is just beginning, and to date has focused on the glycopeptide elicitors released from yeast invertase. The most active of the glycopeptides released from yeast invertase was radiolabeled by derivatization with *t*-butoxycarbonyl-L-[^{35}S]methionine at the amino terminus of the peptide portion of the elicitor [15]. A single class of reversible, high affinity binding sites (apparent K_d = 3.3 nM) for the radiolabeled glycopeptide was detected in membranes prepared from suspension-cultured tomato cells [57]. Saturable binding of the ligand to whole cells (apparent K_d = 0.7 nM) was also observed, suggesting that the glycopeptide elicitor-binding site is located in the plasma membrane. Whether or not the binding site is proteinaceous in nature has not yet been determined. The site concentration is about two orders of magnitude lower (B_{max} = 19 fmol/mg of membrane protein) than observed for the hepta-β-glucoside- and oligochitin-binding sites discussed elsewhere in this paper. This low abundance is likely to hamper purification of sufficient amounts of the glycopeptide elicitor-binding sites for detailed characterization.

The specificity of the membrane-localized binding site has been examined using glycopeptides carrying different glycan structures and with the free glycans [15]. A direct correlation was observed between the ability of various glycopeptides to compete with the labeled ligand for the binding sites and the ability of the glycopeptides to induce ethylene biosynthesis in tomato cells. A radiolabeled peptide obtained by deglycosylation of the most active glycopeptide elicitor did not bind to the membranes, providing evidence that the peptide portion of the glycopeptide elicitors does not contribute significantly to binding of the elicitor. This conclusion is further substantiated by the observation that the ability of free glycans to compete for the binding site is identical to that of the glycopeptides from which the glycans had been released. Thus, it appears that the glycopeptide elicitors and the glycan suppressors interact with the same binding site in the membranes. This is the first example among the oligosaccharide elicitors of an antagonist (free glycan) competing for the same binding site as the agonist (glycopeptide), and presumably thereby preventing the biological response to the elicitor.

6. Oligoglucoside Elicitors

Elicitor-active glucans were first detected in the culture filtrates of *Phytophthora sojae*, a phytopathogenic oomycete [7], and were later also purified from a commercially available yeast extract [66]. These elicitors were shown to be composed of 3-, 6-, and 3,6-linked β-glucosyl residues [7,66], a composition very similar to β-glucans that are important structural components of various mycelial walls [12]. Thus, subsequent research focused on elicitors released from mycelial walls.

6.1 INDUCTION OF PLANT DEFENSE RESPONSES BY OLIGOGLUCOSIDE ELICITORS

Elicitor-active glucans can be released from the mycelial walls of *P. sojae* by either partial chemical [8,9,122] or enzymatic [70,78,139] hydrolysis. The complex mixture of oligo- and polysaccharides released from mycelial walls contains β-glucan fragments ranging in size from short oligomers up to polysaccharides ($M_r > 100{,}000$) [9,122]. Oligoglucans in this mixture have been shown to elicit several plant defense responses. For example, glucan elicitors have been shown to induce phytoalexins in several legumes including soybean [8,122], chickpea [61], bean [33], alfalfa [83], and pea [20], and also in the solanaceous species, potato [33] and sweet pepper [20]. Other glucans present in the mycelial wall hydrolyzates induce the production of hydroxyproline-rich glycoproteins [116] in soybean, and induce resistance to viruses [85] and activate glycine-rich protein gene expression [24] in tobacco. There may be other, as yet unidentified, activities in the oligoglucan mixture as well. The following sections focus on the glucan elicitors of phytoalexin accumulation, and describe the purification of a hepta-β-glucoside elicitor, structure-activity studies that define essential structural elements of this elicitor, and recent research aimed at the identification and characterization of a specific binding protein in soybean for this elicitor.

6.2 OLIGOGLUCOSIDE ELICITOR PURIFICATION

Fractionation of the mixture of oligoglucosides generated by partial acid hydrolysis of mycelial walls of *P. sojae* on a size-exclusion column revealed that elicitor-active oligosaccharides were present in all fractions containing oligomers of DP $\geq$ 6 [122]. The heptamer-enriched fraction was further fractionated on a series of normal- and reversed-phase high-performance liquid chromatography (HPLC) columns [122]. It was estimated that the original heptaglucoside mixture contained >100 structurally distinct heptaglucosides, based on the number of peaks observed in the various chromatographic steps. Homogeneous preparations of the aldehyde-reduced forms (i.e., the hexa-β-glucosyl glucitols) of one elicitor-active hepta-β-glucoside (structure shown in Figure 1) and of seven other elicitor-inactive hepta-β-glucosides were obtained in amounts sufficient to determine their structures [121]. The structure of the elicitor-active hepta-β-glucoside [120] was subsequently confirmed by its chemical synthesis [59,94,102,132].

The ability of the chemically synthesized, unreduced hepta-β-glucoside elicitor to induce phytoalexin accumulation in soybean cotyledons is identical to that of the corresponding hexa-β-glucosyl glucitol purified from fungal wall hydrolyzates [120].

Both are active at concentrations of ~10 nM, making them some of the most active elicitors of phytoalexin accumulation yet observed. The seven other hexa-β-glucosyl glucitols that were purified from the partial hydrolyzates of fungal cell walls had no elicitor activity over the concentration range (≤ 400 μM) tested [122]. These results provided the first evidence that specific structural features are required for an oligo-β-glucoside to be an effective elicitor of phytoalexin accumulation.

Structure-activity studies carried out in our laboratory [30,31] using synthetic oligosaccharides structurally related to the active hepta-β-glucoside have demonstrated that the three non-reducing terminal glucosyl residues are essential for the activity of the elicitor, as is the distribution of the side-chains along the backbone of the molecule. In contrast, the reducing terminal glucosyl residue of the hepta-β-glucoside elicitor does not appear to be essential for activity [31] and provides a suitable site for radiolabeling of the elicitor for use in experiments to identify and characterize glucan elicitor-binding proteins.

6.3 OLIGOGLUCOSIDE ELICITOR PERCEPTION

The first step in the signal transduction pathway induced by the hepta-β-glucoside elicitor is likely to be its recognition by a specific receptor. Indeed, the specificity of the response of soybean tissue to oligoglucoside elicitors of phytoalexin accumulation [31,122] suggests that a specific receptor for the hepta-β-glucoside elicitor exists in soybean cells. Several studies utilizing heterogeneous mixtures of mycelial glucan fragments indicated that binding sites for glucan fragments exist in soybean membranes [38,107,119,138,140]. In particular, binding studies carried out with partially purified elicitor-active glucans from *P. sojae* mycelial walls demonstrated the presence of high affinity glucan-binding sites on soybean root plasma-membranes [119] and protoplasts from suspension-cultured soybean cells [38]. Lower affinity binding sites for a radiolabeled preparation of glucanase-released glucan fragments [140] and radiolabeled laminarin, a β-1,3-glucan [138], have also been identified in soybean membranes. Since the glucan preparations utilized in these binding studies were not homogenous, it is difficult to determine whether the observed binding was specific for the biologically active oligoglucosides.

Subsequent research utilizing homogenous hepta-β-glucoside elicitor coupled either to radio-iodinated aminophenethylamine [37] or to tyramine [32] as the labeled ligand has provided further evidence that membrane-localized glucan elicitor-binding sites exist in soybean cells. The hepta-β-glucoside elicitor-binding sites are present in membranes prepared from every major organ of young soybean plants [32]. These elicitor-binding sites co-migrate with an enzyme marker (vanadate-sensitive ATPase) for plasma membranes in isopycnic sucrose density gradients [30], confirming earlier results obtained with partially purified labeled glucan fragments [119]. Binding of the radiolabeled hepta-β-glucoside elicitor to the root membranes is saturable over a concentration range of 0.1 to 5 nM, which is somewhat lower than the range of concentrations (6 to 200 nM) required to saturate the bioassay for phytoalexin accumulation [31,122]. The root membranes possess a single class of high-affinity hepta-β-glucoside binding sites (apparent $K_d \approx 1$ nM). These binding sites are inactivated by heat or pronase treatment [32], suggesting that the molecule(s) responsible for the binding are proteinaceous. Binding of the active hepta-β-glucoside to the membrane preparation is reversible, indicating that the elicitor does not become covalently attached to the binding protein(s) [32,58]. The membrane-localized elicitor-binding proteins exhibit a high degree of specificity with respect to the oligoglucosides that they bind. More importantly, the ability of an oligoglucoside to bind to soybean root membranes correlates with its ability to induce phytoalexin accumulation in soybean tissues [30,32]. It is this correlation between biological activity and binding affinity that provides the strongest evidence to date that the binding proteins are physiological receptors for the hepta-β-glucoside elicitor.

Solubilization of glucan elicitor-binding proteins from soybean root membranes has been achieved with the aid of several detergents [30,35]. The non-ionic detergent, *n*-dodecylsucrose, and the zwitterionic detergent, *N*-dodecyl-*N*,*N*-dimethyl-3-ammonio-1-propane-sulfonate (ZW 3-12) have been the principal detergents used, and each solubilizes between 40 and 60% of the elicitor-binding activity from soybean membranes. The detergent-solubilized glucan elicitor-binding proteins appear to be fully functional as defined by retention of their affinity and specificity for the hepta-β-glucoside elicitor. Thus, the solubilized binding proteins have an apparent K_d between 1 and 3 nM [30,36,58] and retain the specificity for elicitor-active oligoglucosides characteristic of the membrane-localized proteins [30].

Recently, progress has been made towards identification and purification of hepta-β-glucoside elicitor-binding proteins. Size-exclusion chromatography of detergent-solubilized hepta-β-glucoside elicitor-binding proteins indicates that elicitor binding activity is associated primarily with large detergent-protein micelles [M_r > 200,000; [30,35]], with detergent-protein micelles having a M_r > 669,000 having the highest specific elicitor binding activity. Little or no elicitor binding activity is associated with the smallest detergent-protein micelles [30]. The large elicitor-binding protein-detergent complexes can be disrupted only with a combination of high detergent concentrations and sonication; a treatment that results in the loss of 70% of the binding activity (F. Côté and M.G. Hahn, unpublished results). These data suggest that the elicitor-binding proteins are active in multimeric protein complexes. Indeed, the fraction of solubilized membrane proteins retained on affinity columns carrying either an immobilized mixture of elicitor-active fungal glucans [36,58] or immobilized hepta-β-glucoside elicitor (F. Côté and M.G. Hahn, unpublished results) and subsequently eluted with free ligand contains several proteins as visualized on polyacrylamide gels. Photo-affinity labeling experiments carried out on solubilized soybean membrane preparations suggest that three of these polypeptides (≈ 70, 100, and 170 kDa) carry elicitor-binding domains [36]. Preliminary results indicate that the affinity-purified hepta-β-glucoside binding proteins retain the same specificity for elicitor-active oligoglucosides that was observed for the membrane-localized and crude detergent-solubilized proteins (F. Côté and M.G. Hahn, unpublished results). Further purification of these proteins and assays of the polypeptides, either alone or in combination, for elicitor-binding activity will be required to determine if all of the proteins present in the affinity-purified fraction are essential for elicitor-binding activity. Proof that the hepta-β-glucoside elicitor-binding proteins are functional receptors will require reconstitution of a hepta-β-glucoside elicitor-responsive system using either purified binding proteins or transgenically expressed genes encoding those proteins.

7. Conclusions

The results of extensive research in a number of laboratories over the past 15 years has clearly established that oligosaccharide elicitors constitute an important class of signal molecules that play major roles in the interactions between plants and their pathogens. Progress in elucidating the mechanisms by which these oligosaccharide elicitors exert their effects on plant cells has been hampered by the great molecular heterogeneity of the elicitor preparations. Detailed biochemical investigations of the cellular signaling pathways triggered by oligosaccharide elicitors require the preparation of homogeneous samples of these elicitors in order to assign observed effects unambiguously to single elicitor-stimulated pathways. Recent improvements in the techniques for the purification of oligosaccharides and, more significantly, contributions from synthetic organic chemists, have made available homogeneous preparations of oligosaccharide elicitors, which should greatly facilitate these biochemical studies. Indeed, the availability of chemically synthesized hepta-β-glucoside elicitor and structurally related, less active agonists has led to the tentative identification of a binding protein (putative receptor) for this elicitor. Similarly, research on the other oligosaccharide elicitors has also resulted in the demonstration of specific binding sites for these elicitors in plant cells.

The oligosaccharide elicitors belong to a recently discovered class of signal compounds, the oligosaccharins (oligosaccharides with regulatory properties [3]). This class of compounds includes oligosaccharides isolated from plants, fungi, and most recently bacteria (reviewed in [4,39,40,46]). Oligosaccharins have been shown to play important roles in the regulation of plant defense responses, as briefly reviewed here, plant development, and plant-symbiont interactions. Additional biochemical analysis of the cellular signaling pathways induced by these signal molecules is likely to yield valuable insights into how plants perceive and respond to stimuli from their environment.

8. Acknowledgements

Work in the author's laboratory on oligoglucoside elicitors and hepta-β-glucoside elicitor-binding proteins is supported by a grant from the United States National Science Foundation (MCB-9206882), and in part by the United States Department of Energy-funded Center for Plant and Microbial Complex Carbohydrates (DE-FG09-93ER20097).

9. References

1. Aist JR (1976) Papillae and related wound plugs of plant cells. *Annu Rev Phytopathol* 14:145-163
2. Albersheim P and Anderson AJ (1971) Proteins from plant cell walls inhibit polygalacturonases secreted by plant pathogens. *Proc Natl Acad Sci USA* 68:1815-1819
3. Albersheim P, Darvill AG, McNeil M, Valent BS, Sharp JK, Nothnagel EA, Davis KR, Yamazaki N, Gollin DJ, York WS, Dudman WF, Darvill JE, Dell A (1983) Oligosaccharins: Naturally occurring carbohydrates with biological regulatory functions. In: Ciferri O and

Dure L, III (eds) Structure and Function of Plant Genomes. Plenum Publishing Corp., New York, NY, p 293-312

4. Aldington S and Fry SC (1993) Oligosaccharins. *Adv Bot Res* 19:1-101
5. Anderson AJ (1989) The biology of glycoproteins as elicitors. In: Kosuge T and Nester E (eds) Plant-Microbe Interactions. Molecular and Genetic Perspectives, Vol. 3. McGraw Hill Inc., New York, NY, p 87-130
6. Apostol I, Heinstein PF, Low PS (1989) Rapid stimulation of an oxidative burst during elicitation of cultured plant cells. Role in defense and signal transduction. *Plant Physiol* 90:109-116
7. Ayers AR, Ebel J, Finelli F, Berger N, Albersheim P (1976) Host-pathogen interactions. IX. Quantitative assays of elicitor activity and characterization of the elicitor present in the extracellular medium of cultures of *Phytophthora megasperma* var. *sojae*. *Plant Physiol* 57:751-759
8. Ayers AR, Ebel J, Valent B, Albersheim P (1976) Host-pathogen interactions. X. Fractionation and biological activity of an elicitor isolated from the mycelial walls of *Phytophthora megasperma* var. *sojae*. *Plant Physiol* 57:760-765
9. Ayers AR, Valent B, Ebel J, Albersheim P (1976) Host-pathogen interactions. XI. Composition and structure of wall-released elicitor fractions. *Plant Physiol* 57:766-774
10. Bailey JA (1980) Constitutive elicitors from *Phaseolus vulgaris*; a possible cause of phytoalexin accumulation. *Ann Phytopathol* 12:395-402
11. Barber MS, Bertram RE, Ride JP (1989) Chitin oligosaccharides elicit lignification in wounded wheat leaves. *Physiol Mol Plant Pathol* 34:3-12
12. Bartnicki-Garcia S (1968) Cell wall chemistry, morphogenesis, and taxonomy of fungi. *Annu Rev Microbiol* 22:87-108
13. Basse CW, Bock K, Boller T (1992) Elicitors and suppressors of the defense response in tomato cells. Purification and characterization of glycopeptide elicitors and glycan suppressors generated by enzymatic cleavage of yeast invertase. *J Biol Chem* 267:10258-10265
14. Basse CW and Boller T (1992) Glycopeptide elicitors of stress responses in tomato cells. *N*-linked glycans are essential for activity but act as suppressors of the same activity when released from the glycopeptides. *Plant Physiol* 98:1239-1247
15. Basse CW, Fath A, Boller T (1993) High affinity binding of a glycopeptide elicitor to tomato cells and microsomal membranes and displacement by specific glycan suppressors. *J Biol Chem* 268:14724-14731
16. Bateman DF and Basham HG (1976) Degradation of plant cell walls and membranes by microbial enzymes. In: Heitefuss R and Williams PH (eds) Encyclopedia of Plant Physiology, New Series, Vol. 4, Physiological Plant Pathology. Springer-Verlag, Berlin, p 316-355
17. Baureithel K, Felix G, Boller T (1994) Specific, high affinity binding of chitin fragments to tomato cells and membranes. Competitive inhibition of binding by derivatives of chitin fragments and a nod factor of *Rhizobium*. *J Biol Chem* 269:17931-17938
18. Bell AA (1981) Biochemical mechanisms of disease resistance. *Annu Rev Plant Physiol* 32:21-81
19. Bergmann CW, Ito Y, Singer D, Albersheim P, Darvill AG, Benhamou N, Nuss L, Salvi G, Cervone F, De Lorenzo G (1994) Polygalacturonase-inhibiting protein accumulates in *Phaseolus vulgaris* L. in response to wounding, elicitors and fungal infection. *Plant J* 5:625-634
20. Bhandal IS and Paxton JD (1991) Phytoalexin biosynthesis induced by the fungal glucan polytran L in soybean, pea, and sweet pepper tissues. *J Agric Food Chem* 39:2156-2157

21. Bishop PD, Pearce G, Bryant JE, Ryan CA (1984) Isolation and characterization of the proteinase inhibitor-inducing factor from tomato leaves. Identity and activity of poly- and oligogalacturonide fragments. *J Biol Chem* 259:13172-13177
22. Boller T (1987) Hydrolytic enzymes in plant disease resistance. In: Kosuge T and Nester EW (eds) Plant-Microbe Interactions. Molecular and Genetic Perspectives, Vol. 2. Macmillan Publishing Co., New York, NY, p 385-413
23. Bostock RM, Kuc JA, Laine RA (1981) Eicosapentaenoic and arachidonic acids from *Phytophthora infestans* elicit fungitoxic sesquiterpenes in the potato. *Science* 212:67-69
24. Brady KP, Darvill AG, Albersheim P (1993) Activation of a tobacco glycine-rich protein gene by a fungal glucan preparation. *Plant J* 4:517-524
25. Broekaert WF and Peumans WJ (1988) Pectic polysaccharides elicit chitinase accumulation in tobacco. *Physiol Plant* 74:740-744
26. Bruce RJ and West CA (1989) Elicitation of lignin biosynthesis and isoperoxidase activity by pectic fragments in suspension-cultures of castor bean. *Plant Physiol* 91:889-897
27. Cervone F, De Lorenzo G, Degrà L, Salvi G, Bergami M (1987) Purification and characterization of a polygalacturonase-inhibiting protein from *Phaseolus vulgaris* L. *Plant Physiol* 85:631-637
28. Cervone F, De Lorenzo G, Pressey R, Darvill AG, Albersheim P (1990) Can *Phaseolus* PGIP inhibit pectic enzymes from microbes and plants? *Phytochemistry* 29:447-449
29. Cervone F, Hahn MG, De Lorenzo G, Darvill A, Albersheim P (1989) Host-pathogen interactions. XXXIII. A plant protein converts a fungal pathogenesis factor into an elicitor of plant defense responses. *Plant Physiol* 90:542-548
30. Cheong J-J, Alba R, Côté F, Enkerli J, Hahn MG (1993) Solubilization of functional plasma membrane-localized hepta-β-glucoside elicitor binding proteins from soybean. *Plant Physiol* 103:1173-1182
31. Cheong J-J, Birberg W, Fügedi P, Pilotti Å, Garegg PJ, Hong N, Ogawa T, Hahn MG (1991) Structure-activity relationships of oligo-β-glucoside elicitors of phytoalexin accumulation in soybean. *Plant Cell* 3:127-136
32. Cheong J-J and Hahn MG (1991) A specific, high-affinity binding site for the hepta-β-glucoside elicitor exists in soybean membranes. *Plant Cell* 3:137-147
33. Cline K, Wade W, Albersheim P (1978) Host-pathogen interactions. XV. Fungal glucans which elicit phytoalexin accumulation in soybean also elicit the accumulation of phytoalexins in other plants. *Plant Physiol* 62:918-921
34. Conrath U, Domard A, Kauss H (1989) Chitosan-elicited synthesis of callose and of coumarin derivatives in parsley cell suspension cultures. *Plant Cell Rep* 8:152-155
35. Cosio EG, Frey T, Ebel J (1990) Solubilization of soybean membrane binding sites for fungal β-glucans that elicit phytoalexin accumulation. *FEBS Lett* 264:235-238
36. Cosio EG, Frey T, Ebel J (1992) Identification of a high-affinity binding protein for a hepta-β-glucoside phytoalexin elicitor in soybean. *Eur J Biochem* 204:1115-1123
37. Cosio EG, Frey T, Verduyn R, Van Boom J, Ebel J (1990) High-affinity binding of a synthetic heptaglucoside and fungal glucan phytoalexin elicitors to soybean membranes. *FEBS Lett* 271:223-226
38. Cosio EG, Pöpperl H, Schmidt WE, Ebel J (1988) High-affinity binding of fungal β-glucan fragments to soybean (*Glycine max* L.) microsomal fractions and protoplasts. *Eur J Biochem* 175:309-315
39. Côté F and Hahn MG (1994) Oligosaccharins: Structures and signal transduction. *Plant Mol Biol* (in press)
40. Darvill A, Augur C, Bergmann C, Carlson RW, Cheong J-J, Eberhard S, Hahn MG, Lo V-M, Marfà V, Meyer B, Mohnen D, O'Neill MA, Spiro MD, van Halbeek H, York WS, Albersheim P (1992) Oligosaccharins - oligosaccharides that regulate growth, development and defense responses in plants. *Glycobiology* 2:181-198

41. Davis KR, Darvill AG, Albersheim P (1986) Host-pathogen interactions. XXXI. Several biotic and abiotic elicitors act synergistically in the induction of phytoalexin accumulation in soybean. *Plant Mol Biol* 6:23-32
42. Davis KR, Darvill AG, Albersheim P, Dell A (1986) Host-pathogen interactions. XXIX. Oligogalacturonides released from sodium polypectate by endopolygalacturonic acid lyase are elicitors of phytoalexins in soybean. *Plant Physiol* 80:568-577
43. Davis KR, Darvill AG, Albersheim P, Dell A (1986) Host-pathogen interactions. XXX. Characterization of elicitors of phytoalexin accumulation in soybean released from soybean cell walls by endopolygalacturonic acid lyase. *Z Naturforsch* 41c:39-48
44. Davis KR and Hahlbrock K (1987) Induction of defense responses in cultured parsley cells by plant cell wall fragments. *Plant Physiol* 85:1286-1290
45. De Lorenzo G, Cervone F, Hahn MG, Darvill A, Albersheim P (1991) Bacterial endopectate lyase: evidence that plant cell wall pH prevents tissue maceration and increases the half-life of elicitor-active oligogalacturonides. *Physiol Mol Plant Pathol* 39:335-344
46. Dénarié J and Roche P (1992) Rhizobium nodulation signals. In: Verma DPS (ed) Molecular signals in plant-microbe communications. CRC Press, Boca Raton, p 295-324
47. Diekmann W, Herkt B, Low PS, Nürnberger T, Scheel D, Terschüren C, Robinson DG (1994) Visualization of elicitor binding loci at the plant cell surface. *Planta* (in press)
48. Dixon RA (1986) The phytoalexin response: Elicitation, signalling and control of host gene expression. *Biol Rev* 61:239-291
49. Dixon RA, Dey PM, Lamb CJ (1983) Phytoalexins: Enzymology and molecular biology. *Adv Enzymol* 55:1-136
50. Dixon RA, Jennings AC, Davies LA, Gerrish C, Murphy DL (1989) Elicitor active components from French bean hypocotyls. *Physiol Mol Plant Pathol* 34:99-115
51. Dixon RA and Lamb CJ (1990) Molecular communication in interactions between plants and microbial pathogens. *Annu Rev Plant Physiol Plant Mol Biol* 41:339-367
52. Ebel J (1986) Phytoalexin synthesis: The biochemical analysis of the induction process. *Annu Rev Phytopathol* 24:235-264
53. Ebel J and Cosio EG (1994) Elicitors of plant defense responses. *Int Rev Cytol* 148:1-36
54. El Ghaouth A, Arul J, Grenier J, Benhamou N, Asselin A, Bélanger R (1994) Effect of chitosan on cucumber plants: Suppression of *Pythium aphanidermatum* and induction of defense reactions. *Phytopathology* 84:313-320
55. Farmer EE, Moloshok TD, Saxton MJ, Ryan CA (1991) Oligosaccharide signaling in plants: Specificity of oligouronide-enhanced plasma membrane protein phosphorylation. *J Biol Chem* 266:3140-3145
56. Felix G, Grosskopf DG, Regenass M, Basse CW, Boller T (1991) Elicitor-induced ethylene biosynthesis in tomato cells. Characterization and use as a bioassay for elicitor action. *Plant Physiol* 97:19-25
57. Felix G, Regenass M, Boller T (1993) Specific perception of subnanomolar concentrations of chitin fragments by tomato cells: Induction of extracellular alkalinization, changes in protein phosphorylation, and establishment of a refractory state. *Plant J* 4:307-316
58. Frey T, Cosio EG, Ebel J (1993) Affinity purification and characterization of a binding protein for a hepta-β-glucoside phytoalexin elicitor in soybean. *Phytochemistry* 32:543-550
59. Fügedi P, Birberg W, Garegg PJ, Pilotti Å (1987) Syntheses of a branched heptasaccharide having phytoalexin-elicitor activity. *Carbohydr Res* 164:297-312
60. Grosskopf DG, Felix G, Boller T (1991) A yeast-derived glycopeptide elicitor and chitosan or digitonin differentially induce ethylene biosynthesis, phenylalanine ammonia-lyase and callose formation in suspension-cultured tomato cells. *J Plant Physiol* 138:741-746

61. Gunia W, Hinderer W, Wittkampf U, Barz W (1991) Elicitor induction of cytochrome P-450 monooxygenases in cell suspension cultures of chickpea (*Cicer arietinum* L.) and their involvement in pterocarpan phytoalexin biosynthesis. *Z Naturforsch* 46c:58-66
62. Hadwiger LA and Beckman JM (1980) Chitosan as a component of pea-*Fusarium solani* interactions. *Plant Physiol* 66:205-211
63. Hadwiger LA, Ogawa T, Kuyama H (1994) Chitosan polymer sizes effective in inducing phytoalexin accumulation and fungal suppression are verified with synthesized oligomers. *Mol Plant Microbe Interact* (in press)
64. Hahlbrock K and Scheel D (1987) Biochemical responses of plants to pathogens. In: Chet I (ed) Innovative Approaches to Plant Disease Control. John Wiley & Sons, Inc., New York, NY, p 229-254
65. Hahn MG (1981) Fragments of plant and fungal cell wall polysaccharides elicit the accumulation of phytoalexins in plants. Ph.D. Thesis, University of Colorado, Boulder, CO
66. Hahn MG and Albersheim P (1978) Host-pathogen interactions. XIV. Isolation and partial characterization of an elicitor from yeast extract. *Plant Physiol* 62:107-111
67. Hahn MG, Bucheli P, Cervone F, Doares SH, O'Neill RA, Darvill A, Albersheim P (1989) Roles of cell wall constituents in plant-pathogen interactions. In: Kosuge T and Nester EW (eds) Plant-Microbe Interactions. Molecular and Genetic Perspectives, Vol. 3. McGraw Hill Publishing Co., New York, NY, p 131-181
68. Hahn MG, Cheong J-J, Birberg W, Fügedi P, Pilotti Å, Garegg P, Hong N, Nakahara Y, Ogawa T (1989) Elicitation of phytoalexins by synthetic oligoglucosides, synthetic oligogalacturonides, and their derivatives. In: Lugtenberg BJJ (ed) Signal Molecules in Plants and Plant-Microbe Interactions. NATO ASI Series, Volume H36. Springer Verlag, Heidelberg, FRG, p 91-97
69. Hahn MG, Darvill AG, Albersheim P (1981) Host-pathogen interactions. XIX. The endogenous elicitor, a fragment of a plant cell wall polysaccharide that elicits phytoalexin accumulation in soybeans. *Plant Physiol* 68:1161-1169
70. Ham K-S, Kauffmann S, Albersheim P, Darvill AG (1991) Host-pathogen interactions XXXIX. A soybean pathogensis-related protein with β-1,3-glucanase activity releases phytoalexin elicitor-active heat-stable fragments from fungal walls. *Mol Plant Microbe Interact* 4:545-552
71. Hargreaves JA and Bailey JA (1978) Phytoalexin production by hypocotyls of *Phaseolus vulgaris* in response to constitutive metabolites released by damaged bean cells. *Physiol Plant Pathol* 13:89-100
72. Hargreaves JA and Selby C (1978) Phytoalexin formation in cell suspensions of *Phaseolus vulgaris* in response to an extract of bean hypocotyls. *Phytochemistry* 17:1099-1102
73. Horn MA, Heinstein PF, Low PS (1989) Receptor-mediated endocytosis in plant cells. *Plant Cell* 1:1003-1009
74. Jin DF and West CA (1984) Characteristics of galacturonic acid oligomers as elicitors of casbene synthetase activity in castor bean seedlings. *Plant Physiol* 74:989-992
75. Kauss H (1987) Callose-Synthese. Regulation durch induzierten Ca^{2+}-Einstrom in Pflanzenzellen. *Naturwissenschaften* 74:275-281
76. Kauss H, Jeblick W, Domard A (1989) The degree of polymerization and N-acetylation of chitosan determine its ability to elicit callose formation in suspension cells and protoplasts of *Catharanthus roseus*. *Planta* 178:385-392
77. Keen NT (1975) Specific elicitors of plant phytoalexin production: Determinants of race specificity in pathogens? *Science* 187:74-75
78. Keen NT and Yoshikawa M (1983) β-1,3-endoglucanase from soybean releases elicitor-active carbohydrates from fungus cell walls. *Plant Physiol* 71:460-465

79. Kendra DF, Christian D, Hadwiger LA (1989) Chitosan oligomers from *Fusarium solani* pea interactions, chitinase/β-glucanase digestion of sporelings and from fungal wall chitin actively inhibit fungal growth and enhance disease resistance. *Physiol Mol Plant Pathol* 35:215-230
80. Kendra DF and Hadwiger LA (1984) Characterization of the smallest chitosan oligomer that is maximally antifungal to *Fusarium solani* and elicits pisatin formation in *Pisum sativum*. *Exp Mycol* 8:276-281
81. Kendra DF and Hadwiger LA (1987) Calcium and calmodulin may not regulate the disease resistance and pisatin formation responses of Pisum sativum to chitosan or Fusarium solani. *Physiol Mol Plant Pathol* 31:337-348
82. Kendra DF and Hadwiger LA (1987) Cell death and membrane leakage not associated with the induction of disease resistance in peas by chitosan or *Fusarium solani* f. sp. *phaseoli*. *Phytopathology* 77:100-106
83. Kobayashi A, Tai A, Kanzaki H, Kawazu K (1993) Elicitor-active oligosaccharides from algal laminaran stimulate the production of antifungal compounds in alfalfa. *Z Naturforsch* 48c:575-579
84. Koga D, Hirata T, Sueshige N, Tanaka S, Ide A (1992) Induction patterns of chitinases in yam callus by inoculation with autoclaved *Fusarium oxysporum*, ethylene, and chitin and chitosan oligosaccharides. *Biosci Biotech Biochem* 56:280-285
85. Kopp M, Rouster J, Fritig B, Darvill A, Albersheim P (1989) Host-pathogen interactions. XXXII. A fungal glucan preparation protects Nicotianae against infection by viruses. *Plant Physiol* 90:208-216
86. Köhle H, Jeblick W, Poten F, Blashek W, Kauss H (1985) Chitosan-elicited callose synthesis in soybean cells as a Ca^{2+}-dependent process. *Plant Physiol* 77:544-551
87. Köhle H, Young DH, Kauss H (1984) Physiological changes in suspension-cultured soybean cells elicited by treatment with chitosan. *Plant Sci Lett* 33:221-230
88. Kuchitsu K, Kikuyama M, Shibuya N (1993) *N*-acetylchitooligosaccharides, biotic elictor for phytoalexin production, induce transient membrane depolarization in suspension-cultured rice cells. *Protoplasma* 174:79-81
89. Kuyama H, Nakahara Y, Nukada T, Ito Y, Ogawa T (1993) Stereocontrolled synthesis of chitosan dodecamer. *Carbohydr Res* 243:C1-C7
90. Legendre L, Rueter S, Heinstein PF, Low PS (1993) Characterization of the oligogalacturonide-induced oxidative burst in cultured soybean (*Glycine max*) cells. *Plant Physiol* 102:233-240
91. Lesney MS (1989) Growth responses and lignin production in cell suspensions of *Pinus elliottii* 'elicited' by chitin, chitosan or mycelium of *Cronartium quercum* f.sp. *fusiforme*. *Plant Cell Tiss Organ Cult* 19:23-31
92. Lesney MS (1990) Effect of 'elicitors' on extracellular peroxidase activity in suspension-cultured slash pine (*Pinus elliottii* Engelm.). *Plant Cell Tiss Organ Cult* 20:173-175
93. Liners F, Thibault J-F, Van Cutsem P (1992) Influence of the degree of polymerization of oligogalacturonates and of esterification pattern of pectin on their recognition by monoclonal antibodies. *Plant Physiol* 99:1099-1104
94. Lorentzen JP, Helpap B, Lockhoff O (1991) Synthese eines elicitoraktiven Heptaglucan-saccharides zur Untersuchung pflanzlicher Abwehrmechanismen. *Angew Chem* 103:1731-1732
95. Low PS, Legendre L, Heinstein PF, Horn MA (1993) Comparison of elicitor and vitamin receptor-mediated endocytosis in cultured soybean cells. *J Exp Bot* 44 Suppl.:269-274
96. Masuta C, Van Den Bulcke M, Bauw G, Van Montagu M, Caplan AB (1991) Differential effects of elicitors on the viability of rice suspension cells. *Plant Physiol* 97:619-629

97. Mathieu Y, Kurkdijan A, Xia H, Guern J, Koller A, Spiro M, O'Neill M, Albersheim P, Darvill A (1991) Membrane responses induced by oligogalacturonides in suspension-cultured tobacco cells. *Plant J* 1:333-343
98. Mauch F, Hadwiger LA, Boller T (1984) Ethylene: Symptom, not signal for the induction of chitinase and β-1,3-glucanase in pea pods by pathogens and elicitors. *Plant Physiol* 76:607-611
99. Moloshok T, Pearce G, Ryan CA (1992) Oligouronide signaling of proteinase inhibitor genes in plants: Structure-activity relationships of di- and trigalacturonic acids and their derivatives. *Arch Biochem Biophys* 294:731-734
100. Nothnagel EA, McNeil M, Albersheim P, Dell A (1983) Host-pathogen interactions. XXII. A galacturonic acid oligosaccharide from plant cell walls elicits phytoalexins. *Plant Physiol* 71:916-926
101. Nürnberger T, Nennstiel D, Jabs T, Sacks WR, Hahlbrock K, Scheel D (1994) High-affinity binding of a fungal oligopeptide elicitor to parsley plasma membranes triggers multiple defense responses. *Cell* (in press)
102. Ossowski P, Pilotti Å, Garegg PJ, Lindberg B (1984) Synthesis of a glucoheptaose and a glucooctaose that elicit phytoalexin accumulation in soybean. *J Biol Chem* 259:11337-11340
103. Parker JE, Schulte W, Hahlbrock K, Scheel D (1991) An extracellular glycoprotein from *Phytophthora megasperma* f.sp. *glycinea* elicits phytoalexin synthesis in cultured parsley cells and protoplasts. *Mol Plant Microbe Interact* 4:19-27
104. Pearce RB and Ride JP (1982) Chitin and related compounds as elicitors of the lignification response in wounded wheat leaves. *Physiol Plant Pathol* 20:119-123
105. Peberdy JF (1990) Fungal cell walls - A review. In: Kuhn PJ, Trinci APJ, Jung MJ, Goosey MW, Copping LG (eds) Biochemistry of Cell Walls and Membranes in Fungi. Springer-Verlag, Berlin, FRG, p 5-30
106. Peña-Cortes H, Sanchez-Serrano J, Rocha-Sosa M, Willmitzer L (1988) Systemic induction of proteinase-inhibitor-II gene expression in potato plants by wounding. *Planta* 174:84-89
107. Peters BM, Cribbs DH, Stelzig DA (1978) Agglutination of plant protoplasts by fungal cell wall glucans. *Science* 201:364-365
108. Pospieszny H and Atabekov JG (1989) Effect of chitosan on the hypersensitive reaction of bean to alfalfa mosaic virus. *Plant Sci* 62:29-31
109. Pospieszny H, Chirkov S, Atabekov J (1991) Induction of antiviral resistance in plants by chitosan. *Plant Sci* 79:63-68
110. Ren Y-Y and West CA (1992) Elicitation of diterpene biosynthesis in rice (*Oryza sativa* L.) by chitin. *Plant Physiol* 99:1169-1178
111. Ricci P, Bonnet P, Huet J-C, Sallantin M, Beauvais-Cante F, Bruneteau M, Billard V, Michel G, Pernollet J-C (1989) Structure and activity of proteins from pathogenic fungi *Phytophthora* eliciting necrosis and acquired resistance in tobacco. *Eur J Biochem* 183:555-563
112. Ride JP (1983) Cell walls and other structural barriers in defense. In: Callow JA (ed) Biochemical Plant Pathology. John Wiley & Sons, New York, p 215-236
113. Robertsen B (1986) Elicitors of the production of lignin-like compounds in cucumber hypocotyls. *Physiol Mol Plant Pathol* 28:137-148
114. Robertsen B (1987) Endo-polygalacturonase from *Cladosporium cucumerinum* elicits lignification in cucumber hypocotyls. *Physiol Mol Plant Pathol* 31:361-374
115. Roby D, Gadelle A, Toppan A (1987) Chitin oligosaccharides as elicitors of chitinase activity in melon plants. *Biochem Biophys Res Commun* 143:885-892
116. Roby D, Toppan A, Esquerré-Tugayé M-T (1985) Cell surfaces in plant-microorganism interactions V. Elicitors of fungal and of plant origin trigger the synthesis of ethylene and of cell wall hydroxyproline-rich glycoprotein in plants. *Plant Physiol* 77:700-704

117. Ryan CA (1990) Protease inhibitors in plants: Genes for improving defenses against insects and pathogens. *Annu Rev Phytopathol* 28:425-449
118. Ryan CA and Farmer EE (1991) Oligosaccharide signals in plants: A current assessment. *Annu Rev Plant Physiol Plant Mol Biol* 42:651-674
119. Schmidt WE and Ebel J (1987) Specific binding of a fungal glucan phytoalexin elicitor to membrane fractions from soybean *Glycine max*. *Proc Natl Acad Sci USA* 84:4117-4121
120. Sharp JK, Albersheim P, Ossowski P, Pilotti Å, Garegg PJ, Lindberg B (1984) Comparison of the structures and elicitor activities of a synthetic and a mycelial-wall-derived hexa(β-D-glucopyranosyl)-D-glucitol. *J Biol Chem* 259:11341-11345
121. Sharp JK, McNeil M, Albersheim P (1984) The primary structures of one elicitor-active and seven elicitor-inactive hexa(β-D-glucopyranosyl)-D-glucitols isolated from the mycelial walls of *Phytophthora megasperma* f. sp. *glycinea*. *J Biol Chem* 259:11321-11336
122. Sharp JK, Valent B, Albersheim P (1984) Purification and partial characterization of a β-glucan fragment that elicits phytoalexin accumulation in soybean. *J Biol Chem* 259:11312-11320
123. Shibuya N, Kaku H, Kuchitsu K, Maliarik MJ (1993) Identification of a novel high-affinity binding site for *N*-acetylchitooligosaccharide elicitor in the membrane fraction from suspension-cultured rice cells. *FEBS Lett* 329:75-78
124. Shiraishi T, Saitoh K, Kim HM, Kato T, Tahara M, Oku H, Yamada T, Ichinose Y (1992) Two suppressors, supprescins A and B, secreted by a pea pathogen, *Mycosphaerella pinodes*. *Plant Cell Physiol* 33:663-667
125. Showalter AM and Varner JE (1989) Plant hydroxyproline-rich glycoproteins. In: Marcus A (ed) The Biochemistry of Plants: A Comprehensive Treatise, Vol. 15, Molecular Biology. Academic Press, Inc., New York, NY, p 485-520
126. Spiro MD, Kates KA, Koller AL, O'Neill MA, Albersheim P, Darvill AG (1993) Purification and characterization of biologically active 1,4-linked α-D-oligogalacturonides after partial digestion of polygalacturonic acid with endopolygalacturonase. *Carbohydr Res* 247:9-20
127. Staehelin C, Granado J, Müller J, Wiemken A, Mellor RB, Felix G, Regenass M, Broughton WJ, Boller T (1994) Perception of *Rhizobium* nodulation factors by tomato cells and inactivation by root chitinases. *Proc Natl Acad Sci USA* 91:2196-2200
128. Tani M, Fukui H, Shimomura M, Tabata M (1992) Structure of endogenous oligogalacturonides inducing shikonin biosynthesis in *Lithospermum* cell cultures. *Phytochemistry* 31:2719-2723
129. Tepper CS and Anderson AJ (1990) Interactions between pectic fragments and extracellular components from the fungal pathogen *Colletotrichum lindemuthianum*. *Physiol Mol Plant Pathol* 36:147-158
130. Thain JF, Doherty HM, Bowles DJ, Wildon DC (1990) Oligosaccharides that induce proteinase inhibitor activity in tomato plants cause depolarization of tomato leaf cells. *Plant Cell Environ* 13:569-574
131. Toubart P, Desiderio A, Salvi G, Cervone F, Daroda L, De Lorenzo G, Bergmann C, Darvill AG, Albersheim P (1992) Cloning and characterization of the gene encoding the endopolygalacturonase-inhibiting protein (PGIP) of *Phaseolus vulgaris* L. *Plant J* 2:367-373
132. Verduyn R, Douwes M, Van der Klein PAM, Mösinger EM, Van der Marel GA, van Boom JH (1993) Synthesis of a methyl heptaglucoside: Analogue of the phytoalexin elicitor from *Phytophtora megasperma*. *Tetrahedron* 49:7301-7316
133. Waldmann T, Jeblick W, Kauss H (1988) Induced net Ca^{2+} uptake and callose biosynthesis in suspension-cultured plant cells. *Planta* 173:88-95
134. Walker-Simmons M, Hadwiger L, Ryan CA (1983) Chitosans and pectic polysaccharides both induce the accumulation of the antifungal phytoalexin pisatin in pea pods and

antinutrient proteinase inhibitors in tomato leaves. *Biochem Biophys Res Commun* 110:194-199

135. Walker-Simmons M, Jin D, West CA, Hadwiger L, Ryan CA (1984) Comparison of proteinase inhibitor-inducing activities and phytoalexin elicitor activities of a pure fungal endopolygalacturonase, pectic fragments, and chitosan. *Plant Physiol* 76:833-836
136. Walker-Simmons M and Ryan CA (1984) Proteinase inhibitor synthesis in tomato leaves. Induction by chitosan oligomers and chemically modified chitosan and chitin. *Plant Physiol* 76:787-790
137. Yamada A, Shibuya N, Kodama O, Akatsuka T (1993) Induction of phytoalexin formation in suspension-cultured rice cells by *N*-acetylchitooligosaccharides. *Biosci Biotech Biochem* 57:405-409
138. Yoshikawa M, Keen NT, Wang M-C (1983) A receptor on soybean membranes for a fungal elicitor of phytoalexin accumulation. *Plant Physiol* 73:497-506
139. Yoshikawa M, Matama M, Masago H (1981) Release of a soluble phytoalexin elicitor from mycelial walls of *Phytophthora megasperma* var. *sojae* by soybean tissues. *Plant Physiol* 67:1032-1035
140. Yoshikawa M and Sugimoto K (1993) A specific binding site on soybean membranes for a phytoalexin elicitor released from fungal cell walls by β-1,3-endoglucanase. *Plant Cell Physiol* 34:1229-1237
141. Yoshioka H, Shiraishi T, Yamada T, Ichinose Y, Oku H (1990) Suppression of pisatin production and ATPase activity in pea plasma membranes by orthovanadate, verapamil and a suppressor from *Mycosphaerella pinodes*. *Plant Cell Physiol* 31:1139-1146
142. Young DH and Kauss H (1983) Release of calcium from suspension-cultured *Glycine max* cells by chitosan, other polycations, and polyamines in relation to effects on membrane permeability. *Plant Physiol* 73:698-702
143. Young DH, Köhle H, Kauss H (1982) Effect of chitosan on membrane permeability of suspension-cultured *Glycine max* and *Phaseolus vulgaris* cells. *Plant Physiol* 70:1449-1454

IN VITRO CULTURE AND PLANT REGENERATION IN GRAMINEOUS CROPS

B. V. CONGER and A. I. KUKLIN
Department of Plant and Soil Science
The University of Tennessee
Knoxville, TN 37901-1071
USA

1. Introduction

"Of all the plants of the earth the grasses are of the greatest use to the human race. To the grasses belong the cereals, sugarcane, sorghum and the bamboos; and, since they furnish the bulk of the forage for domestic animals, the grasses are also the basis of animal industry" (Hitchcock 1935). These words are at least as true today as they were when written 60 years ago and the primary reason for the high interest in including biotechnological applications in their improvement. As is well known, techniques for various cellular and molecular manipulations for the Gramineae (Poaceae) have been more difficult and hence slower to develop than for species in various other plant families. This is true in spite of the enormous effort expended by numerous laboratories throughout the world. Limitation of space will not permit a comprehensive review of the voluminous literature in this area; therefore, we will present a historical perspective and what we consider to be some of the significant highlights of *in vitro* culture of species in the Gramineae. Emphasis will be on the cereals and grasses. Additional and more specific information can be obtained from various other presentations in this Congress.

2. Callus Induction and Plant Regeneration

A fundamental requirement for nearly all applications of biotechnology is the regeneration of whole plants from cells and/or tissues cultured *in vitro*. The first papers describing successful regeneration from callus cultures in a gramineous species were authored by Japanese scientists working with rice, *Oryza sativa* L., (Kawata and Ishihara 1968; Nishi et al. 1968; Tamura 1968). Subsequently, regeneration was reported for the other two major cereals, wheat, *Triticum aestivum* L., (Shimada et al. (1969) and maize, *Zea mays* L., (Green and Phillips) 1975). The first successful regeneration of green plants in a forage grass species appears to be from calli derived from triploid ryegrass (*Lolium*) embryos (Ahloowalia 1975). Since then regeneration has been reported for all major cereals, many forage and turf grasses, sugarcane, (*Saccharum* sp.), and from species of various bamboo genera. Reports of regeneration in previously untested gramineous species continue to appear in the literature. For more extensive lists of species in the grass family regenerated from tissue cultures, the reader is referred to Conger (1981), Conger and Gray (1984) and Vasil and Vasil (1986).

A key factor in successful callus induction and plant regeneration in the above and other species in the grass family is the use of meristematic tissues as explants. These include mature

M. Terzi et al. (eds.), Current Issues in Plant Molecular and Cellular Biology, 59-68.

and immature embryos, unemerged inflorescences, and basal leaf tissue. Another essential component is the incorporation of a strong auxin, usually 2,4-dichlorophenoxyacetic acid (2,4-D), into the medium. The requirement for a cytokinin is less clear and results have been variable. For example, in our experiments with orchardgrass (*Dactylis glomerata* L.), we found that zeatin, a natural cytokinin, was inhibitory when high endogenously or added exogenously (Wenck et al. 1988). On the other hand, results with switchgrass (*Panicum virgatum* L.), indicate a much improved response when 6-benzylaminopurine (BAP), a synthetic cytokinin, is added to the medium (Denchev and Conger, 1994).

3. Somatic Embryogenesis

Most of the regeneration observed from callus cultures of gramineous species prior to 1980 was probably by organogenesis. In fact, in the late 1970's, regeneration from tissue cultures of cereals and other grasses was a controversial subject and promoted much discussion (sometimes heated) at the IV IAPTC Congress in Calgary in 1978. The group at the Friedrich Miescher-Institut in Basel considered that "the routine induction of callus cultures (in the classical sense of unorganized, more-or-less uniform population of dedifferentiated, dividing cells) is not possible from cereal explants" (King et al. 1978). Furthermore, they concluded that "most reports of plant regeneration from cultured cereal tissues are no more than derepression of presumptive shoot primordia which proliferate adventitiously in culture and that the capacity for shoot production is usually rapidly lost by dilution of primordia during subculture." Decrease in regeneration capacity by organogenesis over time and additional subcultures is widely accepted and is exemplified by our early results with *Festuca arundinacea* Shreb. (Lowe and Conger 1979).

The clear documentation of somatic embryogenesis in the early 1980's (see reviews by Morrish et al. 1987; Vasil 1987; and Vasil and Vasil 1986) represented a major breakthrough in cell and tissue culture of the Gramineae and provided the potential for genetic manipulation at the cell level. In 1982, we reported somatic embryogenesis in orchardgrass from mature embryos and basal leaf tissue (McDaniel et al. 1982; Hanning and Conger 1982). It was shown that somatic embryos in leaf segments arise directly from the mesophyll (Conger et al. 1983) and histological studies provided strong evidence for a single cell origin (Trigiano et al. 1989).

4. Suspension Cultures

The next logical step is the establishment of regenerable cell suspensions. In general, this has been more difficult than from callus cultures and the list of gramineous species in which successful regeneration has been obtained is much shorter (Vasil and Vasil 1986). In most experiments, calli are initiated from immature embryos, young leaves, or inflorescences. These calli are then separated from the explant and placed in liquid medium. In most cases, somatic embryos develop only to a young proembryo stage in liquid and the suspension must be plated onto solid medium for further embryo development and maturation. An exception is orchardgrass in which embryos develop fully to a germinable stage in a single liquid medium (Gray et al. 1984) and therefore represents a system analogous to that which exists for certain dicot species, e.g., carrot (*Daucus carota* L.). We have documented stages ranging from the initial cell division to embryos which are essentially identical to those in mature caryopses (Conger et al. 1989).

In recent years, suspension cultures have been used in a major way for protoplast isolation and gene transfer experiments. These areas are mentioned below.

5. Protoplast Isolation and Culture

One of the most significant events in the history of plant cell and tissue culture was the demonstration of "parasexual hybridization" between two *Nicotiana* species (Carlson et al. 1972). The most significant contribution of this work may be the high interest and excitement that it created. It provided a stimulus for many to initiate work in this area and it generated financial support from both public and private sources. Unfortunately, the potential applications, especially in terms of creating new hybrids between species which are not closely related, have fallen short of that anticipated at the time.

Again, success in regeneration from protoplasts has been difficult and slow to achieve in species of the grass family. Rice was the first cereal in which plants were regenerated from protoplasts and successfully established in soil (Abdullah et al. 1986; Coulibaly and Demarly 1986; Fujimura et al. 1985; Yamada et al. 1986). See also reviews by Hodges et al. (1993); Lörz et al. (1988); Vasil (1992). This was approximately 15 years after the initial report of whole plant regeneration from tobacco protoplasts (Takebe et al. 1971). Orchardgrass was the first forage grass in which protoplast regeneration was accomplished (Horn et al. 1988a). Since then regeneration of protoplast-derived green plants has been reported for various other forage grasses. Nielsen et al. (1993) have listed these in their paper reporting regeneration from protoplasts of Kentucky bluegrass (*Poa pratensis* L.).

The above mentioned works utilized protoplasts isolated from embryogenic suspension cultures. Regeneration from mesophyll protoplasts in a grass or cereal species was accomplished only very recently. Again the species was rice (Gupta and Pattanayak 1993).

6. Gene Transfer

One of the most exciting areas and one of extremely high interest and activity is that of "gene transfer" or "genetic transformation." The first solid evidence for the transfer, expression and inheritance of foreign genes in higher plants was reported in 1983 and published the following year (De Block et al. 1984; Horsch et al. 1984). These experiments utilized the crown gall inducing bacterium, (*Agrobacterium tumefaciens*), to transfer antibiotic resistance genes into cells of tobacco and petunia. Both of these species are in the family Solonaceae. The fact that infection of gramineous species by *Agrobacterium* is very low or, perhaps, almost nonexistent, has not allowed this method of gene transfer to be a workable procedure in these species.

Early attempts of gene transfer in cereals involved techniques such as germinating seeds in pools of foreign DNA (Ledoux and Huart 1968) or injecting DNA into developing florets (see ref. in Soyfer 1980). Although successes were claimed, the work was subject to criticism (Kleinhofs et al. 1975) and molecular techniques to confirm the transformation, e.g., Southern blotting, were not yet developed.

Maize was the first cereal species to be genetically transformed; however, the plants were morphologically stunted and nonfertile (Rhodes et al. 1988). A few months later, there were three reports of successful transformation in rice which resulted in normal and/or fertile plants (Toriyama et al. 1988; Zhang et al. 1988; Zhang and Wu 1988). Later in the same year, transformation of the first forage grass, orchardgrass, was reported (Horn et al. 1988b). Transformation was obtained in tall fescue (*Festuca arundinacea* Schreb.) four years later (Wang et al. 1992) and very recently transgenic plants were obtained in redtop, *Agrostis alba* L. (Asano and Ugaki 1994). The above examples, utilized direct uptake of DNA by protoplasts and then whole plant regeneration from the altered protoplasts. This remains an important method but difficulties in regeneration from protoplasts continue to be a problem. The pollen-tube method of transformation described in rice (Luo and Wu 1988) has been difficult to repeat in other species (Martin et al. 1992) and therefore, has not found wide acceptance.

A more recent, and currently the most favored method for gene transfer in cereals and other grass species, is bombardment of calli, plated suspensions and other tissues with DNA coated tungsten or gold microprojectiles. This method has resulted in fertile transgenic plants of maize (Fromm et al. 1990; Gordon-Kamm et al. 1990), wheat (Vasil et al. 1992; Weeks et al. 1993), oats, *Avena sativa* L., (Somers et al. 1993), barley, *Hordeum vulgare* L., (Ritala et al. 1994), both indica and japonica rice (Christou et al. 1991; Li et al. 1993; Peng et al. 1992), sugarcane (Bower and Birch 1992) and creeping bentgrass, *Agrostis palustris* Huds., (Zhong et al. 1993). Replacement of the CaMV 35S promotor with the maize ubiquitin gene (Christiansen et al. 1992) has improved transformation rates in various gramineous species.

Examples of gramineous crops in which successful gene transfer (transgenic plants established and grown in soil) has been reported are listed in Table 1. Selected references for each crop are provided. The reader may obtain additional references from these articles and from Hodges et al. (1993).

7. Conclusions

The preceding has been a brief overview of *in vitro* culture in the Gramineae beginning with the first reports of regeneration from callus cultures to the recent reports of genetic transformation. Limitation of space has prevented a more complete and detailed treatment. The areas of anther culture and somaclonal variation were not covered at all. These are covered in other symposia of this Congress and presented in these preceedings. The chairpersons of these symposia have extensive experience with cereal species and therefore, the coverage should be pertinent for those interested in the Gramineae. Additional information on subjects which were presented in this paper, e.g., gene transfer, may also be found elsewhere in these proceedings.

Table 1. Examples of transgenic plants in the family Gramineae. P = direct uptake of DNA by protoplasts. M = microprojectile bombardment.

Genus species	Common name	Transformation method	Selected references
Avena sativa L.	oats	M	Somers et al. 1993
Hordeum vulgare L.	barley	M	Ritala et al. 1994
Oryza sativa L. (indica)	rice	M	Christou et al. 1991
		M	Li et al. 1993
		M	Peng et al. 1992
Oryza sativa L. (japonica)	rice	P	Toriyama et al. 1988
		P	Zhang et al. 1988
		P	Zhang and Wu 1988
		M	Christou et al. 1991
		M	Li et al. 1993
		M	Peng et al. 1992
Triticum aestivum L.	wheat	M	Vasil et al. 1992
		M	Weeks et al. 1993
Zea mays L.	corn or maize	P	Rhodes et al. 1988
		M	Fromm et al. 1990
		M	Gordon-Kamm et al. 1990
		P	Golovkin et al. 1993
Agrostis alba L.	redtop	P	Asano and Ugaki 1994
Agrostis palustris Huds.	creeping bentgrass	M	Zhong et al. 1993
Festuca arundinacea Schreb.	tall fescue	P	Wang et al. 1992
Dactylis glomerata L.	orchardgrass	P	Horn et al. 1988b
Saccharum sp.	sugarcane	M	Bower and Birch 1992

References

Abdullah, R., Cocking, E. C. and Thompson, J. A. (1986) Efficient plant regeneration from rice protoplasts through somatic embryogenesis. Bio/Technology 4, 1087-1090.

Ahloowalia, B. S. (1975) Regeneration of ryegrass plants in tissue culture. Crop Sci. 15:449-452.

Asano, Y. and Ugaki, M. (1994) Transgenic plants of *Agrostis alba* obtained by electroporation-mediated direct gene transfer into protoplasts. Plant Cell Rep. 13, 243-246.

Bower, R. and Birch, R. G. (1992) Transgenic sugarcane plants via microprojectile bombardment. The Plant J. 2, 409-416.

Carlson, P. S., Smith, H. H. and Dearing, R. D. (1972) Parasexual interspecific plant hybridization. Proc. Nat. Acad. Sci. U.S.A. 69, 2292-2294.

Christiansen, A. H., Sharrock, R. A. and Quail, P. H. (1992) Maize polyubiquitin genes: structure, thermal perturbation of expression and transcript splicing, and promotor activity following transfer to protoplasts by electroporation. Plant Molec. Biol. 18, 675-689.

Christou, P., Ford, T. L. and Kofron, M. (1991) Production of transgenic rice (*Oryza sativa L.*) plants from agronomically important indica and japonica varieties via electric discharge particle acceleration of exogenous DNA into immature zygotic embryos. Bio/Technology 9, 957-962.

Conger, B. V. (1981) Agronomic crops. in B. V. Conger (ed.) Cloning Agricultural Plants via *In Vitro* Techniques. CRC Press Inc., Boca Raton, pp. 165-215.

Conger, B. V. and Gray, D. J. (1984) In vitro culture in forage grass improvement. in R. E. Barker and B. L. Burson (eds.) Proc. 28th Grass Breeders Work Planning Conf. USDA-ARS, Mandan, ND. pp. 50-64.

Conger, B. V., Hanning, G. E., Gray, D. J. and McDaniel, J. K. (1983) Direct embryogenesis from mesophyll cells of orchardgrass. Science 221, 850-851.

Conger, B. V., Hovanesian, J. C., Trigiano, R. N. and Gray, D. J. (1989) Somatic embryo ontogeny in suspension cultures of orchardgrass. Crop Sci. 29, 448-452.

Coulibaly, M. Y. and Demarly, Y. (1986) Regeneration of plantlets from protoplasts of rice, *Oryza sativa* L. Z. Pflanzenzüchtg. 96, 79-81.

De Block, M., Herrera, Estrella, L., Van Montagu, M., Schell, J. and Zambrisky, P. (1984) Expression of foreign genes in regenerated plants and their progeny. EMBO J. 3, 1681-1689.

Denchev, P. D. and Conger, B. V. (1994) Plant regeneration from *in vitro* cultures of switchgrass. Crop Sci. (in press).

Fromm, M. E., Morrish, F., Armstrong, C., Williams, R., Thomas, J. and Klein, T. M. (1990) Inheritance and expression of chimeric genes in the progeny of transgenic maize plants. Bio/Technology 8:833-839.

Fujimura, T., Sakurai, M., Akagi, H., Negishi, T. and Hirose, A. (1985) Regeneration of rice plants from protoplasts. Plant Tissue Cult. Lett. 2, 74-75.

Golovkin, M. V., Abraham, M., Mórocz, S., Bottka, S., Fehér, A. and Dudits, D. (1993) Production of transgenic maize plants by direct DNA uptake into embryogenic protoplasts. Plant Sci. 90, 41-52.

Gordon-Kamm, W. J., Spencer, T. M., Mangano, M. L., Adams, T. R., Daines, R. J., Start, W. G., O'Brien, J. V., Chambers, S. A., Adams, W. R. Jr., Willets, N. G., Rice, T. B., Mackey, C. J., Krueger, R. W., Kausch A. B. and Lemaux P. G. (1990) Transformation of maize cells and regeneration of fertile transgenic plants. The Plant Cell 2:603-618.

Gray, D. J., Conger, B. V. and Hanning, G. E. (1984) Somatic embryogenesis in suspension and suspension-derived callus cultures of *Dactylis glomerata*. Protoplasma 122, 196-202.

Green, C. E. and Phillips, R. L. (1975) Plant regeneration from tissue cultures of maize. Crop Sci. 15, 417-421.

Gupta, H. S. and Pattanayak (1993) Plant regeneration from mesophyll protoplasts of rice (*Oryza sativa* L.) Bio/Technology 11, 90-94.

Hanning, G. E. and Conger, B. V. (1982) Embryoid and plant formation from leaf segments of *Dactylis glomerata* L. Theor. Appl. Genet. 63, 155-159.

Hitchcock, A. S. (1935) Manual of the Grasses of the United States. United States Government Printing Office, Washington.

Hodges, T. K., Rathore, K. S. and Peng. J. (1993) Advances in genetic transformation of plants, in Proc. XVII Intern. Grassland Cong., Keeling and Mundy, Palmerston North, New Zealand. pp. 1013-1023.

Horn, M. E., Conger, B. V. and Harms, C. T. (1988) Plant regeneration from protoplasts of embryogenic suspension cultures of orchardgrass (*Dactylis glomerata* L.) Plant Cell Rep. 7, 371-374.

Horn, M. E., Shillito, R. D., Conger, B. V. and Harms, C. T. (1988) Transgenic plants of orchardgrass (*Dactylis glomerata* L.) Plant Cell Rep. 7, 469-472.

Horsch, R. B., Fraley, R. T., Rogers, S. G., Sanders, P. R., Lloyd, A. and Hoffman, N. (1984) Inheritance of functional foreign genes in plants. Science 223, 496-498.

Kawata, S. and Ishihara, A. (1968) The regeneration of rice plant, *Oryza sativa* L., in the callus derived from the seminal root. Proc. Japan Acad. 44, 549-553.

King, P. J., Potrykus, I. and Thomas, E. (1978) *In vitro* genetics of cereals: problems and perspectives. Physiol. Veg. 16, 381-399.

Kleinhofs, A., Eden, F. C., Chilton, M. D. and Bendich, A. J. (1975) On the question of the integration of exogenous bacterial DNA into plant DNA. Proc. Nat. Acad. Sci. USA 72, 2748-2752.

Ledoux, L. and Huart, R. (1968) Integration and replication of DNA of *M. lysodeikticus* in DNA of germinating barley. Nature 218:1256-1259.

Li, L., Qu, R., deKochko, A., Fauquet, C. and Beachy, R. N. (1993) An improved rice transformation system using the biolistic method. Plant Cell Rep. 12, 250-255.

Lörz, H., Göbel, E. and Brown, P. (1988) Advances in tissue culture and progress towards genetic transformation of cereals. Plant Breeding 100, 1-25.

Lowe, K. W. and Conger, B. V. (1979) Root and shoot formation from callus cultures of tall fescue. Crop Sci. 19, 397-400.

Luo, Z. and Wu, R. (1988) A simple method for the transformation of rice via the pollen-tube pathway. Plant Molec. Biol. Rep. 6, 165-174.

Martin, N., Forgeois, P. and Picard, E. (1992) Investigations on transforming *Triticum aestivum* via the pollen tube pathway. Agronomie 12, 537-544.

McDaniel, J. K., Conger, B. V. and Graham, E. T. (1982) A histological study of tissue proliferation, embryogenesis and organogenesis from tissue cultures of *Dactylis glomerata* L. Protoplasma 110, 121-128.

Morrish, F., Vasil, V., and Vasil, I. K. (1987) Developmental morphogenesis and genetic manipulation in tissue and cell cultures of the Gramineae. Adv. Genet. 24, 431-499.

Nielsen, K. A., Larsen, E. and Knudsen, E. (1993) Regeneration of protoplast-derived green plants of Kentucky blue grass (Poa pratensis L.). Plant Cell Rep. 12, 537-540.

Nishi, T., Yamada, Y. and Takahashi, E. (1968) Organ redifferentiation and plant restoration in rice callus. Nature 219, 208-209.

Peng, J., Kononowicz, H. and Hodges, T. K. (1992) Transgenic indica rice plants. Theor. Appl. Genet. 83, 855-863.

Rhodes, C. A., Pierce, D. A., Mettler, I. J., Mascarenhas, D. and Detmer, J. J. (1988) Genetically transformed maize plants from protoplasts. Science 240, 204-207.

Ritala, A., Aspegren, K., Kurten, U., Salmenkallio-Marttila, M., Mannonen, L., Hannus, R., Kauppinen, V., Terri, T. H. and Enari, T.-M. (1994) Fertile transgenic barley by particle bombardment of immature embryos. Plant Molec. Biol. 24, 317-325.

Shimada, T., Sasakuma, T. and Tsunewaki, K. (1969) *In vitro* culture of wheat tissues. I. Callus formation, organ redifferentiation and single cell culture. Can. J. Genet. Cytol. 11, 294-304.

Somers, D. A., Rines, H. W., Gu, W., Kaeppler, H. F. and Bushnell, W. R. (1993) Fertile transgenic oat plants. Bio/Technology 10, 1589-1594.

Soyfer, V. N. (1980) Heredity variability of plants under the action of exogenous DNA. Theor. Appl. Genet. 58, 225-235.

Takebe, I., Labib, G. and Melchers, G. (1971) Regeneration of whole plants from isolated mesophyll protoplasts of tobacco. Naturwissenschaften 58, 318-320.

Tamura, S. (1968) Shoot formation in calli originated from rice embryo. Proc. Japan Acad. 44, 544-548.

Toriyama, K., Arimoto, Y., Uchimiya, H. and Hinata, K. (1988) Transgenic rice plants after direct gene transfer into protoplasts. Bio/Technology 6, 1072-1074.

Trigiano, R. N., Gray, D. J., Conger, B. V. and McDaniel, J. K. (1989) Origin of direct somatic embryos from cultured leaf segments of *Dactylis glomerata*. Bot. Gaz. 150, 72-77.

Vasil, I. K. (1987) Developing cell and tissue culture systems for the improvement of cereal and grass crops. J. Plant Physiol. 128, 193-218.

Vasil, I. K. (1992) Advances in cereal protoplast research. Physiol. Plant. 85, 279-283.

Vasil, V., Castillo, A. M., Fromm, M. E. and Vasil, I. K. (1992) Herbicide resistant fertile transgenic wheat plants by microprojectile bombardment of regenerable embryogenic callus. Bio/Technology 10, 667-674.

Vasil, I.K. and Vasil, V. (1986) Regeneration in cereal and other grass species, in I. K. Vasil (ed.) Cell Culture and Somatic Cell Genetics of Plants V. 3 Plant Regeneration and Genetic Variability, Academic Press Inc., Orlando, pp. 121-150.

Wang, Z., Takamizo, T., Iglesias, V. A., Osusky, M., Nagel, J., Potrykus, I. and Spangenberg, G. (1992) Transgenic plants of tall feseue (*Festuca arundinacea* Schreb.) obtained by direct gene transfer to protoplasts. Bio/Technology 10, 691-696.

Weeks, J. T., Anderson, O. D. and Blechl, A. E. (1993) Rapid production of multiple independent lines of fertile transgenic wheat (*Triticum aestivum*). Plant Physiol. 102, 1077-1084.

Wenck, A. R., Conger, B. V., Trigiano, R. N. and Sams, C. E. (1988) Inhibition of somatic embryogenesis in orchardgrass by endogenous cytokinins. Plant Physiol. 88, 990-992.

Yamada, Y., Yang, Z. Q. and Tang, D. T. (1986) Plant regeneration from protoplast derived callus of rice (*Oryza sativa* L.). Plant Cell Rep. 5, 85-88.

Zhang, H. M.,. Yang, H., Rech, E. L., Golds, T. J., Davis, A. S., Mulligan, B. J. and Cocking, E. C. (1988) Transgenic rice plants produced by electroporation mediated plasmid uptake into protoplasts. Plant Cell Rep. 7, 379-383.

Zhang, W. and Wu, R. (1988) Efficient regeneration of transgenic plants from rice protoplasts and correctly regulated expression of the foreign gene in plants. Theor. Appl. Genet. 76, 835-840.

Zhong, H., Bolyard, M. G., Srinivasan, C. and Sticklen, M. B. (1993) Trangenic plants of turfgrass (*Agrostis palustris* Huds.) from microprojectile bombardment of embryogenic callus. Plant Cell Rep. 13, 1-6.

PROTOPLAST, CALLUS AND SUSPENSION CULTURE OF PERENNIAL RYEGRASS - EFFECT OF GENOTYPE AND CULTURE SYSTEM

A. Olesen, M. Storgaard, M. Folling, S. Madsen & S.B. Andersen
Dept. Agricultural Sciences
Royal Veterinary and Agricultural University
Thorvaldsensvej 41
DK-1871 Frederiksberg C
Denmark

ABSTRACT. Replicated experiments with callus, suspension and protoplast cultures of perennial ryegrass were performed to quantify the genetic component of plant regeneration parameters. In callus culture genotype has by far the greatest influence on plant regeneration. In suspension culture, however, the effect of genotype is less pronounced and the random variation due to uncontrollable variation is larger. Regeneration pattern for protoplasts parallels that of the source suspension. The study shows a positive correlation between plant regeneration in anther culture and in somatic tissue cultures, with upto 30-50 per cent of the genotypic variation for plant regeneration in callus and suspension culture being explained by the correlation with anther culture response parameters. Long-term regenerable suspension cultures can be established in ryegrass, since finely dispersed cell suspensions, maintained for more than 36 months, have retained the ability to regenerate high yield of green plants. In protoplast nurse culture, upto 10 per cent plating efficiency and upto 40.000 green protoplast-derived plantlets per ml sedimented suspension cells was obtained. Such nurse cultures have a stronger positive effect than conditioned medium.

Introduction

During the last decade it has become possible to regenerate viable plants from embryogenic cell suspensions and protoplasts in a number of grass species including perennial ryegrass (*Lolium perenne* L.) (Dalton, 1988; Creemers-Molenaar *et al.*, 1989; Zaghmout & Torello, 1992; Wang *et al.*, 1993.). Protoplast fusion has resulted in cybrid/hybrid callus in perennial ryegrass (Creemers-Molenaar *et al.*, 1992a) and intergeneric hybrid plants have been obtained from fusion between tall fescue and Italian ryegrass protoplasts (Takamizo *et al.*, 1991). However, the general lack of an efficient and reproducible plant regeneration system is still hampering exploitation of protoplast fusion and gene transfer to protoplasts in ryegrass. Rapid decline of regeneration capacity with age in cell suspensions and the occurrence of albino plants are serious problems (e.g. Dalton, 1988; Zagmout and Torello, 1992). A number of studies have revealed the influence of genotype in callus culture (e.g. Creemers-Molenaar *et al.*, 1988). The general magnitude of genetic determination in callus culture and also its possible influence in suspension culture, however, is less well known. This paper focuses on the genetic control of responses in somatic cell culture in order to determine its magnitude and possible correlation with the genetic control of anther culture response. In addition, improvement of the protoplast culture system by means of nurse cultures is studied.

M. Terzi et al. (eds.), Current Issues in Plant Molecular and Cellular Biology, 69-74.

Materials and Methods

GENOTYPIC EFFECTS

The plant material consisted of 10 clones of perennial ryegrass selected for agronomic value and 11 clones with good anther culture response. All clones were evaluated for anther culture response in a separate experiment estimating general combining ability (GCA) for embryo formation, regeneration, green plant formation and green plants per 100 anthers (Madsen *et al.*, 1994). Calli were induced from meristems in three blocks representing three seasonal environments on MS basal medium (Murashige and Skoog, 1962) supplemented with 4 mg/l 2,4-D, 0.2 mg/l BAP, 100 mg/l casein hydrolysate, 3 per cent sucrose and solidified with 0.8 per cent Difco agar. They were subcultured on the same medium and tested for regeneration every 28 days for upto 17, 14 and 11 months, respectively, for the three environments. A number of cell suspensions were initiated from calli from each genotype/environment combination 96 days after plating of the meristems using medium described by Dalton (1988), with the exception that it contained 3 mg/l 2,4-D. Plant regeneration was tested every 14-28 days following methods described by Dalton (1988) and studied for upto 22, 19 and 10 months, respectively, for the three environments. Protoplast culture was performed with cell suspensions 6-9 months old according to the methods established by Dalton (1988), with the exception that the carbohydrate source in the protoplast culture medium (PC4) was 11 per cent glucose. Analysis of variance and calculation of variance components was performed on square root transformed data for somatic cultures using GCA and GCA^2-values for anther culture response as covariables to explain genetic effect on somatic response.

PROTOPLAST NURSE CULTURE

The study utilized four cell suspensions, two derived from mature embryos (89-4C, 26 months old) as described by Dalton (1988), and two from meristems (I 6-8, 20 months old). Cells for protoplast isolation and nurse culture were prepared by subculturing 1 ml sedimented volume of small dense cell colonies 1:80 with fresh cell suspension medium 5 days before protoplast isolation. Conditioned medium was taken from the suspensions 5 days after subculture and the osmolality adjusted to 800 mOsm/kg with glucose. Protoplasts were cultured using a modified agarose drop method (Steinbrenner *et al.*, 1989). Agarose drops (1.2 per cent sea plaque agarose) containing protoplasts with a density of 2×10^5 in culture medium were placed in wells of a 24 well multidish and, following solidification, either 0.5 ml protoplast culture medium (see above) with 30 mg nurse cells or 0.5 ml conditioned medium was added per well. As a control, the protoplasts were cultured in 0.5 ml culture medium solidified with 1.2 per cent agarose. Plating efficiency was determined 27 days after isolation, and the agarose drops were washed and spread onto solid protoplast plating medium (Dalton, 1988). After a further 4-6 weeks, calli were transferred to regeneration medium (Dalton 1988). In all culture systems regeneration frequency was expressed as the frequency of calli that regenerated at least one green or albino plant.

Results

GENOTYPIC EFFECTS

A total of 156 suspensions were established, which regenerated an average of 14.4 plants per 100 cell colonies, with 55.9 per cent of the regenerated plants being green (Table 1). Analysis of variance showed a highly significant effect of genotype on callus induction, the component of which explained about 40 per cent of the total variation for this character. Genotypes had an even

TABLE 1. Regeneration from 156 cell suspensions from 18 genotypes of perennial ryegrass.

	Actual No.	Plants/100 colonies	% Green plants
Total no. cell colonies on reg.medium	34,338		
Total plants	4,951	14.4	
Green plants	2,770	8.1	55.9

larger impact on regeneration frequency and percentage of green plants from callus culture, where this component explained 59 and 83 per cent of the total variation respectively (Figure 1). The main effect of genotype on regeneration from suspension culture was relatively smaller, although still highly significant, and the experimental error relatively larger than in callus culture. Variation between suspensions of the same genotype was large, and explained more of total variation than the genotype itself. The effect of environments on plant regeneration and proportion of green plants was small and in some instances non-significant. Differences between cell suspensions from the same genotype in terms of protoplast yield were significant and again larger than the variation between genotypes. Effect of environments was large and responsible for a greater part of the total variation than genotypes and variation within genotypes (Figure 1). The cell suspensions from the first block (environment) were more finely dispersed than suspensions form the second and third block and this is probably the major reason for the large difference between environments for protoplast yield. Regeneration capacity of the protoplast colonies generally paralleled the regeneration capacity of the donor suspensions.

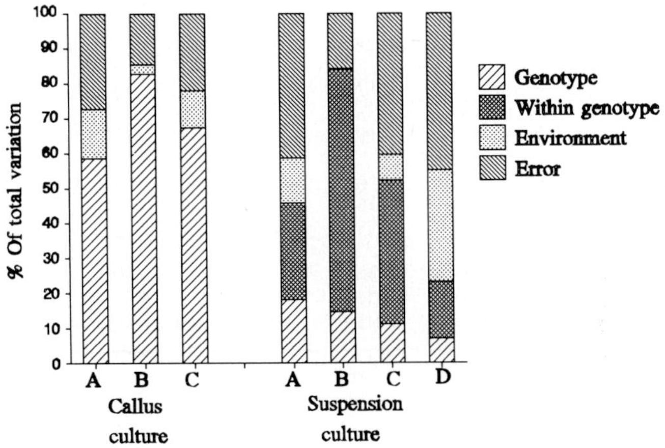

FIGURE 1. Relative contributions of genotype, suspensions within genotypes, environment and error to the total variation in ryegrass somatic tissue culture. A: % Regeneration, B: % Green plants, C: No. of green plants per 100 colonies, D: protoplast yield.

Regeneration frequency in the two somatic tissue culture systems was generally positively correlated when measured over the individual genotypes (data not shown). Furthermore, yield of green plants (green plants/100 colonies) in somatic tissue culture correlated positively with yield of green plants in anther culture (green plants/100 anthers). This relation was partly caused by a correlation between regeneration frequency in somatic tissue culture and anther culture, explaining 18-31 per cent of the genotypic variation for plant regeneration in somatic tissue culture. A correlation of regeneration frequency and green plants/100 colonies in suspension culture with % green plants in anther culture ($r^2 = 0.38$ and 0.52, respectively) contributed further to the overall correlation between the two systems.

PROTOPLAST NURSE CULTURE

Nurse cells had a very positive effect on plating efficiency and regeneration of green plants (Table 2), whereas conditioned medium did not clearly improve plating efficiency. We also investigated conditioned medium mixed 1:1 with protoplast culture medium, but this was inferior to the procedure used here. With nurse cells the plating efficiency of protoplasts was upto 8-10 per cent. A high regeneration frequency and proportion of green plants resulted in a final yield of green plantlets of upto 40,368 when calculated per ml suspension cells. Many of the plants have been established in soil.

TABLE 2. The effect of different culture conditions on the plating effiency and regeneration in protoplast culture of perennial ryegrass. Nurse cells and conditioned medium were taken from the protoplast donor suspension (self-feeding). Data are from two independent experiments separated by 41 days. Each experiment consisted of two replicates.

Suspension line code	Protoplast yield (x 10^6)	Culture method	Plating efficiency (%)	Regeneration			
				Calli per 4 drops (4/24 ml)	Regeneration % (No.)	Green plants % (No.)	Green plants/ ml susp.
I6-8A2	5.7	Control	1.46	151	20.5 (31)	22.5 (7)	1197
		Cond. med.	2.08	383	44.6 (171)	90.1 (154)	26334
		Nurse	10.24	665	36.4 (242)	86.8 (210)	35910
I6-8C	5.8	Control	0.17	3	33.3 (1)	0.0 (0)	-
		Cond. med.	0.33	42	64.3 (27)	63.0 (17)	2958
		Nurse	8.19	819	57.6 (472)	49.2 (232)	40368
89-4C-III	4.3	Control	0.16	11	0.0 (0)	-	-
		Cond. med.	0.16	6	0.0 (0)	-	-
		Nurse	4.55	283	7.8 (22)	100 (22)	2838
89-4C-III G1	7.6	Control	0.36	54	9.3 (5)	100 (5)	1140
		Cond. med.	0.04	0	-	-	-
		Nurse	0.91	122	4.1 (5)	100 (5)	1140

Discussion

Important factors for the application of protoplast technology to grasses relate to the protoplast methodology and to the performance of the donor cell suspension. Yield and plating efficiency of protoplasts isolated from morphogenic ryegrass suspensions are generally fairly low. We found the strong dilution of the cell suspension colonies 5 days before the cells are to be used for protoplast isolation or as nurse beneficial for protoplast yield from low yielding suspensions, and possibly also for the feeding capacity of the nurse (see materials and methods). This may be due to an increased cell division rate which does not last long enough to reduce plant regeneration from the suspension-derived protoplasts. The use of nurse cells has a dramatic effect on plating efficiency and perhaps also on regeneration capacity, since the colonies appear more compact and embryogenic with nurse culture. The major influence of the nurse might be an earlier initiation of cell wall formation, earlier onset of cell division and more rapid initial cell division rate. Creemers-Molenaar *et al.* (1992b) reported addition of conditioning medium to be necessary for the proliferation of microcalli derived from young regeneration-competent suspension cultures. This positive effect of conditioned medium for ryegrass protoplast culture has been confirmed in our study. However, nurse cultures were much better than conditioned medium. The stimulating effect of nurse on ryegrass protoplasts has been reported by others (Zaghmout and Torello, 1992; Wang *et al.*, 1993), but a protoplast culture system of the efficiency found here has not been reported for ryegrass so far. This systems now works so well that protoplast approaches to gene transfer may be an alternative to biolistic procedures for gene transfer.

Quality of the donor suspension is a major determinant of the efficiency of the protoplast system. Ryegrass cell suspensions usually only retain green plant regeneration capacity for upto 25 weeks (e.g. Dalton, 1988). Occurrence of albinos is a major problem in both anther culture and somatic tissue culture of ryegrass (e.g. Olesen *et al.*, 1988, Dalton 1988). For cereal anther culture Day and Ellis (1985) demonstrated that deletions in the chloroplast genome are responsible for albino formation, even if the frequency of green plant formation appears to be controlled by a few major genes on the nuclear chromosomes (Tuvesson *et al.*, 1989; Larsen *et al.*, 1991). The cell suspensions used in the study of protoplast nurse culture were relatively old, but some suspensions are fairly stable in terms of regeneration behaviour when measured over different suspensions of the same origin. We have two genetically different lines of cell suspensions established from mature embryos, which are three years old, finely dispersed and still with very high regeneration frequencies (up to 100 green plants per 100 cell colonies).

We have previously found suspension culture performance to be affected by variety with some varieties being clearly and reproducibly superior to others, some inferior and a large intermediate group of varieties with small differences. To study the effect of genotype further, single-genotype callus and suspension cultures were established. This work identified good callus producing genotypes with a high potential for green plant regeneration, since 59 - 83% of the variation for regeneration parameters was determined by genotype. The relative influence of genotype (although highly significant) is reduced and the uncontrollable variation increased in cell suspension culture. Furthermore, cell suspensions of the same genotype were found to differ in terms of regeneration and proportion of green plants and this intra-genotypic variation was more important than the main effect of genotype. The rapid cell division rate and greater opportunity for random changes and sorting out of variant types when subculturing the suspensions may be the reason for this large variance between suspensions from the same genotype. The difference between environments for plant regeneration was small in these experiments but was more important for protoplast yield.

Regeneration performance of the genotypes in somatic tissue cultures generally correlates positively to their level of regeneration in anther culture. Yield in anther culture is usually measured in green plants per 100 anthers which is obtained as a multiple of the individual components, i.e. embryos per 100 anther, % regeneration and % green plants which seem to be under individual genetic control (Halberg *et al.*, 1990). The correlations between plant regeneration in anther and somatic tissue culture suggests that some of the genes controlling these traits are identical. We utilized the genetic component of anther culture response to construct genotypes of perennial ryegrass with a 10-50 fold improvement in yield of green plants by crossing slightly responsive parents (Halberg *et al.*, 1990). We further showed that these "inducer" types can be used to transfer the ability to produce green doubled haploids to ordinary breeding material (Madsen *et al.*, 1994). The positive genetic correlation between plant regeneration in the two systems and the dominating influence of genotype on plant regeneration in callus culture suggest that anther culture performance can select for genotypes with a higher average *in vitro* ability, of which some may show excellent performance in somatic tissue culture.

The suspensions giving the very high yield of protoplast-derived plants were initiated from such an "inducer" genotype (Table 2, I6-8). Such very responsive genotypes may be used for gene transfer and protoplast fusion and thereby function as "bridges" to ordinary breeding material.

Acknowledgements

This work was partly funded by the Agricultural and Veterinary Research Council. We thank Ms H.Faarup, Ms B. Henriksen and Mr. B. Kastrup for technical assistance and Dr. J. Hill for reading the manuscript.

References

Creemers-Molenaar,J., J.P.M. Loeffen & P. Van der Valk (1988): The effect of 2,4 Dichlorophenoxyacetic acid and donor plant environment on plant regeneration from immature inflorescence-derived callus of *Lolium perenne* L. and *Lolium multiflorum* L.- Plant Science, 57:165-172.

Creemers-Molenaar, J., P.Van der Valk, J.P.M. Loeffen & M.A.C.M Zaal (1989): Plant regeneration from suspension cultures and protoplasts of *Lolium perenne* L.- Plant Science, 63:167-176.

Creemers-Molenaar, J., R.D. Hall & F.A. Krens (1992a): Asymmetric protoplast fusion aimed at intraspecific transfer of cytoplasmic male sterility (CMS) in *Lolium perenne* L..- Theor. Appl.Genet., 84:763-770.

Creemers-Molenaar, J., F.A. Van Eeuwijk & F.A. Krens (1992b): Culture optimization for perennial ryegrass protoplasts.- J. Plant Physiol., 139:303-308.

Dalton, S.J. (1988): Plant regeneration from cell suspension protoplasts of *Festuca arundinacea* Schreb. (tall fescue) and *Lolium perenne* L (perennial ryegrass).- J. Plant Physiol., 132:170-175.

Day, A. & T.H.N. Ellis (1985): Deleted forms of plastid DNA in albino plants from cereal anther culture.- Curr. Genet., 9:671-678.

Halberg N., A. Olesen, I.K.D. Tuvesson & S.B. Andersen (1990): Genotypes of perennial ryegrass (*Lolium perenne* L.) with high anther-culture response through hybridization.- Plant Breeding, 105:89-94.

Larsen,E.T., I.K.D.Tuvesson & S.B.Andersen S.B.(1991): Nuclear genes affecting percentage of green plants in barley (*Hordeum vulgare* L.) anther culture.- Theor. Appl. Genet.,: 82:417-420.

Madsen S, A. Olesen, B. Dennis & S.B. Andersen (1994): Inheritance of anther culture response in perennial ryegrass (*Lolium perenne* L).- Plant Breeding, (submitted).

Murashige, T. & F. Skoog (1962): A revised medium for rapid growth and bioassays with tobacco tissue culture.- Physiol. Plant., 15:473-497.

Olesen A., S.B. Andersen & I.K. Due (1988): Anther culture response in perennial ryegrass (*Lolium perenne* L.).- Plant Breeding, 101:60-65.

Steinbrenner, B., R. Schroeder, B. Knoop and R. Beiderbeck (1989): Viability factors in plant cell suspension cultures - a novel bioassay.- J. Plant Physiol., 134:582-55.

Takamizo T., G. Spangenberg, K. Suginobu & I. Potrykus (1991): Intergeneric somatic hybridization in Gramineae: somatic hybrid plants between tall fescue (*Festuca arundinacea* Schreb.) and Italian ryegrass (*Lolium multiflorum* Lam.).- Mol. Gen. Genet., 231:1-6.

Tuvesson I.K.D., S. Pedersen & S.B. Andersen (1989): Nuclear genes affecting albinism in wheat (*Triticum aestivum* L.) anther culture.- Theor. Appl. Genet., 78:879-883.

Wang Z.Y., J. Nagel, I. Potrykus & G. Spangenberg (1993): Plants from cell suspension-derived protoplasts in *Lolium* species.- Plant Science, 94:179-193.

Zaghmout, O.M.F & W.A. Torello (1992): Plant regeneration from callus and protoplasts of perennial ryegrass (*Lolium perenne* L.).- J. Plant Physiol., 140:101-105.

BREAKTHROUGH IN THE PLANT REGENERATION FROM PROTOPLASTS OF *VICIA FABA* AND *V. NARBONENSIS*

MECHTHILD TEGEDER, OTTO SCHIEDER & THOMAS PICKARDT

Institut für Angewandte Genetik, Albrecht-Thaer-Weg 6, Freie Universität, 14195 Berlin, Germany

Abstract. Protoplasts of ten cultivars of *Vicia faba* and one accession of *V. narbonensis* were isolated from etiolated shoot-tips. Yields of $1.7\text{-}4.7 \times 10^6$ protoplasts/g fresh weight were obtained after a 14 h enzymatic digestion. After washing in a $CaCl_2$ and mannitol solution the protoplasts were embedded in sodium-alginate at a final density of 2.5×10^5 protoplasts/ml. Depending on cultivar cell divisions of 4.4-36.3 % and plating efficiencies up to 0.3 % were obtained. 8 weeks after embedding callusses were transferred to solid media. Somatic embryos were regenerated on auxin media for *V. narbonensis*. In the case of *V. faba* globular structures could be induced on auxin media, but no embryo regeneration was achieved. Shoot regeneration from *V. faba* cv. Mythos and *V. narbonensis* was sucessful on medium containing thidiazuron. Shoots were grafted on seedlings and transferred to soil. Fertile plants of both species were obtained.

Introduction

An efficient protoplast regeneration system is essential for plant cell manipulation such as somatic hybridisation or direct gene transfer. In grain legumes there is a striking discrepancy between the great number of regeneration protocols from explants (Parrot *et al.* 1992) and the few reports on plant regeneration from protoplasts. The regeneration of plants from protoplasts of grain legumes has been gradually overcome in *Glycine clandestina* and *G.canescens* (Newell and Luu 1985, Hammat *et al.* 1988, Myers *et al.* 1989), *Vigna aconitifolia* and *V.unquiculata* (Shekawat and Galston 1983, Sinha *et al.* 1983), *Pisum sativum* (Puonti-Kaerlas and Eriksson 1988, Lehminger-Mertens and Jacobsen 1989) and *Glycine max* (Wei et Xu 1988, Dhir *et al.* 1991).
However, until now plant regeneration from protoplasts in the genus *Vicia* has not been reported. Callus was obtained from protoplast cultures of *V. faba* and related species (Kao and Michayluk 1975, Binding and Nehls 1978, Donn 1978, Röper 1980). Regeneration of plants was only achieved from shoot-tips or nodal-explants (Parrot *et al.* 1992). A *de novo*-regeneration from dedifferentiated tissue has only been demonstrated for *V.narbonensis* (Pickardt *et al.* 1989).
Here we present a regeneration system for protoplasts from etiolated seedlings of *V. faba* and its close relative *V. narbonensis*.

Material & Methods

Plant material: Ten cultivars of *V. faba* and one accession of *V. narbonensis* (see Table 1) were tested. Seeds were surface sterilizied for 1 min in 70% ethanol, 10 min in a 5% sodiumhypochlorit solution and then rinsed 3 times with sterile tap-water. After soaking for 6-8 hours seeds were transferred to water-agar (0.8%) and kept for germination in the dark at 24°C.
Protoplast isolation: Shoot-tips and epicotyls from 7 days old etiolated seedlings were cut in 0.5 mm segments and incubated in enzyme solution (1.5-2g tissue/50ml) containing 5% cellulase TC (Serva), 1%

M. Terzi et al. (eds.), Current Issues in Plant Molecular and Cellular Biology, 75-80.

pectinase (Serva), 1% macerozyme R10 (Yakult), 1% hemicellulase (Sigma), 8 mM $CaCl_2(x2H_2O)$ and 0.2 M mannitol (500 mOsm/kg H_2O, pH 5.5). Digestion was performed in 250 ml flasks for 14 h in the dark at 25°C under continuous rotation in a Cell Production Roller Apparatus (Bellco/USA). After incubation protoplasts were purified in 4 steps: (1) Protoplasts were filtered through two layers of steel sieves (125 and 63 µm mesh size) and centrifuged for 8 min at 150xg (swing-out rotor). (2) Protoplasts were washed twice in 0.2M $CaCl_2(x2H_2O)$ which resulted in the precipitation of undigested tissue and burst protoplasts. (3) Protoplasts were separated from undigested tissue by filtration through a 60 µm nylonfilter. (4) After centrifugation at 150xg for 5 min the protoplast pellet was resuspended in 0.48M mannitol (500 mOsm/kg H_2O)

Protoplast culture: The protoplast solution was adjusted to a density of $5x10^5$ protoplasts/ml with 0.48 M mannitol (500mOsm/kg H_2O). The protoplasts were subsequently embedded in sodium-alginate by mixing equal volumes of the protoplast suspension and a 0.40 M mannitol-solution containing 2.1% sodium-alginate to obtain a final density of $2.5 \cdot 10^5$ protoplasts/ml. 0.5 ml aliquots of this mixture were distributed on agar-plates containing 0.4 M mannitol and 0.02 M $CaCl_2(x2H_2O)$.

After polymerisation the alginate disks containing the embedded protoplasts were transferred to petri dishes (6 cm ∅) containing 4 ml KM medium (Kao and Michayluk 1975) with 0.5 mg/l each of 2,4-D, NAA and BAP. The protoplasts were cultured at 25°C in the dark.

7 days after embedding 0.5 volumes of the medium, were exchanged by fresh medium without mannitol. During the following culture the liquid medium was renewed every 10 days. Cell division frequency was calculated 2 weeks after protoplast isolation. Plating efficiency was calculated after 6 weeks and is defined as the number of protocallusses per number of protoplasts plated.

Regeneration: 6-8 weeks after protoplast isolation protocallusses were released from the alginate by incubation for 30 min in a 20 mM sodium-citrate solution for depolymerisation of alginate matrix. After centrifugation (100xg, 3 min) the callusses were washed twice in distilled water and transferred on solid medium.

For induction of somatic embryos callusses of *V. faba* and *V. narbonensis* were cultivated on MS-medium (Murashige and Skoog 1962) with 3% sucrose, 0.25% Gelrite and 10 mg/l picloram (Pickardt *et al.* 1991). Cultures were kept in the dark at 25 °C. After 4 weeks the medium was exchanged by fresh MS-medium (1.5% sucrose, 0.25 % Gelrite, pH 5.7) containing 1.0 mg/l NAA. The callusses were incubated at 25°C and a 16 h photoperiod (79 µmol m^{-2} s^{-1}). Regenerated somatic embryos were transferred on MS medium with 0.5 mg/l BAP (3 % sucrose, 0.25 % Gelrite, pH 5.7).

For direct shoot regeneration the effect of MS-medium (3 % sucrose, 0.25 % gelrite, pH 5.7) supplemented with different combinations and concentrations of growth regulators was tested. All cultures were incubated at 25°C, 16 h photoperiod and a light intensity of 79 µmol m^{-2} s^{-1}. The callusses were subcultured every 4 weeks. Developing shoots were grafted on seedlings and cultured on hormone free MS medium (1.5 % sucrose, 0.25 % Gelrite, pH 5.7). Growing plantlets were transferred to sterilized soil and kept under "semi-sterile" conditions as described above. After 2-3 weeks the plants were transferred into the greenhouse.

Results

Purified protoplasts of *V. narbonensis* could simply be obtained by filtering and several washing steps in mannitol. For *V. faba*, however, several washes in 20 mM $CaCl_2$ $(x2H_2O)$ were essential to seperate the protoplasts from undigested tissue. The divalent calcium ions produce complexes of undigested tissue and bursted protoplasts. These complexes precipitate and can be removed by filtering which results in a pure protoplast solution.

Protoplasts were obtained from *V. narbonensis* and all tested cultivars of *V. faba*. The protoplast yield ranged from 1.7 to $4.7x10^6$ protoplasts/g fresh weight (Table 1). Incubation in mannitol solution (24 h in the dark at 4°C) followed by embedding in sodium-alginate was essential for successful cultivation of protoplasts. A concentration of 2.1% alginat was optimal for both *Vicia* species. 5-7 days after embedding protoplasts started to divide. Dependent on the cultivar cell division frequencies between 4.4 and 36.3 % were observed for *V. faba*. Plating efficiency up to 0.3 % were achieved (Figures 1 a+b). The division frequency of *V. narbonensis* was 36.0%.

Table 1: Division frequency (%) and plating efficiency (%) of *V. faba* and *V. narbonensis* (plating density 2.5×10^5 protoplasts/ml)

	Protoplast yield/g fw*	Division frequency (%)	Plating efficiency (%)
***V. faba* cv.**			
Albatross	2.3×10^6	13.1±3.7**	0.15±0.05**
Carola	1.9×10^6	7.8±5.3	0.02±0.02
Caspar	2.2×10^6	17.1±5.3	0.08±0.03
Dreifach-Weiße	2.4×10^6	15.0±0.6	0.24±0.01
Herz Freya	3.6×10^6	36.3±14.0	0.33±0.24
Kristall	3.7×10^6	13.6±12.6	0.10±0.10
Mythos	4.7×10^6	20.4±17.7	0.26±0.20
Piccolo	3.3×10^6	28.1±2.2	0.24±0.04
Topas	1.9×10^6	4.9±1.6	0.01±0
Troy	1.7×10^6	4.4±0.7	0.03±0.02
V. narbonensis	2.4×10^6	36.0±28.9	0.14±0.08

* Fresh weight **± Standard deviation (σ_{n-1})

In the case of *V. narbonensis* somatic embryos could be obtained (Table 2). An average number of 2 to 5 somatic embryos were regenerated per embryogenic callus. 20 to 34 % of these embryos showed shoot regeneration. Direct shoot formation was induced on MS media containing thidiazuron (TDZ). After grafting, 80% of the shoots developed into plants. All plants set seeds after self pollination.

Table 2: Somatic embryogenesis and direct shoot formation from protoplasts of *V. narbonensis*

V. narbonensis	Somatic embryogenesis*	Direct shoot formation**
No. of callusses tested	269	609
No. of callusses regenerating	96	79
Regeneration rate (%)	35.7	13.0

*Culture on MS-medium containing 10 mg/l picloram (4 weeks) and subsequent culture (10 weeks) on MS-medium with 1 mg/l NAA. ** Culture on MS-medium containing 2.5 mg/l TDZ, 0.1 mg/l NAA and 50 mg/l CH

Callusses of *V. faba* were cultured on MS-media supplemented with picloram. Globular, embryo-like structures were induced, but no plant regeneration was achieved. Therefore more than 60 different combinations and concentrations of growth regulators were tested for their ability to induce shoot regeneration in *V. faba*. The results of 4 hormone combinations are given in Table 3. Shoots could be regenerated after a culture of 7 months for *V. faba* cv. Mythos on MS medium with TDZ (Figures 1 c+d). Two morphological types of callusses could be clearly distinguished. Shoot formation was only achieved from globular callusses. Up to now 24 protoclones have shown shoot regeneration with a frequency of 7-64 shoots per callus. They were grafted on seedlings of cv. Mythos and transferred to the greenhouse. 25 % of the regenerated plants were fertile (Figures 1 e+f).

Figure 1: Plant regeneration from protoplasts of *V. faba* cv. Mythos. (a) cell division 3 weeks after embedding, (b) protocallusses in alginate matrix, (c) shoot bud, (d) multiple shoot formation, (e) grafted shoot, (f) regenerated flowering plants

Table 3: Shoot regeneration from protocallusses of *V. faba* cv. Mythos

Hormones* (mg/l)	No. of callusses tested	No. of globular callusses	No. of callusses with shoot formation
BAP 0.5 NAA 0.1	207	0	0
BAP 0.5 Zea 0.5 Kin 0.5 NAA 0.1	167	0	0
TDZ 1.0 NAA 0.1 CH 50**	178	24	24
Zea 1.0 NAA 0.1 2,4 D 0.5	113	0	0

* Basal medium MS ** CH casein hydrolysate

Discussion

Etiolated shoot-tips of *V. faba* and *V. narbonensis* seedlings proved to be a suitable source material for protoplast isolation. Meristematic and ontogenetic young tissue seems to be more appropriate for plant regeneration than mesophyll cells especially in recalcitrant families like *Gramineae* and *Fabaceae*. Plant regeneration was achieved for several *Gramineae* using embryogenic suspension culture (Abdulla et al. 1986, Rhodes et al. 1988, Wang et al. 1989, Vasil et al. 1990, Jähne et al. 1991). In grain legumes such as *Pisum sativum* and *Glycine max* seedlings (Puonti-Kaerlas and Eriksson 1988, Lehminger-Mertens and Jacobsen 1989) or immature embryos (Wei and Xu 1988, Dhir et al. 1991) were used for successful plant regeneration from protoplasts. Viable *V. faba* protoplasts could not be isolated without the application of a high enzyme concentration and the $CaCl_2$ purification step. Embedding in sodium alginate was essential for the culture of *V. faba* and *V. narbonensis* protoplasts. This effect has also been described for other species like sugar beet (Schlangstedt et al. 1992), apple (Huancaruna Perales and Schieder 1993) or potato (Schilde-Rentschler et al. 1988). Sodium alginate provides cellular protection against mechanical stress and gradients in enviromental conditions during the critical first days of protoplast culture (Draget et al. 1988). Schnabl et al. (1983) speculated that proteolytic enzymes released from dying cells are absorbed by the alginate matrix. The distance between embedded protoplasts is maintained and the negative effect of metabolites (e.g. phenol derivates) might be reduced. It was shown that the Ca^{2+} ions used for alginate polymerization have a stimulating effect on the cell division of *Arabidopsis thaliana* protoplasts (Park and Wernike 1993).
For *V. narbonensis* a protocol of somatic embryogenesis and regeneration of plants from protoplasts derived from etiolated shoot tips derived protoplasts was established. In the case of *V. faba* only globular embryo-like structures could be obtained on auxin medium. Direct organogenesis could not be induced on media with different combinations and concentrations of cytokinines (BAP, Kin, Zea). The breakthrough in plant regeneration of *V. faba* and *V. narbonensis* via direct organogenesis was achieved by medium with TDZ. Under the commercial name Dropp® it is known as a defoliant for cotton (Arndt et al. 1976). This cytokinin-like substance is the most ef-

fective in woody plant tissue culture (see Huetteman and Preece 1993). Coleman and Estabrooks (1992) supposed an influence of TDZ on cytokinin metabolism. As somatic embryogenesis can also be induced by TDZ (Gill and Saxena 1993, Gray et al. 1993), Visser et al. (1992) suggested on a synergism of endogenous cytokinins and auxins.
A genotype-dependent tissue culture fitness and regeneration ability is known for many legumes (Oelck und Schieder 1983, Gulati and Jaiwal 1992). Up to now we have only achieved plant regeneration from one cultivar of *V. faba* (cv. Mythos). The testing of more cultivars and lines of *V. faba* for their regeneration potential is under way. Apart from this, culture conditions will be improved to shorten the time frame for plant regeneration from protoplasts of *V. faba*.

References

Abdullah, R., Cocking, E., & Thompson, J.A. (1986) Bio/Technology 4, 1087-1090.
Arndt, F.R., Rusch, R., Stillfried, H. V., Hanisch, B. & Martin, W. C. (1976) Plant Physiol 57, 99.
Binding, H., & Nehls, R. (1978) Z. Pflanzenphysiol. 88, 327-332.
Dhir, S.K., Dhir, S., Hepburn, A., & Widholm, J.M. (1991) Plant Cell Rep 10, 106-110.
Donn, G. (1978) Z. Pflanzenphysiol. 86, 65-75.
Draget, K.I., Myhre, S., Skjåk-Bræk, & Østgaard, K. (1988) J Plant Physiol 132, 552-556.
Gray, D.J., Mcolley, D.W. and Compton, M.E. (1993) J Am Soc Hort Sci 118, 425-432.
Gulati, A., & Jaiwal, P.K. (1992) Plant Cell Tiss and Org Cult 29, 199-205.
Hammatt, N., & Davey, M.R. (1988) In Vitro Cell & Dev Biol 24, 601-604.
Huancaruna Perales, E.M. & Schieder, O. (1993) Plant Cell Tiss Org Cult 34, 71-76.
Huetteman, C.A., & Preece, J.E. (1993) Plant Cell Tiss Org Cult 33, 105-119.
Jähne, A., Lazzeri, P.A., Jäger-Gussen, M., & Lörz, H. (1991) Theor Appl Genetics 82, 74-80.
Kao, K.N., & Michayluk, M.R. (1975) Planta 126, 105-110.
Lehminger-Mertens, R., & Jacobsen, H.J. (1989) Plant Cell Rep 8, 379-382.
Murashige, T., & Skoog, F. (1962) Physiol Plant 15, 473-497.
Myers, J., Lazzeri, P., & Collins, G.B. (1989) Plant Cell Rep 8, 112-115.
Newell, C., & Luu, H.T. (1985) Plant Cell Tiss Org Cult 4, 145-149.
Oelck, M., & Schieder, O. (1983) Z. Pflanzenzüchtung 91, 312-321.
Park, H.Y., & Wernicke, W. (1993) J Plant Physiol 141, 376-379.
Parrott, W.A., Bailey, M.A., Durham, R.E. & Mathews, H.V. (1992) In: Moss, J.P. (ed.) Biotechnology and crop improvement in Asia, 115-148, Patancheru, India.
Pickardt, T., Huancaruna Perales, E., & Schieder, O. (1989) Protoplasma 149, 5-10.
Pickardt, T., Meixner, M., Schade, V., & Schieder, O. (1991) Plant Cell Rep 9, 535-538.
Puonti-Kaerlas, J., & Eriksson, T. (1988) Plant Cell Rep 7, 242- 245.
Rhodes, C., Lowe, K., & Ruby, K.L. (1988) Bio/Technology 6, 56-60.
Röper, W. (1980) Z. Pflanzenphysiol. 101, 75-78.
Schilde-Rentschler, L., Boos, G. & Ninnemann, H. (1988) In: Puite, K.J., Dons, J.J. M., Huizing, H.J., Kool, A.J., Koorneef, M. & Kreens, F.A. (eds.) Progress in plant protoplast Research, pp. 195-196, Kluwer Academic Publishers, Dordrecht.
Schlangstedt, M., Hermans, B., Zoglauer, K., & Schieder, O. (1992) J Plant Physiol 140, 339-344.
Schnabl, H., Youngman, R.J. & Zimmermann, U. (1983) Planta 149, 392-397.
Shekhawat, N., & Galston, A.W. (1983) Plant Sci Lett 32, 43-51.
Sinha, R.R., Das, K., & Sen, S.K. (1983) In: Sen, S.K., & Giles, K.L. (eds.) Basic life science. Plant cell culture in crop improvement, pp. 209- 214, Plenum, New York.
Vasil, I.K. (1990) Bio/Technology 8, 797.
Visser, C., Qureshi, J.A., Gill, R., & Saxena, P. K. (1992) Plant Physiol 99, 1704-1707.
Wei, Z., & Xu, Z.H. (1988) Plant Cell Rep 7, 348-351.

PLANT REGENERATION FROM SUSPENSION AND PROTOPLAST CULTURES IN THE TEMPERATE GRASSES *FESTUCA* AND *LOLIUM*

Z.Y. Wang, G. Legris, M.P. Vallés[1], I. Potrykus & G. Spangenberg*
Institute for Plant Sciences, Swiss Federal Institute of Technology,
CH 8092 Zürich, Switzerland; [1]C.S.I.C., Zaragoza, Spain

ABSTRACT. An efficient system for green plant regeneration from protoplasts in different *Festuca* and *Lolium* species: *F. arundinacea, F. pratensis, F. rubra, L. multiflorum, L. perenne* and *L.* x *boucheanum* has been worked out. The protocol is based on established single genotype derived embryogenic cell suspensions, cryopreservation for the long-term storage of established cultures and a protoplast bead-type culture system including fast growing non-morphogenic nurse cells. Conditions required for the recovery of embryogenic suspension cultures upon cryopreservation were established for these recalcitrant graminaceous species. Suspension cultures and their protoplasts obtained from cryopreserved cultures retained their morphogenic capacity. Fully fertile greenhouse-grown plants setting seeds have been reproducibly regenerated from protoplasts in *F. arundinacea, F. pratensis, L. perenne* and *L. multiflorum.* A RAPD analysis on the genetic stability of protoplast- and cell suspension-derived plants in *F. pratensis* and *L. multiflorum* revealed limited newly induced genetic variation at the loci screened with molecular markers.

1. Introduction

Cereal crops and forage grasses are among the most recalcitrant plants to genetically manipulate *in vitro* [1, 2]. The establishment of morphogenic cell suspensions has proven to be an important prerequisite for gene transfer approaches into these graminaceous monocotyledonous species [2, 3]. These cell suspensions are required for the isolation of totipotent protoplasts, which can be subjected to direct gene transfer and fusion, for the production of transgenic plants and somatic hybrids, respectively. In addition, suspension cultures can be directly used as targets for biolistics® transformation, finally leading to the regeneration of transgenic plants.

Festuca (tall, meadow and red fescues) and *Lolium* (Italian, perennial and hybrid ryegrasses) are key forage grasses in the temperate region. For the genetic manipulation of these species, the plant regeneration systems established from cell suspensions and protoplasts need to fulfill following requirements: a) reproducible regeneration of a large number of mature plants; b) regeneration of fertile plants that could be included in further breeding and finally seed production programs; and c) regeneration of largely genetically stable plants.

M. Terzi et al. (eds.), Current Issues in Plant Molecular and Cellular Biology, 81-86.

2. Establishment, Maintenance of, and Plant Regeneration from Embryogenic Cell Suspensions

A screening for the induction of embryogenic callus, appropriate to initiate morphogenic suspension cultures from single genotypes was performed for 5 *F. arundinacea* cultivars, 10 *F. pratensis* cultivated varieties, 3 *F. rubra* varieties, 10 *L. multiflorum* cultivars, 6 *L. perenne* cultivated varieties and 2 breeding lines in *L.* x *boucheanum* [4-6]. Since the *Festuca* and *Lolium* species used are largely self-sterile and out-crossing, seeds within one cultivar may represent different genotypes.

Four varieties of *F. arundinacea* and 7 cultivars out of the 10 tested in *F. pratensis* produced embryogenic calli. In *F. rubra*, embryogenic callus was obtained for the 3 cultivars included in the genotype screening. Depending on the cultivar used, 3% to 20% of the seeds screened produced yellowish friable callus suitable for the initiation of suspension cultures, in all three *Festuca* species evaluated.

In the *Lolium* species considered, single-genotype derived friable yellowish callus was initiated in 5% to 8% of the seeds plated for 5 different cultivars of *L. multiflorum* var. *westerwoldicum* and in a similar range for the 5 cultivars considered in *L. multiflorum* var. *italicum*. Two to 5% of the seeds evaluated in the genotype screening of 6 cultivars in *L. perenne* yielded embryogenic calli. Similar frequency of responsive genotypes was observed for the 2 breeding lines of *L.* x *boucheanum* tested.

Embryogenic callus, obtained from mature embryos of defined single-seed origin, was transferred into liquid culture to initiate suspension cultures [4-6]. For all *Festuca* and *Lolium* species considered, embryogenic cell suspensions showing differences in degree of dispersion and growth rate, but mainly consisting of pro-embryogenic (<2 mm) cell clusters, could be established after 4-6 months.

In the major fescues, regenerable cell suspensions were obtained for independent genotypes in 4 cultivars (cvs. "Roa", "Barcel", "Olga" and "Tacuabé") of tall fescue and for 5 different genotypes in 3 varieties (cvs. "Barmondo", "Belimo" and "Leopard") of *F. pratensis*. Frequencies of green plant regeneration (number of green plantlets/number of plated calli from best responding single-genotype cell suspension within one cultivar) between 50% and 80% were observed in *F. arundinacea*. Corresponding values for *F. pratensis* were in the range of 30% to 80%. The frequency of albinism varied from cultivar to cultivar, being between 2% and 8% for 20 week-old suspensions established from *F. arundinacea*, and in the range of 10% to 20% for equivalent suspensions established for *F. pratensis*. In *F. rubra*, suspension cultures of the cv. "Roland" showed the capacity to regenerate efficiently (over 90% of the plated cell aggregates regenerating plantlets) green plants over a period of 14 months.

For the ryegrasses, plant regeneration from established embryogenic cell suspensions was achieved for 3 cultivars each in *L. multiflorum* var. *westerwoldicum* (cvs. "Andy", "Caramba" and "Limella") and *L. multiflorum* var. *italicum* (cvs. "Axis", "Fedo" and "Lipo"). Depending on the cultivar, green plantlets were regenerated in 5% to 25% of the plated suspension-derived calli for *L. multiflorum*, while albino plantlets were obtained with a frequency up to 15%. In *L. perenne*, embryogenic cell suspensions regenerating green plantlets were established in 5 out of the 6 cultivars tested. The frequency of green plant regeneration in perennial ryegrass varied depending on the cultivar between 8% and 28%. The corresponding green:albino ratio observed for plant regeneration in embryogenic cell suspensions of *L. perenne* was in the range of 0.8:1 to 6:1. In *L.* x *boucheanum* regenerable cell suspensions were recovered for one of the 2 breeding lines and calli plated from the best responding single-genotype cell suspension obtained for the line LH8855 regenerated green and albino plantlets in >30% and >2% of the cases, respectively.

3. Cryopreservation of Suspension Cultures

Established embryogenic suspension cultures retained their potential for regeneration of green plants when evaluated over a period of 8-14 months. However, frequency of plant regeneration as well as green:albino ratio declined over time, for the *Festuca* and particularly *Lolium* species considered, when cell suspensions were maintained under standard conditions. Therefore, a reproducible method for the cryopreservation under liquid nitrogen of 4-6 month-old, single-genotype derived embryogenic cell suspensions was optimized.

Evaluation of different parameters, such as cryoprotectant composition, osmotic pre-freezing treatment of suspension cultures, freezing steps, post-thaw washing, etc. led to species-specific procedures for storage of embryogenic cultures in liquid nitrogen allowing 40% to 60% post-thaw recovery [5, 6]. Four to 6 months after their initiation, embryogenic cell suspensions of *F. arundinacea, F. pratensis, F. rubra, L. multiflorum, L. perenne* and *L.* x *boucheanum* were cryopreserved under the optimal conditions established for each species. Cryopreserved suspension cultures were plated for proliferation after thawing, and their potential for plant regeneration was evaluated after transfer of calli onto regeneration medium. No reduction in plant regeneration frequencies and green:albino ratios from cryopreserved vs. non-frozen original embryogenic cultures was observed for the different grass species tested.

4. Protoplast Culture and Plant Regeneration

4.1. *F. arundinacea, F. pratensis* and *F. rubra*

Protoplasts were readily isolated from 6 to 12 month-old morphogenic cell suspensions initiated from independent genotypes in 4 cultivars (cvs. "Roa", "Barcel", "Olga" and "Tacuabé") of *F. arundinacea*, and in 2 cultivars (cvs. "Barmondo" and "Belimo") of *F. pratensis* [4]. For both major fescues, pure preparations of protoplasts, nearly free of contaminating undigested cells, were routinely obtained with a yield of 0.5 - 1 x 10^6 protoplasts/g fresh weight cells. The overall plating efficiency (number of visible colonies/number of plated protoplasts) in bead-type culture was in the range of 10^{-3} to 10^{-4} provided nurse cells were used during the first week in culture. Colonies growing in the agarose beads were visible after 2 weeks of culture and generated a lawn on the agarose-solidified medium after 3 to 4 weeks.

Plant regeneration from protoplast-derived calli in *F. arundinacea* was possible with frequencies (number of green plantlets/number of protoplast-derived colonies plated onto regeneration medium) up to 50%. In *F. pratensis*, regeneration frequencies lower than for tall fescue, in the range of 20% to 30%, were observed. Few albino plantlets were regenerated from protoplasts of tall and meadow fescues, and their frequency varied with the age of the cell suspension used. Since tall and meadow fescue require cold-treatment for flowering, a representative sample of protoplast-derived plants were vernalized in order to assess their fertility. Flowering protoplast-derived plants in tall fescue and meadow fescue were obtained; they developed normal inflorescences with protruding anthers containing up to 40% viable pollen. Flowering protoplast-derived plants of meadow fescue were crossed to assess their female fertility and seed setting. In addition, pollen from protoplast-derived plants was used for pollination of seed grown meadow fescue plants. Crosses in both directions succeeded and led to production of viable seeds on cytologically normal (2n = 14) plants.

In *F. rubra*, protoplasts were isolated from 4 to 14 month-old highly embryogenic cell suspension of the cv. "Roland"[6]. Protoplast yield was 2 - 5 x 10^5 protoplasts/g fresh weight cells and remained relatively low irrespectively of the age of the cell suspension used within the range tested. Nevertheless, the use of a bead-type culture system, including fast growing non-morphogenic nurse cells during the first week of culture, and the use of 0.5 - 1 x 10^6 protoplasts/ml plating density allowed sustained divisions of the morphogenic red fescue protoplasts. Plating efficiencies in the range of 1 - 3 x 10^{-3} were routinely obtained. More than 85% of the protoplast-derived microcalli grew further on proliferation medium and allowed for differentiation of green plantlets when transferred onto regeneration medium. A representative set of 40 independent protoplast-derived plants was grown under greenhouse conditions.

4.2. *L. multiflorum*, *L. perenne* and *L.* x *boucheanum*

Protoplasts were isolated from 4 - 8 month old morphogenic cell suspensions for representative genotypes within different cultivars of the three *Lolium* species considered [5]. Protoplast yield was between 1 - 5 x 10^5 protoplasts/g fresh weight cells. The use of an agarose bead-type culture system including nurse cells allowed to culture protoplasts from low yielding young suspensions. An overall plating efficiency in the range of 10^{-3} to 10^{-4} was obtained for Italian ryegrass; and was between 4 x 10^{-3} to 10^{-4} for perennial and hybrid ryegrasses.

Plant regeneration from ryegrass protoplast-derived calli was evaluated upon their transfer onto proliferation medium for 3 weeks and later onto regeneration medium. After 3 weeks on regeneration medium, somatic embryos growing on protoplast-derived calli differentiated shoots. Depending on the cultivar, 4% to 30% of the protoplast-derived calli regenerated green plantlets. Albino plantlets were also recovered. Observed green:albino ratio for most cultivars was in the range of 2:1 to 4:1 for Italian ryegrass, 1:2 to 2:1 for perennial ryegrass, and 6:1 for hybrid ryegrass. Over 150 green plantlets regenerated from protoplasts, out of 2 experiments each, in the different ryegrasses were rooted on hormone free MS medium. More than 60 rooted plantlets from protoplasts in Italian, perennial, and hybrid ryegrasses were hardened off and transferred to soil, where all continued growing under greenhouse conditions. A sample of 30 protoplast-derived plants of *L. multiflorum* cv. "Andy", transferred to soil and grown under greenhouse conditions, flowered within 2 to 3 months and produced normal inflorescences with protruding anthers. Anthers formed between 40% to 50% stainable pollen. Crosses for assessing both, male and female fertility of flowering protoplast-derived plants of Italian ryegrass were performed; all plants set seeds and 80% of the collected seeds could be germinated. In the case of protoplast-derived plants of *L. perenne* cvs. "Citadel" and "Bastion" grown in soil, upon vernalization for 8 weeks, first plants were recently brought to flower; they produced normal anthers with 60% to 70% viable pollen.

5. Analysis of Genetic Stability of Plants from Suspensions and Protoplasts

In order to assess the genetic stability of green protoplast-derived plants in *Festuca*, a sample of meadow fescue plants regenerated from protoplasts isolated from a single-genotype derived cell suspension of the cv. "Barmondo" was screened for RAPD (randomly amplified polymorphic DNA) markers using 18 different primers [7]. Similarly, for the ryegrasses, over 30 plants regenerated from protoplasts isolated from a single-seed derived embryogenic suspension of *L. multiflorum* cv. "Andy" were screened with 19 primers [5]. The control RAPD analysis performed allowed to discriminate different fescues (*F. arundinacea*, *F. pratensis* and *F. rubra*) and ryegrasses (*L. multiflorum* vars. *westerwoldicum* and *italicum*, *L. perenne* and *L.* x *boucheanum*) by revealing RAPD markers for most of the primers evaluated. The analysis performed in *F. pratensis*, allowed also discrimination between different genotypes and cell suspensions within one particular cultivar, since different patterns were obtained for 5 independent suspension cultures initiated from different seeds of the cv. "Barmondo" in 4 out of 18 primers tested. Similarly, different patterns of PCR (polymerase chain reaction) amplification products were observed for 6 different cultivars of *L. multiflorum* and even for 3 independent cell suspensions initiated from different genotypes within *L. multiflorum* var. *westerwoldicum* cv. "Andy" for 16 and 9 of the primers used, respectively. This demonstrated the utility of RAPD technology for revealing pre-existing genetic variation among single-seed derived suspension cultures obtained from highly heterozygous out-crossing species, and suggested their potential use for revealing newly induced genetic changes at the loci screened by these molecular markers.

While analyzing representative DNA samples from independent *F. pratensis* plants regenerated from a single-seed derived 20 week-old cell suspension, no variation among the regenerants was revealed. The same held true for the analysis of DNA samples from independent protoplast-derived meadow fescue plants regenerated from the same suspension culture, irrespectively of the primer used. Similarly, most of the plants regenerated from protoplasts isolated from the single-genotype derived suspension of *L. multiflorum* cv. "Andy" included in the RAPD analysis performed, showed identical patterns for most of the primers used. However, variation within the protoplast-derived plants of Italian ryegrass was detectable at loci screened with molecular markers for 3 out of 19 primers tested.

6. Discussion

The results obtained in our laboratory demonstrate that large numbers of mature green plants, surviving transfer to soil and growing under greenhouse conditions, can be reproducibly regenerated from suspensions and protoplasts in different fescues and ryegrasses [4-6]. This point has been for long a bottle-neck to the application of genetic manipulation techniques for *Festuca* and *Lolium* improvement [8-14]. No success in plant regeneration from protoplasts had been previously reported for *F. pratensis*, *F. rubra* and *L.* x *boucheanum*. While no information was available for meadow fescue and hybrid ryegrass; conditions for establishment and maintenance of embryogenic suspensions had been worked out in *F. rubra* [15], but protoplasts isolated therefrom had so far led to the production of morphogenic calli only [8]. In previous reports on ryegrasses, only 5 soil-grown plants from protoplasts were obtained for *L. multiflorum* [9] and only 3 soil-grown plants from protoplasts were recovered for *L. perenne* [9-14].

In many cases, albino plantlets were exclusively [13] or preferentially [9-11, 14] recovered from protoplasts in ryegrasses and fescues. Our results confirm the view that green mature plants can be regenerated from these graminaceous protoplasts if they are isolated from young embryogenic cell suspensions, before the ability of corresponding suspension cultures to regenerate green plants has ceased. In this context, the initiation of highly embryogenic single-genotype derived friable callus for the later establishment of appropriate suspension cultures seemed to be critical. A clear genotype dependence for obtaining adequate callus cultures was observed for different fescues [4, 6, 16] and ryegrasses [5]. In contrast, previous reports on protoplast culture in *Lolium* and in *Festuca*, with one exception [16], have dealt with suspension cultures initiated from unselected mixes of a number of genotypes [9-11, 13], thus being unable: a) to analyze genotype effects, b) to assess possible domination in proliferation of particular genotypes within established cultures, and more importantly, c) to select and maintain highly embryogenic genotypes. This striking difference in the approach followed in our studies [4-6] compared to previous reports [8-11] - particularly for the ryegrasses - is suggested to be responsible for the better performance over time of embryogenic suspension cultures. For example, a rapid (within 25-30 weeks after initiation) loss of green plant regeneration capacity was a reproducible feature in most established suspension cultures using mixtures of genotypes in *L. perenne* and *L. multiflorum* [9-11]. In our studies, regeneration of green ryegrass and fescue plants from suspension cultures and protoplasts derived therefrom was even possible for 8-14 months old cell suspensions [5, 6]. This indicates that the selection of genetic materials with a good response for initiation of single-genotype derived embryogenic suspensions is a key prerequisite to obtain established and long-term regenerating suspension cultures as a source of totipotent protoplasts. In addition, the use of cryopreservation for storage of young embryogenic suspension cultures in liquid nitrogen further extended the useful life of established suspensions in fescues and ryegrasses, thus allowing to perform protoplast culture experiments with particular genotypes repeatedly over a long period of time. Another rationale behind the use of single-genotype derived embryogenic cell suspensions was to allow the evaluation of somaclonal effects. If suspension cultures used for protoplast isolation are composed of different genotypes, as in most previous reports on *Festuca* and *Lolium* protoplast culture [9-11], it is not possible to discriminate between true genotype or somaclonal effects when variation is observed among regenerants. A RAPD analysis performed in a representative set of green plants regenerated from protoplasts isolated from a single-genotype derived embryogenic cell suspension of meadow fescue [7] and of Westerwold's ryegrass [5] revealed no or limited newly generated variation, respectively.

7. References

1 I. Potrykus (1990) Bio/Technology 8: 535-542.
2 I.K. Vasil (1988) Bio/Technology 6: 397-402.
3 I. Potrykus (1991) Ann. Rev. Plant Physiol. Plant Mol. Biol. 42: 205-225.
4 Z.Y. Wang, M.P. Vallés, P. Montavon, I. Potrykus and G. Spangenberg (1993) Plant Cell Rep. 12: 95-100.
5 Z.Y. Wang, J. Nagel, I. Potrykus and G. Spangenberg (1993) Plant Science 94:179-193.
6 G. Spangenberg, Z.Y. Wang, J. Nagel and I. Potrykus (1994) Plant Science 97:83-94.
7 M.P. Vallés, Z.Y. Wang, P. Montavon, I. Potrykus and G. Spangenberg (1993) Plant Cell Rep. 12: 101-106.
8 O.M.F. Zaghmout and W.A. Torello (1990) Plant Cell Rep. 9: 340-343.
9 S.J. Dalton (1988) Plant Cell Tissue Organ Culture 12: 137-140.
10 S.J. Dalton (1988) J. Plant Physiol. 132: 170-175.
11 J. Creemers-Molenaar, P. van der Valk, J.P.M. Loeffen and M.A.C.M. Zaal (1989) Plant Science 63: 167-176.
12 M.G.K. Jones and P.J. Dale (1982) Z. Pflanzenphysiol. 105: 267-274.
13 O.M.F. Zaghmout and W.A. Torello (1992) J. Plant Physiol. 140: 101-105.
14 J. Creemers-Molenaar, F.A. van Eeuwijk and F.A. Krens (1992) J. Plant Physiol 139: 303-308.
15 O.M.F. Zaghmout and W.A. Torello (1992) Plant Cell Rep. 11: 142-145.
16 T. Takamizo, K.I. Suginobu and R. Ohsugi (1999) Plant Science, 72: 125-131.

IN VITRO ORGANOGENESIS, TRANSFORMATION AND EXPRESSION OF DROUGHT-RELATED PROTEINS IN FOREST TREE CULTURES

A. ALTMAN[1,2], A. YA'ARI[2], D. PELAH[2], A. GAL[2], T. TZFIRA[2], W-X. WANG[2]
O. SHOSEYOV[1,2], A. VAINSTEIN[1,2], J. RIOV[2]
[1]The Otto Warburg Center for Biotechnology in Agriculture and [2]The Kennedy-Leigh Centre for Horticultural Research, The Hebrew University of Jerusalem, P.O.B. 12, Rehovot, ISRAEL.

ABSTRACT. A procedure for aseptic in vitro culture of Populus spp. and Pinus halepensis buds from adult trees was developed, and bud cultures have been used to elucidate the hormonal control of bud growth and development. Populus cultures served as well for establishing procedures for in vitro propagation. Axillary bud break was induced, and adventitious buds and roots were formed, resulting in acclimatized plants. Adventitious shoot formation in root cultures of Populus is being studied. Explants of zygotic embryos, seedlings and rejuvenated mature trees of Pinus halepensis are examined for in vitro clonal propagation. Efficient procedures for massive induction of adventitious shoots from mature pine embryos were established, and good rooting was achieved. Efforts are being made to induce somatic embryogenesis. A novel 66 kD boiling-stable protein was highly expressed in Populus tremula shoot cultures and in callus, in response to gradual desiccation, cold and osmotic stress, and ABA application. Populus clones differing in their drought tolerance are characterized by different expression patterns of this protein. This protein shows a high degree of homology with similar proteins form other species. Similar boiling-stable proteins were differentially expressed in germinating pine seeds. Populus shoots and Pinus embryos were transformed using A. rhizogenes, as monitored by root formation and GUS expression, and complete transgenic aspen plants were established.

1. Introduction

The concept of "plant cell culture" has been first conceived by Haberlandt in 1902, stating that "I am not making too bold a prediction if I point to the possibility that ... one could successfully cultivate artificial embryos from vegetative cells". Since then, solid foundations for the art and science of plant tissue culture have been laid by P. White, R. Gautheret, F.C. Steward, F.Skoog, H.E. Street, J.P. Nitsch, G. Morel and many others. Plant tissue culture can be divided into four major periods, as judged by the experimental plant systems employed, representing historical developments towards increased practical utilization: (1) the period of basic studies, using model plants like tobacco, carrot and petunia, (2) the period of in vitro (micro)clonal propagation, especially of ornamental plants (e.g., ferns, orchids, carnation, gerbera), (3) more recently, focus has been turned to application of tissue culture techniques to important crop plants (e.g., cereals, legumes, potato), (4) the period of woody perennials, fruit and - particularly - forest trees, is now evolving as a major discipline.

M. Terzi et al. (eds.), Current Issues in Plant Molecular and Cellular Biology, 87-94.

The importance of forest trees for timber production and in agroforestry, and their unique crucial role in preserving our environment, are well recognized. This calls for intensive, integrated research in forest tree biotechnology and plant tissue culture. Populus and Pinus are two highly important forest tree geni, containing many species, land-races and clones, which comprise a significant proportion of forests worldwide. They are important, both economically for timber production and as major components of the ecosystem, due to their specific growth characteristics and to their partial adaptation to the regional climatic and edaphic conditions.

The ability to clone forest trees, and to study their growth and development under controlled conditions, offers several unique opportunities for forest tree breeding and improvement. In this respect, bud cultures have been used previously by us to elucidate the hormonal control of bud growth and development in trees of Populus trichocarpa, P. tremula and Fagus sylvatica (Nadel et al. 1991a, Nadel et al. 1991b, Nadel et al. 1992). At present, conventional large-scale production of clonal material of Pinus spp., for selection and breeding purposes, is not possible because of considerable difficulties in consistent rooting of cuttings from selected adult trees. Most Populus species root well, but some are recalcitrant. In vitro propagation is an important alternative to traditional propagation by cuttings, whenever a requirement arise for rapid, mass production of selected clones. In contrast to the situation in many ornamental plants, in vitro methods have not been shown as yet economically-efficient in forest trees. Micropropagation has been applied over the last 20 years to an increasing number of conifer species (Aitken-Christie et al. 1988, Gupta and Durzan 1987, Hakman and von Arnold 1988, Thorpe et al. 1991). In some cases, these methods have generated a large number of transplantable plants, however the few studies on Mediterranean species such as P. halepensis and P. brutia, have resulted in a very limited practical success.

The use of novel molecular probes for elucidating the underlying physiological and molecular control mechanisms of drought tolerance is based on recent findings on stress-induced expression of a specific "desiccation-related", or "water-binding" proteins (e.g., Baker et al. 1988, Dure III et al. 1989, Mundy and Chua 1988, Piatkowski et al. 1990). A role in protecting plant structures during water stress is likely for these proteins, with ABA possibly functioning in the stress transduction processes. These proteins share conspicuous physico-chemical properties: they are highly hydrophilic and do not precipitate if boiled in aqueous solution. It was suggested that its specific structure (i.e., a potential amphiphilic helix) enables the protein to associate with membranes, in particular membranal cracks that result from desiccation or cytoplasmic water loss during freezing (Piatkowski et al. 1990).

Successfull genetic transformation of conifers is very limited, and especially in pine species. To the best of our knowledge, transformation via A. rhizogenes was studied in two conifer species only: Larix decidua and Picea glauca (Huang et al. 1991), and the ability of A. rhizogenes strains to incite galls was studied in four pinaceous species, including Pinus ponderosa (Morris et al. 1989). A. tumefaciens-mediated gene transfer was reported in the following Pinus species: P. taeda (Sederoff et al. 1986), P. lambertiana (Loopstra et al. 1990), and P. thunbergii (Choi and Lee 1990).

2. Experimental

2.1. CULTURE CONDITIONS

Shoot explants of aspen (Populus tremula) and other Populus species were routinely subcultured every 5-6 weeks, on 0.5-strength Murashige and Skoog basal medium supplemented with 500 mg/l casein hydrolysate. The shoot cultures formed roots within 1-2 weeks, and were maintained in a growth room at 24^{0}C, under a 16h photoperiod. Rooted aspen plantlets were acclimatized under mist, and the plants were grown in pots. Explants were prepared from embryos and seedlings of Pinus halepensis, plated on MS basal media containing different combinations of auxins (NAA, 2,4-D, and conjugates of phenoxy-auxin with glycine), and cytokinins (BA, Thidiazuron). Explants with newly developed shoots were subcultured in solid or agitated liquid media. Two year old pine plants were sprayed with cytokinins at different concentration (50-500 ppm), to induce fascicular shoot development for further culture in vitro. These were separated, surface-sterilized, and plated on appropriate media. In addition, embryogenic callus cultures are being studied, for induction of pine somatic embryos.

2.2. WATER STRESS TREATMENTS AND PROTEIN ANALYSIS

Excised cultured aspen shoots were wilted at room temperature to 80% or 60% of their fresh weight, then kept enclosed in plastic bags in the laboratory for additional 4 h. Shoots and leaves were removed from the plastic bags at the end of the experiment, and frozen in liquid nitrogen. Detached leaves of the Popularis and the Italica clones, removed from rooted cuttings, were subjected to short or long water stress. In some experiments, shoots were excised from cultured rooted aspen plantlets sprayed with ABA (1mM), and kept enclosed in plastic bags in the laboratory for additional 4 and 18 hours. Frozen plant material was homogenized in an extraction buffer, crude extracts were centrifuged at 10,000 g for 10 min, and total protein in the supernatant was determined. Protein samples were boiled for 10 min at 100^{0}C, and centrifuged at 10,000 g for 10 min. Supernatant heat-stable proteins were precipitated by acetone, centrifuged, and re-dissolved in SDS-PAGE sample buffer. Total and heat-stable proteins were separated by one-dimensional 10% SDS-PAGE, and gels were stained with Coomassie Blue. For western blotting, proteins were electrotransferred to a nitrocellulose membrane, and reacted with the appropriate rabbit antibodies. Similar procedures were used for analysis of pine seed proteins in the course of their germination.

2.3. TRANSFORMATION

Excised cultured aspen shoots were co-cultivated with the A. rhizogenes strain LBA 9402, harboring the kanamycin^{R} gene and p35S GUS INT (Vanacanneyt et al. 1990), using standard procedures. The resulting transgenic roots were excised and cultured in agitated liquid media, where adventitious shoot bud formation occurred at a high rate. Fully developed transgenic aspen plants were recovered, as verified by GUS staining, PCR, and kanamycin resistance. For transformation of Pinus halepensis, the same strain was co-cultivated with excised mature zygotic embryos, or seedlings were innoculated by applying the bacteria onto mildly wounded tissues.

3. Results

3.1. IN VITRO PROPAGATION

Following our previous studies on in vitro propagation and acclimatization of Populus spp. from nodal explants (Nadel et al. 1992), an alternative procedure for large-scale propagation was investigated. In collaboration with M. Ziv and T. Carmi, we studied adventitious shoot bud formation from root cultures. Additional studies were aimed at developing efficient procedures for pine in vitro propagation, using adventitious and fascicular shoot formation from various explants of Pinus halepensis embryos, seedlings and from more mature plants, as outlined in **Table 1**.

Table 1. In vitro propagation of Pinus halepensis and Populus spp., using several types of explants and morphogenetic patterns

---

A. Pinus halepensis

- Organogenesis: adventitious shoot and root formation from explant of -
    * mature embryos
    * seedlings
    * mature trees (cytokinin-induced brachyblasts)
    * transformation with A. rhizogenes
- Somatic embryogenesis and polyembryogenesis: direct and indirect embryo formation from -
    * mature and immature zygotic embryos
    * embryogenic callus and cell suspensions

B. Populus spp. (tremula, trichocarpa):

- Axillary and adventitious shoot proliferation in -
    * nodal stem mini-cuttings
    * shoot tips
- Adventitious shoot bud regeneration in -
    * root cultures

Culture of embryo and seedling explants on a medium with 2,4-D resulted in considerable proliferation of callus, with no further shoot regeneration. Intact embryos, as well as apical shoot explants of germinating seedlings (containing the cotyledons), developed both callus and a large number of shoots on a medium containing the conjugated IAA and a high concentration of BA. Culture of embryos in the inverted position (cotyledons in the medium) resulted in massive regeneration of adventitious shoots (**Fig. 1**), and 50-70 new buds formed on the cotyledons of each one of the embryos. These buds were separated and plated on different media designed to promote shoot elongation. Up to 85 % of these shoots rooted in a sterile peat and vermiculite medium, following IBA pulse treatment. Similar procedures were used with fascicular shoots which were induced by cytokinin spraying of 2 year-old pine seedlings. The induction of fascicular shoots in mature trees, as well as somatic embryogenesis from seed tissues, is being studied.

3.2. EXPRESSION OF DESICCATION-RELATED PROTEINS IN POPLAR AND PINE

Aspen shoots, detached from in vitro cultured plantlets, were exposed to water stress by gradual wilting, and lost 20 or 40 % of their initial fresh weight within 30 or 80 min, respectively, Water loss by aspen shoots was accompanied by a 20-30 % loss of total protein content. However, a water stress-induced boiling-stable protein of a molecular mass of 66 kD, was highly expressed in shoots that lost 20% and 40% of their fresh weight. Expression of the drought-responsive boiling-stable protein occurred as short as 1h after the water stress. The same 66 kD heat-stable protein accumulated in aspen plantlets in response to ABA treatment, as well as in aspen callus cultures subjected to osmotic stress in the medium and to 4^0C. In addition, the expression of this protein was verified by western blots with antibodies raised against the purified protein, and the N-terminal sequencing of the protein was determined. Similar experiments with other poplar clones were initiated, and preliminary studies indicate that drought-tolerant and drought-sensitive Populus clones differed in the rate of water loss and in the expression of the 66 kD boiling-stable protein. Details of these results will be reported elsewhere (Pelah et al. 1994, and unpublished), and are summarized in **Table 2**.

Table 2. Some properties of boiling-stable water stress-induced proteins in Populus

---

* **A 66 kD, boiling-stable protein induced by:**
  - water stress (plant desiccation, low air R.H.)
  - negative water potential of medium, low temperature, ABA
* **Expression of the protein is:**
  - rapid (less than 1 h after stress application)
  - organ and tissue specific (in leaves and callus)
  - genotype-specific (differing in sensitive and tolerant genotypes)
  - highly significant (up to 7 % of total protein content)
* **This boiling-stable protein is:**
  - extremely hydrophilic (possibly an amphiphilic helix)
  - glycosylated
  - shows high homology with wheat germins (gf-2.8/3.8) and barley cold-regulated gene (pa 086)

Preliminary experiments did not reveal specific desiccation-induced boiling-stable proteins in pine seedlings, although many boiling-stable proteins were present, possibly expressed in a constitutive manner. Therefore, an alternative approach was studied: a 300bp DNA fragment was generated using a PCR procedure, with the two DNA primers (degenerated from two conserved sequences in LEA genes) and Pinus halepensis genomic DNA. This insert codes to a very hydrophylic protein, a features characteristic to some of the LEA proteins. These primers are being used in our ongoing studies. In addition, we studied the changes in protein content and profiles during pine seed germination. A decrease in the content of seed total proteins during germination was noted in some seed parts, an increase in others, yet a general decrease in total proteins was found towards the termination of germination. The heat-labile and heat-stable soluble proteins were examined as well, and many changes in the profile of both heat-stable and heat-labile proteins were noted.

3.3. POPLAR AND PINE TRANSFORMATION

Two day co-cultivation of aspen node explants with the A. rhizogenes strain LBA 9402, without a pre-incubation period, resulted in positive GUS expression in the basal ends of ca. 93 % of the stems, and in over 78 % of the adventitious roots which formed on the stem segments. Adventitious shoot buds regenerated readily on those roots, and 91 % of them showed positive GUS staining. Fully developed transgenic aspen plants were recovered, as verified by GUS staining, PCR and kanamycin resistance.

A general procedure for transformation of Pinus halepensis was established. Inoculation at different sites along the stem of 2-3 week old seedlings resulted in both callus and root formation in a considerable number of the seedlings. The induced adventitious roots, as well as the wounded and callused stem tissues, showed extensive GUS expression. Co-cultivation with excised mature zygotic embryos resulted in a high rate of GUS expression in the basal (root) end of the embryo (**Fig. 2**).

4. Concluding remarks

Selection and breeding of forest trees for better performance and for improved tolerance to the major environmental stress conditions is a most important goal. This requires the combination of: (1) conventional breeding and selection techniques, (2) rapid clonal propagation of the selected genotypes, and (3) novel molecular and biotechnological methodologies . All three disciplines should be useful both for determining the physiological and molecular basis of tree tolerance, and for rapid selection and tree improvement programs. Novel procedures, which may help to shorten and improve the current selection and breeding of elite tree genotypes, is of utmost importance for forestry in marginal land and climatic conditions.

Water stress (i.e., drought) and salinity are two major harmful stress conditions which exist in many marginal forest regions. Desiccation stress, due to water shortage, is expressed at the plant level by a series of physiological and molecular changes which adversely affect plant growth and productivity. Previous studies have concentrated on physiological and biochemical responses, and little is known on the molecular control mechanisms of desiccation tolerance in plants in general, and particularly in trees. We have found a unique 66 kD heat stable protein which accumulates in woody plants due to water/osmotic stress, cold stress and ABA application. The function of this protein is still unknown, however its unique physico-chemical properties and its homology with other water stress-responsive proteins may serve as a clue in understanding its function. Furthermore, we hope to use it as a marker for screening for drought and salt stress. Transformation studies in forest trees are very limited, and Populus spp. has been the primary target plant, so far, for successful production of transgenic trees. This was achieved by A. tumefaciens and A. rhizogenes-mediated transformation, and was extended only recently to the use of accelerated particles. The present studies formulated some procedures for transformation of Pinus halepensis and for a high production rate of transgenic aspen plants.

These studies, along with improvements made in large-scale in vitro propagation of Populus and Pinus halepensis, will form the basis for selection, breeding and propagation of water stress-tolerant poplars and pines.

5. Acknowledgments

This research is supported by the German-Israel Agricultural Research Agreement for the Benefit of Developing Countries (GIARA) Project 92-4 and by a grant from the Land Development Authority (Keren Kayemet Le'Israel).

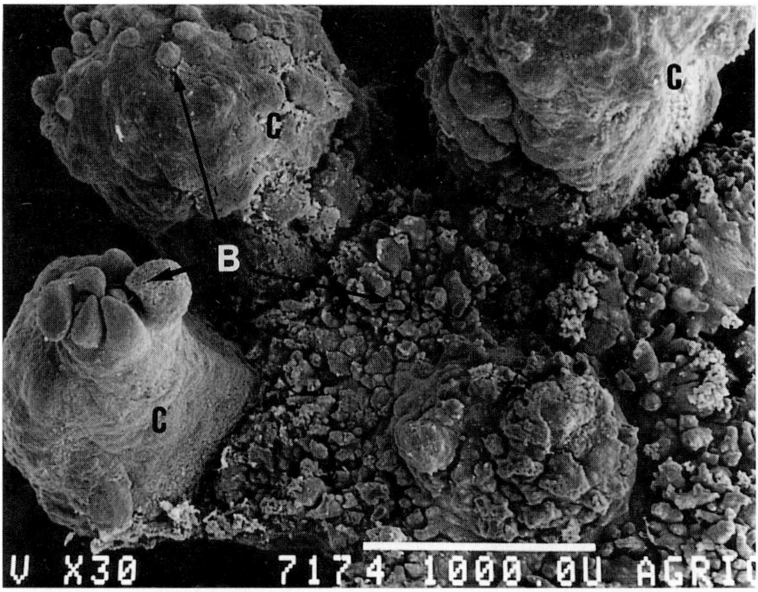

Figure 1. Adventitious shoot formation on cotyledons of inverted Pinus halepensis embryos. The mature embryos were cultured, inverted, in a cytokinin-rich medium for 3 weeks, followed by 3 week culture on a hormone-free medium. Bar = 1 mm. B: adventitious buds, C= embryo cotyledons.

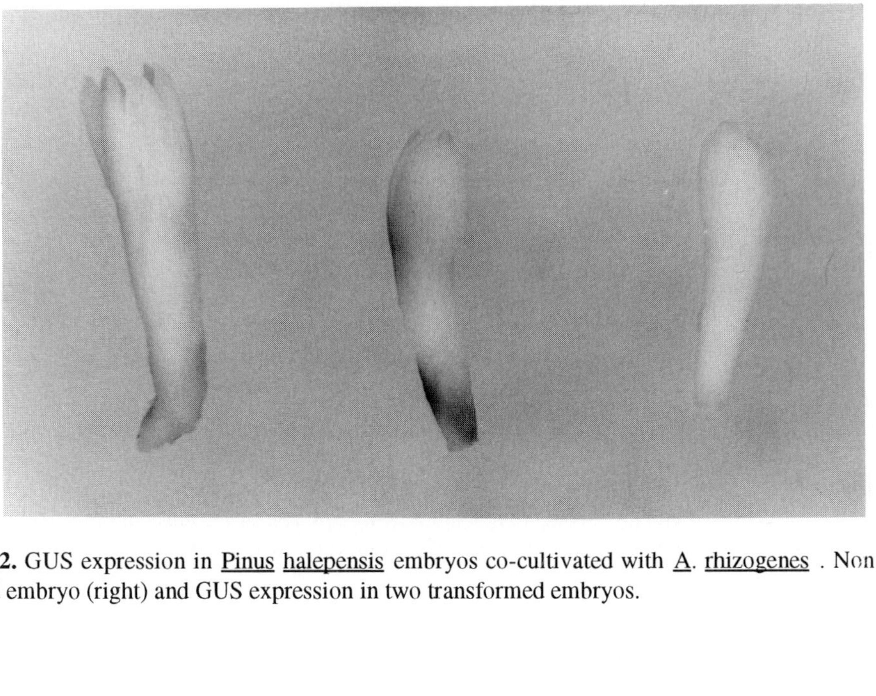

Figure 2. GUS expression in Pinus halepensis embryos co-cultivated with A. rhizogenes . Non infected embryo (right) and GUS expression in two transformed embryos.

6. Literature cited

Aitken-Christie, J., A.P. Singh and H. Davies. 1988. Multiplication of meristematic tissue: A new tissue culture system for radiata pine. In: Hanover JW, Keathley DE (Eds), Genetic Manipulation of Woody Plants (pp 413-432) Plenum Press, New York.

Baker, J.C,, C. Steele and L. Dure III. 1988. Sequence and characterization of 6 Lea proteins and their genes from cotton. Plant Mol. Biol. 11: 277-291.

Dure III, L., M. Crouch J. Harada, T-H. Ho, J. Mundy, R. Quatrano, T. Thomas and Z.R. Sung. 1989. Common amino acid sequence domains among the LEA proteins of higher plants. Plant Mol. Biol. 12: 475-486.

Gupta, P.K. and D.J. Durzan. 1987. Biotechnology of somatic polyembryogenesis and plantlet regeneration in loblolly pine. Bio/Technology 5: 147-151.

Hakman, I. and S. von Arnold. 1988. Somatic embryogenesis and plant regeneration from suspension cultures of Picea glauca (white spruce). Physiol. Plant. 72: 579-587.

Huang, Y., A.M. Diner and D. Karnosky. 1991. Agrobacterium-mediated genetic transformation and regeneration of a conifer: Larix decidua. In Vitro Cell. Develop. Biol. 27P: 201-207.

Loopstra, C,A., A.M. Stomp and R.R. Sederoff. 1990. Agrobacterium mediated DNA transfer in sugar pine. Plant Mol. Biol. 15: 1-10.

Morris, J.W., L.A. Castle and R.O. Morris.1989. Efficacy of different Agrobacterium tumefaciens strains in transformation of pinaceous gymnosperms. Physiol. Molec. Plant Pathol. 34: 451-462.

Mundy, J. and N-H. Chua. 1988. Abscisic acid and water stress induce the expression of a novel rice gene EMBO J. 7: 2279-2286.

Nadel, B.L., A. Altman, S. Pleban and A. Huttermann. 1991a. In-vitro development of mature Fagus sylvatica L. buds. I. The effect of medium and plant regulators on bud growth and protein profiles. J. Plant Physiol. 138: 596-601.

Nadel, B.L., A. Altman, S. Pleban, R. Kocks and A. Huttermann. 1991b. In vitro development of mature Fagus sylvatica L. buds. II. Seasonal changes in the response to plant growth regulators. J. Plant Physiol. 138:136-141.

Nadel, B.L., G. Hazan, R. David, A. Altman and A. Huttermann. 1992. In vitro propagation of Populus species: responses to growth regulators and medium composition. Acta Horticulturae 314: 61-68.

Pelah, D., A. Ya'ari, A. Altman, O. Shoseyov and J. Riov. 1994. Growth, in vitro propagation and desiccation-specific proteins of Populus and Pinus tissues. Proc. IUFRO Int. Workshop :Advances in Tree Development Control and Biotechnique, Beijing Forestry University, 1993 (in press).

Piatkowski, D., K. Schneider, F. Salamini and D. Bartels. 1990. Characterization of five abscisic acid-responsive cDNA clones isolated from the desiccation-tolerant plant Craterostigma plantagineum and their relationship to other water-stress genes. Plant Physiol. 94: 1682-1688.

Sederoff, R., A.M. Stomp, W.S. Chilton and L.W. Moore.1986. Gene transfer into loblolly pine by Agrobacterium tumefaciens. Bio/Technology 4: 647-649.

Thorpe, T.A., I.S. Harry and P.P. Kumar. 1991. Application of micropropagation to forestry. In: Debergh PC, Zimmerman RH (Eds) Microprpagation (pp 311-336) Kluwer Academic Publishers, Dordrecht.

Vanacanneyt, G., R. Schmidt, A. O'Connor-Sanchez, L. Willmitzer and M. Rocha-Sosa. 1990. Construction of an intron-containing marker gene: Splicing of the intron in transgenic plants and its use in monitoring early events in Agrobacterium-mediated plant transformation. Mol. Gen. Genet. 220: 245-250.

EVOLUTION AND AUTOMATION IN MICROPROPAGATION AND ARTIFICIAL SEED PRODUCTION

P.C. DEBERGH
Laboratory of Horticulture
Faculty of Agriculture and Applied Biological Sciences University Gent
Coupure links 653, B-9000 Gent BELGIUM

ABSTRACT. Since the last IAPTC-Congress the whole commercial micropropagation scene changed: production centers moved, plants became cheaper, robotization did not make the expected breakthrough. However, this era also witnessed a more fundamental approach of different physiological and pathological problems. Relevant are: a better understanding of carbohydrate metabolism; the introduction of some new plant growth regulators and improved insight in their metabolism; the sometimes important impact of environmental parameters; rootability and control of dormancy (especially of somatic embryos) received more attention. Liquid media are becoming more and more important, the use of bioreactors is entering a critical stage in its commercial applications.

1. COMMERCIAL MICROPROPAGATION

Since the last IAPTC-Congress in 1990, held in Amsterdam, the commercial micropropagation scene underwent drastic changes. In 1990 the world production was estimated to be 500 million plants, with a majority being produced in W. Europe, followed by N. America (different chapters in Debergh and Zimmerman, 1991). Today the production has probably not increased, but accurate production figures are not available. However, the production centers and prices have changed. Indeed, some low wage countries entered the scene, e.g. Poland and India became major producers. Average prices in 1990 were between 20 and 40 $cent; in 1994 the orientative price is of the magnitude of 10 $cent.

Consequences are: many W. European and N. American labs closed their doors or went bankrupt; the rather important efforts to stimulate automation and robotization of different manipulations in micropropagation are discontinued; there is an overproduction and dumping, which mainly perturbs the ornamental production industry.

To protect their niche in the market we also witness that the interest for somatic embryogenesis, and the related developments for large scale production in bioreactors and the manufacture of artificial seeds, increased in the countries controling the market before 1990.

M. Terzi et al. (eds.), Current Issues in Plant Molecular and Cellular Biology, 95-104.

2. CAPITA SELECTA

2.1. Medium ingredients

More than 30 years after its formulation, the Murashige and Skoog medium (1962) is still top of the bill. The impact of this medium cannot be underestimated, and this is clearly emphasized by the fact it is the first plant science paper that reached the status of science citation classic.

There has been more innovation with respect to the hormonal aspects of tissue culture media. The cytokinin-like activity of thidiazuron was already mentioned in 1982 (Mok et al., 1982), but this product, and some other thiadiazole substitutes, became only commercially available in 1993 (Sigma). In some plants it is more effective than adenine-based compounds for the induction of adventitious and axillary shoots. However, it is still not well documented if thidiazuron is acting as a cytokinin *in se*, or as an inhibitor of cytokinin oxidase (Horgan, 1987). This is also the case for some pesticides which have an cytokinin-like activity [e.g. carbendazim (Debergh et al., 1993)].

N^6-Benzylaminopurine (BA) has long been considered as a synthetic cytokinin, although already in 1973 evidence was provided that BA or its derivatives occured in plants (Horgan et al., 1973; Strnad et al., 1992). Because BA is one of the most often used cytokinins its metabolism has been investigated in more detail by different research groups. In tissue culture, BA itself is quickly metabolised, especially conjugates with ribose or glucose are formed. Depending on the plant species different conjugates can have biological activity (S. Werbrouck, not published).

More than 10 years ago jasmonic acid was identified in plants (Ueda and Kato, 1980). However, only recently it, and related compounds, have been used in vitro, and it received most atttention as promotors of senescence, but also as potent inhibitors of cell division (Ueda and Kato, 1982), and they induce tuberization (Koda and Kikuta, 1991).

Already more than twenty years ago growth inhibitors have been used in tissue culture media (a.o. Wochok and Wetherell, 1971). Nowadays there is a renewed interest, especially in systems using liquid culture media. The major purpose is upscaling, along with a certain degree of automation. Growth inhibitors reduce shoot or leaf development, and can increase the number of propagules (especially with geophytes; Ziv, 1989) or minimize problems with hyperhydricity (leaves are the most susceptible organ for this type of physiological disease; Ziv and Ariel, 1991).

We also witnessed a more widespread use of silver salts ($AgNO_3$ and silverthiosulphate) in culture media to keep the ethylene production in check. In different cases it has proven to be of utmost importance for reliable micropropagation and/or regeneration, especially in those plants which are very susceptible for leaf abscission in vitro.

From a practical point of view control of the development of dormancy in vitro is highly desirable. In different cases it is postulated that abscisic acid (ABA) could be one of the causes for non reactivity. Therefore products have been investigated which interfere in its biosynthesis. A group of herbicides (e.g. fluridone, flurtamone, norflurazone) act upon the phytoene desaturase; this blocks the formation of carotenoids. Interventions in this pathway leads to a decrease in the ABA concentration, because ABA can be produced via a C_{40}-precursor from carotenoids (Rohde and Debergh, 1992; Kim et al., 1994). Those products have proven to be effective to release dormancy in tissue cultured lily corms (Kim et al., 1994), but had no major impact on bud dormancy in some woody species (A. Rohde & P. Debergh, non published results).

2.2. Efficient micropropagation

Increasing the efficiency of micropropagation schemes means that more high quality plants should be produced at competitive prices.

A higher productivity does not necessarely mean that the propagation ratio should be increased, it does also imply that losses should be minimized.

Different angles of incidence exist to increase the propagation ratio.

(1) One of the most obvious interventions is to use higher concentrations of a cytokinin or to change for a more performant one.

(2) It is well documented that lowering the agar concentration, or the use of an agar with a lower gel strength, and in the extreme case liquid media, increase the propagation ratio.

(3) The headspace composition in the container can be responsible for quantitative and qualitative changes. Factors involved are: control of the vapor pressure deficit, the ethylene concentration [positive and negative effects have been reported (Mattys et al., 1994)] as well as the photosynthetic status of the plants (Kozai, 1991a, b).

(4) Most often the shift from solid to liquid medium, does also imply the choice of another propagation unit. Axillary or adventitious shoots are most often used on solid media; bud clusters, nodules or somatic embryos are preferred in liquid media.

Minimizing losses fall into two major categories: physiology and contamination.

(1) An uncontrolled water housekeeping in the container, as well as the use of liquid media can cause hyperhydricity; abusive use of cytokinins can be an evocator of this phenomenon (Debergh et al., 1992). Hyperhydricity is not always visualized by macroscopically visible symptoms. Therefore, losses during acclimatizaton of so-called normal shoots or plantlets can be due to the hyperhydric status.

(2) Nowadays many ornamentals are micropropagated and explanted as clusters. An un-appropriated hormonal balance can, among other symptoms (e.g. bushiness), be responsible for excessive callusing at the base of the cluster. Consequences can be: inhibition of root initiation or root development; delayed and/or very heterogeneous

shoot elongation; attack by weakness parasites (*Sciara* larvae, *Myrothecium, Rhizoctonia*, bacteria), and finally losses during acclimatization (J Van Huylenbroeck and S Werbrouck, unpubl. results).

(3) Endogenous bacteria are still one, if not the major cause of losses in vitro. The lack of an appropriated system for screening cultures remains a serious handicap to fully exploit all the potentials of the tissue culture technology.

(4) Micro-arthropodes are another major source of problems. These insects are not harmful *in se*, but they are as a transport agent for fungal spores. Such a type of infestation does seriously disturb the production planning and can even ruin a tissue culture operation.

(5) To overcome or to prevent losses of tissue cultured plants during the weaning stage in the greenhouse, quite often pesticides are used, just before or just after planting. Different fungicides have a cytokinin-like activity and can be responsible for various kinds of aberrations and problems during the weaning stage.

2.3. Environment

Over the last decennium we witnessed that more and more importance is attached to the three most important parameters of the physical environment controling micropropagation: temperature, humidity and light. Hereafter we just mention a few examples to emphasize their importance.

Most tissue culture laboratories maintain passe-partout temperatures around 20 - 23°C. However, on some occasions it has been important to lower the room temperature, e.g. 12.5°C during the rooting stage of *Paeonia* (Albers and Denissen, 1991).

Exposure of plant tissue culture media to specific light conditions, e.g. to light from fluorescent bulbs changed the growth regulating properties of the media. It caused nutrient medium-dependent photosensitized degradation of the phytohormone indole-3-acetic acid and other media components. Formaldehyde is formed from EDTA and the medium becomes deficient in iron as it becomes unchelated. The use of appropriately filtered light (eliminate wavelengths lower than 450 nm) when culturing plant material can eliminate unnecessary variability by stabilizing the culture media composition (Hangarter and Stasinopoulos, 1991 a, b; Stasinopoulos and Hangarter, 1990).

2.4. Rooting

For many woody species rooting remains a problematic stage, especially when explants are taken from "mature" mother plants. Although some interesting examples of rejuvenation have been published recently, it is obvious that the physiological and biochemical basis of juvenility and maturation remain very incomplete (Jones, 1994).

Moreover, it is too often forgotten that after root induction, there are two phases in adventitious root formation: root initiation, when a short period of exposure to auxin is essential; and root emergence and growth, when auxin is not required or even inhibitory (Went, 1939; Went and Thimann, 1937). Recently more attention has been paid to this point, and this allowed to develop elegant procedures for otherwise recalcitrant species to root. E.g. *Carica papaya*: IBA 10 μM for 3 days before transfer to a hormone free medium; or 2 days in darkness on a medium containing 10 μM IBA + 31 μM riboflavin before transfer to light; or the double layer system is applied as an overlay of 300 μM riboflavin (1 ml over 10 ml of medium) over a root induction medium with 10 μM IBA after 1 day (Drew, 1991).

2.5. Bioreactor production

The possibilities for upscaling in bioreactors are tremendous. Recently many applications have been proposed: for geophytes (Ziv, 1989), nodules (Boxus, 1992) somatic embryo cultures (Texeira et al., 1993).

Especially when shoots or clusters are used for propagation purposes, hyperhydricity becomes a major problem. Different adjuvents, called anti-vitrifying substances have been proposed (o.a. Nairn, 1993; Paques and Boxus, 1987). Another approach is to add growth inhibitors, e.g. paclobutrazol (Ziv, 1989); this reduces the leaf surface, the most susceptible organ for hyperhydricity. The latter approach has been succesfully applied for different ornamentals.

Once embryogenic cultures have been obtained, it is not always evident that this technology can be transfered to the bioreactor scale, this is especially the case for conifers (Aitken-Christie, 1993). This was also examplified by Stuart et al. (1987) with alfalfa, in a bioreactor the conversion rate was 2 to 3% compared to 70 to 90% on semisolid medium.

2.6. Artifical seed

The list of plants that can be regenerated through somatic embryogenesis is still increasing. Since the application of somatic embryogenesis to conifers in 1985 (Chalupa, 1985; Hakman et al., 1985; Nagmani and Bonga, 1985), most research and therefore major progress has been obtained in conifers, mainly for reforestation purposes.

To fully exploit the possibilities of somatic embryogenesis, inexpensive methods need to be developed for synchronous, routine maturation of high quality embryos. Different research teams tackled this problem and best results have sofar been obtained with the following approaches, most often a combination of them (Cantliffe et al., 1993; Fowke and Attree, 1993; Gupta and Pullman, 1993; Gupta et al., 1993):

- adding ABA, which induces maturation, but the frequency is low and the process is not synchronous;

- application of water stress, by growing embryos in the presence of high molecular weight compounds (PEG, dextran), thus inducing a water stress similar to that created by drought conditions;

- charcoal, especially in combination with high levels of ABA, improves embryo quality and reduces the number of transfer steps.

Machine vision based analysis of somatic embryos (Cantliffe et al., 1993) will probably be required, in combination with the forementioned approaches, to select the embryos once they have reached the desired level of maturity, and to scale up the process leading to the production of artificial seeds.

Alginate is still the prefered substrate for encapsulation of somatic embryos. However, the gel is preferably oxygenated by adding to an uncured gel solution a suitably stabilized emulsion of a perfluoracarbon compound or a silicon oil, which compounds are capable of absorbing large amounts of oxygen (Carlson et al., 1993).

Also the shape of artificial seeds can be modified to aid the radicle of a germinating embryo in protrusively rupturing the capsule, thereby facilitating successful germination and minimizing incidence of seedling malformation (Carlson et al., 1993). It has been proposed to include microcapsules into the alginate for sustained, temperature controlled, release of sugar, and to develop self breaking capsules (Onishi and Sakamoto, 1993).

2.7. Acclimatization related physiology, morphology and anatomy

Normal looking tissue cultured plants are quite often anatomically and physiologically aberrant, due to the ambient conditions *in vitro*. For tissue culture, closed containers are used to avoid infections, but the prevailing conditions in such a confined atmosphere are responsible for a different physiological behavior and for morphological and anatomical adaptations. Too often, the latter is not considered in interpreting results. Each tissue culture situation is different, therefore extrapolation and oversimplification should be avoided. This does also imply that, in particular, more attention should be paid in evaluating the role of environmental parameters in the functioning of plants, and their adaptation mechanisms to modified environmental conditions.

There are two approaches to minimize acclimatization problems at the transition from *in vitro* to *in vivo* conditions. Either plants (shoots) are made as photoautotrophic as possible (Kozai, 1991b), or shoots are loaded with carbohydrates while still *in vitro* (Capellades et al., 1990, 1991).

In the lattter case an increase of chloroplastic starch reserves is observed, when the sucrose concentration was increased in the "double layer" medium on top of a final stage II medium (Maene & Debergh, 1985). The starch reserve is acting as it where as the endosperm in a seed, and helps to overcome the difficult transition period, during which the plant (shoot) is not really photoautotrophic.

REFERENCES

Aitken-Christie J (1993) Somatic embryogenesis in conifers and its application o forestry. Abstract n°8.5.3-4 in Book of Abstracts XV International Botanical Congress, Yokohama

Albers MRJ, Denissen CJM (1991) Vermeerdering van Paeonia via weefselkweek, Jaarverslag Proefstation voor de boomkwekerij, Boskoop, Nederland, 93-94

Boxus P (1992) Mass propagation of strawberry and new alternatives for some horticultural crops. In: K Kurata, T Kozai (Eds) Transplant production systems, Kluwer Academic Publishers, Dordrecht, 151-162

Cantliffe DJ, Bieniek ME, Harrell RC (1993) A systems approach to developping an automated seed production model. In: WY Soh, JR Liu, A Komanine (Eds). Advances in developmental biology and biotechnology of higher plants (pp. 160-196) Korean Soc Plant Tissue Culture

Cappelades MQ, Lemeur R, Debergh PC (1990) Kinetics of chlorofyll fluorescence in micropropagated rose shootlets. Photosynthetica 24: 190-193

Cappelades MQ, Lemeur R, Debergh PC (1991) Effects of sucrose on starch accumulation and rate of photosynthesis in Rosa cultured in vitro. Plant Cell Tissue Organ Culture 25: 21-26

Carlson WC, Hartle JE, Bower BK (1993) Oxygenated analogs of botanic seed. United States patent n° 5, 236, 469

Chalupa V (1985) Somatic embryogenesis and plantlet regeneration from cultured immature and mature embryos of Piceae abies (L.)Karst. Comm Inst For 14: 57-63

Debergh PC, De Coster G, Steurbaut W (1993) Carbendazim as an alternative plant growth regulator in tissue culture systems. In Vitro Cell Dev. Biol. 29P:89-91

Debergh P, Aitken-Christie J, Cohen D, Grout B, von Arnold S, Zimmerman R, Ziv M (1992) Reconsideration of the term vitrification' as used in micropropagation. Plant Cell Tissue Organ Culture 30: 135-140

Debergh PC, Zimmerman RH (1991) Micropropagation - Technology and Application, Kluwer Academic Publishers, Dordrecht, pp 484

Drew R (1991) Tissue culture and field evaluation of Papaya (*Carica papaya*). PhD-thesis University Murdoch, Australia

Fowke L, Attree S (1993) Applied and basic studies of somatic embryogenesis in white spruce (Piceae glauca) and black spruce (Piceae mariana). In: WY Soh, JR Liu, A Komanine (Eds). Advances in developmental biology and biotechnology of higher plants (pp. 5 - 17) Korean Soc Plant Tissue Culture

Gupta PK, Pullman GS (1993) Method for reproducing conifers by somatic embryogenesis by stepwise hormone adjustment. United States Patent n° 5,236,841

Gupta PK, Timmis R, Carlson WC (1993) Somatic embryogenesis: a possible tool for large scale propagation of forestry species. In: WY Soh, JR Liu, A Komanine (Eds) Advances in developmental biology and biotechnology of higher plants (pp. 18-37) Korean Soc Plant Tissue Culture

Hakman I, Fowke LC, Von Arnold S, Eriksson T (1985) The development of somatic embryos in tissue cultures initiated from immature embryos of Picea abies (Norway spruce). Plant Science 38: 53-59

Hangarter RP and Stasinopoulos TC (1991a) Repression of plant tissue culture growth by light is caused by photochemical change in the culture medium. Plant Sci 79: 253-257

Hangarter RP and Stasinopoulos TC (1991b) Effect of Fe-catalyzed photooxydation of EDTA on root growth in plant culture media. Plant Physiol 96: 843-847

Horgan R (1987) Plant growth regulators and the control of growth and differentiation in plant tissue cultures. In: CE Green, DA Somers, WP Hackett, DD Biesboer (Eds) Plant Tissue and Cell Culture, Alan R Liss, Inc, New York, pp 135-149

Horgan E, Hewett EW, Purse JG, Wareing PF (1973) A new cytokinin from *Populus robusta*. Tetrahedron Letters 30: 2827-2828

Jones OP ((1994) Physiological change and apparent rejuvenation of temperate fruit trees from micropropagation. In: PJ Lumsden, JR Nicholas, WJ Davies (Eds) Physiology, growth and development of plants in culture. Kluwer Academic Publishers, Dordrecht, 323-331

Kim KS, Davelaar E, De Klerk G-J (1994) Abscisic acid controls dormancy development and bulb formation in lily plantlets regenerated in vitro. Physiol Plant 90: 59-64

Koda Y, Kikuta Y (1991) Possible involvement of jasmonic acid in tuberization of Yam plants. Plant Cell Physiol 32: 629-633

Kozai T (1991a) Micropropagation under photoautotrophic conditions. In: PC Debergh and Zimmerman RH (Eds) Micropropagation - Technology and Application (pp. 484) Kluwer Academic Publishers, Dordrecht

Kozai T (1991b) Photoautotrophic micropopagation. In vitro cell. dev. biol. 27P: 47-51

Maene LJ & Debergh PC (1985) Liquid medium additions to established tissue cultures to improve elongation and rooting in vivo. Plant Cell Tissue Organ Culture 5: 23-33

Matthys D, Gillis J, Debergh P (1994) Ethylene. In: J Aitken-Christie, T Kozai, MAL Smith (Eds) "Automation and environmental control in plant tissue culture" Kluwer Academic Publishers, Netherlands, in press

Mok MC, Mok DWS, Armstrong DJ, Shudo K, Isogai Y, Okamoto T (1982) Cytokinin activity of N-phenyl-N'-1,2,3-thiadiazol-5-ylurea (thidiazuron) Phytochem. 21: 1509-1511

Murashige T, Skoog F (1962) A revised medium for rapid growth and bioassay with tobacco tissue cultures. Physiol Plant 15: 473-497

Nagmani R, Bonga JM (1985) Embryogenesis in subcultured callus of Larix decidua. Can J For Res 15: 1088-1091

Nairn BJ (1992) Commercial micropropagation of radiata pine. In: MR Ahuja (Ed) Micropropagation of woody plants, Chapter 23 (383-394) Kluwer Academic Publishers, Den Haag

Onishi N, Sakamoto Y (1993) Development of cultural systems and encapsulation technology for synthetic seeds. Abstract 8.5.3-5 in Book of Abstracts XV International Botanical Congress, Yokohama

Pâques M, Boxus P (1987) A model to learn "vitrification", the rootstock apple M.26. Present Results. Acta Hort 212: 193-210

Rohde A, Debergh P (1992) Inhibition of carotenoid biosynthesis - An approach to release buds of *Acer palmatum* from dormancy? Med Fac Landbouww Univ Gent 57: 1537-1544

Stasinopoulos TC and Hangarter RP (1990) Preventing photochemistry in culture media by long-pass light filters alters growth of cultured tissues. Plant Physiol 93: 1365-1369

Strnad M, Peters W, Beck E, Kaminek M (1992) Immunodetection and identification of N^6-(o-Hydroxybenzylamino)purine as a new naturally occurring cytokinin in *Populus x canadensis* Moench cv Robusta leaves. Plant Physiology 99: 74-80

Stuart DA, Strickland SG, Walker KA (1987) Bioreactor production of alfalfa somatic embryos. HortScience 22: 800-803

Texeira JB, Sondahl MR, Kirby EG (1993) Somatic embryogenesis from immature zygotic embryos of oil palm. Plant Cell Tissue Organ Culture 34: 227-233

Ueda J, Kato J (1980) Isolation and identification of a senescence-promoting substance from wormwood (Artemisia absinthium L.) Physiol Plant 66: 246-249

Ueda J, Kato J (1982) Inhibition of cytokinin-induced plant growth by jasmonic acid and its methyl ester. Physiol Plant 54: 249-252

Went FW (1939) The dual effect of auxin on root formation. Am J Bot 26: 24-29

Went FW, Thimann KV (1937) Root formation. In "Phytohormones, Macmillan, New York, 183-206

Wochok ZS; Wetherell DF (1971) Suppression of organized growth in cultured wild carrot tissue by 2-chloroethylphosphonic acid. Plant & Cell Physiol 12: 771-774

Ziv M (1989) Enhanced shoot and cormlet proliferation in liquid cultured gladiolus buds by growth retardants. Plant Cell Tissue Organ Culture 17: 101-110

Ziv M, Ariel T (1991) Bud proliferation and plant regeneration in liquid-cultured Philodendron treated with ancymidol and paclobutrazol. J Plant Growth Regul 10: 53-57

A NEW CONCEPT OF PLANT *IN VITRO* CULTIVATION LIQUID MEDIUM : TEMPORARY IMMERSION

C. TEISSON, D. ALVARD
CIRAD / BIOTROP
P.O.Box 5035
34032 Montpellier
France

ABSTRACT
The advantages of *in vitro* culture in a liquid medium are often counterbalanced by technical problems such as asphyxia, hyperhydricity and the need for complex apparatus. A new culture technique has been developed based on temporary immersion that reduces these disadvantages. The technique can be used with a commercial filtration unit with only minor modifications and a laboratory air pump to flush the liquid medium. The system has been successfully employed with different species (*Coffea, Hevea, Musa*) and different multiplication systems (microcutting, somatic embryogenesis, meristem proliferation). The quality of plant development is generally better than that obtained with traditional liquid or semi-solid medium and this can be put down to the close contact with nutrients with no deterioration in gaseous exchanges of the explants. The temporary immersion technique can profitably replace costly biofermentors in both research and mass production.

1. Introduction

The use of liquid medium for *in vitro* culture has many advantages and has been the subject of many studies. It has also frequently been considered an ideal technique for mass production as it reduces manual labor and facilitates changing the medium. Techniques and culture vessels of varying complexity have been developed as a result of these studies (Tisserat & Vandercook ,1986; Aitken-Christie & Davies, 1988; Simonton *et al.* 1991).
Along with other authors we considered that the major disadvantages of a liquid medium could be avoided by exposing the plant to the liquid medium intermittently with periodic floodings rather than continuously. We also wanted to ensure the technique was easy to use so we designed a small, cheap apparatus that is simple to operate. We carried out tests using the apparatus on several species (banana, coffee, rubber) and in several phenomena (microcuttin,somatic embryogenesis and meristem proliferation).

2. Description of the apparatus

The vessel we tested was made from a commercially distributed autoclavable 1-liter filter unit. This was modified by connecting the upper and lower compartment by means of a small glass tube and by placing a screen disc with a selected pore size at the bottom of the upper compartment. This apparatus functioned in the opposite way to usual: plants were placed in the

M. Terzi et al. (eds.), Current Issues in Plant Molecular and Cellular Biology, 105-110.

upper vessel and the nutrient solution (generally 250 ml) in the lower one. When put under pressure by means of a small air pump, the liquid flowed up through the glass tube and immersed the plants in the upper vessel.During the immersion period the flow of air renewed the atmosphere inside the culture vessel. A home timer was used to control the frequency and length of the pressure increase and thus the frequency and duration of immersion. When pressure dropped, the medium returned to the lower part of the apparatus by gravity. All the air flows were sterilized through 0.2μ hydrophobic filters which ensured passive venting of the vessel. Depending on the model of air pump used it may be necessary to add a three-way solenoid valve to the entry tube to allow the pressures to adjust rapidly and the nutrient liquid to return rapidly to the lower vessel.

3. Examples of use

3.1. COFFEE MICROCUTTINGS

In coffee (*C.arabica* or *canephora*) microcuttings multiplication on a semi-solid medium is relatively slow, the multiplication coefficient being 6 to 7 every 3 months (Sondhal *et al.* 1989). With temporary immersion the same coefficient can be obtained in 5 weeks. Optimum results were obtained with an immersion rate of 4 times 15 minutes per 24 hours, shorter intervals and durations were not as efficient and longer immersion led to vitrification. In this system a brief treatment by gibberellic acid is also easy to apply. Gibberellic acid leads to lengthening of internodes which are then easier to section, but treatment must be brief as gibberellic acid is toxic for coffee. After final division, a simple change in the composition of the medium allows rooting to take place in the same vessel. This simplifies final handling before the plants are transferred to the nursery for hardening.

The yields we obtained indicate mass propagation of *Coffea Arabica* is possible in more competitive conditions.

3.2. COFFEE EMBRYOGENESIS

Somatic embryogenesis in *Coffea* sp. on liquid medium has been mastered only with calluses showing good friability (Berthouly, 1990). Irrespective of the frequency and duration of immersion our system produces embryos much more rapidly than with agitated liquid medium. In addition to the speed of embryo production, it is also important to note the difference in the modes of expression of embryogenesis as a function of the different types of immersion. With 4 immersions of 15 minutes per 24 hours the embryos continue to develop up to germination and may then be transferred directly to the greenhouse. On a medium of identical composition but with only one immersion of 1 minute per 24 hours primary embryos stop developing and start producing adventitious embryos. Subsequently one series of 4 immersions of 15 minutes per 24 hours leads to development and germination of the secondary embryos.

Thus a simple modification in the programme of immersions which requires neither a change in the medium nor the culture vessel means two successive stages of embryogenic expression can be achieved.

3.3. RUBBER EMBRYOGENESIS

Somatic embryogenesis in *Hevea brasiliensis* is induced after callogenesis on inner integuments of young fruits (Carron *et al* 1992). The expression of somatic embryogenesis on friable callus became possible when the medium was completely devoid of growth regulators and was tested both on a solid medium and by temporary immersion .Only temporary immersion at a frequency of 2 times 1 minute per day (fig 1) resulted in embryos that were satisfactory both from the point of view of quality and quantity. In rubber, this is the first time that such a degree of control of somatic embryogenesis has been achieved after long-term callogenesis.

The conversion of embryos into plantlets requires a three weeks maturation step during which abscisic acid is supplied together with desiccation (Etienne *et al.* 1993). Temporary immersion during this stage resulted in the best germination rate and in the best conversion rate with an extremely brief immersion period: 1 minute per week !

True germination is also more satisfactory using temporary immersion; however in this case a much higher immersion rate is required : 4 times 15 minutes per day. The physical appearance of the plantlets obtained by temporary immersion is much closer to seedlings in the development of cotyledons, the quality of the root system and the absence of vitrification .

3.4. BANANA MICROPROPAGATION

The banana is one of the most intensively multiplied major food crops in the world using *in vitro* techniques. The classic procedure is based on meristem proliferation (Vuylsteke ,1989). In our experiments shoot clusters (cv.Grande Naine) were used as initial material for the comparison of five different liquid medium cultures (permanent immersion with or without bubbling, partial immersion, cellulose support and temporary immersion) with semi-solid medium culture. Temporary immersion produced the best results both in proliferation rate and in increase in dry weight (Alvard *et al* ., 1993).

More recently we developed a somatic embryogenesis procedure following callogenesis on very young male flowers (Escalant *et al.*, 1994.). Primary embryos formed on this callus can be used to establish cell suspensions or may be placed in a temporary immersion system in the presence of Picloram. In this case adventitious somatic embryogenesis took place from epidermal cells of the primary embryos. The phenomenon was repeated and in 6 months the number of somatic embryos increased 40 fold. With these embryos and after transfer to germination medium recovery frequencies were 60 to 70%. Primary embryos placed on a semi-solid medium of the same composition were transformed into compact, white, callus-like structures.

4. Description of physical parameters

The originality of this system lies in the physical conditions in which the plants are placed and this is why these conditions should be described in some detail.

Relative humidity inside the compartment containing the plants always remains near saturation irrespective of the frequency and duration of immersion. Figure 2 shows variations in relative humidity and temperature during culture with 4 immersions of 1 minute distributed regularly over 24 hours. During the dark period relative humidity was permanently saturated, during the light period it remained over 98%. In these conditions only the temperature varied by approximately $5^{\underline{0}}$ C due to a greenhouse effect during the light period (30 µmol.m2.s).

The high values for relative humidity are caused both by the water retained at the surface of the plant tissue and on the sides of the culture vessel by capillarity and by the opening that communicates with the lower vessel, which always contains liquid medium. The quantity of water retained by capillarity will of course depend on the surface area of the explants; with equal mass, the smaller the explants, the greater the quantity. Figure 3 shows the quantity of water covering the explants as a percentage of their mass. Over a 24-hour period and although there was absolutely no immersion, the quantity of water remained relatively high.

The reduced volume of the vessel and the hydrophobic character of the event explains the constantly high relative humidity and thus the absence of damage to the plants even in the case of very rare and brief immersion. The persistence of a capillary film of water on the surface of the tissues not only enables nutrition outside immersion periods but also disturbs gaseous exchanges much less than total immersion.

Gaseous exchanges inside and outside the vessel mainly take place during the immersion stage. Apart from the immersion stage, there is only passive gaseous exchange caused by the event, and in this case it depends to a great extent on the concentration gradient inside and outside the vessel and on the type of event and is usually very low. During immersion gaseous exchanges are much higher and are caused both indirectly by movement of the liquid and directly by the air pump. In our most frequent experimental conditions this corresponds to complete renewal of the culture atmosphere after 5 minutes of immersion.

Dissolved oxygen concentration in the liquid medium reachs saturation after only one minute of bubbling when it does not exceed 95 %, in the absence of plant tissue, in an agitated Erlenmeyer flask.

5. Conclusion

The experiments described here represent only a small selection of those in which we use this apparatus. Other satisfactory results were achieved in somatic embryogenesis in oil palm, *Elaeis guineensis*,and in *Citrus deliciosa*

This alternative micropropagation system makes it possible to supply the culture with both nutrients and aeration without sophisticated technology such as bioreactors. It combines the advantages of being cheap and easy to use, and allows control of morphogenetic responses.

Its small size could be considered a disadvantage for mass production. However, considering how frequently it is necessary to manage different populations following distinct clones and taking into account the risk of contamination, we feel the size is justifiable and that even in the case of industrial production many small vessels are preferable to one big one. In any event, the volume of the filter unit described here is large enough to create a batch of 20 coffee microcuttings or 2 to 3 thousand banana somatic embryos, to cite but two examples.

The use of a gas to displace the liquid, in our case compressed air, means the use of other gases can be envisaged that could have an effect on the behavior of the plants: for example $C0_2$ to facilitate photosynthesis (Kozai & Iwanami 1987) or air poor in oxygen for the maturation of somatic embryos (Carman ,1988).

The ease with which the screen in the bottom of the upper compartment can be changed means the appropriate pore size can be selected to guarantee optimum drainage while at the same time retaining plant tissue. In our different experiments the size varied from 50µ for embryogenic clusters to 500µ for coffee microcuttings.

Another aspect of our work that is worth underlining concerns the frequency and duration of immersion. During preliminary trials we used frequencies and immersion times much higher than those described here. We were consequently extremely surprised by the success we achieved with flooding times as short as 1 minute per week. However the most surprising results of all were the effects on morphogenesis of different rates of immersion such as those described here in connection with somatic embryogenesis in coffee. This implies that a new tool is available for experimenters to control *in vitro* plant reactions.

Generally speaking plant behavior resembles more closely the behavior of 'normal' plants, i.e. *ex vitro* than the disturbed behavior often encountered in *in vitro* culture. This is probably because the supply of nutrient elements is more effective than on a semi-solid medium and there is minimum disturbance of gaseous exchanges through the capillary film covering the tissues.Growth and morphogenetic responses of plants either equaled or exceeded that of plants on concurrent agar or liquid media and so in our laboratory different researchers who have had the opportunity to test this system are to a larger and larger extent using it to replace traditional plant tissue culture systems and at the present time more than two hundred units are operating.

References

AITKEN-CHRISTIE J., DAVIES H., 1988. Development of a semi-automated micropropagation system. Acta Hort. 230 : 81-87.

ALVARD D.,COTE F.,TEISSON., 1993. Comparison of methods of liquid medium culture for banana micropropagation .Plant Cell Tiss. Org. Cult.32:55-60

BERTHOULY M., 1991. Cell suspension and somatic embryogenesis regeneration in liquid medium in coffea. 14° International Plant Biotechnology Congress CIPNAET, San Jose Costa Rica, January 14-18

CARMAN J.G 1988 .Improved somatic embryogenesis in wheat by partial simulation of the *in-ovulo* oxygen,growth regulator and desiccation environments.Planta ,175:417-424.

CARRON M.P., D'AUZAC J., ETIENNE H., EL HADRAMI I., HOUSTI F., MICHAUX-FERRIERE N., MONTORO P., 1992. Biochemical and histological features of somatic embryogenesis in rubber (*Hevea brasiliensis* Müll. Arg.). Indian journal of Natural Rubber Research, 5 (182) : 7-17 .

ESCALANT J.V,TEISSON C ,COTE F 1994 .Amplified somatic embryogenesis from male flowers of triploid banana and plantain cultivars (*Musa sp.*) In Vitro .in press

ETIENNE H., MONTORO P., MICHAUX-FERRIERE N., CARRON M.P., 1993. Effects of desiccation, medium osmolarity and abscisic acid on the maturation of *Hevea brasiliensis* somatic embryos. J. Exp. Bot., 44, (267): 1613-1619.

KOZAI T,et IWAMANI Y 1987.Effects of CO^2 enrichment on the plantlet growth during the multiplication stage.Plant Tissue Culture Letters,4(1):22-26

SIMONTON W., ROBACKER* C., KRUEGER S., 1991. A progammable micropropagation apparatus using cycled liquid medium. Plant Cell Tiss.Org. Cult.27 : 211-218.

SONDAHL M.P., NAKAMICRA F., MEDINCE FILHO M.P., CARVOLLHO A., FAZUOLI L.C., and CASLO W.M., 1989. Coffee Handbook of Plant Cell Culture. Volumen III chapter 21

TISSERAT B., VANDERCOOK C.E., 1986. Computerized longterm tissue culture for orchids. Amer. Orchid. Soc. Bull. 55 (1) : 35-42.

VUYLSTEKE DR 1989 Shoot-tip culture for the micropropagation,conservation and exchange of Musa germplasm.I B P G R,Rome (pp 1-56). ISBN 92-9043-140-7.

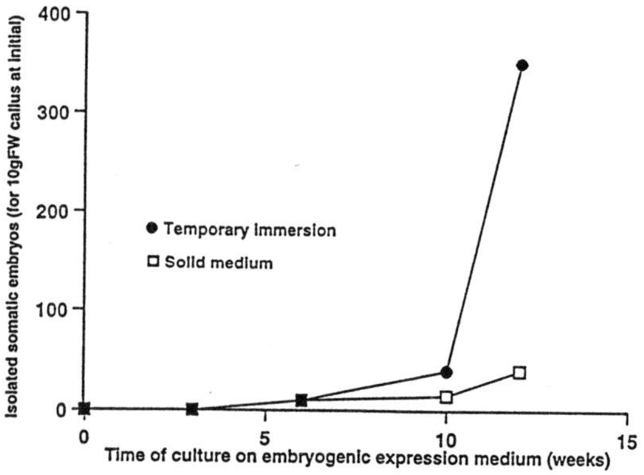

Fig 1 Effect of temporary immersion on Hevea somatic embryos production

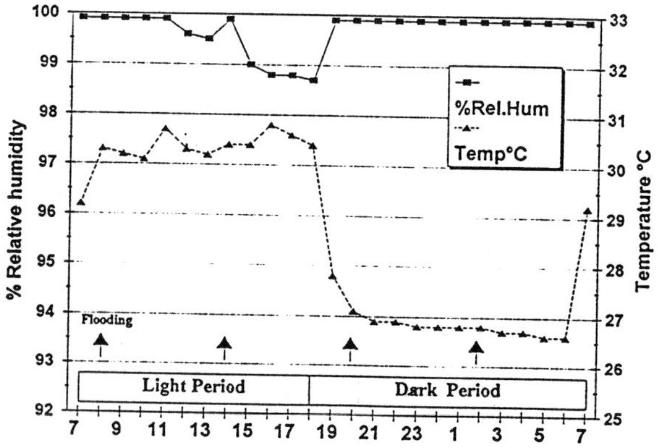

Fig. 2 : Relative humidity and temperature in the upper compartment
Four immersions/24H . Light intensity : 30 µmol . m-2 . s-1
External temperature : 27 °C

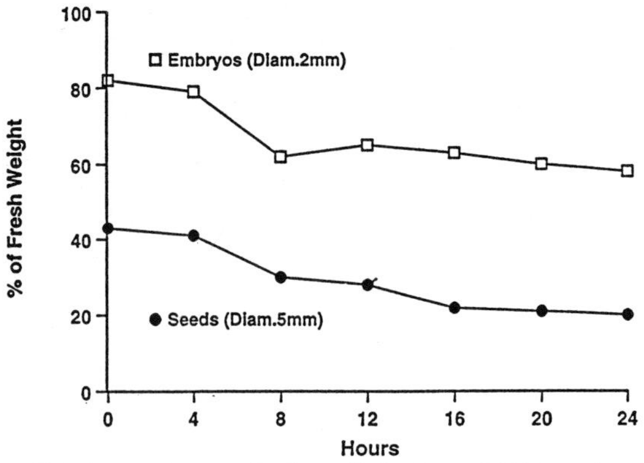

Fig 3 Amount of water retained by capillarity after one initial immersion.

HORMONE REQUIREMENTS DURING THE SUCCESSIVE PHASES OF ROOTING OF *MALUS* MICROCUTTINGS

GEERT-JAN DE KLERK
Centre for Plant Tissue Culture Research
PO Box 85, 2160 AB Lisse
The Netherlands

ABSTRACT. Just as other regeneration processes, adventitious root formation may be dissected into three phases (dedifferentiation, induction, and morphological differentiation). The timing of these phases was established by giving pulses with the anti-auxin *p*-chlorophenoxyisobutyric acid (PCIB) during the rooting period. This compound is expected to block rooting specifically during the time of action of auxin, *i.e.*, during the induction phase. The results matched our previously reported data with cytokinin-pulses. This determination of the timing of the three phases enables an examination of the differential effects of hormonal and other factors during each phase. Here we examine the effects of various auxins and related compounds. The various auxins had very different effects in the successive phases. For example, indolebutyric acid (IBA) was very promotive in the induction phase but was only a weak inhibitor during morphological differentiation; 2,4-dichlorophenoxyacidic acid (2,4-D) was only little effective in the induction phase, but was a very strong inhibitor during morphological differentiation. Among the various auxins that have been examined, naphthaleneacetic acid (NAA) had the highest activity in root induction. However, NAA had also a strong inhibitory effect during morphological differentiation. Phloroglucinol, a phenolic compound that has been frequently reported to enhance rooting, only acted during the first 24h, indicating that it was not active during the induction phase.

1. Introduction

Rooting of microcuttings is a crucial step in micropropagation because the rooting system determines performance after planting in soil (Van Telgen *et al.*, 1992). Rooting is usually achieved by a simple treatment with auxins. Our research aims at the development of new rooting protocols based on differential hormonal sensitivity of the successive phases in the rooting process. Previously, we have established the timing of the phases in the rooting process of *Malus* 'Jork 9' microcuttings by giving during the rooting treatment pulses with cytokinin (benzylaminopurine, BAP) or auxin (IBA). Analogous to the phases described by Christianson and Warnick (1983) in shoot regeneration from *Convulvus* leaf explants, we distinguished **(1) 0 - 24h: dedifferentiation** during which the cells develop competence to respond to the rhizogenic signal; **(2) 24 - 96h: induction** during which the cells become determined to form roots, and **(3) after 96h: morphological differentiation** during which the roots develop (De Klerk & Ter Brugge, 1992). Now we examine the differential effects of hormones and other factors during the various phases. The current paper reports on the effects of various auxins and related compounds (anti-auxin and phenolic compounds) during the three phases.

M. Terzi et al. (eds.), Current Issues in Plant Molecular and Cellular Biology, 111-116.

2. Materials and Methods

Shoot production of *Malus* 'Jork 9' was maintained as described previously (De Klerk *et al.*, 1990). Rooting was examined in 1-mm disks excised from stems of microcuttings (10 disks from each stem) as described by Van der Krieken *et al.* (1993). The disks were cultured with the apical side down on a nylon mesh (4 x 4 cm) in a 9-cm Petri dish on 25 ml rooting medium to which auxins, PCIB, 2,3,5-triiodobenzoic acid (TIBA), phloroglucinol or BAP had been added as shown for each experiment. The Petri dish was incubated upside down in the dark in a culture room at 25°C. The nylon mesh, with the disks attached, was transferred at the indicated times to Petri dishes with medium of another hormonal composition. At transfer, the mesh was first put for 5 sec on filter paper to remove liquid. At the time of transfer from medium with to medium without auxins, the disks were also transferred to the light. In each Petri dish, 30 disks were cultured originating from 6 shoots. For each determination, 90 disks were used. Rooting was also examined in shoots as described previously (De Klerk *et al.*, 1990). For each determination, 30 shoots were used.

3. Results and Discussion

Timing of the induction phase by pulses with the anti-auxin PCIB

PCIB is considered to be a genuine anti-auxin as opposed to auxin-transport inhibitors like TIBA (McRae & Bonner, 1953). Figure 1 shows that PCIB inhibits rooting at a low IBA-concentration, but promotes rooting at a supraoptimal IBA-concentration. This demonstrates that it is indeed a genuine anti-auxin. In a similar experiment with TIBA (at 0.1, 1 or 10 μM), we observed an inhibition of rooting by TIBA at low and optimal IBA-concentrations, but no stimulation at a supraoptimal IBA-concentration (data not shown).

Disks were given 24h pulses with PCIB during a rooting treatment of 5 days at 3 μM IBA. There was a significant decrease of rooting from the 0-24h pulse to the 24-48h and 48-72h pulses ($p<0.05$) (Fig. 2). The pulses of 72-96h and 96-120h showed an increase of rooting to the value of the control. These data demonstrate that the induction phase during which auxins exert their rhizogenic action, is from 24 to 72h. This agrees with previous

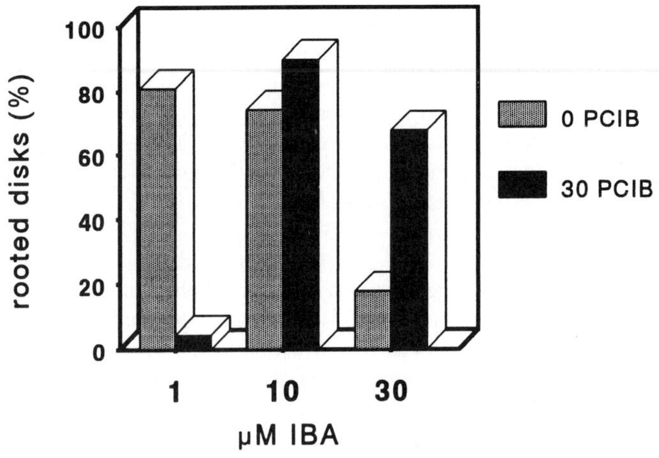

Figure 1. Effect of the anti-auxin PCIB on rooting. Disks were cultured for 5 days with addition of 1, 3 or 10 μM IBA. The medium did or did not contain 30 μM PCIB.

Figure 2. Effect of 24h PCIB-pulses during the rooting period. Disks were cultured for 5d on medium containing 3 μM IBA. Then they were transferred to hormone-free medium. PCIB (200 μM) was added during the indicated periods. (During the PCIB-pulse 3 μM IBA was also present in the medium.)

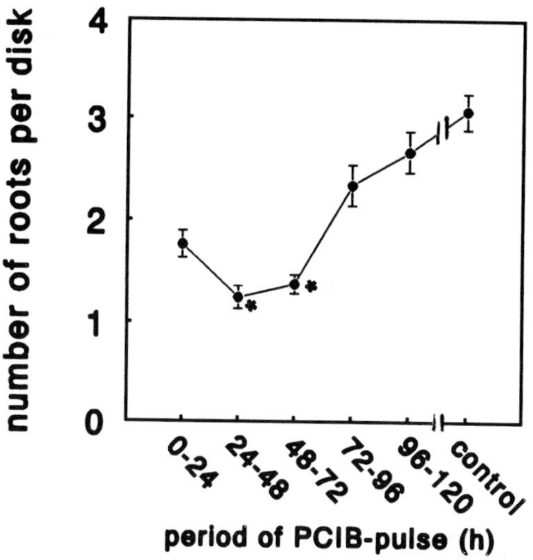

findings with BAP- and IBA-pulses (De Klerk & Ter Brugge, 1992). In contrast to the effect of BAP-pulses, the effect of PCIB-pulses is only small. This may be explained by the different ways of action of these compounds. BAP redirects the process. Just like withholding IBA, addition of PCIB only has a small, redirecting effect, and has as a major effect that the process comes to a standstill up to the removal of PCIB (De Klerk & Ter Brugge, 1992).

Effect of 2,4-D on root induction and root outgrowth

In adventitious root formation, two major effects of auxins occur: auxins promote root induction, but are inhibitory once the root primordia have been formed (a.o. De Klerk *et al.*, 1990). We examined whether IBA and 2,4-D had differential effects in these phases. When supplied as 24h pulses at 0-24h, 2,4-D was much less effective than IBA in inducing roots (Fig. 3). This may be due to a poor induction *per se*, or to persistence of 2,4-D in the tissue resulting in blocking of the outgrowth of the root meristemoids. To examine this, we gave disks that had been treated with 25 μM IBA at 0-24h, at various times 24h pulses with 3 μM 2,4-D. As a control, we also gave pulses with 0.5 μM BAP. Figure 4 shows characteristic low rooting for the 24-48h and 48-72h BAP-pulses (De Klerk & Ter Brugge, 1992). The pulses with 2,4-D had a large effect when given at 48-72h or 72-96h. When 2,4-D-pulses were given early (0-24h or 24-48h), they had no effect. This shows that the poor rooting after application of 2,4-D was not due to blocking of outgrowth because of persistence of 2,4-D in the tissue, but to poor induction. When IBA was given at various concentrations at 72-96h, it had only a minor inhibitory effect (data not shown).

Figures 3 and 4 demonstrate that the effects of 2,4-D and IBA are very different from one another in the induction and morphological-differentiation phases: induction is strongly promoted by IBA, but only slightly by 2,4-D; in contrast, IBA is only slightly inhibitory during outgrowth, but 2,4-D strongly. The different effects of these auxins may be explained by the occurrence of two receptors: the first is related to the induction of root meristems and has a high affinity to IBA and a low to 2,4-D; the second is related to the inhibition of root growth and has a high affinity to 2,4-D and a low to IBA.

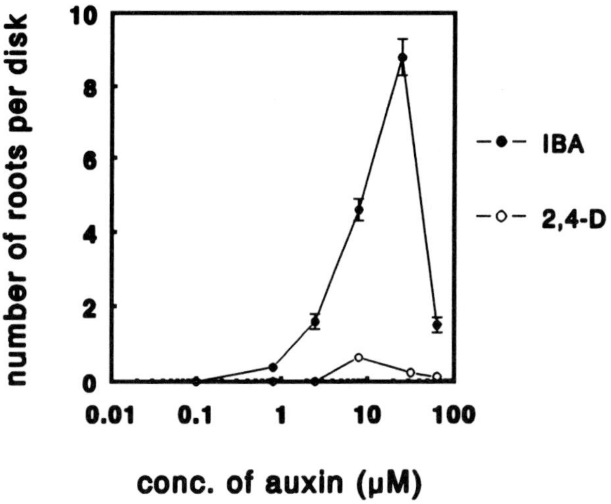

Figure 3. Effect of the auxins IBA and 2,4-D on rooting. Disks were cultured from 0 to 24h with addition of increasing concentrations of IBA or 2,4-D. Then they were transferred to hormone-free medium.

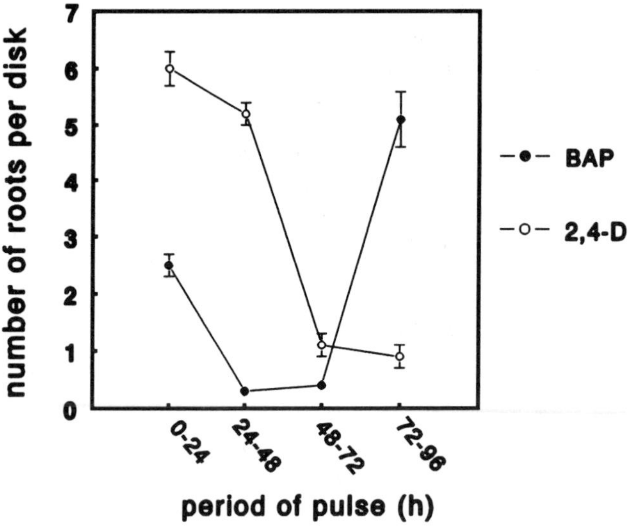

Figure 4. Effect of 24h pulses with BAP or 2,4-D during the rooting period. Disks were treated from 0 to 24h with 25 μM IBA and then transferred to medium without IBA. During the rooting period, they were given pulses with 0.5 μM BAP or 3 μM 2,4-D. Thus, for the 0-24h BAP pulse, the disks were cultured on medium with both BAP and IBA, and after 24h transferred to hormone-free medium. For the 24-48h BAP-pulse, they were first cultured for 24h on medium with 25 μM IBA, after that transferred to medium with only BAP, and after another 24h to hormone-free medium.

Effect of a series of auxins on the induction and morphological differentiation
A series of auxins was given from 0 to 24h to examine their effect on root induction. NAA was the most active auxin. IAA, 2,4-D, picloram and PAA hardly induced any roots. To examine the effect of the various auxins on outgrowth, disks were first incubated for 24h with 25 μM IBA and then on hormone-free medium. From 72 to 120h, they were given 48h

	induction	morphological differentiation
IAA (indoleacetic acid)	±	-
IBA (indolebutyric acid)	++++	- -
NAA (naphthaleneacetic acid)	+++++	- - - -
2,4,5-T (2,4,5-trichlorophenoxyacidic acid)	+++	- - - -
2,4-D (2,4-dichlorophenoxyacidic acid)	±	- - - - -
picloram (4-amino-3,5,6-trichloropicolinic acid)	±	- - - - -
PAA (phenylacetic acid)	0	0

Table 1. The effect of some auxins on two phases in the rooting process: induction and morphological differentiation. To determine their effect on root induction, the auxins were applied at 1, 3 or 10 μM from 0-24h. To determine their effect on morphological differentiation, the disks were treated from 0 to 24h with 25 μM IBA. After that, they were treated from 72-120h with the various auxins at 1, 3 or 10 μM.

pulses with the various auxins. Picloram and 2,4-D showed the strongest inhibition. It should be noted that addition of NAA and 2,4,5-T resulted in both strong promotion of induction and strong inhibition of morphological differentiation. In these experiments, IAA had a very low activity at the applied concentrations. To achieve high root induction, the concentration of IAA had to be ten times higher (100 μM; data not shown). PAA had a slight effect when applied at concentrations of 100 or 300 μM.

Effect of phloroglucinol in the various phases of the rooting proces
Phenolic compounds may, depending on their type, promote or inhibit root formation. It has been suggested that this is due to their effect on peroxidase activity (Wilson & Van Staden, 1990). Phloroglucinol, a triphenol, has been reported to enhance rooting (Jones, 1976). We also found an increase in rooting with continuous addition (data not shown). When phloroglucinol was given as 24h pulses during the rooting period, it only enhanced rooting when applied during the first 24h (Fig. 5). The other pulses had no or only a small effect. Since the main period of auxin action is the induction period (24-96h), it is unlikely that phloroglucinol acts by reducing peroxidase activity and in that way auxin degradation. The time of action of phloroglucinol corroborates the hypothesis of Wilson & Van Staden (1990), that phenolic compounds promote rooting by enhancing stress or injury.

4. General Discussion and Conclusions

The timing of the three phases of adventitious root formation has been established with BAP-, IBA-, and PCIB-pulses (Fig. 2, and De Klerk & Ter Brugge, 1992). We now examine the specific effects of hormonal and other factors during the three phases. The current paper

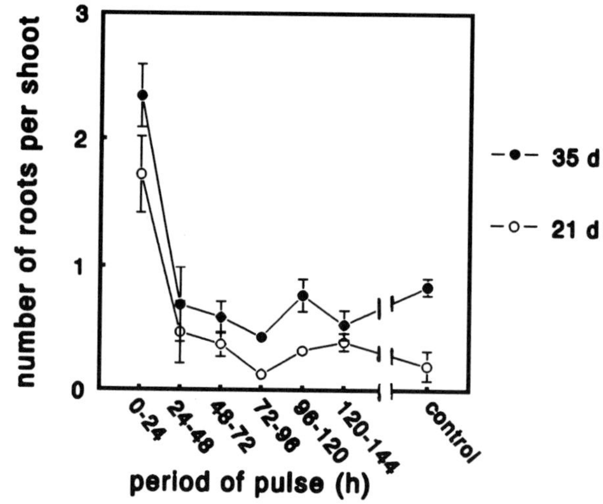

Figure 5. Effect of phloroglucinol pulses on rooting. Shoots were cultured on hormone-free medium and given pulses with 1 mM phloroglucinol at the indicated times. Rooting was scored after 21 and 35 days.

reports on different auxins applied during the first two phases (dedifferentiation + induction) or during the third phase. We found that various auxins had very different effects in these phases. NAA was the most effective auxin in root induction. However, because NAA also strongly inhibited the third phase, it is not the optimal auxin for rooting. IBA, being very effective in root induction and only slightly inhibitory during the third phase, is the preferable auxin for rooting. The time of action of added compounds may explain their mode of action. Thus, the time of action of phloroglucinol (0-24h) shows that this compound unlikely acts by protecting auxin from degradation by peroxidase. Now, the mode and time of action of other compounds (polyamines, ethylene) are being examined.

Acknowledgements. The results presented in this paper have been obtained in co-operation with Jolanda ter Brugge, Emmanuelle Caillat, Elisa Catenaro, and Marina Keppel. This research was carried out in the framework of the COST 87 working group 'Root Regeneration'.

References

Christianson, M.L., and Warnick, D.A., 1983. Competence and determination in the process of in vitro shoot organogenesis. *Dev. Biol.*, **95**, 288-293.

De Klerk, G.J., Ter Brugge, J., Smulders, R., and Benschop, M., 1990. Basic peroxidases and rooting in microcuttings of *Malus*. *Acta Hortic.*, **280**, 29-36.

De Klerk, G.J., and Ter Brugge, J., 1992. Factors affecting adventitious root formation in microcuttings of *Malus*. *Agronomie*, **12**, 747-755.

Jones, O.P., 1976. Effect of phloridzin and phloroglucinol on apple shoots. *Nature*, **262**, 392-393.

McRae, D.H., and Bonner, J., 1953. Chemical structure and antiauxin activity. *Physiol. Plant.*, **6**, 485-510.

Van der Krieken, Breteler, H., Visser, M.H.M., and Mavridou, D., 1993. The role of the conversion of IBA into IAA on root regeneration in apple: introduction of a test system. *Plant Cell Rep.*, **12**, 203-206.

Van Telgen, H.J., Van Mil, A., and Kunneman, B., 1992, Effect of propagation and rooting conditions on acclimatization of micropropagated plants. *Acta Bot. Neerl.*, **41**, 453-460.

Wilson, P.J., and Van Staden, J., 1990. Rhizocaline, rooting co-factors, and the concept of promotors and inhibitors of adventitious rooting - A review. *Ann. Bot.*, **66**, 479-490.

PROPAGATION OF *MAGNOLIA* AND *LIRIODENDRON* VIA SOMATIC EMBRYOGENESIS

S. A. MERKLE and H.D. WILDE
D.B. Warnell School of Forest Resources
University of Georgia,
Athens, GA 30602 USA

ABSTRACT. The family Magnoliaceae contains an number of trees valued for both horticultural and forestry purposes. Some species, particularly in the genus *Magnolia*, are quite rare, and vegetative propagation by conventional means has been difficult. We have developed embryogenic regeneration systems for a number of rare *Magnolia* species, such as *M. cordata* and *M. pyramidata*, as well as for *Liriodendron tulipifera* and hybrids between *L. tulipifera* and *L. chinense*. All of these systems use the same basic protocol for culture initiation, which consists of culturing immature zygotic embryo or seed explants at very early stages of development on a medium supplemented with 2,4-dichlorophenoxyacetic acid (2,4-D) and benzylaminopurine (BAP) to induce repetitive embryogenesis. However, optimal treatments vary for somatic embryo production from cultures of each species. Some species must be maintained on 2,4-D in order to continue to proliferate as proembryogenic masses (PEMs). Others lose embryogenic capacity if maintained on 2,4-D and instead proliferate indefinitely via repetitive embryogenesis following transfer to hormone-free medium. Somatic embryogenesis has made possible the production of virtually unlimited numbers of rare *Magnolia* trees, some of which have produced flowers after only 2 seasons in the field. Somatic embryo-derived hybrid *Liriodendron* trees have shown rapid growth rates and other signs of hybrid vigor. Of the species cultured to date, *L. tulipifera*, which has the ability to proliferate as PEMs in suspension culture, has the best potential for scale-up for mass propagation, using a size-fractionation/plating protocol. Field tests of thousands of somatic embryo-derived trees of this species have already been established, and transgenic trees produced from the PEMs following microprojectile-mediated gene transfer have maintained expression of two marker genes for over 3 years.

1. Introduction

The genera *Magnolia* and *Liriodendron* are both members of the Magnoliaceae, which is one of the most primitive families of flowering plants. Fossil evidence of their presence on earth dates to the tertiary period (2-65 million years ago). Magnolias, of which there are approximately 80 species today, are distributed mainly in two distinct temperate and tropical regions of the world, eastern America and eastern Asia. Trees and shrubs included in this group are particularly well known on account of their flowers, which vary in size (up to 36 cm in diameter in some species), in fragrance, and in color, from a pure white to royal purple (Treseder 1978). They constitute one of the most widely-employed groups of ornamental woody perennials in the world.

The genus *Liriodendron* includes only two species, *L. tulipifera* (yellow-poplar) of eastern North America, and *L. chinense* (Chinese tuliptree), native to central mainland China. Yellow-poplar is one of the most distinctive and valuable hardwood trees in the eastern United States and is characterized by rapid height growth on good sites, reaching up to 40 m

M. Terzi et al. (eds.), Current Issues in Plant Molecular and Cellular Biology, 117-122.

in 50 years. Chinese tuliptree, which is morphologically similar to yellow-poplar, is found across a wide geographical area covering most of the Chinese provinces of Anhwei, Kiangsi, Fukien, Hupeh, Szechuan, Kweichow, Kwangsi and Yunnan. It is also found in Thapa, Vietnam (Gardiner 1989). Allozyme data, sequence divergence in the plastid genomes of the two species and paleobotanical evidence have been used to estimate the time since divergence from a common ancestor at 10-16 million years (Parks and Wendel 1990). Hybrids between the two species have been synthesized which are male fertile, although most are female sterile (Santamour 1972, He and Santamour 1983, Parks et al. 1983). Highly heterotic growth rates have been reported for the hybrid trees.

2. Yellow-poplar somatic embryogenesis and plantlet regeneration

Somatic embryogenesis in yellow-poplar tissue cultures was first reported by Merkle and Sommer (1986), and the protocol for culture initiation remains basically unchanged since being optimized by Sotak et al. (1991). The optimal developmental stage of zygotic embryo explants for initiation of embryogenic cultures is the globular stage, 0.2 mm or less in diameter, which appears 7-9 weeks following pollination. Based on these results, immature aggregates of samaras are collected approximately 8 weeks following pollination and dissected into individual samaras using a grafting knife. Samaras are dewinged and surface-sterilized using the following sequence: 70% ethanol (20 s), 10% Roccal (National Laboratories; 2 min.), repeat ethanol and Roccal steps, 100% Clorox (5.25% sodium hypochlorite; 5 min), water rinse (3 min), 0.01 N HCl rinse (3 min), and three additional water rinses (3 min each). Following surface-sterilization, dewinged samaras are dissected aseptically and embryos and endosperm are placed on a semisolid induction medium consisting of Blaydes' (Witham et al. 1971) major salts, Brown's minor salts (Sommer and Brown 1980), Murashige and Skoog's (1962) iron, Gresshoff and Doy's (1972) vitamins, 40 g/l sucrose, 2 mg/l 2,4-D, 0.25 mg/l BAP, 8 g/l Phytagar (Gibco) and 1 g/l casein hydrolysate (enzymatic; CH). The medium is dispensed in 60 mm plastic Petri dishes. Petri dishes are sealed with Parafilm and incubated in the dark at 22° C. Explants are transferred to fresh induction medium monthly. Proembryogenic masses (PEMs) begin to appear from the explants within 2 months and their proliferation can be maintained by monthly passage to fresh induction medium.

To initiate development of somatic embryos, PEMs can be simply transferred to semisolid basal medium (same as induction medium, but lacking growth regulators). Somatic embryos begin to appear about one month following transfer to basal medium. However, over the past few years, we have developed a protocol for mass production of synchronous populations of embryos from suspension cultures of PEMs. Yellow-poplar PEMs proliferate rapidly as suspension cultures. Embryogenic suspension cultures are grown in 125 ml Erlenmeyer flasks containing 40 ml of liquid induction medium on a gyratory shaker at 90 rpm and maintained by transferring approximately 1 g of PEMs to fresh medium every 3 weeks. Various protocols involving size fractionation of embryogenic suspension cultures have been tested for their potential to produce synchronous populations of mature somatic embryos (Merkle et al. 1990). In the protocol we currently use, two weeks following transfer to fresh induction medium, 1 g of PEMs is sieved on a 140 μm stainless steel screen and the fraction passing through is resieved on a 38 μm screen. The fraction remaining on the 38 μm screen is rinsed with basal medium and then backwashed from the screen onto a single layer of filter paper in a Buchner funnel. PEMs are again rinsed with basal medium while under mild vacuum, which serves to spread PEMs into a single layer. When excess liquid medium has been drawn off, the filter paper with PEMs is placed on semisolid basal medium and incubated under fluorescent light (16 hr/day) at 30° C. As PEMs develop into somatic embryos on the filter paper, mature torpedo-stage embryos with well-developed cotyledons are selected and transferred to Petri plates containing germination medium (basal medium lacking CH). This modification of the basal medium was made following our discovery that CH inhibited somatic embryo germination, probably by raising the osmotic potential of the medium, and that eliminating it promoted vigorous germination and cotyledon greening within one week (Merkle et al. 1990). Using this method, a 60-70% synchronous population of embryos can be produced within two weeks following plating of PEMs. Following transfer to GA7 vessels (Magenta Corp.) containing 100 ml of plantlet development medium, which is a modified Risser and White's (1964) medium with 2% sucrose and no CH, an average of 32% of these somatic embryos convert to plantlets (Merkle et al. 1990). This protocol was applied to 9 embryogenic lines to produce over 5500 somatic embryo-derived plantlets for field testing (Merkle et al. 1991). Recently, we have found that newly-germinated somatic embryos with only a single non-cotyledon leaf can be transferred to potting mix and successfully acclimatized in a humidifying chamber, by gradually lowering the relative humidity from 100% to ambient conditions over 6-8 weeks. (An experiment comparing in vitro and ex vitro development of newly germinated pyramid magnolia somatic embryos is detailed below). During this time plantlets are fertilized weekly with 1 ml of a modified Hoagland's solution.

3. Hybrid *Liriodendron* somatic embryogenesis

Initiation of hybrid *Liriodendron* embryogenic cultures follows the same protocol as for yellow-poplar culture initiation, except that the explant material is immature hybrid zygotic embryos. To generate hybrid seeds, flowers are collected from a Chinese tuliptree and the anthers used for controlled pollinations of yellow-poplar mother trees. Aggregates of samaras resulting from the pollinations are harvested 8 weeks post-pollination and embryos and endosperm excised to initiate cultures on the yellow-poplar induction medium described above. In contrast to yellow-poplar embryogenic cultures, hybrid *Liriodendron* cultures tend not to maintain growth as PEMs while exposed to 2,4-D. Instead, they proliferate via repetitive embryogenesis as globular- or even later-stage embryos. Thus, they are not as amenable to suspension culture as yellow-poplar embryogenic cultures and efforts to use the fractionation/plating protocol described above have been unsuccessful. However, clusters of immature hybrid somatic embryos transferred to basal medium will develop into mature embryos after 4-6 weeks. Mature embryos germinate vigorously following transfer to basal medium lacking CH and can be immediately planted in potting mix as described for yellow-poplar. Hybrid *Liriodendron* plantlets have shown evidence of hybrid vigor by their rapid growth rates and leaves that are 2-3 times the size of leaves of the parent species. We have verified the hybrid nature of all of these embryogenic lines and plants derived from them by Southern hybridization analysis with a species-specific DNA marker (Merkle et al. 1993).

4. Magnolia somatic embryogenesis and plantlet regeneration

Using our experience with yellow-poplar as a basis, we initiated experiments to establish embryogenic cultures of several magnolia species native to the southeastern United States, some of which are quite rare: *M. virginiana* (sweetbay magnolia), *M. fraseri* (fraser magnolia), *M. cordata* (yellow cucumbertree) (Merkle and Wiecko 1990), *M. macrophylla* (bigleaf magnolia) (Merkle and Watson-Pauley 1993) and *M. pyramidata* (pyramid magnolia) (Merkle and Watson-Pauley, in press). Culture initiation protocols and media are the same as for yellow-poplar, with one main distinction: The optimal developmental stage of zygotic embryo for initiation of embryogenic cultures is reached much sooner following pollination in magnolias than in yellow-poplar. We have found that 3-5 weeks post-pollination magnolia seeds have the highest embryogenic potential. To initiate embryogenic magnolia cultures, developing fruits (aggregates of follicles) are collected, washed for 5 minutes in 10% Roccal and dissected using a grafting knife to remove the seeds. Seeds are surface-sterilized using the same sequence as is used with yellow-poplar samaras. Following surface-sterilization, seeds are bisected longitudinally with a scalpel and the halves were placed cut surface downward in 60 mm plastic Petri dishes containing semisolid yellow-poplar induction medium. Cultures are maintained in darkness at 22° C and transferred to fresh induction medium monthly.

Each magnolia species responds somewhat differently to culture on yellow-poplar induction medium. Sweetbay magnolia and bigleaf magnolia in general fail to produce PEMs on induction medium, and instead, embryogenic cultures first appear as clumps of somatic embryos at the micropylar ends of the explanted seeds. If transferred to fresh induction medium, sweetbay magnolia embryogenic cultures darken and embryo production declines, and bigleaf magnolia cultures produce round nodules which grow in diameter but do not proliferate. Embryogenic cultures of each of these species are best maintained by transfer to basal medium following a few months on induction medium. Monthly transfer to fresh basal medium maintains repetitive embryo production for years. Fraser magnolia, yellow cucumbertree and pyramid magnolia all produce PEMs in response to culture on induction medium, which are easily maintained by monthly transfer to fresh induction medium. As with yellow-poplar, somatic embryos are produced following transfer of PEMs to basal medium. However, all embryogenic magnolia cultures differ from yellow-poplar cultures in that they cannot be maintained as suspensions of PEMs. Thus prospects for large-scale propagation of these magnolias via synchronization of embryo development are not promising currently.

As with yellow-poplar, germination of mature somatic embryos of all magnolia species is accomplished by transferring them to basal medium lacking CH and incubating in the light at 30° C. Germinants transferred to yellow-poplar plantlet development medium in test tubes or GA7 vessels continue to grow, but both root and shoot development are slow. Because of this lack of plantlet vigor in vitro, we set up an experiment to compare development of newly germinated pyramid magnolia somatic embryos in vitro versus that ex vitro. We found that frequency of plantlet production (conversion) from germinants was higher and the resulting plantlets were more vigorous when germinants were transferred directly to potting mix and grown in the humidifying chamber instead of being maintained in vitro on plantlet development medium. Ex vitro-converted pyramid magnolia somatic embryos had, on the average, leaves that were 3 times larger and root systems that were 9 times longer than in vitro-converted somatic embryos of the same age (Merkle and Watson-Pauley, in press). Based on these results, our standard protocol for newly germinated somatic embryos of all of our magnolia and *Liriodendron* cultures is to transfer them directly to potting mix for conversion in the humidifying chamber.

Although no organized field testing of magnolia somatic embryo-derived plantlets is planned, those trees which have been planted out resemble seedling-derived trees in form, growth rate and bud phenology. Surprisingly, somatic embryo-derived trees of both sweetbay magnolia and yellow cucumbertree have produced flowers as early as 2 seasons following outplanting (unpublished).

5. Use of yellow-poplar embryogenic suspensions for gene transfer

One application for which we have found embryogenic yellow-poplar suspension cultures to be highly amenable is gene transfer via microprojectile bombardment (Klein et al. 1988). Using a DuPont Biolistics PDS-1000 particle gun, we (Wilde et al. 1992) have stably transformed embryogenic cells of yellow-poplar, from which transgenic plantlets have been regenerated via somatic embryogenesis. Briefly, our transformation protocol is as follows: Plasmid DNA (pBI121.1) containing genes encoding the reporter ß-glucuronidase (GUS; Jefferson et al. 1987) and the selectable marker neomycin phosphotransferase (NPTII) is precipitated onto 1.1 μm tungsten or gold particles. The particles are used to bombard individual cells and small cell clusters that have been isolated by size fractionation of an embryogenic yellow-poplar suspension, and collected on a filter paper disk. After 2 days of incubation on semisolid induction medium, filters are transferred to induction medium containing 100 μg/ml kanamycin. After 5-6 weeks, clusters of kanamycin-resistant PEMs, approximately 1 mm in diameter, are transferred individually to fresh antibiotic-containing plates. Suspension cultures are initiated from the resulting lines by transferring PEMs to liquid induction medium containing 50 μg/ml kanamycin, which is sufficient to inhibit growth of nontransformed yellow-poplar suspensions.

The ability to grow these cultures as suspensions is particularly desirable as it allows stringent selection pressure to be applied, thereby reducing the likelihood of regenerating chimeras. Southern analysis indicates that independently transformed embryogenic sublines carry from 3 to 30 full-length copies of the GUS gene. A histochemical assay for GUS activity employing the substrate 5-bromo-4-chloro-3-indolyl-ß-D-glucuronic acid (X-Gluc) reveals a heterogeneous staining pattern in PEMs, which seems to indicate that they are chimeric for the GUS gene. However, extracts from cell clusters reacting positively (blue) or negatively (white) with this substrate both demonstrate GUS activity when a fluorometric assay is performed using the substrate 4-methylumbelliferyl-ß-D-glucuronide (MUG). Furthermore, somatic embryos regenerated from transformed sublines are uniformly GUS positive, using the histochemical assay. Plantlets derived from transformed somatic embryos and grown in the greenhouse continue to express GUS and NPT II in leaves and roots over 3 years following transformation (unpublished).

6. References

Gardiner, J.M. (1989) Magnolias: Their Care and Cultivation. Cassell, London. 143 p.

Gresshoff P.M., and Doy, C.H. (1972) Development and differentiation of haploid *Lycopersicon esculentum* (tomato). Planta 107:161-170.

He, S.A., and Santamour, F.S., Jr. (1983) Isoenzyme verification of American-Chinese hybrids of *Liquidambar* and *Liriodendron*. Ann. Missouri Bot. Gard. 70:748-749.

Jefferson, R.A., Kavanaugh, T.A., and Bevan, M.W. (1987) GUS fusions: ß-glucuronidase as a sensitive and versatile gene fusion marker in higher plants. EMBO J. 6:3901-3907.

Klein, T.M., Gradziel, T., Fromm, M.E., and Sanford, J.C. (1988) Factors influencing gene delivery into *Zea mays* cells by high-velocity microprojectiles. Bio/technology 6:440-444.

Merkle, S.A., and Sommer, H.E. (1986) Somatic embryogenesis in tissue cultures of *Liriodendron tulipifera*. Can. J. For. Res. 16:420-422.

Merkle, S.A., and Wiecko, A.T. (1989) Somatic embryogenesis in three magnolia species. J. Amer. Soc. Hort. Sci. 115(5):858-860.

Merkle, S.A., Wiecko, A.T., Sotak, R.J., and Sommer, H.E. (1990) Maturation and conversion of *Liriodendron tulipifera* somatic embryos. In Vitro Cell. Dev. Biol. 26:1086-1093.

Merkle, S.A., Schlarbaum, S.E., Cox, R.A., and Schwarz, O.J. (1991) Mass propagation of somatic embryo-derived plantlets of yellow-poplar for field testing. *In* Proceedings of the 21st Southern Forest Tree Improvement Conference, June 17-20, 1991, Knoxville, TN. pp 56-68.

Merkle, S.A., and Watson-Pauley, B.A. (1993) Regeneration of bigleaf magnolia by somatic embryogenesis. HortScience 28:672-673.

Merkle, S.A., Hoey, M.T., Watson-Pauley, B.A., and Schlarbaum, S.E. (1993) Propagation of *Liriodendron* hybrids via somatic embryogenesis. Plant Cell Tissue Organ Cult. 34:191-198.

Merkle, S.A., and Watson-Pauley, B.A. Ex vitro conversion of pyramid magnolia somatic embryos. HortScience (in press).

Murashige, T., and Skoog, F. (1962) A revised medium for rapid growth and bioassays with tobacco tissue culture. Physiol. Plant. 15:473-497.

Parks, C.F., Miller, N.G., Wendel, J.F., and McDougall, K.M. (1983) Genetic divergence within the genus *Liriodendron* (Magnoliaceae). Ann. Missouri Bot. Gard. 70 658-666.

Parks, C.R., and Wendell, J.F. (1990) Molecular divergence between Asian and North American species of *Liriodendron* (Magnoliaceae) with implications for interpretation of fossil floras. Amer. J. Bot. 77(10):1243-1256.

Risser, P.G., and White, P.R. (1964) Nutritional requirements spruce tumor cells *in vitro* Physiol. Plant. 15:620-635.

Santamour, F.S., Jr. (1972) Interspecific hybrids in *Liriodendron* and their chemical verification. For. Sci. 18(3):233-236.

Sommer, H.E., and Brown, C.L. (1980) Embryogenesis in tissue cultures of sweetgum. For. Sci. 26(2):257-260.

Sotak, R.J., Sommer, H.E., and Merkle, S.A. (1991) Relation of the developmental stage of zygotic embryos of yellow-poplar to their somatic embryogenic potential. Plant Cell Rep. 10:175-178.

Treseder, N.G. (1978) Magnolias. Faber and Faber, London. 243 p.

Wilde, H.D., Meagher, R.B., and Merkle, S.A. (1992) Expression of foreign genes in transgenic yellow-poplar plants. Plant. Physiol. 98:114-120.

Witham, F.H., Blaydes, D.F., and Devlin, R.M. (1971) Experiments in Plant Physiology. Van Nostrand-Reinhold Co, New York. 245 p.

TUBERIZATION IN VITRO AND DORMANCY OF POTATO (SOLANUM TUBEROSUM L.) MICROTUBERS

D. Levy, Edna Fogelman, Alexandra Levine & Y. Itzhak

Department of Vegetable Crops, The Volcani Center, ARO, Bet Dagan 50 250, Israel.

ABSTRACT. Tuberization capacity in vitro of S. tuberosum cultivars and clones and of S. phureja clones was assessed. All genotypes produced tubers in vitro and varietal differences in the rate of tuberization were observed.
Temperatures affected tuberization in vitro. Microtubers up to 6 mm in diameter were produced by single node custtings, and up to 11 mm in diameter by whole plantlets. The incubation media contained 5 ppm of the growth retardant ancymidol (α-cyclopropyl-4-methoxy-α-primidin-5yl-benzyl alcohol), which might have had an effect on dormancy. However, sprouting of the microtubers occured between 78-135 days after 'harvest' in S. tuberosum cultivars, and between 49-110 days after 'harvest' in S. phureja clones. At this stage, the microtubers were planted in pots containing a mixture of peat, vermiculite and volcanic tuff and placed in a greenhouse. Growth and development of plants was monitored. In micrsotubers larger than 3 mm in diameter, there were no noticeable differences in emergence and growth rates, but in the smaller microtubers (less than 2 mm in diameter) emergence and growth were much retarded. There was a tendency to obtain greater yields of minitubers from larger microtubers.

INTRODUCTION. Small tubers (microtubers) produced in vitro, usually 3-15 mm in diameter, are usefull for potato production, storing, shipping, and for planting in the field for certified seed production or other purposes. Thus, it became advantageous to develop a mass in vitro microtuber production (Tovar et al. 1985; Wang and Hu, 1982). Studies of potato in vitro tuberization, and of the behaviour of the in vitro tubers are important, and we present our recent data on the effects of temperature on tuberization, on dormancy of microtubers and on their performance.

MATERIALS AND METHODS. Propagation of in vitro plantlets and the procedures for in vitro tuberization were described earlier (Levy et al. 1993; Seabrook et al. 1993).
The cultivars and clones included in the present studies were: Aquila, Atica, Atzimba, Baronesa, Cara, Claustar, Desiree, Huinkul, Idit, Norchip, Ori, Spunta, Superior, DTO-28, LT-1, LT-2, LT-8, NT-8 and few S. Phureja clones.
For in vitro tuberization the plant material was held in darkness mostly at 20°C unless otherwise stated. The tuberization medium contained 80 g/l sucrose and 5 mg/l of the growth retardant ancymidol in Murashige and Skoog (1962) salts medium.

M. Terzi et al. (eds.), Current Issues in Plant Molecular and Cellular Biology, 123-125.

RESULTS AND DISCUSSION. The effect of temperature during the incubation period on tuberization of single nodes was investigated. Tuberization occured in the temperatures tested, 15°C, 20°C, 25°C, and a combination of 12 hours at 30°C and 12 hours at 20°C. Tubers at 25°C tended to be of greater diameter than those formed at 15°C and incubation at high temperatures (30°C during 12 hours) caused growth of the apical buds ('eyes'). This resembles the effect of high temperatures on field grown potatoes which interfer with tuber bud dormancy (Levy, 1986). We also noticed that in the late maturing cultivar Cara, tuberization was slower as compared to earlier cultivars or clones such as LT1, resembling the behaviour of early and late cultivars under field conditions. Similar trends were observed in whole plantlets: the greater numbers of tubers per plantlet and the greater tuber diameters occured in the earlier maturing cultivars.

Microtubers less than 2 mm in diameter tended to loose volume when kept at room temperature, and their emergence and growth were much retarded. Sprouting of microtubers of S. tuberosum held at room temperature (ca 25°C) started 78-135 days after they were removed from the culture ('harvest'). In S. phureja clones, sprouting started 49-110 days after harvest. Microtubers were planted in pots and grown in a greenhouse at 20-25°C. Microtubers larger then 3 mm in diameter emerged and grew well and no noticeable differences were noted among microtubers of 3-9 mm. However, higher yields of tubers (minitubers) were obtained from microtubers of ca. 7 mm in diameter. These observations should be repeated before an 'optimal' microtuber size is determined. The present results are means of many cultivars and clones, and differences among genotypes in microtuber performance may be expected.

Efforts to develop efficient technology for mass production of microtubers and the evaluation of microtuber performance are on going. Storage, handling and preparation of microtubers for planting will also be studied.

REFERENCES

Levy, D. 1986. Tuber yield and tuber quality of several potato cultivars as affected by seasonal high temperatures and by water deficit in a semi-arid environment. Potato Research 29: 95-107.

Levy, D., Seabrook, J.E.A. and Coleman, S. 1993. Enhancement of tuberization of axillary shoot buds of potato (Solanum tuberosum L.) cultivars cultured in vitro. J. Exp. Bot. 44: 381-386.

Murashige, T. and Skoog, F. 1962. A revised medium for rapid growth and bioassays with tobacco tissue culture. Physiol. Plant. 15: 473-497.

Seabrook, J.E.A., Coleman, S. and Levy, D. 1993. Effect of photoperiod on in vitro tuberization of potato (Solanum tuberosum L.). Plant Cell Tiss. Org. Cult. 7: 3-10.

Tovar, P., Estrada, R., Schilde-Rentschler, L. and Dodds, J.H. 1985. Induction and use of in vitro potato tubers. CIP Circular 13(4), 1-5.

Wang, P.J. and Hu, C.Y. 1982. In vitro mass tuberization and virus-free seed-potato production in Taiwan. Am. Potato J. 5 9: 33-37.

HAPLOIDS - AN INTEGRAL PART OF APPLIED AND BASIC RESEARCH

G. WENZEL[1,2], U. FREI[1,2], A. JAHOOR[2], A. GRANER[1],
B. FOROUGHI-WEHR[1]
1. Federal Centre for Breeding Research on Cultivated Crops,
Institute for Resistance Genetics, D- 85461 Grünbach
2. Technical University of Munich, Institute for Agronomy and Plant Breeding,
D-85350 Freising
Germany

ABSTRACT. On the example of barley it will be demonstrated that the use of haploids is a common step in modern applied breeding programs. It is irrelevant whether these haploids are produced partheno- or androgenetically but normally parthenogenesis is more laborious than androgenesis since it demands both, intensive greenhouse work and in vitro equipment. In most cases microspore androgenesis has been improved to such an extent that the number of green haploid plants and derived DH lines necessary for applied work (normally about 100 DH-lines per donor genotyp) can be produced. This is particularly true for self-fertile cereals and for Brassicaceae. In crops where there are still difficulties, like most outbreeders, a concentrated input might solve the problem. In those crops where the technique works, the question is no longer how to produce haploids but rather how to use them most efficiently. When the aim is rapid incorporation of a monogenic trait the time gain is beneficial, but much more important are strategies which allow the combination of quantitatively inherited characters. Here haploids have a big potential. Further, with increasing progress in somatic hybridization, e.g. in potato, the production of haploids becomes more essential resulting in a broad haploid population as fusion parents. A rather new field of haploid usage is the molecular genome identification. Here, haploids due to their simpler segregation patterns are an important prerequisite to identify genes. Most RFLP maps of barley are based on such DH-populations. Particularly for QTL analysis unselected DH-populations are an important tool to produce reproducible DNA-polymorphisms.

1. Introduction

A plant breeder who is dependent on genetic variability seeks to circumvent the diploidy of higher plants, at least in several parts of a breeding program. Of course, the final breeding product must be diploid again, resulting in doubled haploid lines (DH). The principle ideas about how to make use of the simpler haploid genomes are more than 50

M. Terzi et al. (eds.), Current Issues in Plant Molecular and Cellular Biology, 127-135.

years old, and very few new concepts have been added since Blakeslee et al. listed them in 1922. It is no longer in doubt that the gametes are the most appropriate material for the reproducible production of haploids. For their production two principal methods exist, either androgenesis, in which the male gamete develops into the haploid or doubled haploid plantlet, or parthenogenesis, when the female gives rise to the haploid. Since normally more microspores are produced, androgenesis offers at least theoretically the more efficient system. In a number of cases, however, parthenogenesis is less genotype dependent and consequently advantageous when only a limited number of haploids is needed. Parthenogenesis often demands both classical combination breeding and tissue culture in form of embryo rescue. If this is the case, the technique becomes more expensive than androgenesis which normally needs only the tissue culture steps.

2. The Technique

Figure 1 gives an example of androgenetic DH production in Germany, and shows how the methods developed from basic research to application during the last 20 years. It is difficult to summarize such an extensive topic. It is even more difficult, if this topic concentrates rather on a tool than on a real research area. Thus, I will restrict myself on plants and research of the groups at Grünbach and Weihenstephan. I like to compare the tool tissue culture in plant breeding with the invention of the fork for selecting solid and liquid food as an improvement of the spoon used exclusively before. For barley anther culture the medium once developed by Clapham (1973) with some minor alterations (Foroughi-Wehr 1993) is still the most efficient one. Under an economic aspect the price for one liter of basic medium is rather modest with about 10US cents. What makes the medium expensive but also functional are the additions. In Figure 2 the effect and the price of such additions is demonstrated. There are probably chances to improve also a medium without maltose, thus circumventing the patent of Shell.

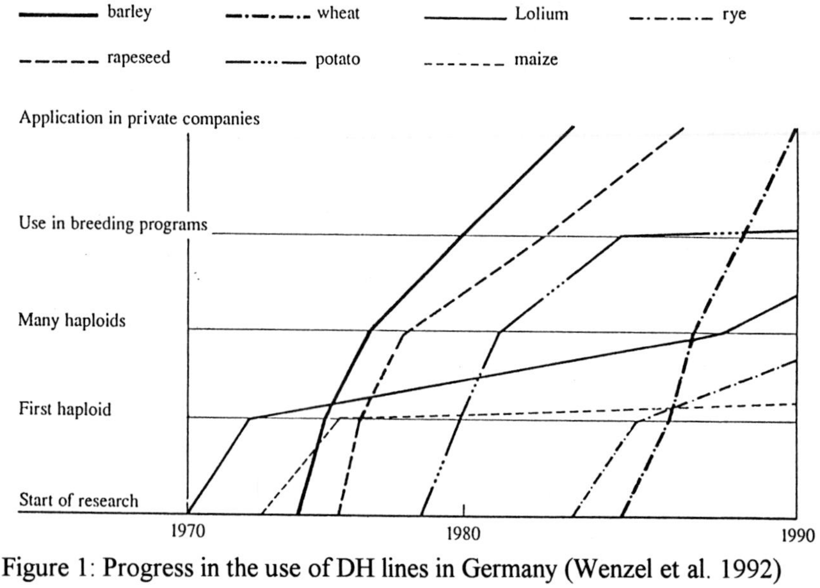

Figure 1: Progress in the use of DH lines in Germany (Wenzel et al. 1992)

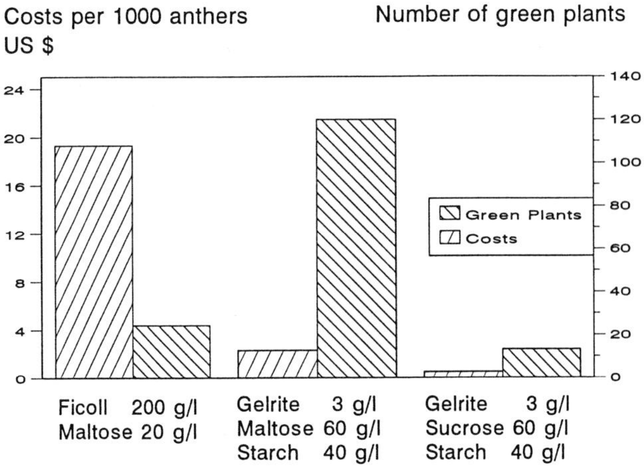

Figure 2: Price and effect in green plantlet formation in media with different additions (average over many genotypes)

3. Use of Haploids

Presently the question is no longer how to make haploids but rather how to use them. In a first attempt they can easily be used to quickly demonstrate how many genes are responsible for a specific character.

3.1. Haploids in Barley Breeding

Table 1 gives an example of the production of barley lines resistant to BaYMV. It is obvious that e.g. in the line Bizen Wase 5 several dominant genes are responsible for the resistance. Depending on the type of parents to be combined, one should know how many recombinations are probably necessary to guarantee the new genome combination. If characters from related varieties are to be combined, one haploid step followed by selection is the most efficient procedure. The advantage is that the highly selected genomes of both parents undergo only one recombination and big genome areas will be kept undisturbed. If more recombinations are needed, a combination of recurrent selection that guarantees variability followed by haploid selection that guarantees secure selection, has been developed. The breeding technique is called recurrent selection alternating with haploid steps, shown in Fig. 4. The necessary offspring size of the DH population for quantitatively inherited characters depends on the degree of linkage and on the number of genes involved.

3.2. Microspore Selection

3.2.1 *In Wheat.* Microspore populations descending from F_1 hybrids of wheat varieties with different levels of resistance to *Fusarium culmorum* or *F. graminearum* were

Table 1. Estimation of gene effects due to the segregation in DH populations

Resistant parent	Number of susceptible cultivars used as 2. parent	DH lines produced	segregation susc.	:	res.
Kobinkatangi	4	166	1.2	:	1
Cebada	4	210	1	:	1.2
Smooth	2	136	1.2	:	1
Chikurin	6	508	1	:	1.1
Pl 329037	1	71	1	:	2.4
Kersho	4	749	1		1.7
Cl 9346	3	279	1	:	3
Bizen Wase	2	129	1	:	4
Bizen Wase 5	3	485	1	:	4.4

screened with a phytotoxin of *Fusarium* before and during regeneration. Two selection methods were compared: either embryoids and calli were first initiated from anthers in toxin-free medium and then grown on medium with the toxin, or anthers were immediately cultured on regeneration medium containing the toxin. Microspores from donor hybrids which where produced from two susceptible cultivars were killed by lower toxin

Table 2. Estimation of the toxin concentration which kills about 90% of regenerating microspores compared to the growth of untreated controls (Fadel and Wenzel 1993)

Anther donors (F_1 hybrid)	Number of anthers/ regenerants - toxin	Toxin concentration ml/l	Number of anthers/ regenerants + toxin
Florida (5) x Carisuper (3)*	8250/5057	2.5	10300/722
Florida (5) x Falke (5)	9955/4496	2.5	9165/285
Florida (5) x Kraka (6)	8230/4850	2.5	9165/510
Florida (5) x Boxer (9)	9510/4290	2	8850/221
Carisuper (3) x Falke (5)	1175/ 245	2	7060/141
Carisuper (3) x Kraka (6)	6085/2192	2	7995/160
Carisuper (3) x Boxer (9)	3330/ 885	2	7010/140
Kraka (6) x Boxer (9)	1220/ 200	1	5490/109

* 1 = very resistant; 9 = very susceptible

concentrations than microspores from hybrids of parents being both rather resistant. F_1 microspores descending from combinations resistant x susceptible showed an intermediate reaction (Tab. 2). The principle of this selection is not the formation of new variability via mutation but rather the use of recombination. The selection should uncover a gene complex which should be present in the microspore population. The lines selected from the differing parents showed under greenhouse tests the resistance level of the better parent (Fadel and Wenzel 1993). The field tests under natural infection conditions are not yet finished.

3.2.2 *In Potato*. In contrast to these encouraging experiments in the wheat/*Fusarium* system, the situation is different in the system potato/*Phytophthora*. Again, regeneration of microspores is rather reproducible, allowing selection within microspores of the heterozygous outbreeder potato. In the presence of *Phytophthora* toxins, however, no correlation was found between toxin resistance of the microspores and resistance of the plants. In some clones the toxin was even increasing the regeneration frequency of the isolated microspores (Möllers et al. 1992).

3.3. Haploids as Basic Breeding Material for Protoplast Fusion in Potato

Dihaploid (2n = 2x = 24) potato clones can be used to fasten the combination of qualitative and especially quantitative traits via cell fusion. Field evaluation of intraspecific somatic hybrids revealed new variability within the hybrids of a single fusion combination which theoretically should be uniform (Thach et al. 1993). A final understanding of this unexpected variability has not been given but the differences between hybrids can be ascribed to several reasons such as somaclonal variation, aneuploidy, the composition of the cytoplasm, and interaction with the nucleus.

The RFLP analysis of the regenerants showed deviating banding patterns, e.g. missing and additional bands. Nearly all chromosomes were affected but the percentage of deviation differed, probably depending on the duration ot the regeneration. With very rare exceptions the plastids segregated in all fusion combinations 1 : 1 into the two parental types. Recombinations of the plastome could not be observed. This was in contrast to the findings for the mitochondria. In 10% of the hybrids totally new restriction fragments appeared, indicating recombination of the mitochondrial DNA. Independent of the fusion combination mtDNA rearrangements were detected in about 75% of the somatic hybrids (Lössl et al. 1994).

Some mt-probes differentiated significantly between genotypes with high and low yield. The mt-type of the one parent seems to be superior in starch content to the other parent. According to total yield, genotypes with more homogeneous mt-genomes were superior to more heterogeneous groups. Mixtures of mitochondria or recombined mtDNA did not express a favourable role in the genotypes screened up till now (Lössl et al. 1994).

4. Haploids and Molecular Work in Barley

With the detection and application of restriction enzymes a molecular knife was added to the biotechnological fork. It became pretty soon clear that most informative for polymorphic markers are segregating F_2 families or inbred lines. Inbred lines are equivalent to DH´s, where no limitation for the material is given. With DH's work on single genotypes became possible, what is of particular importance in the area of QTL´s. Androgenetic DH lines showed varying degrees of distorted segregation compared to *Hordeum bulbosum* derived parthenogenetic lines. This did not interfere, however, with the map construction. The Igri x Franka DH-lines used in early mapping experiments (Graner et al. 1990) were produced already several years ago by a procedure which has since undergone improvements (e.g. instead of organogenesis, regeneration via an embryogenic pathway). Analysis of DH-progenies which were produced according to the improved protocols reveals less or even no distorted segregation.

With the use of such DH-lines several groups mapped DNA markers which in the meantime could be integrated (Tab. 3). To the markers phenotypes could be correlated, in the beginning only monogenic ones, now increasingly polygenic ones. A quantitative trait has to be dissected into several loci and the contribution of each individual locus to the over-all variation can be determined. Similar to the analysis of monogenic traits, the use of DH lines greatly reduces the genetic complexity of the progeny, since only homozygous genotypes occur. Using DH's the likelyhood to recover a distinct homozygous genotype is

Table 3. Marker numbers of some of the presently most comprehensive RFLP maps of barley (Graner 1994)

Cross	Progeny type	Total number of markers	Reference
Proctor x Nudinka	androgenetic DH	154	Heun et al. 1991
Igri x Franka	androgenetic DH	376	Graner et al. 1994
Varda x H. spontaneum	F_2/F_3	163	Graner et al. 1991
Steptoe x Morex	parthenogen. DH	423	Kleinhofs et al. 1993
Harrington x TR306	parthenogen. DH	177	Kasha and Kleinhoffs 1994
Vogelsanger Gold x Alf	parthenogen. DH	80	Giese et al. 1994

1 per 2n, however, it is 1 per 4n in an F_2 progeny (n = number of unlinked genes). The size of a progeny which is screened for a desired genotype can be reduced accordingly. In order to select a specific recombination of five genes with a probability of 95%, about 3000 F_2 plants have to be tested compared to 100 F_1 derived DH-lines (Graner and Wenzel 1992). Those DH-lines can be propagated allowing repeated testing in several locations for several years, which forms the basis for accurate phenotypic analysis. Using this approach, already several QTLs have been mapped (Backes et al. 1994). In order to facilitate the practical application of marker assisted selection, large efforts have been undertaken during recent years to locate genes of agronomic interest and to identify closely linked markers. Linked markers are identified either by analysis of pairs of near isogenic lines (NILs) or by bulked segregation analysis. While the production of NILs is a time consuming process, bulked segregation analysis can be performed in any type of segregating progeny, e.g. a F_1 derived DH-progeny. Based on the phenotype of the trait of interest, two contrasting bulks are formed. Given a close genetic linkage between a

marker and the gene of interest, each bulk will show e.g. the resistant or the susceptible pattern (Fig. 3).

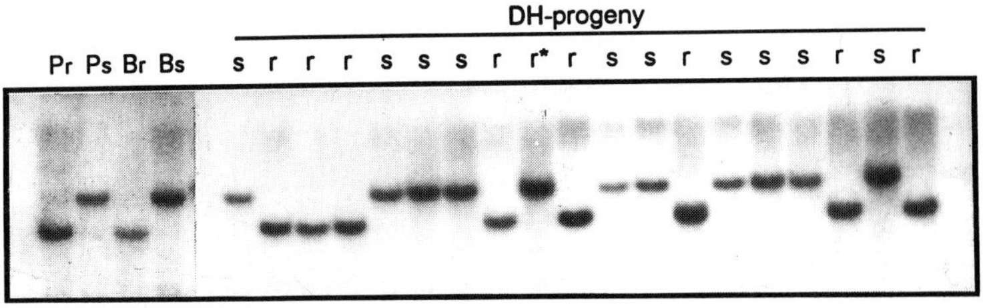

Figure 3: Parents, bulks and DH progeny with and without resistance to BaYMV resistance

5. Conclusion

In terms of practical plant breeding the combination of molecular markers, haploid techniques and classical breeding steps will facilitate the identification and combination of appropriate germplasm and the application of improved selection strategies to combine both qualitative and quantitative traits. However, despite the strength of each technology

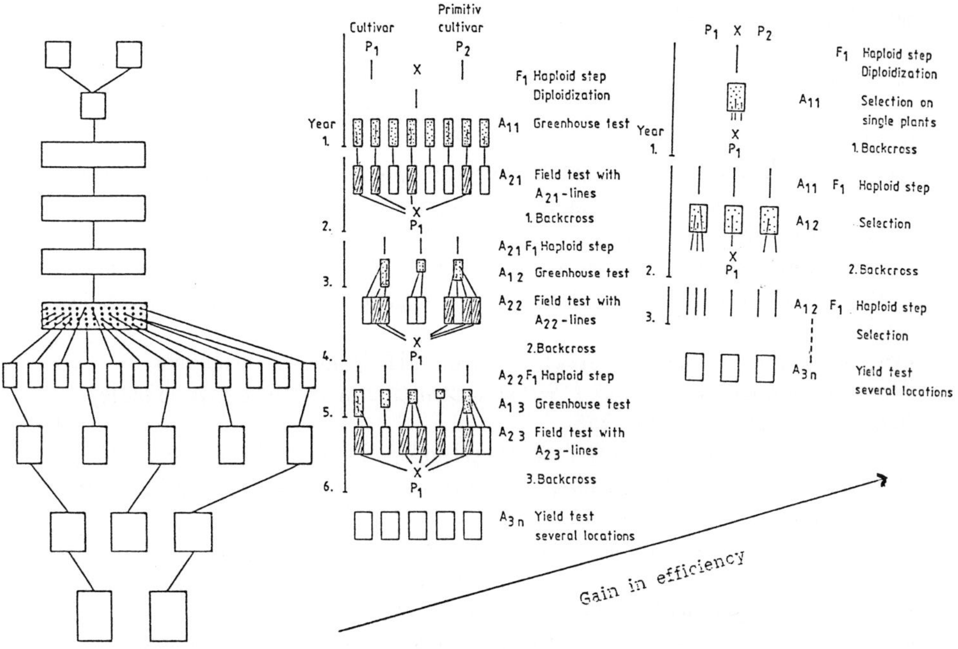

Figure 4: Summary of the gain in efficiency of partial bulk selection, recurrent selection with repeated haploid steps (Foroughi-Wehr and Wenzel 1990) and the combination of the last procedure with marker assisted selection

on its own, only the deliberate combination of molecular marker techniques, biotechnology and classical cross breeding will maximize the efficiency of any individual breeding concept. Spoon, fork and knife build a unit. In Fig. 4 the possible gain in efficiency in barley breeding programs using partial bulk, DH or marker assisted selections are summarized. If the food becomes however, really difficult to eat, we often even use fingers and finally there are also completely different approaches as the chop sticks.

6. References

Backes, G., Graner, A., Foroughi-Wehr, B., Fischbeck, G., Wenzel, G. and Jahoor A. (1994) Localization of quantitative trait loci (QTL) for agronomic important characters by the use of a RFLP map in barley (*Hordeum vulgare* L.), Theor. Appl. Genet. (in press).

Clapham, D (1973) Haploid Hordeum plants from anthers in vitro, Z.Pflanzenzüchtg 69, 142-155.

Fadel, F. and Wenzel, G. (1993) In vitro selection for tolerance to Fusarium in F_1 microspore populations of wheat. Plant Breeding 110, 89-95.

Foroughi-Wehr, B. (1993) Protocol for the efficient production of doubled haploid wheat and barley plants, Nachrichtenbl. Deut.Pflanzenschutzd. 45, 263-267.

Foroughi-Wehr, B. and Wenzel, G. (1990) Recurrent selection alternating with haploid steps - A rapid breeding procedure for combining agronomic traits in inbreeders, Theor. Appl. Genet. 80, 564-568.

Giese, H., Holm-Jensen, A.G., Mathiassen, H., Kjaer, B., Rasmussen, S.K., Bay, H. and Jensen, J. (1994), Distribution of RAPD markers on a linkage map of barley Hereditas (in press).

Graner, A (1994) RFLP-mapping the haploid genome of barley (*Hordeum vulgare* L.), in S.M. Jain (ed.), In vitro Haploid Production in Higher Plants, Kluwer Academic Publishers, Dordrecht.

Graner, A., Bauer, E., Kellermann, A., Kirchner, S., Muraya, J.K., Jahoor, A. and Wenzel, G. (1994) Progress of RFLP-map construction in winter barley, BGN 23 (in press).

Graner, A., Jahoor, A., Schondelmaier, J., Hiedler, H., Pillen, K., Fischbeck., G., Wenzel, G. and Herrmann, R.G. (1991) Construction of an RFLP map in barley, Theor. Appl. Genet., 83, 250-256.

Graner, A. and Wenzel, G. (1992) Towards an understanding of the genome - New molecular markers increase the efficiency of plant breeding, Agro-Food-Industry Hi Tech. 3, 18-23.

Heun, M., Kennedy, A.E., Anderson, J.A., Lapitan, N., Sorrells, M.,E. and Tanksley, S.D. (1991) Construction of a restriction fragment length polymorphism map for barley (*Hordeum vulgare*), Genome 34, 437-447.

Kasha, K.J. and Kleinhofs, A. (1994) Mapping of the barley cross Harrington x TR306, BGN 23 (in press).

Kleinhofs, A., Kilian, A. and Kudrna, D. (1994) The NABGMP Steptoe x Morex mapping progress report, BGN 23 (in press).

Lössl, A., Frei, U. and Wenzel G. (1994) Interaction between cytoplasm composition

and yield parameters in somatic hybrids of *S. tuberosum* L., Theor. Appl. Genet. (in press).
Möllers, C., Zitzlsperger, J. and Wenzel, G. (1992) The effects of a toxin preparation from *Phytophthora infestans* on potato protoplasts and microspores, Physiological Molec. Plant Pathol. 41, 427-4325.
Thach, N.Q., Frei, U. and Wenzel, G (1993) Somatic fusion for combining virus resistances in *Solanum tuberosum* L. Theor. Appl. Genet. 85,863-867.
Wenzel, G., Graner, A., Fadel, F., Zitzlsperger J. and Foroughi-Wehr, B. (1992) Production and use of haploids in crop improvement, in J.P. Moss (ed.) Biotechnology and Crop Improvement in Asia, pp. 169-179 ICRISAT, Patancheru.

UTILIZATION OF ANTHER CULTURE AS A BREEDING TOOL IN RICE IMPROVEMENT

M. S. Alejar[1], F. J. Zapata[2], D. Senadhira[1], G. S. Khush[1], and S. K. Datta[1]

[1]Plant Breeding, Genetics and Biochemistry Division, The International Rice Research Institute, P. O. Box 933, 1099 Manila, Philippines
[2]Present Address: Department of Genetics, College of Agriculture, Sao Paolo University, C. P. 83, 13400 Piracicaba, Sao Paolo, Brazil

ABSTRACT. The general breeding objective of IRRI's anther culture program is to improve anther culture efficiency, specifically that of indica rice cultivars, and to apply this technique in rice varietal improvement. The materials for anther culture use consisted of F_1 crosses generated by plant breeders for different ecosystems or biotic and abiotic stresses. These include materials for irrigated, rainfed, upland, and tidal wetland ecosystems as well as materials for cold tolerance, blast resistance and hybrid rice heterotic lines. As of December 1993, a total of 4143 dihaploid lines were regenerated for various in-house (IRRI) and international collaborative research. Six anther culture-derived lines were utilized as donor parents for breeding programs in six countries. A high-yielding, salt-tolerant anther culture line, IR51500-AC11-1 from the cross IR5657-33-2/IR4630-22-2-5-1-3, was approved by the Rice Varietal Improvement Group of the Philippine Seed Board in 1993 as one of the two prerelease varieties intended for saline soil conditions.

INTRODUCTION

The main advantages of using haploids in a breeding program are the production of homozygous lines in the shortest possible time, and more reliable and effective selection (Morrison and Evans 1988; Foroughi-Wehr and Wenzel 1989).

For many years, IRRI plant breeding effort was geared towards field evaluation of anther culture-derived (AC) lines (Zapata et al., 1991). International testing of AC lines was also conducted in different test sites in East, Southeast, South and West Asia through the International Network for Genetic Evaluation of Rice (INGER) field testing program, to evaluate the performance of these lines under different biotic and abiotic stresses.

Results of field evaluation showed very good performance of AC lines. It also confirms the possibility of transferring through anther culture the high level of salt tolerance from traditional cultivars into improved rice cultivars.

MATERIALS AND METHODS

Anther Culture Materials

F_1 seeds of single and multiple crosses involving japonica/indica, japonica/japonica, indica/japonica, and indica/indica parental combinations are used for anther culture of different breeding materials. IRRI rice breeders who work on different rice ecosystems select the F_1 crosses for anther culture and for international collaboration, scientists sent by their respective country or research system to learn the anther culture technique bring with them F_1 seeds bred for their own requirements.

M. Terzi et al. (eds.), Current Issues in Plant Molecular and Cellular Biology, 137-142.

The indica varieties , IR43 and IR58, and the japonica variety Taipei 309, are used in basic studies on media improvement.

Selection and Pretreatment of Materials

The primary panicles used in anther culture are collected (from the screenhouse or from the F_1 nursery) when the distance of the flagleaf's auricle to that of the first subtending leaf is between 5 and 10 cm, depending on the genotype or variety. This distance would have the microspores in the mid-uninucleate to early binucleate, the most responsive stage for anther culture. For accurate determination of pollen stage, staining the anthers with 2% acetocarmine is required. Panicles are sprayed with 70% ethanol and cold-shocked for 8 days at $8^{o}C$ prior to anther plating.

Callus Induction

Selected florets are surface-sterilized with 20% solution (commercial bleach containing 5.2% NaOCl) for 20 minutes before plating. The 90 anthers are plated aseptically in a 60x15 mm Falcon petri dish containing callus induction medium and kept in the dark at $25^{o}C$. The plates are examined periodically 2-8 weeks from plating for callus formation.

Plant Regeneration and Care of the Plants

Embryogenic calli of at least 2 mm in size are transferred in flasks containing plant regeneration medium and kept for 4 weeks under a 12-h photoperiod of 3000 lux supplied by cool white fluorescent lamps at $25^{o}C$.

Plantlets are transferred in flasks (Fig. 1) containing the same regeneration medium for further root and shoot development, and incubated under light at 3000 lux intensity for 12 hours/day. Plantlets with vigorous roots are transferred in styrofoam boards with holes and placed over a plastic tray filled with culture solution. The plants are grown in the Phytotron at $21/29^{o}C$ night/day temperature under normal daylength (9-10 hours/day, 14-15 hours/night). A plant or regenerant is given a code number (indicating the cross and callus source) before transplanting. Each plant is transplanted to an 8-inch diameter pot with soil and fertilizer added at the rate of 5 g $(NH_4)SO_4$, 2.5 g P_2O_5 and 1.5 g K_2O. Panicles are individually bagged before the onset of anthesis to prevent cross-pollination.

RESULTS AND DISCUSSION

Plant Regeneration in Various Collaborative Projects

In-House and International Collaboration. We used 292 F_1 crosses for anther culture during the period 1989 wet season (WS) to 1993 dry season (DS), 213 (72.9%) produced dihaploid lines. A total of 4143 fertile plants were regenerated from the 213 crosses. The remaining 1889 anther culture lines (for rainfed, upland and cold tolerance) will be ready for field evaluation by June 1994. Anther culture-derived lines from various sources are listed in Table 1.

Evaluation of AC Lines

National Testing. The National Cooperative Tests Trials were conducted from 1990 WS-1993 DS at five different saline-prone irrigated lowland sites in the Philippines. The anther culture line IR51500-AC11-1 from the cross IR5657/IR4630 had good salinity tolerance, and had a comparatively higher season average grain yield than other test lines and check varieties (Fig. 2). This line was approved by the Rice Varietal Improvement Group of the Philippine Seed Board in 1993 as one of the two prerelease varieties for saline soil conditions.

Figure 1. Mass production of anther culture-derived plants.

Figure 2. Salt tolerant AC line IR51500-AC11-1 showing very good phenotypic acceptability and high yield potential.

Table 1. Plants regenerated through anther culture (1989WS-1993DS)

Materials for different conditions	Crosses for anther culture (No.)	Crosses with response (No.)	Anthers plated (No.)	Diploid lines (No.)
IRRI				
Irrigated	77	51	174,368	621
Cold tolerance	57	40	53,616	524
Salt tolerance	38	29	76,840	413
Rainfed	54	35	173,130	232 (985)[a]
Upland	3	3	55,900	- (473)
Hybrid rice	6	5	28,860	107
Blast resistance	8	6	146,000	- (431)
Genetic studies	1	1	16,110	69
International				
Vietnam[b]	30	30	110,000	1,434
Egypt[c]	10	9	55,380	529
Nepal[d]	11	7	54,810	214
Total	292	213	889,114	4,143 (1,889)

[a]Data in parentheses are number of regenerated green plants as of December 1993.

[b]Egypt - Rice Research and Training Center, Sakha.

[c]North Vietnam - The Food Crops Research Institute, Hai-Hong Province.

[d]Nepal - Lumle Agricultural Center, Pokhara, Gandaki Amchal.

International Testing. Results of the 15th International Rice Salinity Tolerance Observational Nursery (IRSTON) conducted at 16 sites in East, Southeast, South, and West Asia, and Latin America, showed that AC lines outperformed salt-tolerant varieties and other breeding lines.

The AC line IR51500-AC11-1 has improved plant type and very good salinity tolerance at maturity, with an average rating of 5.7 in three locations: Hwasong and Kyewa Korea, and Panvel in India. The salt-tolerant check varieties Nona Bokra and Pokkali have average salinity ratings of 6.0 and 6.5, respectively.

AC lines IR51500-AC9-7, IR51500-AC11-9, and IR51500-AC11-3 have a good phenotypic acceptability; they have an average score of 4.0 in saline soils in Hwasong and Kyewa in Korea, where electrical conductivity is 10 ds/m.

The anther culture line IR51500-AC11-1 has a good phenotypic acceptability score of 5.0 in alkaline soils in Dalip Nagar in India. In saline-alkaline soils in Karnal in India and in Kala Shah Kaku in Pakistan, the average phenotypic acceptability score of IR51500-AC11-1 is 6.0 compared with 9.0 for Nona Bokra and 7.0 for Pokkali.

UTILIZATION OF AC LINES IN VARIOUS BREEDING PROGRAMS

Thirty three AC lines were utilized as donor parents for different IRRI breeding programs from 1989 WS up to 1990 DS. Fourteen AC lines were used as parents because of very good phenotypic acceptability, cold tolerance, and high yield potential. Anther culture line IR66467-AC5-3 from Puntenu/Djembel was used as donor parent for irrigated condition because of very good phenotypic acceptability.

Six AC lines were likewise used as donor parents for breeding programs in seven countries. In Egypt, anther culture line IR51500-AC11-1 was used as a parent because of tolerance for salinity, excellent phenotypic acceptability, high yield potential and resistance to blast. In Thailand, two anther culture lines, IR51491 and IR51500 from the crosses IR4630/Pokkali and IR5657/IR4630, respectively, had very good phenotypic acceptability, and were utilized as parents in the breeding program for acid-lowland conditions. Anther culture lines IR51500-AC9-7 and IR53649-AC3 from the crosses IR5657/IR4630 and IR8192/Nona Bokra, respectively, were used as parents in the breeding programs in Myanmar. For salinity tolerance, anther culture line IR51491-AC10 was used in the hybridization programs in India and Sri Lanka.

Improvement of Technique

Our attention is now focused on media manipulation (Zapata et al., 1983) to increase plant regeneration efficiency in indicas. These are: (1) use of "nurse cell" or cross feeding technique to enhance callus regeneration capacity; (2) use of liquid medium with ficoll to enhance the growth and regenerative capacity of microspore-derived embryogenic callus (Datta et al., 1990); and (3) change of carbohydrate or use of colchicine to enhance higher frequency dihaploid fertile plants.

CONSTRAINTS AND PROSPECTS

The problems being faced by rice scientists who utilize anther culture in their basic and applied research include: (1) low culturability in indicas; (2) significant increase of haploids in relation to green plants regenerated in some F_1 crosses; and (3) need for cooperation among rice breeders and tissue culture scientists.

Our results show that with proper manipulation of the culture media (such as the use of maltose and optimizing the ammonium-nitrate ratio) and possibly nongenetic factors, the anther culture response of indicas could be highly improved. Genetic manipulation such as the transfer of culturability from japonicas into indicas could be initiated to increase the response of the latter.

Rice breeders usually require at least 150 dihaploid lines per cross for field evaluation, which is not possible in many cases.

In some instances, however, promising lines were selected in spite of the limited number of regenerated plants. A possible reason is that the anther culture process exerts a selection pressure on the microspores, allowing only the superior recombinants to proceed toward androgenesis and eventually plant regeneration.

To solve the problem of high number of haploids in relation to green plants regenerated, we are conducting a study using antimicrotubule herbicides (i.e., amiprophosmethyl and pronamide) to enhance the growth and regenerative capacity of callus along with the ploidy and seed set of regenerated plants.

The success of anther culture as an effective breeding tool in rice improvement depends to a large extent on the mutual interests and cooperation of tissue culturists and rice breeders. The former should accelerate their research on the improvement of conditions for maximum anther culture response, while the latter should actively push for the evaluation of AC lines under their respective breeding programs. At IRRI, we are working together to fulfill the objective in dihaploid rice breeding for varietal improvement.

ACKNOWLEDGEMENTS

The authors wish to thank Florencia Aquino, Reynaldo Garcia and Daniel Pasuquin for their careful work and help.

REFERENCES

Datta, S.K., Datta, K. and Potrykus, I. (1990) Embryogenesis and plant regeneration from microspores of both indica and japonica rice (*Oryza sativa*). Plant Sci. 67, 83-88.

Foroughi-Wehr B., Wenzel G. (1989) Androgenetic haploid production. IAPTC News Letter 58, 11-18.

Morrison, R.A., Evans, D.A. (1988) Haploid plants from tissue culture: new plant varieties in shortened time frame. Bio/Technology 6, 684-690.

Zapata, F.J., Alejar, M.S., Torrizo, L.B., Novero, A.U., Singh, V.P. and Senadhira, D.S. (1991) Field performance of anther culture-derived lines from F_1 crosses of indica rices under saline and non-saline conditions. Theor. Appl. Genet. 83, 6-11.

Zapata, F.J., Khush, G.S., Crill, J.P., Heu, M.H., Romero, R.O., Torrizo, L.B. and Alejar, M.S. (1983) Rice anther culture at IRRI In Cell and Tissue Culture Techniques for Cereal Crop Improvement, Science Press, Beijing, pp. 27-46.

APPLICATION OF HAPLOIDY IN GENETIC MANIPULATION OF CANOLA

M.B.Westecott & B.Huang
Pioneer Hi Bred Production Ltd
12111 Mississauga Rd
Georgetown, Ontario L7G 4S7
Canada

ABSTRACT. Haploidy through anther and microspore culture has long been used in breeding programs of canola and other grain and oil crops. More recently, application of haploidy in genetic manipulation has been explored in canola. (1) Uninucleate microspores and microspore-derived embryos were transformed via *Agrobacterium tumefaciens*. Transformation efficiency was higher for embryos (ca. 0.5%) than for the thick-walled microspores. For most efficient haploid transformation, young embryos (21-28 days after microspore culture) were physically wounded and inoculated with disarmed *A.tumefaciens* (10^8/ml) harboring a binary vector containing genes of interest. When a NPTII gene was used as the selectable marker gene, kanamycin selection @ 50mg/l was applied 2 to 7 days after co-cultivation. Shoots or secondary embryos developed under selection were uniform and were hence presumably of single-cell origin. Transgenic canola plants thus obtained were homozygous after colchicine doubling. They did not segregate for the transgene. (2) Microspore culture was successfully used to recover homozygous doubled-haploid plants from hemizygous, primary transgenic canola plants. Level and tissue/stage specificity of gene expression in homozygous transgenics are comparable to that in hemizygous transgenic plants.

1. INTRODUCTION

This manuscript summarizes the methodology and results from our research on canola (*Brassica napus*) haploidy, including microspore culture and haploid transformation.

In addition to being an important oil crop, *Brassica* species also represent a model system of microspore embryogenesis. *Brassica* species are an exception to the rule in that isolated microspore culture is more efficient in yielding microspore-derived embryos than anther culture. In most *B. napus* genotypes tested, thousands of haploid embryos can be produced from cultured microspores. Isolated microspores are haploid single cells amenable to genetic manipulations such as mutagenesis (Swanson et al. 1989) and transformation (Pechan, 1989, Huang, 1992). In general, microspores are much more responsive to *in vitro* culture than their female counterpart. Products from microspore culture, such as embryos, plants, as well as explants or protoplasts derived from them, have also been successfully used as recipients in transformation experiments of canola (Swanson and Erickson, 1989) as well as other species (Sangwan et al.

M. Terzi et al. (eds.), Current Issues in Plant Molecular and Cellular Biology, 143-148.

1993). In spring *B.napus*, over 80% of the plants whick regenerate from the microspore-derived embryos remain haploid until chromosome doubling treatments are applied. Consequently, a majority of the transgenic plants developed from the microspore culture system are homozygous for the transgene after chromosome doubling.

2. MICROSPORE CULTURE METHODOLOGY

The detailed procedure of *B.napus* microspore culture, developed and optimized by a number of laboratories in Europe and Canada, has been described by Huang and Keller (1989). Briefly, microspore donor plants of *B.napus* maybe grown in the field or under a controlled environment at 10-20°C, with a regular supply of fertilizers and routine watering. For best response of microspore embryogenesis, it is desirable to grow donor plants at 10°C/5°C (day/night temperature, 16 hour photoperiod) after bolting.

Flower buds containing microspores at the late uninucleate to early bicellular stages range from 3-4mm in length for most genotypes tested. They are surface-sterilized for 15 minutes in 5% sodium hypochlorite (commercial bleach) and given three five-minute rinses in sterile water. Buds are then homogenized in B5 medium (Gamborg et al. 1968) supplemented with 13% sucrose either in a blender (suitable for $\geq$ 30 buds, Swanson et al. 1987) or in a sterile beaker with a glass rod or syringe barrel for smaller samples (Huang and Keller 1989). The homogenate is then passed through a 48 micron nylon mesh and the filtrate is centrifuged at 100x*g* for 3 to 5 minutes. The microspore pellet is resuspended in B5 medium and centrifuged again. This step is repeated twice for a total of three washes.

Before the last centrifugation, the microspore number is determined using a hemacytometer. The washed microspore preparation is resuspended @ 20,000-50,000/ml NLN medium (Gland et al. 1988), or NLN with the concentration of four major salts reduced by 50%, supplemented with 13% sucrose. Microspores are incubated at 32-35°C in darkness for the first 3-7 days of culture, after which standard culture temperatures of 20-25°C (16 hour photoperiod) can be used. The elevated temperature is a key inducing factor for microspore embryogenesis in *B.napus* as well as other *Brassica* species tested.

Embryos developed from the isolated microspores are visible 10-14 days after culture. For secondary embryogenesis or plant regeneration, two to five week old embryos are transferred to agar-solidified B5 medium supplemented with 2% sucrose. It has been reported that higher temperature (25°C) during the regeneration period stimulates secondary embryogenesis, whereas lower culture temperature (20°C) or cold treatment at 5-10°C for 28 days favours direct plant regeneration from microspore-derived embryos (Huang et al. 1991). Another factor influencing embryo development on the regeneration media is the age of embryos at transfer. The frequency of embryos undergoing secondary embryogenesis increases with the delay of transfer of embryos from liquid culture to the regeneration media.

Chromosome doubling can be induced by immersing cleaned roots of well-established haploid plants in 0.1-0.5% aqueous colchicine solution for 2-3 hours. This can be done either prior to or after transplanting of plantlets into soil.

3. HAPLOIDY IN GENETIC MANIPULATION OF CANOLA

3.1. TRANSFORMATION TECHNOLOGIES

All transformation technologies applicable to diploid explants, cells or protoplats can be used for transformation of their haploid counterparts. These include introduction of DNA into cells or protoplast through *Agrobacterium tumefaciens* or *A. rhizogenes*, through chemical means such as PEG-mediated DNA uptake, through physical means such as microprojection and microinjection, as well as the combination of any of the biological, chemical, and physical approaches such as electroporation and Agrobacterium-mediated transformation in conjunction with chemical or physical treatments.

Limitations likely to be encountered in haploid transformation do not differ from that in the diploid transformation systems. They can arise from inadequate regeneration, inefficient DNA delivery and integration (transformation), and the de-coupling of regeneration and transformation. However, for most plant species, effort required for induction and maintenance of cell or tissue cultures is greater for haploid than for diploid cells. In some plant species where large numbers of plants can be regenerated from diploid cell or tissue culture, plant regeneration from the haploid microspores has either been unsuccessful or extremely inefficient. Consequently, reports on successful transformation of haploid explants and cells have been few in number, compared to the vast pool of literature on transformation of diploid explants or cells.

3.2. *AGROBACTERIUM*-MEDIATED TRANSFORMATION OF HAPLOID CANOLA EXPLANTS

Amongst all the transformation technologies available and tested to date, *Agrobacterium*-mediated transformation is by far the most efficient system for canola transformation for tissues of both haploid and diploid origin. This also appears to be true for other dicot species. While any haploid explant or cell can be used for Agrobacterium-mediated transformation, we have focused on microspores or embryos at the cotyledonary stage. The former have the advantage of being single cells, whereas the latter possess high regenerative potential through secondary embryogenesis.

3.2.1. *Microspores.* Freshly isolated microspores (10^5/ml NLN containing $20\mu M$ acetosyringone) are co-cultivated with overnight cultures of *A.tumefaciens* cells (10^{7-8}cells/ml) for 30-60 minutes at 32°C. Microspores are collected after centrifugation at 100xg for 5 minutes, resuspended in NLN medium at 10^5/ml, and cultured according to standard protocol (session 2 of this paper) thereafter. Carbenicillin, or other appropriate antibiotics, is added 1-2 days later at 200mg/l to eliminate *Agrobacterium* growth. When neomycin phosphotransferase II gene (NPTII) is used as the selectable marker gene for plant cells, kanamycin at 10mg/l is added to the culture medium 7 days after the initiation of co-cultivation. In general, antibiotics levels used for inhibiting bacteria or selection of transformants are lower for microspores. For shoots, 500mg/l carbenicillin and 50mg/l kanamycin are used.

Electron micrographs of *B.napus* microspores during co-cultivation show tight association of *Agrobacterium* and microspore exine through an elaborate network of cellulose fibrils. Due to the strong bacterial attachment to microspores, it is often impossible to dissociate the bacteria from the microspores by washing. Antibiotics at levels tolerable by the microspores offer insufficient inhibition to prevent bacterial regrowth. For *Agrobacterium*-mediated transformation of isolated microspores, cellulose-minus mutant strains are highly recommended.

The frequency of transformation has been low at about 1 transformant/2 x 10^7 microspores, using the A6 cellulose-minus strain (A6/ce$^-$, Matthysse, 1983). This efficiency may be acceptable, considering the fact that 10^7 microspores can be obtained from 100-200 buds in one experiment. Integration of the transgene may or may not occur at the single cell stage. During the course of technology development for microspore transformation, two transgenic *B.napus* CV Topas plants were obtained from a total of 5 x 10^7 microspores co-cultivated with a A6/ce$^-$ containing pBI121 (Jeffeson 1987). Seeds from these transgenic plants did not segregate for kanamycin resistance, suggesting both homozygosity and homogeneity of the plants.

3.2.2. *Microspore-derived embryos. B.napus* embryos, approximately 28 days after microspore culture, are scratched with forceps, mixed with *A.tumefaciens* at ca. 20 embryos and 10^8 bacterial cells per ml NLN medium containing 20μM acetosyringone, and incubated in a 50ml tube at 25-30°C for 30 minutes. After removal of liquid medium, embryos are cultured on agar-solidified B5 medium supplemented with 0.1mg/l 2,4-dichlorophenoxyacetic acid, 0.05mg/l 6-benzyladenine(BA) and 20μM acetosyringone for 2 days (100-200 embryos per 9cm Petri dish). Embryos are then transferred to solid B5 medium containing 0.05mg/l BA and 250mg/l carbenicillin for 5 days (50-100 embryos per 9cm Petri dish), and finally to the same medium supplemented with appropriate selection agent (50mg/l kanamycin for NPTII gene) for continuous culture with subculture intervals of 2-3 weeks (20-50 embryos per 9cm Petri dish). Shoots develop from the surface of embryo axis within 4 weeks after co-cultivation. These are grown on B5 medium containing carbenicillin and kanamycin and transplanted to soil after roots are well established. The protocol outlined above has been used successfully to produce homozygous canola transgenics in several genotypes of B.napus, at an average frequency of 0.5% (5 out of 1,000 embryos co-cultivated produce transgenic shoots as confirmed by NPTII dot blot assays). Although this is a lower frequency than that routinely observed using the cotyledonary petioles as explants (Moloney et al. 1989), the efficiency of embryo transformation is quite high as embryos do not need to be handled individually during co-cultivation. We have produced a large number of canola transgenics through *Agrobacterium*-mediated transformation of microspore-derived embryos. About 50% of these transgenics contain single integration of the transgene, most of the remainder have two or three integrations. More than 90% of the transgenics thus produced did not segregate for the transgene, as confirmed by seed germination tests on medium containing 100mg/l kanamycin. For transgenics that produced both green and bleached seedlings on kanamycin, segregation could have resulted from either chimerism or hemizygosity.

This technology is applicable to all the genotypes tested so far, the only limitation appears to be the ability to produce microspore-derived embryos in sufficient quantity for transformation purposes. Another factor to be considered is the time required for production of transgenic seeds. While transformation experiments can be initiated using the cotyledonary petioles a few days after seeds become available, it would take at least 10 weeks from the time when seeds are available to co-cultivation of embryos.

Both octopine (LBA4404) and nopaline strains of *A.tumeficiens* have been used to produce canola transgenics from microspore-derived embryos. We routinely use the nopaline strain GV3101 (C58C1, Rif^r) harboring a disarmed Ti plasmid pMP90 which is a deletion of pTi58 (Koncz and Schell, 1986), and a binary vector containing a selectable marker gene and gene(s) of interest. The plant selectable marker gene we have used is the NPTII gene, which when driven by a CaMV 35S promoter gives sufficient resistance to 100mg/l kanamycin in shoots and seedlings.

4. PRODUCTION OF DOUBLED HAPLOIDS FROM TRANSGENIC MICROSPORES

Alternatively, homozygosity may be achieved by recovering doubled haploids from culturing microspores from primary transgenic plants which are hemizygous for the transgene(s). Microspore culture is particularly valuable when the primary transgenic plant has more than one unlinked insertions of the foreign gene, in which case plants homozygous for all, or any combination of, the transgene insertions can be produced in one step. Minimal times required to produce homozygous transgenic seeds for alternative approaches are estimated as follows.

Explant source	Starting point	Time to seed	
		Heterozygous	Homozygous
Stem segment	Seed	12 months	16 months
	Bolting plant	8 months	12 months
Cotyledons	Seed	8 months	12 months
Microspores or embryos	Seed	NA	14 months
	Bolting plant	NA	12 months

5. CONCLUSION

Microspore culture of B.napus is one of the most efficient plant regeneration systems available. For the development of genetically modified plants, the microspore culture system has demonstrated its value in transformation efficiency as well as accelerating homozygosity.

6. REFERENCES

Gamborg, O.L., Miller, R.A., and Ojima, K. (1968) 'Nutrient requirements of suspension cultures of soybean root cells. Exp Cell Res 50, 151-158.

Gland, A., Lichter, R. and Schweiger, H.G. (1988) 'Genetic and exogenous factors affecting embryogenesis in isolated microspore cultures of Brassica napus L. J Plant Physiol 132, 613-617.

Huang, B. (1992) 'Genetic manipulation of microspores and microspore-derived embryos', In Vitro Cell Dev Biol, 28P, 53-58.

Huang, B., Bird, S., Kemble, R., Miki, B. and Keller, W. (1991) 'Plant regeneration from microspore-derived embryos of Brassica napus: effect of embryo age, culture temperature, osmotic pressure, and abscisic acid. In Vitro Cell Dev Biol, 27P, 28-31.

Huang, B. and Keller, W.A. (1989) 'Microspore culture technology', J Tissue Culture Methods,

12, 171-178.
Jardinaud, M-F., Souvrë, A. and Blibert, G. (1993) 'Transient GUS gene expression in *Brassica napus* electroporated microspores'. Plant Science, 93: 177-184.
Jefferson, J.A. (1987) Assaying chimeric genes in plants: The GUS gene fusion system. Plant Molecular Biology Reporter, 5, 387-405.
Koncz, C. and Schell, J. (1986) The promoter of T_L-DNA gene 5 controls the tissue specific expression of chimeric genes carried by a novel type of *Agrobacterium* binary vector. Mol Gen Genet, 204, 383-396.
Matthysse, A.G. (1983) Role of bacterial cellulose fibrils in *Agrobacterium tumefaciens* infection. J Bacteriology, 154, 906-915.
Moloney, M.M., Walker, J.M. and Sharma, K.K. (1989) High efficiency transformation of *Brassica napus* using *Agrobacterium* vectors. Plant Cell Reports, 8, 238-242.
Pechan, P.M. (1989) 'Successful co-cultivation of *Brassica napus* microspores and pro-embryos with *Agrobacterium*'. Plant Cell Reports, 8:387-390.
Sangwan, R.S., Ducrocq, C. and Sangwan-Norreel, B. (1993) '*Agrobacterium*-mediated transformation of pollen embryos in *Datura innoxia* and *Nicotiana tabacum*: production of transgenic haploid and fertile homozygous dihaploid plants'. Plant Science, 95: 99-115.
Swanson, E.B. and Erickson, L.R. (1989) 'Haploid transformation in *Brassica napus* using an octopine-producing strain of *Agrobacterium tumefaciens*'. Theor Appl Genet, 78: 831-835.
Swanson, E.B., Herrgesell, M.J., Arnoldo, M., Sippell, D.W. and Wong, R.S.C. (1989) 'Microspore mutagenesis and selection: Canola plants with field tolerance to the imidazolinones'. Theor Appl Genet, 78: 525-530.

ACKNOWLEDGEMENT

We wish to thank our former colleagues Drs Larry Erickson, Eric Swanson and Ms Liane Cooke for their effort in developing the haploid transformation system. Thanks are also due to MaryAnne Arnoldo and Roger Kemble, both at Pioneer Hi Bred, for their contribution and leadership to the canola transformation program.

DIHAPLOID CHICORY (*CICHORIUM INTYBUS* L.) VIA MICROSPORE CULTURE

R. THEILER-HEDTRICH
Swiss Federal Research Station
CH-8820 Waedenswil
Switzerland

C.S. HUNTER
Department of Biological Sciences
University of the West of England
Bristol BS16 1QY, UK

ABSTRACT. Pollen of selected Witloof, Robin and Treviso chicory genotypes was isolated and cultured in a modified MS medium supplemented with 0.5 mg/l 2,4-dichlorophenoxy-acetic acid (2,4-D), 0.5 mg/l indole-3-acetic acid (IAA) and 2.0 mg/l zeatin (Z). During culture periods up to many (6-12) months, the pollen cells emerged and divided, were transferred to MS medium + 0.5 mg/l IAA and 0.5 mg/l 6-benzylaminopurine (BA) for further callus growth. Shoots were induced on the calli following transfer to a low salt medium containing 0.2 mg/l IAA + 0.4 mg/l kinetin (KN); shoots were rooted on a half-concentrated Lepoivre medium + 0.2 mg/l indole-3-butyric acid (IBA). The plants were maintained either *in vitro* or transferred *ex vitro* to soil. A range of ploidy levels was measured by DNA fluorescence analysis *via* flow cytometry in the microspore-derived progeny: some plants were predominantly haploid in a diploid background. Genetic fingerprinting by RAPD markers of some of these plants revealed clear differences compared to the parent plants.

1. Introduction

One commercial objective in chicory breeding is to obtain Witloof chicory with red leaves: this has been achieved by crossing Witloof with either Radicchio or Treviso chicory (Kiss 1963) or by developing inbred lines to obtain F1 hybrid seed (de Coninck and Bannerot 1989). Progeny from these crossings and later generations were heterozygous for anthocyanin expression, showing green, red and colour-speckled leaves. As a potential source of homozygous plants, dihaploids needed construction and evaluation. In the work reported here, microspore cultures from selected plants of Witloof, Treviso and Robin (a commercial hybrid of Witloof x Treviso) were established and dihaploid plants regenerated.

2. Materials and Methods

Single plant selections of Witloof (WI-BO) a green-leaf variety, Robin (R-4) a commercial red-leaf variety, and a genotype of Treviso (TR-1), with fully-red coloured leaves were grown in a greenhouse for flowering. Capitula, 3-4 mm in length, were harvested for microspore isolation,

M. Terzi et al. (eds.), Current Issues in Plant Molecular and Cellular Biology, 149-154.

surface sterilised in NaOCl solution (0.75% available chlorine) for 20 min on a shaker and then washed three times in sterile distilled water.

From a single capitulum, florets at the same microspore development stage were used. For the release of microspores, single florets were laid at the edge of a drop of culture medium (50-100 µl) in a sterile Petri dish (50 mm diameter), the base of the floret was cut off (in the drop of medium) and the microspores squeezed out from the floret. For one drop of medium, microspores from 3-8 florets from one capitulum were prepared, before being sieved and rinsed through a 45 µm mesh sieves and diluted in culture medium (7.5-10 ml/Petri dish). Only tetrad, microspore and uninuclear pollen stages were used for cultures.

For the growth of microspores, cell and colony formation medium R92.01, a modified MS liquid medium was used (Vermeulen *et al.* 1992) supplemented with 0.5 mg/l 2,4-dichlorophenoxy-acetic acid (2,4-D), 0.5 mg/l indole-3-acetic acid (IAA) and 2.0 mg/l zeatin (Z). Calli were subcultured to medium R92.25 (the same basal medium as above) plus 0.5 mg/l IAA and 0.5 mg/l 6-benzylaminopurine (BA) for further callus growth. Shoots developed after transfer of nodular calli to a "low salt" medium D91.1 (Dubois *et al.* 1991) containing 0.2 mg/l IAA plus 0.4 mg/l kinetin (KN) and were rooted on medium PM-109, a half-concentrated Lepoivre medium (Quoirin *et al.* 1977) containing 0.2 mg/l indole-3-butyric acid (IBA). Rooted plants *ex vitro*, approximately 5-10 cm tall, were transferred to commercial peat mixture (Brill No 3) in the greenhouse and further cultivated in the field.

Initially, microspore cultures were kept either in darkness or dim light at 25°C±1°C for 10-14 days. For the growth of callus, plant regeneration and rooting, cultures were incubated in a 16 h photoperiod at 20 µmol $m^{-2}.s^{-1}$ (Durotest True lite) at 22°±1°C. Rooted plants were kept under high humidity (95±5% RH) and in a 16 h photoperiod (40 µmol $m^{-2}.s^{-1}$) for 10-14 days prior to transfer to normal greenhouse conditions .

Ploidy of plants obtained either from seeds, clonal propagation or microspore-derived plants was detected by flow cytometry (Sharma *et al.* 1983).

Plants growing vegetatively in the nursery bed were indexed according to their anthocyanin expression in the leaves: *viz*; green leaves (-); mainly green leaves with only a few cells with anthocyanin on leaf lamina or red coloured midribs (-/+); green-red speckled leaves with 25-75% anthocyanin (+/-); and red leaves with >75% of the upper leaf lamina evenly anthocyanin coloured (+).

3. Results

Microspores grown in the Petri dishes exhibited slow rates of cell division during periods up to many (6-12) months. Within a Petri dish, cell cultures originating from single microspores could not be separated from other colonies, consequently subcultures were made non-selectively with respect to the progeny from single microspores.

After isolation of microspores and their culture in liquid medium (plate 1a) a gradual degradation of the exine was observed, leading to high quantities of debris in the Petri dishes. Microspores emerged from the pollen cell wall revealing a gametoplast (plate 1b) whose volume slowly expanded up to 3-6-fold before cell division (plate 1c). From divided pollen cells small colonies (plate 1d) and calli formed. Calli approximately 0.5-1.0 mm diameter were then transferred to gelled regeneration medium (R92.25), on which further growth was induced before meristematic nodules and leaves formed. For further shoot growth, these nodular calli were transferred to medium D91.1 and to medium PM-109 for rooting.

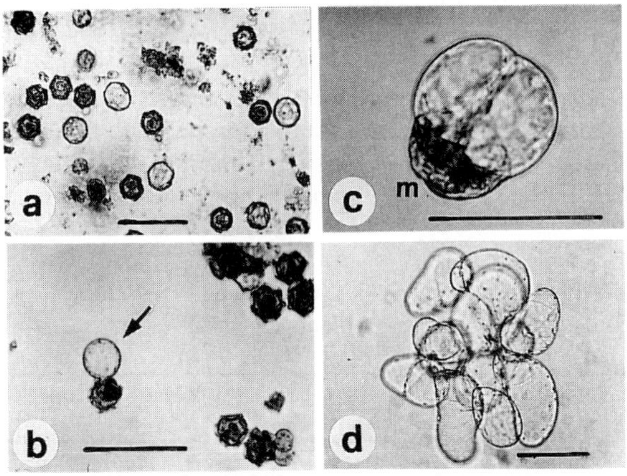

Plate 1: culture and development stages of chicory microspore cultures *in vitro*; 1a)typical sample of microspores; 1b) gametoplast emerging from pollen grain (arrow); 1c) first mitotic division of gametophytic cell, m = remains of microspore; 1d) microspore-derived cell colony. Size bar = 100 μm.

From the three genotypes, Witloof (WI-BO), Robin (R-4) and Treviso (TR-1), microspores from 93 capitula were isolated and cultured, but only microspores from florets from 6 capitula formed colonies and calli which developed into plants (table 1).

TABLE 1. Number of capitula per genotype for microspore culture, number of capitula with microspores that divided and number of plants raised and investigated

year	genotype	No. of capitula excised for microspore culture	No of capitula with microspores that divided (capitulum No.)	No. of regenerated plants tested by flow cytometry
1991	Witloof	13	1 (WI-BO/2)	80
	Treviso	11	2 (TR-1/16)	50
			(TR-1/22)	120
	Robin	10	1 (R-4/3)	205
1992	Robin	59	5 (R-4/11)	68
			(R-4/20)	40
			(R-4/33)	39
			(R-4/66)	3
			(R-4/69)	13

After transfer of calli and colonies to regeneration medium R92.25, high rates of cell divisions occurred, leading to nodular callus. Shoots from these calli must have ariven either from single microspores or from more than one microspore, in which case the shoot could be chimeric. Calli were subcultured for three to four pasages to raise as many plants per callus as possible.

During subculture on regeneration medium R92.25 and shoot induction on medium D91.1, up to 50% of the plants formed inflorescences *in vitro* : of these flowering plants, some were transferred to soil and grew on to flower in the greenhouse, but many of those plants transferred to soil died. For the ploidy analysis (figure 1) only the results from vegetative plants are given (table 2).

TABLE 2. Ploidy determination by flow cytometry of chicory progeny derived from microspore cultures. Number of plants at various ploidy states

genotype	capitulum number	haploid 1n*)	diploid 2n	tetraploid 4n	mixoploid **)	sum
Witloof	WI-BO/2	2	74	-	4	80
Treviso	TR-1/16	2	47	-	1	50
	TR-1/22	24	85	5	6	120
Robin	R-4/3	16	98	83	8	205
	R-4/11	3	44	18	3	68
	R-4/20	-	28	9	3	40
	R-4/33	1	32	4	2	39
	R-4/66	3	-	3	-	3
	R-4/69	1	10	1	1	13

*) haploid plants with a clear haploid peak (figure 1a) or as in the control (figure 1b), where the number of 1n and 2n nuclei were approximately equal.

**) mixoploid: leaf samples in which different ploidy levels could be detected eg 2n+3n, 2n+4n, 2n+4n+6n.

In most cases the haploid plants contained diploid cells indicating that they were chimeric, originating from haploid cells of which some had diploidized spontaneously. The ploidy state of plants with haploid cells (figure 1a) was compared with a diploid standard either from Witloof or Treviso plants to confirm that haploid cells were present (figure 1b).

Microspore-derived plants revealed some differences in leaf phenotype, either in lamina size or form. The most obvious differences were in leaf coloration (anthocyanin expression) which could be seen as a genetic marker. For progeny derived from microspores of the two fully-red donor plants Robin (capitulum No. R-4/3) and Treviso (capitulum No. TR-1/22), leaf-colour was indexed and grouped according to ploidy level (table 3).

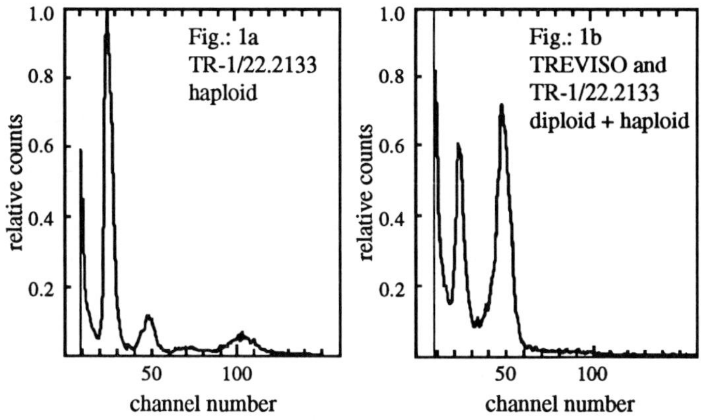

Figure 1: Flow cytometric histograms for chicory leaves; 1a: haploid microspore-derived plant from capitulum No. TR-1/22, plant No. 2133; 1b: control of diploid and haploid tissue from Treviso (diploid standard) and TR-1/22.2133 (haploid).
Horizontal axis: channel number, showing relative intensity of nuclei fluorescence, for G1 cells of haploid nuclei at channel 25, for G1 cells of diploid nuclei (standard) at channel 50. Vertical axis: relative counts of impulses per channel. Counts below channel 20 are neglectable background impulses from debris.

TABLE 3. Leaf colour index of microspore-derived plants from Robin (capitulum No. R-4/3) and Treviso (capitulum No. TR-1/22): numbers of plants

microspore-derived plants from capitulum number	ploidy level	leaf lamina mainly green with a few anthocyanin containing cells (-/+)	green-red speckled leaves (25-75%) anthocyanin (+/-)	red leaves (+)
R-4/3 (+)	haploid	3	11	2
	diploid	4	93	1
	tetraploid	28	55	0
TR-1/22 (+)	haploid	3	15	6
	diploid	4	48	33
	tetraploid	3	2	0

4. Discussion

Ploidy determination of regenerated plants supported their microspore origin. Most plants had become polyploid spontaneously *in vitro* leading to 1n+2n, 2n, 2n+4n and other mixoploid plants. This phenomen is well known from microspore cultures in which plant organogenisis occurs from callus (Mix *et al.* 1978): this leads to a difficulty in distinguishing between different genotypes originating from single microspores. During the callus phase and further plant propagation, clones of the same genotype could develop. Therefore methods to identify differnt genotypes need to be tested, for example by random amplified polymorphic DNA (RAPD) markers (Williams *et al.* 1990). Our prelimnary RAPD analysis showed clear differences in DNA banding between parent plants and microspore-derived progeny. From the DNA samples analysed, one haploid and one tetraploid plant showed the same DNA pattern, indicating that they originated from the same microspore genotype. Further investigations are necessary to confirm these results. For the results presented here, anthocyanin expression was used as a genetic marker and here, too, clear differences were shown (table 3).

With the above protocol for dihaploid chicory plant regeneration, a considerable improvement in chicory breeding should be achievable either for increased root inulin content (Rambaud *et al.* 1992), or improved F1-hybrids for Roodloof type chicons. A further improvement might also be achieved by raising dihaploid plants which are completely self-incompatable.

Acknowledgements

We thank Iris Finger, Andrea Bzonkova and H.U. Bisang (Swiss Federal Research Station, Wädenswil) and P. Fisher, (UWE, Bristol) for their assistance with aspects of the work.

References

de Coninck, B. and Bannerot, H. (1989) 'Creation d'hybrides F1 de chicoree de Bruxelles (*Cichorium intybus* L.) a limb rouge adaptes au forage hydroponique', Acta Horticulturae **242**, 191-192.

Dubois,T., Guedira, M. Dubois, J. and Vasseur, J. (1991) 'Direct somatic embryogenesis in leaves of *Cichorium* a histological and SEM study of early stages', Protoplasma **162**, 120-127.

Kiss, P. D., (1963) 'Genetische Untersuchungen zur Züchtung einer rotblättrigen Treibzichorie', Dissertationsschrift aus der Eidg. Versuchsanstalt für Obst-, Wein- und Gartenbau, Wädenswil.

Mix, G., Wilson, H.M. and Foroughi-Wehr, B. (1978), 'The cytological status of plants of *Hordeum vulgare* L. regenerated from microspore callus', Zeitschrift für Pflanzenzüchtung **80**, 89-99.

Quoirin, M., Lepoivre, P. and Boxus, P. (1977) 'Un premier bilan de 10 annees de recherches sur les cultures de meristemes et la multiplication in vitro de fruitiers ligneux', C.R. Rech. 1976-1977 et Rapports de Synthese, Stat. Cult. Fruit. et Maraich. Gembloux, pp. 93-117

Rambaud, C., Dubois, J. and Vasseur, J. (1992), 'The induction of tetraploidy in chicory (*Cichorium intybus* L. var. Magdebourg) by protoplast fusion', Euphytica **62**: 63-67.

Sharma, D. P., Firoozabady, E., Ayres N. M. and Galbraith, D. W. (1983), 'Improvement of anther culture in Nicothiana: Media cultural conditions and flow cytometric determination of ploidy levels', Z. Pflanzenphysiol. **111**, 441-451.

Vermeulen, A., Vaucheret, H., Pautot, V. and Chupeau, Y. (1992) '*Agrobacterium* mediated transfer of a mutant *Arabidopsis* acetolactate synthase gene confers resistance to chlorsulfuron in chicory (*Cichorium intybus* L.)', Plant Cell Reports **11**: 243-247.

Williams J.G.K., Kubelik, A.R., Livak, K.J., Rafalski, J.A., and Tingey, S.V. (1990) 'DNA polymorphisms amplified by arbitrary primers are useful as genetic markers', Nucleic Acids Research **18**, 6531-6535.

USE OF ANTHER CULTURE METHOD IN PEPPER BREEDING: FACTORS AFFECTING PLANTLETS PRODUCTION

NERVO G., CARANNANTE G., AZZIMONTI M.T. and ROTINO G.L.
Istituto Sperimentale per l'Orticoltura Sezione di Montanaso L. Via Paullese, 28 - 20075 Montanaso Lombardo (MI) Italy.

ABSTRACT. The aim of this work was to incorporate anther culture in a pepper breeding program for disease resistance. Anthers of twelve pepper genotypes were cultured following the method of Dumas de Vaulx et al (1981) with modifications in the induction medium regarding: growth regulators, gelling agents, sugar, silver nitrate and activated charcoal. From 8169 cultured anthers an overall of 3310 embryoids (40.5%) and 512 mature plantlets (6.3%) were obtained. All the anther donor genotypes produced androgenetic plantlets with a frequency ranging from 32.8 to 1%. The presence of silver nitrate (5 mg/l), glucose (30 g/l), and activated charcoal (0.05%) in the induction media gave a favorable effect on androgenetic plants production. The results suggest that anther culture is a useful tool for practical purpose.

INTRODUCTION

Homozygous plants obtained through in vitro culture are useful for fundamental studies and practical purposes. Haploids from several monocotyledon and dicotyledon species have been successfully induced by in vitro male or female gametophyte culture. Nevertheless, in some cases, the low efficiency impedes the use in plant breeding (1).

Several authors described results on in vitro androgenesis in pepper (2, 3, 4), however none of the methods described produced a great enough amount of haploids.

The frequency of microspore-derived plants increased consistently by heat treatments to the cultured anthers (5) and the protocol seems now promising for a large amount of doubled haploid plants production. More recently the importance of donor plant growth condition, medium composition and genotype were investigated (6,7). In this report we describe the effect of induction media on the response of twelve pepper genotypes used in a breeding program for disease resistance.

M. Terzi et al. (eds.), Current Issues in Plant Molecular and Cellular Biology, 155-160.

MATERIALS AND METHODS

As plant material, eleven breeding genotypes and the hybrid "PM687 x Yolo Wonder", as control, were utilized (table 1).

Flower buds of greenhouse-grown plants were collected when the corolla was slightly longer than the calyx and were surface sterilized. At this stage microspore were at the late mononucleate or early binucleate phase. Anthers were aseptically excised from flower buds and cultured on eleven induction media (table 2). We used the basal medium C, already described by Dumas de Vaulx et al. (5), with modifications regarding growth regulators, gelling agent, sugar, silver nitrate and activated charcoal. Anther cultures were incubated at 35°C in the dark for eight days and placed at 25°C under light at a 16h photoperiod for an additional period of four days. After twelve days, anthers were transferred to medium R, as reported by Dumas de Vaulx et al (5), supplemented with the same cytokinin (KIN or BAP) but without 2,4D. After 2-4 weeks of culture embryoids with globular or torpedo shape were removed from anther and recultured to fresh medium R. Well developed embryoids were transferred to hormone-free medium and rooted plantlets transplanted to soil under high humidity. The ploidy level of the androgenetic plants was determined by counting the number of chloroplast in guard cells (8) and confirmed by checking the number of the chromosomes in root tip cells.

RESULTS AND DISCUSSION

An overall of 3310 embryoids (40.5%) and 512 mature plantlets (6.3%) were obtained (table 3). The average number of androgenetic embryoids and plantlets obtained per 100 cultured anthers ranged from 10.2 to 185.3 and from 1.0 to 32.8, respectively.

All the twelve tested genotypes produced androgenetic embryoids; the hybrid "PM687 x Yolo Wonder" confirmed its good behaviour and achieved the highest number of embryoids and plantlets. High frequencies were also obtained with the accession "SCM 334", the hybrid "H. Taltoz x Perennial" and the hybrid "Cuneo x SCM 334".

Most of the anthers did not give any embryoids, while the responding ones produced many embryoids at various stages of development. The low convertion rate of embryoids into plantlets (15,5%) could be explained as a kind of inhibitory effect of the first well developed embryoid on the later globular or torpedo embryoids of the same anther or as an early senescence of the tissues.

In table 4 are reported the embryoids and plantlets frequencies obtained with the eleven induction media, calculated over the seven genotypes with the higher number of cultured anthers (PM687 x Yolo Wonder, Osir F_1, P374 F_1, Perennial, SCM 334, Osir x SCM 334, Osir x Perennial). The embryoids and plantlets yield ranged from 26.9 to 96.2

and 1.5 to 14.6 per 100 cultured anthers, respectively. The presence of 5 ppm silver nitrate in the induction media showed generally a favourable effect on embryoids (74.7%) and plantlets (13.5%) regeneration with variation among the genotypes. Glucose stimulated androgenesis both at the level of 30 and 100 g/l, whereas high amount of sucrose (100 g/l) caused a strong decrement in plantlets formation (1.5%). Among the gelling agent agar-agar showed higher frequencies in plantlets induction (10.1%) while the differences between gelrite and agarose were not evident. The presence of activated charcoal gave the lowest yield of embryoids (26.9%) but the highest convertion rate to mature plants (40%). These effects could be explained by a lower availability of growth regulators in the induction medium that reduced the embryoid formation. Subsequently, the induced embryoids developed in a better condition due to the absorption of inhibitory factors. Generally KIN gave higher induction in most of the pepper genotyes, even if in the line "SCM 334" and in "P 374 F_1" BAP performed better (tab.5).

In any case the high variability observed in the embryoids and plants frequencies suggest that it is better to consider the behaviour of each genotypes on different induction media. So, for breeding purpose we used induction medium containing KIN, glucose and silver nitrate for the anther culture of the hybrid "PM 687 x Yolo Wonder", and BAP both for "SCM 334" and "P 374" (table 4).

The ploidy level of the androgenetic plants, determined using chloroplast counts, gave almost an equal proportion of haploids and diploids, with differences between the genotypes.

CONCLUSION

The results presented here indicates that anther culture can be successfully applied to practical breeding.

Genotype and induction medium showed to be important for haploid plants production. Although all the genotypes used gave androgenetic plants, the frequencies varied highly according to the induction media. Thus, in order to apply anther culture to pepper breeding we suggest to search the best combination between genotypes and induction media. This work point out the importance of silver nitrate, activated chaorcoal and type of cytokinine (KIN, BAP) in the induction medium when new genotypes have to be used.

Considering the high frequencies of embryoids obtained in all the genotypes, the convertion rate into mature plants represents the "bottle neck" of pepper anther culture technique. For this reason further research on the factors involved in the embryoid development may permit to achieve an higher efficiency.

Acknowledgment: This research was in part supported by the Italian Ministry of Agriculture and Forestry, Project: Development of Plant Biotechnology.

REFERENCES

1. Bajaj Y.P.S. (1990) Biotechnology in Agriculture and Forestry Vol.12., Springer Verlag. Berlin, pp. 3-44.
2. George L. and Narayanaswamy S. (1973) Protoplasma 78, 467-470.
3. Novàk F.J. (1974) Z. Pflanzenzücht 72, 46-54.
4. Sibi M., Dumas De Vaulx R. and Chambonett D. (1979) Ann. Amélior. Plant. 29, 583-606.
5. Dumas De Vaulx R., Chambonett D. and Pochard E. (1981) Agronomie 1, 859.
6. Morrison R.A., Koning R.E. and Evans D.A. (1986) J. Plant Physiol. 126, 1-9.
7. Kristiansen K. and Andersen S.B. (1993) Euphytica 67, 105-109.
8. Qin X. and Rotino G.L. (1993) Capsicum and Eggplant Newsletter 12, 59-62.

TABLE 1. Genotypes of pepper utilized as anther donor; origin and main useful characteristics.

Genotypes	Origin	Useful characteristics(*)
Osir	F1 commercial hybrid	t. Ph. capsici and CMV
P 374	" " "	t. " " "
"	" " "	R. TMV and PVY
Perennial	accession	R. CMV
Serrano Criollo	"	R. Ph.capsici
Morelos(SCM334)		R. TMV and PVY
H.Taltoz x SCM334	F1 breeding hybrid	R. Ph. capsici
Osir x SCM 334	" " "	R. "
Cuneo x SCM334	" " "	R. "
P374 X SCM334	" " "	R. " ,TMV,PVY
H.Taltoz x Perennial	" " "	R. CMV
Osir x Perennial	" " "	R "
P374 x Perennial	" " "	R. TMV,PVY
PM687 x Yolo W.	F1 control	High anther response

(*) Resistant to (R), tolerant to (t).

TABLE 2 - Composition of the elevan induction media employed having macroelements, microelements and vitamins described by Dumas de Vaulx et al. (5).

Media code	Growth regulators	Sugars %		Others	Gelling agents		
1	KIN	Sucrose	3	-	Agar	0.9	
2	"	"	3	-	Gelrite	0.3	
3	"	"	3	-	Agarose	0.7	(*)
4	"	"	3	$AgNO_3$ 2,5 ppm	Agar	0.9	
5	"	"	3	" 5 ppm	"	0.9	
6	"	"	10	-	"	0.9	
7	"	Glucose	3	-	"	0.9	
8	"	"	10	-	"	0.9	
9	"	Sucrose	3	-	Agarose	0.8	(**)
10	"	"	3	A.charcoal 0.05%	Agar	0.9	
11	BAP	"	3	-	"	0.9	

(*) from Calbiochem; (**) from Sigma

TABLE 3. Number of cultured anthers, number and frequency (%) of embryoids and plantlets obtained from twelve pepper genotypes regardless of the induction media employed.

Genotypes	anther number	embryoids			plantlets		
		number	%	±s.e.	number	%	±s.e.
PM687 x Y.Wonder	530	982	185.3	48.1	174	32.8	7.3
Osir F_1	1012	288	28.5	10.9	28	2.8	1.4
P374 F_1	1109	442	39.9	11.3	11	1.0	0.3
Perennial	1079	48	4.4	1.6	12	1.1	0.5
SCM334	1185	883	74.5	18.7	145	12.2	2.4
H.Taltoz x SCM334	171	70	40.9	16.3	10	5.8	1.8
Osir x SCM334	1164	219	18.8	6.2	62	5.3	2.1
Cuneo x SCM334	75	46	61.3	80.3	9	12.0	20.0
P374 x SCM334	222	47	21.2	51.1	5	2.2	9.8
H.Taltoz x Perennial	98	66	67.3	48.6	17	17.3	9.8
Osir x Perennial	1092	175	16.0	6.3	34	3.1	1.3
P374 x Perennial	432	44	10.2	6.6	5	1.2	1.1
Total	8169	3310	40.5		512	6.3	

TABLE 4. Frequency (%) of embryoids and plantlets obtained on the eleven induction media. Data concerning the seven genotypes with the higher number of cultured anthers.

Media	Embryoids		Plantlets	
	%	±s.e.	%	±s.e.
1	61.8	48.4	10.1	7.9
2	55.2	27.8	6.1	2.7
3	79.1	29.5	8.0	4.5
4	43.3	13.5	8.8	3.5
5	74.7	37.8	13.5	8.5
6	41.6	21.3	1.5	1.0
7	96.2	76.6	14.6	8.1
8	62.8	57.8	12.1	10.9
9	73.4	39.7	5.9	3.9
10	26.9	9.3	11.2	4.9
11	44.4	18.8	8.0	4.2

TABLE 5. Frequency (%) of embryoids and plantlets from three pepper genotypes cultured on the eleven induction media.

Media	PM 687 x Yolo Wonder		SCM 334		P 374 F1	
	embryoids	plants	embryoids	plants	embryoids	plants
1	351.4	57.1	32.8	9.4	23.0	2.6
2	167.9	16.9	156.4	10.8	12.3	0.0
3	185.3	32.8	167.8	9.3	1.4	0.0
4	107.1	17.8	45.8	18.7	40.0	0.7
5	291.4	63.7	42.9	12.9	87.0	0.9
6	133.9	6.7	9.0	3.6	111.5	0.0
7	552.2	60.8	75.5	16.3	0.0	0.0
8	409.1	77.2	0.0	0.0	3.2	0.8
9	286.7	26.6	129.7	13.5	62.7	0.0
10	78.2	38.4	23.2	10.8	17.2	1.0
11	18.6	9.3	139.6	31.3	59.0	4.0

PROTOPLAST-FUSION DERIVED CYBRIDS IN SOLANACEAE

E. GALUN
Department of Plant Genetics
The Weizmann Institute of Science
Rehovot 76100
Israel

ABSTRACT. The term 'cybrid' is neat and concised but ill-defined. For this review this term shall include protoplast-fusion derived cells and plants in which most or all the nuclear genome of one (organelle-donor) fusion partner is eliminated while this partner does contribute some or all the organelle's (chloroplast and/or mitochondria) genomes. There exist efficient procedures to suppress the contribution of the nuclear genome from the donor fusion-partner as well as a choice of procedures to select fusion-derivatives containing the plastome or chondriome components of either fusion-partner. Thus, there exists experimental tools to strongly shift the range of cybrids towards the desired one. This implies not only to cybrids but also to asymmetric hybrids. Solanaceae genera (e.g. *Nicotiana, Petunia, Lycopersicon & Solanum)* and species are especially amenable to protoplast-fusion. Hence there exist ample data in respect with the compatibilities between a given nuclear genome and organelles of an alien genus or species, that provide us with important theoretical and practical information.

1. Introduction

The purpose of this presentation is to furnish an overview on cybridization in the Solanaceae.Emphasis will be put on intra-and intergeneric cybridizations in which *Solanum* served as one of the protoplast-fusion partners. An introspective overview on plant cybrids was published about a year ago (Galun,1993) that provided ample citations on cybridizations by others and in our own laboratory. Here I shall deal mainly with three questions: (1) what are the purposes of cybrid formation; (2) what are the methodologies for cybrid formation, selection and identification; (3) what are the main results of cybridization and asymmetric hybridization that involve Solanaceae protoplasts. This presentation is a *keynote* lecture, it should therefore also serve as a *key* to open other presentations in this symposium, that are provided in this Congress as either oral lectures or posters.

This overview is not an authoritative review on cybridization and assymmetric hybrids in Solanaceae. Time and space are too limited for it. Those interested in additional information are referred to recent reviews such as in two theses (Derks, 1992; Xu, 1993), the review of Medgyesy (1994) and our reviews (Breiman and Galun, 1990; Galun and Aviv, 1991; Galun, 1993).

On the otherhand, it *is* intended to provide here some critical remarks on purposes and methodologies of cybridization. My current research work shifted from cybridization and organelle-transfer to other topics. Therefore,my treatment of cybridization, while still introspective, will probably not lack objectivity.

2. The Aims of Cybridization and Asymmetric Hybridization

We previously detailed the aims of cybrid production(e.g. Galun, 1981; Galun and Aviv, 1983, 1986) in several plant genera as well as specifically in *Solanum* (Galun et al., 1994). Generally the aims are in two different fields of investigation. First, there is the interest to obtain biological

M. Terzi et al. (eds.), Current Issues in Plant Molecular and Cellular Biology, 161-169.

information on intra-and intergeneric transfer of organelles. It includes the study of the interaction of the plastome and/or the chondriome of a given species with the nuclear genome of another species. What is the phylogenetic limit of compatibility between organelles and nucleus. The other field of investigation is an applied one: to replace, in a given genotype, either or both the chloroplasts and the mitochodnria by organelles of another genotype - - usually of a different species or of a specific organelle-mutant donor. In practical terms mitochodnria were replaced to cause cytoplasmic male sterility that is useful for the production of hybrid cultivars (see Breiman and Galun, 1990 for details). Chloroplasts were replaced (especially in Brassica) to confer resistance to specific herbicides (e.g. atrazin).

Assymmetric hybridization does not differ substantially from cybridization. Actually it is doubtful whether or not pure cybrids were ever produced (see: Galun, 1993). Thus the purposes of asymmetric hybridization are rather similar to those of cybridization. But there are some differences. In asymmetric hybridization it is frequently intended to transfer specific nuclear coded traits such as resistance to viral diseases (e.g. Valkohen et al. 1994). Moreover asymmetric hybridization may furnish a tool to reveal which nuclear components of the organelle donor, are essential for maintaining nuclear-organelle compatibility (e.g. Wolters et al., 1991, 1993).

3. Methodologies: Techniques and Considerations

The main methods for the establishment of cybrids and asymmetric hybrids are summarized in Fig. 1. These shall not be detailed here because they were described by us previously. Basically these methods are frequently based on two procedures: irradiation of the donor protoplasts (Zelcer et al., 1978) and transient metabolic inhibition of the recipient protoplasts (Sidorov et al., 1981; Aviv et al., 1986). Isolation of protopalsts and their fusion are now well established procedures and need no reiteration. Protoplasts are commonly isolated from either leaf-mesophyll or from *in vitro* cultured cells (e.g. calli or cell-suspensions). Frequently investigators fused either mesophyll protoplasts of one species with mesophyll protoplasts of another species, or used mesophyll protopalsts to fuse with suspension-culture cells. There is a considerable difference between these two fusion pairs. Fig. 2 demonstrates the roles of cell source and treatment of fusion-partners, on the pre-fused protoplasts and, consequently, on the fusion-derived protoplasts. Mesophyll-cell derived protoplasts have ample plastomes and condriomes. When these protopalsts are used as γ-or x-irradiated organelle-donors their organelle genomes are probably not affected by this treatment but the nuclear genome is vastly damaged. This damage is apparently correlated with the level of irradiation and thus it is quite common that after low and medium doses of irradiation at least fractions of the nuclear genome are retained and can then integrate into the genome of the fusion product. When mesophyll-cell derived protoplasts are used as recipients and are treated with a metabolic inhibitor (e.g. iodoacetate) their nuclear genome stays intact but their organelles are probably damaged. Thus, a fusion prodcut between such donor and recipient protoplasts will have the nuclear genome of the recipient with the possibility of some nuclear-genome components of the irradiated donor The chondriomes and the plastomes of both fusion partners will be included in the fusion product. But, most of these will be from the donor because those of the recipient were damaged by the prefusion treatment. Very frequently there will be a recombination between the DNAs of the chondriomes of the recipient and the donor. This could lead, after sorting out, to fusion products with the recipient's chondriome, with the donor's chondriome, or with a new type of recombinant chondriome.

When the recipient protoplast is derived from a suspension-culture cell, it probably has a much lower number or plastome copies (but see Li and Sink, 1992). Iodoacetate treatment may further reduce the copies of the plastome and the chondriome. When such a treated recipient protoplast is fused with an irradiated donor protoplast that was derived from a mesophyll-cell, the fusion product will have a different popualtion of plastomes than in the previously mentioned fusion. Most or all the chloroplasts will be from the mesophyll-derived donor. The nuclear-genome and the chondriomes will be about the same as those of the previously described fusion-product. Sorting out could frequently lead to plants containing only the donors chloroplasts and having either the pure chondriome of the donor or a novel chondriome, containing components from both the donor

MAIN METHODS TO ESTABLISH CYBRIDS AND ASYMMETRIC HYBRIDS

Treatment of Fusion - Partner Protoplasts

Recipient of organelles & nuclear components

- Iodoacetate Treatment
- Albino Plastome Mutants
- Albino Nuclear - genome Mutants
- Thodamine 6 -B Treatment

Donor of organelles & nuclear components

- Gamma or X - Irradiation
- Plastome Resistance to Antibiotics (Strep., Spec.)
- Chondriome Resistance to Antibiotics (Oligomycin)
- Nuclear Resistance to drugs (Methotrexate, Kanamycin)

Selection of Fusion Derivatives

- No Selection - All Regenerated Plants are Putative Cybrids
- Selection on Plastome - Affecting Antibotics (Strep. Spec.)
- Selection on Chondriome - Affecting Antibiotics (Oligo.)
- Selection of Green Regenerants
- Selection on Nuclear - Genome Affecting Drugs (Methotr.,Kana.)

Fig. 1.

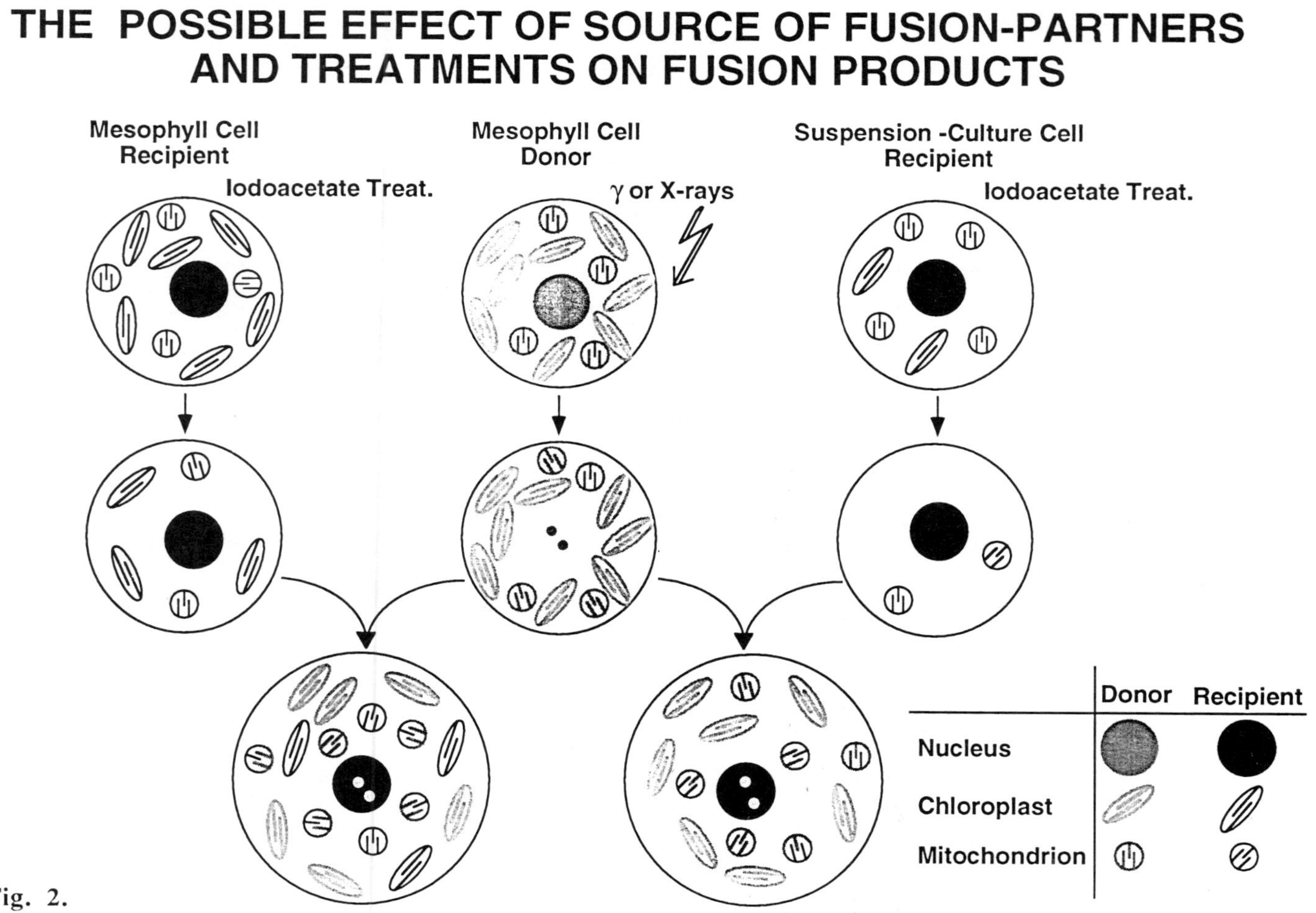
THE POSSIBLE EFFECT OF SOURCE OF FUSION-PARTNERS
AND TREATMENTS ON FUSION PRODUCTS
Mesophyll Cell
Recipient
Iodoacetate Treat.
Mesophyll Cell
Donor
γ or X-rays
Suspension -Culture Cell
Recipient
Iodoacetate Treat.
Donor
Recipient
Nucleus
Chloroplast
Mitochondrion

Fig. 2.

and the recipient. Only in rare cases the resulting plants will have the pure chondriome of the recipient fusion-partner.

Selection of fusion-products by the use of nuclear mutants.- Up to now we described the scenarios of fusions where no selection is imposed during culture and regeneration of the fusion-products. To "force" the inclusion of nuclear-genome components from the donor into the fusion product a donor with a selectable nuclear-coded trait can be used. One example is to use a kanamycin-resistant donor (e.g. Bates, 1990). Such a donor may be a result of transformation with *Agrobacterium* harboring a kanamycin-resistance DNA fragment in its plasmid. If the respective fusion product is cultured in a kanamycin containing medium - - only those plantlets that contain the nuclear components of the donor, with the kanamycin resistance, will survive selection; especially during rooting on kanamycin. A selectable nuclear-coded resistance to methatrexate was used in a similar manner (Dudits et al., 1987). Selection of nuclear components can be performed also by using a recipient with nuclear-coded chlorophyll deficiency (Gupta et al., 1984). Complementation of such a mutant by nuclear components from an irradiated donor protoplast is a possible method although the chances of such a complementation are probably low.

The role of organelle-genome coded mutants - - Such mutants are coded in the plastome. They may be resistances to antibiotics as streptomycin (e.g. Menczel et al., 1981) or spectinomycin, that can easily be produced (e.g. Fluhr et al., 1985). If one of the fusion-partners contains such a plastome mutant then selection of fusion products in the presence of the respective antibiotics will retain only chloroplasts containing the mutated plastome. There are not many useful chondriome mutants. One of them is oligomycin resistnace in tobacco (Aviv and Galun, 1988; Perl et al., 1991a), and it was used to select cybrids having chondriome components of the donor.

4. Achievements

The available arsenal of methods mentioned above can lead to many combinations of organelles and nuclear components in the fusion derived plants. But there is one great limitation. Only systems in which plants can be regenerated from cultured protoplasts are useful. The following summary of results is based on recent publications on cybridization and asymmetric hybridization in Solanaceae by other authros (e.g. Bonnema et al., 1991; Derks et al., 1992; Hinnisdaels et al., 1991; Li and Sink, 1992; Melchers et al., 1992; Melzer and O'Connell, 1992; Parokonny et al., 1994; Wolters et al., 1991, 1993; Xu et al., 1993) as well as on our own studies (e.g. Perl et al., 1990a, 1990b, 1991a, 1991b).

1. Cybridization and asymmetric hybridization by the donor-recipient protoplast-fusion is an efficient method to transfer chloroplasts among closely related species. When the phylogenetic distance is extensive pigmentation deficiencies may occur. Extensive phylogenetic distances require severe selection of the required plastome and this could lead to plastome recombination. (e.g. Sidorov et al. 1994; Thanh and Medgyesy, 1989).But plastome recombination is generally rare in fusion derivatives. It also seems that the transfer of some nuclear components from the plastome-donor to a distant recipeint are required to obtain functional asymmetric hybrids (see detailed discussion in Breiman and Galun, 1990) .

2. Cybridization and asymmetric hybridization is also an efficient method for intrageneric transfer chondriome components. Unlike plastomes, chondriomes from two species frequently recombine in the protoplast-fusion product. Transfer of chondriome components can induce cytoplasmic male sterility in many interspecific cybridizations. (see detailed discussion in Breiman and Galun, 1990). As with plastomes, extensive phylogenetic distances between a given nuclear genome and an alien chondriome can be bridged by an appropriate selectable chondriome mutation. It seems that phylogenetic distance is a less severe limitation for the transfer of chondriome components than for the transfer of plastomes.

3. Sorting-out of plastomes, and even more so the stabilization of the chondriome, in the fusion derivaties, may require many cell-divisions and thus not be achieved in the respective cybrid or asymmetric-hybrid plant. Selective pressure during sorting-out may therefore affect the final organelle composition. Because in asymmetric hybrids sorting out takes place also in respect with

nuclear-genome components the stabilization of the latter components may also require many cell divisions. These sorting-outs and fixations may even not reach completion in the regenerated plant and require one or more sexual cycles.

Cybridization and asymmetric hybridization also resulted, in several cases, in breeding lines that have agronomic value. These were in three areas. First, in transfer of male-sterility into crop species, then inducing resistance to herbicides and more recently, conferring disease resistance in asymmetric somatic hybrids. The transfer of chondriome components among species by the donor-recipient protoplast fusion was already revealed 16 years ago (e.g.Zelcer et al., 1978; Aviv et al., 1984) in *Nicotiana*. Cybridization also resulted in male-sterile potato plants that could be useful for the production of true-potato-seed hybrids (Perl et al., 1990a, Melchers et al., 1992). Protoplast-fusion mediated transfer of male sterility was amply reported also in another plant family - in the genus *Brassica* (see: Pelletier et al., 1983; Sundberg et al., 1991; Walters et al., 1992; Walters and Earle, 1993 and see previous literature in these authors' publications). Several publications by Earle and collaborators as well as by others (e.g. Thomzik and Hain, 1988) showed that atrazine and triazine resistant chloroplasts could be transferred into target cultivars of *Brassica* by protoplast-fusion.

The usefulness of asymmetric hybridization in crop improvement was advocated already several years ago; actual examples are beginning to be reported: Bates (1990) imposed TMV resistance in tobacco and Valkonen et al. (1994) transferred resistances to several viral diseases of potato from *Solanum brevidens* to *S. tuberosum.*

5. Epilogue

Is trying to understand the concerted interaction of plasmones and nuclear genomes or establishing useful plant genotypes the sole reasons for the hundreds of cybridization and asymmetric hybridization studies? I doubt it. The manifold efficient tools of cybridization (and asymmetric hybridization), in conjunction with the introdcution of selectable markers by genetic transformation, can lead to the production of exotic genotypes. Feeling as a "Creator" of novel plants probably gives a considerable boost to the investigator.

Acknowledgement
I am grateful to Drs. E.D. Earle, Y.Y. Gleba, E. Jacobson, P. Medgyesy, E. Pehu and K.C. Sink for generously furnishing me with reprints and preprints; I apologize that not all of these could be cited in this presentation. Thanks are also due to Dr. Dvora Aviv and Dr. Avihai Perl who collaborated with me in the studies on *Nicotiana* and *Solanum* cybrids and conducted with me numerous discussions on the subject of this presentation.

6. References

Aviv, D., Chen, R. and Galun, E. (1986) 'Does pretreatment, by Rhodamine 6-G affect the mitochondrial composition of fusion-derived *Nicotiana* cybrids', Plant Cell Rep. 5: 227-230.

Aviv, D and Galun, E. (1988) 'Transfer of cytoplasmic organelles from an oligomycin resistant *Nicotiana* cell suspension into tobacco protoplasts yielding oligomycin resistant cybrid plants', Mol. Gen. Genet. 215: 128-133.

Bates, G.W. (1990) 'Asymmetric hybridization between *Nicotiana tabacum* and *N. repanda* by donor-recipeint protoplast fusion: transfer of TMV resistance', Theor. Appl. Genet. 80: 481-487.

Bonnema, A.B., Melzer, J.M. and O'Connell, M.A. (1991) 'Tomato cybrids with mitochodnrial DNA from *Lycopersicon pennelli',* Theor. Appl. Genet. 81: 339-348.

Breiman, A. and Galun, E. (1990) 'Nuclear-mitochondrial interrelation in angiosperms', Plant Science 71: 3-19.

Derks, F.H.M. (1992) 'Organelle transfer by protoplast fusion in Solanaceae', Thesis, Free University of Amsterdam, Amsterdam, The Netherlands, 120p.

Derks, F.H.M., Hakkert, J.C., Verbeek, W.H.J. and Colijn-Hooymans, C.M. (1992) 'Genome composition of asymmetric hybrids in relation to phylogenetic distance between parents. Nucleus-chloroplast interaction',. Theor. Appl. Genet. 84: 930-940.

Dudits, D., Maroy, E., Praznovsky, T., Olah, Z., Gyorgyey, J. and Cella, R. (1987) 'Transfer of resistance traits from carrot into tobacco by asymmetric somatic hybridization: regeneration of fertile plants,' Proc. Natl. Acad. Sci. (USA) 84: 8434-8438.

Fluhr, R., Aviv, D., Galun, E. and Edelman, M. (1985) 'Efficient induction and selection of chloroplast-coded antibiotic-resistant mutants in *Nicotiana*', Proc. Natl. Acad. Sci. (USA) 82: 1485-1489.

Galun, E. (1981) 'Plant protoplasts as physiological tools', Ann. Rev. Plant Physiol. 32: 237-266.

Galun, E. (1993) 'Cybrids - An introspective overview', IAPTC Newsletter 70:2-10.

Galun, E. and Aviv, D. (1983) 'Cytoplasmic hybridization - genetic and breeding applications', in D.A. Evans et al. (eds), Handbook of Plant cell Culture, Vol. 1, Macmillan, N.Y., pp. 358-392.

Galun, E. and Aviv, D. (1986) 'Organelle Transfer', Methods in Enzymol. 118: 595-611.

Galun, E. and Aviv, D. (1991), 'Cybrid production and selection', in K. Linsey (ed.) Plant Tissue Culture Manual pp. D4 1-17, Kluwer Acad Publ. Dordrecht, The Netherlands.

Galun, E., Perl., A. and Aviv, D. (1994) 'Cybridization in potato', in Y.P.S. Bajaj (ed), Biotechnology in Agriculture and Forestry, vol. 27, Springer Verlag Heidelberg (in press).

Gupta, P.P., Schieder, O. and Gupta, M. (1984) 'Intergeneric nuclear gene transfer between somatically and sexually, incompatible plants through asymmetric protoplast fusion', Mol. Gen. Genet. 197: 30-35.

Hinnisdaels, S., Bariller, L., Mouras, A., Sidorov, V., Del-Favero, J. Veuskens, J., Negrutiu, I. and Jacobs, M. (1991) 'Highly asymmetric intergeneric nuclear hybrids between *Nicotiana* and *Petunia:* evidence for recombinogenic and translocation events in somatic hybrid plants after "gamma"-fusion', Theor. Appl. Genet. 82: 609-614.

Li, Y. and Sink, K.C. (1992) 'Cell type determines plastid transmission in tomato intergeneric somatic hybrids', Curr. Genet. 22: 167-171.

Medgyesy, P. (1994) 'Transfer of chloroplast traits between sexually incompatible Solanaceae species through protoplast fusion', in Y.P.S. Bajaj (ed.) Biotechnology in Agriculture and Forestry Vol. 27, Springer Verlag, Heidelberg (in press).

Melchers, G., Mohri, Y., Watanabe, K., Wakabayashi, S. and Harada, K. (1992) 'Onestep generation of cytoplasmic male sterility by fusion of mitochondrial-inactivated tomato protopalsts with nuclear-inactivated tomato protoplasts', Proc. Natl. Acad. Sci. (USA) 89: 6832-6836.

Melzer, J.M. and O'Connell, M.A. (1992) 'Effect of radiation dose on the production of and the extent of asymmetry in tomato asymmetric somatic hybrids', Theor. Appl. Genet. 83: 337-344.

Menczel, L., Nagy, F., Kiss, Zs. R. and Maliga, P. (1981) 'Streptomycin resistant and sensitive somatic hybrids of *Nicotiana tabacum* + *N.knightiana:* correlation of resistance to *N. tabacum* plastids', Theor. Appl. Genet. 59: 191-195.

Pelletier, G., Primard, C., Vedel, F., Chetrit, P., Remy, R., Rouselle, P. and Penard, M. (1983) 'Intergeneric cytoplasmic hybridization in Crucifera by protoplast fusion', Mol. Gen. Genet. 191: 244-250.

Parokonny, A.S., Kenton, A., Gleba, Y.Y. and Bennet, M.D. (1994) 'Genome interaction and the fate of recombinant chromosomes in *Nicotiana* asymmetric somatic hybrids and their sexual progeny', Theor. Appl. Genet. (in press).

Perl, A., Aviv, D. and Galun, E. (1990a) 'Protoplast-fusion-derived CMS potato cybrids: potential seed-parent for hybrid, true-potato-seeds,' J. Heredity 81: 438-442.

Perl, A., Aviv, D. and Galun, E. (1990b) 'Protoplast-fusion-derived *Solanum* cybrids: application and phylogenetic limitations', Theor. Appl. Genet. 79: 632-640.

Perl, A., Aviv, D. and Galun, E. (1991a). 'Protoplast fusion mediated transfer of oligomycin resistance from *Nicotiana sylvestris* to *Solanum tuberosum* by intergeneric cybridization', Mol. Gen. Genet. 225: 11-16.

Perl, A., Aviv, D. and Galun, E. (1991b) 'Nuclear-organelle interaction in *Solanum:* interspecific cybridizations and their correlation with a plastome dendogram', Mol. Gen. Genet. 228: 193-200.

Sidorov, V.A., Evtushenko, D.P., Shakhovsky, A.M. and Gleba, Y.Y. (1994) 'Cybrid production based on mutagenic inactivation of protoplasts and rescue of mutant plastids in fusion products: potato with plastome from *S. bulbocastanum* and *S. pinnatisectrum'*,. Theor. Appl. Genet. (in press).

Sidorov, V.A., Menczel, L., Nagy, F., and Maliga, P. (1981) 'Chloropalst transfer in *Nicotiana* based on metabolic complementation between irradiated and iodoacetate-treated protoplasts', Planta 152: 341-345.

Sundberg, E., Lagercrantz, U., and Glimelius, K. (1991) 'Effect of cell type used for fusion on chromosome elimination and chloroplast segregation in *Brassica oleracea* + *Brassica napus* hybrids', Plant Sci. 78: 89-98.

Thanh, N.D. and Medgyesy, P. (1989) 'Limited gene transfer via recombination overcomes plastome-genome incompatibility between *Nicotiana tabacum* and *Solanum tuberosum',* Plant Mol. Biol. 12: 87-93.

Thomzik, J.E. and Hain, R. (1988) 'Transfer and segregation of triazine tolerant chloroplasts in *Brassica napus'*, Theor. Appl. Genet. 76: 1650171.

Valkonen, J.P.T., Xu, Y.-S., Rokka, V.-M., Pulli, S. and Pehu, E. (1994) 'Transfer of resistance to potato leafroll virus, potato virus Y and potato virus X from *Solanum brevidens* to *S. tuberosum* through symmetric and designed asymmetric somatic hybridization', Ann. Appl. Biol. 124 (in press).

Walters, T.W. and Earle, E.D. (1993) Organelle segregation, rearrangement and recombination in protoplasts fusion derived *Brassica oleracea* calli. Theor. Appl. Genet. 85: 761-769.

Walters, T.W., Mutscher, M.A. and Earle, E.D. (1992) 'Protoplast fusion derived Ogura male sterile cauliflower with cold tolerance', Plant Cell Report. 10: 624-628.

Wolters, A.M.A., Koornneef, M. and Gilissen, L.J.W. (1993) 'The chloroplast and mitochondrial DNA type are correlated with the nuclear composition of somatic hybrid calli of *Solanum tuberosum* and *Nicotiana plumbaginifolia', Curr. Genet.* 24: 260-267.

Wolters, A.M.A.,, Schoenmakers, H.C.H., van der Meulen-Muisers, J.J.M., Van der Knaap, E., Derks, F.H.M., Koornneef, M. and Zelcer, A. (1991) 'Limited DNA elimination from the irradiated potato parent in fusion products of albino *Lycopersicon esculentum* and *Solanum tuberosum*,' Theor. Appl. Genet. 83: 225-232.

Xu, Y.-S. (1993) 'Production and characterization of somatic hybrids between *Solanum tuberosum* and *S. brevidens.* ' Thesis, University of Helsinki, Helsinki, Finland, Publ. no. 38, 56p.

Xu, Y.S., Murto, M., Dunckley, R., Jones, M.G.K. and Pehu, E. (1990) 'Production of asymmetric hybrids between *Solanum tuberosum* and irradiated *S. brevidens'*, Theor. Appl..Genet. 85: 729-734.

Zelcer, A., Aviv, D. and Galun, E. (1978) 'Interspecific transfer of cytoplasmic male sterility by fusion between protoplasts of normal *Nicotiana sylvestris* and X-ray irradiated protoplasts of male-sterile *N. tabacum'*, Z. Pflanzenphysiol. 90: 397-407.

BRASSICA OLERACEA CYBRIDS FOR HYBRID VEGETABLE PRODUCTION

E.D. EARLE[1] and M.H. DICKSON[2]

[1]*Dept. of Plant Breeding*
Cornell University
Ithaca, NY 14853-1902
United States of America

[2]*Dept. of Horticulture*
New York State Agricultural Experiment Station
Geneva, NY 14456
United States of America

ABSTRACT. Horticulturally improved vegetable lines were obtained by fusion of cauliflower (*Brassica oleracea*) protoplasts carrying the Ogura type of cytoplasmic male sterility (CMS) with protoplasts from fertile lines of cauliflower, *B. rapa*, or *B. napus*. Molecular, phenotypic, and flow cytometric analysis of more than 100 fusion-derived plants identified some diploid cauliflower cybrids that were CMS but contained chloroplasts from the fertile fusion partner. These no longer showed the chlorosis seen in the original Ogura CMS material at low temperatures. After field and greenhouse selection for good female fertility and horticultural characters, cold-tolerant CMS broccoli, cauliflower, and cabbage lines were developed. These are now being tested for use in commercial hybrid production at over 20 seed companies worldwide. Current issues of interest include stability of the fusion-derived materials, comparison of sexual and protoplast-based methods for transfer of the improved cytoplasms to additional vegetable lines, and "finger-printing" of cytoplasms from different sources. This work illustrates how somatic hybridization can contribute to applied plant breeding.

Introduction

Availability of cytoplasmic male-sterile (CMS) lines greatly facilitates production of hybrid seed. Use of the radish-derived Ogura male sterile cytoplasm for hybrid production in *Brassica* species has been limited because leaves of Ogura CMS plants show chlorosis at low temperatures [1]. This chlorosis is related to the presence of *Raphanus sativus* chloroplasts in Ogura CMS *Brassica* materials [2]. Protoplast fusion provides a way to replace the *Raphanus* chloroplasts with *Brassica* ones, while retaining the CMS associated with the *Raphanus* mitochondria. Many groups have used this approach in attempts to obtain plants suitable for applied breeding programs. This report describes how cold-tolerant Ogura CMS *B. oleracea* vegetable lines with good horticultural characters were developed at Cornell University by means of protoplast fusion followed by conventional breeding.

Materials and Methods

Three types of fusion experiments were done (see [3], [4] for details). The first [3] involved fusions of leaf protoplasts from an Ogura CMS cauliflower inbred (7642A) with hypocotyl or leaf protoplasts of male-fertile *B. rapa* (cv. Candle) carrying chloroplast-encoded resistance to the herbicide atrazine. Division of the cauliflower line, which regenerated well from protoplasts, was inhibited by treatment with 5mM iodoacetate. The *B. rapa* protoplasts produced callus but not plants. In some cases, the *B. rapa* protoplasts were treated with γ-irradiation (22-66 krad) prior to polyethylene glycol-induced fusion.

M. Terzi et al. (eds.), Current Issues in Plant Molecular and Cellular Biology, 171-176.

The plants recovered were examined by flow cytometry to determine nuclear DNA content, cellulose acetate electrophoresis of isozymes, and molecular analysis of organellar composition using hybridization of total DNA with probes that distinguished the parental lines. Phenotypic characters related to the organelles (CMS, atrazine-resistance, cold tolerance) were also assessed. Cauliflower plants that were diploid, CMS, and cold-tolerant were evaluated further in the greenhouse and the field. They were initially crossed with a white-flowered broccoli and then backcrossed to either broccoli or cauliflower. At each generation, plants were selected for good seed set as well as for horticultural characters.

The second fusion experiment [4] was similar, except that the male-fertile partner was an atrazine-sensitive cauliflower line (NY 3317) closely related to 7642A instead of *B. rapa.*

A third fusion experiment combined leaf protoplasts of the Ogura CMS cauliflower with irradiated (66 krads) leaf protoplasts from a male-fertile atrazine-resistant *B. napus* somatic hybrid (Temple and Earle, unpublished). This somatic hybrid (4420) was produced by Jourdan et al. [5] by fusing protoplasts of the Ogura CMS cauliflower and male-fertile atrazine-resistant *B. rapa*. The experiment could thus be described as a "back-fusion".

Results

OGURA CMS CAULIFLOWER + ATRAZINE-RESISTANT FERTILE *B. RAPA*

Over 190 shoots were recovered from 62 calli from seven fusion experiments of this type. About 40% of the plants examined had close to the diploid nuclear DNA content for cauliflower and only cauliflower isozyme patterns as well as some cytoplasmic components from *B. rapa*; these were considered cauliflower cybrids. Some of these were cold-tolerant (and also atrazine-resistant) and produced no pollen. After further greenhouse and field trials, plants from three calli (Table 1) were selected as the most promising. Two of the cybrid calli were recovered after irradiation of the *B. rapa* fusion partner, but the third (3142) was produced without irradiation. All three calli came from fusions in which the *B. rapa* protoplasts were isolated from hypocotyls rather than leaves. Plants from two of the calli (3142 and 3105) showed only radish mitochondrial DNA (mtDNA) regions for the probes tested. Plants from callus 3125 had mtDNA regions from both fusion partners.

TABLE 1. Characteristics of selected cauliflower cybrids recovered from fusions of protoplasts from Ogura CMS cauliflower and atrazine-resistant fertile *B. rapa*

				Organellar probes							
		No. of	pg DNA/	cp	mitochondrial					Male	Cold
Callus	krad[a]	Plants[b]	nucleus	s8	p5.0	p5.2	p9.7	b2.3	D23	fertility	tolerance
	Cauliflower		1.24	*rph*[c]	*ogu*[c]	*ogu*	*ogu*	*ogu*	*ogu*	-	-
	B. rapa		1.03	*rapa*[c]	*rapa*	*rapa*	*rapa*	*rapa*	*rapa*	+	+
3142	0	13[d]	1.34	*rapa*	*ogu*	*ogu*	*ogu*	*ogu*	*ogu*	-	+
3105	66	5[d]	1.34	*rapa*	*ogu*	*ogu*	*ogu*	*ogu*	*nd*[d]	-	+
3125	66	10[d]	1.38	*rapa*	*ogu*	*ogu*	*rapa*	*rapa*	*nd*	-	+

[a] γ-irradiation to the *B. rapa* protoplasts prior to fusion. [b] not all plants from each callus were tested in each assay. [c] *rph*: chloroplasts from *R. sativus*; *rapa*: from *B. rapa*; *ogu*: from Ogura CMS *R. sativus.* [d] not determined, but presumably *ogu* since this is the region correlated with the Ogura CMS phenotype in *Brassica* cybrids [6, 7, 8]

TABLE 2. Seed set on cold-tolerant Ogura CMS broccoli in the field (first backcross)

Pedigree			No. of plants with seed set			Line no.[d]
			Poor	Medium	Good	
89-3142[a]-9	x WF[b]	x 2393[c]	9	0	3	NY7865
89-3142-11	x WF	x 2403[c]	5	4	6	NY7866
89-3142-9	x WF	x 2403	6	0	6	NY7867
89-3142-9	x WF	x 2403	5	4	7	NY7868
89-3142-9	x WF	x 2393	7	1	8	NY7869

[a] 89-3142-9 and 89-3142-11 are plants from cauliflower cybrid callus 3142 (see Table 1). [b] a white flowered broccoli. [c] heat-tolerant broccoli inbred lines. [d] lines NY7865-7869 are bulks of seed from plants that set well in both field and greenhouse cages and backcrossed to 2393, 2403, or 5415 (another heat-tolerant broccoli).

Progeny of two plants from callus 3142 were backcrossed to heat-tolerant broccoli lines to produce broccoli lines designated NY7865, 7866, 7867, 7868 and 7869. When these were grown in the field with natural bee pollination, seed set varied substantially, from poor to good (Table 2). When cuttings from the plants with good seed set were pollinated by flies in greenhouse cages, all set well. Female fertility of subsequent generations has continued to be good.

Backcrosses of the original cybrids or their progeny to various good cauliflower lines produced cauliflower lines NY7720 and 7760 (from plants 10 and 11 from callus 3142), line 7721 (from plant 5 from callus 3105) and NY7719 (from plant 1 from callus 3125). NY7719 and 7720 showed some erratic reversion to male fertility in subsequent field generations and may not be suitable material for commercial use. Preliminary genetic studies suggest that the male fertility seen is not due to a cytoplasmic reversion event.

OGURA CMS CAULIFLOWER + FERTILE CAULIFLOWER

Fusions of leaf protoplasts from Ogura CMS and fertile cauliflower also produced several cold-tolerant Ogura CMS cybrids [4]. Plants from the three selected calli showed only Ogura CMS hybridization patterns for the five mitochondrial probes used (Table 3).

TABLE 3. Characteristics of selected plants from fusions of protoplasts from Ogura CMS cauliflower and atrazine-sensitive male-fertile cauliflower (modified from [4])

			Organellar probes							
	No. of	pg DNA/	cp	mitochondrial					Male	Cold
Callus	Plants	nucleus	s8	s10.1	p5.0	p5.2	b2.3	D23	fertility	tolerance
CMS cauliflower		1.24±.03	*rph*[b]	*ogu*[b]	*ogu*	*ogu*	*ogu*	*ogu*	-	-
Fertile cauliflower[a]		1.24±.03	*ole*[b]	*ole*	*ole*	*ole*	*ole*	*ole*	+	+
DF2.1	2	1.25-1.37	*ole*	*ogu*	*ogu*	*ogu*	*ogu*	*ogu*	-	+
DF2.2	1	1.35	*ole*	*ogu*	*ogu*	*ogu*	*ogu*	*ogu*	-	+
DF2.3	1	1.26	*ole*	*ogu*	*ogu*	*ogu*	*ogu*	*ogu*	-	+

[a] protoplasts from the fertile line were treated with 20.2 krad of γ-irradiation prior to fusion. [b] *rph*: from *R. sativus*; *ole*: from *B. oleracea*; *ogu:* from Ogura CMS *R. sativus*.

These cybrids were used in sexual crosses to produce atrazine-sensitive cold-tolerant Ogura CMS lines of cauliflower (NY8424, 8425, 8426, 8427) and broccoli (NY8464, 8465). These lines have good seed set, although the first generation of progeny from the cybrids varied greatly in female fertility.

OGURA CMS CAULIFLOWER + ATRAZINE-RESISTANT *B. NAPUS* SOMATIC HYBRID

Among the eight plants recovered from this fusion was one cold-tolerant CMS cauliflower cybrid (2829). It contained atrazine-resistant chloroplasts from the *B. rapa* line that had been used to create the original somatic hybrid. After pollination with several different cabbage lines, the cybrid cauliflower showed very good seed set. Further backcrosses have produced a number of different cold-tolerant Ogura CMS cabbage lines, some of which carry additional desirable features such as resistance to blackrot (*Xanthomonas campestris* pv. *campestris*), *Fusarium* yellows disease, and insects (by incorporation of the Glossy leaf character). The lines varied in pod length and curling; the latter reflects the male parent used.

Discussion

In this work several different fusion combinations produced some desirable *B. oleracea* vegetable cybrids from which lines suitable for hybrid production were developed by appropriate greenhouse and field selection and crosses. It was necessary to evaluate many fusion-derived plants to obtain ones that combined all the desired characteristics. Efficient systems for protoplast fusion and culture and evaluation of plant phenotypes were therefore important.

Although most of the cybrids were produced using irradiation to eliminate the nucleus of the male fertile fusion partner, it is notable that one cybrid callus (3142) developed without prior irradiation. In this case, the *B. rapa* nucleus was spontaneously eliminated, probably during fusion or at an early stage of protoplast culture.

Some of the cold-tolerant vegetable lines carry atrazine-sensitive chloroplasts (from *B. oleracea*), while other contain atrazine-resistant chloroplasts (from *B. rapa*). Atrazine-resistant chloroplasts cause a yield penalty in *B. napus* rapeseed materials (9), but they do not appear to be detrimental in *B. oleracea* vegetables, either at the seedling stage [10] or in mature plants [11]. Although triazine herbicides are not registered in the U.S. for use on *Brassica* vegetables and are no longer considered environmentally appropriate, the atrazine-resistant CMS lines might have some benefits for planting in soils with high levels of triazine residues.

Most of the cybrids advanced in the breeding program showed only Ogura CMS mtDNA hybridization bands with the probes used. However, more detailed analysis of a population of fertile and CMS *B. napus* somatic hybrids with a broader of probes and restriction enzymes showed that each had a unique mtDNA pattern [6]. This was true even for different plants regenerated from the same callus. It is therefore likely that the cybrids described here could be distinguished from each other and from ones produced elsewhere through use of sufficient probe/enzyme combinations.

Many of the cybrids that appeared to have the desired nuclear and organellar combinations were eliminated on the basis of poor initial seed set in the field tests. The variable seed set seen among progeny of a given cybrid (Table 2) may be due to segregation/elimination of extra nuclear DNA initially present in the cybrids. The fact that the nuclear DNA content of many of the cybrids was somewhat higher than that of seed-grown cauliflower material (Tables 1 and 3) is consistent with this concept. Fortunately, plants selected for good seed set continued to show good female fertility in later generations. With the exception of cauliflower lines NY7719 and 7720, whose progeny showed erratic male fertility in some trials, the male sterility of the selected lines has also been stable over at least four generations.

Some of the same fusion combinations used to obtain the cauliflower cybrids also produced *B. napus* somatic hybrids that were cold-tolerant and Ogura CMS. The latter were sexually backcrossed to Chinese cabbage and pak choi to produce *B. rapa* vegetable lines suitable for hybrid production [12]. The improved Chinese cabbage line (NY8481) is also resistant to blackrot.

The various lines described above are available for testing by signing a biological materials agreement. They have already been released to over 20 seed companies worldwide. Licensing for commercial use is also possible.

Now that cold-tolerant CMS *Brassica* vegetable lines are available, additional biological and legal issues require attention. One is the best strategy for transferring the improved cytoplasms to additional commercially valuable lines. In the case of annual crops, sexual backcrosses may be the most efficient approach; however, for biennials such as cabbage many years are required to return to the initial type. For that reason, fusion-mediated transfer is worth considering, particularly if the line of interest regenerates well from protoplasts. Direct transfer of the mtDNA sequence associated with the Ogura CMS [13, 14] may become an option in the future. Regulation of plants with novel combinations of organelles as genetically manipulated organisms in some countries will probably slow adoption of CMS-based hybrid systems there. In view of the patent issues related to fusion-derived *Brassica* materials, "finger-printing" of cytoplasms from different sources may also be important.

Acknowledgments

This work was supported in part by a grant from the Cornell Center of Advanced Technology in Biotechnology, which is sponsored by the New York State Science and Technology Foundation, a consortium of industries and the National Science Foundation, by the NSF/DOE/USDA Plant Science Center, a unit in the Cornell Biotechnology Program, and by USDA grant No. 85-CRCR-1-16.

References

1) Bannerot, H., Boulidard, L., Cauderon, Y., and Tempe, J. (1974) 'Cytoplasmic male sterility transfer from *Raphanus* to *Brassica*', Eucarpia Cruciferae Newsl. 2, 16.

2) Pelletier, G., Primard, C., Vedel, F., Chétrit, P., Rémy, R., Rouselle, P., and Renard, M. (1983) 'Intergeneric cytoplasmic hybridization in Cruciferae by protoplast fusion', Mol. Gen. Genet. 191, 244-250.

3) Stephenson, J.C. (1991) 'The effect of gamma irradiation on protoplast fusion between atrazine-resistant *Brassica campestris* and Ogura male sterile *B. oleracea*', M.S. Thesis, Cornell University, Ithaca, NY.

4) Walters, T.W., Mutschler, M.A., and Earle, E.D. (1992) 'Protoplast fusion-derived Ogura male sterile cauliflower with cold tolerance', Plant Cell Rep. 10, 624-628.

5) Jourdan, P.S., Earle, E.D., and Mutschler, M.A. (1989) 'Synthesis of male-sterile, triazine-resistant *Brassica napus* by somatic hybridization between cytoplasmic male sterile *B. oleracea* and atrazine-resistant *B. campestris*', Theor. Appl. Genet. 78, 445-455.

6) Temple, M., Makaroff, C.A., Mutschler, M.A., and Earle, E.D. (1992) 'Novel mitochondrial genomes in *Brassica napus* somatic hybrids', Curr. Genet. 22:243-249.

7) Bonhomme, S., Budar, F., Ferault, M., Pelletier, G. (1991) A 2.5 kb *Nco* I fragment of Ogura radish mitochondrial DNA is correlated with cytoplasmic male-sterility in *Brassica* cybrids', Curr. Genet. 19, 121-127.

8) Earle, E.D., Temple, M., Walters, T.W. (1992) 'Organelle assortment and mitochondrial DNA rearrangements in *Brassica* somatic hybrids and cybrids', Physiol. Plant. 85, 325-333.

9) Reboud, X., and Till-Bottraud, I. (1991) 'The cost of herbicide resistance measured by a competition experiment', Theor. Appl. Genet. 82, 690-696.

10) Christey, M.C., Makaroff, C.A., and Earle, E.D. (1991) 'Atrazine-resistant cytoplasmic male-sterile-*nigra* broccoli obtained by protoplast fusion between cytoplasmic male-sterile *Brassica oleracea* and atrazine-resistant *Brassica campestris*', Theor. Appl. Genet. 83, 201-208.

11) Christey, M.C., and Earle, E.D. (1994) 'Field testing of atrazine resistant cytoplasmic male sterile-Nigra broccoli obtained by protoplast fusion', J. Genet. & Breed. (in press)

12) Heath, D.W., Earle, E.D., and Dickson, M.H. (1994) 'Introgressing cold-tolerant Ogura cytoplasms from rapeseed into pak choi and Chinese cabbage', HortScience 29, 202-203.

13) Krishnasamy, S., and Makaroff, C.A. (1993) 'Characterization of the radish mitochondrial *orf* B locus: possible relationship with male sterility in Ogura radish', Curr. Genet. 24, 156-163.

14) Bonhomme, S., Budar, F., Lancelin, D., Small, I., Defrance, M.C., Pelletier, G. (1992) 'Sequence and transcript analysis of the *NcoI* Ogura-specific fragment correlated with cytoplasmic male sterility in *Brassica* cybrids', Mol. Gen. Genet. 235, 340-348.

RECOVERY OF CITRUS SOMATIC HYBRIDS TOLERANT TO *PHOMA TRACHEIPHILA* TOXIN, COMBINING SELECTION AND IDENTIFICATION BY RAPD MARKERS

Z.N. DENG, A. GENTILE, F. DOMINA, E. NICOLOSI, E. TRIBULATO AND A. VARDI[1]
Istituto di Coltivazioni arboree, University of Catania, Italy
[1]*Institute of Horticulture, The Volcani Center, Israel*

ABSTRACT. *Phoma tracheiphila* is a pathogen that causes a serious tracheomycotic disease in commercial lemon cultivars. Somatic hybridization offers the possibility to transfer the trait for tolerance into susceptible cultivars. One of the limitations of this approach is selection and identification of nuclear hybrids at an early stage. Protoplasts isolated from calli of 'Murcott' tangor (tolerant) and 'Messina' lemon (susceptible) were fused (PEG treatment). Prior to fusion, 'Murcott' protoplasts were treated with 0.25 mM iodoacetate. Fusion treated protoplasts were plated in liquid medium for 40 days and then exposed to 0.6 μM of *P. tracheiphila* toxin. Total DNA was extracted from the selected calli and corresponding embryos and analyzed by RAPDs. With the selected primers, hybrids were successfully identified. In the survey of coculture with 1 μM toxin, the hybrid callus lines showed tolerance to the toxin. Immunoblot analysis of the recovered hybrid calli indicated the presence of chitinase among the extracellular proteins. The combined methods, selection and identification by RAPD markers, resulted in 90% recovery of nuclear hybrids tolerant to the *P. tracheiphila* toxin.

Introduction

The severe tracheomycotic disease, mal secco, caused by the fungus *Phoma tracheiphila* (Petri) Kanc. et Ghik., seriously damages some citrus species in the Mediterranean region. One of the most important species growing in the south part of Italy is lemon [*Citrus limon* (L.) Burm. f.] and no genetic source of tolerance to mal secco has been found among its high quality commercial varieties. Up to now the available practices failed to control the disease. The two lemon cultivars that are tolerant to the disease, 'Interdonato' and 'Monachello', have poor bio-agronomic characteristics. Tolerant sources exist among other *Citrus* species such as sweet orange [*C. sinensis* (L.) Osbeck], mandarin (*C. reticulata* Blanco) and some hybrids like 'Murcott' tangor. Conventional breeding aiming to transfer this important trait to the high quality lemon varieties is restricted because of high polyembryonic characteristic of lemon and the above mentioned tolerant species, which leads to small F_1 zygotic population. Moreover, crosses between lemon and the other *Citrus* species have not produced lemon hybrids with high commercial value (Russo and Tribulato, 1986).

Somatic hybridization is an alternative approach for gene transfer. Since the first regeneration of citrus embryos from protoplasts (Vardi et al., 1975), many intergeneric and interspecific somatic hybrids have been obtained between leaf protoplasts, which are not able to divide, and callus protoplasts (Oghawara et al., 1985; for review see Gmitter et al., 1992). Selection and identification of nuclear hybrids, at an early stage, are the major limitations of somatic

M. Terzi et al. (eds.), Current Issues in Plant Molecular and Cellular Biology, 177-183.

hybridization in many plants. Various physical and chemical methods, as well as biochemical mutants have been used to separate heterokaryons (Bhojwani and Razdan, 1983). Sjödin and Glimelius (1989) using the toxin of *Phoma lingam* were able to obtain asymmetric somatic hybrids tolerant to the fungus. In citrus no special selection methods or agent has been utilized (Gmitter et al., 1992).

Identification of citrus nuclear hybrids was usually performed on plant morphological characteristics, isozyme and RFLP analyses (Grosser and Gmitter, 1990). Isozyme analysis in *Citrus* usually generated low level of polymorphisms, therefore in many cases more than one enzyme system is required. In the case of RFLP technique, relatively large quantity of plant material is needed to extract suffucient DNA for the analysis.

In previous studies we showed differential responses of citrus calli and protoplasts, originated from tolerant and susceptible genotypes, to the partially purified toxin produced by *P. tracheiphila* (Gentile et al., 1992a). Toxin - tolerant cell line was selected (Gentile et al., 1992b) by exposing susceptible nucellar callus of 'Femminello' lemon to sublethal toxin concentration. Moreover, the selected cell line and other tolerant genotypes were found to over-secrete pathogenesis related (PR) proteins as compare to the susceptible ones (Gentile et al., 1993; Gentile et al., in litteris). Recently we were able to identify in vitro and in vivo lemon mutants by RAPD (random amplified polymorphic DNA) markers. This technique has been used by Baird et al. (1992) and Xu et al. (1993) to characterize potato somatic hybrids.

The aim of the present study was to improve the fusion efficiency between protoplasts derived from embryogenic calli, to select somatic hybrid cell lines tolerant to *P. tracheiphila* toxin and to verify the nuclear hybrid nature of the selected colonies, at an early possible stage, by RAPD analysis.

Materials and Methods

PROTOPLAST ISOLATION AND FUSION TREATMENT

Protoplasts were isolated from embryogenic callus of 'Messina' lemon, susceptible, and 'Murcott' tangor, tolerant to mal secco, according to Vardi et al. (1982). 'Murcott' protoplasts were treated with 0.25 mM iodoacetate for 25 min followed by three washes with BH3 medium. The fusion was performed with polyethylene glycol (PEG) and the fusion treated protoplasts were plated in BH3 liquid medium in dark at 27 $\pm$ 1°C as described by Grosser and Gmitter (1990).

SELECTION WITH TOXIN TREATMENT

The filter sterilized partially purified toxin was isolated from *P. tracheiphila* as described in previous report (Gentile et al., 1992a). Three toxin concentrations (0.3, 0.6 and 0.8 μM) were added into the liquid protoplast culture medium immediately after the fusion treatment and 40 days after fusion when small cell masses were formed. In the control Petri dishes only liquid medium was added. The successive culture was according to Grosser and Gmitter (1990) and the induction of embryos and regeneration of plants were performed as described by Vardi et al. (1982).

RAPD ANALYSIS

DNA isolation. Total DNA was isolated according to Doyle and Doyle (1987) with some modifications. About 100 mg callus were taken from separated callus colonies and one piece of tissue (about 50-100 mg) was cut off from the embryos formed from corresponding callus colonies. Callus and leaves were also sampled from two parents. All the samples were homogenized with 800 μl warm (60°C) CTAB extraction buffer [3% CTAB, 1.4 M NaCl, 0.2% β-mercaptoethanol, 20 mM EDTA, 100 mM Tris-HCl (pH 8.0), 1% PVP-40]. The sample mixture was incubated at 60°C for 30 min and extracted with chloroform-isoamyl (24:1). Following a centrifugation, the upper phase was incubated with RNase A (10 μg/ml) at 37°C for 30 min. Then DNA was precipitated with 660 μl cold ethanol and 2.5 M ammonium acetate and resuspended with sterile distilled water. The DNA concentration was determined with TKO-100 DNA fluorometer (Hoefer Scientific).

Amplification. The amplification was carried out according to Williams et al. (1990). Reactions were performed in a Perkin Elmer Cetus GeneAmp PCR System 9600 programmed as follows: 5 min at 94°C for initial strand separation, 45 cycles of 15 sec at 94°C, 15 sec at 36°C and 80 sec at 72°C, followed by 10 min at 72°C for final extension. Fifty 10-mer primers (Operon Technologies Inc., Alameda, California) were tested to detect polymorphisms between two parents. The primers that generated distinct markers were used for identifying nuclear hybrids.

IN VITRO TESTS FOR TOLERANCE

Toxin tolerance. The hybrids verified by RAPD markers were further tested for tolerance to *P. tracheiphila* toxin compared with the parents. A small piece of callus was subcultured on MT medium containing 1μM toxin. After one month, the callus growth was estimated.

Immunoblotting. Callus of 3 hybrids and the corresponding parents were cultured in suspension for determining the release of chitinase to the medium. The experiment was carried out as described previously (Gentile et al., 1993).

Results

SELECTION OF FUSION PRODUCTS

'Murcott' protoplasts treated with 0.25 mM iodoacetate were plated at the same density and conditions as the fusion treated protoplasts. In all the iodoacetate treated control plates a complete inhibition of cell division was recorded.

a b c d

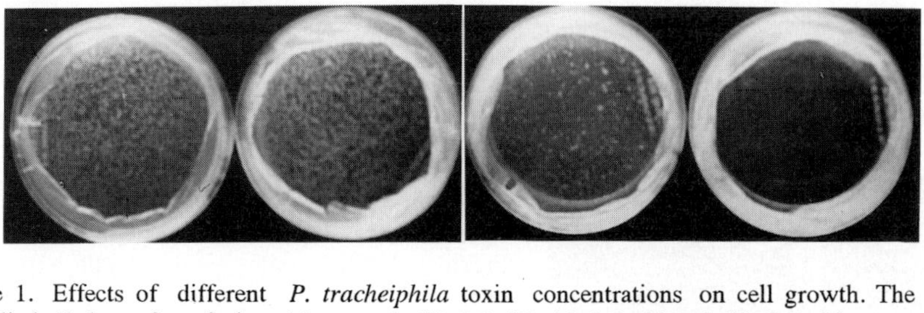

Figure 1. Effects of different *P. tracheiphila* toxin concentrations on cell growth. The toxin was applied 40 days after fusion. (a) control; (b) 0.3 μM; (c) 0.6 μM and (d) 0.8 μM.

The selection effect of *P. tracheiphila* toxin strongly depends on the concentration and the time application. When the toxin (0.3, 0.6 and 0.8 μM) was added immediately after the fusion treatment, all the protoplasts died within a week. The best selection effect was obtained when the toxin was added to the culture medium 40 days after the fusion, when masses of 20-30 cells had already formed. In 0.8 μM toxin treatment very few colonies were observed, while in 0.3 μM cells grew as well as those of the control. In 0.6 μM toxin treatment about 25% of the total treated cell masses formed colonies (Fig.1).

According to these results, the suitable selection was to treat cell masses with 0.6 μM toxin 40 days after fusion.

HYBRIDS SCREENING BY RAPD ANALYSIS

Polymorphism detection between two parents. All the 50 primers generated amplification products ranging from 150 - 3500 bp. Each primer produced a profile composing from 2 (A20) to 15 bands (A07). High level of polymorphisms were obtained between 'Messina' and 'Murcott' with 45 of the tested primers (Fig. 2).

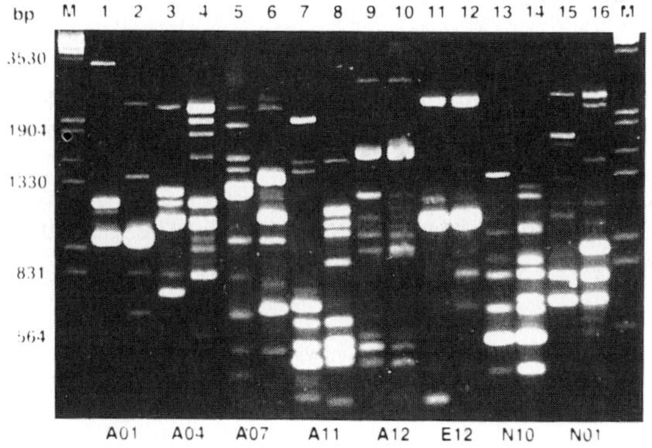

Figure 2. RAPD profiles generated by 8 primers showing high level of polymorphisms between 'Murcott' (odd-numbered lanes) and 'Messina' (even-numbered lanes). *M* DNA size markers (*EcoR* I/*Hind* III - digested DNA).

As RAPDs usually reveal dominant - null alleles (Williams et al., 1990), a primer useful for identification of hybrids should be able to generate at least one dominant amplification product from each parent. From the 45 primers which generated polymorphisms, 10 useful primers were selected for screening hybrids.

Identification of somatic hybrids by RAPD markers. Among the 10 selected primers, A20 was particularly useful for identification of hybrids in the present combination because it produced

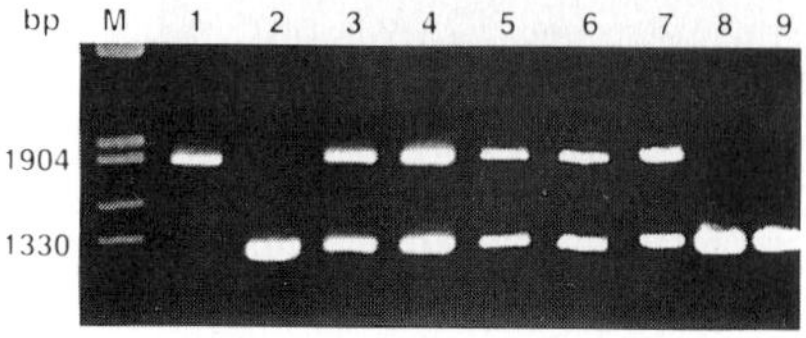

Figure 3. RAPD profiles generated with A20 primer. *Lane 1* 'Murcott', *lane 2* 'Messina', *lane 3-7* somatic hybrids, *lane 8-9* homokaryon types of 'Messina', *M* DNA size markers (*EcoR* I/*Hind* III - digested DNA).

one unique fragment (A20-1330) for 'Messina' and another (A20-1900) for 'Murcott'. With this primer, 20 - 30 callus colonies from 0.3, 0.6 μM toxin treatments and the control were analyzed (callus from 0.8 μM treatment were not enough for the analysis). Two out of 20 and 1 out of 25 were identified as hybrids from 0.3 μM treatment and from control, respectively, while 90 % of the colonies (26 out of 30 tested) resulted in hybrids. The hybrids showed the combination RAPD profile of the two donors (Fig.3). To ascertain the hybrid character, verified with primer A20, other 9 primers were further used to analyze 10 hybrids. Combination patterns were obtained with all the primers except two (A01 and N20), with which one band from 'Murcott' was lost in the hybrid profiles (Fig.4). Successively, DNA was extracted from embryos derivated from 20 corresponding hybrid callus lines for RAPD analysis and same results were obtained as in the callus DNA analysis.

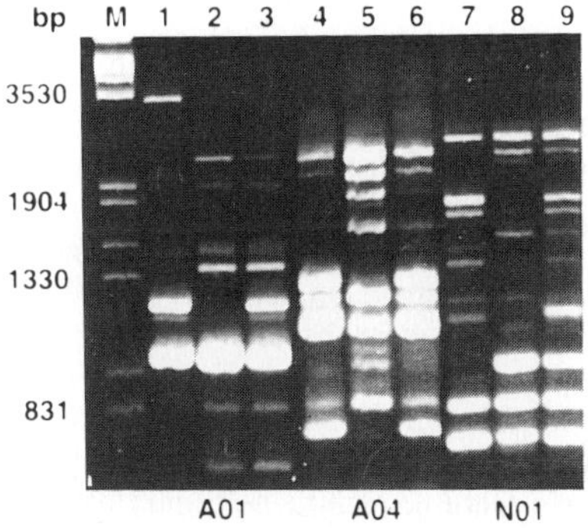

Figure 4. RAPD profiles obtained with 3 primers. The hybrid showed the combination pattern of the two parents. One fragment (A01-3500) presented in 'Murcott' was absent in the somatic hybrid. *Lanes 1,4, 7* 'Murcott', *lanes 2, 5, 8* 'Messina', *lanes 3, 6, 9* somatic hybrid, *M* DNA size markers (*EcoR* I/*Hind* III - digested DNA).

IN VITRO TESTS FOR TOLERANCE

The 'Murcott' and hybrid calli grew well on MT + 1 μM toxin, while those of 'Messina', including the homokaryons from the fusion, were inhibited from growing on the same toxin medium (Fig.5).

In the western blotting of extracellular proteins from suspension cultures of three hybrids and 'Murcott', using antichitinase specific for tobacco, immunoreactivity was observed at molecular weight of 34 kDa, while in the analysis of proteins from the susceptible parent 'Messina' lemon, no cross reactivity was detected (Fig.6).

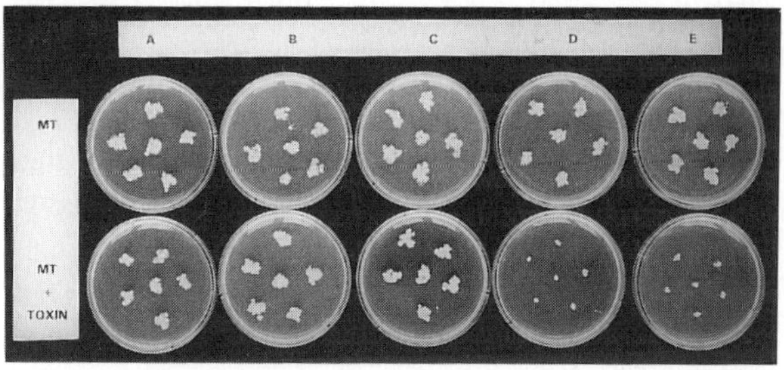

Figure 5. Coculture test, with 1 μM toxin, of calli of two somatic hybrids (B and C), one homokaryon of 'Messina' genotype (D) compared with the parents 'Murcott' (A) and 'Messina' (E).

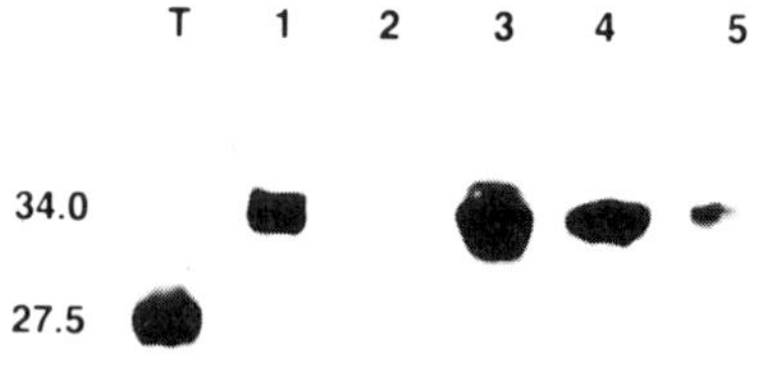

Figure 6. Western blot, using tobacco antichitinase, of extracellular proteins collected from suspension culture. *Lane T* TMV, *lane 1* 'Murcott', *lane 2* 'Messina', *lane 3-5* somatic hybrids.

Discussion

Fusion with PEG treatment resulted, generally, a very small proportion (0.5 - 10%) of heterokaryon cells in the culture mixture (Bhojwani and Razdan, 1983). An effective selection and identification scheme of somatic hybrids, at an early stage, is of key importance for successful fusion experiments. Most of the selection methods, which have been used in other plants, have not been successfully applied in citrus cell fusion. Citrus somatic hybrids were obtained by using embryogenic callus protoplasts as one partner and leaf protoplasts, unable to divide, as another. The inhibition of the embryogenesis of unfused and self fused protoplasts by 0.6 M sucrose (Ohgawara et al., 1985) or by a long period of suspension culture (Grosser and Gmitter, 1990), have enabled the discrimination between homokaryons and heterokaryons. However, these methods could not inhibit the embryogenesis of lemon and 'Tarocco' orange protoplasts isolated from embryogenic callus (data not shown). Utilization of iodoacetate treatment (to the tolerant partner), when both protoplast partners derived from calli, and selection of the fusion products with *P. tracheiphila* toxin were proven to be highly effective tools for heterokaryons recovery.

With RAPD technique, somatic hybrids could be identified at an early stage. The high level of polymorphisms, revealed between the two genotypes employed, allowed quick screening of somatic hybrids with just one primer. This technique seems more advantageous over isozyme and RFLP analysis and, in addition, it saves space and labour .

One band presented in 'Murcott' was absent in the RAPD profiles of somatic hybrids when the template DNA was amplified with 2 out of 10 primers. This probably resulted from an amplification competition between the DNAs of the two partners and, in this case, DNA from one partner might be ineffectively amplified, which led to the lack of visualization of the given fragment in the hybrids. The same phenomenon was observed using 1:1 mixture of the two parental DNAs for the amplification.

In previous studies it was shown that the responses of calli and protoplasts to *P. tracheiphila* toxin and release of chitinase in the culture medium were in full agreement with the in vivo behaviour of citrus genotypes to the mal secco disease (Gentile et al., 1992a, 1993). Even if it still remains enigmatic why selection for tolerance to a toxin, whose mode of action is presently unknown, would result in cell cultures that over-secrete PR proteins, the use of such selection system to recover toxin tolerant somatic hybrids seems to be very useful to achieve the main goal of the present study. Moreover, the system presented here will allow the identification of somatic hybrids within the lemon group and the reduction of backcrosses required to select a high quality lemon tolerant to mal secco.

References

Baird, B., Cooper-Bland, S., Waugh, R., DeMaine, M., and Powell, W. (1992) 'Molecular characterization of inter - and intraspecific somatic hybrids of potato using randomly amplified polymorphic DNA (RAPD) markers', Mol. Gen. Genet. 233, 469-475.

Bhojwani, S.S. and Razdan, M.K. (1983) 'Plant Tissue Culture: Theory and Practice', Elsevier, Amsterdam.

Doyle, J.J. and Doyle, J.I. (1987) 'A rapid DNA isolation procedure for small quantities of fresh leaf tissue', Phytochemical Bulletin 19, 11-15.

Gentile, A., Tribulato, E., Continella, G., and Vardi, A. (1992a) 'Differential responses of citrus calli and protoplasts to culture filtrate and toxin of *Phoma tracheiphila*', Theor. Appl. Genet. 83, 759-764.

Gentile, A., Tribulato, E., Deng, Z.N., and Vardi, A. (1992b) 'In vitro selection of nucellar lemon callus and regeneration of plants tolerant to *Phoma tracheiphila* toxin', Adv. Hort. Sci. 4, 151-154.

Gentile, A., Tribulato, E., Deng, Z.N., Galun, E., Fluhr, R., and Vardi, A. (1993) 'Nucellar callus of 'Femminello' lemon, selected for tolerance to *Phoma tracheiphila* toxin, shows enhanced release of chitinase and glucanase into the culture medium', Theor. Appl. Genet. 86, 527-532.

Gmitter, F.G.Jr., Grosser, J.W., and Moore, G.A. (1992) 'Citrus', in F.A. Hammerschlag and R.E. Litz (eds.), Biotechnology of Perennial Fruit Crops, C.A.B. International, pp.335-369.

Grosser, J.W. and Gmitter, F.G.Jr. (1990) 'Protoplast fusion and citrus improvement', Plant Breed. Rev. 8, 339-374.

Ohgawara, T., Kobayashi, S., Ohgawara, E., Uchimiya, H., and Ishii, S. (1985) 'Somatic hybrid plants obtained by protoplast fusion between *Citrus sinensis* and *Poncirus trifoliata*', Theor. Appl. Genet. 71, 1-4.

Russo, F. and Tribulato, E. (1986) 'La ricerca e il miglioramento genetico degli agrumi', Il Recente Contributo della Ricerca allo Sviluppo dell'Agrumicoltura Italiana, Cagliari pp.23-29.

Sjödin, C. and Glimelius, K. (1989) 'Transfer of resistance against *Phoma lingam* to *Brassica napus* by asymmetric somatic hybridization combined with toxin selection', Theor. Appl. Genet. 78, 513-520.

Vardi, A., Spiegel-Roy, P., and Galun, E. (1975) 'Citrus cell culture: isolation of protoplasts, planting densities, effect of mutagens and regeneration of embryos', Plant Sci. Lett. 4,231-236.

Vardi, A., Spiegel-Roy, P., and Galun, E. (1982) 'Plant regeneration from *Citrus* protoplasts: variability in methodological requirements among cultivars and species', Theor. Appl. Genet. 62,171-176.

Williams, K.G.K., Kubelik, A.R., Livak, K.J., Rafalski, J.A., and Tingey, S.V. (1990) 'DNA polymorphisms amplified by arbitrary primers are useful as genetic markers', Nucleic Acid Res. 18, 6531-6533.

Xu, Y.S., Clark, M.S., and Pehu, E. (1993) 'Use of RAPD markers to screen somatic hybrids between *Solanum tuberosum* and *S. brevidens*', Plant Cell Reports 12, 107-109.

Towards producing bloat-safe *Medicago sativa* L. through protoplast fusion

Y.-G. Li[1,2], G. J. Tanner[1], and P. J. Larkin[1]

[1]CSIRO, Division of Plant Industry, Canberra, ACT 2601, Australia
[2]Xinjiang Institute of Biology and Soil Sciences, CAS, Wulumuqi, Xinjiang 830011, China

Abstract. Intergeneric somatic hybrid plants have been produced for the first time between two sexually incompatible legume species. Iodoacetamide-inactivated *Medicago sativa* (lucerne, alfalfa) leaf protoplasts were fused with gamma-irradiated *Onobrychis viciifolia* (sainfoin) suspension-cell protoplasts. Sainfoin is bloat-safe because of foliar condensed tannins. Following optimised electrofusion, functional complementation permitted only the heterokaryons to survive. A total of 570 putatively heterokaryon-derived plantlets survived transplantation to soil. Southern analysis using an improved total-genomic probing technique showed low levels of sainfoin specific DNA in 43 out of 158 tested regenerants. Chromosome counting revealed aneuploidy (2n = 30, 33-78) in the majority of the hybrids. Pollen germination tests indicated that about 60% of the hybrids were fertile. Tannin determination by using dimethylaminocinnamaldehyde showed a hybrid containing 0.03% condensed tannins in dry matter of leaves. The feasibility of using tannin-positive asymmetric somatic hybrids for bloat-safe lucerne is under investigation.

Introduction

Bloat is a world-wide problem for ruminants, particularly for beef cattle and dairy cows grazing on bloat-causing forage species such as lucerne and white clover. It is caused by the formation of stable proteinaceous foams in the rumen which prevent gas escape and result in sickness and death [4]. Experimental evidence both *in vitro* and *in vivo* has indicated that condensed tannins (CT) play a substantial role in preventing bloat. CT can suppress and destabilise the foams by protein precipitation [6] and microbial inhibition [9].

Foliar CT is a common trait in bloat-safe legume species such as sainfoin and *Lotus* species. Extensive surveys within the *Medicago* genus have not identified any plant possessing foliar CT [e.g. 5, 11, 16]. Transfer of the genes encoding CT biosynthesis to CT-negative species may confer bloat-safety. The formation of tannin-protein complex may also increase the fraction of dietary protein that escapes rumen degradation. Up to 40% total dietary protein can be wasted due to microbial degradation and ammonia formation in the rumen [3, 15].

To date there have been only two reports on hybrid calli produced by intergeneric somatic hybridisation of *Leguminosae* species, *Lotus corniculatus* with *Medicago sativa* [13] or with *Glycine max* [7]. Total-genomic probing has been applied only in grass species for detection of sexual hybrids [e.g. 8, 17]. We report here the production of intergeneric somatic

M. Terzi et al. (eds.), Current Issues in Plant Molecular and Cellular Biology, 185-190.

hybrid plants in the *Leguminosae*. This is the first application of the total genomic probing technique in legume species for detection of somatic hybrids. We also report a highly sensitive method using dimethylaminocinnamaldehyde (DMACA) for rapid CT screening. This is a preliminary report on the first CT-positive lucerne somatic hybrid plant.

Materials and Methods

Plant materials, protoplast isolation, fusion and regeneration, DNA extraction and total-genomic probing, chromosome and pollen analyses were as described in Li et al. [10].

Tannin assay

Green and healthy leaves were collected, freeze-dried, ground into powder and sieved through 0.6 mm mesh. The powder was extracted four times (30 min for three times, then once overnight) with 70% acetone containing 0.1% ascorbic acid at 4°C in the dark. The residue was retained for bound-CT assay. The extracts were combined, mixed with 0.75 vol. of diethyl ether, then set at 4°C in the dark for separation of the two phases. Only the lower phase contained free-CT and was mixed with 0.5 vol. of methanol and 0.3 vol. of DMACA solution, reacting at room temperature for 20 min then measured at 643 nm.. The DMACA solution contained 1% (w/v) dimethylaminocinnamaldehyde (DMACA) in a cold mixture of ethanol and 6 M HCl (1:1 v/v). Lucerne was used as the blank. The butanol/HCl-SDS method [18] was used in the residue analysis.

A qualitative CT assay was also applied to the fresh leaves. The leaves were decolourised by soaking overnight in Carnoy's solution containing glacial acetic acid and ethanol (1:3 v/v), and stained with the DMACA solution as above. The CT-rich cells stain dark blue.

Results and Discussion

Fusion treatments and plant regeneration

When lucerne leaf protoplasts were treated with 0.3-10 mM iodoacetamide (IOA) or sainfoin suspension cell protoplasts were gamma-irradiated at 600-1,000 Gy, neither could initiate sustained divisions, though both remained viable for a short time.

The optimum electrofusion conditions were determined as follows: AC: 150 V_{p-p} /cm, 2.3 MH_z, 30 s; DC: 1,400 V/cm, one pulse, 10 μs; protoplast density: 3.5×10^5/ml; fusion buffer: 1 mM $CaCl_2$, 530 mM glucose, 0.5 mM MES, pH 5.9. The heterokaryon frequency was 4.8 ± 1.2% (mean ± SE, n = 5), which stands for the heterokaryon number 1 h after fusion versus the parental protoplast number before fusion. Similarly the viable heterokaryon frequency was 1.6 ± 0.3% as estimated using fluorescein diacetate 24 h after fusion. Since an average of 3.1×10^6 protoplasts were fused in each fusion experiment, about 50,000 viable heterokaryons were produced, which were predominantly from binary fusions.

No colonies were found in 5×10^6 pretreated, but unfused, control protoplasts, indicating the prefusion treatments had successfully prevented parental escapes, while the functional complementation did allow heterokaryons to survive.

In total 706 putatively heterokaryon-derived plantlets were regenerated, of which 570 survived transplantation into soil. Most of these regenerant plants have shown various degrees of morphological differences from the fusion parents. However, all the plants, as expected from the gamma treatment of the sainfoin protoplasts, resembled the fusion "recipient", lucerne, which has pinnately trifoliolate leaves. Fig. 1 shows one confirmed asymmetric hybrid. The hybrid leaf shape appeared to be an intermediate type between the two parents.

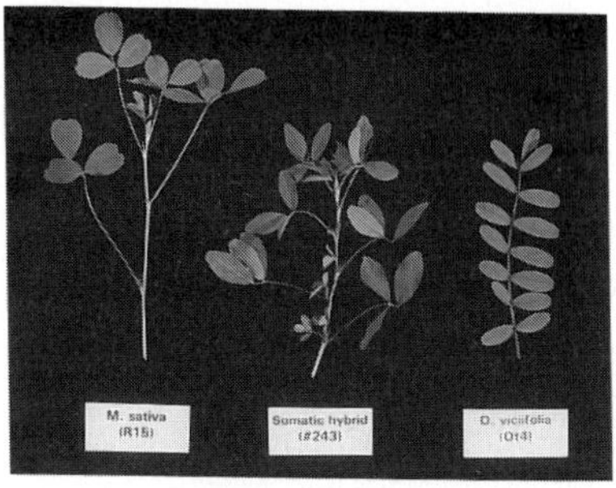

Fig. 1 Morphological comparison between vegetative shoots of an asymmetric somatic hybrid (middle) and the parents, lucerne (left) and sainfoin (right).

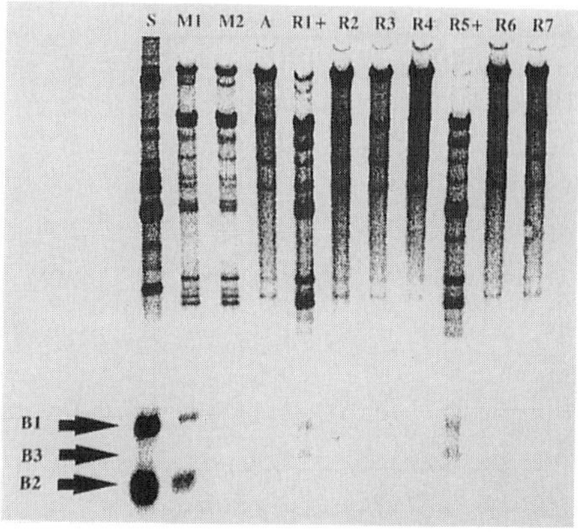

Fig. 2 Southern analysis using the total genomic probing method* to confirm hybridity of the putative hybrid plants regenerated following fusion of lucerne (R15) and sainfoin (Ot4) protoplasts.
(lanes) A: R15 total genomic DNA 2 μg. M1 and M2: R15 total genomic DNA 2 μg plus 10 ng and 1 ng Ot4 total genomic DNA respectively. S: Ot4 total genomic DNA 100 ng. R1-R7: Regenerants tested, 2 μg total genomic DNA each.
(arrows) B1-2: Ot4 specific DNA bands 1-2. B3: a novel DNA band.
* *Hinc* II digests of total genomic DNA on Hybond N+ charged nylon membrane were hybridised at 65°C with 50% formamide and washed with 0.1 × SSC at 72°C for 1 h, probe concentration 20 ng/ml

DNA analysis

When Southern-transferred filters were probed with labelled sainfoin genomic DNA and blocked with 400 fold excess of unlabelled lucerne genomic DNA, the very high stringency (65°C with 50% formamide in blocking and probing; 0.1 × SSC at 68-72°C for 1 h in washing) enhanced the detection sensitivity to about 0.001. That is, one part (2 ng) of sainfoin total genomic DNA is discernible among one thousand parts (2 μg) of lucerne total genomic background. Under lower stringency control the detection sensitivity decreased remarkably.

The lower part (around 300-1,000 base pairs) of the lanes consistently showed no hybridisation if only lucerne DNA was loaded but two bands if sainfoin DNA was present (B1, B2; Fig. 2). Those hybridisation signals were also proportional to the amount of sainfoin DNA in lucerne/sainfoin mixtures (Table 1). That part of the lane was therefore chosen as the diagnostic area for hybrid detection in the DNA analysis of the putative hybrid plants.

Of the 570 regenerated putative hybrids 158 were analysed using total-genomic probing. An example of such DNA analysis is shown in Fig. 2, in which two hybrids were confirmed. R1 and R5 showed detectable levels of at least one sainfoin specific DNA band (B1) and perhaps a novel band (B3) that is not found in either the fusion parental species.

Hybridisation signals corresponding to approximately 6 ng and 9 ng of sainfoin genomic DNA were detected by the PhosphorImager in hybrids R1 and R5 respectively (Table 1). In Table 1 some signal was also detected in R2, R3, R4, and R6 with the PhosphorImager but this was not discernible on X-ray film and not significantly above the background noise (lane A). Signals of sainfoin genomic DNA of one nanogram or so were not significantly above background noise (e.g. M2 compared to A, Table 1).

Table 1. Sainfoin total genomic probe hybridising to the diagnostic area (Band 1 to Band 2 of Fig. 2) as quantitated by signal volume integration with PhosphorImager.

Lane	Sainfoin DNA loaded (ng)	Signals relative to S lane (%)
S	100	100
M1	10	10.4
M2	1	0.9
A	0	0.8
R1+	-	6.2
R2	-	2.2
R3	-	1.7
R4	-	1.7
R5+	-	9.6
R6	-	1.4
R7	-	0.6

Table 2. Chromosome analysis of lucerne (R15), sainfoin (Ot4) and some confirmed hybrids.

	Mitotic chromosome number (2n)	Number of hybrids
R15	32	-
Ot4	48	-
Group 1	32	6
Group 2	30, 33-44	7
Group 3	54-78	23

Among 158 tested regenerants 43 were confirmed asymmetric hybrids containing low levels of sainfoin specific DNA.

Extrachromosomal DNA generally accounts for only 0.1-0.01% of the total DNA of higher plants [14] and it is highly conserved between species. The excess lucerne total genomic DNA would have blocked the extrachromosomal DNA and made it undetectable.

Tannin assay

Among 40 tested hybrid plants, one hybrid, #487, contained CT in foliage. The free-CT content of this hybrid was 0.02% and the bound-CT content was 0.01% in dry matter of leaves. Detailed characterisation of this hybrid and CT is in progress and the feasibility of using such a hybrid in bloat prevention is under investigation.

The DMACA qualitative assay was rapid and convenient, but it seemed not as sensitive as the quantitative assay. The hybrid did not show obvious CT-containing cells.

Other analyses of the regenerants

Of the 43 hybrid plants obtained, 36 were analysed for their mitotic chromosome numbers. A dramatic variation in the chromosome number was observed (Table 2).

The chromosomes of lucerne and sainfoin are not readily distinguishable morphologically. However one average sainfoin chromosome in the presence of 32 average lucerne chromosomes would be the equivalent of about 30 ng sainfoin genomic DNA in 1 μg lucerne genomic DNA sample. This would give a hybridisation signal stronger than any of the asymmetric hybrids in this study. The signals of the confirmed hybrids could be generated by the inclusion of about 0.1 to 0.6 of an average sainfoin chromosome in a lucerne genome (2n = 32). The extra-chromosomes, e.g. in Group 3 (Table 2), would therefore be of lucerne origin. This is consistent with the lucerne-like morphology of all the hybrid plants. So in Table 2 the hybrids with 32 chromosomes (Group 1) might be due to translocation(s) between lucerne and sainfoin chromosome(s). The hybrids with 30 to 44 chromosomes (Group 2) might result from fusion of one or two lucerne protoplasts with one sainfoin protoplast followed by random deletion of chromosomes. The hybrids with 54 to 78 chromosomes (Group 3) might result from fusion of two or three lucerne protoplasts with one sainfoin protoplast followed by extensive loss of chromosomes. Quite a few plants in the last group also resembled octoploid lucerne in general morphology.

Thirty seven of the 43 confirmed hybrid plants were analysed for male fertility. The control lucerne plants had pollen germination above 65%, which is regarded as normal fertility. Twenty four plants were of normal fertility; 7 were of low fertility (20 ± 15% pollen germination) and 6 were effectively male sterile (below 5% germination).

Conclusion

The protoplast fusion described above has produced a number of highly asymmetric somatic hybrid plants which resemble lucerne in general morphology but carrying small amounts of sainfoin DNA. The majority of these plants were male fertile. One hybrid also contained CT in foliage. Although there was only a trace level of CT, it is important evidence of the transfer of the desired character.

The breeding strategy reported in this paper has four distinctive advantages: firstly, the protoplasts of donor species are not required to be regenerable, and this non-regenerability may even favour the production of highly asymmetric hybrids; secondly, selective markers are not required in the fusion parents for hybrid selection, since the pretreatments using IOA and gamma rays efficiently prevented parental escapes while functional complementation allowed survival of heterokaryons; thirdly, cloned species-specific probes are not required for the hybrid confirmation, since the Southern analysis using total-genomic probing is versatile and sensitive enough to detect asymmetric hybrids containing small amounts of donor DNA; and fourthly, a simple and rapid detection method is available for screening of the desired character, foliar CT. DMACA offers the specificity of vanillin to flavanols but without the attendant disadvantages of vanillin tests [12]. The DMACA quantitative assay presented in this paper is highly sensitive and can detect 0.01% CT in dry matter. This is about 100 times more sensitive than the commonly used vanillin/leaf-squash assay, which explains why CT was not detected in the hybrid, #487, in the past [10].

Compared with employing cloned species-specific probes, the total-genomic probing is more simple and economical in terms of time and labour. Moreover, a cloned probe may be only homologous to a few chromosomes or chromosome segments [1]. A total-genomic probe involves all the species-specific sequences and using such a probe is similar to using a mixture of all the possible cloned probes. The detection sensitivity of total-genomic probing was greatly improved by using stringent hybridisation as well as stringent washing. This is different from conventional stringency control which uses permissive hybridisation followed by stringent washing [2]. However, the sensitivity needs to be further improved if detection of extremely asymmetric hybrids is required.

Acknowledgments. The financial support to Y.-G. Li by the Australian International Development Assistance Bureau and the Australian Dairy Research & Development Cooperation. We thank Drs. Warren McNabb, Rudi Applels, Paul Whitfeld, John Watson, and T. J. Higgins for helpful discussions of DNA analysis.

References

1. Anamthawat-Jónsson K, Schwarzacher T, Leitch AR, Bennett MD, Heslop-Harrison JS (1990) Discrimination between closely related *Triticeae* species using genomic DNA as a probe. Theor Appl Genet 79:721-72
2. Anderson MLN, Young BD (1985) Quantitative filter hybridization. In: Harris BD, Higgins SJ (eds) Nucleic acid hybridization. IRL Press, Oxford Washington DC, pp 73-111
3. Barry TN, Manley TR (1983) The role of condensed tannins in the nutritional value of *Lotus pedunculatus* for sheep. British J Nutrition 51:493-504
4. Clarke RTJ, Reid CSW (1974) Foamy bloat of cattle. A review. Journal of Dairy Science 57:753-785
5. Goplen BP, Howarth RE, Sarkar SK, Lesins K (1980) A search for condensed tannins in annual and perennial species of *Medicago, Trigonella,* and *Onobrychis.* Crop Sci 20:801-804
6. Jones WT, Mangan JL (1977) Complexes of the condensed tannins of sainfoin (*Onobrychis viciifolia* Scop.) with fraction 1 leaf protein and submaxillary mucoprotein, and their reversal by polyethylene glycol and pH. J Sci Food Agric 28:126-136
7. Kihara M, Cai K-N, Ishikawa R, Harada T, Niizeki M, Saito K (1992) Asymmetric somatic hybrid calli between leguminous species of *Lotus corniculatus* and *Glycine max* and regenerated plants from the calli. Japan J Breed 42:55-64
8. Le HT, Armstrong KC, Miki B (1989) Detection of rye DNA in wheat-rye hybrids and wheat translocation stocks using total genomic DNA as a probe. Plant Mol Biol Rep 7:150-158
9. Lees GL (1984) Tannins and legume pasture bloat. In: Report of the 29th Alfalfa Improvement Conf, Lethbridge, Alberta, Canada, 1984 p 68
10. Li Y-G, Tanner GJ, Delves AC, Larkin PJ (1993) Asymmetric somatic hybrid plants between *Medicago sativa* L. (alfalfa, lucerne) and *Onobrychis viciifolia* Scop. (sainfoin). Theor Appl Genet 87:455-463
11. Marshall DR, Broue P, Grace J, Munday J (1981) Tannins in pasture legumes. 2. The annual and perennial *Medicago* species. Aust J Exp Agric Anim Husb 21:55-58
12. McMurrough I, McDowell J (1978) Chromatographic separation and automated analysis of flavanols. Analytical Biochemistry 91:92-100
13. Niizeki M, Saito K-i (1989) Callus Formation from protoplast fusion between leguminous species of *Medicago sativa* and *Lotus corniculatus.* Japan J Breed 39:373-377
14. Pelletier G, Vedel F, Belliard G (1985) Cybrids in genetics and breeding. Hereditas Suppl 3:49-56
15. Pilgrim AF, Gray FV, Weller RA, Belling CB (1970) Synthesis of microbial protein from ammonia in the sheep's rumen and the proportion of dietary nitrogen converted into microbial nitrogen. British J Nutrition 24:589-598
16. Rumbaugh MD (1979) A search for condensed tannins in the genus *Medicago.* In: Agron Abstr p 75
17. Schwarzacher T, Leith AR, Bennett MD, Heslop-Harrison JS (1989) *In situ* localization of parental genomes in a wide hybrid. Ann Bot 64:315-324
18. Terrill TH, Rowan AM, Douglas GB, Barry TN (1992) Determination of extractable and bound condensed tannin concentrations in forage plants, protein concentrate meals and cereal grains. J Sci Food Agric 58:321-329

SELF AND CROSS POLLINATION OF OVULES IN TEST TUBES

M.ZENKTELER
Laboratory of General Botany, Faculty of Biology, Adam Mickiewicz University, 61-713 Poznań, Poland

ABSTRACT. Direct in vitro pollination of ovules can be successfully applied to those species which ovaries have large placentas covered with many ovules. This method enables to overcome some pre- and postzygotic barriers of incompatibility. In vitro pollination of single ovules (cut off from the placenta) is a much more complicated procedure and it was successfully applied to only few species. The present paper contains the results concerning: a) development of embryos and plants after self and cross pollineted placentas and single ovules of some species representing Brassicaceae, Caryophyllaceae, Fabaceae, Liliaceae, and Solanaceae families; b) germination of pollen grains and entrance of pollen tubes into the embryo sacs after distant crosses.

1. Introduction

Success of in vitro fertilization of ovules is determined by pollen viability, ovules receptivity and environmental influences. In vitro pollination of ovules and subsequently development of embryos have been obtained in 57 species representing 14 families (1,2,3,4). It is best to experiment with those species in which the ovaries are large and contain many ovules. Therefore, the best results were obtained with species representing the following families: Brassicaceae, Caryophyllaceae, Liliaceae, Papaveraceae, Primulaceae and Solanaceae. Direct in vitro pollination of ovules, irrespectively of the applied method, enables to overcome pre-fertilization barriers which include: failure of pollen germination, slow pollen tube growth and arresting of pollen tube in the style and ovaries. The post-fertilization barriers are much more difficult to overcome as those are responsible for the abortion of the zygote or immature embryo which usually starts to degenerate a few days following fertilization. Mostly the barriers of hybrid embryo survival occur during the mid to late stages of embryogenesis.
This paper reviews some results of in vitro self and cross pollination of placentas and single ovules in species representing 5 families.

2. The general applied techniques

In vitro fertilization can takes place in one of the following ways: 1) pollination of stigmas on the pistil (e.g. Trifolium species); 2) direct pollination of ovules through the cut opened upper part of ovary, or a perforation made on the side ovary wall - placental pollination (e.g. Brassicaceae and Fabaceae species); 3) ovary wall removed and ovules are directly pollinated - - placental pollination (e.g. Caryophyllaceae or Solanaceae species); 4) pollination of single ovules cultured directly on medium (e.g. Fabaceae species); 5) grafted style method in which pollen grains first germinate in a compatible style which is cut after several hours and later

M. Terzi et al. (eds.), Current Issues in Plant Molecular and Cellular Biology, 191-199.

attached to the ovary of an incongruent mother plant (e.g.Lilium species); 6) fertilization of single isolated egg cells (up to date only Zea mays).

This paper concerns only results obtained on placental pollination and pollination of single ovules of some species representing Brassicaceae, Caryophyllaceae, Fabaceae, Liliaceae and Solanaceae families. Fresh flower buds (1-2 days prior to anthesis) were sterilized in 70% ethanol followed by chlorine water and later on washed with sterile water. The next step was to cut off the perianth, stigma and style, however, in same cases, depending on the species, only a part of the corolla and calyx were removed. Pollen grains extracted from florets having dehisced anthers were transferred directly on the ovules of the opened ovaries. Pollinated placentas were cultivated on Murashige and Skoog medium (5) supplemented with various concentrations of witamins, cytokinins, auxins, 3% sucrose and 0.8 - 0.9% Difcobacto agar. Embryos of the torpedo stage were dissected from the enlarged ovules and transferred to MS medium constaining only IAA in the concentration of 0.5 mg/l.

Single ovules of Fabaceae species were cut off from the opened ovaries and transferred onto the agar medium. In our experiments we have usually used the same media as for placental pollinated ovules with the only exception that they were composed of a higher concentration of sucrose, usually from 6 - 8%. Several hours after in vitro pollination, ovules were removed from culture, stained in aniline blue stain (Merck Anilinblau WC, 0.1% in aqueous 0.1M K_3PO_4) and examined for pollen tube growth using epifluorescence microscope. For embryological studies the enlarged ovules were fixed, embedded in paraffin, sectioned and stained in haematoxylin or crystal fiolet.

3. Results

3.1. SELF POLLINATION OF PLACENTAS

Kanta et. al. (6) from the University of Delhi were the first to report the successful in vitro pollination, fertilization and development of embryos in ovules cultured jointly with a segment of placenta of Papaver somniferum. Placental pollination enables germination of pollen on naked ovules, pollen tubes entrance into the embryo sac, fertilization and the development of embryo and endosperm. Thus in vitro the process of embryogenesis may occur similarly as in vivo, however the size of placentas and the composition of the medium may play an important role in this process. The role of the medium is especially important in those species which contain poor placentas or when single ovules are cultured directly in the medium. The nature of the medium plays also a very important role in the culture of ovules at the stage of early embryogenesis, especially after distant crosses. This is still a very complicated problem, especially in finding the most suitable environmental conditions for the induction of hybrid globular embryos. As reviewed in the literature (2,4,7,8) fully formed seeds can be obtained easily in some species of such families as Caryophyllaceae, Brassicaceae, Liliaceae, Primulaceae and Solanaceae. In those species placentas are usually cultured with portion of the ovary wall and with a part of the calyx. In some plants, particularly those representing Brassicaceae, ovary wall and the calyx have a beneficial nutritive effect on the embryo development.

During the past 3 years intensive investigations have been undertaken on various species of the Brassicaceae. Some of these results have been already published (9) but some more ones have been obtained only recently. Pollen grains were put on the cut opened upper part of the ovary where during the next few hours they started to germinate, pollen tubes entered inside the embryo sacs and later on endosperm and embryo development was noticed. The number of developing seeds varied in different species e.g. in Brassica napus, B.oleracea var.sabellica, B.cretica, B.pekinensis, Sinapis arvensis usually several seeds have been found in each ovary. In other species e.g. Arabis caucassica, Sinapis alba and Lunaria officinalis seeds containing mature embryos were developed sporadically, usually 1-2 seeds in certain ovaries. Histological analysis of those ovaries revealed that in some ovules small globular embryos degenerated shortly after fertilization. Successes on self pollination of placentas of Brassicaceae are shown in Table 1.

Table 1. Self pollination of placentas of Brassicaceae*

Species	Immature embryos	Viable seeds
Brassica napus cv. Bronowski		+
B. oleraceae var. sabellica		+
B. campestris		+
B. cretica		+
B. pekinensis		+
Arabis caucassica		+
Alliaria officinalis		+
Diplotaxis tenuifolia		+
Moricandia arvensis		+
Sinapis arvensis		+
Sinapis alba		+
Hesperis matronalis	+	
Bunias orientalis	+	
Lunaria officinalis	+	
Camelina sativa	+	

* List of species representing other families is included in paper already published (2).

Attempts at placental pollination have been made also with species of Fabaceae (10) but usually no seeds were set. In soybean, fertilization of ovules has been described but the development of seeds was not reported (11). An interesting contribution is that of Leduc et al. (12) in which viable seeds of Trifolium repens were obtained by placing pollen grains on the perforation made in the ovary wall or in the base of the style. This method appears to be efficient in by-passing the gametophytic system of self-incompatibility. Similarly achievements on overcoming self-incompatibility barriers had been obtained earlier in Petunia hybrida (13), P.axilars (14) and Lilium (15).

3.2. FERTILIZATION OF SINGLE OVULES

There are only a few reports describing succes in culturing in vitro pollinated ovules. Kameya et al.(16) inoculated pollinated ovules of Brassica oleracea and obtained two seeds. Stewart (8) obtained interspecific hybrid embryos between Gossypium hirsutum and G.arboreum after placing pollen grains on the micropylar end of ovules. We were attending to fertilize single ovules of several species of Fabaceae, mainly the following ones: Trifolium pratense, T.rubens, T.medium, T.repens, Galega officinalis, Lathyrus odoratus, Mellilotus albus, Cytisus albus, Coronilla coronata, Pisum sativum. The media used were mainly composed of MS macro- and microelements but with 6-8% of sucrose and with various concentrations of witamins and aminoacids. Pollen grains of T.pratense, T.repens, T.medium and Galega officinalis when selfed directly on the ovules germinated sporadically and no clear evidence was obtained of the entrance of pollen tubes into the micropyles. Pollen of some other species e.g. Lathyrus odoratus, Mellilotus albus, Cytisus albus, Coronilla coronata, Trifolium rubens germinated abundantly when selfing or even on ovules of Trifolium species (Figs.1-2). Several days after self pollinating of ovules of T.rubens and cross pollinating ovules of T.pratense with Cytisus albus some of them enlarged (Figs.3-4) and in those ones globular embryos were found. These experiments are still in progress as it seems that the technique of in vitro pollination of single ovules is possible. The list of pollination of single ovules in Fabaceae is depicted in Table 2.

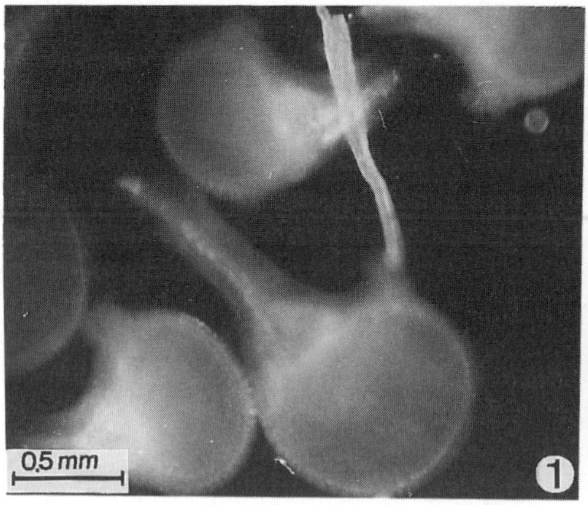

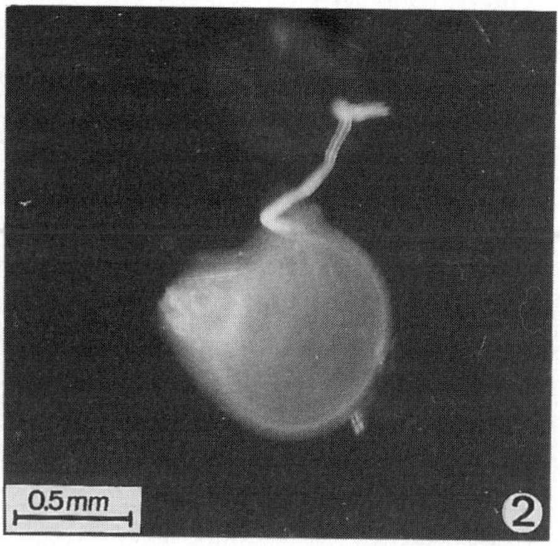

Fig.1. Pollen tube of Lathyrus odoratus entering the micropylar region of ovule of Trifolium pratense. **Fig.2.** Pollen tube of Cytisus albus entering the micropylar region of ovule of Trifolium medium.

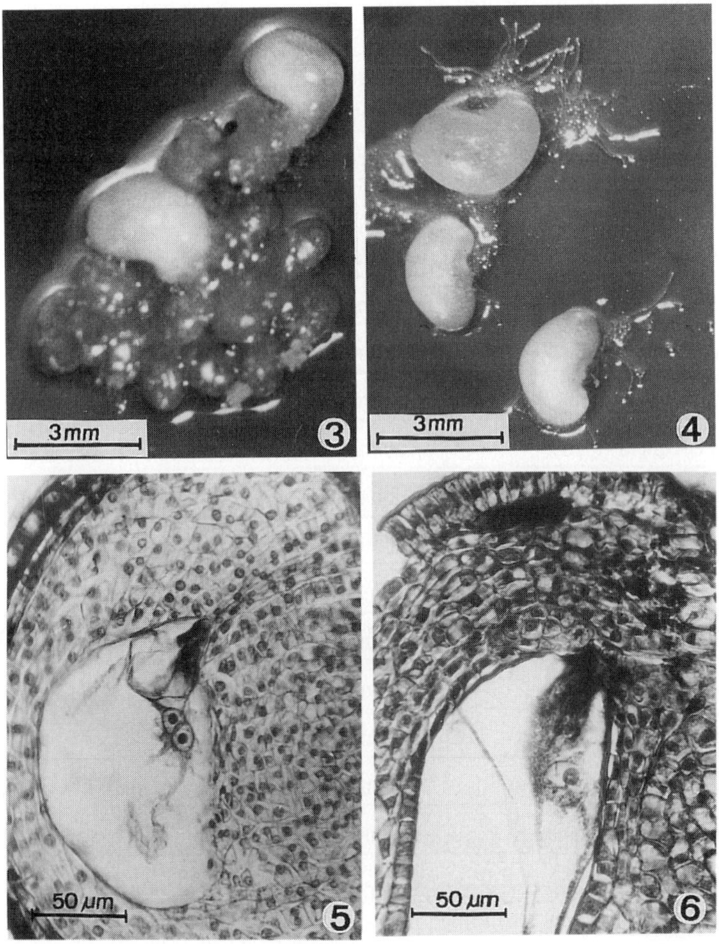

Fig.3. Enlarged ovules of T. rubens cultured directly on medium, 10 days after self pollination. **Fig.4.** Enlarged ovules of T. pratense cultured directly on medium, 11 days after pollination with pollen of Cytisus albus. **Fig.5.** Entrance of pollen tube of Nicotiana tabacum inside the embryo sac of Melandrium album, 8 hours after placental pollination. **Fig.6.** Entrance of pollen tube of Coronilla coronata inside the embryo sac of Trifolium medium, 14 hours after pollination of single ovules.

Table 2. Self and cross pollination of single ovules in Fabaceae

Species	Pollen germination	Globular embryo
Trifolium pratense (self)	sporadic	-
T. medium (self)	sporadic	-
T. repens (self)	sporadic	-
T. rubens (self)	abundant	+
Mellilotus albus (self)	abundant	-
Cytisus albus (self)	abundant	-
Coronilla coronata (self)	abundant	-
Lathyrus odoratus (self)	abundant	-
Pisum sativum (self)	abundant	-
Galega officinalis (self)	sporadic	-
T. rubens x Lathyrus odoratus	abundant	-
T. rubens x Cytisus albus	abundant	+
T. pratense x Pisum sativum	abundant	-
T. pratense x Mellilotus albus	abundant	-
T. medium x T. pannonicum	sporadic	-

3.3. CROSS POLLINATION OF PLACENTAS

The main significance of the method of in vitro pollination of ovules is that it provides a mean of bypassing the pollen/style rejection response present in self-incompatible or inter-specifically incompatible plants. Applying a similar technique as for self pollinating ovules, interspecific, intergeneric and even interfamiliar crosses were tried in 68 combinations, of these in 26 combinations viable seeds forming hybrid seedlings were obtained (17). The most satisfactory development of hybrid embryos and plants were found among species of Caryophyllaceae, Nicotiana, Brassica, Lilium and Gossypium. It is worth pointing out that the entrance of pollen tubes inside the embryo sacs occurs even between distant crosses e.g. Melandrium album x Nicotiana tabacum (Fig.5) or Trifolium repens x Coronilla coronata (Fig.6). In some combinations of crosses male gametes were noticed inside the egg cells e.g. M.album x Petunia hybrida. In some other crosses e.g. M.album x Cucubalus baccifer, Brassica napus x Bunias orientalis (Fig.7) or B.oleracea var. sabellica x Bunias orientalis only proembryos developed, while the endosperm was absent. Conversely, in ovules of M.album x Datura stramonium, M.album x Cerastium arvense only few endosperm nuclei developed. In some other crosses e.g. B.napus x Synapis alba both globular embryos and endosperm developed (Fig.8) and in some other ones as e.g. B.pekinensis x S.alba the full process of embryogenesis and endosperm formation was noticed (Figs.9a,b). Our recent observations on distant pollinations of ovules have also shown that germination of pollen grains may occur even between species representing various families e.g. abundant pollen grain of Lathyrus odoratus (Fabaceae) germinated on ovules of Nicotiana tabacum (Solanaceae) or pollen of Petunia hybrida (Solanaceae) on ovules of Pisum sativum (Fabace). It is worth to notice that no germination of pollen among those species occurred after stigma pollination. The list of interspecific and intergeneric placental pollination in various families is in press (17).

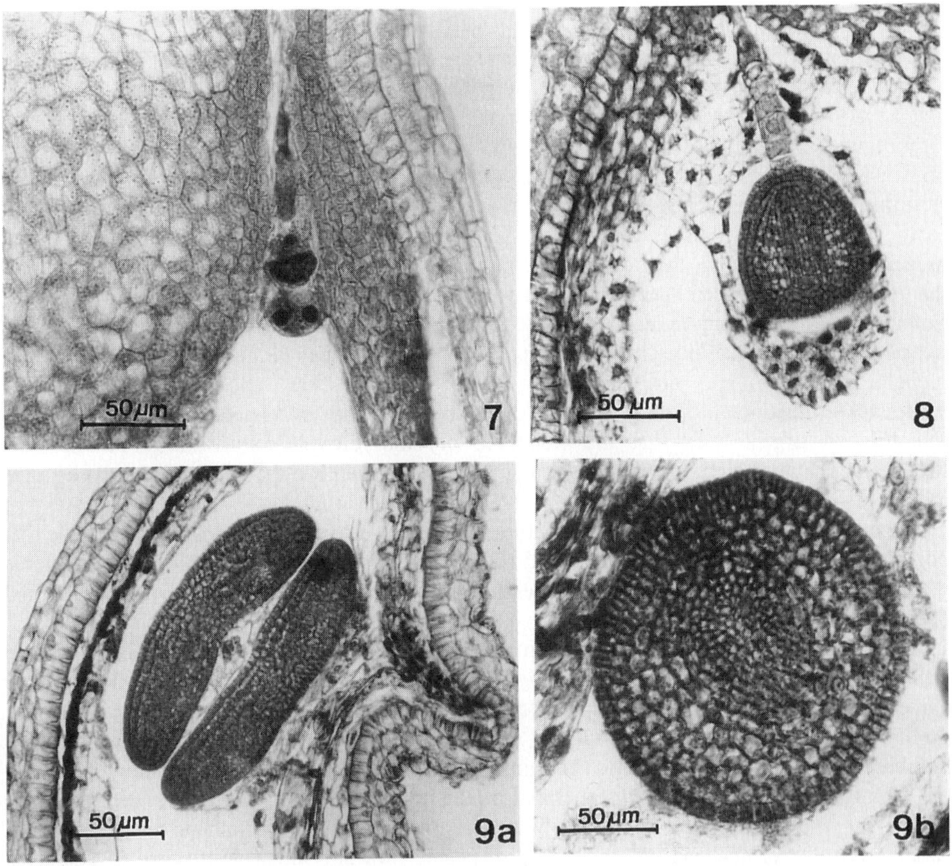

Fig.7. Proembryo of Brassica napus x Bunias orientalis, 5 days after placental pollination, endosperm not present. **Fig.8.** Globular embryo and endosperm of B. napus x Sinapis alba, 8 days after placental pollination. **Fig.9a-9b.** Longitudinal (a) and cross (b) sections of torpedo shaped embryo of B.oleracea var. pekinensis x Sinapis alba, 18 days after placental pollination.

4. Discussion

The method of application of pollen grains onto the surface of ovules can lead to fertilization. Since the stigma and style are by-passed, pollen tubes of one species can, therefore have a direct access to the micropyle of another from which in nature they would be excluded. From the work to date concerning the application of placental pollination, it is appearent that this technique enables to overcome some self and cross incompatibility barriers. As shown in our paper barriers hindering the fusion of gametes between distant species are not ovewhelming and fertilization can occur between species belonging to different genera. Much more dificult is to overcome the postfertilization barriers, particularly in those cases when the two gametes derived from different parents, a situation referred to as "genomic incompatibility". Once fertilization has occurred, postfertilization barriers may restrain hybrid embryo growth. In several instances as e.g. in Caryophllaceae, Brassicaceae, Solanaceae and Liliaceae, these barriers can be overcome using embryo or ovule rescue methods. Unfortunately, the number of hybrids obtained in test tubes is still very limited as there are no suitable methods for culturing hybrid globular embryos. Often the oveles which were fertilized in test tubes lacked endosperm and hybrid embryos stopped their development at early globular stage. Successful embryo rescue depends upon developmental stage at which breackdown occurs and the availability of culture medium. The nutritional requirements of the embryo vary between species and between different stages of embryo development. Embryo culture techniques are the most significant experimental procedure to overcome hybrid embryo inviability in wide crosses. Although the results on culture hybrid globular embryos are not satisfactory, this approach has enormous potential for successful utilization of distant relatives of cultivated plants in various breeding programs. There is an urgent need that the technique of embryo/ovule culture should be improved to rescue the hybrid embryos at very young stage. Only then can the value of method of wide crosses in test tubes be appreciated.

The method of placental pollination enables to pollinte immature ovules. It is propable that pollination of those ovules can evoke certain deviations from the normal developmental patterns of megasporogenesis and megagametogenesis. Thanks to this procedure it is possible to find out answers on the following questions:
1) do pollen grains are capable to germinate on immature ovules; 2) does a pollen tube is capable to enter inside an ovule which is at the stage of megasporogenesis; 3) do the male gametes are capable to fuse with the female nuclei inside the immature embryo sac.Our preliminary experiments carried out on some species of Solanaceae and Caryophyllaceae families show that pollen grains germinate on immature ovules and that pollen tubes enter the immature female gametophytes. Based on the up to date observations we are not able to find out whether the male gametes fuse with the nuclei of an immature embryo sac. It appears that deeper analysis of the specific pecularities of pollination in vitro of immature ovules will provide answers on the questions mentioned above.

References

1. Sladky, Z. Griga, M. and Juroch, J. (1982) "Verification of some fertilization methods in vitro in further plants", Scripta Fac.Sci.Nat.Univ.Purk.Brun. 12,8 (Biologia) 371-376.
2. Zenkteler, M. (1990) "In vitro fertilization and wide hybridization in higher plants", Crit.Rev.Plant Sci. 9, 267-279.
3. Zenkteler, M. (1990) "In vitro fertilization of some species of Brassicaceae", Plant Breeding 105, 221-228.
4. Zenkteler, M. (1991) "Ovule culture and test tube fertilization", Med. Fac. Landbouww. Rijksuniv. Gent 56 (4a), 1403-1410.
5. Murashige, T. Skoog, F. (1962) "A revised medium for rapid growth and bioassay with tobacco tissue culture", Physiol.Plant. 15, 473-497.
6. Kanta, K. Rangaswamy, N.S. and Maheshwari, P. (1962) "Test-tube fertilization in flowering plant", Nature (London) 194, 1214-1217.
7. Zenkteler, M. (1980) "Intraovarian and in vitro pollination", Int.Rev.Cytol.Suppl. 11B, 137-156.
8. Stewart, J. McD. (1981) "In vitro fertilization and embryo rescue", Environ.Exp.Bot. 21, 301-315.
9. Zenkteler, M. (1992) "Wide hybridization in higher plants by applying the method of test tube pollination of ovules", in Y.Datte., C.D.Dumas and A.Gallais (eds.), Reproductive Biology and Plant Breeding, Springer Verlag, pp. 205-213.
10. Zubkova, M. and Sladky, Z. (1975) "The possibility of obtaining seeds following placental pollination in vitro", Biol. Plant. 17, 276-280.
11. Tilton, V.R. and Russel, S.H. (1983) "In vitro pollination and fertilization of soybean Glicine max (L) Herr.(Leguminoseae)", in Mulcahy, D.L. and Ottaviano, E. (eds.) Pollen: biology and implication for plant breeding. Elsevir Sci. Publ. Amsterdam pp.281-286.
12. Leduc, N., Dougls, G.C. and Monnier, M. (1992) "Methods for nonstigmatic pollination in Trifolium repens (Papilionaceae): seed set with self-and-cross-pollination in vitro", Theor. Appl. Genet. 83, 912-918.
13. Niimi, Y. (1970) "In vitro fertilization in self-incompatible plant of Petunia hybrida", J. Jpn. Soc. Hort. Sci. 39, 345-352.
14. Rangaswamy, N.S. and Shivanna, K.R. (1967) "Induction of gametic compatibility and seed formation in axenic culture of diploid self-incompatibile species of Petunia", Nature 216, 237-239.
15. Van Tuyl, J.M. Straathof, T.P. Bino, R.J. and Kwakkenbos, A.A.M. (1988) "Effect of three pollinatiom methods on embryo developemnt and seedset in intro and interspecific crosses between seven Lilium species", Sex.Plant.Reprod. 1, 119-123.
16. Kameya, T. Hinata,K. and Mizushima,U. (1966) Fertilization in vitro of excised ovules treated with calcium chloride in Brassica oleraceae L.", Proc. Jap. Acad. 42, 165-167.
17. Zenkteler, M. (1994) "In vitro pollination and embryo rescue", in K.L.Giles and J.Prakash (eds), Current Plant Sciences and Biotechnology in Agriculture, Kluwer Academic Publishers (in press).

EMBRYOGENESIS AND PLANT REGENERATION FROM *IN VITRO* FUSED ISOLATED GAMETES OF MAIZE

ERHARD KRANZ AND HORST LÖRZ
Institute for General Botany, Applied Molecular Biology of Plants (AMP II)
University of Hamburg
Ohnhorststrasse 18
D-22609 Hamburg, Germany

ABSTRACT. *In vitro* fertilization techniques using single isolated gametic protoplasts of maize are described. Pairs of isolated sperm and egg cell protoplasts were fused electrically. Unequal distribution of cell organelles was observed followed by unequal first cell division in most of the zygotes. The individually cultured zygotes developed via direct embryogenesis to globular structures, proembryos, transition-phase embryos and finally to phenotypically normal and fertile maize plants. These plants were examined karyologically and morphologically, and the segregation ratios of the kernel color of the selfed fusion plants were investigated, indicating their hybrid nature.

The fusion of the female and male gametic protoplasts of maize is mediated also by a high calcium concentration and a high pH followed by growth to multicellular structures. The non-electrical fusion method enables to study adhesion, recognition and fusion of higher plant gametes. However, it has to be investigated, whether gamete specific receptors are involved in these processes. The likelyhood of nongamete specific fusion is indicated by the observation of spontaneous fusion of two egg cell protoplasts as well as by the fusion of two sperm cell protoplasts.

Possible applications of the methods of *ex ovulo* fertilization for fundamental and applied studies are discussed.

1. Introduction

Biotechnological procedures have been applied in addition to sexual crossings for a number of years. Cell and tissue culture methods such as somatic cell genetics, anther and microspore culture are extensively used for cell manipulation. Methods of ovule culture and *in vitro* pollination/fertilization of flower explants, ovaries and ovules have also been used to overcome cases of self-and cross-incompatibility (Zenkteler 1990; 1992). An extension of these approaches is the technique of *in vitro* fertilization (IVF) with isolated single gametes (Kranz et al., 1990; Kranz et al., 1991a). Pairs of single maize (*Zea mays* L.) sperm and egg cell protoplasts were fused electrically. The individually cultured zygotes developed via direct embryogenesis to globular structures, proembryos, transition-phase embryos and finally to phenotypically normal and fertile maize plants. These plants were examined karyologically and morphologically, and the segregation ratios of the kernel color of the selfed fusion plants were investigated, indicating their hybrid

M. Terzi et al. (eds.), Current Issues in Plant Molecular and Cellular Biology, 201-206.

nature (Kranz and Lörz, 1993). Additionally, karyogamy was observed in the fusion products (Faure et al., 1993; Tirlapur et al., 1994). Recently evidence was presented for chemically induced fusion of isolated maize gametic protoplasts (Kranz, 1993; Faure et al., 1994). The fusion of the gametic protoplasts, mediated by a high calcium concentration and a high pH was followed by the formation of multicellular structures (Kranz and Lörz, 1994). Herein details of the IVF technique are described and its relevance for basic research and breeding is discussed.

2. Isolation and selection of male and female gametes, synergids and central cells

Maize sperm cells were isolated from pollen, collected from freshly dehisced anthers and used immediately or after storage in a moistened atmosphere for some hours (Kranz and Lörz, 1990; Kranz et al., 1991a). Sperm cells were obtained from pollen grains by osmotic shock. Viable egg cells and nongametic cells of the embryo sac were isolated using methods of enzymatic degradation of cell walls in a combination with microdissection. The embryo sac containing ovular tissues were treated with an enzyme mixture (0.75% pectinase, 0.25% pectolyase Y23, 0.5% hemicellulase and 0.5% cellulase Onozuka RS) for about 30 min and 24°C (Kranz and Lörz, 1993). The yield of isolated embryo sacs and egg cells could be improved by reducing the enzyme concentrations and by shortening the time of incubation of ovule-pieces in the enzyme mixture and especially by manual dissection using microneedles (Kranz et al., 1990; Kranz, 1992). Glass needles, prepared with a microforge, were used for the dissection. The yield was finally determined by the manual isolation step. Routinely about 5 egg cells per 20 ovule pieces can be obtained. Depending on the "quality" of the plant material, up to 40 egg cells can be isolated by microdissection within 2 to 3 hours. The "bottleneck" to get a high number of isolated egg cells for performing IVF by mass fusion can be circumvented by the use of single cell techniques which include individual selection, transfer and fusion of gametes.

Egg cell protoplasts are taken up by a microcapillary with a tip opening of about 200 μm, sperm cells by a capillary with an opening of about 20 μm. The two sperm cell protoplasts can be differentiated from the smaller vegetative nucleus. Sperm, egg and central cell protoplasts are selected by the microcapillaries and transferred by use of a hydraulic system and a computer-controlled dispenser/dilutor (Koop and Schweiger, 1985a; Kranz, 1992).

3. *In vitro* fusion of male and female gametes

IVF was performed either electrically by applying an electrical pulse, or chemically, e.g. mediated by a high calcium concentration and high pH (Kranz et al., 1990, 1991a; Kranz, 1993; Kranz and Lörz, 1994). Gametes are used as protoplasts. IVF might be performed therefore by any other method, available for the fusion of somatic protoplasts. Maize sperm cells are able to fuse after osmotic shock mediated isolation without any further treatment with other sperm, egg and nongametic protoplasts (Kranz et al., 1991a,b; Kranz and Lörz, 1994). Comparable with somatic protoplast fusion, the osmolarity of the fusion medium is an important factor for the fusion of the protoplasts of the egg cell, the sperm cell and of the nongametic cells of the embryo sac. This has been observed for the electrical fusion as well as for the fusion in a fusiogenic medium. For the electrical fusion, 540 - 570 mosmol/kg H_2O and for the high calcium/ high pH mediated fusion 400 - 430 mosmol/kg H_2O was used. Pairs of an egg and a sperm cell protoplast as well as of a sperm cell and a central cell protoplast of maize were fused mainly electrically, as described by Kranz et al. (1991a,b) and Kranz (1992). The fusions were performed on a coverslip in microdroplets which are overlayered by mineral oil. Each microdroplet consists of 2µl mannitol solution. The individual electrical fusion was performed under microscopic observation. The electrodes were fixed to a support mounted under the condensor of an inverted microscope. Cell fusion is induced by single or multiple negative dc-pulses after dielectrophoretic alignment on one of the electrodes (Koop and Schweiger, 1985b). The frequencies of electrofusion of egg and sperm cell protoplasts are high (mean frequency 85% of about 2000 fusion products). The individual gamete fusion can also be performed in a microdroplet of a fusiogenic medium, e.g. containing a high calcium concentration and at a high pH (0.01 - 0.05 M $CaCl_2$; pH 11.0) (Kranz, 1993; Kranz and Lörz, 1994). This medium was developed originally for the fusion of somatic tobacco protoplasts (Keller and Melchers, 1973). Microneedles were used to accomplish the alignment of the two gametic protoplasts. The mechanical alignment, however, leading to adhesion of the gametic protoplasts, was more time consuming than the dielectrophoretic alignment procedure. As with somatic protoplasts, no deleterious effects in the gametic cells could be observed after the treatment. Furthermore, fusion products of the gametic protoplasts were able to develop to microcalli. It has to be investigated, whether gamete specific receptors are involved in the adhesion and fusion or if these processes are occurring only in a nongamete specific manner, comparable with the fusion of somatic protoplasts. The likelyhood of nongamete specific fusion is indicated by the observation of fusion of two egg cell protoplasts as well as the fusion of two sperm cell protoplasts (Kranz and Lörz, 1994). IVF methods performed with fusiogenic media, e.g. mediated by sodium nitrate (Power et al., 1970), calcium (Keller and Melchers, 1973), or even spontaneously induced fusion (Ito and Maeda, 1973), may allow studies of adhesion, fusion and possible mechanisms to prevent polyspermy. However, compared to the procedure using a fusiogenic medium in obtaining artificial zygotes, electrically performed IVF was much more efficient.

4. Culture of *in vitro* produced zygotes and plant regeneration

Zygotic embryogenesis and plant regeneration from *in vitro* produced zygotes was achieved by electrofusion mediated IVF (Kranz and Lörz, 1993). High frequency multicellular structure formation was obtained using a wide range of different maize lines as donor plants for sperm and egg cells. A prerequisite for the successful culture of the zygotes is the use of a suitable microculture system. A "Millicell/feeder suspension"-system proved to be capable of achieving sustained growth of the zygotes. The non-morphogenic feeder suspension cells have derived from excised zygotic maize embryos (Kranz et al., 1991a). The fusion products were cultivated in a liquid media on a semipermeable membrane of "Millicell-CM" dishes, which were placed in a fast growing feeder suspension. Sustained growth of the multicellular structures could only be achieved with feeder cells that were cytoplasmically rich. Modified MS-media (Murashige and Skoog, 1962) containing 1.0 mg/l 2,4-D and 0.02 mg/l kinetin or 2.0 mg/l 2,4-D without kinetin as well as modified N_{6ap}-media (Rhodes et al., 1988) with 1.0 mg/l 2,4-D were found to be suitable.

Karyogamy was demonstrated within one hour after gamete fusion (Faure et al., 1993; Tirlapur et al., 1994) followed by a distinct unequal distribution of the cell organelles towards the cell periphery. The first cell division occurred between 40 - 60 hours after cell fusion. The early development of the zygotes, e.g. the unequal first division of most zygotes, the formation of multicellular structures, proembryos and transition-phase embryos resemble those observed *in vivo*

(Van Lammeren, 1981, 1986; Schel et al., 1984). The development of the early stages of the *in vitro* maize zygote following the first unequal cell division and a decrease of cell size, were found to occur without detectable orientation pattern, comparable with the observations made *in vivo* (Randolph, 1936). Embryo structures are formed with a pronounced axis, consisting of meristematic tissue at the top and a suspensor at the base. They were transferred onto solid regeneration media, usually 10 - 12 days after gamete fusion, when they reach a size of about 0.4 x 0.5 mm. Coleoptiles and plantlets developed about 22-34 days after gamete fusion. The further development of *in vitro* created early embryos was accompanied by an overgrowth and occasionally by polyembryony. These structures were comparable to those obtained by somatic embryogenesis (e.g. Lu et al., 1982). Phenotypically normal and fertile maize plants were obtained within 99 - 171 days after gamete fusion. In four independent experiments 11 fertile plants were regenerated from 28 fusion products. The hybrid nature of the fusion plants has been confirmed (Kranz and Lörz, 1993). The F_2 plants were phenotypically normal and no mutational effects have been observed.

In order to optimize the growth of later embryo stages, modified regeneration media may have to be developed, e.g. special maturation media. Whereas the maize zygotes are developing under comparable conditions with a high frequency of cell division, egg cells never divide in culture (Kranz et al., 1991a). Fusion products of a sperm cell with a central cell developed in culture, thus endosperm might be produced *in vitro* by this fusion combination (Kranz et al., 1991b; E. Kranz, unpublished data).

5. Concluding comments

The gamete fusion system fills a gap between sexual crossing and somatic hybridization. IVF, first developed using different maize lines, may also be applied to other plant species. The usefulness of this technique in plant breeding, especially in combinations with remote species has still to be demonstrated. Intra- or interspecies incompatibilities, located on the stigma, in the style or ovary might be overcome by this method. Zygotic incompatibility might be circumvented by the use of special tissue culture conditions. High rates of microcallus formation were observed after electrofusion of sperm cell protoplasts of *Coix lacryma-jobi* L. and egg cell protoplasts of *Zea mays* L. and sperm cell protoplasts of *Sorghum bicolor* (L.) Moench and egg cell protoplasts of *Zea mays* L. (E. Kranz and A. Jahnke, unpublished results).

Somatic protoplast fusion holds considerable promise for new combinations of genomes and plastomes which cannot be produced by other methods. Numerous hybrid and cybrid plants have been obtained by this method. Any somatic protoplasts, irrespectively of the taxonomic distances, can be fused. This may also be true for gametic and nongametic protoplasts of the embryo sac. Fertile triploid hybrids via gametosomatic hybridization were produced (e.g. Pirrie and Power, 1986). Asymetric somatic hybrids were obtained and used for sexual hybridization. However, in wide distance somatic hybridizations regeneration capacity often is absent, reduced, or results in sterility, caused by polyploidization, aneuploidy or elimination of parts or entire genomes (Schoenmakers et al., 1994). The karyotypical instability was widely studied and also observed in hybrids after conventional crossing (e.g. O'Donoughue and Bennett, 1994). Thus, it can be expected that novel hybrids might be produced by IVF using isolated gametes rather with more closely related species.

With the access to isolated higher plant gametes, experiments can be carried out to study for example adhesion, fusion, recognition, and the prevention of polyspermy, comparable to animal and lower plant systems. Furthermore, the individual culture of the produced zygotes offers new possibilities of investigating zygotic embryogenesis *in vitro*, especially very early events.

The technique of IVF allows experimentation with the possibilities of gamete manipulation before fertilization as well as providing access to very early events, for example egg cell activation and zygote polarity, without any influence of cells of the mother plant. The experimental system described herein is used for developmental studies at the molecular level, such as the detection of zygote-specific genes from very early stages of zygote development. As a first step, a cDNA library of isolated maize egg cells was constructed (Dresselhaus et al., 1994).

References

Dresselhaus, T., Lörz, H. and Kranz, E. (1994) Representative cDNA libraries from few plant cells. *Plant J.* 5, 605-610.

Faure, J.E., Mogensen, H.L., Dumas, C., Lörz, H. and Kranz, E. (1993) Karyogamy after electrofusion of single egg and sperm cell protoplasts from maize: cytological evidence and time course. *Plant Cell* 5, 747-755.

Faure, J.E., Digonnet, C., and Dumas, C. (1994) An in vitro system for adhesion and fusion of maize gametes. *Science* 263, 1589-1600.

Ito, M. and Maeda, M. (1973) Fusion of meiotic protoplasts in liliaceous plants. *Exptl. Cell Res.* 80, 453-456.

Keller, W.A. and Melchers, G. (1973) The effect of high pH and calcium on tobacco leaf protoplast fusion. *Z. Naturforsch.* 28 c, 737-741.

Koop, H.-U. and Schweiger H.-G. (1985a) Regeneration of plants from individually cultivated protoplasts using an improved microculture system. *J. Plant Physiol.* 121, 245-257.

Koop, H.-U. and Schweiger H.-G. (1985b) Regeneration of plants after electrofusion of selected pairs of protoplasts. *Eur. J. Cell Biol.* 39, 46-49.

Kranz, E. (1992) *In vitro* fertilization of maize mediated by electrofusion of single gametes, in K. Lindsey (ed.), *Plant Tissue Culture Manual,* Supplement 2, E1, Kluwer Academic Publishers, Dortrecht, pp. 1 - 12.

Kranz, E. (1993) *In vitro* Befruchtung mit einzelnen isolierten Gameten von Mais (*Zea mays* L.). Habilitationsschrift. Universität Hamburg.

Kranz, E. and Lörz H. (1990) Micromanipulation and in vitro fertilization with single pollen grains of maize. *Sex. Plant Reprod.* 3, 160-169.

Kranz, E. and Lörz, H. (1993) In vitro fertilization with isolated, single gametes results in zygotic embryogenesis and fertile maize plants. *Plant Cell* 5, 739-746.

Kranz, E. and Lörz, H. (1994) *In vitro* fertilization of maize by single egg and sperm cell protoplast fusion mediated by high calcium and high pH. *Zygote* 2, in press.

Kranz, E., Bautor, J. and Lörz, H. (1990) *In vitro* fertilization of single, isolated gametes, transmission of cytoplasmic organelles and cell reconstitution of maize (Zea mays L.), in H.J.J. Nijkamp, L.H.W. Van der Plas, & J. Van Aartrijk (eds.), *Progress in Plant Cellular and Molecular Biology*, Proc. of the VIIth International Congress on Plant Tissue and Cell Culture, Amsterdam, The Netherlands, 24-29 June 1990, Kluwer Academic Publishers, Dordrecht, pp. 252-257.

Kranz, E., Bautor, J. and Lörz, H. (1991a) *In vitro* fertilization of single, isolated gametes of maize mediated by electrofusion. *Sex. Plant Reprod.* 4, 12-16.

Kranz, E., Bautor, J. and Lörz, H. (1991b) Electrofusion-mediated transmission of cytoplasmic organelles through the in vitro fertilization process, fusion of sperm cells with synergids and central cells, and cell reconstitution in maize. *Sex. Plant Reprod.* 4, 17-21.

Lee, C.H. and Power, J.P. (1988) Intraspecific gametosomatic hybridization in *Petunia hybrida. Plant Cell Rep.* 7, 17-18.

Lu, C., Vasil, I.K. and Ozias-Akins P. (1982) Somatic embryogenesis in *Zea mays* L. *Theor. Appl. Genet.* 62, 109-112.

Murashige, T. and Skoog, F. (1962) A revised medium for rapid growth and bioassays with tobacco tissue cultures. *Physiol. Plant.* 15, 473-497.

O'Donoughue, L.S. and Bennett, M.D. (1994) Comparative responses of tetraploid wheats pollinated with *Zea mays* L. and *Hordeum bulbosum* L. *Theor. Appl. Genet.* 87, 673-680.

Pental, D., Mukhopadhyay, A., Grover, A. and Pradhan, A.K. (1988) A selection method for the synthesis of triploid hybrids by fusion of microspore protoplasts (n) with somatic cell protoplasts (2n). *Theor. Appl. Genet.* 76, 237-243.

Pirrie, A. and Power, J.B. (1986) The production of fertile, triploid somatic hybrid plants (*Nicotiana glutinosa* (n) + *N. tabacum* (2n) via gametic: somatic protoplast fusion. Theor. Appl. Genet. 72, 48-52.

Power, J.B., Cummins, S.E. and Cocking, E.C. (1970) Fusion of isolated protoplasts. *Nature* 225, 1016- 1018.

Randolph, L.F. (1936) Developmental morphology of the caryopsis in maize. *J. Agric. Res.* 53, 881-916.

Rhodes, C.A., Lowe, K.S. and Ruby, K.L. (1988) Plant regeneration from protoplasts isolated from embryogenic maize cell cultures. *Bio/Technol.* 6, 56-60.

Schel, J.H.N., Kieft, H. and Van Lammeren A.A.M. (1984) Interactions between embryo and endosperm during early developmental stages of maize caryopses (*Zea mays*). *Can. J. Bot.* 62, 2842 - 2853.

Schoenmakers, H.C.H., A.-M.A. Wolters, A. de Haan, A.K. Saiedi and Koorneef, M. (1994) Asymmetric somatic hybridization between tomato (*Lycopersicon esculentum* Mill) and gamma-irradiated potato (*Solanum tuberosum* L.): a quantitative analysis. *Theor. Appl. Genet.* 87, 713-720.

Tirlapur, U.K., Kranz, E. and Cresti, M. (1994) Characterization of isolated egg cells, in vitro fusion products and zygotes of *Zea mays* L. using the technique of image analysis and confocal laser scanning microscopy. *Zygote*, submitted.

Van Lammeren, A.A.M. (1981) Early events during embryogenesis in Zea mays L. *Acta Soc. Bot. Pol.* 50, 289-290.

Van Lammeren, A.A.M. (1986) Developmental morphology and cytology of the young maize embryo (*Zea mays* L.). *Acta Bot. Neerl.* 35, 169-188.

Ward, M., Davey, M.R., Mathias, R.J., Cocking, E.C., Clothier, R.H., Balls, M. and Lucy, J.A. (1979) Effects of pH, Ca^{2+}, temperature, and protease pretreatment in interkingdom fusion. *Somatic Cell Genetics* 5, 529-536.

Zenkteler, M. (1990) *In vitro* fertilization and wide hybridization in higher plants. *Crit. Rev. Plant Sci.* 9, 267-279.

Zenkteler, M. (1992) Wide hybridization in higher plants by applying the method of test tube pollination of ovules, in Y. Dattée, C. Dumas and A. Gallais, (eds.) *Reproductive biology and plant breeding*, The XIIIth Eucarpia congress from the 6th to the 11th Jyly 1992, Angers, France, Springer, Heidelberg, pp. 205-214.

FERTILE BARLEY PLANTS CAN BE REGENERATED FROM MECHANICALLY ISOLATED PROTOPLASTS OF FERTILIZED EGG CELLS

P.B. HOLM[a], S. KNUDSEN[a], P. MOURITZEN[a], D. NEGRI[b], F. L. OLSEN[a] AND C. ROUÉ[a,c]

[a]*Carlsberg Research Laboratory, 10, Gamle Carlsberg Vej, DK-2500 Copenhagen, Denmark*
[b]*Secobra Recherches S.A., Centre de Bois Henry, F-78580 Maule, France*
[c]*Tepral, Brasseries Kronenbourg, 68, route d'Oberhausbergen, F-67200 Strasbourg, France*

ABSTRACT. The present paper summarizes techniques that have allowed the isolation and regeneration of barley zygote protoplasts. A purely mechanical procedure was developed where viable protoplasts of unfertilized barley eggs and zygotes could be isolated with a frequency of 75%. Regeneration of zygote protoplasts was achieved by cocultivation with barley microspores undergoing microspore embryogenesis. About 75% of the zygotes developed into embryo-like structures and fully fertile plants were regenerated from about 50% of the embryo-like structures. Zygotes, grown in medium only, divided a few times and formed microcalli that degenerated. Protoplasts of unfertilized eggs were never observed to divide irrespective of the culture conditions. Regeneration required cocultivation for more than 3 weeks and the development of the zygote-derived structures and their regeneration potential related directly to the age of the microspore culture. The isolation-regeneration frequencies appeared to be largely genotype independent. Protoplasts of unfertilized eggs and zygotes of wheat were isolated by the same procedure and a fully fertile wheat plant was regenerated by cocultivation with barley microspores.

1. Introduction

Experimental studies on egg development, fertilization and early embryogenesis in plants require the development of techniques for the isolation and culture of the female gametes and zygotes. Simple and efficient techniques are available for the isolation of viable sperm cells from a variety of plant species (for review, see Chaboud and Perez, 1992) but it has been difficult to obtain viable egg cells. Protoplasts of the cells of the embryo sac have in a few cases been isolated by enzyme maceration and micromanipulation of ovules, but in general the viability of the cells was poor, possibly due to the enzyme treatment (for review, see Theunis et al., 1991; Huang and Russell, 1992; van der Maas et al., 1993).

In maize, viable protoplasts of unfertilized eggs have been obtained from enzymatically digested maize ovules (Kranz et al., 1991; Faure et al., 1992). The protoplasts could be electrofused to isolated sperm cells with a high frequency (Kranz et al., 1991; Faure et al., 1993) and regeneration of the fusion products to normal, hybrid plants was achieved by cocultivation with a nonmorphogenic maize cell suspension (Kranz and Lörz, 1993). Spontaneous fusion of sperm cells and unfertilized egg protoplasts has also recently been documented (Faure et al. 1994).

M. Terzi et al. (eds.), Current Issues in Plant Molecular and Cellular Biology, 207-212.

Isolation and regeneration of *in vivo* fertilized eggs has so far only been achieved in barley (Holm et al. 1994a). It was shown that highly viable protoplasts of the unfertilized egg and the zygote could be isolated with a high frequency by a simple dissection of the ovule. Normal, fully fertile plants were regenerated by cocultivation with barley microspores undergoing embryogenesis. It is the intention of the present paper to summarize the techniques developed and briefly discuss the potential of this approach for basic studies and biotechnology.

2. Materials and methods

Donor plants of the winter barley cultivar Igri, the spring malting barley cultivar Alexis and the wheat cultivar Walter were grown in growth cabinets. A detailed description of the media used (Kao90, a modified Kao medium with 90 g/l maltose), the processing of the ovaries and ovules, and the isolation and regeneration of the egg protoplast has been given in Holm et al. (1994a, 1994b). For regeneration the protoplasts were transferred to Transwell tissue culture inserts and cocultured with microspores undergoing embryogenesis.

3. Results

3.1. ISOLATION OF PROTOPLASTS

The ovaries were dissected to expose the micropylar tip of the ovule. Removal of the two integuments allowed that the egg apparatus could be seen with a high resolution (Figure 1A). The initial step in the isolation of the protoplasts was to puncture the central cells giant vacuole. This resulted in a partial collapse of the central cell. The collapse of the central cell caused the egg cell and often also the persistent synergid to loose their adhesion to the nucellus layer and their own cell wall and become spherical (Figure 1B). A gentle tabbing on the micropylar end of the ovule further eased the liberation of the protoplasts. After removing the few nucellus cell layers in the micropylar region, the egg protoplasts could be extruded by applying a gentle pressure on the ovule with the forceps (Figures 1C and 1D). Occasionally synergid and central cell protoplasts were also isolated.

Protoplasts of unfertilized eggs and zygotes were isolated with the same frequencies. On average protoplasts were isolated from 75% of the dissected ovaries and about 10 protoplasts could be isolated per hour. The isolation frequency was the same in the two cultivars analyzed. Zygote protoplasts could be isolated until shortly before the first division of the zygote but dyads did not liberate protoplasts.

Calcofluor-staining did not reveal any fluorescent wall material around the protoplasts. Preliminary electron microscopic studies have also failed to reveal a wall or wall remnants around the egg protoplasts (H.L. Mogensen and P.B. Holm, unpublished observation).

The protoplasts of fertilized and unfertilized eggs were remarkable stable in the Kao90 medium (350-375 mosmol/kg H_2O). Viable and regenerable protoplasts could also be isolated at higher and lower osmotic pressures although they responded to the changes in osmolarity by a slight swelling at the lower osmolarity and a shrinkage at the higher osmotic pressure. In Kao 90 medium without Ca^{2+} and Mg^{2+}, protoplasts could still be isolated, but they appeared to lack a plasmalemma and degenerated shortly after isolation. Agarose-embedded protoplasts could readily be microinjected. Preliminary studies showed that about 70% of the protoplasts survived the injection, and of these more than 50% continued their development.

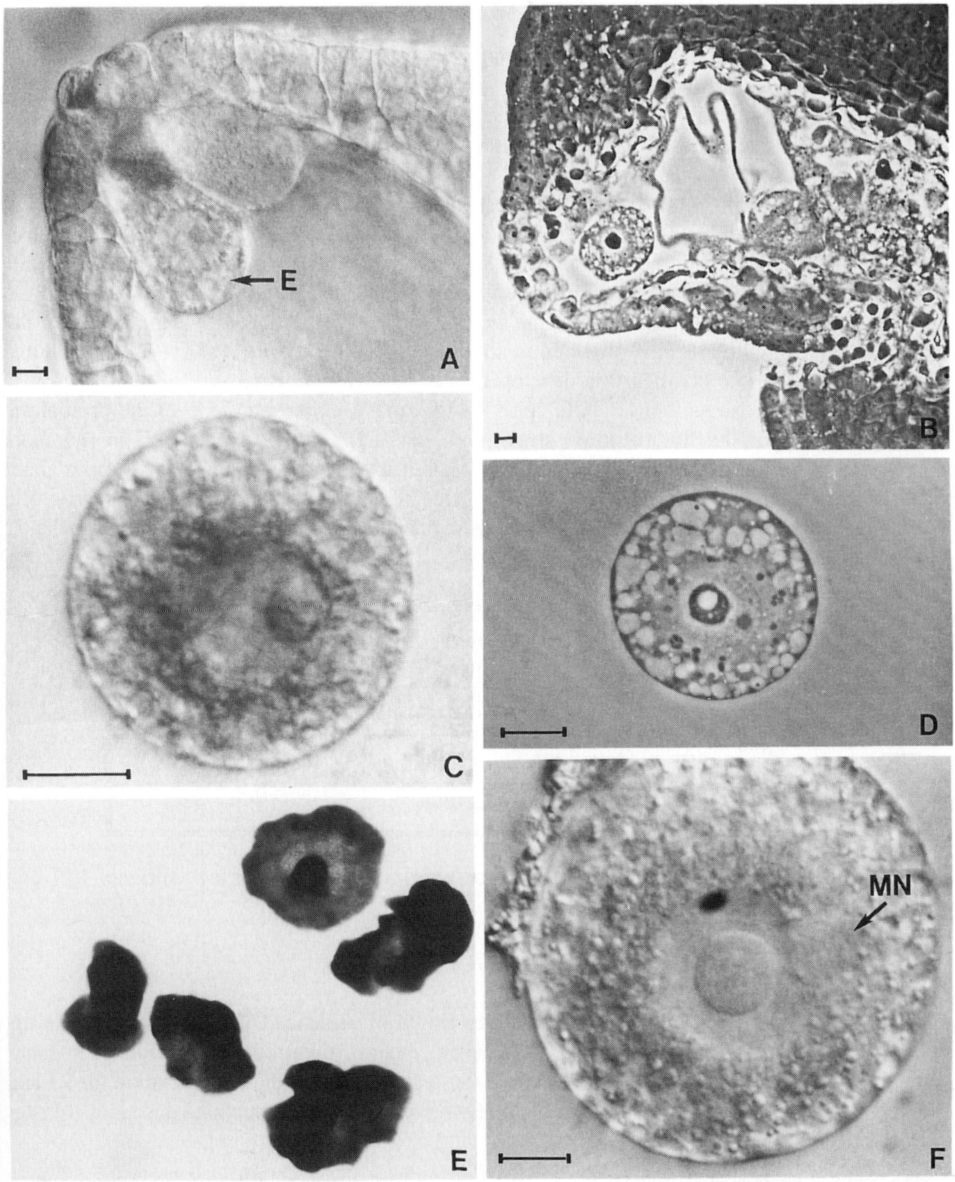

Figure 1. Photographs of intact and sectioned ovules, protoplasts and derived structures. **1A)** Micropylar tip of dissected barley ovary where both integuments are removed; E, egg cell. **1B)** Egg protoplast in ovule where the vacuole of the central cell has been punctured. **1C)** Intact barley zygote protoplast. **1D)** Sectioned barley zygote protoplast. **1E)** Embryo-like structures after 27 days of cocultivation. **1F)** Intact protoplast of wheat zygote; MN, male nucleus. Bar $= 10\ \mu m$.

3.2. REGENERATION BY COCULTIVATION WITH MICROSPORES

Zygote protoplasts, cultured in medium only, divided a few times and formed a spherical microcallus, after which the cells developed large vacuoles and degenerated. In no case did protoplasts of unfertilized eggs divide and they degenerated after a few days in culture.

Regeneration was achieved by cocultivation with microspores undergoing embryogenesis. The regeneration potential of the zygote appeared to be the same from immediately after fertilization until shortly before the first division. Three to 4 days after isolation, multicellular globular microcalli had formed that developed into bipolar structures. After 2 to 3 weeks, the protoplast-derived structures often attained an embryo-like shape with a sheet-like scutellum (Figure 1E). Sometimes the scutellum-like structures were very irregular, and globular proembryos often formed along the rim of the scutellum; in particular, this occurred when culturing the protoplasts in Kao90 medium supplied with one mg/l 2,4-D. The embryo-like structures germinated readily into plantlets that developed into completely normal and fully fertile plants. The data in Table 1 show that for both cultivars Igri and Alexis about 75% of the protoplasts developed beyond the microcallus stage and plantlets were regenerated in numbers corresponding to the number of developing structures. On average plants formed from about 50% of the plantlets. Fiftyfive Igri and 13 Alexis plants were grown to maturity. The regenerated plants were fully fertile and indistinguishable from seed grown control plants. Albino plants were never observed.

Cultivars	No. of Protoplasts	Developing Structures		No. of Plantlets	
		No.	%	No.	%
Igri	56	44	79	46	105
Alexis	40	29	73	22	76

Table 1. Regeneration frequencies for zygote protoplasts of the barley cultivars Igri and Alexis.

3.3. FACTORS ESSENTIAL FOR ZYGOTE REGENERATION

Zygote regeneration showed dependence on the length of the cocultivation. Cocultivation for 9 days resulted in a very limited growth and development of the protoplasts, while protoplasts cocultivated for 16 days developed into small, loose callus structures. Cocultivation for 23 and 30 days resulted in a high frequency of plant regeneration, but the structures germinated faster into plantlets with 30 days of cocultivation.

Microspore preparations, ranging from freshly isolated microspores to 6-week-old cultures in which embryoids had formed, all supported growth of the zygote beyond the microcallus stage. However, when cocultivated with up to 3-day-old microspore cultures, the protoplasts developed into small, loose aggregates where multiple globular proembryo-like structures formed. Cocultivation with 5-to 19-day-old microspore preparations resulted in the formation of embryo-like structures. Microspore cultures older than 21 days supported only the formation of small spherical structures. It was not possible to regenerate plantlets from egg protoplasts in acceptable frequencies using microspore preparations older than 21 days.

3.4 ISOLATION AND REGENERATION OF WHEAT PROTOPLAST

Protoplasts of the unfertilized wheat egg were readily obtained with the methods described above while it was more difficult to obtain protoplasts of the zygote (Figure 1 F). The diameter of these protoplasts isolated in Kao 90 exceeded that of barley by a factor of 2. As for barley, the unfertilized protoplasts were unable to divide while the zygote protoplasts divided a few times and formed spherical, multicellular structures that subsequently degenerated. A few zygote protoplasts were cocultivated with barley microspores and in one case a completely normal, fully fertile wheat plant was regenerated.

4. Discussion

The study showed that it is possible by simple dissections and puncturing of the giant vacuole of the central cell to mechanically isolate protoplasts of unfertilized egg cells and zygotes of barley and wheat. This can be attributed to the fact that unfertilized plant egg cells and zygotes *in vivo* largely lack surrounding walls, at least in the chalazal region (for review and discussion see Huang and Russell, 1992, Holm et al. 1994a). The absence of a wall is probably essential for penetration of the sperm nucleus into the egg and perhaps also for sperm and egg cell recognition and interaction. Wall formation in the chalazal part of the egg appears thus to be a postfertilization event that may be triggered by the fertilization process itself. The number of plasmodesmata between the egg cell and its neighbouring cells may be a second essential factor for the mechanical isolation of undamaged egg protoplasts. In some species plasmodesmata are present in all walls except the external embryo sac wall, and in others plasmodesmata are entirely lacking or confined to the walls between synergids (for review, see Huang and Russell, 1992).

There were no indications for a strong genotype effect on the regeneration of plants from protoplasts of the fertilized egg cell. Likewise, protoplasts isolated immediately after fertilization until the first division appeared to have the same regeneration potential. This is also the case when culturing zygotes *in situ* using ovule culture (Holm et al. 1994b).

Unfertilized egg cells of barley and wheat did not divide while the zygote protoplasts formed microcalli when cultured in medium only. Cocultivation with barley microspores undergoing embryogenesis proved to be very efficient for supporting the formation of embryo-like structures. Formation of embryo-like structures with a high regeneration capacity required cocultivation with microspores less than 19 days old, while cocultivation with older structures gave rise to globular, nonregenerable structures. One possible explanation is that the microspore cultures at all stages excrete compounds promoting a basic type of undifferentiated growth, whereas embryogenesis in the protoplast-derived structures requires an additional set of stimuli from the microspores at an early stage of embryogenesis. Microspore cocultivations have also allowed regeneration of fertile barley plants from protoplasts isolated from 36-day-old microspore-derived embryoids and promoted development of microspores undergoing embryogenesis from other barley cultivars (unpublished observation). Preliminary investigations have also suggested that microspore cocultivation is superior to cocultivation with morphogenic and in particular non-morphogenic suspensions of barley.

The technology developed may facilitate experimental studies on fertilization and early embryogenesis in barley and other plants. It is considered likely that mechanically isolated egg protoplasts are superior to egg protoplasts liberated by enzyme treatment if the objective is to study gamete surface receptors or other plasmalemma factors essential for gamete recognition and fusion. Cultivation of egg protoplasts may also be used as a very sensitive bioassay for the identification of embryogenesis-promoting compounds. As already documented in maize, *in vitro* fertilization can be achieved which later may allow the generation of new hybrids or cybrids. Protoplasts of the egg, fertilized *in vivo* or *in vitro*, are also ideal targets for transformation by microinjection. Provided that sufficiently high transformation frequencies can be achieved, selectable markers may not be needed. In the cereals only a few useable selectable genes are available which may constitute a problem if several different genes are to be introduced successively into the same host (Yoder and Goldsbrough 1994). Foreign genes, conferring herbicide or antibiotic resistance, may also be unwanted in transgenic plants to be used for human consumption. Finally, microinjection may be superior to other techniques if large DNA fragments such as YAC chromosomes are to be introduced.

5. Acknowledgements

We are indebted to Ann Sofi Steinholz for producing the photographs. The work was supported by Eureka grant EU 270 and is ABIN (Adaptation of Barley for Industrial Needs) publication No. 140.

6. References

Chaboud, A., and Perez, R. (1992) 'Generative cells and male gametes: isolation, physiology, and biochemistry', Int. Rev. Cytol. 140, 205-232.

Faure, J.-E., Mogensen, H.L., Kranz, E., Digonnet, C., and Dumas, C. (1992) 'Ultrastructural characterization and three-dimensional reconstruction of isolated maize (*Zea mays* L.) egg cell protoplasts', Protoplasma 171, 97-103.

Faure, J.-E., Mogensen, H.L., Dumas, C., Lörz, H., and Kranz, E. (1993)' Karyogamy after electrofusion of single egg and sperm cell protoplasts from maize: Cytological evidence and time course', Plant Cell 5, 747-755.

Faure, J.-E., Digonnet, C. and Dumas, C. (1994) 'An in vitro system for adhesion and fusion of maize gametes', Science 263, 1598-1600.

Holm, P.B., Knudsen, S., Mouritzen, P., Negri, D., Olsen, F.L., and Roué, C. (1994a) 'Regeneration of fertile barley plants from mechanically isolated protoplasts of the fertilized egg cell', Plant Cell (in press).

Holm, P.B., Knudsen, S., Mouritzen, P., Negri, D., Olsen, F.L., and Roué, C. (1994b).'Regeneration of the barley zygote in ovule culture', Sex Plant Reprod. (in press).

Huang, B.-Q., and Russell, S.D. (1992) 'Female germ unit: Organization, isolation and function', Int. Rev. Cytol. 140, 233-293.

Kranz, E., and Lörz, H. (1993) 'In vitro fertilization with isolated, single gametes results in zygotic embryogenesis and fertile maize plants', Plant Cell 5, 739-746.

Kranz, E., Bautor, J., and Lörz, H. (1991) 'In vitro fertilization of single, isolated gamets of maize mediated by electrofusion', Sex Plant Reprod. 4, 12-16.

Maas van der, H.M., Zaal, M.A.C., Dejong, E.R., Krens, F.A., and van Went, J.L. (1993)' Isolation of viable egg cells of perennial ryegrass (*Lolium perenne* L)', Protoplasma 173, 86-89.

Theunis, C.H., Pierson, E.S., and Cresti, M. (1991) 'Isolation of male and female gametes in higher plants', Sex. Plant Reprod. 4, 145-154.

Yoder, J.I. and Goldsbrough, A.P. (1994) 'Transformation systems for generating marker-free transgenic plants', Bio/Technology 12, 263-267.

EXPERIMENTS ON INTER-GENERA AND INTER-FAMILIES POLLEN GRAINS PROTOPLAST FUSION

J. M. MONTEZUMA-DE-CARVALHO AND M. C. TOMÉ
Department of Botany
University of Coimbra
3049 Coimbra Codex
Portugal

ABSTRACT. Here we demonstrate that isolation of protoplasts from pollen grains, after enzymatic treatment, depends on their type of aperture. Besides the 1-sulcate type, other related types as the 2--sulculate and zonisulcate also give protoplasts. Up to now, we have isolated pollen protoplasts from 31 species belonging to 13 angiosperm families. Considering the total number of species with these types of pollen grains apertures, it is expected that several thousands of angiosperm plants can easily give viable pollen protoplasts. We have used these protoplasts for fusion experiments using a PEG protocoll. Preliminary results of fusion between species of different families show that fusion products are of the homoplasmic or heteroplasmic type, and two or more protoplasts can be involved. Attempts to cultivate these fusion products have not yet given complete satisfactory results.

1. Introduction

Recently it was reported the isolation of pollen protoplasts from a few species belonging to the families Liliaceae, Amaryllidaceae and Iridaceae [1, 2, 3]. However, none of the authors has given a complete rationale explanation for this phenomenon. Our hypothesis was that the phenomenon would be related with a special type of pollen grain aperture, namely the 1-sulcate type. Knowing that this type of aperture is the usual type in the Liliaceae, Amaryllidaceae and Iridaceae [4], we tested further our hypothesis [5], with success, in other angiosperm families, either from Dicotyledons or Monocotyledons, using species of the 1-sulcate pollen type.

Although pollen protoplast culture has already been studied by some authors, as far as we know no fusion experiments have been done. We report here some preliminary results on pollen protoplast fusion.

2. Materials and Methods

2.1. PLANT MATERIAL

Mature pollen grains from dehisced anthers were used, except in a few cases where pollen was obtained from squashing undehisced anthers before flowering. Thirty-one species belonging to 13 families were studied, most of them growing at the Botanical Garden of Coimbra University. For protoplast fusion experiments it was used pollen stored in the cold for about one year.

2.2. SEM OBSERVATION

In order to test the type of aperture, pollen of all species was acetolyzed [6] and dehydrated with

M. Terzi et al. (eds.), Current Issues in Plant Molecular and Cellular Biology, 213-218.

ethanol. Some samples were dryed with a critical point dryer through ethanol-CO_2, sputter coated with gold with a *magnetron* head and observed on a JEOL T330 scanning electron microscope.

2.3. PROTOPLAST ISOLATION

Experiments were not carried out in aseptic conditions. 300 μl aliquots of a solution with 1% Cellulase RS, 1% Macerozyme R-10 and seven different mannitol concentrations (550 - 0 mM) in a modified V-KM medium [7], pH 5.5, were placed in each well of Costar® 3424 clusters. Pollen grains incubation was carried out at 25 °C with different degrees of agitation (400 - 0 rpm).

2.4. PROTOPLAST FUSION AND CULTURE

Digested pollen solutions of two species were mixed together in a total volume of 600 μl. After 5 min. protoplasts have settled down, excess liquid was taken out and 500 μl of a solution with 40% polyethylene glycol (PEG) 1500, 0.3 M glucose and 50 mM $CaCl_2.2H_2O$, pH 7.0 was added for 40 min. Addition of 500 μl of a solution containing 0.1 M $CaCl_2.2H_2O$ and 14% sorbitol, pH 7.0 followed and, after 10 min., 1 ml of the same solution was poured in. Protoplasts were washed several times in this solution and then cultured in a modified V-KM medium with the appropriate mannitol concentration, pH 5.8. Staining of protoplasts and fusion products was accomplished with the carbol fuchsin [8] and fluorochrome 4',6-diamidino-2-phenylindole (DAPI) [9] methods.

3. Results and Discussion

As can be seen from Table 1 and 2, most of the protoplasts are released between 10 - 120 minutes. In these tables, of the seven concentrations of mannitol used, we present only the concentration that for each plant species gives more than 50% of undamaged protoplasts. Due to

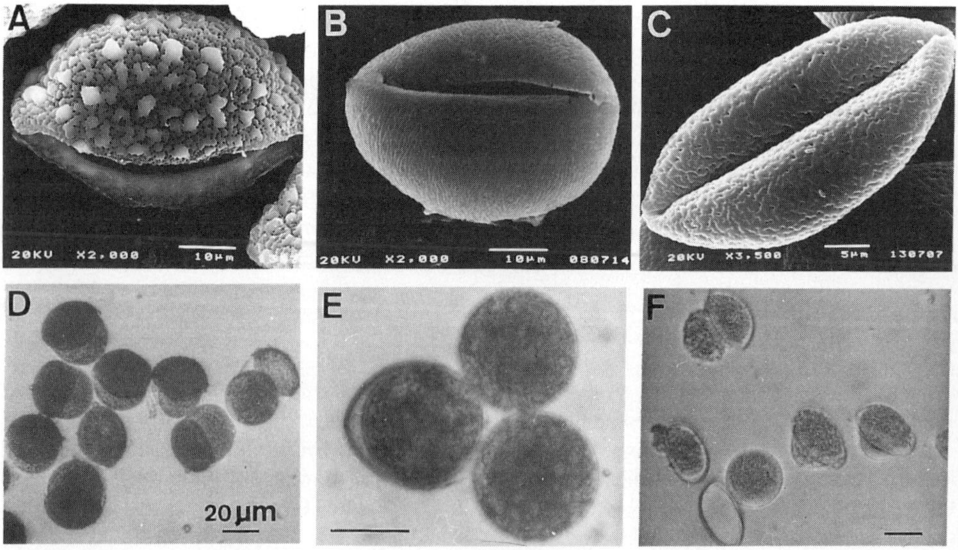

Figure 1. **A-C**: SEM pollen observation; **D-E**: pollen protoplasts released in enzymatic solution. **A**, **D**: *Liriodendron tulipifera* (Dicotyledons); **B**, **E**: *Tacca chantrieri* (Monocotyledons); **C**, **F**: *Cocos eriospatha* (Monocotyledons).

TABLE 1. Conditions for the release of protoplasts from pollen of several **dicotyledonous** species.

Family (nº genera / nº species)	Species studied	[mannitol] (mM)	rpm	Time of release (min.)
Magnoliaceae (12 / 230)	*Magnolia soulangeana*	367	100	150
	Liriodendron tulipifera	367	250	90
Aristolochiaceae (7 / 400)	*Aristolochia sempervirens*	550	0	120
Euryalaceae (2 / 3-4)	*Victoria amazonica*	183	200	150
	Euryale ferox	183	0	150

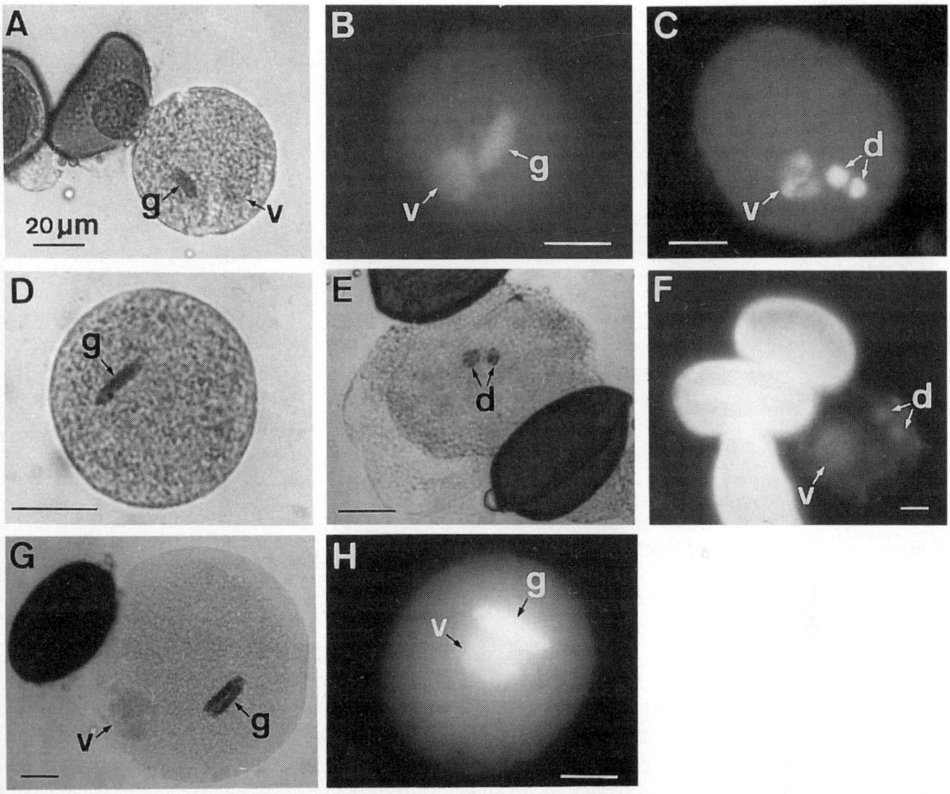

Figure 2. Pollen protoplast culture. **A-C**: *Agapanthus praecox* (Alliaceae); **D-F**: *Jubaea chilensis* (Palmae); **G, H**: *Hippeastrum reginae* (Amaryllidaceae). **A, D, E, G**: carbol fuchsin staining; **B, C, F, H**: DAPI staining. **A, B, D, E, F**: fresh isolated protoplasts; **C, E, F**: culture after 27, 34 and 34 hours, respectively. **v** - vegetative nucleus; **g** - generative nucleus; **d** - division of the generative nucleus.

the large exposure surface (Figure 1. A-C) of the pectocellulosic intine in a 1-sulcate pollen grain, its digestion by the cellulase-pectinase mixture is quite rapid. According to the quickness of digestion, protoplasts can be released dumbbell-shaped, oval or even spherical (Figure 1. D-F). It is noteworthy that besides the 1-sulcate type pollen grain, other related types as the 2-sulculate (*Pontederia cordata*; *Euryale ferox*), trichotomosulcate (*Simethis planifolia*; *Phormium tenax*)

TABLE 2. Conditions for the release of protoplasts from pollen of several **monocotyledonous** species.

Family (nº genera / nº species)	Species studied	[mannitol] (mM)	rpm	Time of release (min.)
Liliaceae (250 / 3700)	*Simethis planifolia*	367	400	35
	Ornithogalum unifolium	550	250	80
Alliaceae (30 / 600)	*Agapanthus praecox*	550	100	330
	Nothoscordum inodorum	458	200	90
	Allium roseum	275	250	60
Amaryllidaceae (85 / 1100)	*Narcissus pseudonarcissus*	550	0	120
	Pancratium maritimum	458	0	90
	Hippeastrum reginae	550	200	120
	Clivia miniata	275	100	120
	Clivia gardenii	550	250	40
Agavaceae (20 / 670)	*Cordyline australis*	550	250	35
	Phormium tenax	367	0	20
	Agave americana	367	0	10
Pontederiaceae (7 / 30)	*Pontederia cordata*	458	200	120
Dioscoreaceae (5 / 750)	*Tamus communis*	550	100	60
Bromeliaceae (44 / 1400)	*Ananas comosus*	550	100	60
	Cryptanthus acaulis	183	200	120
Commelinaceae (38 / 500)	*Tradescantia palludosa*	183	400	60
	Rhoeo discolor	550	200	120
Taccaceae (2 /31)	*Tacca chantrieri*	550	0	30
Palmae (217 / 2500)	*Arecastrum romanzoffianum*	550	0	40
	Jubaea chilensis	183	0	60
	Livistona australis	275	0	1020
	Cocos eriospatha	458	200	20
	Cocos weddeliana	458	400	120
	Chamaerops humilis	0	200	240

and zonisulcate (*Victoria amazonica*) also give rise to protoplasts. Up to now we have studied 31 species, belonging to 13 angiosperm families. Considering however the number of genera and species [4] in these families (see Tables 1 and 2) and also considering that there are other families [10] where the 1-sulcate type pollen grain also prevails (Aponogetonaceae, Apostasiaceae, Butomaceae, Canellaceae, Haemodoraceae, Mayaceae, Myristicaceae, Piperaceae, Rapataceae, Velloziaceae, Xiridaceae) we can confidently expect that several thousands of angiosperm plant species can easily give rise to pollen protoplasts.

Cultured protoplasts of the three species used showed division of the generative nucleus after 27-34 hours (Figure 2. C, E, F). Considering that, in *in vitro* cultured intact pollen, the generative nucleus is programmed to enter mitosis inside the pollen tube in about the same time (24 hours) after germination, the occurrence of mitosis in pollen protoplasts is perfectly predictable. In fact, division of the generative nucleus in cultured pollen protoplasts has been observed before [1, 2]. However, since numerous studies have demonstrated *in vitro* embryogenesis either through anther or pollen culture, it is also expected that pollen protoplasts can enter in an embryogenic pathway.

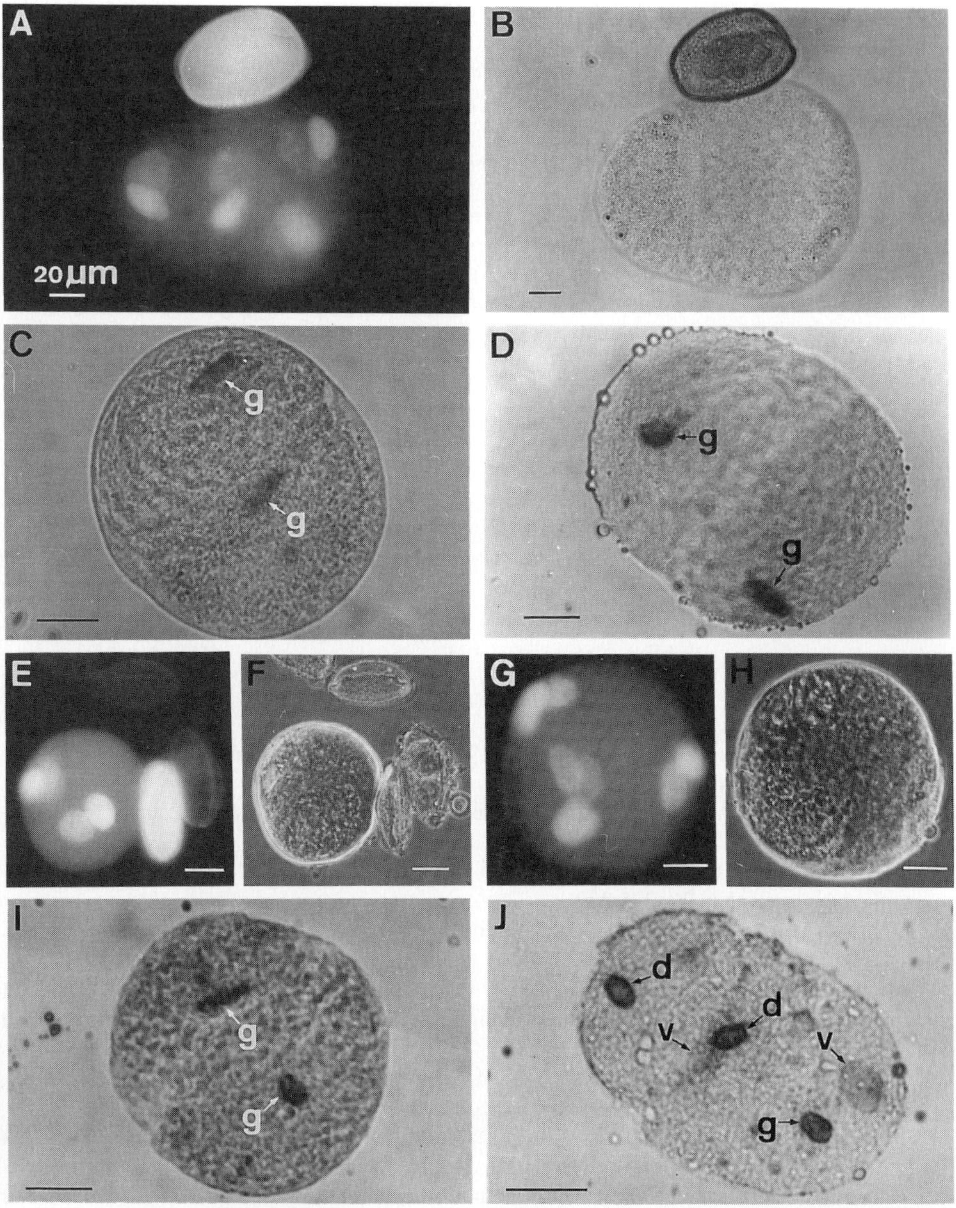

Figure 3. PEG fusion of pollen protoplasts. **A-D**: fusion between *Agapanthus praecox* and *Hippeastrum reginae*; **E-J**: fusion between *Agapanthus praecox* and *Jubaea chilensis*. **A**, **B**: multiple fusion stained with DAPI and without staining, respectively; **C**, **D**: a probable homoplasmic and heteroplasmic fusion, respectively, carbol fuchsin stained. **E**, **F**: fusion between two protoplasts, DAPI stained and in phase contrast, respectively; **G**, **H**: triple fusion, DAPI stained and in phase contrast, respectively; **I**: a probable heteroplasmic fusion, carbol fuchsin stained; **J**: division of the generative nucleus after 34 hours of culture. **v** - vegetative nucleus; **g** - generative nucleus; **d** - division of the generative nucleus.

Identification of fused protoplasts was not easy in our experiments, and could only be based on the differences in size of the generative nuclei of the three species involved: *Jubaea chilensis* has a quite long nucleus, the nucleus of *Hippeastrum reginae* is a little shorter and in *Agapanthus praecox* is the shortest. Therefore, in Figure 3. C one could see a homoplasmic fusion while Figure 3. D shows a heteroplasmic fusion.

Until now, fused protoplasts in culture have not shown evidence of nuclear divisions, except in one probable case (Figure 3. J). Experiments are in progress to find the most suitable culture medium and to evaluate the probable toxic effect of PEG in division. Based on the fact that a pollen protoplast differs greatly from a somatic protoplast, since it does not contain a nucleus but instead two cells (vegetative and generative cell), it is expected that the chance of nuclear fusion is remote. Therefore, in a heteroplasmic fusion product, the most probable embryogenic alternative will be either the formation of two genetically independent haploid embryoids or, the much more interesting situation, the formation of an embryoid with a haploid chimaeric structure.

4. Acknowledgements

We thank Professor Vasco Bairos, Faculty of Medicine, for help on gold coating of pollen samples and their visualization in SEM. This work was supported by grants from J.N.I.C.T. and Fundação Oriente, Portugal.

5. References

1 Tanaka, I., Kitazume, C., and Ito, M. (1987) 'The isolation and culture of lily pollen protoplasts', Plant Sci. 50, 205-211.

2 Zhou, C. (1989) 'A study on isolation and culture of pollen protoplasts', Plant Sci. 59, 101--108.

3 Fellner, M., and Havránek, P. (1992) 'Isolation of *Allium* pollen protoplasts', Plant Cell Tissue Organ Cult. 29, 275-279.

4 Erdtman, G. (1971) Pollen Morphology and Plant Taxonomy, Hafner Publishing Company, New York.

5 Montezuma-de-Carvalho, J.M., and Tomé, M.C. (1993) 'Plant species with 1-sulcate pollen grains as a source for isolation of pollen protoplasts', Abstracts of The XV International Botanical Congress, Yokohama, p. 548.

6 Erdtman, G. (1960) 'The acetolysis method - a revised description', Svensk. Bot. Tidskrift. 54, 561-564.

7 Bokelmann, G., and Roest, S. (1983) 'Plant regeneration from protoplasts of potato (*Solanum tuberosum* cv. Bintje)', Z. Pflanzenphysiol. 109, 259-265.

8 Kao, K.N. (1975) 'A nuclear staining method for plant protoplasts', in O.L. Gamborg and L.R. Wetter (eds.), Plant Tissue Culture Methods, Saskatoon, pp. 60-62.

9 Coleman, A.W., and Goff, L.J. (1985) 'Applications of fluorochromes to pollen biology. I.Mithramycin and 4',6-diamidino-2-phenylindole (DAPI) as vital stains for quantitation of nuclear DNA', Stain Technol. 60, 145-154.

10 Willis, J.C. (1973) A Dictionary of The Flowering Plants and Ferns, Eigth Edition, Cambridge University Press, Cambridge.

MITOCHONDRIAL REGULATION OF PETAL AND STAMEN DEVELOPMENT IN CYTOPLASMIC MALE STERILE CULTIVARS OF NICOTIANA TABACUM.

K. GLIMELIUS, M. HERNOULD AND P. BERGMAN
Dept of Plant Breeding Research Uppsala Genetic Center, Swedish University of Agricultural Sciences, Box 7003, 750 07 Uppsala, Sweden

Abstract

Different alloplasmic and cytoplasmic male-sterile, CMS, cultivars of *Nicotana tabacum* have been investigated regarding the interactions between mitochondrial and nuclear genes and their regulatory role on development of floral organs. According to our hypothesis the different morphologies that characterize the different CMS cultivars are indicative of variation in activity of several mitochondiral genes. The genes which might be a result of intra- or intermolecular recombination events, are "phenotypically silent" in the original nuclear background, but when transferred to a different nuclear background lacking restorer genes, the male-sterile phenotypes are obtained. In support of our hypothesis new transcripts or changes in existing transcripts which could be correlated with male sterility have been found when mitochondria from the wild *Nicotiana* species have been introduced to the nuclear environment of *N. tabacum*. Restoration to fertility either by restorer genes or via somatic cybridization, which in the latter case resulted in recombination between the mitochondrial genomes, altered or completely abolished the CMS specific expression. Gel blot hybridizations of mtDNA isolated from the fertile cybrids, using different mitochondrial genes as probes, revealed that DNA-sequences correlated to the CMS-cultivars were deleted.

A detailed molecular investigation of the restored fertile cybrids revealed that a chimeric gene was associated with one of the CMS cultivars, *N. tabacum* with *N. biglovii* cytoplasm. The chimeric gene corresponds to an urf coding for 38 amino acids fused in frame with a truncated *atp*A gene sequence corresponding to the 220 C-terminal amino acids. We have recently shown that this chimeric gene is a target for post-transcriptional editing, in which 7 cytosines are replaced by uridines in the mt mRNA.

1. Introduction

By introducing the nucleus of one species into the cytoplasm of another, but related species, via interspecific hybridization followed by consecutive backcrossings, cytoplasmic male-sterile (CMS) cultivars can be obtained (Kaul, 1998). The alloplasmic combinations between different species, e.g. different *Nicotiana* species (Bonnett et al. 1991), frequently lead to functional disorders, and thus inhibited pollen production and male-sterile plants, but also developmental disorders which results in production of abnormal floral organs. These abnormal features are maternally inherited and have been associated with the mitochondrial genome (Beillard et al. 1978, Aviv and Galun 1980, Glimelius et al. 1981). During evolution a cooperation and cordinated gene expression between the nuclear and the organellar genomes within a species have evolved into a functional unit. Thus, processes like photosynthesis, respiration and certain stages in plant development which are regulated both by genes in the nucleus and in the organelles, function without problems in a cooperative interaction between the genomes. However, when combining a nucleus from one species with mitochondria and chloroplasts from another, a disturbance in the cooperation between intracellular genomes may be the result.

M. Terzi et al. (eds.), Current Issues in Plant Molecular and Cellular Biology, 219-224.

Usually, the incompatibility between the nuclear and the mitochondrial genomes only affects the reproductive parts and not the vegetative development and growth. (Hosfield and Vernsman, 1974).

2. Morphological characterization of cytoplasmic male-sterile tobacco cultivars and certain cybrids produced between some of these cultivars.

The genus *Nicotiana* is represented by a large number of species which can be crossed with *N. tabacum*, which after backcrossings to tobacco as the male parent will result in alloplasmic male-sterile cultivars (Chaplin 1969, Gerstel 1980, Breiman and Galun 1990). The alloplasmic cultivars display a large variation in the floral structures, especially in the stamen development and corolla morphology (Bonnett et al. 1991). However, only the floral whorls from which the petals and stamens differentiate are affected. The modifications and defects obtained are similar to homeotic mutants which demonstrates that such disorders also can be regulated by nuclear-mitochondrial interacting genes. The disorders obtained affect different stages of stamen development, all the way from an early inhibition of further development of stamen primordia, over stages where anther filaments without and with anther bags are formed to more final stages where fully developed anthers with microspores, but no fertile pollen are formed. (Kofer et al. 1992). Depending on the cytoplasmic donor different modifications of the anthers are obtained which in some cases even lead to development of petaloid structures or into filaments tipped with stigmatoids (Bonnett et al. 1991). The alterations obtained in different stages of stamen development indicate that there are consecutive steps in the anther development that require a cooperative interaction of the nuclear and mitochondrial genomes. Since the nuclear genome in all these cases is from *N. tabacum* and the genome that varies is the mitochondrial, our hypothesis is that several mitochondrial genes are involved in the regulation of flower development (Rosenberg and Bonnett, 1983; Bonnett el al. 1991, Kofer et al. 1992). To test this hypothesis protoplast fusions were performed between sexually produced alloplasmic male-sterile cultivars (Kofer et al. 1990, 1991 a,b). The results obtained from these experiments support this hypothesis. Cybrid plants displaying floral morphologies in a recombined biparental male-sterile pattern as well as novel male-sterile stamen morphologies were obtained, as were also cybrids exhibiting restored pollen production. In the following presentation we will focus on the molecular investigation performed of the cybrids produced between the male-sterile cultivars of tobacco with *N. undulata* cytoplasm, N tab (und) S and tobacco with *N. biglovii* cytoplasm N tab (big) S which resulted in plants with restored stamen development and pollen production (cybrids 29-3) as well as novel male-sterile plants (cybrid 28).

3. A molecular characterization of the mitochondrial DNA and RNA in the pollen producing and novel male sterile cybrids

From a battery of mitochondrial (mt) genes (*atp 6, atp 9, atp A, Cox I, Cyt 6, orf25, rrn8/rrn5, rrn26)* available as heterologous probes, only the probe corresponding to the region of *atp*A from *Zea mays* (Braun and Levings, 1985) revealed a difference in hybridization pattern between sterile and pollen-producing materials. The 5'-end of the *atp*A probe hybridized to one fragment for each parent (fig.1a). In the progeny from the restored hybrids from 29-3 hybridization to the two parental-specific fragments of about 4 kb was obtained. However, when hybridized to the 3' part of *atp*A. the novel male-sterile cybrids (28) hybridized, besides to the two parental specific fragments of about 4 kb also to an additional fragment of about 5 kb, also found in the CMS parent Nta big)S. This fragment could be correlated to the split male-sterile flower type of Nta(big)S as it was only found in the novel male-sterile offspring of cybrid 28 and in the male-sterile parent.

From RNA blot hybridization to the 5'- and 3'- ends of the *atp*A probe from *Zea mays* a correlation of the transcription pattern to the mt-DNA pattern was obtained. The probe covering the 5'-part of *atp*A visualized two major transcripts in the CMS cultivar containing *N. undulata* cytoplasm and three transcripts in the cultivar with *N. biglovii* cytoplasm. In the pollen producing progeny from cybrid 29-3 a sum of these transcripts was found (fig.1b). Regarding hybridization to the 3'-end of *atp*A(fig. 2) additional transcripts of approximately 1500 and 7000 nucleotides were found in the CMS cultivars. In the pollen-producing progeny from 29-3 both of these transcripts were absent, while the novel male-sterile cybrids progeny from callus 28 revealed hybridization to the additional transcript of 7000 nt's. Thus, a clear correlation to the mt DNA pattern was found.

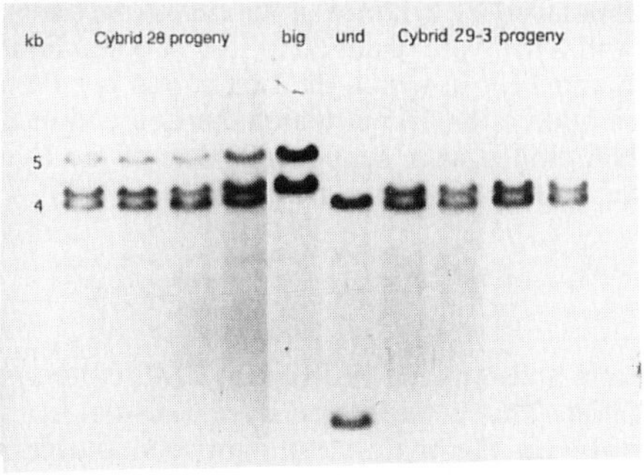

Fig1a Gel blot hybridization to total-DNA of progeny of cybrids from callus 28, progeny from cybrid 29-3 as well as the sterile parents Nta(big)S and Nta(und)S marked as big and und respectively, using a probe homologous to BLSC1.

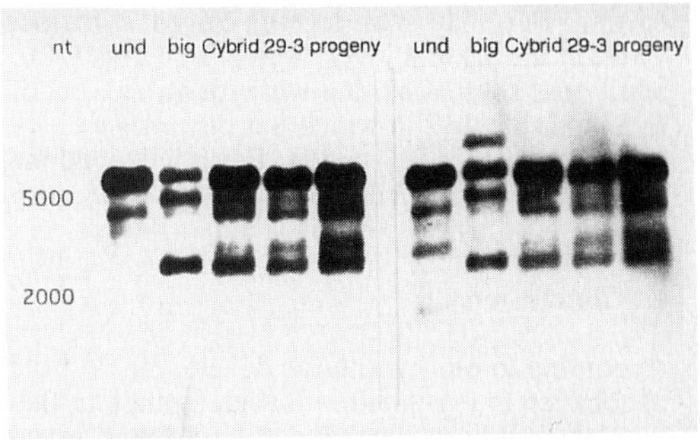

Fig1b mtRNA gel blot hybridizations to progeny of cybrid 29-3 and male-sterile parents. Nta(und)S is marked as und, and Nta(big)S is marked as big.
(A) Hybridization with a *Zea mays* probe covering the 5'-region of the *atp*A gene. (B) Hybridization with probe BLSC1 covering the 3'-flanking part of the coding region of *atp*A.

4. Cloning and sequencing of the 3' specific region of *atp*A correlated to the CMS cultivar of Nta (big) S.

The 5 kb DNA fragment which hybridized specifically to the 3'-end of *atp*A from *Zea mays* was cloned from mt-DNA of Nta (big)S. A *Pst* I- *Hind* III fragment, of 1.7 kb was then subcloned and sequenced (Bergman et al. 1994a).
The sequence from this clone revealed the presence of a chimeric gene, *orf 38/220,* corresponding to an in frame fusion of an *urf* coding for 38 amino acids and the sequence

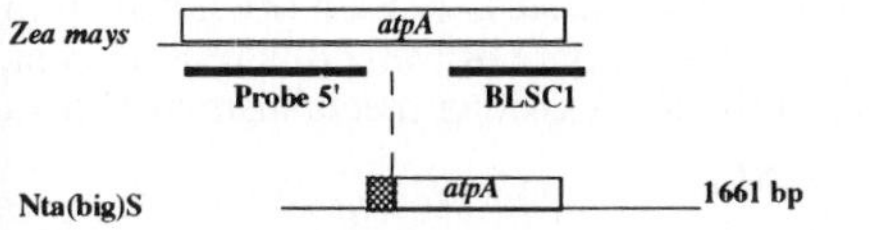

Fig 2 A comparison of the sequenced *atp*A-gene region in Nta(big)S and the *atp*A-gene from *Zea mays.* Open boxes represent *atp* A-gene open reading frames. The hatched vertical line indicates the recombination point in Nta(big)S *atp*A. The grey box marks an unidentified open reading frame fused in frame to the truncated *atp*A-sequence of Nta(big)S. The positions of *Zea mays* Probe 5' (Braun and Levings, 1985) and BLSC1 (Small et al., 1987) are shown by horizontal black bars.

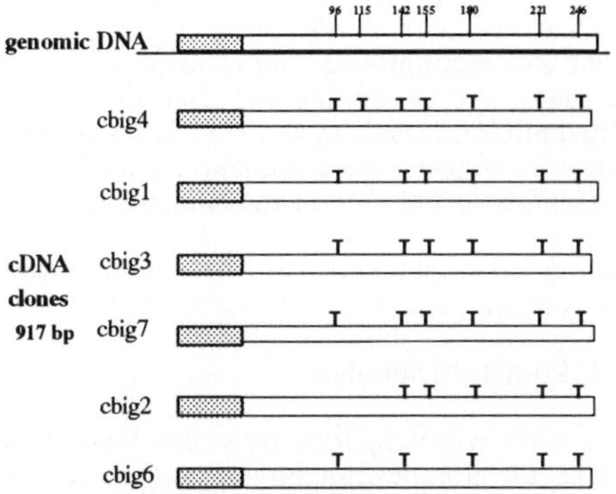

Fig. 3 Comparison of the RNA- editing events in the different *orff30/220* cDNA molecules. Edited codon positions are shown on the genomic DNA.

coding for the carboxy-terminal 220 amino acids of the *atp*A subunit (fig.2). This CMS-correlated *atp*A sequence shows 95% homology to the *Zea mays atp*A gene at the nucleotide level and corresponds to about half of the length of the *atp*A gene.

5. Post-transcriptional editing of the chimeric *orf38/220* gene.

Analysis of cDNAs of *orf38/220* revealed that the gene is a target for post-transcriptional editing (Bergman et al. 1994b). Seven cytosines are replaced by uridines in the mt RNA (fig.3). Thus, at the protein level three Pro codons (180, 22 and 246) were changed to Leu codons. Four additional sites of editing were identified, codons 96 and 115, in which the Leu codon is edited to Ser, codon 142 in which Ser becomes a Leu codon and codon 155 in which the Leu codon is transformed to a Phe codon. Eigth independent cDNA clones were sequenced. One of the five completely sequenced clones is fully edited, while the rest of the clones are partially edited.

6. Conclusions.

According to our hypothesis development of the corolla and stamens in different CMS cultivars of tobacco is regulated by several genes in the mtDNA. This is supported by the results from this work where we have focused our investigation on the male-sterile tobacco cultivar containing mitochondria from *N. bigelovii*. We found a chimeric gene that was transcribed and could be correlated to the male-sterile traits found in the Nta(big)S cultivar.

The finding that the chimeric gene is post-transcriptionally edited supports the speculation that this gene codes for a gene product. RNA editing has been proposed as a mechanism involved in amino acid sequence conservation (Covello and Gray, 1989; Gualberto et al. 1989, Begu et al. 1990). As reported recently by Hernould et al. (1993) RNA editing could be of utmost importance to obtain a functional mitochondrial protein. In agreement with this the editing of the chimeric *orf38/220* gene results in an increased similarity of the part corresponding to the *atp*A carboxy-terminus sequence carrying the active site. The *atp*A C-terminus is involved in the insertion of the functional *atp*A protein in the F_1-ATPase complex. In addition, the N-terminus of the funtional *atp*A which contributes to ADP binding is lost in the *orf38/220* gene (Walker et al. 1984).

Thus, RNA editing of the chimeric gene could be required for the insertion of the chimeric protein in the F_1-ATP ase complex. Due to the absence of the protein region involved in ADP binding, the ORF38/220 protein probably would be inactive even though it might, if correctly folded, compete with the functional *atp*A protein, thus resulting in a malfunction of the ATP synthase complex. Furthermore, the ORF38/220 is nearly identical to the corresponding region of the *Zea mays* protein, and identical to the *N. plumbaginifolia atp*A C-terminus (Chaumont et al. 1988), and the editing increased the homology to the *Beta vulgaris* protein predicted from edited mRNA (Senda et al 1993). RNA editing in the mitochondrial ORF38/220 chimeric gene transcript suggest also a sequence-specific regulation of the RNA editing mechanism as shown by Gualberto et al. (1991) and Kumar and Levings, (1993).

Studies are in progress to show if the chimeric gene indeed causes the abnormal flower development, but also to elucidate the genetic regulation of other CMS cultivars in tobacco.

7. Acknowledgements.

This work was supported by grants from the Swedish Natural Science Council, Lars Hiertas Minne, the Nilsson-Ehle Foundation and the Carl Tryggers Foundation.

8. References.

Aviv, D. and Galun, E. Restoration of fertility in cytoplasmic male sterile (CMS) *Nicotiana sylvestris* by fusion with X-irradiated *N. tabacum* protoplasts. Theor. Appl. Genet., 58 (1980) 121-127.

Bergman, P., Kofer, W., Håkansson, G., and Glimelius K. (1994a). A chimeric and truncated mitochondiral *atp*A-gene is correlated with male sterility in alloplasmic tobacco with *Nicotiana bigelovii* mitochondria. Plant Journal (submitted).

Bergman, P., Hernould, M. and Glimelius, K. (1994b). The chimeric mitochondrial *ORF38/220* gene associated with cytoplasmic male-sterility in tobacco is post-transcriptionally edited. Physiol. Plant. (in press).

Begu, D., Graves PV., Domec C., Arselin G., Litvak S., Araya A. (1990) RNA editing of wheat mitochondrial ATP synthase subunit 9: direct protein and cDNA sequencing. Plant Cell 2: 1283 1290

Belliard, G., Pelletier, F., Vedel and Quetier, F. Morphological characteristics and chloroplast DNA distribution in different cytoplasmic parasexual hybrids of *Nicotiana tabacum.* Mol. Gen. Genet., 165 (19789 231-237.

Bonnett, H.T., Kofer, W., Håkansson, G. and Glimelius, K. (1991). Mitochondrial involvement in petal and stamen development studied by sexual and somatic hybridization of *Nicotiana* species. Plant Science 80, 119-130.

Braun, C.J. and Levings, C.S. III. (1985). Nucleotide sequence of the F_1-ATPase alpha-subunit gene from maize mitochondria. Plant Physiol. 79, 571-577.

Breiman, A. and Galun, E. Nuclear-mitochondrial interrelation in Angiosperms. Plant Sci., 71 (19909 3-19.

Chaplin, J.F., Use of male-sterile tobaccos in the production of hybrid seed. Tob. Sci., 8 (1964) 105-109.

Chaumont, F., Boutry, M., Briquet, M. and Vassarotti, A. (1989. Sequence of the gene encoding the mitochondrial F_1 ATPase alpha subunit from *Nicotiana.*

Covello PS, Gray MW (1989). RNA editing in plant mitochondria. Nature 341:662-666

Gerstel, D.U. Cytoplasmic male sterility in *Nicotiana* (a review). NC Agric. Res. Serv. Tech. Bull., 263 (1980) 1-31.

Glimelius, K. Chen, K. and H.T. Bonnett, Somatic hybridization in *Nicotiana:* Segregation of organellar traits among hybrid and cybrid plants. Planta 153 (1981) 504-510.

Gualberto JM, Lamattina L., Bonnard G., Weil JH., Grienenberger JM. (1989) RNA editing in wheat mitochondria results in the conservation of protein sequences. Nature 341:660-662

Hernould M., Suharsono S., Litvak S., Araya A., Mouras A. (1993) Male-sterility induction in transgenic tobacco plants with an unedited *atp9* mitochondrial gene from wheat. Proc. Nat. Acad. Sci. 90:2370-2374

Hosfield, G.L. and Wernsman, E.A. Effect of an alien cytoplasm and fertility restoring factor on growth, agronomic characters, and chemical constituents in a male-sterile variety of flue-cured tobacco. Crop Sci., 14 (1974) 575-577.

Kaul, M.L.H. (1988). Male sterility in higher plants, Chapter 3. In *Monographs on Theoretical and Applied Genetics,* volume 10 (Frankel, R., Grossman, M., Linskens, H.F., Maliga, P. and Riley, R., eds). Berlin: Springer-Verlag, pp. 97-192.

Kofer, W., Glimelius, K. and Bonnett, H.T. (1990). Modifications of floral development in tobacco induced by fusion of protoplasts of different male-sterile cultivars. Theor. Appl. Genet. 79, 97-102.

Kofer, W., Glimelius, K. and Bonnett, H.T. Restoration of normal stamen development and pollen formation by fusion of different cytoplasmic male-sterile cultivars of *Nicotiana tabacum. Theor. Appl. Genet., 81 (1991a) 390-396*

Kofer, W., Glimelius, K. and Bonnett, H.T. Modifications of mitochondrial DNA cause changes in floral development in homeotic-like mutants of tobacco. The Plant Cell, (1991b) 759-767.

Kofer, W., Glimelius, K. and Bonnett, H.T. (1992). Fusion of male-sterile tobacco causes modifications of mtDNA leading to changes in floral morphology and restoration of fertility in cybrid plants. Physiologia Plantarum 85:334-338.

Rosenberg, S.M. and Bonnett, H.T., Floral organogenesis in *Nicotiana tabacum:* a comparison of two cytoplasmic male-sterile cultivars with a male-fertile cultivar. Am. J. Bot., 70 (19839 266-275.

Senda M., Mikami T., Kinoshita, T., (1993) The sugar beet mitochondrial gene for the ATPase alpha-subunit: sequence, transcription and rearrangements in cytoplascmi malesterile plants. Curr. Genet. 24:164-170.

Small, I.D., Isaac, P.G. and Leaver, C.J. (1987). Stoichiometric differences in DNA molecules containing the *atp*A gene suggest mechanisms for the generation of mitochondrial genome diversity in maize. EMBO J. 6, 865-969.

Walker JE, Saraste M., Gay NJ (1984) The unc operon. Nucleotide sequence, regulation and structure of ATP-synthase. Biochim. Biophys. Acta 768:164-200

EXPLOITING SOMACLONAL VARIATION - ESPECIALLY GENE INTROGRESSION FROM ALIEN CHROMOSOMES

Larkin, P.J. and Banks, P.M.[1]

CSIRO Division of Plant Industry, P.O.Box 1600, Canberra 2601, Australia

[1] present address: Queensland Wheat Research Institute, Toowoomba, Australia 4350

ABSTRACT The following example is discussed of using somaclonal variation to achieve transfer of valuable genes from an alien chromosome to a crop plant. Barley yellow dwarf virus (BYDV) resistance has been transferred to wheat (*Triticum aestivum*) from a disomic addition line with a *Thinopyrum* (*Agropyron*) *intermedium* chromosome. The resistance locus is on the long arm of the *7Ai-1* chromosome. BYDV resistant recombinant lines were identified after selection against a short arm marker (red coleoptile) and for BYDV resistance. Resistance in seven of the cell culture induced recombinants has been inherited with Mendelian segregation ratios for eight generations. Heterozygote meiosis indicates that the alien chromatin in the cell culture induced recombinants is small enough to allow regular meiotic behaviour. A probe, pEleAcc2, hybridising to dispersed repetitive sequences, was used to detect *Th. intermedium* chromatin. The translocations involved a portion of alien chromatin less than the long arm of *7Ai-1*. RFLP analysis confirmed the loss of a short arm marker (probe psr152) but the retention of two long arm markers (probes psr129 and csIH81-1) which assort with the BYDV resistance. A third long arm alien marker (probe psr690), was absent in one of the recombinants.

Introduction

Barley Yellow Dwarf Virus (BYDV) is a phloem-limited luteovirus and one of the most economically significant viral pathogens of small grain cereals. The disease has been identified in all cereal growing regions of the world. In wheat, partial tolerance has been identified in several lines, but resistance to BYDV, as substantially reduced virus accumulation, has not been reported in wheats despite extensive screening.

A number of perennial grasses related to wheat, have been reported to be resistant or immune to BYDV (Sharma *et al.* 1984; Comeau and Plourde 1987; Larkin *et al.* 1990a; Banks et al, 1991; Xu *et al.* 1993). *Thinopyrum (Agropyron) intermedium* has been suggested as an amenable source of BYDV resistance for wheat (Sharma *et al.* 1984; Xin *et al.* 1988).

We identified a major BYDV resistance factor conferred by the ß arm (long arm) of a homoeologous group 7 *Thinopyrum intermedium* chromosome (Brettell *et al.* 1988), also called 7Ai-1L (Friebe et al, 1992). 7Ai-1L is the long arm of the added chromosome in line L1 (2n=44)(Cauderon, 1966). The resistance has been demonstrated to be effective against Australian RPV and PAV isolates (Banks *et al.* 1992; Brettell *et al.* 1988), as well as Chinese GPAV, GAV and GPV isolates (Qian, Zhou, Zhou and Cheng, unpublished), and European PAV

M. Terzi et al. (eds.), Current Issues in Plant Molecular and Cellular Biology, 225-234.

and MAV isolates (Banks and Jahier, unpublished). No resistance breaking isolate has been discovered.

We have utilised two methods to induce genetic exchange between the group 7 *Thinopyrum* chromosome and wheat chromosomes: (i) non-homologous exchange associated with cell culture (Larkin and Scowcroft 1981; Davies *et al.* 1986); and (ii) homoeologous pairing induced by the *ph* mutant gene and deleted chromosome 5B (Wang *et al.* 1977).

Materials and methods

CELL CULTURE

Cultures were initiated either from (a) 12 to 14 day old F_1 embryos (2n=43) of L1 (2n=44) x wheat (2n=42); or (b) immature inflorescences of approximately 5-35 mm length from F_1 hybrid plants of L1 x wheat. All media were as described in Larkin *et al*, 1984.

RED COLEOPTILE SCREENING

Seeds were surface sterilised in bleach and germinated for 2 to 3 days at 20°C. When the shoots were 5 to 10 mm long, they were transferred to conditions of 15°C and 200 $\mu E.m^{-2}.s^{-1}$. After 24 hours, the colour of the coleoptiles was assessed. All individual seedlings expressing the red coleoptile character were culled. The remaining seedlings were screened for BYDV resistance.

BYDV RESISTANCE TESTING

Germinated seeds were space planted into seedling trays and grown for 1 week at 15 to 18°C prior to inoculation. Viruliferous aphids were produced by propagating *Rhopalosiphum padi* on barley and then feeding the aphids for 3 days on oats which had previously been infected with a BYDV-PAV isolate. After more than five viruliferous aphids were transferred to each, seedlings were individually covered with a perspex tube (covered at the top with fine mesh) and maintained in lighted conditions at 15°C for 3 to 4 days. The tubes were then removed, the aphids killed with pyrethrins and the trays transferred to growth cabinets at 18°/16°C and a 12 hour photoperiod of 300 to 400 $\mu E.m^{-2}.s^{-1}$. After 2 weeks growth, sap extracts were prepared from approximately 0.5 g samples of the youngest fully expanded leaves and virus titres assessed by ELISA. The double antibody sandwich ELISA for detecting BYDV has been described previously (Sward and Lister 1987; Xin *et al.* 1988).

INDUCING RECOMBINATION BETWEEN THE GROUP 7 *THINOPYRUM* CHROMOSOME AND WHEAT

Genetic exchange between the group 7 *Thinopyrum* chromosome and wheat chromosomes was promoted by two methods: (i) non homologous exchange associated with cell culture-induced breakage and fusion; or (ii) homoeologous meiotic pairing induced by the *ph* mutant gene and deleted 5B chromosome. The cell culture procedure, described above, was applied to immature embryos or inflorescences which were genetically monosomic for the alien chromosome (2n=43) and produced by backcrossing disomic addition plants (2n=44) to wheat.

Eighty five families utilising the ph mutant approach were created by Dr R.A. McIntosh, University of Sydney, Australia and kindly provided to us.

SELECTING BYDV RESISTANT TRANSLOCATION FAMILIES

Individual seedlings which expressed the red coleoptile character in the first generation of screening and progeny lines in the second generation of screening were culled. This culling was designed to eliminate from the selection program most of the material which carried the entire 7Ai-1 chromosome or its short arm. Screening for BYDV resistance was conducted each generation and only resistant plants propagated. In the third generation of screening, when the numbers of plants in each progeny line could be increased to more than 25, only families in which the segregation ratio of resistance was consistent with Mendelian ratios were maintained. The chromosome number of resistant plants in the remaining families was determined from root tip squashes and all families with more than 42 chromosomes were culled. The 42 chromosome, BYDV resistant families were backcrossed to several commercial wheats. Backcross material was used for meiotic chromosome pairing analysis. Following two generations of selfing, DNA from homozygous resistant plants was analysed by hybridisation to the DNA probes in Table 1.

DNA ANALYSIS

DNA was extracted according to the procedure of Appels and Moran (1984) and DNA hybridisation conducted according to Lagudah *et al* (1991). A number of clones which hybridise to loci on group 7 chromosomes and a couple of dispersed repetitive probes which assay the *Th. intermedium* chromatin in a wheat background were used. These clones are listed in Table 1. When required, the intensity of hybridisation was quantified using either phosphoImager analysis of the hybridisation membrane or densitometer analysis of the X-ray film.

Table 1. DNA probes used in the analysis of the translocations

Clone	Source	Location	Reference
psr152	*T.aestivum*	7AL, 7BL, 7DL	Sharp *et al.* 1989
psr129	*T.aestivum*	7AL, 7BL, 7DL, 7RL	Sharp *et al.* 1989
csIH81-1	*T.tauschii*	7DL, 7H	Lagudah *et al.* 1991
psr690	*T.aestivum*	7AL, 7BL, 7DL, 7RL	Rognli *et al.* 1992
pEleAcc2	*A. elongatum*	dispersed	Hohmann *et al.* 1994
A600	*A. elongatum*	dispersed	"

Results

RECOMBINANT SELECTION

BYDV resistant recombinant lines were identified after three or more generations of selection against the red coleoptile marker and for BYDV resistance. From 1200 plants regenerated from cultured monosomic addition material, 14 families were identified in which BYDV resistance was inherited consistently with Mendelian segregation ratios in early generations, coleoptiles of progeny were green and the chromosome number was 42. Seven of the 14 families were deleted from the backcrossing program because the proportion of resistant plants in pollen cross progeny from heterozygous resistant plants was less than 50% and/or meiotic chromosome pairing in heterozygous resistant backcross plants was irregular. Six of the seven TC lines were derived from regenerated plants from cultured immature inflorescence material and one, TC9, from a cultured immature embryo.

Individuals with red coleoptiles were identified and culled in the early generations of some families. Only sub-families in which all resistant individuals had green coleoptiles were propagated.

From the 85 families derived from Sears' *ph1b* mutant, one family, called 5395, was identified with green coleoptiles, 42 chromosomes and BYDV resistance segregation ratios which have been consistent with Mendelian ratios for eight generations. The 85 families do not represent a random population because they were previously selected for promising segregation ratios for stem rust resistance, which resides on the short arm of 7Ai-1.

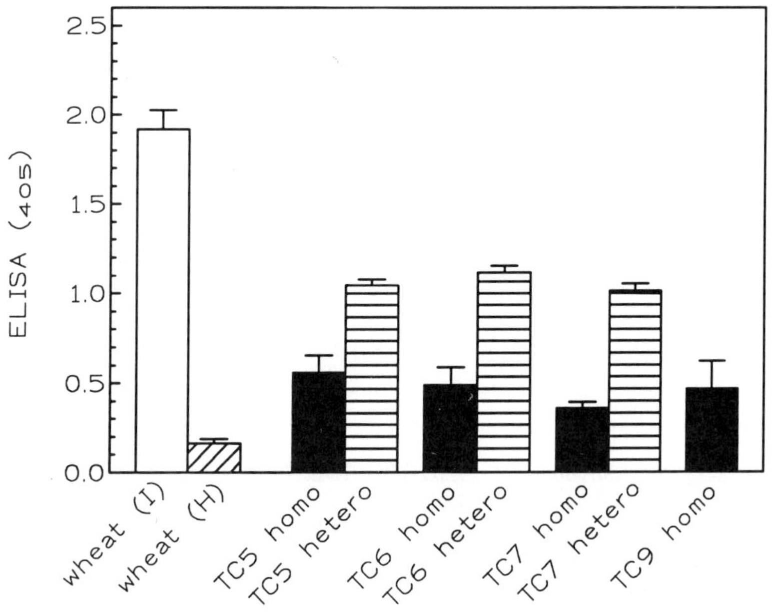

Figure 1. ELISA values for barley yellow dwarf virus in wheat, with (I) and without (H) infection, and in the lines TC5, TC6, TC7 and TC9 where the resistance gene is in either heterozygous (horizontal stripes) or homozygous (solid fill) condition.

Additive genetic reduction in virus titre

The BYDV resistance of the L1 derived recombinant families, appears to be controlled by an additive genetic factor. The suppression of virus multiplication was found to be greater in homozygous resistant plants compared to heterozygous resistant plants with a similar wheat background (Fig. 1).

SYMPTOM SUPPRESSION

The leaf yellowing symptoms of BYDV were only occasionally observed in the screening program in the Canberra glasshouses. However symptom expression was noticable in the field in both Canberra and Beijing where also the resistance gene gave dramatic suppression of symptoms.

REGULAR MEIOTIC PAIRING OF HETEROZYGOTES OF THE TC LINES

Regular meiotic pairing (21 bivalents) was observed in heterozygous resistant plants of TC5, TC6, TC7, TC9 and TC10. The analysis was conducted with F1 cross plants to wheat cultivars. PMCs with 21 ring bivalents were observed, suggesting pairing can be complete and involve all arms. By contrast, meiotic pairing in the ph1b mutant/L1 derived family, 5395, was usually irregular. In F1 plants of 5395 crossed and backcrossed to wheat, the usual meiotic configuration was 20 bivalents + 2 univalents. The two unpaired chromosomes appeared to differ slightly in length and arm ratios. In about 5% of PMCs of the six F1 plants, 21 bivalents were observed.

RFLP MARKERS

Table 2 Association of disease, seedling, RFLP and repetitive sequence markers with the parental and recombinant lines

	7Ai-1 short		7Ai-1 long				Acc2 dispersed probe
	red coleoptile	psr152	psr690	$BYDV^R$	psr129	csIH81-1	
TAF46 (2n=56)	+	+	+	+	+	+	+
wheat	-		-	-	-	-	-
L1 addition (2n=44)	+	+	+	+	+	+	+
7Ai-1L telo addition	-	-	+	+	+	+	+
7Ai-1S telo addition	+	+	-	-	-	-	+
TC5	-	-	+	+	+	+	+
TC6	-	-	+	+	+	+	+
TC7	-	-	+	+	+	+	+
TC8	-	-	+	+	+	+	+
TC9	-	-	+	+	+	+	+
TC10	-	-	+	+	+	+	+
TC14	-	-	-	+	+	+	+
5395	-	-	+	+	+	+	+

Two 7Ai-1 short arm markers, red coleoptile and the RFLP marker detected using probe psr152, are absent in all the TC lines and the 5395 recombinant (Table 2). This, together with the cytological data, confirms that these lines are not whole chromosome substitutions. In addition two 7Ai-1 long arm markers, using RFLP probes psr129 and csIH81-1, both assorted with the BYDV resistance in all the recombinants. A third 7Ai-1 long arm RFLP marker, using probe psr690, assorted with BYDV resistance in all but TC14 (Table 2). It is not known how closely linked any of these loci are to each other on 7Ai-1L. In view of the broken linkage of psr690 with TC14, it is interesting to note that in wheat the psr690 marked locus is less than 2 map units from the centromere (Rognli *et al.* 1992).

All homozygous recombinant families, except TC7, appear to have involved loss of wheat 7DL loci, evidenced with the psr129 and psr690 probe. When the DNA is from a plant hemizygous for the recombinant chromosomes, except TC7, the 7D wheat band is of reduced intensity relative to the 7A and 7B bands. When the DNA is from a homozygous plant the 7D band is missing.

ASSAYING ALIEN CHROMATIN WITH DISPERSED REPETITIVE PROBE

The dispersed repeated DNA sequence probes pEleAcc2 and A600, hybridise to a wheat band as well as a *Thinopyrum* band. The ratio of hybridisation signals between the two bands gives a measure of the relative amount of *Thinopyrum* chromatin in the lines. This is confirmed by the dose response observable with the monotelosomic and ditelosomic long arm additions and also by comparing the short arm to the long arm additions (Table 3). All of the recombinants appear to have less *Thinopyrum* chromatin than the 7Ai-1L telosomic addition. Although the repetitive sequence being analysed may not be uniformly dispersed along the chromosomes, the comparisons in Table 3 are with the long arm from which the recombinants derive. Therefore we infer that there is less than a whole long arm present in the recombinants.

Table 3. Ratio of *Thinopyrum* to wheat band in Southerns hybridised to the dispersed repetitive probe Acc2

	Ratio of Ti/wheat band with Acc2 (± SD)	
	heterozygote	**homozygote**
telosomic 7Ai-1S		0.30
telosomic 7Ai-1L	0.30	0.67 ± 0.06
TC5	0.19	0.33 ± 0.05
TC6	0.19	0.30 ± 0.04
TC7	0.12 ± 0.04	0.39 ± 0.01
TC8		0.36 ± 0.08
TC9	0.24	0.39
TC10	0.25	0.55
TC14		0.31
5395	0.27	0.40 ± 0.06

Discussion

BYDV resistance has been transferred to bread wheat from the long arm of a group 7 *Thinopyrum intermedium* chromosome (7Ai-1). The resistance has been transferred in several independent chromosomal translocations or recombinants. The wheats produced in this introgression program are the first true bread wheats with a gene which reduces BYD virus multiplication. The resistance is additive and is inherited as a simple Mendelian factor in the recombinant families. Because of the effectiveness of the resistance against all of the isolates of BYDV tested in Australia, China and Europe, and because of its simple inheritance, we expect that the resistance will be widely used.

Isolation of the desirable translocation families from material regenerated from cell cultures supports the proposal of Larkin and Scowcroft (1981) that cell culture-induced chromosome rearrangement can be used to translocate desirable genes from alien chromosomes - an idea also suported by others (eg Sharma and Baenziger, 1986; Lapitan et al, 1984; McCoy et al, 1982).

The frequency of translocations conferring BYDV resistance was about 1% (14 translocation families from 1200 regenerated plants). In simple backcross material from L1 to commercial wheats, recombination was not identified (X Chen, pers. comm.) when no tissue culture phase was included, emphasising the importance of the cell culture step in the procedure.

Mutations have been discovered at the homoeologous pairing inhibitor (Ph) loci in wheat (review Sears, 1984). For a very limited number of alien genomes of the Triticeae there is little recombination barrier with the wheat genomes (A, B and D). For others, such as the *Agropyron* E genome, mutations at the Ph1 locus have enabled a sufficient relaxation of the suppression of homoeologous pairing, that recombinants with wheat have been recovered. However it appears that there is a biological ceiling for homoeologous pairing which the present mutations have reached (Sears, 1984). Not all Triticeae genomes are able to recombine with the wheat genomes even in the presence of the most effective mutation, ph1b. The rye genome is sufficiently unresponsive to ph1b that it is usually considered unlikely to be of practical use for exchanging genes from rye to wheat (Jouve and Giorgi, 1986; Naranjo et al, 1988). The chromosomes of the barley genome do not seem to respond at all to the ph1b effect.

The question therefore arises whether the cell culture approach has advantages over the use of the ph mutants which genetically relax the suppression of homoeologous pairing and thereby increase the possibility of recombination between wheat and alien chromosomes. Despite our early ambitions to furnish an appropriate comparison, our research does not supply adequate data on the frequency of ph-induced recombination with the 7Ai-1 chromosome, and no conclusion would be justifiable on this question. Irregular pairing in heterozygotes of the *ph*-mutant induced recombinant, 5395, suggests that it possesses a larger portion of the alien chromatin than the cell culture derived recombinants. This conclusion however is not strongly supported by the ratio of *Thinopyrum* to wheat bands using the Acc2 and A600 dispersed probes (Table 3).

We have previously described evidence for cell culture induced recombination between wheat and rye chromosome 6R (Larkin et al, 1990b). Similarly, following cell culture of wheat x rye hybrids, exchanges have been observed between 1R and 4D, and 3R and 2B (Lapitan et al, 1984; Lapitan et al, 1986). At the very least we may conclude that the cell culture approach will be a valuable alternative to *ph*-induced recombination in wheat. X- and gamma rays have also been used to induce translocations between rye and wheat (Friebe et al, 1990, 1993)

In most species, of course, there are no mutations available analagous to the ph system. In these the cell culture approach to alien chromosome recombination may prove particularly valuable.

Despite these results in wheat and the promising indications for breeding other species, a word of caution is necessary. The evidence is only circumstantial that the recombinations conferring BYDV resistance were induced during the cell culture procedure. It is conceivable that they are the result of meiotic recombination. Yachevskaya et al (1988) reported that crosses between *Th. intermedium* and wheat could overcome the expression of the homoeologous suppression genes on chromosome 5B. Zhang *et al.* (1992) similarly concluded from meiotic studies of F1 and BC1 plants of *Th. intermedium* x wheat, that the *Th. intermedium* had genes which promote homoeologous chromosomes to pair. However if this were the case for the addition line, L1, we would expect it to be meiotically unstable. We have not observed such instability and it has not been reported by any of those who have worked with this line. It therefore seems more likely that the recombinations reported here, and which were identified in regenerants after advancing through only one meiosis, are a result of the cell culture step.

The mechanism of cell culture induced somatic recombination is unknown. Nuti Ronchi (1990) has observed chromosome association and chiasmata-like chromosome interactions in somatic cells during cell culture. If these observations prove valid and general, a type of somatic meiosis may explain the high rate of cell culture induced recombination. Alternatively, rapid and continuing somatic cell divisions in cultured cells may cause chromosome breakage and fusion.

Acknowledgements

Much of the work was done with close collaboration of colleagues in the Chinese Academy of Agricultural Sciences, Chen Xiao, Xu Huijun, Xin Zhiyong, Qian Youting, Zhou Zhiming, Cheng Zhuomin and Zhou Guanghe. This work was conducted with the support of grants from the Australian Centre for International Agricultural Research (ACIAR) and the Grains Research and Development Corporation (GRDC).

References

Appels, R. and Moran, L.B. (1984). Molecular analysis of alien chromatin introduced into wheat. *In* Gene Manipulation in Plant Improvement. *Edited by* J.P.Gustafson. Stadler Genet. Symp. 16th, 1984. pp.529-557.

Banks, P.M., Waterhouse, P.M. and Larkin, P.J. (1992). Pathogenicity of three RPV isolates of Barley Yellow Dwarf Virus on barley, wheat and wheat alien addition lines. Ann. Appl. Biol. **121**: 305-314.

Banks, P.M., Xu, S.-J., and Larkin, P.J. (1991). Sources of resistance to BYDV in wheat-related grasses. *In* Aphid-Plant Interactions: Populations to Molecules. Edited by D.C. Peters, J.A. Webster and C.S. Chouber. USDA/ARS, Oklahoma State Uni, p.324.

Brettell, R.I.S., Banks, P.M., Cauderon, Y., Chen, X., Cheng, Z.M., Larkin, P.J. and Waterhouse, P.M. (1988). A single wheatgrass chromosome reduces the concentration of barley yellow dwarf virus in wheat. Ann. Appl. Biol. **113**: 599-603.

Cauderon, Y. (1966). Etude cytogénétique de l'évolution du matériel issu de croisement entre *Triticum aestivum* et *Agropyron intermedium*. I. Création de types d'addition stables. Ann. Amélior. Plant. 16: 43-70.

Comeau, A. & A. Plourde, (1987). Cell tissue culture and intergeneric hybridization for barley yellow dwarf virus resistance in wheat. Can. J. Plant Path. 9: 188-192

Davies, P.A., Pallotta, M.A., Ryan, S.A., Scowcroft, W.R. and Larkin, P.J. (1986). Somaclonal variation in wheat : genetic and cytogenetic characterisation of alcohol dehydrogenase 1 mutants. Theor.Appl.Genet. 72: 644-653.

Friebe, B., Hatchett, J.H., Sears, R.G., Gill, B.S. (1990) Transfer of Hessian fly resistance from "Chaupon" rye to hexaploid wheat via a 2BS/2RL wheat-rye chromosome translocation. Theor. Appl. Genet. 79: 385-389

Friebe, B., Mukai, Y., Gill, B.S. and Cauderon Y. (1992). C-banding and *in situ* hybridization analyses of *Agropyron intermedium*, a partial wheat x *Ag.intermedium* amphiploid, and six derived chromosome addition lines. Theor. Appl. Genet. 84: 899-905.

Friebe, B., Jiang, J.., Gill, B.S. and Dyck, P.L. (1993). Radiation-induced nonhomoeologous wheat-*Agropyron intermedium* chromosome translocations conferring resistance to leaf rust. Theor. Appl. Genet. 86: 141-149.

Hohmann, U., Appels, R., Ohm, H. and Hogie, L. (1994). Amplification of DNA sequences in wheat and its relatives: the Acc2 family of *Agropyron* (syn. *Thinopyrum*) amplified sequences. Genome (in press)

Jouve, N. and Giorgi, B. (1986) Analysis of induced homoeologous pairing in hybrids between 6X triticale *ph1* mutant and *Triticum aestivum* L. Can. J. Genet. Cytol. 28:696-700.

Lagudah, E.S., Appels, R., Brown, A.D.H. and McNeil, D. (1991). A molecular genetic analysis of *Triticum tauschii* - the D genome donor to hexaploid wheat. Genome 34: 375-386.

Lapitan, N., Sears, R. and Gill, B. (1984) Translocations and other karyotypic structural changes in wheat x rye hybrids regenerated from tissue culture. Theor.Appl.Genet. 68:547-554.

Lapitan N, Sears R, Rayburn A, Gill B (1986) Wheat-rye translocations. J.Heredity 77:415-419.

Larkin, P. J. (1985). In vitro culture and cereal breeding. IN Cereal Tissue and Cell Culture (Eds. SWJ Bright, MGK Jones). pp 273-296.

Larkin, P.J. and Scowcroft, W.R. (1981). Somaclonal variation - a source of variability from cell cultures for plant improvement. Theor. Appl. Genet. 60:197-214.

Larkin, P.J., Ryan, S.A., Brettell, R.I.S. and Scowcroft, W.R. (1984). Heritable somaclonal variation in wheat. Theor. Appl. Genet. 78: 443-456.

Larkin, P.J., R.I.S. Brettell, P. Banks, R. Appels, P.M. Waterhouse, Z.M. Cheng, G.H. Zhou, Z.Y. Xin & X. Chen, (1990a). Identification, characterization, and utilization of sources of resistance to barley yellow dwarf virus. *In* Burnett, P.A. (ed.), World Perspectives on Barley Yellow Dwarf, pp 415-420, CIMMYT, Mexico, D.F., Mexico.

Larkin, P.J., Spindler, L.H. and Banks, P.M. (1990b) 'The use of cell culture to restructure plant genomes for introgressive breeding', in Kimber, G. (ed.), Proceedings of the Second International Symposium on Chromosome Engineering in Plants, Uni of Missouri-Columbia, pp. 80-89.

McCoy TJ, Phillips RL, Rines HW (1982) Cytogenetic analysis of plants regenerated from oat (*Avena sativa*) tissue cultures; high frequency of partial chromosome loss. Can. J. Genet. Cytol. 24:37-50.

Naranjo T, Roca A, Giraldez R, Goicoechea PG (1988) Chromosome pairing in hybrids of ph1b mutant wheat with rye. Genome 30:639-646.

Nuti Ronchi, V. (1990). Cytogenetics of plant cell cultures. In Bhojwani, S.S. and Novak, F.J. (eds.), Plant Tissue Culture: Applications and Limitations, pp 276-300, Elsevier, Amsterdam.

Rognli, O.A., Devos, K.M., Chinoy, C.N., Harcourt, R.L., Atkinson, M.D. and Gale, M.D. (1992). RFLP mapping of rye chromosome 7R reveals a highly translocated chromosome relative to wheat. Genome 35: 1026-1031

Sears ER (1984) Mutations in wheat that raise the level of meiotic chromosome pairing. In: Gene Manipulation in Plant Improvement, 16th Stadler Genetics Symposium, JP Gustafson (Ed.) pp.295-300. Plenum Press, New York

Sharma HC, Baezinger PS (1986) Production, morphology, and cytogenetic analysis of *Elymus caninus* (*Agropyron caninum*) X *Triticum aestivum* F1 hybrids and backcross-1 derivatives. Theor Appl Genet 71:750-756.

Sharma, H.C., Gill, B.S. and Uyemoto, J.K. (1984). High levels of resistance in Agropyron species to barley yellow dwarf and wheat streak mosaic viruses. Phytopath. Z. 110: 143-147

Sharp, P.J., Chao, S., Desai, S. and Gale, M.D. (1989). The isolation, characterization and application in the Triticeaea of a set of wheat RFLP probes identifying each homoeologous chromosome arm. Theor. Appl. Genet. 78: 342-348

Sward, R.J. & R.M. Lister, (1987). The incidence of barley yellow dwarf viruses in wheat in Victoria. Aust. J. Agric. Res. 38: 821-828

Wang, R. C., Liang, G. H. and Heyne, E. G. (1977). Effectiveness of ph gene in inducing homoeologous chromosome pairing in Agrotricum. Theor. Appl. Genet. 5: 139-142.

Xin, Z.Y., R.I.S. Brettell, Z.M. Cheng, P.M. Waterhouse, R. Appels, P.M. 'Banks', G.H. Zhou, X. Chen & P.J. Larkin, (1988). Characterization of a potential source of barley yellow dwarf virus resistance for wheat. Genome 30: 250-257

Xu, S.J., Banks, P.M., Dong, Y.S., Zhou, R.H. and Larkin, P.J. (1993) Evaluation of Chinese Triticeae for resistance to barley yellow dwarf virus (BYDV). Genetic Resources and Crop Evolution (in press)

Yachevskaya, G.L., Pukhal'skii, V.A., Lapochkina, I.F., Dovnar, T.I. and Volkova, G.A. (1988) Cytogenetic characteristics of wheat-Agropyron and wheat-Aegilops F1 and F1BC1 hybrids. English abstract in Wheat, Barley and Triticale Abstracts 5(5): abstract 4197. Original in Referativnyi Zhurnal (1987) 7.65.62

Zhang, X., Li, Z. and Chen, S. (1992). Production and identification of three 4Ag(4D) substitution lines of *Triticum aestivum - Agropyron*: relative transmission rate of alien chromosomes. Theor. Appl. Genet. 83: 707-714.

ACETOHYDROXYACID SYNTHASE GENE AMPLIFICATION INDUCES CLORSULFURON RESISTANCE IN DAUCUS CAROTA L.

S. CARETTO, M.C. GIARDINA, C. NICOLODI and D. MARIOTTI
Institute of Plant Biochemistry and Ecophysiology - CNR
Via Salaria km 29,300 - 00016 Monterotondo Scalo (Rome) ITALY

ABSTRACT. *In vitro* cell selection was used to isolate *Daucus carota* L. cell cultures resistant to increasing concentrations of the sulfonylurea herbicide chlorsulfuron (CS). The stepwise scheme produced cells 300-fold more resistant to CS than the wild type. CS resistance was stably maintained after several passages in the absence of selective pressure. During the selection the specific activity of acetohydroxyacid synthase (AHAS), the sulfonylurea target enzyme, increased along with resistance. Dot-blot analysis carried out on the genomic DNA of the selected cultures, using a heterologous AHAS probe, revealed that AHAS gene copy number also increased gradually. Gene amplification was considered responsible for the overexpression of AHAS activity and the consequent CS resistance. Morphogenetic ability of the selected cultures was not affected since plants were regenerated from the resistant cell lines. In several cases the resistant trait was maintained at the whole plant level.

INTRODUCTION

The achievement of herbicide resistance in crop species represents one of the expected major applications of plant biotechnology. At present two approaches are available and can be used alternatively to produce plants resistant to a wide range of herbicide molecules. The first one consists of transferring into crops either genes encoding herbicide insensitive forms of the target enzymes or genes encoding enzymes able to degrade or detoxify the herbicide molecules (Botterman and Leemans, 1988). In the second approach the wide genetic variability characteristic of *in vitro* cultures, extensively studied as somaclonal variation, is exploited to select herbicide resistant cell lines from which plants can be regenerated. Recently, *in vitro* cell selection has been extensively used since it offers the advantage of isolating cell mutants characterized by different mechanisms of herbicide resistance which can be further investigated.

Gene amplification is a common mechanism of drug resistance that has been largely investigated over the last 15 years mostly in animal cell systems. It has been reported that when inhibitors that kill mammalian cells or arrest their growth were used to select resistant mutants a substantial increase of the activity of a protein, the target of the inhibitor, was observed and in several cases gene amplification was the basis of resistance (Stark and Wahl, 1984). To our knowledge, the only cases of gene amplification that have been so far reported in plant systems

M. Terzi et al. (eds.), Current Issues in Plant Molecular and Cellular Biology, 235-240.

are referred to a selective response of tissue cultures to herbicide treatments; no information has been given about such mechanism at the whole plant level. An overproduction of the target enzyme 5-enolpyruvylshikimate 3-phosphate synthase (EPSPS) due to gene amplification has been shown in different plant cell lines selected with the herbicide glyphosate (Shah et al., 1986; Hauptmann et al., 1988; Goldsbrough et al., 1990; Shyr et al, 1992; Shyr et al., 1993). Such mechanism was also the basis of phosphinotricin resistance in alfalfa cells where a substantial increase in the gene copy number of the target enzyme glutamine synthetase was found (Donn et al., 1984).

Chlorsulfuron (CS, commercial name "Glean", Dupont) is a sulfonylurea herbicide known to act by inhibiting the enzyme acetohydroxyacid synthase (AHAS,EC 4.1.3.18) which catalyses the first step in the biosynthesis of leucine, isoleucine and valine (Chaleff and Mauvais, 1984; Ray, 1984). Up to date, a large number of reports have been published on the isolation of cell lines and plants resistant to sulfonylurea herbicides (Chaleff and Ray, 1984; Haughn and Somerville, 1986; Jordan and McHughen, 1987; Sebastian and Chaleff, 1987; Saxena and King, 1988; Swanson et al., 1988; Saxena et al., 1990; Harms and Di Maio, 1991; Saunders et al., 1992; Caretto et al., 1993). Differently from the herbicides mentioned above, the mechanism shown to underly sulfonylurea resistance is mostly a mutation in the nuclear gene encoding AHAS. Gene amplification related to sulfonylurea resistance has been described only in a tobacco cell line characterized by one AHAS gene mutated and amplified (Harms et al., 1992) and in the carrot cell line resistant to CS in which we have shown the presence of different AHAS amplified sequences (Caretto et al., 1994).

We describe here the isolation, through a stepwise selection, of a *Daucus carota* L. callus culture which is 300-fold more resistant to the herbicide CS than the wild type for a mechanism of gene amplification. The selected callus cultures have proven to be highly morphogenetic and plants that maintained the CS resistant traits have been regenerated from cell lines isolated at different steps.

MATERIALS AND METHODS

Callus cultures, initiated from aseptically grown *Daucus carota* L. var. Scarlet Nantes seedlings were selected stepwise for CS resistance. Small pieces of callus tissue were transferred at 1 to 2 months intervals onto agarized (1% w/v) Gamborg's B5 medium (Gamborg et al., 1968) supplemented with 0.5 mg/L 2,4-dichlorophenoxyacetic acid (2,4-D), 0.25 mg/L 6-benzylaminopurine (6-BAP) and gradually increasing CS concentrations. The plates were maintained at 23 °C under cool fluorescent light (3000 lux) on a 12 h photoperiod. CS was kindly provided by DuPont (Wilmington, DE); the stock solution 10 µM was filter sterilized.

The callus cultures were induced to regenerate plants after transferring onto herbicide-free B5 medium supplemented with 2,4-D and subsequently on B5 without hormones.

Resistance test was carried out by foliar spray application with 0.5 µM CS and 0.2% Tween 20 (v/v) for one month. The level of resistance was estimated by evaluating the typical toxic effects on plant growth (extreme growth inhibition, chlorosis, necrosis). Plants were classified in three groups characterized by: lack of resistance (-), intermediate degree of resistance (+) and high degree of resistance (++).

AHAS activity was measured as previously described (Caretto et al., 1993).

Dot-blot analysis has been carried out using "Bio Dot Microfiltration Apparatus (BIORAD). Hybridization has been performed using a *Brassica napus* AHAS cDNA clone pBI-284, kindly provided by W.L. Crosby (Bekkaoui et al., 1991).

RESULTS

When a *Daucus carota* wild type cell line designated SC0 was exposed to increasing CS concentrations a 50 % inhibition of cell growth was obtained with 10 nM, while no growth was observed in the presence of 50 nM CS. The SC0 cells able to grow on medium supplemented with 10 nM CS were used to start a stepwise selection by transferring to a medium containing progressively higher CS concentrations at intervals of few months. While no resistant cells could be revealed by a single treatment using a CS concentration as high as 1 µM, after 10 steps, cells able to grow on a medium containing 5 µM CS (designated SC5000) were obtained.

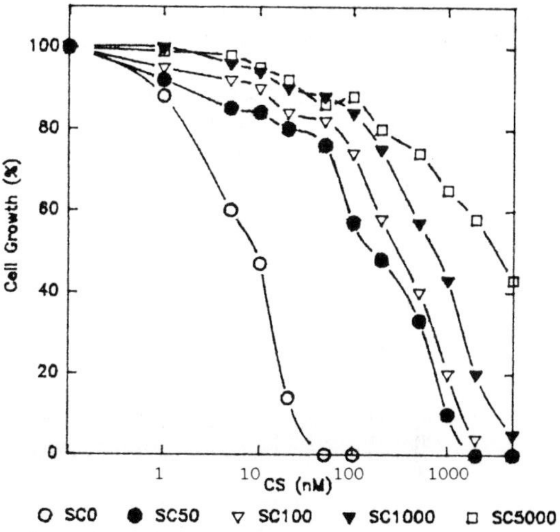

Figure 1. Growth inhibition of different carrot cell lines by increasing chlorsulfuron concentrations.

The stability of resistance was tested by transferring the selected cells on herbicide-free medium for several months. After this period, without selective pressure, calluses were subcultured onto medium containing increasing CS concentrations and cell fresh weight was determined to evaluate cell growth. The results of the experiment carried out using not only SC5000 but also SC50, SC100 and SC1000, which represent some intermediate selection steps, are shown in Figure 1: the CS concentration able to inhibit by 50% the cell growth of the selected lines resulted progressively higher up to the SC5000 value which was about 300-fold higher with respect to the wild type cells.

In order to characterize the CS resistance, the AHAS enzymatic activity was measured in crude extracts from the selected cell lines in the absence and in the presence of increasing CS concentrations. The AHAS activity level measured as nmoles acetoin/mg protein/h rose gradually during the selection from 248 in the wild type to 610 in SC50, 923 in SC100, 1380 in SC1000, reaching in the last step a 7-fold

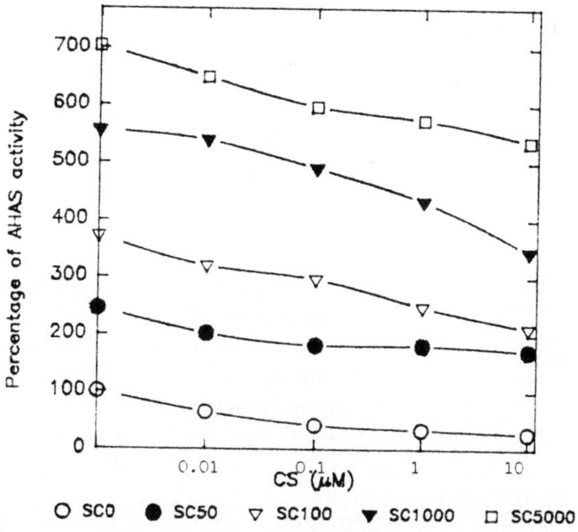

Figure 2. Chlorsulfuron-dependent inhibition of AHAS activity in crude extracts from wild type and resistant carrot cell lines. Activity is expressed as a percentage of the activity of the wild type in the absence of herbicide.

increase with a value of 1874 (Fig. 2). As a result, AHAS enzyme activity in the selected cell lines remained very high even in the presence of CS concentrations which severely inhibited AHAS from the wild type. At the highest CS concentration used (10 μM) the enzyme activity from the selected lines SC1000 and SC5000 were inhibited respectively by 38 and 29 % comparing to a 72 % inhibition of the wild type; the lowest CS concentration used (0.01 μM) did not inhibit the enzyme from the above mentioned selected lines but inhibited the enzyme from SC0 by 38%.

Genomic DNA was isolated from SC0, SC50, SC100, SC1000 and SC5000 and a dot-blot analysis was carried out. When the membrane was hybridized using as probe a AHAS cDNA from *Brassica napus* signals of higher intensity appeared in correspondence to the DNA samples from the selected lines with respect to the wild type SC0. Furthermore the hybridization signals observed from SC50 to SC5000 increased gradually as the resistance increased, thus indicating a gradual AHAS gene amplification. The AHAS gene copy number of SC5000 has been estimated as 13 times the wild type (Fig.3).

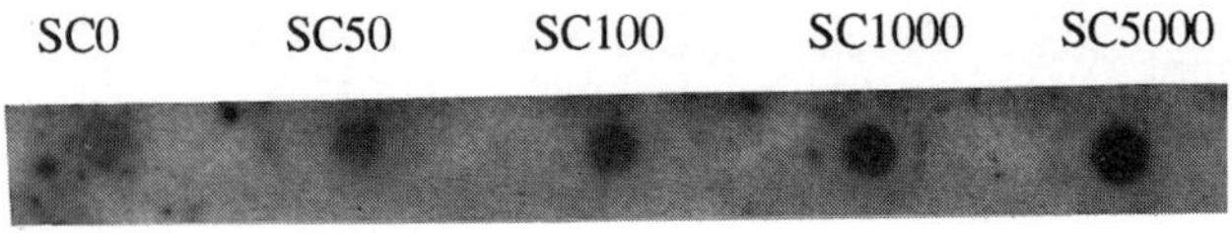

Figure 3. Dot-blot analysis of DNAs from different selected cell lines probed with a AHAS cDNA from *B.napus*

Regeneration experiments were performed to analyze if the CS resistance obtained in the cell lines was maintained at the whole plant level. Several plantlets were regenerated from SC50, SC100 and SC1000 thus showing that the selection procedure did not affect the morphogenetic ability of the callus cultures. CS resistance was then tested by foliar spray application using a concentration of 0.5 μM. Table 1 shows that the percentage of highly resistant plants regenerated from the selected lines ranged from 1.7 to 13%, while no resistant plants were observed among the plants regenerated from SC0.

TABLE 1. Effects of foliar application of chlorsulfuron to SC0 control and putatively resistant SC50, SC100 and SC1000 carrot plants. The values represent the percentage of plants characterized by: (-) lack of resistance; (+)intermediate degree of resistance; (++) high degree of resistance

	Number of tested plants	(-)	(+)	(++)
SC0	53	100	0	0
SC50	58	81	17	1.7
SC100	51	53	39	7.8
SC1000	53	45	41	13

DISCUSSION

A valuable source of variability (somaclonal variation) useful in crop improvement is represented by the genetic changes observed in response to *in vitro* cell growth. In the present study we have successfully used *in vitro* cell selection to obtain carrot cell lines highly resistant to a widely used sulfonylurea herbicide and gene amplification has resulted the molecular mechanism involved.

Our data show that a stepwise scheme is necessary in our cell system to obtain a high level of CS resistance since a single treatment with high CS concentrations could not reveal any mutant. From the cell growth data two important pieces of information derive: 1) the selected CS resistance is stable, since it is maintained during several cell divisions in the absence of the selective agent; 2) it is acquired gradually during the selection. In a previous report (Caretto et al., 1994) the production of a cell line resistant to a concentration of 1 μM was shown, in the present paper we demonstrate that further selection steps have been able to achieve a 5-fold higher resistance.

During the stepwise selection in our cell system a linear correspondence among CS resistance, AHAS enzyme activity and AHAS gene copy number has been detected, thus indicating that the overproduction of AHAS enzyme due to AHAS gene amplification is responsible for the observed CS resistance. Even if gene amplification after cell selection has proven to be the basis of phosphinotricin and glyphosate resistance, many reports have shown sulfonylurea resistance related to a mutation in the nuclear gene encoding AHAS. Recently, Harms et al. (1992) reported in tobacco a case of CS resistance due to amplification of the mutated gene coding for AHAS, in this situation it was not possible to discriminate the source of resistance.

Up to date there is a lack of information about the maintainance at the whole plant level of the gene amplification obtained after *in vitro* cell selection. This is due to the frequent loss of regenerative ability of cells submitted to a stringent selection. In the present study we show that the selected carrot cultures have not lost their ability to regenerate plants and, more interestingly, have maintained in a considerable percentage the resistant trait. These results suggest that the presence of molecular alterations in the nuclear genes does not affect the morphogenetic pattern of carrot cells selected *in vitro* for CS resistance.Preliminary results indicate that in the highly resistant plants an increased AHAS specific activity can be observed and experiments are in progress to ascertain whether gene amplification is still present.

REFERENCES

Bekkaoui, F., Condie, J.A., Neustaedter, D.A., Moloney, M.M. and Crosby, W,L, (1991) 'Isolation,structure and expression of cDNA for acetolactate synthase from *Brassica napus*', Plant Mol. Biol. 16,741-744.

Bottermann, J. and Leemans, J. (1988) 'Engineering herbicide resistance in plants', Trends in Genet. 4, 219-222.

Caretto, S., Giardina, M.C., Nicolodi, C. and Mariotti, D. (1993) 'In vitro cell selection: production and characterization of tobacco cell lines and plants resistant to the herbicide chlorsulfuron', J. Genet. Breed. 47,115-120.

Caretto, S., Giardina, M.C., Nicolodi, C. and Mariotti, D. (1994) 'Chlorsulfuron resistance in *Daucus carota* cell lines and plants: involvement of gene amplification', Theor. Appl. Genet. (in press).

Chaleff, R.S. and Mauvais, C.J. (1984) 'Acetolactate synthase is the site of action of two sulfonylurea herbicides in higher plants', Science 224,1443-1445.

Chaleff, R.S. and Ray, T.B. (1984) 'Herbicide-resistant mutants from tobacco cell cultures',

Science 223, 1148-1151.
Donn, G., Tischer, E., Smith, J.A. and Goodmann, H.M. (1984) 'Herbicide-resistant alfalfa cells: an example of gene amplification in plants', J. Mol. Appl. Genet. 2, 621-635.
Gamborg, O.L., Miller, R.A. and Ohyama, K. (1968) 'Nutrient requirements of suspension cultures of soybean root cells', Exp. Cell Res. 50, 151-158.
Goldsbrough, P.B., Hatch, E.M., Huang, B., Kosinski, W.G., Dyer,W.E., Hermann, K.M. and Weller, S.C. (1990) 'Gene amplification in glyphosate tolerant tobacco cells', Plant Sci. 72, 53-62.
Harms, C.T., Armour, S.L., DiMaio, J.L., Middlesteadt, L.A., Murray, D., Negrotto, D.V., Thompson-Taylor, H., Weymann, K., Montoya, A.L., Shillito, R.D. and Jen, G.C. (1992) 'Herbicide resistance due to amplification of a mutant acetohydroxyacid synthase gene', Mol. Gen. Genet. 233, 427-435.
Harms, C.T. and DiMaio, J.J. (1991) 'Primisulfuron herbicide-resistant tobacco cell lines. Application of fluctuation test design to *in vitro* mutant selection with plant cells', J. Plant Physiol. 137, 513-519.
Haughn, G.W. and Somerville, C. (1986) 'Sulfonylurea-resistant mutants of *Arabidopsis thaliana*', Mol. Gen. Genet. 204, 430-434.
Hauptmann, R.M., della-Cioppa, G., Smith, A.G., Kishore, G.M. and Widholm, J.M. (1988) 'Expression of glyphosate resistance in carrot somatic hybrid cells through the transfer of an amplified 5-enolpyruvylshikimate-3-phosphate synthase gene', Mol. Gen. Genet. 211, 357-363.
Jordan, M.C. and McHughen, A. (1987) 'Selection for Chlorsulfuron resistance in Flax (*Linum usitatissimum*) cell cultures', J. Plant Physiol. 131, 333-338.
Ray, T.B. (1984) 'Site of action of Chlorsulfuron', Plant Physiol., 75, 827-831.
Saunders, J.W., Acquaah, J., Renner, K.A. and Doley, W.P. (1992) 'Monogenic dominant sulfonylurea resistance in sugarbeet from somatic cell selection', Crop Sci. 32, 1357-1360.
Saxena, P.K. and King, J. (1988) 'Herbicide resistance in *Datura innoxia*', Plant Physiol. 86, 863-867.
Saxena, P.K., Williams, D. and King, J. (1990) 'The selection of Chlorsulfuron-resistant cell lines of independent origin from an embryogenic cell suspension culture of *Brassica napus* L.', Plant Sci. 69, 231-237.
Sebastian, S.A. and Chaleff, R.S. (1987) 'Soybean mutants with increased tolerance to sulfonylurea herbicides', Crop Sci. 27, 948-952.
Shah, D.M., Horsch, R.B., Klee, H.J., Kishore, G.M., Winter, J.A., Tumer, N.E., Hironaka C.M., Sanders, P.R., Gasser, C.S., Aykent, S., Siegel, N.R., Rogers, S.G. and Fraley, R.T. (1986) 'Engineering herbicide tolerance in transgenic plants', Science 233, 478-481.
Shyr, YY. J., Caretto, S. and Widholm, J.M. (1993) 'Characterization of the glyphosate selection of carrot suspension cultures resulting in gene amplification', Plant Sci. 88, 219-228.
Shyr, YY. J., Hepburn, A.G. and Widholm, J.M. (1992) 'Glyphosate selected amplification of the 5-enolpyruvylshikimate-3-phosphate synthase gene in cultured carrot cells', Mol. Gen. Genet. 232, 377-382.
Stark, G.R. and Wahl, G.M. (1984) 'Gene amplification', Annu. Rev. Biochem. 53, 447-491.
Swanson, E.B., Coumans, M.P., Brown, G.L., Patel, J.D. and Beversdorf, W.D. (1988) 'The characterization of herbicide tolerant plants in *Brassica napus* L. after in vitro selection of microspores and protoplasts', Plant Cell Rep. 7, 83-87.

RAPD ANALYSIS OF SUGARCANE DURING TISSUE CULTURE

P. W. J. TAYLOR[1], T. A. FRASER[1], H-L. KO[2] and R. J. HENRY[2]
1. Bureau of Sugar Experiment Stations, PO Box 86, Indooroopilly, Queensland, Australia 4068; 2. Queensland Agricultural Biotechnology Centre, Gehrmann Laboratories, The University of Queensland, Australia 4072

ABSTRACT. The sensitivity of random amplified polymorphic DNA using arbitrary 10-mer oligonucleotide primers (RAPD) to detect genetic change in sugarcane (*Saccharum* spp. hybrid) plants during tissue culture was assessed. RAPD analysis of sugarcane plants regenerated from embryogenic callus revealed very few polymorphisms. However, RAPDs detected gross genetic change in protoplast-derived callus of sugarcane. This callus had been in tissue culture for over two years and had lost the ability to differentiate shoots and regenerate plants. RAPD assessment of the genetic integrity of sugarcane plants from an *in vitro* germplasm collection, regenerated from apical buds and maintained in storage at 18°C for 3, 6 and 12 months, revealed no significant genetic change. These results suggest that RAPDs are suitable for detecting gross genetic change but may not be sensitive enough to detect small changes and that minimal genetic change appears to take place in sugarcane plants regenerated from embryogenic callus or apical buds.

1. Introduction

Genetic variability generated during tissue culture is defined as somaclonal variation (Larkin and Scowcroft 1981). Tissue cultured plants lose totipotency with length of time in culture which is probably due to loss or mutation of genes that are essential for plant regeneration (Scowcroft 1984).

Transgenic sugarcane plants were produced from embryogenic callus that was transformed using microprojectile bombardment technique (Bower and Birch 1992). An *in vitro* germplasm collection has been established for *Saccharum* spp. hybrid clones by regenerating plants from apical meristems (Taylor and Dukic 1993). Transgenic sugarcane plants and plants from an *in vitro* germplasm collection need to have minimum genetical changes to be used directly as commercial cultivars or in breeding programs. Hence, methods are required for detecting genetic change in plants during tissue culture.

Chowdhurry and Vasil (1993) found no DNA variation in sugarcane plants regenerated from callus, cell suspension cultures and protoplasts, analysed by DNA restriction fragment length polymorphism (RFLP). Similarly, RFLP techniques have failed to detect variation among plants regenerated from cell suspension cultures and protoplasts of *Festuca* (Vallés *et al.* 1993) and rice (Saleh *et al.* 1990). The lack of detection of somaclonal variation in the above studies, using RFLP markers, may have been due to the use of unsuitable probes or the fact that tissue culture plants were regenerated from embryogenic cells which are less prone to genetic changes (Vasil 1988).

M. Terzi et al. (eds.), Current Issues in Plant Molecular and Cellular Biology, 241-246.

Polymerase chain reaction (PCR) techniques were developed in the late 1980s (Saiki *et al.* 1985) and have been applied widely in the genetic identification of biological samples. The method requires only a small amount of DNA, of which certain regions are amplified, to produce a specific marker profile for identification of the sample (Welsh and McClelland 1990). Random amplified polymorphic DNA (RAPD) assays utilise arbitrary 10-mer oligonucleotide sequences as primers (Williams *et al.* 1990). Primers hybridise to two nearby sites in the template DNA that are complimentary to the primer sequence. Deletions or insertions in the amplified regions or base changes altering primer binding sites will result in polymorphisms. This paper assesses the sensitivity of arbitrary (RAPD) primers to detect genetic change in sugarcane plants during tissue culture.

2. Materials and Methods

2.1. ESTABLISHMENT OF *IN VITRO* PLANTS FROM APICAL BUDS

Apical meristems and surrounding 2 to 3 whorls of developing leaves that contained apical buds in the leaf axils were cultured on MS medium with 0.2 mg/L BAP and 0.1 mg/L kinetin (Taylor and Dukic 1993). Individual plantlets were transferred to rooting medium (MS with 60 g/L sucrose and 1 mg/L NAA). *In vitro* plants were then transferred to half strength MS medium and maintained in storage at 18°C. Field trials were established with plants of five genotypically diverse *Saccharum* spp. hybrid clones that were stored *in vitro* for three, six and 12 months. Single-bud vegetative cuttings of field-grown clones (same five as above) were also planted adjacent to the *in vitro* plants.

2.2. ESTABLISHMENT OF CALLUS CULTURES AND PLANT REGENERATION

Callus was initiated from immature leaf explants of sugarcane (*Saccharum* spp. hybrid) cultivar Q63 cultured on MS medium with 3 mg/L 2,4-D and embryogenic callus was selectively subcultured, as described by Taylor *et al.* (1992a). After 4.5 months of culture, embryogenic callus was transferred to solid MS medium without plant growth regulators and incubated under diffuse light with a 12-h photoperiod at 27°C. Individual plantlets were then transferred to rooting medium. Eight plants were established, one per 4.5 L pot, in the glasshouse. Field-plants of source clone Q63 were established in the field from vegetative cuttings.

2.3. PROTOPLAST CALLUS

Non-morphogenic callus was regenerated from protoplasts isolated from a homogeneous cell suspension culture established from Q63 SP cell line (Taylor *et al.* 1992a; 1992b). This callus was maintained by subculture onto fresh MS medium with 3 mg/L 2,4-D every three weeks.

2.4. DNA ISOLATION

DNA was extracted from the spindle leaves of field-grown plants and plants regenerated from embryogenic callus and apical buds, and from protoplast-derived callus of Q63 SP using the modified CTAB method of Graham *et al.* (1994). Leaf material or callus (0.5g fwt) was ground in liquid nitrogen and DNA extracted using 2% (w/v) hexadecyltrimethylammonium bromide (CTAB, BDH, Australia), 0.1 M Tris-HCL pH 8.0/ 1.4 M sodium chloride/ 0.02 M EDTA at 55°C for 20 min, followed by centrifugation (5 min at 12,000 *g*). Genomic DNA was quantified spectrophotometrically by measuring absorbances at 260 and 280 nm (Sambrook *et al.* 1989).

2.5. POLYMERASE CHAIN REACTION

A total of 25 arbitrary 10-mer oligonucleotide sequences (Operon Technologies, USA) were screened for amplification of plant genomic DNA fragments. Conditions for DNA amplification were standardised for all primers. Each 25 µl reaction volume containied: 30 ng of DNA template, 0.8 units of Taq DNA polymerase (Boehringer Mannheim), 0.24 mM each of dATP, dGTP, dTTP and dCTP (Promega Corporation), 0.05 µM primer and PCR buffer with a final concentration of 10.0 mM Tris-HCl pH 8.3, 50.0 mM KCl, 3.0 mM $MgCl_2$, 0.1 mg/ml gelatin.

Amplification was performed in a Perkin Elmer 9600 GeneAmp PCR System and was initiated by a denaturation of 1 min at 94°C, followed by 33 cycles of 10 sec at 94°C, 30 sec at 40°C and 1 min at 72°C. The amplification was completed with one cycle of 5 min at 72°C.

Reaction products were resolved by electrophoresis on 1.5% (w/v) agarose gels, stained with ethidium bromide and visualised under UV illumination. PCR reactions were repeated to establish reproducibility of results.

3. Results

Analysis of the genomic DNA of sugarcane plants based on RAPD markers using arbitrarily chosen oligonucleotide primers resulted in the resolution of 67 scorable markers from 13 primers out of 25 primers screened. Each primer produced between 2 and 10 amplification products that ranged in size between approximately 400 and 3800 bp. The primers that did not produce scorable markers either produced faint or non-consistent amplification product, or no amplification product was produced.

No genetic change was detected in plants of the five clones that were regenerated from apical buds and stored *in vitro* for three, six and 12 months. Consistent RAPD markers were produced for the 3 replicates of each treatment for each clone (Fig. 1).

PCR amplification products for the eight somaclones of Q63 regenerated from embryogenic callus were similar to the field-grown plants except with primer M6 where the 950 bp fragment was missing from two of the somaclones (Fig. 2). The polymorphism in the two somaclones was consistent in three DNA extractions over a five month period and after the plants were vegetatively propagated.

Protoplast-derived callus of Q63 SP cell line expressed a high degree of polymorphism relative to field-grown source cultivar with six of the primers tested (Fig. 3).

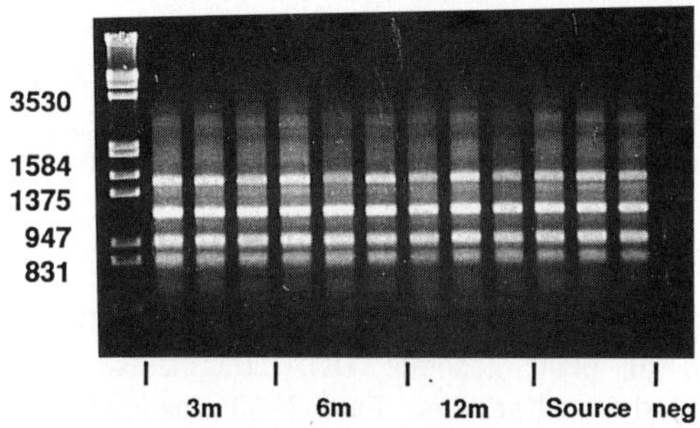

Figure 1. Amplification with primer OPC-14 of genomic DNA of sugarcane clone TS67-74 after plants were regenerated from apical buds and stored *in vitro* for 3, 6 and 12 months. Size markers are Lambda/Eco RI/Hind III.

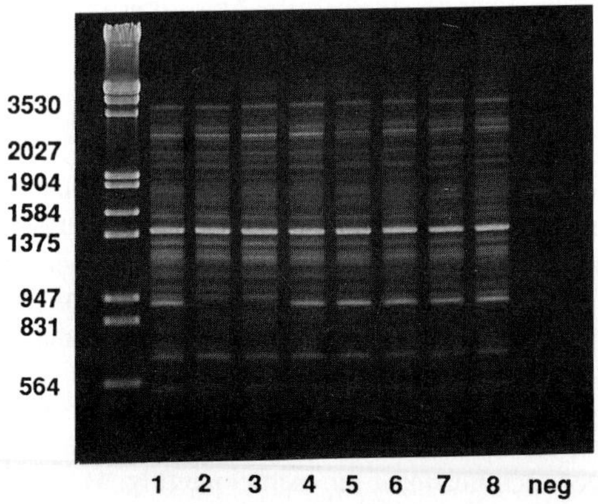

Figure 2. Amplification with primer OPM-06 of genomic DNA of 8 somaclones of Q63 regenerated from embryogenic callus. Size markers are Lambda/Eco RI/Hind III.

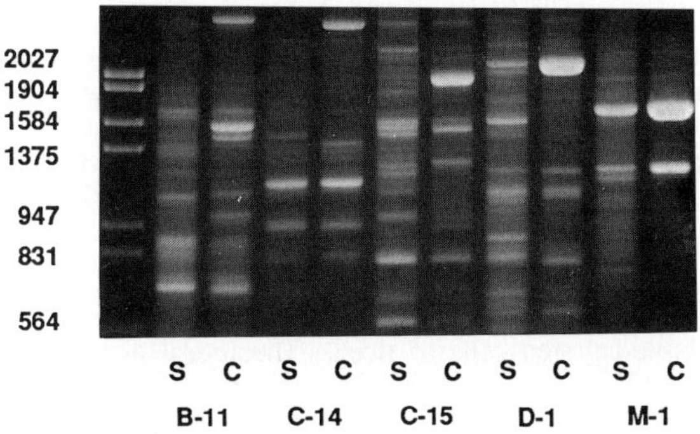

Figure 3. Amplification with OP-primers of genomic DNA of sugarcane clone Q63 (S) and protoplast-derived callus of clone Q63 (C). Size markers are Lambda/Eco RI/Hind III.

4. Discussion

The sensitivity of RAPDs may not be sufficient to detect small and irregular genetic changes in tissue cultured plants that result from somaclonal variation. No genetic change was detected in plants that were regenerated from apical buds and stored *in vitro* for three, six and 12 months. Only one polymorphic marker (with primer OPM-06) from the 67 scorable markers was resolved in plants regenerated from 4.5-month-old embryogenic callus. The lack of detection of genetic change in these experiments may have been due to too few primers screened, the use of unsuitable primers or the fact that tissue culture plants regenerated from apical buds or embryogenic cells had minimal genetic change. Chowdhurry and Vasil (1993) also found no DNA variation in sugarcane plants regenerated from *in vitro* cultures, using RFLP analysis. They claimed that this was most likely due to the use of embryogenic cultures which are less prone to genetic changes because of stringent selection favour of normal cells during somatic embryo formation.

Nevertheless, RAPDs revealed gross polymorphisms in protoplast-derived callus of cultivar Q63 SP, with four out of the six primers revealing polymorphisms. This callus had been in tissue culture for over two years and had lost the ability to differentiate shoots and regenerate plants.

5. References

Bower, R. and Birch, R.G. (1992) `Transgenic sugarcane via microprojectile bombardment', The Plant Journal 2, 409-416.

Chowdhury, M.K.U. and Vasil, I.K. (1993). `Molecular analysis of plants regenerated from embryogenic cultures of hybrid sugarcane cultivars (*Saccharum* spp.)', Theoretical and Applied Genetics 86, 181-188.

Graham, G.C., Mayers, P. and Henry, R.J. (1994). 'Simple and rapid method for the preparation of fungal genomic DNA for PCR and RAPD analysis'. BioTechniques 16, 2-3.

Larkin, P.J. and Scowcroft, W.R. (1981). 'Somaclonal variation - a novel source of variability from cell cultures for plant improvement'. Theoretical and Applied Genetics 60, 197-214.

Saiki, R.K., Gelfand, D.H., Stoffel, S., Scharf, S. J., Higuchi, R., Horn, G.T. and Erlich, H.A. (1985). 'Primer-detected enzymatic amplification of DNA with a thermostable DNA polymerase'. Science 239, 487-491.

Saleh, N.M., Gupta, H.S., Finch, R.P., Cocking, E.C. and Mulligan, B.J. (1990). 'Stability of mitochondrial DNA in tissue-cultured cells of rice'. Theoretical and Applied Genetics 79, 342-346.

Sambrook, J., Fritsch, E.F. and Maniatis, T. (1989). Molecular cloning. A laboratory manual, Cold Spring Harbor Laboratory Press, Cold Spring Harbor, N.Y.

Scowcroft, W.R. (1984). 'Genetic variability in tissue culture: impact on germplasm conservation and utilization'. International Board for Plant Genetic Resources, Rome.

Taylor, P.W.J. and Dukic, S. (1993). 'Development of an *in vitro* culture technique for conservation of *Saccharum* spp. hybrid germplasm'. Plant Cell, Tissue and Organ Culture 34, 217-222.

Taylor, P.W.J., Ko, H-L., Adkins, S.W., Rathus, C. and Birch, R.G. (1992a). 'Establishment of embryogenic callus and high protoplast yielding suspension cultures of sugarcane (*Saccharum* spp hybrids)'. Plant Cell, Tissue and Organ Culture 28, 69-78.

Taylor, P.W.J., Ko, H-L. and Adkins, S.W. (1992b). 'Factors affecting protoplast isolation and the regeneration of shoots from protoplast-derived callus of sugarcane (*Saccharum* spp hybrids)'. Australian Journal of Botany 40, 863-876.

Vallés, M.P., Wang, Z.Y., Montavon, P., Potrykus, I. and Spangenberg, G. (1993). 'Analysis of genetic stability of plants regenerated from suspension cultures and protoplasts of meadow fescue (*Festuca pratensis*)'. Plant Cell Reports 12, 101-106.

Vasil, I.K. (1988). 'Progress in the regeneration and genetic manipulation of cereal crops'. Bio/Technology 6, 397-402.

Welsh, J. and McClelland, M. (1990). 'Fingerprinting genomes using PCR with arbitrary primers'. Nucleic Acids Research 18, 7213-7218.

Williams, J.G.K., Kubelik, A.R., Livak, K.J., Rafalski, J.A. and Tingey, S.V. (1990). 'DNA polymorphisms amplified by arbitrary primers are useful as genetic markers'. Nucleic Acids Research 18, 6531-6535.

EVIDENCE OF SOMACLONAL VARIATION IN SOMATIC EMBRYO-DERIVED PLANTLETS OF WHITE SPRUCE (*PICEA GLAUCA* (MOENCH) VOSS.)

N. ISABEL, R. BOIVIN, C. LEVASSEUR, P. M. CHAREST [1], J. BOUSQUET AND F. M. TREMBLAY

Centre de recherche en biologie forestière, Faculté de foresterie et géomatique; and [1]Département de phytologie, Faculté des sciences de l'agriculture et de l'alimentation, Université Laval, Sainte-Foy, Québec G1K 7P4, Canada

ABSTRACT. Unlike angiosperms (especially crop plant species), conifers are considered to be genetically stable following somatic embryogenesis. However, we have been able to identify four different *variegata* phenotypes among 2270 somatic embryo-derived white spruces. The four types of variegated plants differ from each other with respect to the extent and distribution of their chlorophyll-deficient needles. Microscopy shows that certain leaves of a selected variant are formed of a chimeral mixture of green and white cells. Cells in completely white needles of this variegated plant are characterized by large nuclei with predominant euchromatin, absence of large cytoplasmic vacuoles, and vacuolized plastids with aberrant morphologies. Various observations suggest that the recovered *variegata* phenotypes reflect some kind of genetic instability of either chloroplastic or nuclear genomes. Molecular approaches, including the use of RAPD markers, are currently employed to find out whether DNA rearrangements are involved in conferring these *variegata* phenotypes.

1. Introduction

Somatic embryogenesis represents a promising method for the propagation of genetically improved conifer trees. Although available data are less extensive than for angiosperms, conifers have been considered to be genetically stable following somatic embryogenesis. For instance, no somaclonal variation of embryogenic tissues or somatic embryos has been detected by flow cytometry analyses for Norway spruce (Mo et al. (1989)), by culture morphology and isozyme analyses for interior spruce (Eastman et al. (1991)), nor by RAPD markers for black spruce (Isabel et al. (1993)).

In the case of angiosperms, there have been numerous reports on somaclonal variation where plants derived from tissue culture differ from the parental type (see Karp (1991), Meins (1983)). One of the most frequent and noticeable manifestations of somaclonal variation is the production of partially (variegated) or completely (albino) achlorophyllous plants. Variegation, defined as the presence of spots, streaks, or sectors of tissue of various colors on an organ or an organism, may have a genetic, physiological, morphological, or pathological basis (Marcotrigiano et al. (1990)). Variegated phenotypes are classified as either pattern or nonpattern according to the predictable or random distribution, respectively, of the color pattern (Marcotrigiano et al. (1990)).

The present study aims at characterizing the first *variegata* phenotypes to be reported for conifers derived from somatic embryogenesis. The study of these variants might provide clues to the nature of somaclonal variation in conifers and might lead to a better understanding of the long term effects of their *in vitro* culturing.

M. Terzi et al. (eds.), Current Issues in Plant Molecular and Cellular Biology, 247-252.

2. Material and Methods

2.1. SOMATIC EMBRYOGENESIS AND PLANT REGENERATION

Seeds of white spruce were obtained from three controlled crosses among six selected genotypes. These parents are part of a large breeding population maintained by the Canadian Forest Service of Quebec (Natural Resources Canada). Embryogenic tissue lines were obtained from excised mature embryos and maintained as suspensions by subculturing every week on liquid HLM-1 medium (Tremblay (1990)). Somatic embryos were produced using the protocol described by Isabel et al. (1993), except that concentration of abscisic acid was between 45 and 60 μM, depending on the tissue line. Germination and acclimatization of the somatic embryos-derived plantlets were done using unpublished procedures (to be reported elsewhere). For the acclimatization, plantlets were transferred to soil under greenhouse conditions after epicotyls had reached 1 cm in length.

2.2. MICROSCOPY

Needle specimens were fixed in 3% glutaraldehyde-0.1 M cacodylate buffer for three days at 4°C and post-fixed in 1% osmium tetroxyde-0.1 M cacodylate buffer for 1.5 h at room temperature. Samples were dehydrated with ethanol and embedded in Jembed 812 resin. Thin sections (approximately 90 nm) were stained with 2% uranyl acetate and then, with a lead citrate preparation (Reynolds 1963). Numerous sections for each sample were examined using a JEOL 1200 EX transmission electron microscope.

2.3. RAPD ANALYSES

DNA samples were isolated from 2-year old plants of clones nos. 39, 40, 41, and 45, derived from the same controlled cross, using a procedure modified from Ziegenhagen et al. (1993) and Bousquet et al. (1990). In addition, DNA was extracted from embryogenic tissue lines nos. 40 and 41. In the case of variegated plants, DNA was obtained from totally white needles, while control DNA was isolated from green needles of normal plants produced at the same time from the same clone. More than sixty primers (10-mers or 11-mers) were used for RAPD-PCR amplifications following methods previously described by Roy et al. (1992).

3. Results

The *variegata* phenotype of somatic embryo-derived plantlets was observed six to eight months after their transfer to soil. One *variegata* phenotype, presented in Figure 1, is characterized by the presence of white and green to completely white needles, mostly found at the extremities of about two third of the branches. The variegation pattern displayed by another individual obtained from tissue line no. 41 is restricted to smaller sectors of the plant, including the alternation of green and white needles on a same branch (not shown). These two variegated plants are of the nonpattern type, since the white parts appeared distributed randomly.

The four different types of variants were all obtained, at a relatively low frequency, from tissue lines nos. 40 and 41, whereas no variegated individual was obtained from tissue line no. 39, derived from the same controlled cross, and from which was produced the largest number of plants (Table 1).

Figure 1. Distribution of white needles for a variegated white spruce recovered from embryogenic tissue line no. 40, one year after transfer in soil

TABLE 1. Occurrence of the *variegata* phenotypes for seven tissue lines used in somatic embryogenesis of white spruce

Tissue line no.	No. of plantlets	No. of variegated plants	Frequency
39	924	0	-
40	459	2	0.44 %
41	648	2	0.31 %
42	76	0	-
43	40	0	-
44	9	0	-
45	114	0	-
Total	2270	4	0.18 %

Structural characterization of the variants was done by light and electron microscopy. Preliminary observations showed that the white and green leaves of the individual shown in Figure 1 are formed of a chimeral mixture of green- and white-coloured cells. Electron microscopy on completely white needles of this variant revealed a cellular morphology rather typical of meristematic cells and characterized by large round nuclei with predominant euchromatin as well as absence of large cytoplasmic vacuoles (not shown). In comparison, the normal nucleus of green cells appeared smaller and asymetrical, while it harboured mainly zones of heterochromatin. The amorphous and vacuolized plastid-like organelles in white needles present aberrant morphologies and contain ubiquitous lipid droplets (right side; Figure 2). In contrast, chloroplasts in needles of control plant contain large starch granules (not shown), while chloroplasts of the green leaves of the variant seldom contained such reserve material (left side; Figure 2). As shown in Figure 2, normal chloroplasts with well-defined grana sometimes coexist in the same cell as chloroplasts showing more or less disorganized thylakoid membranes.

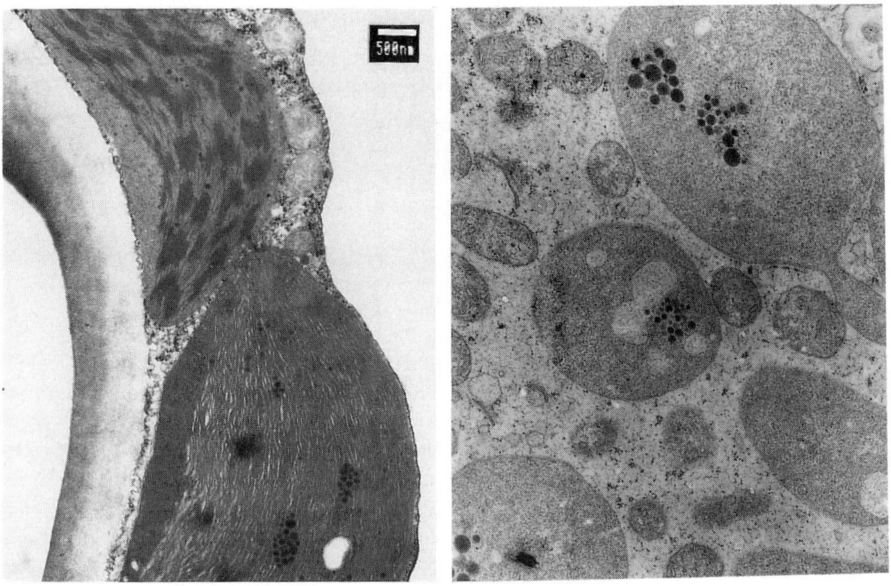

Figure 2. Electron micrographs showing plastid morphology in green (left) and white needles (right) of the same variegated plant shown in Figure 1. A scale bar represents 500 nm. Largest structures are chloroplasts (left) or plastid-like organelles (right), while smaller structures are mitochondria.

Some of the morphological observations suggest that these *variegata* phenotypes reflect some kind of genetic instability of either chloroplastic or nuclear genome. Among molecular approaches, RAPD analysis was first used to test these hypotheses. More than sixty primers, generating at least two hundred genetic markers showing various levels of polymorphism, were employed to screen the DNA of these variants. These markers allowed us to distinguish easily between clones originated from the same controlled cross. Almost all markers were invariant within a given clone and tissue line, indicating a relative stability of the sampled DNA regions. However, the few markers that yielded differences in RAPD patterns, associated or not with the *variegata* phenotype, are currently under investigation.

4. Discussion

We report here the morphological and preliminary molecular characterization of different variegated white spruces derived from somatic embryogenesis. The observed *variegata* phenotypes are not consistent with alterations due to pathological or physiological factors.

No sign of viral infection was observed, whereas the quite improbable physiological hypothesis could be ruled out by grafting of the variegated branches.

While somaclonal variation can either be of genetic or epigenetic origin (Meins (1983)), especially difficult to distinguish for trees having a long-life cycle (Karp (1991)), many reports on tissue culture-associated variations suggest underlying alterations at the DNA level (Brown et al. (1993), Cecchini et al. (1992), Müller et al. (1990), Shirzadegan et al. (1991)). Mutations affecting levels of pigment production in angiosperms occur quite frequently: some of these are coded by the nuclear genome, while others are coded by the chloroplast genome (Aviv and Galun (1985)). In the case of white spruce, the assignment of these variations by conventional crossing methods to either the chloroplast or the nucleus is hindered by the long-life cycle of this conifer tree. However, it has been demonstrated, at least in some cereals derived from anther culture, that the plastid genome of most chlorophyll-deficient plants had undergone deletion or alteration of specific restriction fragments (Day and Ellis (1985), Dunford and Walden (1991)). On the other hand, some forms of variegation are caused by differences in genetic expression among genetically identical cells (Peary et al. (1988)). Although the genetic status of the variations in white spruce has not been established, the type of somatic sectoring observed in variegated leaves rather suggests a juxtaposition of genetically dissimilar cells and the segregation of mutant plastids. For most of the white spruce *variegata* phenotypes, the conversion from green to white cells seemed irreversible, since subsequent growth on the variegated branches remained achlorophyllous. No matter if the events leading to this phenotypic instability are the result of genetic or epigenetic changes, the variegation pattern of the white spruces argues in favor of repeated events occurring at a relatively high frequency at different places and stages throughout growth and development.

5. Acknowledgements

We are grateful to Alain Goulet for skillful work in light and electron microscopy, Dr. Hélène Chamberland for helpful discussions, Laurence Tremblay in charge of the plant inventories, and Julie Bélanger responsible for observations and maintenance in greenhouses. This work was financially supported by a Synergie grant of the Ministère de l'Enseignement Supérieur et de la Science (Québec) to F. T.

6. References

Aviv, A. and Galun, E. (1985) 'An *in vitro* procedure to assign pigment mutations in *Nicotiana* to either the chloroplast or the nucleus', J. Heredity 76, 135-136.

Bousquet, J., Simon, L., and Lalonde, M. (1990) 'DNA amplification from vegetative and sexual tissues of trees using polymerase chain reaction', Can. J. For. Res. 20, 254-257.

Brown, P.T.H., Lange, E.D., Kranz, E., and Lörz, H. (1993) 'Analysis of single protoplasts and regenerated plants by PCR and RAPD technology', Mol. Gen. Genet. 237, 311-317.

Cecchini, E., Natali, L., Cavalli, A., and Durante, M. (1992) 'DNA variations in regenerated plants of pea (*Pisum sativum* L.)', Theor. Appl. Genet. 84, 874-879.

Day, A.and Ellis, T.H.N. (1985) 'Deleted forms of plastid DNA in albino plants from cereal anther culture', Curr. Genet. 9, 671-678.

Dunford, R. and Walden, R.M. (1991) 'Plastid genome structure and plastid-related transcript levels in albino barley plants derived from anther culture', Curr. Genet. 20, 339-347.

Eastman, P.A.K., Webster, F.B., Pitel, J.A., and Roberts, D.R. (1991) 'Evaluation of somaclonal variation during somatic embryogenesis of interior spruce (*Picea glauca engelmannii* complex) using culture morphology and isozyme analysis', Plant Cell Rep. 10, 425-430.

Isabel, N., Tremblay, L., Michaud, M., Tremblay, F.M., and Bousquet, J. (1993) 'RAPDs as an aid to certify genetic integrity of somatic embryogenesis-derived populations of *Picea mariana* (Mill.) B.S.P.', Theor. Appl. Genet. 86, 81-87.

Karp, A. (1991) 'On the current understanding of somaclonal variation', in B.J. Miflin (ed.) Oxford Surveys of Plant Molecular and Cell Biology. Vol. 7, pp. 1-58.

Marcotrigiano, M., Boyle, T.H., Morgan, P.A., and Ambach, K.L. (1990)' Leaf color variants from coleus shoot cultures', J. Amer. Soc. Hort. Sci. 115, 681-686.

Meins, F. (1983) 'Heritable variation in plant tissue culture', Ann. Rev. Plant Physiol. 34, 327-346.

Mo, L. H., von Arnold, S., and Lagercrantz, U. (1989) 'Morphogenic and genetic stability in long term embryogenic cultures and somatic embryos of Norway spruce (*Picea abies* (L.) Karst)', Plant Cell Rep. 8, 375-378.

Müller E., Brown, P. T. H., and Lörz, S. H. (1990) 'DNA variation in tissue culture-derived rice plants', Theor. Appl. Genet. 80, 673-679.

Peary, J.S., Lineberger, R.D., Malinich, T.J., and Wertz, M.K. (1988) 'Stability of leaf variegation in *Saintpaulia ionantha* during *in vitro* propagation and during chimeral separation of a pinwheel flowering form', Amer. J. Bot. 75, 603-608.

Reynolds, E.S. (1963) 'The use of lead citrate at high pH as an electron-opaque stain in electron microscopy', J. Cell Biol. 17, 208.

Roy, A., Frascaria, N., MacKay, J., and Bousquet, J. (1992) 'Segregating random amplified polymorphic DNAs (RAPD) in *Betula alleghaniensis*', Theor. Appl. Genet. 85, 173-180.

Shirzadegan, M., Palmer, J.D., Christey, M., and Earle, E.D. (1991) 'Patterns of mitochondrial DNA instability in *Brassica campestris* cultured cells', Plant. Mol. Biol. 16, 21-37.

Tremblay, F.M. (1990) 'Somatic embryogenesis and plantlet regeneration from embryos isolated from store seeds of *Picea glauca*', Can. J. Bot. 68, 236-242.

Ziegenhagen, B., Guillemaut, P., and Scholz, F. (1993) 'A procedure for mini-preparations of genomic DNA from needles of silver fir (*Abies alba* Mill.)', Plant Mol. Biol. Rep. 11, 117-121.

GENE TECHNOLOGY FOR DEVELOPING COUNTRIES: GENETIC ENGINEERING OF INDICA RICE

I. Potrykus, P.K. Burkhardt, S.K. Datta, J. Fütterer, G.C. Ghosh-Biswas, A. Klöti, G. Spangenberg, J. Wünn
Institute of Plant Sciences, Swiss Federal Institute of Technology (ETH), ETH-Centre, CH-8092 Zurich, Switzerland

Abstract:

Indica-type rice provides the staple food for 2 billion people in Third World Countries. We have established gene transfer to IRRI breeding lines to explore the contributions of genetic engineering to sustained and stable production of high quality food. Experiments are in progress on the development of resistance towards Yellow Stem Borer, towards Rice Tungro Virus, towards fungal pests, and towards accumulation of provitamin A in the endosperm.

Key words: *Oryza sativa*, Indica-type rice, genetic engineering, vitamin A endosperm, insect-resistance, virus-resistance, fungus-resistance.

Introduction

Indica-type rice feeds more than two billion people, predominantly in developing countries. In humid and semihumid Asia where rice is the basic food, the population is expected to increase by 58 percent over the next 35 years. In thirty years the world will need 70 percent more rice than it requires today. And these ca. 800 million tons of rice will have to be grown with considerable reduction in the input of agrochemicals under sustainable conditions (IRRI, 1993). This immense task requires that traditional plant breeding is supported by every possible contribution from novel technical developments. Genetic engineering, applied with consequence and care, has the potential to contribute to the sustainable production of affordable food for the increasing population in developing countries. For maximum benefit for developing countries, application of gene technology should focus on important problems for which solutions by conventional approaches are not available. For Indica-type rice such problems have been identified and described in a joint study of the International Rice Research Institute (IRRI), Manila, and the Rockefeller Rice Biotechnology Program, New York (Khush and Toenniessen, 1991). Among the problems to be solved with high priority are a) resistance to fungal diseases, b) resistance to Tungro virus, c) resistance to Yellow stem borer, d) stable supply of provitamin A, and e) improvement of nutritional quality. The problems mentioned are especially severe for people depending on Indica-type rice. It is, to date, relatively easy to genetically engineer Japonica-type rice, and relatively difficult to do the same with Indica-type rice. The tasks mentioned can be solved only by focusing work on Indica-type rice. The International Rice Research Institute, Manila (IRRI) has the world

M. Terzi et al. (eds.), Current Issues in Plant Molecular and Cellular Biology, 253-262.

mandate to develop breeding lines to the benefit of small rice farmers, and has already released numerous successful varieties to the rice growing countries (IRRI, 1992). The following research projects are, therefore, performed in collaboration with IRRI and using IRRI breeding lines, to assure direct transfer of experimental success to the target population. Best possible solutions can be found only with the tight involvement of the international scientific community. This is achieved by collaboration with the Rockefeller Rice Biotechnology Program (Toenniessen, G.H. et al., 1989).

Gene transfer for integrative transformation

Plant breeding with transgenic characters requires populations of independent transgenic and fertile plants to chose the most stable and best expressing lines for subsequent traditional breeding. This in turn requires routine and efficient gene transfer protocols leading to fertile transgenic plants. Although recovery of transgenic Indica-type rice plants and offspring have been described from our laboratory (Datta et al., 1990, 1992) and from others (Christou et al., 1991), gene transfer to IRRI breeding lines is not yet routine and efficient enough, and recovery of fertile transgenic plants is still rather inefficient and requires further optimisation. Gene transfer to Indica-type rice is, to date, possible via direct gene transfer to protoplasts (Datta et al., 1990, Ghosh-Biswas et al., 1993) and our group is using this technique still routinely with IR43, IR72 and other Indica rice varieties. With IR72, the at present most advanced breeding line, we still face, however, severe fertility problems; with IR43 we see, so far, better chances to raise sufficient numbers of independent and fertile transgenic plants in the near future. Transgenic Indica rice plants have also been recovered from biolistic treatment of immature embryos (Christou et al., 1991) and from electroporation to split embryos from mature seeds (Xhu and Li, 1994). We have a series of transgenic clones and plants from IR43 and IR72 under investigation and hope to be able at the time of the symposium to present data which allow to judge which of both techniques can be recommended. We also applied electroporation to cells of immature embryos for transformation of cereal cells. Although gene transfer to scutellum cells of wheat was routine and efficient (Klöti et al., 1993), it was not possible, so far, to repeat these results with rice. *Agrobacterium*-mediated transformation of rice has been attempted by other laboratories (e.g. Rainieri et al., 1990). However, *Agrobacterium*-mediated transgenic rice plants have, so far, not been recovered from any laboratory.

Approach towards Yellow Stemborer resistance

Insect damage is one of the major factors for yield loss in rice farming all over the world. In Southeast Asia alone the value of forgone production caused by insect damage reaches more than 600 million US $ per year. More than two third of these losses are caused by two insect species, the rice brown planthopper (BPH; *Delphax oryzae*, Homoptera) and the yellow stemborer (YSB; *Scirpophaga incertulas*, Lepidoptera, Herdt, 1991). Unlike the case of BPH there is no wild type rice variety known to contain resistance genes against YSB, which could be used as a source for getting YSB resistant rice by conventional breeding methods.

The entomocidal sporeforming soil-bacterium *Bacillus thuringiensis* offers a promising variation of genes which encode for specific endotoxins. Until now more than 40 nucleotide sequences of such genes have been determined. They are clearly related to each other and classified in 17 distinctly different crystal protein genes, the so called *cry*-genes (Peferoen, 1991). These genes encode for proteins either of some 130-140 kDa or some 70 kDa which are first dissolved and then proteolytically cleaved in the midgut of the insects to small toxic fragments of approximately 60 kDa (Faust and Bulla, 1982). After cleavage these toxins bind to specific proteins on the brush-border membranes of the insect gut (Hofmann et al., 1988), followed by disrupting the epithelium, disturbing the ionic balance and thereby paralysing the gut.

Different B.t. formulations have been used as biological insecticides for many years. Disadvantages like poor persistence and short duration of effect under tropical conditions especially during the rainy season, could be overcome by a transgenic approach, the expression of Bt-genes in the rice plant itself.

First success has been reported from tobacco (Vaeck et al., 1987), potato (Peferoen et al., 1991), tomato (Fischhoff et al., 1987) cotton (Perlak et al., 1990) maize (Koziel et al., 1993) and Japonica rice (Fujimoto et al., 1993). These plants showed a high level of insect resistance and it is expected that some of these crops will be released on the market in the next few years (Peferoen, 1991).

The aim of our project is the transformation of advanced Indica rice breeding lines with genes conferring resistance to YSB. Therefore we are using the lepidopteran specific *cry* I A(b)-gene, which has been shown to be effective against YSB as well as rice leaffolder (Lepidoptera, D. Bottrell, personal communication).

Since it is known that the highly conserved C-terminal half of the Bt-toxin is not necessary for the toxicity of the protein, truncated forms of the *cry*-genes have been cloned (Fischhoff et al., 1987). These truncated genes encode only for the N-terminal half of the crystal proteins and show significantly enhanced expression levels of the B.t. toxins in transgenic plants (Fischhoff et al., 1987; Delannay et al., 1989; Perlak et al., 1990; Koziel et al., 1993; Fujimoto et al., 1993).

As plants in general show a different codon usage than bacteria, a synthetic *cry* I A(b)-gene was constructed (Koziel et al., 1993) which shows a high G-C content in the coding region. As the maize codon usage resembles the codon usage pattern of monocots in general (Murray et al., 1989), this should also lead to a sufficient expression of the *cry* I A(b)-gene in rice.

As a first step towards YSB resistant rice plants we have transformed a truncated version (645 codons) of a wild type and a synthetic *cry* I A(b)-gene to elite Indica rice breeding lines. Resistant clones were analysed by Southern Blotting and showed a clear integration of the *cry* I A(b)-gene in the rice genome. Plants were regenerated, transferred to soil and are growing under greenhouse conditions. For further analysis R_0 and R_1 plants will be checked by Southern, Northern, and Western Analysis and by insect feeding studies as well.

To minimise the possible development of resistance in insects we put this gene under control of a tissue specific promoter which directs the expression of the *cry* I A(b)-gene only to the leaf sheath, the primary target site of the YSB . Furthermore we are planning to use a second Bt-gene, the *cry* II A-gene which is known to bind to a different receptor site in the brush-border membrane of the insect gut, as resistance to Bt-toxins could be due to a change in such a receptor site (Ferré et al., 1991).

However, it has to be emphasised that such transgenic rice plants should be planted under field conditions only according to the concept of Integrated Pest Management, IPM, to keep the Bt-toxins as an effective and sustainable tool not only for the next few years (McGaughey et al., 1992).

Approach towards tungro virus resistance

The rice tungro disease is caused by a complex of two viruses, rice tungro spherical virus (RTSV) and rice tungro bacilliform virus (RTBV, Hibino et al., 1978). The severe symptoms are caused by RTBV (Dasgupta et al., 1991). RTBV is a member of the newly assigned group of badnaviruses (Hay et al., 1991; Qu et al., 1991) and is related to the better studied caulimoviruses in its life cycle and genome organisation (Hohn and Fütterer, 1992). While engineered virus resistance has been achieved for a great number of RNA plant viruses, for the DNA containing badna- and caulimoviruses no successful strategy has been reported so far (Wilson, 1993). Since these viruses have very different replication cycles it is not clear whether approaches that worked with RNA viruses (i.e. constitutive expression of viral coat proteins or wild-type or mutated replicases) will also work for RTBV. It is, however, to be expected that also for RTBV, expression of functional viral proteins already at the onset of virus infection could interfere with an ordered progression through the viral life cycle and that expression of mutated viral proteins might interfere with the function of normal viral proteins by competition. We therefore have introduced into indica rice breeding lines (IR43, IR72) constructs designed to express RTBV proteins 1, 3 and 4. The protein 3 is a precursor from which by proteolytic processing the viral coat protein, reverse transcriptase and probably a variety of other proteins are generated. The locations of these proteins within the precursor can at present be deduced only from sequence homologies to related viruses. A variety of constructs for direct expression of the coat protein or the reverse transcriptase have been prepared on the basis of such estimations. Transgenic plants will be screened for expression of the proteins with antisera obtained from R. Hull (John Innes Institute, Norwich) or with antisera against a protein tag that has been incorporated into the expression constructs.

To create mutant proteins that might act as competitive inhibitors for critical viral functions, mutations have been introduced into the coat protein and the reverse transcriptase:

In the coat protein region, a sequence motif containing three invariable cysteines and one histidine is conserved between almost all viruses using reverse transcriptase in their replication cycle (retro- and plant pararetroviruses; Covey, 1986). This motif, which is involved in several RNA binding steps, has been mutated to glycine since such a mutation has been shown to preserve the structure (and thus part of the function) of the remaining part of the coat protein but to abolish infectivity of a retrovirus (De Rocquigny et al., 1992; Morellet et al., 1992).

In the reverse transcriptase region, a mutation was introduced into a highly conserved sequence motif containing two aspartates (Argos, 1988). In addition, subfragments of the polyfunctional reverse transcriptase have been cloned in analogy to results from some RNA viruses where subfragments of the polymerase gene produced high levels of protection (Wilson, 1993).

Since functions for the remaining RTBV proteins are unknown, sensible mutations are difficult to design. In the protein 4 we localised a leucine zipper similar to those that are involved in protein-protein interactions in many other proteins (Gruissem, 1990). In a yeast system (Fields and Song, 1989) we found that the protein 4 indeed has the capacity to dimerise but the dimerisation domain resides outside the leucine zipper which, however, may interact with another protein. We have cloned subfragments of the protein 4 coding region containing either the leucine zipper or the dimerisation domain to express proteins that lack one of the interaction domains and thus probably will not have a complete function but still will interact with one of the original partners and therefore will act as competitive inhibitor.

In addition to these approaches involving expression of a protein, we also have introduced a construct expressing antisense RNA against the leader sequence of the RTBV pregenomic RNA. Antisense RNAs had little effect on RNA viruses (Wilson, 1993), but viruses like RTBV with a nuclear phase might be more susceptible.

Most of these strategies follow the work performed with RNA viruses. Approaches that would more specifically use the particular molecular biology of RTBV require a detailed knowledge of the viral life cycle. Therefore, we also study viral gene expression mechanisms and the potential function of viral gene products in transient expression systems (Fütterer et al., 1993).

Approach towards provitamin A accumulation in rice endosperm

According to UNICEF statistics world-wide, over 124 million children are estimated to be vitamin A deficient (Humphrey et al., 1992). Improved vitamin A nutrition would be expected to prevent approximately 1--2 million deaths annually among children aged 1--4 years. An additional 0.25-0.5 million deaths may be avoided if improved vitamin A nutriture can be achieved during the later childhood. Improved vitamin A nutriture alone therefore could prevent 1.3--2.5 million of nearly 8 million late infancy and preschool-age child deaths that occur each year in the highest-risk countries (West Jr. et al., 1989).

Rice in its milled form, as it is consumed by most people, in South East Asia is characterised by the complete absence of provitamin A. The milled rice kernel consists exclusively of the endosperm The embryo and the aleuron layer have been removed during processing of the rice grain.

The aim of this project is to initiate the carotenoid biosynthesis in the rice endosperm tissue to increase the daily vitamin A uptake of people predominantly feeding on rice.

It is known for maize and sorghum that cereal endosperm cells can produce and accumulate carotenoids (Buckner et al., 1991). Furthermore the starch storage tissues of potato and cassava (Pentedao and Almeida 1988), accumulate carotenoids in considerable amounts.

To provide the minimum requirements of relevant carotenoids to young infants, and assuming rice as the sole dietary source, 1--2 μg β-carotene per gram uncooked rice would be needed in rice endosperm (The Rockefeller Foundation, 1993). This is roughly 1/4--1/2 of the amount produced in maize endosperm, and enough to turn the rice noticeably but not dark yellow.

The carotenoid pathway is a branch of the central isoprenoid pathway which is characterised by 4 key enzymes that are necessary for carotenoid biosynthesis. These are the phytoene synthase, the phytoene desaturase, the ζ-carotene desaturase and the lycopene cyclase. The genes encoding for these enzymes are available both from higher plants (Fray and Grierson, 1993, Ray et al., 1987, Linden et al., 1993, Bartley et al., 1991) and bacteria (Armstrong et al., 1989).

Our strategy is to produce transgenic indica rice varieties which contain either single genes or several genes in combination. Experiments to produce plants that contain a phytoene synthase cDNA from daffodil (Beyer et al., unpublished results) under the control of endosperm specific promoters (Okita et al., 1989) to ensure exclusive expression in the endosperm are underway. The analysis of these plants will be done in close collaboration with Dr. P. Beyer, Freiburg (FRG). The biochemical basis for the production of carotenes in rice endosperm have been examined

Furthermore we have established a system for high level transient in endosperm tissues which allows us to asses the capacity of genes and promoters which we will use for transformation. Based on this system we want to bombard immature endosperm of maize, wheat and rice with a particle inflow gun (Finer et al., 1992). Via visualisation of the gene products or by HPLC we want to test the constructs we use for stable transformation. The first construct to test will be the phytoene synthase under the control of a CaMV 35S promoter as well as under the control of two different rice glutelin promoters (Okita et al., 1989). We also will complement a phytoene deficient maize mutant (*y1*), described by Buckner et al. (1991), thus showing the proper function of our constructs. This work will be continued with the other necessary key enzymes which will be tested in appropriate maize endosperm mutants and afterwards also be transformed to indica rice varieties.

Approach towards fungal disease resistance:

Rice blast (*Magnophorthe grisea*) and sheath blight (*Rhizoctonia solani*) are fungal diseases of rice that cause significant yield losses (Reissig et al. 1986; Toenniessen, 1991). It has been estimated that important productivity gains could be possible if these challenges - in the case of rice blast particularly in association with upland drought - would be overcome (Herdt, 1991). In addition, conventional approaches to improve these traits in rice have been scored as ineffective even with substantial research (Herdt, 1991).

Multiple natural host response mechanisms, including the accumulation of defensive enzymes (e.g. chitinases, β-1,3-glucanases, etc.) are involved in plant resistance to phytopathogenic fungi (Boller, 1988). Chitinase preparations, especially in combination with β-1,3-glucanases, inhibit fungal growth *in vitro* (Mauch et al. 1988; Arlorio et al. 1992; Sela-Buurlage et al. 1993). Chitinases have also been shown to accumulate around invading hyphae *in planta* (Benhamou et al. 1990; Collinge et al. 1993). Transgenic approaches based on the constitutive expression of a bean endochitinase gene in tobacco (Broglie et al. 1991) and canola (Benhamou et al. 1993), or based on the wound-inducible expression of a barley seed ribosome inactivating protein (RIP) in tobacco (Logemann et al. 1992) have been reported to lead to increased protection over *Rhizoctonia solani*. In addition, osmotin-like proteins inducible by osmotic stress, such as tobacco AP24 (Melchers et al. 1993), have been shown to be pathogen-induced proteins with inhibitory activity toward fungal pathogens (Woloshuk et al. 1991).

The development of gene transfer systems for Indica and Japonica rice opened up possibilities for testing the effects of expression of these candidate anti-fungal (and stress-response) genes as strategies to overcome sheath blight and upland drought/blast constraints. Putative transgenic rice clones are under selection for: a) barley RIP cDNA under control of the inducible rice chitinase promoter RCH10 (Zhu et al. 1993), and b) β-1,3-glucanase and chitinase driven by the Ti-1'2' dual promoter (collaboration J. Mundy, Copenhagen, Denmark). In addition, research aimed at expressing in transgenic rice plants: a) tobacco osmotin-like AP24, b) bean endochitinase, and c) tobacco β-1,3-glucanase, individually and in a concerted manner, has recently been initiated (collaboration G. Selman-Housein, Havana, Cuba) Progress on these lines will be reported

Concluding remarks

The goal of our scientific work is to contribute to future sustained production of affordable and high quality food in developing countries. We tried to organise this step out of the ivory tower of pure science into application in such a way it will not end in an academic exercise. We tried to make sure to work on problems which are a heavy burden on a great number of poor people and we are trying to organise our science in such a way that it complements traditional plant breeding. We can reach our goal only, if the novel characters we introduce into Indica rice will be used in breeding programmes; this is guaranteed through our collaboration with IRRI. The novel characters will be successful only in breeding if they are stable and effective. This requires that we can provide the breeders with a collection of many transgenic plants for every novel character to select the best possible case for his breeding programme. This in turn requires more efficient gene transfer protocols than those available to date.

Success or failure of our goal will, however, not only depend on success or failure of our experiments and subsequent breeding programmes. It will also depend on political, social, and psychological circumstances in those countries in which the novel, genetically engineered varieties are supposed to help solving problems. Risk assessment will be an integral part of the projects, however, the judgement of scientists and national biosafety committees on the security of transgenic plants or food gained from transgenic plants will not necessarily lead to an acceptance of these plants or food by the local population . If we were e.g. able to recover a transgenic rice which accumulates sufficient provitamin A to stop vitamin A deficiencies, there is no guarantee, that people would be willing to eat this rice. Therefore, there is much educational and political work ahead of us in addition to what we are trying to achieve.

References

Argos, P., 1988. A sequence motif in many polymerases. Nucleic Acids Res. 16: 9909-9916.

Arlorio, M., A. Ludwig, T. Boller & P. Bonfante, 1992. Inhibition of fungal growth by plant chitinases and β-1,3-glucanases. A morphological study. Protoplasma 171: 34-43.

Armstrong, G.A., M. Alberti, F. Leach & J.E. Hearst, 1989. Nucleotide sequence, organization, and nature of the protein products of the carotenoid gene cluster of *Rhodobacter capsulatus*. Mol. Gen. Genet. 216: 254-268.

Bartley, G.E., P.V. Vitanen, I. Pecker, D. Chamovitz, J. Hirschberg & P.A. Scolnik, 1991. Molecular cloning and expression in photosynthetic bacteria of soybean cDNA coding for phytoene desaturase, an enzyme of the carotenoid biosynthesis pathway. Proc. Natl. Acad. Sci. USA, 88: 6532-6536.

Benhamou, N., M.H.A.J. Joosten & P.J.G.M. de Wit, 1990. Subcellular localization of chitinase and of its potential substrate in tomato root tissues infectd by *Fusarium oxysporium* f.sp. *radicis-lycopersici*. Plant Physiol. 92: 1108-1120.

Benhamou, N., K. Broglie, I. Chet & R. Broglie, 1993. Cytology of infection of 35S-bean chitinase transgenic canola plants by *Rhizoctonia solani*: cytochemical aspects of chitin breakdown *in vivo*. The Plant J. 4: 295-305.

Boller, T., 1988. Ethylene and the regulation of antifungal hydrolases in plants. Oxf. Surv. Plant Mol. Cell. Biol. 5: 145-174.

Bryan, J.K., 1980. Synthesis of the aspartate family and branched-chain amino acids. In: B.J. Miflin (Ed.), The Biochemistry of Plants Volume 5, pp. 403-452. New York Academic Press, New York.

Broglie, R., I. Chet, M. Holliday, R. Cressman, P. Biddle, S. Knowlton, C.J. Mauvais & R Broglie, 1991. Transgenic plants with enhanced resistance to the fungal pathogen *Rhizoctonia solani*. Science 254: 1194-1196.

Buckner, B., T.L. Kelson & D.S.Robertson, 1991. Cloning of the *y1* locus of maize, a gene involved in the biosynthesis of carotenoids. The Plant Cell 2: 867-876.

Christou, P., L.F. Tameria & M. Kofron, 1991. Production of transgenic rice (*Oryza sativa*) plants from agronomically important Indica and Japonica varieties via electric discharge particle acceleration of exogenous DNA into immature zygotic embryos. Bio/Technology 9: 957-962.

Collinge, D.B., K.M. Kragh, J.D. Mikkelsen, K.K. Nielsen, U. Rasmussen & K. Vad, 1993. Plant chitinases. The Plant J. 3: 31-40.

Covey, S.N. 1986. Amino acid sequence homology in the gag region of reverse transcribing elements and the coat protein gene of cauliflower mosaic virus. Nucleic Acids Res. 14: 623-633.

Dasgupta, I., R. Hull, S. Eastop, C. Poggi-Pollini, M. Blakebrough, M.I. Boulton & J.W. Davies, 1991. Rice tungro bacilliform virus DNA independently infects rice after agrobacterium-mediated transfer. J. Gen. Virol. 72: 1215-1221.

Datta, S.K., K. Datta & I. Potrykus, 1990. Fertile Indica rice plants regenerated from protoplasts isolated from microspore derived cell suspension. Plant Cell Reports 9: 253-256.

Datta, S.K., A. Peterhans, K. Datta & I. Potrykus, 1990. Genetically engineered fertile Indica-rice recovered from protoplasts. Bio/Technology 8: 736-740.

Datta, S.K., K. Datta, N. Soltanifar, G. Donn & I. Potrykus, 1992. Herbicide-resistant Indica rice plants from IRRI breeding line IR72 after PEG-mediated transformation of protoplasts. Plant Mol.Biol. 20: 619-629.

Datta,K., I. Potrykus & S.K. Datta, 1992. Efficient fertile plant regeneration from protoplasts of the Indica rice breeding line IR72 (Oryza sativa L.). Plant Cell Reports 11: 229-233.

De Rocquigny, H., C. Gabus, A. Vincent, M.C. Fournie-Zaluski, B. Roques & J.L. Darlix, 1992. Viral RNA annealing activities of HIV-1 nucleocapsid protein require only peptide domains outside the zinc fingers. Proc. Natl. Acad. Sci. USA 89: 6472-6476.

Delannay, X., B.J. La Valle, R.K. Proksch, R.L. Fuchs, S.R. Sims, J.T. Greenplate, P.G. Marrone, R.B. Dodson, J.J. Augustine, J.G Layton. & D.A. Fischhoff, 1989. Field performance of transgenic tomato plants expressing the *Bacillus thuringiensis* var. *kurstaki* insect control protein. Bio/Technology 7: 1265-1269

Faust, R.M. & L.A. Bulla jr., 1982. Bacteria and their toxins as insecticides. In E. Kurstak (Ed.), Microbial and Viral Pesticides. Dekker, New York, pp. 75-208.

Ferré, J., M.D. Real, J. Van Rie, S. Jansens & M. Peferoen, 1991. Resistance to the *Bacillus thuringiensis* bioinsecticide in a field population of *Plutella xylostella* is due to a change in a midgut membrane receptor. Proc. Natl. Acad. Sci. USA 88: 5119-5123.

Fields, S. & O.-K. Song, 1989. A novel genetic system to detect protein-protein interactions. Nature 340: 245-246.

Finer, J.J., P. Vein, M.W Jones. & M.D. McMullen, 1992. Development of the particle inflow gun for DNA delivery to plant cells. Plant Cells Reports 11: 323-328.

Fischhoff, D. A., K.S. Bowdish, F.J. Perlak, P.G. Marrone, S.M. McCormick, J.G. Niedermeyer, D.A. Dean, K. Kusano-Kretzmer, E.J. Mayer, D.E. Rochester, S.G. Rogers & R.T. Fraley, 1987. Insect tolerant transgenic tomato plants. Bio/Technology 5: 807-813.

Fray, R.G. & D. Grierson, 1993. Identification and genetic analysis of normal and mutant phytoene synthase of tomato by sequencing, complementation and co-suppression. Plant Mol. Biol. 22: 589-602.

Fujimoto, H., K. Itoh, M. Yamamoto, J. Kyozuka & K. Shimamoto, 1993. Insect resistant rice generated by introduction of a modified δ-endotoxin gene of *Bacillus thuringiensis*. Bio/Technology 11: 1151-1155.

Fütterer, J., I. Potrykus, M.P. Valles Brau, I. Dasgupta, R. Hull & T. Hohn, 1993. Splicing in a plant pararetrovirus. Virology (in press.)

Ghosh Biswas, G.C., V.A. Iglesias, S.K. Datta & I. Potrykus 1993. Transgenic indica rice (*Oryza sativa* L.) plants obtained by direct gene transfer to protoplasts. J. Biotechnology (in press).

Gruissem, W, 1990. Of fingers, zippers and boxes. Plant Cell 2: 827-828.

Hay, J.M., M.C. Jones, M.L. Blakebrough, I. Dasgupta, J.W. Davies & R. Hull, 1991. An analysis of the sequence of an infectious clone of rice tungro bacilliform virus, a plant pararetrovirus. Nucleic Acids Res. 19: 2615-2621.

Herdt, R.W., 1991. Research priorities for rice biotechnology. In: G.S. Khush & G.H. Toenniessen (Eds.), Rice Biotechnology, pp. 19-54. C·A·B· International & IRRI, Wallingford/Manila.

Hibino, H., M. Roechan & S. Sudarisman, 1978. Association of two types of virus particles with penyakit habang (tungro disease) of rice in Indonesia. Phytopathology 68: 1412-1416.

Hoffmann, C., H. Vanderbruggen, J. Van Rie, S. Jansens. & H. Van Mellaert, 1988. Specifity of *Bacillus thuringiensis* δ-endotoxins is correlated with the presence of high-affinity binding sites in the brush border membrane of target insect midguts. Proc. Natl. Acad. Sci. USA 85: 7844-7848.

Hohn, T. & J. Fütterer, 1992. Pararetroviruses and retroviruses: a comparison of expression strategies. Seminars in Virology 2: 55-69.

Humphrey, J.H., K.P. West Jr. & A. Sommer, 1992. Vitamin A deficiency and attributale mortality among under-5-year-olds. Bulletin of the World Organization 70(2): 225-232.

IRRI: Annual Report of the International Rice Research Institute, 1992. IRRI, Los Banos, The Philippines.

Khush, G.S. & G.H. Toenniessen (Eds), 1991. Rice Biotechnology. C·A·B· International & IRRI, Wallingford/Manila.

Kortt, A.A., J.B. Caldwell, G.G. Lilley & T.J. Higgins, 1991. Amino acid and cDNA sequences of a methionine -rich 2S protein from sunflower seed (Helianthus annus L.). Eur. J. Biochem. 195:329-344.

Koziel, M.G, G.L. Beland, C. Bowman, N.B. Carozzi, R. Crenshaw, L. Crossland, J. Dawson, N. Desai, M. Hill, S. Kadwell, K. Launis, K. Lewis, D. Maddox, K. McPherson, M.R. Meghji, E. Merlin, R. Rhodes, G.W. Warren, M. Wright & S.T. Evola, 1993. Field performance of elite transgenic maize plants expressing an insecticidal protein derived from *Bacillus thuringiensis*. Bio/Technology 11: 194-200.

(*Oryza sativa*) protoplasts. Planta 178: 325-333.

Linden, H., A. Vioque & G. Sandmann, 1993. Isolation of a carotenoid biosynthesis gene coding for ζ-carotene desaturase from *Anabena* PCC 7120 by heterologous complementation. FEMS Microbiol. Lett. 106: 99-104.

Logemann, J., G. Jach, H. Tommerup, J. Mundy & J. Schell, 1992. Expression of a barley ribosome-inactivating protein leads to increased fungal protection in transgenic tobacco plants. Bio/Technology 10: 305-308.

Mauch, F., B. Mauch-Mani & T. Boller, 1988. Antifungal hydrolases in pea tissue. II. Inhibition of fungal growth by combinations of chitinase and β-1,3-glucanase. Plant Physiol. 88: 936-942.

McGaughey, W. H. & M. E. Whalon, 1992. Managing Insect resistance to *Bacillus thuringiensis* toxins. Science 258: 1451-1455.

Melchers, L.S., M.B. Sela-Buurlage, S.A. Vloemans, C.P. Woloshuk, J.S.C. Van Roekel, J. Pen, P.J.M. van den Elzen & B.J.C. Cornelissen, 1993. Extracellular targeting of the vacuolar tobacco proteins AP24, chitinase and β-1,3-glucanase in transgenic plants. Plant Mol. Biol. 21: 583-593.

Morellet, N., N. Jullian, H. De Rocquigny, B. Maigret, J.-L. Darlix & B.P. Roques, 1992. Determination of the structure of the nucleocapsid protein NCp7 from the human immunodeficiency virus type 1 by ^{1}H NMR. EMBO J. 11: 3059-3065.

Murray, E.E., J. Lotzer & M. Eberle, 1989. Codon usage in plant genes. Nucleic Acids Research 17: 477-493.

Okita, T.W., Y.S. Hwang, J. Hnilo, W.T. Kim, A.P. Aryan, R. Larson & H.B. Krishnan, 1989. Structure and expression of the rice glutelin multigene family. J. Biol. Chem. 264: 12573-12581.

Peferoen, M. 1991. Engineering of insect-resistant plants with *Bacillus thuringiensis* crystal protein genes. In: Biotechnology in Agriculture 7, pp. 135-153.

Penteado, M.V.C. & L.B. Almeida, 1988. Occurence of carotenoids in roots of five cultivars of cassava (*Manhiot esculenta*, Crantz) from São Paulo. Rev. Farm. Bioquim. Univ. S. Paulo 24: 39-49.

Perlak, F.J., R.W. Deaton, TA. Armstrong, R.L. Fuchs, S.R. Sims, J.T. Greenplate & D.A. Fischhoff, 1990. Insect resistant cotton plants. Bio/Technology 8: 939-943.

Potrykus, I., 1990. Gene transfer to plants: assessment of published approaches and results. Bio/technology 8: 535-542.

Potrykus, I., 1993. Gene transfer to plants: approaches and available techniques. In: Hayward, M.D., N.O.Bosemark & I.Romagosa (Eds.), Plant Breeding: Principles and Prospects. Chapmann & Hall, London, pp. 126-137.

Qu, R., M. Bhattacharyya, G.S. Laco, A. DeKochko, B.L. Subba Rao, M.B. Kaniewska, J.S. Elmer, D.E. Rochester, C.E. Smith & R.N. Beachy, 1991. Characterization of the genome of rice tungro bacilliform virus: comparison with commelina yellow mottle virus and caulimoviruses. Virology 185: 354-364.

Raineri, D.M., P. Bottino, M.P. Gordon & E.W. Nester, 1990 Agrobacterium-mediated transformation & rice (*Oryza sativa* L.) Bio/Technology 8: 33-38.

Ray, J., C.R. Bird, M.J. Maunders, D. Grierson & W. Schuch, 1987. Sequence of pTOM5, a ripening related cDNA from tomato. Nucl. Acids Res. 15: 1057.

Reissig, W.H., E.A. Heinrichs, J.A. Litsinger, K. Moody, L. Fiedler, T.W. Mew & A.T. Barrion, 1986. Illustrated guide to integrated pest management in rice in tropical Asia, IRRI, The Philippines.

Sela-Buurlage, M.B., A.S. Ponstein, S.A. Bres-Vloemans, L.S. Melchers, P.J.M. van den Elzen & B.J.C. Cornelissen, 1993. Only specific tobacco (*Nicotiana tabacum*) chitinases and β-1,3-glucanases exhibit antifungal activity. Plant Physiol. 101: 857-863.

Shaul, O. G. Gaalili, 1992. Increased lysine synthesis in tobacco plants that express high levels of bacterial dihydrodipicolinate synthase in their chloroplasts. The Plant J. 2(2):203-209.

Toenniessen, G.H., R.W. Herdt & L.A. Sitch, 1989. Rockefeller Foundations international network on rice biotechnology. In: A. Mujeed-Kai & L.A. Sitch (Eds.) Genetic manipulation in crops pp. 265-274. CIMMYT, Mexico, IRRI, Philippines.

Toennissen, G.H., 1991. Potentially useful genes for rice genetic engineering. In: G.S. Khush & G.H. Toenniessen (Eds.), Rice Biotechnology, pp. 253-280. C·A·B· International & IRRI, Wallingford/Manila.

Vaeck, M., A. Reynarts, H. Höfte, S. Jansens, M. De Beuckeleer, C. Dean, M. Zabeau, M. Van Montague & J. Leemans, 1987. Transgenic plants protected from insect attack. Nature 327: 33-37.

Wandelt, C.I., M.R.I. Khan, S. Craig, H.E. Schroeder, D. Spencer & T.J.V. Higgins, 1992. Vicillin with carboxy-terminal KDEL is reteined in the endoplasmic reticulum and accumulates to high levels in the leaves of transgenic plants. The Plant J. 2:181-192.

West Jr., K.P., G.R. Howard & A. Sommer, 1989. Vitamin A and infection: public health implications. Annual review of nutrition 9: 63-86.

Wilson,T.M.A., 1993. Strategies to protect crop plants against viruses: Pathogen-derived resistance blossoms. Proc. Natl. Acad. Sci. USA 90: 3134-3141.

Woloshuk, C.P., J.S. Meulenhoff, M. Sela-Buurlage, P.J.M. van den Elzen & B.J.C. Cornelissen, 1991. Pathogen-induced proteins with inhibitory activity toward *Phytophthora infestans*. The Plant Cell 3: 619-628.

Xhu, X., & B. Li, 1994. Fertile transgenic Indica rice plants obtained by electroporation. Plant Cell Reports (in press).

PRODUCTION OF TRANSGENIC CEREAL CROPS

BECKER, D., JÄHNE, A., ZIMNY, J.*, LÜTTICKE, S. and LÖRZ, H.
Zentrum für Angewandte Molekularbiologie der Pflanzen, AMP II,
Institut für Allgemeine Botanik der Universität Hamburg
Ohnhorststrasse 18, D-22609 Hamburg, Germany
**Plant Breeding and Acclimatization Institute - Radzikow*
P.O.Box, 1019, PL-00950 Warszawa, Poland

ABSTRACT. Applied aspects set cereals as an attractive goal for transformation. Many efforts have been made during recent years towards the establishment of reliable transformation techniques. We report here the development of three transformation systems based on micro-projectile-mediated gene transfer and discuss alternative methods.

A reproducible transformation method for hexaploid wheat as well as hexaploid triticale (x *Triticosecale* Wittmack) based on particle bombardment of scutellar tissue of immature embryos has been developed. A DNA construct containing the ß-glucuronidase gene (*uidA, gus*) and the selectable marker gene *bar* giving resistance to phosphinothricin (PPT, herbicide BASTA) was introduced into scutellar tissue. In the case of wheat, bombarded embryos were transferred to callus induction media supplemented with phosphinothricin after 2 weeks. The selection conditions were also applied during plant regeneration. From a total of 1050 bombarded immature embryos 12 transgenic plants were regenerated which showed enzyme activity for both introduced genes.

In triticale transformation experiments more than 4000 plantlets were regenerated, without any selection during the callus induction phase and the first step of plant regeneration. All regenerated plants were transferred to selection media and 300 plants survived selection. Thirty plants from 14 independent transformation experiments show enzyme activity for one or both introduced marker genes. Southern blot analysis showed stable integration of foreign genes into the genomic DNA of putative wheat and triticale transformants as well as their progeny.

Bombardment of barley microspores was performed using the same plasmid construct as described for wheat. Under optimal bombardment conditions about 1% of the bombarded microspores show transiently GUS activity. During regeneration of plants GUS activity remains in developing tissue and up to now 12 fertile transgenic plants were regenerated on selection media supplemented with 5 mg/l phosphinothricin. The integration and enzyme activity of the introduced marker genes was analyzed by Southern blot and enzyme assays, for primary transformants and progeny recovered from selfing of R_0 plants. Particle bombardment provides a reliable tool for gene transfer to cereal crops and all major cereals have been transformed recently with this technique using different target cells.

1. Introduction

The development of plant transformation techniques during the past decade has made it possible to improve crop plants by introduction of cloned genes. For most dicotyledoneous species, the *Agrobacterium*-mediated transformation system can be used to generate many transformants

M. Terzi et al. (eds.), Current Issues in Plant Molecular and Cellular Biology, 263-269.

while for the most monocotyledoneous species, especially the agronomically important cereals, current transformation systems still need to be improved. Of the various approaches to gene transfer, three transformation methods have led to the production of transgenic plants:
- protoplast based direct gene transfer,
- tissue electroporation,
- microprojectile-mediated gene transfer.

In the past considerable progress has been made in establishing reliable and efficient *in vitro* culture systems for most cereals. However, thus far embryogenic suspension cultures are the only reliable source for totipotent protoplasts. Nevertheless, it is very difficult and time-consuming to start and maintain these cultures. Furthermore, regeneration capability has been observed to gradually decline during cultivation in cereal suspension cultures. The direct DNA transfer into isolated protoplasts, induced by polyethylen glycol (PEG) or electric pulses is a successful and routinely used method to obtain transformed cell lines, but the regeneration of transgenic plants remains difficult. Only in rice and maize which reproducible give rise to protoplast-derived, fertile transgenic plants, it has been possible to obtain transgenic plants by this method (Shimamoto et al. 1989, Datta et al. 1990, Donn et al. 1992). The fundamental problem of this transformation method is the continous loss of embryogenic capacity of the suspension cultures during long time culture (Jähne et al. 1991), occurrence of somaclonal variation (Wang et al. 1992), and high expenditure of labour and energy. As an alternative microprojectile-mediated gene transfer (Sanford et al. 1987) or tissue electroporation (D`Halluin et al. 1993) have the potential to overcome these limitations. The essence of microprojectile systems for plant genetic transformation is to use high velocity particles to penetrate cell walls and to introduce DNA into intact cells thus circumventing the host range limitation of *Agrobacterium* and the problems of plant regeneration from protoplasts.

The transfer of DNA into cells and tissues with embryogenic capacity takes place with high efficiency. The choice of appropriate target cells is of major importance as there are only few tissues and cells capable of plant regeneration. Using embryogenic suspension cells and embryogenic callus cultures, successful transformation and regeneration of cereals, such as maize (Gordon-Kamm et al. 1990), rice (Cao et al. 1992), wheat (Vasil et al. 1992), oat (Somers et al. 1992) and sugarcane (Bower and Birch 1992) could be achieved. However the morphogenetic competence of cells is significantly reduced during long term maintenance and the phenomenon of somaclonal variation limits the suitability of these cells for transformation.

These limitations could be overcome by directly targeting tissues or cells which can be obtained easily and manipulated *in vitro*. In cereals, scutellar tissue of immature embryos, immature inflorescences or microspores are suitable primary explants for bombardment or tissue electroporation as it was demonstrated in maize (D`Halluin et al. 1992). The time necessary for preparation of the target cells is comparatively low and the risk of somaclonal variation is neglectable as the period in *in vitro* culture is reduced to a few weeks. Another advantage of microprojectile bombardment of primary explants is that even genotypes which are recalcitrant in protoplast culture can be transformed easily. Up to now, scutellar tissue of rice, maize, hexaploid triticale, and wheat (Christou et al. 1991, Koziel et al. 1993, Zimny et al. submitted, Weeks et al. 1993, Vasil et al. 1993, Becker et al. 1994), immature inflorescences of tritordeum (Barcelo et al. 1994) and barley microspores (Jähne et al. in press) have been used successfully to obtain fertile transgenic plants.

We report here the progress made in our laboratory towards biolistic transformation of cereals using scutellar tissue of wheat and triticale, and barley microspores as targets for particle bombardment. The inheritance of the introduced marker genes *bar* and *uidA* was studied in the R_1 generation.

2. Material and methods

Details of material and methods are described by Becker et al. (1994), Jähne et al. (1994) and Zimny et al. (submitted).

3. Results and Discussion

The primary requirement for an optimal target is that the tissue or cells receiving exogenous DNA are culturable *in vitro*, actively dividing and capable of giving rise to fertile plants. In our experiments we used scutellar tissue of immature wheat and triticale embryos and barley microspores as targets for particle bombardment. In barley a spontaneous autoendoreduplication of the genome during the first cell division of the microspore leads to homozygous, dihaploid regenerants. Therefore, this haploid target makes the regeneration of homozygous R_0 plants possible. There are existing well established *in vitro* culture systems in our laboratory, from which fertile plants can be regenerated in a high frequency.

Each target tissue has been subject of individual optimization experiments to improve conditions of particle bombardment. Optimization was perfomed by transient transformation experiments using the plasmid pDB1 (Figure 1), containing the *uidA* (*gus*) and *bar* marker gene. The aim of these experiments was to enhance transient transformation by minimizing tissue damage, which is correlated with a reduced regeneration capability. In our experiments we observed that the degree of tissue damage depends on the type of explant or cells, the particle density and the acceleration pressure used for bombardment. For example, high particle densities (116 µg/bombardment) caused a severe tissue damage and a reduced regeneration capability in scutellar tissue of wheat and triticale. In comparison, the viability of barley microspores or immature inflorescences of tritordeum (Barcelo et al. 1994) has not been influenced by the same amount of particles. Results from these optimization processes demonstrate that the optimal particle densities have to be determined for each cell type, whereas the acceleration pressure can be relatively wide ranged without having a negative effect on regeneration capability or the number of transient transformation events (Jähne et al. in press, Becker et al. 1994). Under optimal culture and bombardment conditions an average number of 100 transient GUS-signals per embryo has been counted in wheat. In barley about 1% of the bombarded microspores transiently expressed the *uidA* gene.

3.1. SELECTION, ANALYSIS AND PHENOTYPE OF R_0 PLANTS

An important component of transformation systems involves selection of transgenic tissue cultures and plants. In cereal transformation antibiotics such as kanamycin, G 418, and hygromycin were sucessfully used to obtain transgenic plants (Vasil et al. 1992, Barcelo et al. 1994, Walters et al. 1992). In recent years, herbicides such as chlorsulfuron, phosphinothricin or bialaphos were predominantly used as selective agents in cereal transformation experiments. The advantage of herbicide selection is the possibility to select plants *in vitro* and *in vivo* by a simple spray test, whereas the identification of putative transformants obtained from transformation experiments using antibiotics as selective agent was more complicated and labour-intensive. For our transformation experiments we used the *bar* gene as a selectable marker and phosphinothricin as selective agent to identify transgenic plants. The plasmid pDB1 used in these transformation experiments contained in addition the non-selectable marker gene *uidA*.

In seven independent transformation experiments 59 wheat plants were regenerated and 12 of these independently regenerated plants showed enzyme activity for both marker genes. This corresponds to a transformation frequency of 1 transgenic plant per 83 immature embryos bombarded. This frequency is substantially higher than the 1-2 plants per 1000 bombarded embryos reported by Weeks et al. (1993) after particle bombardment of 5 days old callus cultures recovered from scutellar tissue of wheat. The transformation frequency may be influenced by the different genotypes used in both experiments. It is possible that scutellar tissue of various wheat genotypes contains different numbers of cells competent to integrate DNA and to divide and to form somatic embryos. It is also possible that *in vitro* preculture of scutellar tissue has positive or negative influence on transformation frequency. To answer these questions, further experiments with other genotypes and different preculture treatments prior to bombardment are necessary.

In triticale we were able to obtain a high transformation frequency of at least 3.3%. From a total number of 30 plants from 14 independent transformation experiments which survived Basta spraying, 25 show also enzyme activity for the second marker gene *uidA*.

In barley, in total 12 plants were regenerated from 9 successful transformation experiments. Only plants showing phosphinothricin acetyltransferase (PAT) activity survived a selection pressure of 5 mg/l PPT used for selection. In all plants, enzyme activity for one or both marker genes was detectable as described below.

In barley, independent transformation events led to the recovery of about one plant per 2,8 x 10^6 bombarded microspores on average (Jähne et al., 1994). The time from embryo excision or microspore isolation and transfer of putative transformants to soil was between 15 to 17 weeks in wheat, 8 to 11 weeks in triticale, and 7 to 8 weeks in barley. This underlines that the use of primary explants as targets for transformation had several advantages over previously published protocols using suspension or callus cultures as targets for bombardment. The time of *in vitro* culture is highly reduced, in contrast to the time-consuming and labour-intensive process needed to establish competent cell suspension or callus cultures as targets.

Using the pDB1 construct for stable transformation experiments, it was possible to screen regenerants by spraying with the herbicide Basta containing PPT, or by histochemically assaying the activity of the *uidA* gene. All regenerants were first tested for GUS activity after transfer into the greenhouse.

In wheat, 12 out of 59, in triticale 23 out of 30 and in barley 4 out of 12 regenerants showed GUS activity in leaf pieces. The intensity of staining varied between individual transformants. Only in a few cases, regenerants showed dark blue staining at all injured surfaces. In most cases intensive staining could only be observed in vascular tissue. Histological GUS assays of leaf pieces of the GUS positive plants showed that all cells of the leaves contained blue cristals, but in varying concentrations. Differences in the staining intensity of various cell typs of the leaves has been described previously in transgenic maize and papaya plants (Gordon-Kamm et al. 1990, Fitch et al. 1992). This may be attributed to differential expression of the *uidA* gene or to variable penetration of the GUS substrate.

Furthermore, the regenerants have been sprayed in a later stage of development with an aqueous solution of the herbicide Basta. One week after herbicide application plants were scored and in wheat only plants which were also GUS positive remained green, confirming functional activity of PAT. In barley 12 regenerants were resistant against herbicide applications and all 30 triticale transformants survived herbicide application.

Barley, triticale and wheat plants which showed enzyme activity for one or both introduced marker genes were analyzed by Southern blot analysis. Southern blot analysis of R_0 plants indicated that these plants contained restriction endonuclease digestion fragments corresponding to intact copies of the coding region of one or both introduced genes. Furthermore, larger or smaller fragments can be observed in most cases, suggesting that deletions, rearrangements and/or methylations at restriction sites have occured. The integration pattern was in most cases very complex. Only in a few cases transgenic plants contain single copy integration of both introduced markers. The integration of only one copy is desirable for applied genetic engineering to avoid potential problems of co-suppression. However, the integration of multiple copies does not necessarily preclude functional levels of gene expression. For example, in our experiments a correlation between integrated copy number and level of resistance against the herbicide or GUS protein accumulation in transgenic wheat, triticale and barley was not observed.

All transformed plants were phenotypically normal. Their development and flowering was comparable with seed derived plants. The use of primary explants for transformantion experiments has the advantage that they are highly regenerative and the probability that abnormal plants are produced is very low in comparison to transgenic plants obtained from long-time suspension or callus cultures (Gordon-Kamm et al. 1990, Somers et al. 1992, Vasil et al. 1992).

In barley transformation experiments we never regenerated albino plants as described by Wan and Lemaux (1994) after particle bombardment of calli derived from microspores of the same cultivar. The frequency of regeneration of albino plants in cereals often increases with the length of time of *in vitro* culture (Kott and Kasha 1984, Vasil 1987, Jähne et al. 1991). In our experiments, time of *in vitro* culture was highly reduced (7-8 weeks), whereas the regeneration of transgenic plants from bombarded microspore-derived embryos takes approximately 7 months (Wan and Lemaux 1994).

The segregation of the introduced marker gene *uidA* has been visualized histochemically in pollen grains of R_0 plants. A 1:1 segregation as well as segregation in a non mendelian fashion has been observed in wheat and triticale.

In barley, all pollen grains showed GUS activity, indicating the homozygous state of the transformed plants.

3.2. INHERITANCE OF THE INTRODUCED MARKER GENES

Progeny from wheat and triticale transformants show mendelian as well as non mendelian segregation of one or both introduced marker genes. In barley all progeny plants show enzyme activity for one or both marker genes. Wheat and triticale progeny plants which show a 3:1 segregation and progeny from homozygous primary barley transformants have been further analyzed by Southern-blot analysis. Wheat, triticale and barley progeny showing enzyme activity for one or both marker genes contained DNA fragments corresponding to the correct size of the coding region of the genes. Furthermore, these plants had the same integration pattern as the parental line. This indicates that the introduced marker genes must be closely linked and inherited to progeny as a genetic unit.

4. Conclusion

Microprojectile-mediated transformation of scutellar tissue of wheat and triticale, and microspores of barley, allowed us to generate fertile transgenic wheat, triticale and barley plants. The development of microprojectile-mediated transformation systems made the rapid progress towards recovery of transgenic cereals possible. The first reports described the regeneration of transgenic plants from bombarded embryogenic suspension or callus cultures. Recent results show that primary explants seem to be more advantageous for the routine production of fertile transgenic cereals, as the time in culture is comparatively short and the risk of somaclonal variation is significantly reduced. Now the routine production of transgenic wheat, triticale and barley plants is possible. The ability to transform cereals is of great importance for approaching fundamental questions as well as for crop improvement.

5. Acknowledgements

The wheat transformation was part of a Ph.D. project at the Institut für Allgemeine Botanik der Universität Hamburg. This work was supported by the Gesellschaft zur Förderung der privaten Deutschen Pflanzenzüchtung e.V. (GFP, Bonn). Barley transformation was supported by the Pflanzenzuchtbetrieb W. von Borries-Eckendorf (Leopoldshöhe, Germany).

6. References

Barcelo, P., Hagel, C., Becker, D., Martin, A. and Lörz, H., 1994. Transgenic cereal (Tritordeum) plants obtained at high efficiency by microprojectile bombardment of inflorescence tissue. Plant Journal 5(4): 583-592

Becker, D., Brettschneider, R. and Lörz, H., 1994. Fertile transgenic wheat from microprojectile bombardment of scutellar tissue. Plant Journal, 5(2): 299-307

Bower , R. and Birch, R.G., 1992 Transgenic sugarcane plants via microprojectile bombardment. The Plant Journal, 2(3), 409-416

Cao, J., Duan, X., McElroy, D. and Wu, R., 1992. Regeneration of herbicide resistant transgenic rice plants following microprojectile mediated transformation of suspension culture cells. Plant Cell Rep. 11, 586-591

Christou, P., Ford, T.L. and Kofron, M., 1991. Production of transgenic rice (*Oryza sativa* L.) plants from agronomically important indica and japonica varieties via electrical discharge particle acceleration of exogenous DNA into immature zygotic embryos. Bio/Technol., 9, 957-962

D`Halluin, K., Bonne, E., Bossut, M., De Beuckeleer, M. and Leemans, J., 1993. Transgenic maize plants by tissue electroporation. Plant Cell 4, 1495-1505

Datta, S.K., Peterhans, A.,Datta, K. and Potrykus, I., 1990. Genetically engineered fertile indica-rice recovered from protoplasts. Bio/Technol., 8, 736-740

Donn, G., Eckes, P. and Müllner, H., 1992. Genübertragung auf Nutzpflanzen. BioEngineering 5+6, 40-46

Fitch, M.M.M., Mansquhardt, R.M., Gonsalves, D., Slightom, J.L. and Sanford, J.C., 1992. Virus resistant papaya plants derived from tissues bombarded with the coat protein gene of papaya ringspot virus. Bio/Technol., 10, 1466-1472

Gordon-Kamm, W.J., Spencer, T.M., Mangano, M.L., Adams, T.R., Daines, R.J., Start, W.G., O'Brian, J.V., Chambers, S.A., Adams, J.W.R., Willetts, N.G.,Rice, T.B., Mackey, C.J., Krueger, .W., Kausch, A.P. and Lemaux, P.G. 1990. Transformation of maize cells and regeneration of fertile transgenic plants. Plant Cell, 2, 603-618

Jähne, A., Lazzeri, P.A. und Lörz, H., 1991. Regeneration of fertile plants from protoplasts derived from embryogenic cell suspensions of barley (*Hordeum vulgare* L.). Plant Cell Rep. 10, 1-6

Jähne, A., Becker, D. and Lörz, H., 1994. Regeneration of transgenic, microspore-derived, fertile barley. Theor. Appl. Genet., in press

Kott, L.S. and Kasha, K.J., 1984. Initation and morphological development of somatic embryoids from barley cell cultures. Can. J. Bot., 62, 1245-1249

Koziel, G.M., Beland, G.L., Bowman, C., Carozzi, N.B., Crenshaw, R., Crossland, L., Dawson, J., Desai, N., Hill, M., Kadwell, S., Launis, K., Lewis, K., Maddox, D., McPherson, K., Meghji, M.R., Merlin, E., Rhodes, R., Warren, G.W., Wright, M. and Evola, S.V. 1993. Field performance of elite transgenic maize plants expressing an insecticidal protein derived from *Bacillus thuringiensis*. Bio/Technol., 11, 194-200

Mordhorst, A.P. and Lörz, H., 1993. Embryogenesis and development of isolated barley (*Hordeum vulgare* L.) microspores are influenced by the amount and composition of nitrogen sources in culture media. J. Plant Physiol., 142, 485-492

Sanford, J.C., Klein, T.M., Wolf, E.D. and Allen, N., 1987. Delivery of substances into cells and tissues using a particle bombardment process. J. Part. Sci. Technol. 5, 27-37

Shimamoto, K., Terada, R., Izawa, T. and Fujimoto, H., 1989. Fertile transgenic rice plants regenerated from transformed protoplasts. Nature 338, 2734-276

Somers, D.A., Rines, H.W., Gu, W., Kaeppler, H., F. and Bushnell, W. R., 1992. Fertile, transgenic oat plants. Bio/Technol. 10, 1589-1594

Vasil, I.K., 1987. Developing cell and tissue culture systems for the improvement of cereal and grass crops. Plant Physiol., 128, 193-218

Vasil, V., Castillo, A.M., Fromm, M.E. and Vasil, I.K., 1992. Herbicide resistant fertile transgenic wheat plants obtained by microprojectile bombardment of regenerable embryogenic callus. Bio/Technol. 10, 667-674

Vasil, V., Srivastava, V., Castillo, A.M., Fromm, M.E. and Vasil, I.K. 1993. Rapid production of transgenic wheat plants by direct bombardment of cultured immature embryos. Bio/Technol. 11: 1553-1558

Walters, D.A., Vetsch, C.S., Potts, D.E. and Lundquist, R.C., 1992. Transformation and inheritance of a hygromycin phosphotransferase gene in maize plants. Plant Mol. Biol., 18, 189-200

Wan, Y. and Lemaux, P.G., 1994. Generation of large numbers of independently transformed fertile barley plants. Plant Physiol. 104: 37-48

Wang, X.-H. , Lazzeri, P.A. and Lörz, H., 1992. Chromosomal variation in dividing protoplasts derived from suspensions of barley (*Hordeum vulgare* L.). Theor. Appl. Genet., 85, 181-185

Weeks, J.T., Anderson, O.D. and Blecchl, A.E., 1993. Rapid production of multiple independent lines of fertile transgenic wheat (*Triticum aestivum*). Plant Physiol., 102, 1077-1084

Zimny, J., Becker, D., Brettschneider, R. and Lörz, H.. Fertile, transgenic *Triticale* (x *Triticosecale* Wittmack). Plant Mol. Biol., submitted

THE EMBRYO AS A TOOL FOR GENETIC ENGINEERING IN HIGHER PLANTS

R.S. SANGWAN, F. DUBOIS, C. DUCROCQ, Y. BOURGEOIS, B. VILCOT, N. PAWLICKI, B.S. SANGWAN-NORREEL,
Laboratoire Androgenèse et Biotechnologie, Université de Picardie Jules Verne, 33, rue Saint-Leu, Ilot des Poulies, 80039 Amiens Cédex, France.

1. Introduction

It is only during the last decade that genetic engineering of plants has been developed only recently that it has become an important tool for crop improvement. Several molecular techniques are widely used to introduce recombinant DNA molecules into a variety of plant species (1-3) however, one of the most effective means of gene transfer into plants is the utilization of soil bacteria from genus *Agrobacterium* and essentially *A. tumefaciens* or *A. rhizogenes*, as vectors. The fundamental problem, in the production of fertile transgenic plants, especially in recalcitrant species, does not seem to lie so much in the delivery or perhaps integration of the introduced DNA, but rather in the regeneration of transgenic plants from the transformed cells. In most transformation experiments, *in vitro* regeneration of plants has become the main limiting factor for success. At present the major source of highly regenerative cells has been immature zygotic embryo or meristematic tissues but a routine *Agrobacterium*-mediated transformation protocol for these tissues is lacking. Circumvention of the difficulties inherent in *in vitro* regeneration by transforming the easily regenerable explants such as zygotic embryos would therefore be highly advantageous. Although meristematic and embryogenic cells were thought to be insensitive to *Agrobacterium* infection (4), it has been demonstrated recently that they are susceptible to Agro-infection (5) and under certain *in vitro* conditions e.g. with phytohormone treatment prior to infection by *Agrobacterium*, the embryonic cells become competent for infection (6-8).

We have studied the feasibility of zygotic and pollen embryo transformation in higher plants (6-8). In 1988, as a first model system we choose to transform the zygotic embryos of *Arabidopsis* then the pollen embryos of *Datura* and *Nicotiana* and lastly we have exploited the direct somatic embryogenesis system from zygotic embryos in *Datura* (Fig.1). In this paper we provide a summary of these results and thus report on 3 transformation systems : 1) routine zygotic embryo transformation and regeneration of fertile transgenic plants in *Arabidopsis thaliana*. 2) transformation of cotyledonary stage pollen embryos in *Datura* and *Nicotiana*. 3) a successful combination of the technique of direct somatic embryogenesis with *Agrobacterium*-mediated gene transfer in *Datura*. This later method does not require wounding nor a callus phase. We will also outline the possible impact on transgenic plant technology of our results.

M. Terzi et al. (eds.), Current Issues in Plant Molecular and Cellular Biology, 271-277.

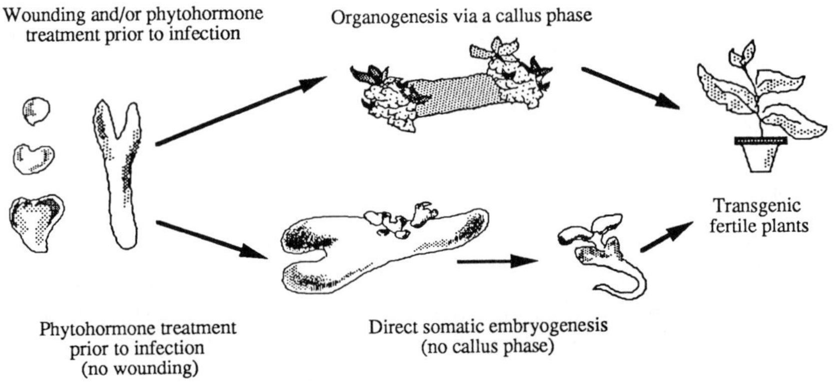

Figure 1

(Dedifferentiation of differentiated cells is required for transformation)

2. Procedures

Only a brief description of the procedures used are described below, as details have been reported elsewhere (6-9).

2.1 Bacterial strains

For transformation the binary (pGS Gluc 1, pGS 943, pBI 121 and p35SGUSINT) and cointegrate (pGV 246) vector systems (Fig.2) were used (6-9).

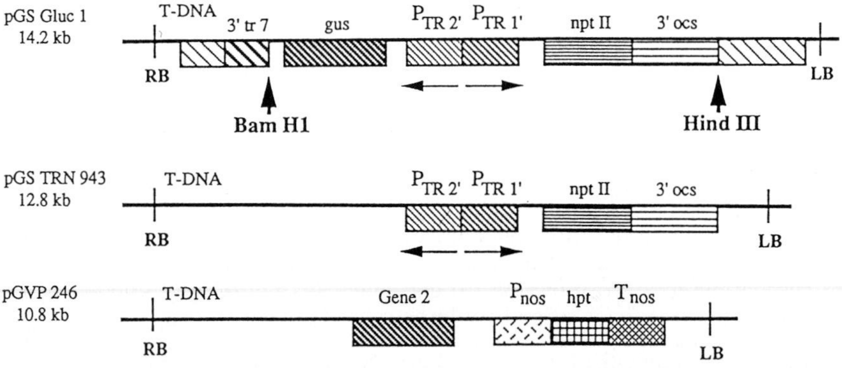

Fig. 2. Schematic representation of chimæric gene constructs used for transformation experiments. RB, right border; LB, left border; gus, ß-glucuronidase; nptII, neomycin phosphotransferase; ocs, octopine synthase; nos, nopaline synthase; hpt, hygromycin phosphotransferase; P, promoter; T, terminator

2.2 Zygotic embryo transformation and regeneration procedure in *Arabidopsis* see Sangwan et al (6).

2.3 Pollen embryo transformation and regeneration procedure in *Datura* and *Nicotiana* see Sangwan et al (7).

2.4 Zygotic embryo transformation in *Datura* (8) : Entire or injured zygotic embryos were incubated with gentle shaking, for 10-15 min in a Petri dish containing 20 ml MS liquid medium and 1 ml overnight bacterial solution (approximatively 10^7 Agrobacteria/ml). Explants were blotted on to a sterile filter paper and placed on **EIM** medium Lin and Staba macro and micro-elements (10), vitamins of Nitsch and Nitsch (11) 12% sucrose, 1 mg/l IAA, 8 g/l agar. After 3 days of cocultivation, embryos were washed with MS liquid medium containing 500 mg/l cefotaxime, blot dried and placed on **EIM2** selective medium (EIM supplemented with 200 mg/l kanamycin and 300 mg/l cefotaxime).
Somatic embryos which appeared on embryonic explants within 3-4 weeks were isolated and subcultured on **MM2** medium (MM medium containing 2% sucrose, 0.1 mg/l IAA, 200 mg/l kanamycin and 300 mg/l cefotaxime) and then on **RM2** medium (RM medium 1/2 LS containing 1% sucrose, o.1 mg/l IAA and 200 mg/l kanamycin).

3. Results

3.1 Parameters affecting zygotic embryo transformation efficiency in *Arabidopsis*

Zygotic embryos of *Arabidopsis* were found to be convenient explants for the initiation of *in vitro* culture and showed considerable potential for shoot regeneration and whole plant formation after 5-6 weeks of culture. Preliminary assays showed that cultured mature and immature *Arabidopsis* embryos were very sensitive to kanamycin and hygromycin. No growth of calli and no escapes were observed on media containing 50 mg/l kanamycin or 25 mg/l hygromycin. The transformation process and subsequent regeneration of transgenic plants was followed using Ti plasmid pGS Gluc 1. It was observed that the nature of the preculture treatment prior to infection was a crucial step in the zygotic embryo transformation of *Arabidopsis*. Hence, the effect of preculture duration was studied. The results showed that the transformation frequency depended on the duration of preculture of embryo explants on callus-inducing medium (BM_2) (6). At least 3 days of preculture were required for successful transformation. In addition to pGS Gluc 1, we confirmed the above results with several other vectors. For example, in the case of embryos transformed with the pGS Gluc 1 binary vector, 76% of the explants produced Km^R calli after 6-7 days of preculture as compared to 34% after 3 days of treatment. Similarly, for embryos transformed with the pGV 246 cointegrate-type vector, transformation rates of 79% and 34% were obtained after a 7-day and 3-day pretreatment, respectively. Although preincubation of more than 5 days resulted in high transformation frequencies, only 25% of the resistant calli were able to regenerate shoots. We also tested whether addition of acetosyringone during the infection period could substitute for hormonal pretreatment of embryos. No transformation was obtained in the presence of acetosyringone without prior application of a hormone pretreatment. However, the duration of co-cultivation did influence the transformation frequency. Whereas 65% of the embryo explants co-cultured for 2 days produced Km^R calli, the mean transformation rate observed after co-cultivation of embryo explants for 4 days was only 45%. Co-cultivation for more than 5 days led to explant loss (about 50%) due to bacterial overgrowth, and a considerable decrease in the yield of Km^R calli (< 25%). Interestingly, the direction of cutting the embryo explants and genotype also had a significant effect on transformation frequency. Cutting the embryos longitudinally resulted in a transformation frequency more than twice that of transversely cut embryos, under optimal conditions (5 days preculture and 2 days co-culture). The two ecotypes, Landsberg-erecta and Columbia, showed transformation frequencies of 15% and 20%, respectively, lower than that obtained with C24 (70%) under the same conditions. In summary, preculture (> 3 days) on a phytohormone-containing medium (BM_2 or BM_3) was absolutely indispensible for transformation of the embryo explants, whereas duration of co-culture, plane of sectioning and genotype had a pronounced effect on transformation efficiency. Analysis for kanamycin-resistance and ß-glucuronidase activity showed that the transgenes were inherited as dominant Mendelian traits. The T-DNA integration pattern after Southern analysis corresponded well to the classical patterns when *A. tumefaciens* was used as a vector (Fig.3).

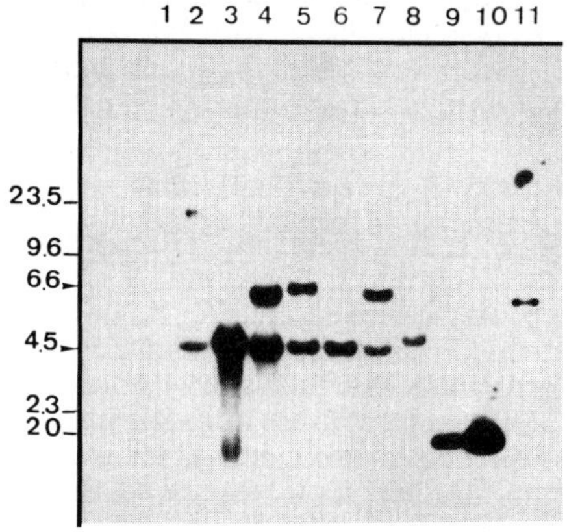

Fig. 3. Southern blot analysis of *A. thaliana* C24 R1 progenies. All transgenic plants were obtained by transformation with the binary plasmid pGS Gluc 1. DNA was isolated from GUS-positive plants, digested with *Bam*H1 and *Hind*III, separated by agarose gel electrophoresis, blotted to a membrane and hybridized with a ^{32}P-labelled fragment of GUS coding sequence. Lane 1, DNA from an untransformed (control) plant. Lanes 2-8, DNA from transformed R1 families. All these are true transformants as indicated by the 4.5 kb band hybridizing with the GUS probe. Lanes 9-10, 1 copy and 10 copies of GUS coding (1.8 kb) sequence used as hybridization control

3.2 Parameters effecting pollen embryo transformation efficiency in *Datura* and *Nicotiana*

Since no transformants were obtained when isolated pollen grains or proembryos enclosed within the pollen wall were used, factors such as the developmental stage of embryos i.e., late globular to cotyledon stage, inoculation method and bacterial strains were investigated. Of the several inoculation methods tested, wounding or cutting the embryo explants prior to infection was most successful for obtaining transformation. Transformation and subsequent regeneration of transgenic plants was studied by using the vectors pGS Gluc 1 and p35S GUSINT. The transformation frequency depended on the stage of embryo explant. For example, no transformation was obtained using globular embryos of *Datura* while the transformation frequency was > 81% in the cotyledonary stage cut embryo explants. A similar pattern was found in *Nicotiana*, although transformation frequencies were slightly lower when compared to *Datura*. Wounding by glass pieces greatly reduced both the embryo survival and transformation frequency. For example, in *Nicotiana* only 6% of cotyledonary embryos were transformed as compared with 72% in cut embryos. Approximately 60% of the *Datura* shoots rooted and 95% of the *Nicotiana* shoots rooted after 10 days. The rooted plants were transferred to the greenhouse, and grown to maturity. About 50% of the regenerants in *Datura* and 5% in *Nicotiana* appeared to be spontaneous diploids. In addition to pGS Gluc 1, we confirmed the above results with several other vectors. For example, in *Nicotiana* with pGV 246, a cointegrate vector, transformation frequency was 69% while with pGS TRN 943 it was 75%. This indicates that the binary and cointegrate vectors are equally suitable for obtaining high transformation frequencies in *Nicotiana* and *Datura*.

3.3 Parameters affecting zygotic embryo transformation in *Datura*

Cocultivation with *Agrobacterium* did not affect somatic embryogenic potential of zygotic embryos. The number of somatic embryos and plants produced on non-selective regeneration medium were similar with or without cocultivation. The transformation and subsequent regeneration of transgenic plants was investigated using two vectors containing the GUS reporter gene (pGS Gluc 1 and p35SGUSINT). Entire zygotic embryos were cocultured for 3 days and transferred to selection medium (EIM2) containing 200 mg/l kanamycin. After 2 to 3 weeks of culture, kanamycin resistant somatic embryos were observed. Somatic embryos only developed from the epidermal cells. While GUS activity (blue coloration) was observed in the epidermal cells (without wounding) and also at the cut surface (after wounding), 3 days after infection. The kanamycin-resistant embryos also showed GUS activity indicating their transformation. Uninoculated controls or explants inoculated with a strain lacking the GUS gene did not produce blue staining when tested.

It has been reported that wounding prior to infection was a crucial step in *Agrobacterium*-mediated transformation of higher plants (12, 13). Hence, the effect of wounding of zygotic embryos on transformation was studied. No transformation or kanamycin resistant somatic embryos were obtained after wounding. Zygotic embryos were cut into two halves or injured with a razor. Wounded zygotic embryos turned brown on the selective medium, and died after 2-3 weeks of culture. In contrast, the infected entire embryos gave a transformation frequency of 54%. This clearly shows that wounding is not required for transformation of zygotic embryos.

Zygotic embryos were precultured on EIM medium (8) for different durations (0, 1, 2, 4 and 8 days) and then cocultured for 3 days with bacteria. Maximum transformation (76%) was obtained after a 1 day preculture and decreased to 40% and 52% after 2 and 8 days of preculture, respectively. Similarly, the effect of prolonged coculture was also investigated. On coculture with the vector pGS Gluc 1 for 3 days, transformation frequency was 70%. However, a further increase in coculture duration decreased transformation frequency. For example, after 4 days of coculture, only 40% of explants produced transgenic somatic embryos and about 50% explants were lost because of bacterial overgrowth. Thus, in all subsequent transformation experiments, one day preculture and a 3 day coculture period was used to maximise the percentage of somatic embryos capable of regeneration. Following this protocol, Km^R somatic embryos were obtained after 3 weeks of culture on EIM2 medium. For further growth, the Km^R somatic embryos were then transferred to MM medium (8) for 10 days and then to germination medium for 2-3 weeks. The rooted plants were grown to maturity in a greenhouse. Under identical experimental conditions, explants cocultured with or without *Agrobacterium* carrying the disarmed Ti plasmid pGV 2260 did not produce Km^R somatic embryos. Others parameters such as genotype and vector were also tested for their influence on the frequency of embryo transformation. All vectors tested produced Km^R somatic embryos but with variable frequency. The frequency of transformation was significantly different for p35SGUSINT (76%) and pGS TRN 943 (15.8%). However, no significant difference between the 3 genotypes tested were observed.

Discussion

Our goal has been to develop reliable and routine transformation systems in higher plants using zygotic and pollen embryos as explants so that these systems (6-8) like the leaf disc system (14) can be efficiently used for transformation of plant species in which regeneration is possible only from cultured embryonic tissues. Above we have described three protocols for regeneration of morphologically normal and fertile transgenic plants of *Arabidopsis, Datura* and *Nicotiana* from embryonic tissues. Among the parameters examined, pretreatment of explants with phytohormones prior to infection and embryo stage appear to be very important. This suggests that wounding of the explant alone does not cause cell activation in the dormant *Arabidopsis* zygotic embryonic cells and that it is not sufficient for efficient transformation. Moreover, a certain degree of activation or division, is required for transformation. This is in accordance with the observation that actively dividing cells are more prone to transformation than non-dividing cells (15) and that only the metabolically active plant cells produce molecules which activate the vir region of the Ti plasmid (16). In agreement with several previous publications we found that both the zygotic embryos of *Arabidopsis* and pollen embryos of *Datura* and *Nicotiana* required wounding and a callus phase for obtaining transgenic plants. Although successful transformation has been reported in *Brassica* using young pollen embryos (17) we were unsuccessful in transforming pollen and young pollen embryos of *Datura* and *Nicotiana*. In contrast, we observed a direct correlation between developmental stage and the competence of pollen embryos for transformation e.g. transformation frequency was the highest (> 75%) with injured or cut cotyledonary-stage pollen embryos. Moreover, GUS assays showed that only dedifferentiating and dividing cells of the embryo explants at the cut surface were transformed. For a productive conjugation, *Agrobacterium* appears to bind specifically to the dividing cells at the wounded surface (4, 16, 18), corroborating our earlier observations in *Arabidopsis* in which dedifferentiation was essential for transformation (6, 9). Apparently, the embryo explants allow *Agrobacterium* transformation to be targeted to highly regeneration-competent cells. Several hundred haploid and spontaneous dihaploid transgenic plants were obtained. The spontaneous dihaploids or colchicine-treated dihaploids were fertile, and set seeds.

In the last protocol we have described a novel method to obtain transgenic somatic embryos directly from infected zygotic embryos of *Datura* without wounding and without an intermediate callus phase, and subsequent regeneration of fertile transgenic plants from embryos. Evidence is presented to show that the epidermal cells of hypocotyl of zygotic embryos are competent both for transformation and regeneration, and that the competency for transformation and regeneration is highly specific to the embryogenic stage. Nearly all of the previously published reports state that a callus phase is essential for *Agrobacterium* based transformation. However, our findings show that T-DNA can be delivered by the *Agrobacterium* into embryogenic competent cells in the preconditioned immature zygotic embryos, and that the transformed cells developed into plants via somatic embryogenesis without an intervening callus phase. We also found that the transformation competence in zygotic embryos was stage specific. A similar type of competence for *Agrobacterium* has been reported in germinating pinto bean (4) developing pollen embryos (7) *Asparagus* zygotic embryos (19) and maize zygotic embryos (20). Both our cytological and molecular genetic analyses showed that the embryogenic competent cells of zygotic embryos, transformed with *Agrobacterium*, regenerate plant from single cells through somatic embryogenesis (8).

Hence, a direct somatic embryogenesis system has a great advantage over protoplast, leaf root, cotyledon explants etc. for normal, fertile transgenic plant production. Moreover, this system was not genotype dependent, since transgenic plants were obtained from all the genotypes tested. Southern analysis performed on transgenic plants confirmed independent insertion events. The transformation protocol reported in this study is simple and rapid, because it involves the use of simple media in sequence and distinct transgenic somatic embryos were visible within three weeks. In conclusion, the above described procedures should facilitate the transformation of recalcitrant but economically important plant species including soybean, pea, sugarbeet etc. where only infrequent and sporadic transgenic plants are obtained often after a very long *in vitro* culture period using an *Agrobacterium* system. This is because in these plants, regeneration from immature zygotic embryos is routine.

Acknowledgements
We thank Dr. M. Hodges for critically reading the manuscript, F. Flandre for drawing figure 1, M. Poiret and G. Vasseur for technical assistance.

References

1) Hooykass, P.J.J. and Schilperoort, R.A. (1992) Plant Mol. Biol. 19, 15-38.
2) Kahl, G. and Weising, K. (1993) In Rehm H.J. and Reed, G. (eds.) Biotechnology, V.C.H. Weinheim, pp. 344-547.
3) Potrykus, I. (1991) Ann. Rev. Plant Physiol. Plant Mol. Biol. 42, 205-225.
4) Lippincott, J.A. and Lippincott, B. (1978) Science 199, 1075-1078.
5) Grimsley, N., Hohn, B., Ramos, C., Kado, C., Rogowsky, P. (1989) Mol. Gen. Genet. 217, 309-316.
6) Sangwan, R.S., Bourgeois, Y., Sangwan-Norreel, B.S. (1991) Mol. Gen. Genet. 230, 475-485.
7) Sangwan, R.S., Ducrocq, C. and Sangwan-Norreel, B.S. (1993) Plant Science 95, 99-115.
8) Ducrocq, C., Sangwan, R.S., Sangwan-Norreel, B.S. (1994) Plant Mol. Biol. Mol. Breeding (In press).
9) Sangwan, R.S., Bourgeois, Y., Brown, S., Vasseur, G. and Sangwan-Norreel, B.S. (1992) Planta 188, 439-456.
10) Lin, M.L. and Staba, J. (1961) Lloydia 24, 139-145.
11) Nitsch, J.P. and Nitsch, C. (1965) Ann. Physiol. Vég. 7, 251-258.
12) Braun, A.C. (1947) Am. J. Bot. 34, 234-240.
13) De Cleene, M. and De Ley, J. (1976) Bot. Rev. 42, 389-466.
14) Horsch, R.B., FRY, J.E., Hoffmann, N.L., Wallroth, M., Eichholtz, D., Rogers, S.G. and Fraley, R.T. (1985) Science 227, 1229-1231.
15) An, G. (1985) Plant Physiol. 79, 568-570.
16) Zambryski, P. (1992) Ann. Rev. Plant Physiol. Plant Mol. Biol. 43, 465-490.
17) Pechan, P.M. (1989) Plant Cell Rep. 8, 387-390.
18) Binns, A.N. and Thomashow, M.F. (1988) Ann. Rev. Microbiol. 42, 575-606.
19) Delbreil, B., Guerche, P. and Jullien, M. (1993) Plant Cell Rep. 12, 129-132.
20) Schaläppi, M. and Hohn, B. (1992) Plant Cell 4, 7-16.

PRODUCTION OF FERTILE TRANSGENIC MAIZE PLANTS BY SILICON CARBIDE WHISKER-MEDIATED TRANSFORMATION

Bronwyn R. Frame[1], Paul R. Drayton[2], Susan V. Bagnall[1], Carol J. Lewnau[1], W. Paul Bullock[1], H. Martin Wilson[1], James M. Dunwell[2], John A. Thompson[2], and Kan Wang[2]
[1] ICI Seeds, Research Department, 2369 330th Street, Box 500, Slater, IA 50244, USA.
[2] ZENECA Seeds, Plant Biotechnology Department, Jealott's Hill Research Station, Bracknell, Berkshire, RG12 6EY, UK.

ABSTRACT. A simple and inexpensive system for the generation of fertile, transgenic maize plants has been developed. Cells from embryogenic maize suspension cultures were transformed using silicon carbide whiskers to deliver plasmid DNA carrying the bacterial *bar* and *uidA* (gus) genes. Transformed cells were selected on medium containing the herbicide bialaphos. Integration of the *bar* gene and activity of the enzyme phosphinothricin acetyl transferase (PAT) were confirmed in all bialaphos-resistant callus lines analysed. Fertile transgenic maize plants were regenerated. Herbicide spraying of progeny plants revealed that the *bar* gene was transmitted in a Mendelian fashion.

1. INTRODUCTION

Fertile transgenic maize plants have been produced through bombardment of embryogenic suspension cultures (Gordon-Kamm *et al.*, 1990, and Fromm *et al.*, 1990) and immature zygotic embryos (Koziel *et al.*, 1993). Wounding of such embryos followed by electroporation (D'Halluin *et al.*, 1992), as well as direct gene transfer to protoplasts (Golovkin *et al.*, 1993) have also resulted in fertile transgenic maize plants.

In contrast to these technically demanding methods the simple mixing of cells in liquid medium with whiskers and plasmid DNA has also resulted in DNA delivery. Stable transformation of maize cultures was achieved (Kaeppler *et al.*, 1992) but the cell line used - Black Mexican Sweet corn - was non-regenerable.

In this paper we report the production of fertile maize plants using the silicon carbide whisker transformation method.

2. MATERIALS AND METHODS

2.1 CELL SUSPENSION CULTURES

These were prepared exactly as described in Register *et al.*, (1994). Suspension cultures recovered from cryopreservation were used in the transformations. The suspensions were cryopreserved according to Shillito *et al.*, (1989).

M. Terzi et al. (eds.), Current Issues in Plant Molecular and Cellular Biology, 279-284.

2.2 PLASMID DNA

Plasmid DNA was purified using a QIAGEN plasmid Maxi kit (QIAGEN Inc., Chatsworth, CA, USA).

2.3. TRANSFORMATION PROCEDURE

2.3.1 *Whisker Preparation.* A sterile 5% (w/v) suspension of silicon carbide whiskers (Silar SC-9 whiskers, Advanced Composite Materials Corp., Greer, SC, USA) in water was prepared immediately prior to use (Kaeppler *et al.*, 1990).

2.3.2 *Tissue Preparation.* Two hundred and fifty μl of packed cell volume (pcv) of suspension cells one day after subculture were dispensed into a 1.5 ml Eppendorf tube. The cells were then dispersed in 1 ml of liquid S/M medium (N6 with 0.25M sorbitol and 0.25M mannitol) and maintained for 30 minutes at room temperature. After treatment, 750 μl of S/M medium was withdrawn leaving 250 μl of medium with the cells.

2.3.3 *DNA delivery.* Forty μl of 5% whisker suspension and 25 μl of plasmid DNA (1 μg/μl) were added to S/M-treated suspension cells. Tube contents were first finger tapped to mix and then placed either upright in a multiple sample head on a Vortex Genie II mixer (Scientific Industries Inc., Bohemia, NY, USA) or horizontally in the holder of a Mixomat dental amalgam mixer (Degussa Canada Ltd., Burlington, Ontario, Canada). Transformation was carried out by mixing at full speed for 60 seconds (Vortex Genie II) or shaking at fixed speed for 1 second (Mixomat).

2.4 SELECTION OF TRANSFORMED TISSUE AND PLANT REGENERATION

After treatment, cells were plated on filter paper (Whatman no.4, 5.5 cm) overlaying N6 medium (2 mg/l 2,4-D, 3% sucrose, 0.3% Gelrite, pH 6.0) in 60 x 20 mm plastic petri dishes and wrapped with Urgopore tape (Sterilco, Brussels, Belgium). One week later, the filter paper and cells were transferred to the surface of N6 selection media with 1 mg/l bialaphos. Subsequent handling of the cells and plant regeneration was exactly as described in Register *et al.* (1994).

2.5 HERBICIDE SPRAY ASSAYS

Screening of R2 segregating progeny was accomplished by spraying 9-12 day-old seedlings with Ignite herbicide. Response was recorded 6 days later.

2.6 ANALYSIS OF TRANSFORMED CALLUS AND PLANTS

Histochemical GUS assays were conducted following published protocols (Jefferson, 1987, McCabe *et al.*, 1988). PAT activity was determined using a thin layer chromatographic assay (De Block *et al.*, 1987, Spencer *et al.*, 1990). Genomic DNA was isolated from leaf tissue for Southern analysis essentially as described by Saghai-Maroof *et al.* (1984), and as described by Edwards *et al.* (1991) for polymerase chain reaction amplification. The probe used in Southern analysis was produced by gel purification of a BamH1 restriction enzyme fragment from pBARGUS and was labelled with digoxigenin (DIG)-dUTP via random primed labelling (the Genius System, Boehringer Mannheim Corp.). Standard procedures were used for DNA blot analysis (Sambrook *et al.*, 1989).

3. RESULTS AND DISCUSSION

The components of the plasmid used for transformation, pBARGUS (Fromm *et al.*, 1990), are shown in Figure 1.

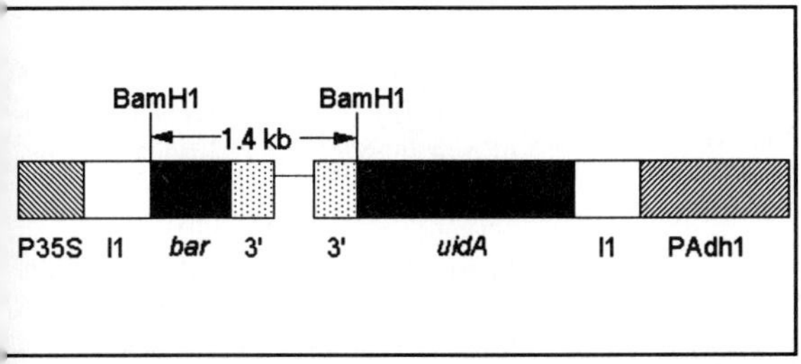

Figure 1. Schematic representation of pBARGUS (Fromm *et al.*, 1990). The 1.4 kb BamHl fragment was used as a probe for Southern analysis. P35S, cauliflower mosaic virus 35S promoter; I1, intron 1 of maize alcohol dehydrogenase 1 (Adhl); *bar*, bar coding region; 3', nopaline synthase polyadenylation region; *uidA, uidA* (*gus*) coding region; PAdhl, promoter of maize alcohol dehydrogenase 1.

Table 1 summarises the results from 9 independent stable transformation experiments. The efficiency of stable clone recovery ranged from 0.2 to 1.7 clones per ml PCV of treated cells. A total of 311 plants were regenerated from 22 independently transformed callus lines. Both the phenotype and fertility of the plants regenerated from bialaphos-resistant callus were similar to plants regenerated from non-transformed callus.

Table 1. Summary of clone recovery from 9 whisker-mediated transformation experiments

Exp. No.	Cells treated (mls of pcv)	Clones recovered	PCR (+/total)	Ratio (clone/ml of pcv)
1	5	4	4/4	0.80
2	8	3	2/3	0.38
3	20	4	4/4	0.20
4	10	4	4/4	0.40
5	6	3	3/3	0.50
6	10	17	17/17	1.70
7	5	1	1/1	0.20
8	4	1	1/1	0.25
9	7	3	3/3	0.43
Total	75	40	39/40	0.53

The transformed status of callus and plants was confirmed by PAT assay (Fig. 2), PCR, or leaf painting with Ignite herbicide (data not shown). Of the nine putative transformants included in

Fig. 2, eight were PAT positive.

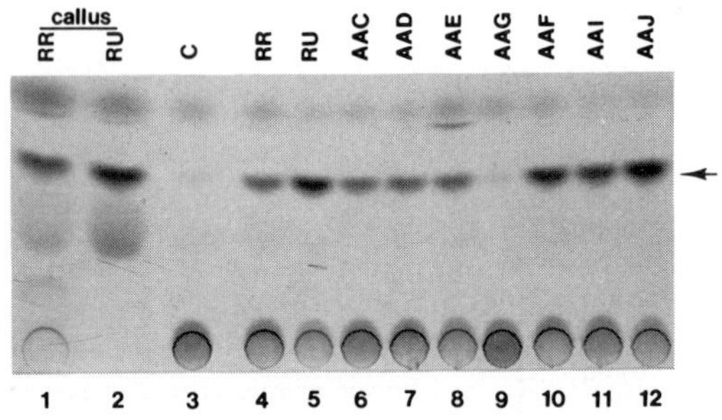

Figure 2. PAT activity in a non-transformed control (Lane 3) and 9 independent transformants (Lanes 4 to 12). PAT activity, as indicated by acetylated phosphinothricin (arrow), is seen in the calli and leaves of clones RR (Lanes I and 4) and RU (Lanes 2 and 5), leaves of clones AAC, AAD, AAE, AAF, AAI and AAJ (Lanes 6-8 and 10-12). Twenty-five micrograms of protein extract were loaded per lane.

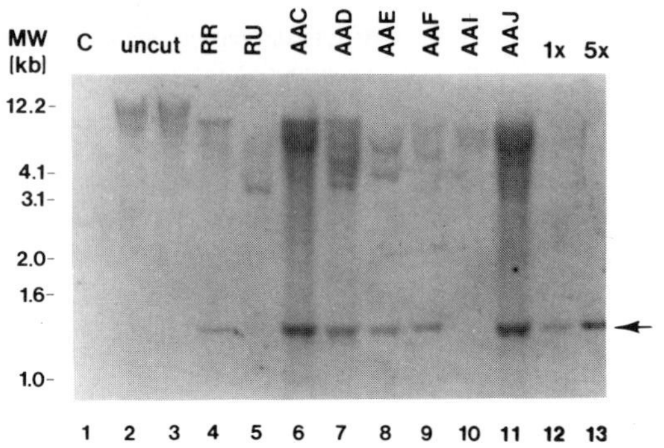

Figure 3. Autoradiograph of Southern blot of genomic DNA extracted from leaf tissue of a nontransformed control (Lane 1) and 8 independent transformants. Lanes 2 and 3, undigested DNAs from transformants RR and RU; Lanes 4 to 12, the BamHl digested DNAs (12 mg per lane) from transformed lines RR, RU, AAC, AAD, AAE, AAF, AAI and AAJ. The blot was hybridized with a digoxigenin (DIG) -dUTP labelled 1.4 kb *bar* gene fragment from pBARGUS. Lanes 12 and 13, one copy and five copies of the *bar* gene fragment controls, refer to the diploid genome and contain 5.2 pg and 26 pg, respectively, of the 1.4 kb *bar* expression unit released from pBARGUS with BamHl. The arrow indicates the 1.4 kb BamHl *bar* gene fragment. Molecular markers (kb) are shown on the left.

Fig. 3 shows the Southern hybridisation analysis of several primary transformants (R_0). Genomic DNA was digested with BamH1 and hybridised with a 1.4 kb BamH1 fragment containing the *bar* gene (Fig. 1). The non-transformed plant control showed no hybdridisation (Lane 1), whereas the digested DNA samples from six of eight transformants (Lanes 4, 6 - 9, and 11) showed a 1.4 kb hybridising band co-migrating with the plasmid control (Lanes 12 and 13), indicating the presence of the *bar* gene. The estimated copy number ranged from one to five, although in some cases more than ten copies were observed. Clones RU (Lane 5) and AAI (Lane 10) showed hybridising fragments at positions higher than 1.4 kb. Since both clones exhibited PAT activity (Fig. 3, Lanes 5 and 11), this suggested that the *bar* gene remained intact. The hybridising fragments greater than 1.4 kb in the two clones may be the products of plasmid/plant junctions which could be due to the loss, during integration, of the BamH1 site between the *uidA* gene and the 3' region (Fig. 1).

Progeny from several independently transformed R_0 plants were analysed to determine the inheritance of the *bar* gene. Its expression was assessed by the spraying of herbicide on 7 to 9 day-old plantlets. The plants expressing the *bar* gene developed no symptoms, whereas the non-transformed plants showed necrosis and died within two weeks of treatment. The herbicide spray permitted monitoring of the segregation of PAT expression in large numbers of progeny. For example, Table 2 presents segregation data from the R_1 progeny of three transformed R_0 plants. The results were not significantly different from a 1 to 1 segregation in reciprocal crosses and a 3 to 1 segregation in selfings, as estimated by Chi-Square analysis, and indicated that the PAT activity was encoded by, and transmitted as, a single dominant gene.

Table 2. Segregation of *bar* gene activity based on leaf spraying assays in R_1 progeny of transgenic maize plants[a]

Population	Cross	Herbicide sensitive	Herbicide resistant	χ^{2b}	P[c]
RR2	Self	9	30	0.008	0.93
	Male	50	37	1.65	0.19
RU2	Self	23	56	0.51	0.47
	Male	34	36	0.014	0.90
	Female	37	30	0.53	0.46
UI2	Self	24	57	0.695	0.40
	Male	46	50	0.094	0.76
	Female	43	52	0.673	0.41

[a] Bialaphos resistant R1 plants were either selfed or used as male or female parent in crosses with non-transformed B73 plants.
[b] χ^2 = chi-square values with Yates (continuity) correction.
[c] P = χ^2 probability with 1 degree of freedom.

This paper represents the first report of transgenic plant production, in any higher plant species, using silicon carbide whiskers as the DNA delivery system. As with clones derived from particle bombardment, ease of plant regeneration and fertility of regenerants were cell line and clone dependent. Generally, plants regenerated from independently transformed clones were similar in phenotype to non-transformed control plants. The reduced seed set sometimes observed in these plants was most likely due to the effects of cell culture age and in vitro selection rather than the transformation process itself. This was confirmed by the complete fertility of transgenic plants in subsequent generations.

Transformation efficiency in these experiments, based on treated cell volume, is estimated at one fifth to one tenth that achieved in our laboratories with microprojectile bombardment of A188xB73 cell suspension cultures. We believe, however, that the efficiency of whisker transformation can be further improved through experimentation and that even at its current efficiency it can be considered as a practical alternative transformation method for maize, requiring no expensive equipment or consumables, especially for those laboratories which do not have access to other more sophisticated technologies.

4. REFERENCES

De Block, M., Botterman, J., Vandewiele, M., Dockx, J., Thoen, C., Gossele, V., Rao Movva, N., Thompson, C., Van Montagu, M. and Leemans, J. (1987) "Engineering herbicide resistance in plants by expression of a detoxifying enzyme", The EMBO Journal 6, 2513-2518.

D'Halluin, K., Bonne, M., de Beuckleer, M. and Leemans, J. (1992) "Transgenic maize plants by tissue electroporation", The Plant Cell 4, 1495-1505.

Edwards, K., Johnstone, C. and Thompson, C. (1991) "A simple and rapid method for the preparation of plant genomic DNA for PCR analysis", Nucelic Acids Res. 19, 1349.

Fromm, M.E., Morrish, F., Armstrong, C., Williams, R., Thomas, J. and Klein, T. (1990) "Inheritance and expression of chimeric genes in the progeny of transgenic maize plants", Bio/Technology 8, 833-839.

Golovkin, M.V., Abraham, M., Morocz, S., Bottka, S., Feher, A. and Dudits, D. (1993) "Production of transgenic maize plants by direct DNA uptake into maize protoplasts", Plant Science 90, 41-52.

Gordon-Kamm, W., Spencer, T., Mangano, M., Adams, T., Daines, R., Start, W., O'Brien, J., Chambers, S., Adams, W., Willetts, N., Rice, T, Mackey, C., Krueger, R., Kausch, A., and Lemaux, P. (1990) "Transformation of maize cells and regenration of fertile transgenic plants", The Plant Cell 2, 603-608.

Jefferson, R. (1987) "Assaying chimeric genes in plants. The GUS gene fusion system", Plant Molec. Biol. Rep. 5, 387-405.

Kaeppler, H., Somers, D., Rines, H. and Cockburn, A. (1990) "Silicon carbide fiber-mediated DNA delivery into plant cells", Plant Cell Reports 9, 415-418.

Kaeppler, H., Somers, D., Rines, H. and Cockburn, A. (1992) "Silicon carbide fiber-mediated stable transformation of plant cells", Theoretical and Applied Genetics 84, 560-566.

Koziel, M., Beland, G., Bowman, C., Carozzi, N., Crenshaw, R., Crossland, L., Dawson, J., Desai, N., Hill, M., Kadwell, S., Launis, K., Lewis, K., Maddox, D., McPherson, K., Meghji, M., Merlin, E., Rhodes, R., Warren, G., Wright, M. and Evola, S. (1993) "Field performance of elite transgenic maize plants expressing an insecticidal protein derived from *Bacillus thuringiensis*", Bio/Technology 11, 194-200.

McCabe, D., Swain, W., Martinelli, B. and Christou, P. (1988) "Stable transformation of soybean (*Glycine max*) by particle acceleration", Bio/Technology 6, 923-926.

Register, J., Peterson, D., Bell, P., Bullock, P., Evans, I., Frame, B., Greenland, A., Higgs, N., Jepson, I., Jiao, S., Lewnau, C., Sillick, J. and Wilson, M. (1994) "Structure and function of selectable and nonselectable transgenes in maize following introduction by particle bombardment", Plant Mol. Biol. (in press).

Saghai Maroof, M., Soliman, K., Jorgenson, R. and Allard, R. (1984) "Ribosomal DNA spacer-length polymorphisms in barley: Mendelian inheritance, chromosomal location, and population dynamics", Proc. Natl. Acad. Sci. USA 81, 8014-8018.

Sambrook, J., Fritsch, E. and Maniatis, T. (1989) "Molecular cloning: A laboratory Manual, Cold Spring Harbor Laboratory Press, Cold Spring Harbor.

Shillito, R., Carswell, G., Johnson, C., Di Mais, J and Harms, C. (1989) "Regeneration of fertile plants from protoplasts of elite inbred maize", Bio/Technology 7, 581-587.

Spencer, T., Gordon-Kamm, W., Daines, R., Start, W. and Lemaux, P. (1990) "Bialaphos selection of stable transformants from maize cell culture", Theoretical and Applied Genetics 79, 625-631.

STUDIES ON THE STABILITY OF FOREIGN GENES IN THE PROGENY OF TRANSGENIC LINES OF *VICIA NARBONENSIS*

M. MEIXNER, U. SCHNEIDER, O. SCHIEDER AND T. PICKARDT
Institut für Angewandte Genetik der Freien Universität Berlin,
Albrecht-Thaer-Weg 6,
14195 Berlin
Germany

ABSTRACT. In order to study the presence and the stable inheritance of genes introduced into the grain legume *Vicia narbonensis*, we have analysed the progeny of three independent transgenic lines that were obtained by transformation with *Agrobacterium*-strain C58C1pGV 3850Hygro. Southern analyses revealed that each transformant carries a single copy of the T-DNA. Progeny analysis was done up to the fourth generation by analysing the expression of the nopaline synthase gene, testing leaf explants in respect to their hygromycin resistance and showing the presence and stable inheritance of the right and left border regions as well as the hygromycin phosphotransferase- and nopalin synthase gene in part of the progeny from these three lines by southern blot hybridization.

In all three lines the foreign genes were generally inherited by a Mendelian transmission resulting in a 3:1 ratio in the R1 and a segregation in homozygous and hemizygous plants in the following generations. However, the offspring of some plants showed a non-Mendelian ratio.

1. Introduction

The recent development in recombinant DNA technology together with efficient plant transformation methods in principal allows the introduction of foreign genes into a broad spectrum of plant species. Within the last few years in an increasing number of species transgenic plants became available. Unfortunately, up to now grain legumes remaine a group of plants that is difficult to transform. Although some progress has been reported in soybean (Christou 1992, Falco et al. 1994, Townsend & Thomas, 1994) and pea (Puonti-Kaerlas et al. 1992, Davies et al. 1993, Schroeder et al. 1993), a general applicability of the published protocols is limited up to now.

In our group the grain legume *Vicia narbonensis*, a close relative of the faba bean was successfully transformed by *Agrobacterium* mediated gene transfer (Pickardt et al. 1991). Meanwhile many transgenic lines containing different constructs were produced and the progeny is available for most of these lines (Pickardt et al. 1992, Saalbach et al. 1994)

The results from many investigations in the last decade showed, that in general foreign genes can be introduced into the plant genome of a recipient plant and can be stable inherited to the offspring. However, it is also well known, that transgenic lines may become unstable by inactivation or loss of the foreign genes.

M. Terzi et al. (eds.), Current Issues in Plant Molecular and Cellular Biology, 285-290.

In order to obtain some knowledge concerning the situation in a transgenic grain legume, we started to investigate the progeny of three transgenic lines, all harbouring a single copy insertion of the T-DNA from the plasmid pGV3850Hygro.

2. Materials and Methods

AGROBACTERIUM STRAIN: Inoculations were performed using the *Agrobacterium tumefaciens* strain C58C1 that harbours the disarmed vector pGV3850Hygro (Kreuzaler & Kato, unpublished). A partial map, showing the organization of the T-DNA and hybridization probes used in this study, is presented in Fig. 1.

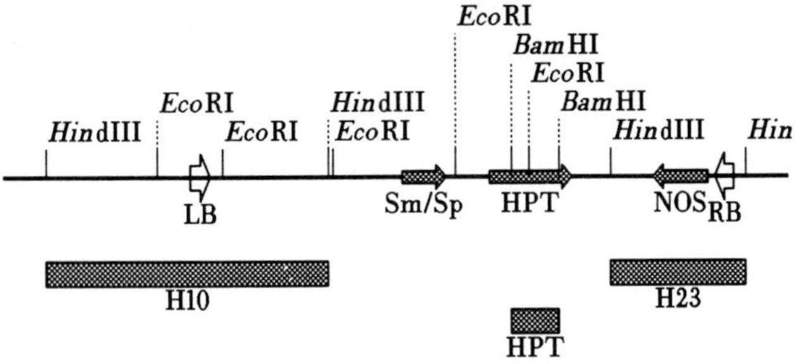

Fig. 1: T-DNA region from the plasmid pGV3850Hygro. Abbreviations: LB, RB, left and right border sequences of the T-DNA; Sm/Sp, bacterial streptomycin/spectinomycin resistance; HPT, plant gene cassette for the hygromycin B resistance; NOS, nopaline synthase gene; H10, 6.6 kbp HindIII fragment, containing the left border and neighbouring sequences; H23, 3.2 kbp HindIII fragment harbouring the right border and neighbouring sequences.

TRANSFORMATION OF *VICIA NARBONENSIS*: The protocol for transformation of *Vicia narbonensis* via *Agrobacterium* mediated gene transfer is described in Pickardt et al. (1991).
NOPALIN TEST: Nos activity was determined as described by Reynaerts et al. (1988). Nontransformed plant and callus were used as a negative control and synthetic nopaline/arginine as standard.
TEST FOR HYGROMYCIN RESISTANCE: Sterilized leafs from transgenic plants and from nontransformed controls were placed on MS medium containing 1 mg l^{-1} Picloram and 25 mg l^{-1} hygromycin. After a 10 d incubation explants expressing the hygromycin phosphotransferase could be clearly distinguished from controls and hygromycin sensitive explants.
SOUTHERN BLOT ANALYSIS: Southern blot analysis of total plant DNA was performed as discribed previously (Pickardt et al. 1991; Schmidt-Rogge et al. 1993) by using 5 x SSC, 0.5 % skim milk powder and 0.5% SDS at 62-65°C as hybridization buffer.

3. Results and Discussion

3.1. ANALYSIS OF THREE PRIMARY TRANSFORMANTS

Transgenic *in vitro*-shoots of *Vicia narbonensis* transformed with *Agrobacterium*-strain C58C1/3850Hygro (Kreuzaler & Kato, unpublished) which contains the hygromycin resistance gene controlled by the CaMV-35S promoter and the nopaline synthase gene within its T-DNA (see Fig. 1) had been previously selected (Pickardt et al. 1991). Southern analyses using genomic DNA isolated from several *in-vitro* shoot clones were performed with three different probes (the left- and right border fragments H10 and H23 and the 1.1 kbp BamHI fragment with the coding region of the HPT-gene, see Fig. 1). The hybridization pattern of the right border fragment H23 in three R0-clones (4, 9 and 18) shows that each clone contain a single copy of the T-DNA region (Fig. 2a). When reprobing the same blot with the fragment H10, which contains the left border and neighbouring DNA sequences, a unique signal appears at 6,5 kbp in all three R0-clones (data not shown for R0, see Fig. 2c for R2-R4). Since H10 (like H23) represents a border-probe, it is expected to hybridize to (junction-) fragments of different length within each transformant (as H23 does). Until now it remains unclear if these unique hybridization signals are fortuitous or if there is some unexpected event, e.g. deficient homologous recombination in *Agrobacterium* or incorrect recognition and transfer of the T-DNA.

After grafting of *in vitro*-shoots (rooting was unsuccessful) and generating fertile plants the corresponding R0-lines 4.1, 9.1 and 18.1 were chosen for further progeny analysis.

3.2. ANALYSIS OF THE PROGENY IN RESPECT TO THE PRESENCE AND EXPRESSION OF THE NOS AND THE HPT GENE

From the R0 plants 4.1, 9.1 and 18.1 we obtained 7, 28 and 21 seeds (respectively) which were capable of germination. The R1 was grown up in the greenhouse and a segregation analysis was performed in respect to the hygromycin resistance and the nopaline production (Tab. 1). Both tests showed 100% conformity. The deviations from the expected 3:1 ratio are not significant, as revealed by chi^2-test. The segregation in all three lines therefore followed a Mendelian transmission.

TABLE 1: Segregation of three transgenic R0-lines after self pollination, based on hygromycin resistance and nopalin production

R0-plant	R1-seeds		ratio (3:1 expected)	χ^2	P (1 d.f.)
	hyg^r:hyg^s	nop^+:nop^-			
4.1	6:1	6:1	6:1	0,429	>0,5
9.1	23:5	23:5	4.6:1	0,761	>0,3
18.1	17:4	17:4	4.2:1	0,397	>0,5

In the next step hygromycin resistant and nopaline producing R1 plants were chosen for further investigations. A scheme representing the progeny analysis of the three transgenic lines on the basis of the nopaline assay is given in Table 2.

Especially for the line 9.1 homozygous candidates were selected in the R2 generation. Because of the small number of the siblings from one line or subline, segregation was also tested within the R3 and R4 respectively. From these results it seems to be evident, that the R1 plants 9.1.19 and 18.1.29 are homozygous in respect to the nopaline synthase gene. Using such homozygous lines, hygromycin resistance tests were performed in addition to the nopaline assay. These studies showed that the entire part of the T-DNA was maintained and stably inherited for at least four generations. In the case of the subline 9.1.19, also the fifth generation showed a 10:0 segregation of the foreign genes.

TABLE 2. Segregation analysis in the selfed progeny of the transgenic lines 4.1, 9.1 and 18.1 based on nopaline assays.

R0	R1	R2	R3	R4	R5
			4.1.6.1 → (3*:0)		
4.1 →	(6:1)	4.1.6 → (4*:0)	4.1.6.2 → <u>(4:10)</u>	**4.1.6.2.7** → (24:0)	
		4.1.9 → (5*:3)	4.1.9.1 → <u>(6:8)</u>	4.1.9.1.10 → (18*:11)	
9.1 →	(23:5)	9.1.12 → (3:0)		**9.1.19.1.2** → (7:0)	
		9.1.13 → (7:0)	**9.1.19.1** → (18*:0)	**9.1.19.1.4** → (4:0)	
		9.1.19 → (25:0)		**9.1.19.1.8** → (5:0)	
		9.1.20 → <u>(4:8)</u>	**9.1.19.2** → (13:0)	**9.1.19.2.9** → (14:0)	
		9.1.21 → (10*:0)		**9.1.19.2.12** → (16*:0)	**9.1.19.2.12.2** → (10:0)
18.1 →	(17:4)	18.1.9 → (3*:0)			
		18.1.25 → (16:4)			
		18.1.29 → (18:0)	**18.1.29.13** → (44:0)		
		18.1.30 → (16:6)	18.1.30.12 → (18*:7)	18.1.30.12.5 → (6:1)	
				18.1.30.12.6 → <u>(3:6)</u>	

Only plants which were nopaline-positive were further analysed. Since the primary transformants contain a single T-DNA insertion, a 3:1 segregation is expected in the offspring of selfed hemizygous plants. Numbers of nopaline positive versus negative individuals from the next generation are given in parenthesis. Unexpected segregations are underlined. Segregations from putative homozygous plants are typed in bold letters. Progenies from which individual plants were subjected to southern analysis (see Fig. 2b/c) are marked by asterisks.

In most experimental work published until now only R1 and R2 progenies or backcrosses were investigated. Because of the relative small number of seeds which were obtained in *V. narbonensis* when growing up in the greenhouse (see Table 2), we followed transmission of the foreign genes up to the fourth or fifth generation. Moreover, we are interested to analyse the transgene stability and continued expression over a long period.

3.3. INVESTIGATION OF THE PROGENY IN VIEW TO THE PRESENCE AND STABILITY OF THE T-DNA BORDER REGIONS

Biochemical and southern blot analyses revealed that the major part of the T-DNA (if not complete) of pGV3850Hygro, including the hygromycin resistance gene as well as the right and left border regions are present in the initial transformants. In order to monitor the stability and the behaviour of the foreign genes in the offspring of the corresponding three lines (4.1, 9.1, 18.1), southern blot experiments were performed. We randomly selected individual R2, R3 and R4 plants (marked by an asterisk in Table 2) for isolation of total DNA, which was digested with the restriction endonuclease HindIII (in the same manner as the initial transformants, see Fig. 2a). The filters shown in Fig. 2b+c were hybridized to the H23-probe, representing the DNA region containing the right border (2b) and subsequently reprobed with the H10-left border probe (2c). All members of the progeny from one line showed an identical pattern in respect to the occurence of the hybridization signals and these patterns are identical to those of the initial transformants.

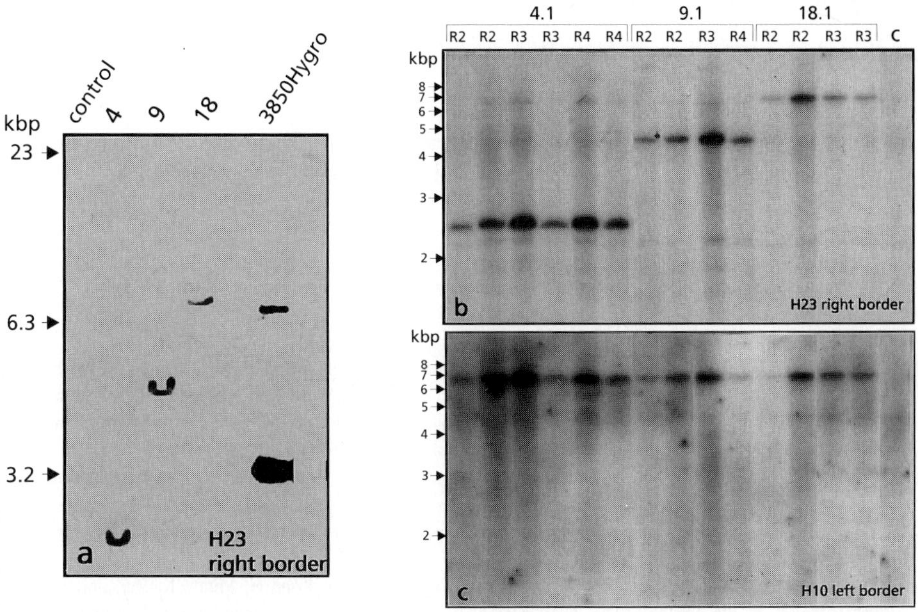

Fig. 2: Southern blot analysis of transgenic *Vicia narbonensis*. **a)** Analysis of three initial transformants, clone 4; 9 and 18, probed with the H23 probe, a 3.2 kbp HindIII right-border fragment (see also Fig. 1). Control: DNA from an untransformed plant, 3850Hygro: DNA from *A. tumefaciens* C58C1 pGV3850Hygro. **b)** Southern blot analysis of individual plants from the progeny of the three initial transformants shown in Fig. 2a and probed with the same fragment. **c)** The same blot as in Fig. 2b was reprobed with the 6.6 kbp H10 left-border fragment (see Fig. 1). C: DNA from an untransformed plant. In all cases total DNA was digested with the restriction endonuclease HindIII.

There are many reports for the analysis of transgenic progenies, especially (a) for plants which are easy to transform (e.g. solanaceous species), (b) for plant species with a short reproduction cycle or (c) for plants where large progenies can be obtained. Generally, both, stable inheritance of the trangenes to the progeny together with a segregation in the Mendelian fashion and instability or loss of the trangenes in the sexual offspring and/or unusual segregation patterns were found (for examples see Karesch et al. 1992, Spencer et al. 1992). Only a few informations are available concerning data on the stable inheritance and progeny analysis in the recalcitrant grain legumes (e.g. Puonti-Kaerlas et al. 1992). To put some light on the inheritance of foreign genes in a grain legume, we started this work.

In conclusion, the results obtained until now show a stable inheritance for the integrated T-DNA in all three transgenic lines regarding the expression of the marker genes NOS and HPT. However, the progeny analysis also showed some unexpected segregations.From the data presented in Table 2 it is obvious that only progenies from hemizygous plants tend to produce an unexpected high number of negative plants. Are transgenes less stable in a hemizygous situation because they miss a homologous sequence during chromosome pairing in meiosis? (see Potrykus et al. 1985). This phenomenon needs further elucidation.

4. Acknowledgements

This work was supported by the Bundesministerium für Forschung und Technologie/Germany. The authors wish to thank Lothar Willmitzer for providing us with constructs, containing the H23 and H10 fragments, and Thilo Schmidt-Rogge for critical reading of the manuscript.

5. References

Christou P: Genetic transformation of crop plants using microprojectile bombardment. Plant Journal 2: 275-281 (1992)

Davies DR, Hamilton J, Mullineaux P: Transformation of peas. Plant Cell Rep 12: 180-183 (1993)

Falco SC, Beaman C, Chui CF, Guida T, Hirata L, Jones T, Keeler S, Knowlton S, Kostow C, Locke M, Mauvais J, McAdams S, Reiter K, Rice J, Sanders C, Schreiner R, Wandelt C, Ward RT, Webber EI: Transgenic crops with improved amino acid composition. J. Cell Biology, Supplement 18A (Keystone Symposium "Improved crop and plant products through biotechnology", Abstract X1-017) (1994)

Karesh H, Bilang R, Mittelsten Scheid O, Potrykus I: Direct gene transfer to protoplasts of *Arabidopsis thaliana*. Plant Cell Rep 9: 571-574 (1991)

Pickardt T, Meixner M, Schade V, Schieder O: Transformation of *Vicia narbonensis* via *Agrobacterium*-mediated gene transfer. Plant Cell Rep 9: 535-538 (1991)

Pickardt T, Schieder O, Saalbach I, Tegeder M, Machemehl G, Saalbach G, Kohn H, Müntz K: Regeneration and transformation in the legume *Vicia narbonensis*. In: Plant Tissue Culture and Gene Manipulation for Breeding and Formation of Phytochemicals. Eds.: Oono K et al., NIAR Japan: 89-94 (1992)

Potrykus I, Paszkowski J, Saul MW, Petruska J, Shillito RD: Molecular and general genetics of a hybrid foreign gene introduced into tobacco by direct gene transfer. Mol Gen Genet 199: 169-177 (1985)

Puonti-Kaerlas J, Eriksson T, Engstrom P: Inheritance of a bacterial hygromycin phosphotransferase gene in the progeny of primary transgenic pea plants. Theoretical and Applied Genetics 84: 443-450 (1992)

Reynaerts A, De Block M, Hernalsteens JP, Van Montague M: Selectable and screenable markers. in: Plant Mol Biol Manual (Eds. Gelvin S, Schilperoort RA, Verma DPS), Kluwer Academic Publishers/Belgium, A9: 1-16 (1988)

Saalbach I, Pickardt T, Machemehl F, Saalbach G, Schieder O, Müntz K: A chimeric gene encoding the methionine-rich 2S albumin of Brazil nut (*Bertholletia excelsa* H.B.K.) is stably expressed and inherited in transgenic grain legumes. Mol Gen Genet 242: 226-236 (1994)

Schmidt-Rogge T, Meixner M, Srivastava V, Guha-Mukherjee S, Schieder O: Transformation of haploid Datura innoxia protoplasts and analysis of the plasmid integration pattern in transgenic plants. Plant Cell Rep 12: 390-394 (1993)

Schroeder HE, Schotz AH, Wardley-Richardson T, Spencer D, Higgins TJV: Transformation and regeneration of 2 cultivars of pea (*Pisum sativum* L). Plant Physiology 101: 751-757 (1993)

Spencer TM, O'Brien JV, Start WG, Adams TR, Gordon-Kamm WJ, Lemaux PG: Segragation of transgenes in maize. Plant Mol Biol 20: 201-210 (1992)

Townsend JA, Thomas LA: Factors which influence the *Agrobacterium-mediated* transformation of soybean. J Cell Biology, Supplement 18A (Keystone Symposium "Improved crop and plant products through biotechnology", Abstract X1-014) (1994)

MOLECULAR ANALYSIS OF CHLOROPLAST DIVISION

R. RESKI, K. REUTTER, B. KASTEN, M. FAUST, S. KRUSE, G. GORR, R. STREPP, W.O. ABEL
University of Hamburg
Institute of General Botany
Ohnhorststr. 18
22609 Hamburg
Germany

ABSTRACT. The molecular events underlying chloroplast division are studied with a mutant of a moss, *Physcomitrella patens*, which is defective in chloroplast division thus possessing one giant lobed chloroplast per cell. This macrochloroplast is severed by the enlarging cell plate during cytokinesis. Its division can be induced by cytokinin and by blue light. Concomitantly, maturation of complex plastid transcripts and a transient occurence of plastid polypeptides can be detected. Southern-analyses revealed methylation of the mutants plastid DNA around an open reading frame (ORF), possibly encoding a zinc-finger protein. This ORF is conserved from cyanobacteria to the plastids of archegoniates but is absent from the plastid DNA of monocots. Somatic hybridization were performed to allocate the mutations either to nuclear or to plastid DNA. Four cytokinin-modulated cDNAs representing novel genes were isolated by molecular subtraction. Transformants with the bacterial *ipt*-gene were generated, one of which has lost sensitivity towards cytokinin and blue light in the chloroplast division process.

Introduction

Since Schimper (1883) suggested the continuity of plastids, numerous studies on the number, orientation and division of chloroplasts have been undertaken to unravel the mechanisms which provide for their nearly equal distribution into daughter cells during cytokinesis. Favourable systems for study are unicellular, have a tip-growing apical cell or are monoplastidic.

The haploid protonemata of mosses are well established plant systems for studying the morphogenic potential of different stimuli such as cytokinin and light. Furthermore, the availability of differentiation mutants and protoplast fusion techniques facilitate the genetic analysis of morphogenesis (Bopp 1981, Cove and Knight 1993). In order to identify genes involved in chloroplast division, we analyse *Physcomitrella patens* with special emphasis on the mutant PC22, which is defective in chloroplast division. Thus, the majority of cells contain only one giant chloroplast. This mutant macrochloroplast is a mechanical obstacle which is divided during cell division in a hitherto unknown manner: It is severed by the enlarging cell plate, resulting in abnormal cross walls, less orderly branching of the protonema, and retarded growth. Additionally, mutant PC22 is impeded in bud formation, that is the transition from an

M. Terzi et al. (eds.), Current Issues in Plant Molecular and Cellular Biology, 291-296.

apical cell to a three faced apical cell (Abel et al. 1989). We report here on our analyses of these differentiation processes.

Materials and Methods

PLANT MATERIAL AND CULTURE CONDITIONS

Physcomitrella patens (Hedw.) B.S.G. wild type as well as the cytokinin-sensitive differentiation mutant PC22, which was obtained after X-ray treatment of wild type spores, have been described in detail (Reski and Abel 1985, Abel et al. 1989, Ye et al. 1989). Plants were grown under sterile conditions in agitated liquid medium under defined conditions (Reski et al. 1994). Protoplasts were isolated after Driselase maceration of cell walls by floating in sucrose; fusion of protoplasts were facilitated with PEG (Rother et al. 1994).

ISOLATION OF ORGANELLES AND ANALYSIS OF THEIR NUCLEIC ACIDS

Series of protocols for the isolation of intact nuclei, chloroplast and mitochondria and their respective DNA have been established (Marienfeld et al. 1989). Southern- and Northern-analyses were performed according to Reski et al. (1991), cloning and sequencing of organellar DNA according to Marienfeld et al. (1991) and to Kasten et al. (1992). Flow cytometry, chromosome staining, denaturation kinetics and cDNA-cloning have been described recently (Reski et al. 1994).

Results

GENOME ANALYSIS

Molecular analysis of mosses is scarce; even the chromosome number of *Physcomitrella patens* has been a matter of debate over decades. The genome size of the strain currently under investigation was estimated to be about 600 Mbp on 27 chromosomes (Reski et al. 1994). The first nucleotide sequences of plastid DNA of a moss (Kasten et al. 1991, Kasten et al. 1992) and the first mitochondrial sequence of an archegoniate (Marienfeld et al. 1991) had been reported with *Physcomitrella.*

SOMATIC HYBRIDIZATION

Moss protonemata show a sequence of two cell types (chloronema and caulonema) growing by apical cells and a subsequent transition to a three faced apical cell (bud), which give raise to leafy shoots. Bud induction is under control of cytokinins. An X-ray derived *Physcomitrella*-mutant, designated PC22, is defective in two differentiation steps: It is impeded in chloroplast division as well as in bud induction, thus forming solely protonema with cells containing at the most one giant chloroplast. Both defects can be compensated for by exogenous cytokinin, additionally, blue light induces division of macrochloroplasts as well (Abel et al. 1989).

Obviously, the defect in cellular differentiation (bud induction) is due to a nuclear mutation. However, to investigate, whether the lack in chloroplast division is due to a nuclear or to a plastid mutation we performed crossing experiments. As mosses are self-fertile haplonts and, moreover, PC22 is not only a chloroplast mutant, but also a sterile differentiation mutant, conventional genetic analysis is impossible. Therefore, one-to one-protoplast fusions with wild type and mutant protoplasts were performed. Anticipations were the following: i) All somatic hybrids would have wild-type plastids, if the mutant macrochloroplast is due to a recessive nuclear mutation. ii) A mixture of plastids would occur in the fusants with a subsequent sorting out of the macrochloroplast, if the mutant plastid is due to a plastome mutation. Most of the somatic hybrids showed wild type characteristics. However, one had wild type chloroplast but showed no budding, while two formed leafy gametophores but contained mutant macrochloroplasts (Rother et al. 1994). Thus, the defects in differentiation are two uncoupled genetic traits in this particular mutant. Furthermore, these data favour a plastid mutation as cause for the defect in chloroplast division, although partial inactivation of the wild type genome can not be excluded. Therefore, the molecular analysis of chloroplast division in this mutant concentrates on four items: 1) RFLP-analysis, 2) plastid proteins, 3) cytokinin-modulated cDNAs, and 4) transformants.

RFLP-ANALYSIS

Although RFLP-analysis with several nuclear, mitochondrial and plastid *loci* were performed (Reski et al. 1991, Marienfeld et al. 1992, Reski et al. 1994), the only difference between mutant PC22 and the wild type revealed a methylation of plastid DNA around the *rbc*L gene in the mutant. Molecular cloning and sequence analysis of this region generated the first nucleotide data of moss plastid DNA (Kasten et al. 1991) and revealed an open reading frame possibly encoding a zinc finger protein with homology to the regulatory GAL4-family (Kasten et al. 1992). This ORF was designated *zfp*A. Sequence comparison of all *zfp*A genes published so far reveal that this ORF is conserved from cyanobacteria to the plastids of archegoniates. In dicots it gained additional sequences whilst it is truncated or totally absent from monocot plastid DNA. In the moss, *zfp*A contains procaryotic promotor elements and three eucaryotic type TATA-sequences. In Northern-blots a gene specific probe detects three transcripts larger than the ORF. Hybridization with strand-specific probes proved that these are immature transcripts of the *zfp*A operon. Compared to the wild type in mutant PC22 the largest transcript is accumulated. Addition of cytokinin promotes processing of these transcripts. Nucleotide sequences in this region do not differ between wild type and mutant. The plastid ORF from moss was expressed in *E. coli* and the respective protein purified to homogeneity.

PLASTID PROTEINS

Addition of cytokinin to five day old mutant protonemata induces massive chloroplast division after three and budding after seven days. This induced chloroplast division is accompanied by a transient occurence of several plastid polypeptides which can not be detected in control plants nor in plants treated for seven days with cytokinin. When plants were grown in blue light, they possess already divided chloroplasts. Likewise, addition of cytokinins neither affects chloroplast division nor alters plastid protein composition (Reski et al. 1991). Thus, these transient proteins appear to be associated with chloroplast division in this particular mutant.

During isolation, chloroplasts form two distinct bands in percoll-gradients. The upper band contains stripped and partial defective plastids whilst from the lower band intact chloroplasts with visible double membranes can be collected (Marienfeld et al. 1989). As the mutant macrochloroplast can not be isolated intactly, the ratio between protein contents from stripped and from intact plastids band was taken as a degree of chloroplast division in this mutant. In wild type this ratio equals one and does not change during cytokinin treatment. In untreated mutant plants this ratio equals eight and decreases to 2.8 after three days of cytokinin treatment. Taking this measure cytokinin-induced chloroplast division can already be detected after hours of cytokinin treatment. Plastid proteins from defined stages were isolated and subjected to two-dimensional protein gel electrophoresis (IEF/SDS-PAGE). Proteins occuring early in this cascade were identified.

CYTOKININ-MODULATED cDNAs

In order to identify nuclear genes that may be involved early in cytokinin-induced differentiation, large cDNA-libraries (more than one million independent clones) have been established. Two of these libraries were subjected to molecular subtractive hybridization and the resulting clones were analysed. Four out of these proved to be responsive to exogenous cytokinin within at least one hour. Three of the genes are developmentally expressed. The cytokinin-modulated cDNAs have been sequenced partially. Yet there are no significant homologies to sequences in the EMBL-data bank.

TRANSFORMANTS

Physcomitrella patens can be genetically transformed via the PEG method (Schaefer et al. 1991) or ballistic methods (Sawahel et al. 1992). We used PEG-mediated DNA uptake to transform moss protoplasts with the bacterial *ipt*-gene under its own promotor. This gene encodes a protein involved in the first step of cytokinin biosynthesis and mediates elevated cytokinin contents in transgenic higher plants (Schmülling et al. 1992). Likewise, *Physcomitrella* transformed with this construct exhibited the expected cytokinin-overproducing phenotype: Number of buds were enhanced, they appeared earlier and the resulting gametophores appeared to be enlarged and swollen. One stable *ipt*-transformant of mutant PC22 (tPC22*ipt*1) was an exception regarding chloroplast division, as it possesses a macrochloroplast that remained unable to divide, even when plants were treated with blue light or exogenous cytokinin. Southern analysis revealed that two copies of the plasmid had been integrated independently into the genome of this transformant.

Expression of four cytokinin-responsive cDNAs during development of the wild type, mutant PC22 and of tPC22*ipt*1 was analysed in Northern blot experiments. Differences in RNA accumulation between these genotypes were detected.

Discussion

Since decades moss protonemata have been developed as model systems to analyse the morphogenetic potential of different stimuli such as light and phytohormones. However, molecular analysis of these processes remained scarce. Meanwhile, all essential techniques of modern plant molecular biology have been adopted for *Physcomitrella patens*. Genome analysis have demonstrated a fairly small genome (eight times *Arabidopsis*) as well as differences in the organisation of mitochondrial and plastid genes when compared to higher plants. Therefore, the genetic dissection of *Physcomitrella* may contribute to understanding the evolution of plant developmental processes as already suggested by Goldberg in 1988.

In these studies special interest is directed to mutant PC22, which is defective in chloroplast division. Although occassionally mutants with enlarged chloroplasts have been reported, this particular mutant seems to be unique. As this defect can partially be compensated for by exogenous cytokinin and by blue light, molecular analysis of genome/plastome-interaction during cellular differentiation and chloroplast division seems feasible (Reski 1994).

Currently, isolated *zfpA*-protein is used in gel-shift assays, South/Western- and North/Western-blots to establish its nucleic acids binding function and to describe its binding characteristics. Transient plastid polypeptides which are associated with chloroplast division were isolated and will be subjected to microsequence analysis to establish the role of known genes in this process, and, hopefully, to find new genes which may be involved in chloroplast division. Four cytokinin-responsive cDNAs represent novel genes, which are developmentally regulated and responsive to cytokinin. Their biological significance will be tested in transformation experiments with anti-sense constructs. Transformation techniques have been established and *ipt*-genes under control of inducible and organ-specific promoters will provide insight into a molecular physiology of cytokinin action. One of these PC22-transformants has lost its sensitivity to cytokinin and to blue light in the induction of chloroplast division. The most obvious explanation for this phenotype is, that by chance one of the two plasmid integrations has inactivated and tagged a gene which is essential in the signal transduction pathway that leeds to chloroplast division.

Acknowledgements

We are grateful to the Commission of the European Union, the Deutsche Forschungsgemeinschaft, Fonds der Chemischen Industrie and the City of Hamburg for financial support. The work described here is based in part on the current Ph.D. studies of K.R., B.K., M.F., S.K. and G.G. performed at the Faculty of Biology, University of Hamburg.

References

Abel, W. O., Knebel, W., Koop, H.-U., Marienfeld, J. R., Quader, H., Reski, R., Schnepf, E.,and Spörlein, B. (1989) 'A cytokinin sensitive mutant of the moss, *Physcomitrella patens*, defective in chloroplast division', Protoplasma 152, 1-13.

Bopp, M, (1981) 'Entwicklungsphysiologie der Moose', in W. Schultze-Motel (ed.), Advances in Bryology., Cramer-Verlag, Braunschweig, pp 11-77.

Cove, D. J., and Knight, C. D. (1993) 'The moss *Physcomitrella patens*, a model system with potential for the study of plant reproduction', Plant Cell 5, 1483-1488.

Goldberg, R. B. (1988) 'Plants: novel developmental processes', Science 240, 1460-1467.

Kasten, B., Wehe, M., Reski, R., and Abel, W. O. (1991) '*trnR*-CCG is not unique to the plastid DNA of the liverwort *Marchantia*: gene identification from the moss *Physcomitrella patens*', Nucl. Acids Res. 19, 5074.

Kasten, B., Wehe, M., Kruse, S., Reutter, K., Abel, W. O., and Reski, R. (1992) 'The plastome-encoded *zfp*A gene of a moss contains procaryotic as well as eucaryotic promoter consensus sequences and its RNA abundance is modulated by cytokinin', Curr. Genet. 22, 327-333.

Marienfeld, J. R., Reski, R., Friese, C., and Abel, W. O. (1989) 'Isolation of nuclear, chloroplast and mitochondrial DNA from the moss *Physcomitrella patens*', Plant Sci. 61, 235-244.

Marienfeld, J. R., Reski, R., and Abel, W. O. (1991) 'The first analysed archegoniate mitochondrial gene (cox3) exhibits extraordinary features', Curr. Genet. 20, 319-329.
Marienfeld, J. R., Reski, R., and Abel, W. O. (1992) 'Chondriom analysis of the moss *Physcomitrella patens*', Crypt. Bot. 3, 23-26.
Reski, R (1994) 'Plastid genes and chloroplast biogenesis', in D. W. S. Mok and M. C. Mok (eds.), Cytokinins. Chemistry, Activity, and Function, CRC Press, Boca Raton, Ann Arbor, London, Tokyo, pp 179-195.
Reski, R, and Abel, W. O. (1985) 'Induction of budding on chloronemata and caulonemata of the moss, *Physcomitrella patens*, using isopentenyladenine', Planta 165, 354-358.
Reski, R., Wehe, M., Hadeler, B., Marienfeld, J. R., and Abel, W. O. (1991) 'Cytokinin and light quality interact at the molecular level in the mutant PC22 of the moss *Physcomitrella*', J. Plant Physiol. 138, 236-243.
Reski, R., Faust, M., Wang, X.-H., Wehe, M., and Abel, W. O. (1994) 'Genome analysis of a moss, *Physcomitrella patens* (Hedw.) B.S.G.', Mol. Gen. Genet., in press.
Rother, S., Hadeler, B., Orsini, J. M., Abel, W. O., and Reski, R. (1994) 'Fate of a mutant macrochloroplast in somatic hybrids', J. Plant Physiol. 143, 72-77.
Sawahel, W., Onde, S., Knight, C., and Cove, D. (1992) 'Transfer of foreign DNA into *Physcomitrella patens* protonemal tissue by using the gene gun', Plant Mol. Biol. Rep. 10, 314-315.
Schaefer, D., Zryd, J.-P., Knight, C. D., and Cove, D. J. (1991) 'Stable transformation of the moss, *Physcomitrella patens*', Mol. Gen. Genet. 226, 418-424.
Schimper, A. F. (1883) 'Über die Verteilung der Chlorophyllkörner und Farbkörper', Bot. Ztg. 41, 105-162.
Schmülling, T., Schell, J., and Spena, A. (1992) 'Construction of cytokinin overproducing mutants in higher plants', in M. Kaminek, D. W. S. Mok, and E. Zazimalova (eds.), Physiology and biochemistry of cytokinins in plants, SPB Academic Publishing bv, The Hague, pp 87-94.
Ye, F., Gierlich, J., Reski, R., Marienfeld, J., and Abel, W. O. (1989) 'Isoenzyme analysis of cytokinin sensitive mutants of the moss *Physcomitrella patens*', Plant Sci. 64, 203-212.

INTEGRATION OF *SOLANUM* DNA INTO THE *NICOTIANA* PLASTOME THROUGH PEG-MEDIATED TRANSFORMATION OF PROTOPLASTS

P.J. DIX, N.D. THANH[1], T.A.KAVANAGH[2] AND P.MEDGYESY[1]
Biology Dept., St. Patrick's College, Maynooth, Co. Kildare, Ireland
[1]*Biological Research Center, Szeged, Hungary*
[2]*Genetics Dept., Trinity College, Dublin, Ireland*

ABSTRACT. A new spectinomycin plus streptomycin resistant double mutant of *Solanum nigrum* has base substitutions in the 16S rRNA and rps12(3') genes on the chloroplast DNA. A cloned fragment containing both these sites was used to transform *Nicotiana plumbaginifolia* protoplasts, using polyethylene glycol in a protocol previously developed for efficient chloroplast transformation. Selection of transformants was on the basis of greening of protoplast-derived colonies on antibiotic containing medium, and was for either simultaneous resistance to both antibiotics, or resistance to spectinomycin followed by screening for streptomycin resistance. Shoots have been recovered from a number of double resistant colonies and homologous integration of *S. nigrum* DNA confirmed by PCR using primers diagnostic for both species.

1. Introduction

The chloroplast genome encodes a number of products essential for the normal growth of plants. These include (Dyer 1985) the large subunit for ribulose-1,5-biphosphate carboxylase, the 32 kDa quinone binding (Q_B) protein, cytochromes, ATP synthase, and a number of ribosomal and envelope proteins. Additional components of the plastid protein synthesis machinery, such as rRNAs and tRNAs, are also plastome encoded. Quite detailed genetic maps are now available, as are complete sequences for the plastome of the liverwort *Marchantia polymorpha* (Ohyama *et al.*), and tobacco (Shinozaki *et al.* 1986). There remain, however, a number of open reading frames of unknown function, and there is a great deal still to learn, for example about the complex interactions between nuclear and plastid genes in fundamental processes such as photosynthesis.

Studies on the genetics of higher plant chloroplasts have lagged behind elucidation of the structure and sequence of their DNA, because of the physical barrier (exclusive maternal inheritance of organelles) most plants have to plastome recombination, which limits the use of one of the most fundamental tools for genetic analysis. A further complication is the large plastome copy number per cell. Several developments in the past decade have improved the

M. Terzi et al. (eds.), Current Issues in Plant Molecular and Cellular Biology, 297-301.

prospects for us to move towards a fuller understanding of the action of chloroplast genes. A number of well characterised chloroplast genetic markers are now available for *Nicotiana* species, conferring resistance to the antibiotics streptomycin (Etzold *et al.* 1987, Fromm *et al.* 1989, Galili *et al.* 1989), spectinomycin (Fromm *et al.* 1987, Svab and Maliga 1991) and lincomycin (Cséplö *et al.* 1988). Recently, a similar set of mutants have been isolated (McCabe *et al.* 1989) and characterised (Kavanagh *et al.* 1994) for *Solanum nigrum*. Protoplast fusion has allowed the generation of a heteroplasmic state, not usually accomplished during gamete fusion, which, together with the improved genetic markers, allows analysis of recombination and segregation in a mixed plastid population (reviewed by Medgyesy 1990). Finally, it has become possible to genetically transform chloroplasts with foreign DNA. This was first accomplished by Svab *et al.* (1990) who obtained transplastomic tobacco plants by microprojectile bombardment of leaf tissue with a pUC based plasmid containing a 16S rRNA gene of tobacco carrying mutations conferring resistance to spectinomycin and streptomycin and a novel *Pst*I restriction site. Initial selection of transformants was for spectinomycin resistance, followed by screening for the flanking markers, streptomycin resistance and the *Pst*I site.

Subsequently, there have been significant developments in the chloroplast transformation procedures, most notably the use of polyethylene glycol induced DNA uptake into protoplasts as an alternative to biolistic delivery systems (O'Neill *et al.* 1993), and of dominant heterologous antibiotic resistance genes as a selection strategy (Svab and Maliga 1993, Carrer *et al.* 1993). A critical evaluation of the procedures currently available for delivery of DNA into plastids, and for the selection of the resulting transformants, has recently been published (Dix and Kavanagh 1994). To date, the integrated DNA has been homologous (ie. gene replacement), *Nicotiana* into *Nicotiana*, or heterologous (ie. gene insertion). We believe the current report is the first to describe intergeneric transplastomic plants, in which homologous recombination has been achieved by transformation, rather than by protoplast fusion (Thanh and Medgyesy 1989). The delivery system used was PEG-mediated DNA uptake, and the recipient protoplasts were *Nicotiana plumbaginifolia*. The donor DNA was a fragment cloned from the plastome of a double mutant of *Solanum nigrum*. This has an alteration at nucleotide 1140 in the 16S rRNA gene, conferring spectinomycin resistance, and another at codon 87 of the 3' *rps*12 gene, converting a lysine to glutamine in this ribosomal protein (Kavanagh *et al.* 1994).

2. Materials and methods

Procedures for the transformation of protoplasts from shoot cultures of diploid wild type *Nicotiana plumbaginifolia* (DNP), or *N. tabacum* var. Petit Havana, were as described previously (O'Neill *et al.* 1993). The transforming DNA was pUC19 plasmid containing a cloned 7.7 kb *Hind* III fragment of the plastid DNA of the *Solanum nigrum* double mutant StSp1, covering the 16S rRNA and 3' *rps*12 genes in the inverted repeat, where spectinomycin and streptomycin resistance mutations, respectively, are located (Kavanagh *et al.* 1994).Selection of putative transformant colonies, was as described previously (O'Neill *et al.* 1993), involving either selection for spectinomycin resistance, followed by screening for resistance to streptomycin, or simultaneous selection for resistance to both antibiotics. Antibiotics were used at a concentration of 500 mg l^{-1}.

Polymerase chain reaction (PCR) was performed on DNA from regenerated plants, using the custom-synthesised oligonucleotide primers shown in Table 1. One of the upstream primers is diagnostic for the *S. nigrum* plastome (CPS); while the other (based on *N. tabacum* sequences) is diagnostic for the *N. plumbaginifolia plastome* (CPN). The downstream primer (CPC) is based on common sequences.

TABLE 1. Primers for analysis of putative transplastomic plants

All three primers lie between *rps*12 (3') and the 5' end of the 16S rRNA gene.

Solanum nigrum-specific (CPS)	
5'- ACA GTT CAT CAC GGA AG	Position: 101849 - 101833
Nicotiana-specific (CPN)	
5' - CAA CAA TTC ATC AGA CT	Position: 101851 -101835
Common primer (CPC)	
5' - GGA CCA ATT TAG TCA CG	Position: 101027- 101011

3. Results

3.1. PRODUCTION OF PLASTOME TRANSFORMANTS

Both selection procedures yielded double antibiotic resistant (green) colonies at a frequency of 1 in 10^4 viable colonies growing on Petri plates, for both *N. plumbaginifolia* and *N. tabacum*. By contrast no double resistant colonies were found in 8-12 x 10^4 plated colonies from non- transformed protoplasts of both species, and spontaneous frequencies of forward mutation to spectinomycin resistance alone are 1-3 x 10^{-5} per plated colony.

Shoots have now been regenerated from more than thirty spectinomycin plus streptomycin resistant putative transplastomic colonies.

3.2. ANALYSIS OF PUTATIVE TRANSPLASTOMIC PLANTS.

Shoot cultures of four double resistant *N. plumbaginifolia* plants were subjected to PCR analysis, together with wild type shoot cultures of both species, using the primers described in Table 1. Primers CPS + CPC, gave a PCR product with *S. nigrum*, but not *N. plumbaginifolia*, wild type DNA, while the converse was true for primers CPN + CPC. Three of the four putative transformants gave products with CPS + CPC, which stained intensely on gels, revealing the presence of *S. nigrum* plastome sequences. These three also gave a less intensively stained product with the other primer pair. The fourth plant only gave a product with the *Nicotiana* specific primers.

4. Conclusion

The results support the use of PEG-mediated DNA uptake, as an efficient procedure for the production of stable chloroplast transformants in higher plants. Furthermore, it is demonstrated that the use of the procedure for gene replacement studies by homologous recombination is not restricted to intrageneric species combinations, and should be more widely applicable to the genetic analysis of the higher plant plastome. In the present work, at least three transformants have been identified and these will be further examined, together with a number of additional lines, by restriction and sequence analysis, to assess the borders of the integrated sequences.

The presence of *N. plumbaginifolia* specific PCR products in the transgenic plants, in addition to the *S. nigrum* introduced sequence, runs counter to previous experience on the selection of plastid recombinants (e.g.. Thanh and Medgyesy 1989, Fejes *et al.* 1990), implying the retention of some non-recombinant plastome copies in the transplastomic *N. plumbaginifolia* shoots. A more thorough analysis of these plants should clarify this unexpected observation.

5. References

Carrer, H., Hockenberry, T.N., Svab, Z., and Maliga, P. (1993) 'Kanamycin resistance is a selectable marker for plastid transformation in tobacco', Mol. Gen. Genet. 241, 49-56.

Cséplö, A., Etzold, T., Schell, J., and Schreier, P. (1988) 'Point mutations in the 23S rRNA genes of four lincomycin resistant *Nicotiana plumbaginifolia* mutants could provide new selectable markers for chloroplast transformation', Mol. Gen. Genet. 214, 295-299.

Dix, P.J., and Kavanagh, T.A. (1994) 'Transforming the plastome: Genetic markers and DNA delivery systems', Euphytica (in press).

Dyer, T.A. (1985) 'The chloroplast genome and its products', Oxford Surveys of Plant Molecular and Cell Biology 2, 147-177.

Etzold, T., Fritz, C.C., Schell, J., and Schreier, P.H. (1987) 'A point mutation in the chloroplast 16S rRNA gene of a streptomycin resistant *Nicotiana tabacum*', FEBS Lett. 219, 343-346.

Fejes, E., Engler, D., and Maliga, P. (1990) 'Extensive homologous chloroplast DNA recombination in the pt14 *Nicotiana* somatic hybrid', Theor. Appl. Genet. 79, 28-32.

Fromm, H., Edelman, M., Aviv, D., and Galun, E. (1987) 'The molecular basis for rRNA-dependent spectinomycin resistance in *Nicotiana* chloroplasts', EMBO J. 6, 3233-3237.

Fromm, H., Galun, E., and Edelman, M. (1989) 'A novel site for streptomycin resistance in the '530 loop' of chloroplast 16S ribosomal RNA', Plant Mol. Biol. 12, 499-505.

Galili, S., Fromm, H., Aviv, D., Edelman, M., and Galun, E. (1989) 'Ribosomal protein S12 as a site for streptomycin resistance in *Nicotiana* chloroplasts', Mol. Gen. Genet. 218, 289-292.

Kavanagh. T.A., O'Driscoll, K.M., McCabe, P.F., and Dix, P.J. (1994) 'Mutations conferring lincomycin, spectinomycin and streptomycin resistance in *Solanum nigrum* are located in three different chloroplast genes', Mol. Gen. Genet. 242, 675-680.

McCabe, P.F., Timmons, A.M., and Dix, P.J. (1989) 'A simple procedure for the isolation of streptomycin resistant plants in *Solanaceae*' Mol. Gen. Genet. 216, 132-137.

Medgyesy, P. (1990) 'Selection and analysis of cytoplasmic hybrids', in P.J. Dix (ed.), Plant Cell Line Selection, VCH Publishers, Weinheim, pp. 287-316.

Ohyama, K.,Fukuzawa, H., Kohchi, T., Shirai, H., Sano, T., Sano, S., Umesono, K., Shiki, Y., Takeuchi, M., Chang, Z., Aota, S.-I., Inokuchi, H.. and Ozeki, H. (1986) 'Chloroplast gene organization deduced from complete sequence of liverwort *Marchantia polymorpha* chloroplast DNA', Nature 322, 572-574.

O'Neill, C., Horváth, G.V., Horváth, E., Dix, P.J., and Medgyesy, P. (1993) 'Chloroplast transformation in plants: polyethylene glycol (PEG) treatment of protoplasts is an alternative to biolistic delivery systems', Plant J. 3, 729-738.

Shinozaki, K., Ohme, M., Tanaka, M., Wakasugi, T., Hayashida, M., Matsubayashi, T., Zaita, N., Chungwongse, J., Obokata, J., Yamaguchi-Shinozaki, K., Ohto, C., Torozawa, K., Meng, B.Y., Sugita, M., Deno, H., Kamogashira, T., Yamada, K., Kusuda, J., Takaiwa, F., Kato, A., Tohdo, N., Shimada, H., and Sugiura, M. (1986) 'The complete nucleotide sequence of the tobacco chloroplast genome: its gene organization and expression', EMBO J. 5, 2043-2049.

Svab, Z., and Maliga, P. (1991) 'A mutation proximal to the tRNA binding region of the *Nicotiana* plastid 16S rRNA confers resistance to spectinomycin', Mol. Gen. Genet. 228, 316-319.

Svab, Z. and Maliga, P. (1993) 'High-frequency plastid transformation in tobacco by selection for a chimeric *aad*A gene', Proc. Natl. Acad. Sci. USA 90, 913-917.

Svab, Z., Hajdukiewitz, P., and Maliga, P. (1990) 'Stable chloroplast transformation in higher plants', Proc. Natl., Acad., Sci. USA 87, 8526-8530.

Thanh, N.D. and Medgyesy, P. (1989) 'Limited chloroplast gene transfer via recombination overcomes plastome-genome incompatibility between *Nicotiana tabacum* and *Solanum tuberosum*', Plant Mol. Biol. 12, 87-93.

PHYSICAL MAPS OF MITOCHONDRIAL GENOMES FROM MALE-FERTILE AND MALE-STERILE SUGAR BEETS

T. KINOSHITA, T. MIKAMI AND T. KUBO
Plant Breeding Institute,
Faculty of Agriculture,
Hokkaido University,
Sapporo, 060, Japan

ABSTRACT. A complete physical map of the normal mitochondrial genome (mtDNA) was constructed as 358 kbp in size and contained three copies of a recombination repeat.

Maps of S, S-2 and S-3 mtDNAs were also made and then aligned with the N map on the basis of the hybridization results.

In addition, the transcriptional alterations were detected in clones carrying *coxII*, *atpA*, *atp6* and *rps13* genes in the male-sterile cytoplasms.

1. Introduction

The use of cytoplasmic male sterility (CMS) is of great importance in the commercial production of hybrid seed, because it provides an alternative to troublesome hand or mechanical emascultaion.

As in all other species so far investigated, the mitochondrial genome (mtDNA) of the CMS sugar beet exhibited altered organizational properties when compared with that of the male-fertile genotype, whereas the chloroplast genome was found to be collinear in the two genotypes (Kinoshita 1990).

Recently, alterations in the transcription patterns of the three mitochondrial genes *coxII*, *atpA*, *atp6* have been observed in CMS sugar beet, but no definite proof that these modifications are responsible for CMS has ever been presented (Senda et al. 1991, 1993). Moreover, the determination of the coding capacity and the detailed organization of sugar beet mtDNA remains incomplete. In this study, we compared the organization and expression of the mitochondrial genomes from a pair of near-isonuclear male-sterile and male-fertile sugar beet lines to identify mutation that might be involved in CMS.

2. Organization of the male-fertile (N) mitochondrial genome

2.1 *Physical map of N mtDNA*

A detailed genetic and physical map of the mitochodrial genome from male-fertile sugar beet is a prerequesite to understanding the structural and/or organizational alterations associated with CMS. The genome walking approach was adopted to determine the physical organization of male-fertile mtDNA. A library of mtDNA in the bacteriophage lambda vector was constructed by partial digestion with *MboI* and size fractionation. The starting points for genome walking were produced by isolating lambda clones that

M. Terzi et al. (eds.), Current Issues in Plant Molecular and Cellular Biology, 303-308.

hybridized to probes for the eleven mitochondrial genes conserved among plant mitochondrial genomes as shown in Table 1. Based upon the combined data of 36 overlapping clones, the entire sequence complexity with a content of 358 kbp can be arranged on a single circular master chromosome (Fig.1). Two rRNA and nineteen polypeptide genes as well as the restriction sites of four enzymes *Xho*I, *Sal*I, *Xba*I and *Sma*I were mapped in this manner (Fig.2).

2.2 *Repeated sequences*

It is known that most of the plant mitochondrial genomes contain recombination repeats (Hanson 1991). There are sequences of a few hundred to a few thousand nucleotides which are present in more than two loci per master chromosome. This was also true for the sugar beet N mtDNA in which four families of repeated sequences were included.

The largest repeat occurred in three copies on the master chromosome and contained *rrn26* gene. Although the region common to all three members was 4.7 kbp in length, the homologous region between two copies out of the three extended to a minimum of 20 kbp. If this repeated sequence functions as an active recombination repeat, nine genomic environments of the three repeat copies will result from intra-molecular recombination. Actually we isolated cosmid clones representing the nine genomic contexts of the repeat indicating that *rrn26* harboring repeat sequences are substrates for frequent recombination events.

Other repeated sequences were less than 0.5 kbp in length, so that recombination across these sequences, if it occurrs, must be very rare. One of the families consisted of an intact *coxII* gene and a pseudo-copy containing the 281 nucleotides of the *coxII* 5' coding region. The *coxII* 5' flanking sequence and *orfB* 5' flanking sequence (COXFS and ORFFS) families had five and six repeats, respectively. Four members of multicopy set of COXFS resided in the 5' flanking regions of the *nad9*, *atp6* and *coxII* loci of sugar beet. ORFFS members were also found in the 5' flanking region of higher plant mitochondrial genes according to a data base search.

2.3 *Chloroplast DNA-homologous sequences*

DNA transfer from chloroplast to mitochondrion has been revealed as a general phenomenon in higher plants. To determine whether an interorganellar exchange has also occurred in sugar beet cpDNA clones from sugar beet, tobacco and rice were hybridized to the clone bank of the male-fertile sugar beet mitochondrial genome. Consequently, six regions of a chloroplast-homologous sequence were found.

3. Mitochondrial DNA rearrangements and transcriptional alteration in the male sterile (S) cytoplasm

3.1 *Physical map of S mtDNA*

First, Brears and Lonsdale (1988) physically mapped sugar beet CMS mtDNA. Their data are limited to *Sma*I and XhoI cleavage sites, and a clone bank is not available. It is thus necessary to construct a detailed physical map of S mtDNA, and to compare it with that of N mtDNA.

Two mapping strategies were adopted. The first approach included RFLP analysis using a set of lambda clones covering the entire N mtDNA as probes. This analysis identified fifteen regions (designated as linkage groups) as shown in Fig.2.

The second approach was based upon the cloning of altered regions of S mtDNA. Probes flanking the rearrangement break points were used to screen the mtDNA library of S cytoplasm line. The positive clones were used for RFLP mapping. A mtDNA was thus mapped by these two strategies.

S mtDNA had three sets of repeated sequences (called repeat-1, -2, and -3), two of which were apparently active in recombination. Repeat-1 consisted of three members which included *rrn26* locus. Two out of three members shared an additional sequence containing *coxII*.

Two members of repeat-3 were arranged in the inverted orientation. The detailed analysis of repeat-3-containing clones failed to reveal the four configurations corresponding to the reciprocal exchange of the 5' and 3' sequences flanking the repeat. This indicates that much of S mtDNA is in the form of linear mitochondrial chromosomes with repeats-1 or -3 residing at the termini of the linear molecule.

3.2 *Genome rearrangement*

A restriction map of S mtDNA was then aligned with that of N mtDNA based upon the hybridization results. This alignment demonstrated numerous changes in the order and orientation of the fifteen linkage groups. It is noteworthy that some members of COXFS and ORFFS families were mapped near the rearranged points. It can be supposed that COXFS and/or ORFFS are substrates for a rare recombination event(s) that is involved in the generation of S mtDNA.

N and S mtDNAs were also characterized by their own unique sequences: they have common sequences of about 250 kb, but about 25% to 28% of the total sequence complexity is unique to each cytoplasm (Fig.2).

3.3 *Transcriptional alterations*

The important question was whether or not the genome rearrangements can affect the transcription of individual mitochondrial genes. To address this issue, a complete transcriptional map of N mtDNA was constructed as shown in Fig 3. Thirty-two independent lambda clones were selected to cover the entire N mtDNA. These clone DNAs were digested with *XhoI, SalI, XbaI* or *SmaI*, and hybridized to cDNA probes derived from total mtRNA of male-fertile sugar beet. Twenty-three regions gave positive hybridization signals. Known mitochondrial genes account for 18 of the 23 hybridization signals and 15 regions were situated in close proximity to the rearrangement points.

The correlation of altered transcription concurrent with the genome reorganization was examined by Northern hybridization. Included in this analysis were 20 clones of S mtDNA and 16 clones of N mtDNA that cover all the altered regions. The clones carrying *coxII, atpA* and *atp6* genes detected different transcripts between male-fertile and male-sterile lines.

3.4 *CMS-unique open reading frame*

The *rps13* locus and its flanking regions have been characterized from both N and S mtDNAs. The N and S segments in question shared a common 875 bp region bearing *rps13* coding region, the exon b of the *nad1* gene and their intergenic spacer region. However, the nucleotide sequences diverged completely 56 bp upstream of *rps13* initiation codon. Upstream of S *rps13* was 300-codon open reading frame (*orf300*) that comprises part of *orfB* gene and 676 bp sequence from unknown sources. Northern blotting indicated the transcription of *orf300*. On the other hand, an *orf300*-homologous transcript was not detected in mtRNA from normal sugar beet which was consistent with the absence of ORF from S mtDNS.

4. Structure of mitochondrial genomes in S-2 and S-3 cytoplasms derived from wild beet

Previously, cytoplasmic polymorphism was identified at both molecular and plant levels and designated as S-2, S-3 and S-4 cytoplasms related to male-sterility. (Kinoshita et al., 1990). The mtDNAs from N and four kinds of S cytoplasms gave five different restriction profiles (Mikami et al., 1985), indicating that mitochondrial polymorphism.

MtDNAs from S-2 and S-3 cytoplasm lines were partially digested with *Sau*3A or *MboI* and the resulting fragments were used to produce a clone library in the bacteriophage vector lambda. We identified the families of overlapping phages by probing the clone library with the 11 mitochondrial genes. Twenty five phage clones of N mtDNA were also included for the screening of mtDNA libraries. Following the genome walking technique, the sequence complexity with a content of 393kbp has so far been arranged with *XhoI, SalI XbaI* and *SmaI*. Our preliminary results indicate that a single large circular master chromosome can be constructed both in S-2 and S-3 mtDNAs although there is an insertion of 13kbp in S-3 mtDNA.

Alignment of the restriction map for S-2 mtDNA with that for N mtDNAs allowed the identification of at least five linkage groups within which restriction maps are conserved between the two genomes. However, the order and orientation of these linkage groups were also different between the two genomes.

Thus it was demonstrated that the structural polymorphism which existed among N, S, S-2 and S-3 mtDNAs is deeply related with the expression of male sterility at plant level suggesting complexity in the relations between nuclear and cytoplasmic genomes.

5. References

Brears, T. and Lonsdale, D. M. (1988) 'The sugan beet mitochondrial genome: a complex organization generated by homologous recombination'. Mol. Gen. Genet. 214:514-522.

Hanson, M.R. (1991) 'Plant mitochondrial mutations and male sterility'. Annu. Rev. Genet., 25, 461-486.

Kinoshita, T. (1990) 'Cytoplasmic engineering on the male sterility of sugar beet'. The Nucleus, 33, 44-57.

Kinoshita, T. and Mikami, T. (1990) 'Classification of male sterile cytoplasmic types in sugar beet (*Beta vulgaris* L.)'. J. Fac. Agr. Hokkaido Univ. 64, 219-228.

Mikami, T., Kishima, Y., Sugiura, M. and Kinoshita, T. (1985) 'Organelle genome diversity in sugar beet with normal and different sources of male sterile cytoplasms'. Theor. Appl. Genet. 71, 166-171.

Senda, M., T. Harada, T. Mikami, M. Sugiura and T. Kinoshita (1991) 'Genomic organization and sequence analysis of the cytochrome oxidase subunit II gene from normal and male-sterile mitochondria in sugar beet'. Current Genetics 19, 175-181.

Senda, M., T. Mikami and T. Kinoshita (1993) The sugar beet mitochondrial gene for the ATPase alpha-subunit: sequence, transcription and rearrangements in cytoplasmic male-sterile plants. Current Genetics. 24, 164-170.

Table 1. Hybridization probes used in gene identification

Gene	Product	Source	Reference
cox I	Cytochrome oxidase subunit I	Sugarbeet	Senda *et al*. (1991)
cox II	Cytochrome oxidase subunit II	Sugarbeet	Senda *et al*. (1991)
cox III	Cytochrome oxidase subunit III	*Oenothera*	Hiesel *et al*. (1987)
atpA	ATPase subunit alpha	Pea	Morikami and Nakamura (1987a)
atp6	ATPase subunit 6	*Oenothera*	Schuster and Brennicke (1987)
atp9	ATPase subunit 9	Pea	Morikami and Nakamura (1987b)
cob	Apocytochrome b	Wheat	Boer *et al*. (1985)
rrn26	26S rRNA	Pea	Morikami and Nakamura (1982)
rrn18	18S rRNA	Pea	Morikami and Nakamura (1982)
nad1 (exon b)	NADH dehydrogenase subunit 1	*Oenothera*	Wissinger *et al*. (1991)
orf192	unknown	Sugarbeet	Kubo *et al*. (unpublished)

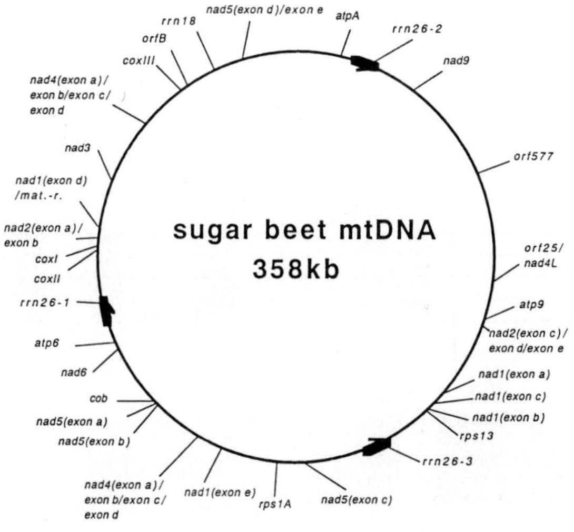

Figure 1. Master circle of mitochondrial genome from male-fertile sugar beet. The 4.7 kb repeats are shown by arrows.

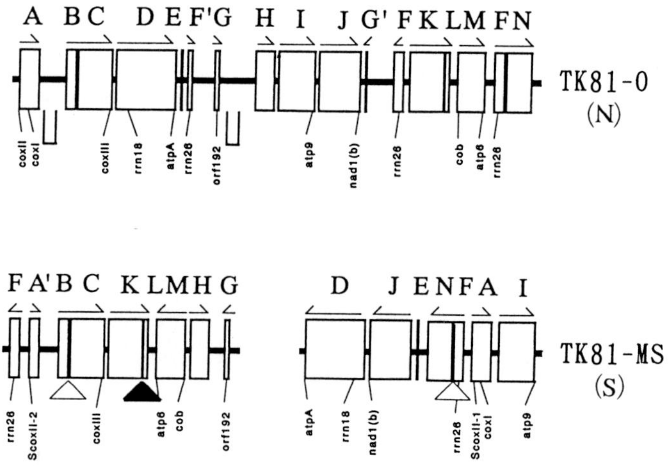

Figure 2. Relative arrangement of linkage groups in N (TK81-0) and S (TK81-MS) mitochondrial genomes. Small deletions or insertions in S vs. N mtDNA are indicated by open and filled triangles, respectively.

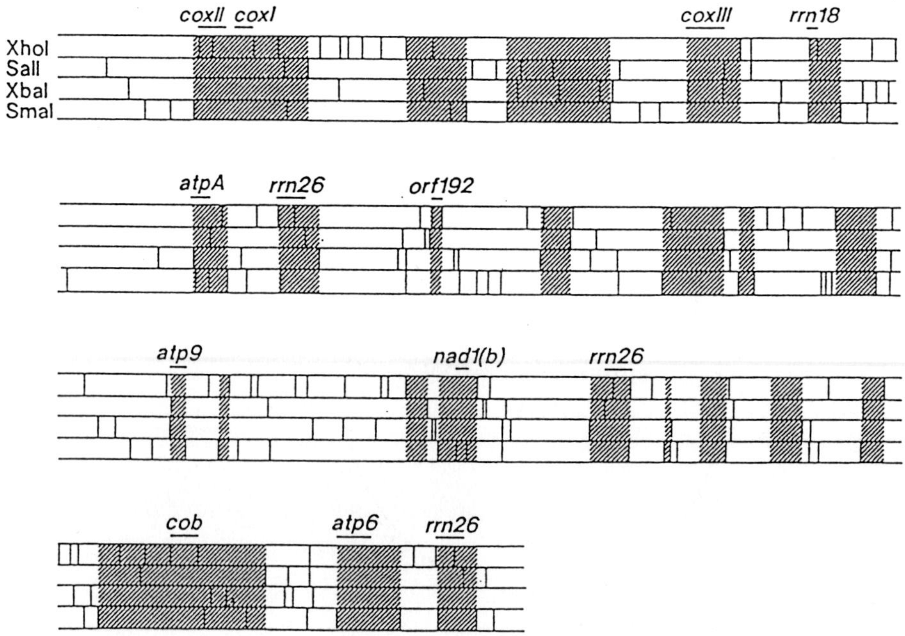

Figure 3. Transcriptional map of N mtDNA. Hatched regions on *Sma*I map indicate the segments to which cDNA probes derived from total mtDNA hybridized.

A tRNA GENE MAPPING WITHIN THE CHLOROPLAST rDNA CLUSTER IS DIFFERENTIALLY EXPRESSED DURING THE DEVELOPMENT OF DAUCUS carota

F. MANNA°, D.R. MASSARDO°, K. WOLF*, G. LUCCARINI^,
M. S. CARLOMAGNO', F. RIVELLINI', P. ALIFANO',
L. DEL GIUDICE°.
° I.I.G.B., C.N.R., Via G. Marconi 12, I-80125 Napoli, ITALY
* Inst. für Biol. IV (Mikrobiol.), R. - W.T.H., Worringer Weg, D-52056 Aachen, GERMANY
^ Ist. Mutagenesi e Differenziamento, C.N.R., Via Svezia 10, I-56124 Pisa, ITALY
' C.E.O.S. - C.N.R., Uni. Napoli, Via S. Pansini 5, I-80131 Napoli, ITALY

ABSTRACT. *In vivo* analysis of expression of the chloroplast rDNA cluster during somatic embryogenesis of *Daucus carota (D. carota)* was performed by Northern-blot analysis with different DNA probes, spanning both the 16S rRNA gene, the 16S-23S rRNA spacer, which contains the two spliced tRNA genes $tRNA_{Ile}$ and $tRNA_{Ala}$, and the region upstream of the 16S rRNA gene, where a $tRNA_{Val}$ maps. In this paper we show that expression both of the spacer tRNAs $tRNA_{Ile}$ and $tRNA_{Ala}$ is not significantly regulated during development whereas the amount of the transcript corresponding to $tRNA_{Val}$, whose gene maps upstream of the 16S rRNA gene, is not detectable during early embryonic stages (undifferentiated cells and globular embryoids), and progressively accumulates during late phases (heart-shaped, torpedo-shaped and early plantula).
Multiple transcription start sites have been identified upstream of the $tRNA_{Val}$ gene by S_1 mapping analysis, which are activated late during the embryogenesis.
These data indicate that during carrot embryogenesis developmental control mechanisms act on plastid gene expression.

INTRODUCTION

In developing plants, chloroplasts are derived from small proplastids, which are the undifferentiated plastids present in meristematic cells.

At present little is known about the mechanisms that modulate the differential gene expression in plastids during the development. The control of chloroplast gene expression operates on several stages, namely transcription, post-transcription, translation and post-translation during chloroplast gene expression (1).

Recent studies suggest that the role of transcription in controlling gene expression is rather limited and that gene expression is more tightly controlled at the post-transcriptional level (1).

The ribosomal RNA genes (rDNAs) in higher-plant chloroplasts are clustered and arranged like in *Escherichia coli (E. coli)* in the order 16S-23S-4.5S-5S rRNA with an interspersion of transfer RNA (tRNA) genes within this cluster (1).

The transcription of plastid rDNA clusters was shown to be regulated differently than the transcription of protein coding genes during chloroplast development (2).

However, at the present time no experimental data at the molecular level has been obtained which could explain such differential gene expression in chloroplast.

In order to understand the regulatory mechanisms which affect rRNA synthesis in chloroplasts during plant embryogenesis, we have cloned and analyzed the promoter region of the carrot

M. Terzi et al. (eds.), Current Issues in Plant Molecular and Cellular Biology, 309-314.

chloroplast rDNA cluster. This DNA clone, which is presented here, encodes the chloroplast $tRNA_{Val(GAC)}$ as well as the start of the 16S rRNA gene and contains sequences similar to the Pribnow box and the -35 region found in *E. coli* promoters (3). *In vivo* transcription of the DNA fragment containing this region reveals that the $tRNA_{Val(GAC)}$ gene is differentially expressed during somatic embryo development of *D. carota.*

MATERIALS AND METHODS

Plant material and culture conditions

Were as previously described (4)

Bacterial strains and growth conditions

Bacterial strains, media and growth conditions were as described by Del Giudice (5).

DNA procedures

Plant DNA was isolated either from dark grown 5 to 6 days old seedlings or from growing calli, according to (6). Preparation of DNA fragments, nick translation, 5' end-labelling with polynucleotide kinasi, colony hybridization and DNA sequencing were performed as described in Sambrook et al. (7).

Construction of a carrot *(D.carota)* genomic library

High molecular-weight DNA, obtained from carrot, was partially digested with *Hind* III, fractionated on a 10 to 40% sucrose gradient and used to construct a genome library in pGEM-3Zf(+) Hind III digested vector DNA.

RNA procedures

Total plant RNA was isolated from carrot embryonic stages by a LiCl small-scale procedure (8).
In vitro 5'-^{32}P-labelling of total RNA was performed according to the procedure of (7).
Northern-blot and RNA slot-blot analyses were performed according to (7).
RNA/DNA-hybridization, S_1 nuclease digestion, and analysis of the hybrids on polyacrylamide denaturing gels were performed according to the procedure described by Alifano et al. (9) .

In vitro transcription assay

Supercoiled plasmid DNA was transcribed *in vitro* according to Alifano et al. (9).
The unlabelled transcripts were analyzed by primer extension according to Sambrook et al. (7).

RESULTS

Cloning and characterization of the chloroplast rDNA cluster from *D. carota*

Our aim was to isolate genes differentially expressed during the somatic embryogenesis. The different embryoid stages: globular, heart-shaped and torpedo-shaped were obtained according to (4). A *Hind* III genomic library from *D. carota* was constructed in the vector pGEM3Zf(+). A replica plating experiment was performed to screen the library with different radiolabelled total RNA probes extracted from undifferentiated cells, from different stages of carrot embryoids, and from plantula.

The restriction map of a positive clone is depicted in Figure 1A. The genetic map was deduced on the basis of the nucleotide sequence homology with the chloroplast rDNA clusters of *Nicotiana tabacum* and *Zea mays (10).*

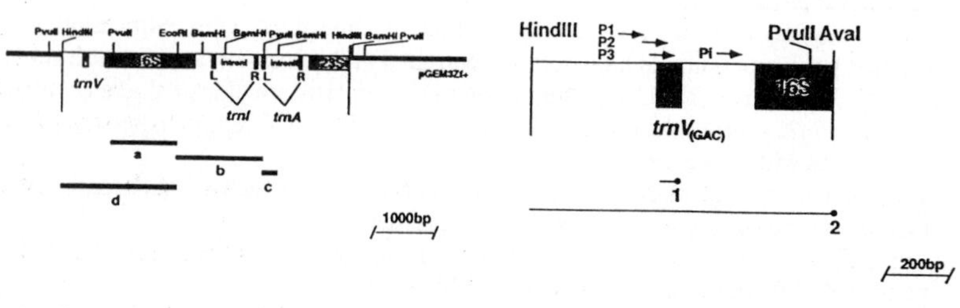

Figure 1. Genetical and physical map of the cloned region of the D. carota chloroplast rDNA cluster .

Northern blot analysis of the specific transcripts from the chloroplast rDNA cluster during the different phases of somatic embryogenesis

Total RNA was extracted from the different developmental stages and analyzed by Northern blot hybridization with the DNA probes indicated in Figure 1A. Probes d, a, b, and c were used in Northern experiments in Figures 2, 3A, 3B and 3C, respectively.

In Figure 2 a signal for the $tRNA_{Val(GAC)}$, which was undetectable in the cellular stage, increased from the globular phase to the plantula when compared with the amount of 16S rRNA detected with the same probe d (Figure 2), and with probe a in Figure 3A. In Figures 3B and 3C hybridization was carried out with probes b and c to show that the embryo-regulated effect did not extend to the spacer tRNAs within the rDNA cluster. These results were confirmed by a quantitative slot-blot analysis (data not shown) using labelled synthetic oligonucleotides as probes which were complementary to the nucleotide sequence of $tRNA_{Val(GAC)}$, $tRNA_{Ile(GAU)}$ and $tRNA_{Ala(UGC)}$ (Figure 1A). In Figure 3B two additional faint bands are visible. The slower migrating band very likely corresponds to the intron I; the faster migrating band corresponds in size to a non-functional maturation product located between the 16S rRNA 3'-and $tRNA_{Ile(GAU)}$ 5'- maturation sites (see also Figure 1A).

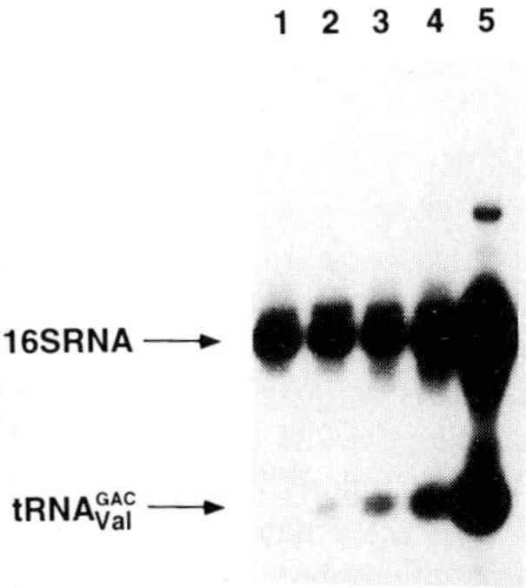

Figure 2. Northern-blot analysis of trnV-16S rDNA region

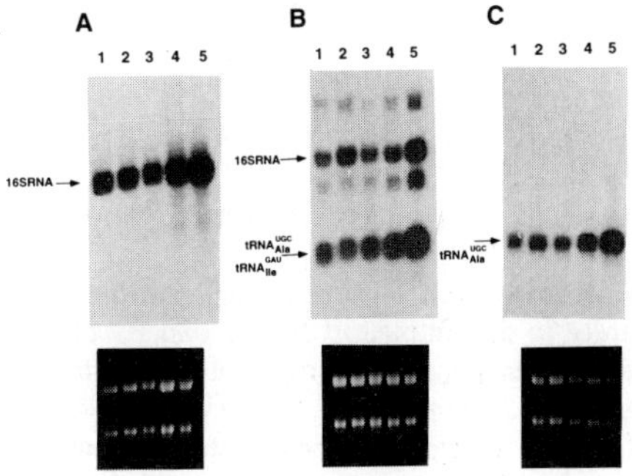

Figure 3. Northern-blot analysis of the 16S rDNA-trnI-trnA- 23SrDNA region

High resolution mapping of the specific transcripts from the chloroplast rDNA cluster during embryo development

In order to investigate the possible molecular mechanisms responsible for the embryo-regulated expression of the $tRNA_{Val(GAC)}$, a detailed map of transcripts in the region spanning the genes for $tRNA_{Val}$ and the 16S rRNA was obtained by S_1 nuclease mapping and primer extension experiments (Figures 4 and 5).

A 781 bp *Hin*dIII-*Ava*I DNA fragment, 5'end-labelled at the *Ava*I site (Figure 1B and lane 7 of Figure 4), was used as a probe in the S_1 mapping experiment shown in Figure 4. The RNAs used in this experiment were extracted from the different developmental phases. In addition to a signal corresponding to the mature 16S rRNA, two faint bands were observed. The faster migrating one (*Pi*) was present in the same amount during all developmental phases, whereas the slower migrating band ($tRNA_{Val(GAC)}$) was only detectable in the early plantula. The sizes of these two bands were determined on the basis of the sequencing ladder which was run in parallel (lanes 8 and 9).

The *Pi* transcript starts from the promoter in the $tRNA_{Val(GAC)}$-16S rDNA intergenic region (Figure 6) previously identified in *Z. mays* (2,11). The transcript indicated as $tRNA_{Val(GAC)}$ has the 5'-end coincident with the 5'-end of the coding region of the *trnV* gene (Figure 1A). This transcript possibly represents an intermediate in the 5'-end maturation of the $tRNA_{Val(GAC)}$-16S rRNA precursor molecule.

To analyze the putative promoter driving transcription of the *trnV* gene, a primer extension experiment was performed with a labelled oligonucleotide internal to the *trnV* sequence (Figure 1B). Three major transcripts were detected (P1, P2 and P3 in Figure 5) when total rRNA extracted from early plantula was used in (lane 6). These transcripts were not detected in the undifferentiated cells (lane 5). The 5'-ends of these species were mapped using a Sanger sequencing ladder obtained by priming the sequencing reactions with the same oligonucleotide (Figure 5, lanes 1-4, and Figure 6). The same results were obtained by S_1 mapping analysis (data not shown).

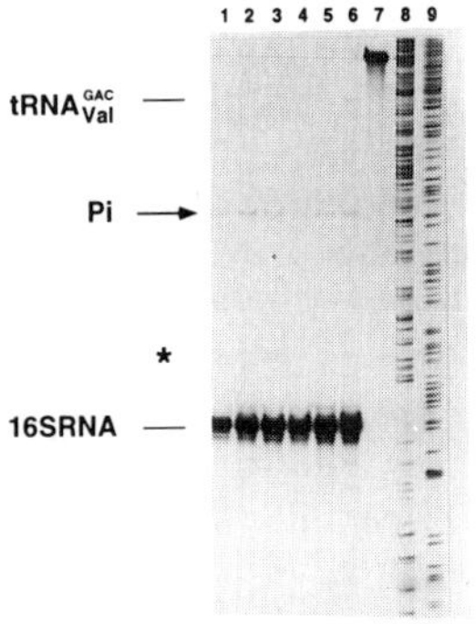

Figure 4. S1 mapping analysis of the trnV-16S rDNA region.

***In vitro* transcription of the 5'-region of the *trnV* gene by *E. coli* RNA polymerase**

In order to discriminate if P1, P2 and P3 transcripts were the results of processing events or directly promoter, we performed an *in vitro* transcription assays, using the 781 bp *Hin*dIII-*Ava*I fragment as a template and purified *E. coli* RNA polymerase. Previous reports have shown that this enzyme recognizes promoter sequences upstream to the *trnV* gene (2). Putative transcription start points in the region upstream of the *trnV* gene of carrot chloroplast DNA could be demonstrated by this method.

The result of this experiment is shown in Figure 5, lane 7.

A heterogeneity of spurious transcription start points was identified, the more prominent of which coincide in size with the transcripts P1, P2 and P3 detected in the early plantula *in vivo*.

Two minor RNA species (* and **), possibly processed products, were only detected *in vivo*.

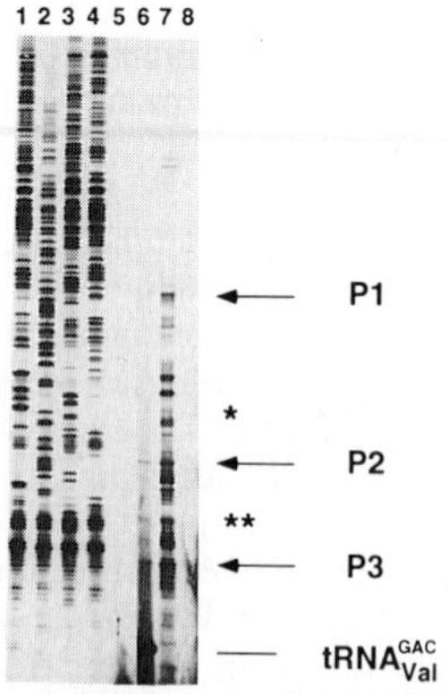

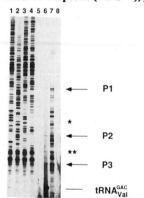

Figure 5. Primer extention analysis of the the trnV-16S rDNA region.

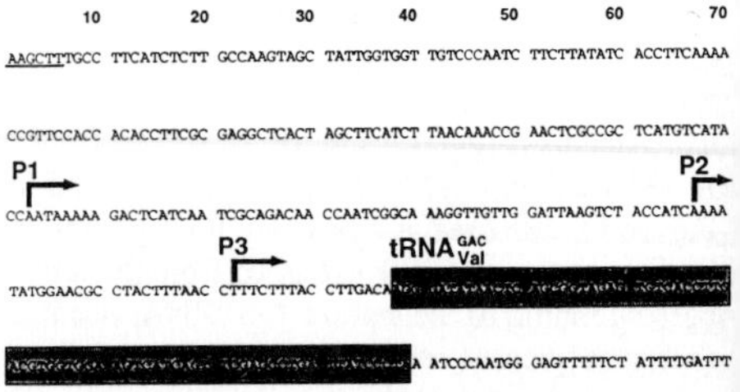

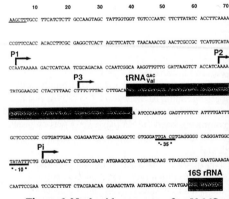

Figure 6. Nucleotide sequence of trnV-16S rDNA region and mapping of the putative promoter sequences.

DISCUSSION.

The gene for $tRNA_{Val(GAC)}$ is approximately 300 bp upstream of the start of the 16S rRNA gene in maize , tobacco and carrot (1). It was shown that there is one strong transcription initiation site in the chloroplast rDNA leader sequence of tobacco, and that this site is 20 to 26 bp upstream of the prior to start of the $tRNA_{Val(GAC)}$ gene (1). It is just downstream of sequences resembling the consensus of E.coli promoters (3).

Our data suggests that the differential expression of the $tRNA_{Val(GAC)}$ during the embryogenesis of *D. carota* is dependent on the differential activity of upstream promoter sequences. Both transcriptional competence of the template and limiting RNA polymerase levels could play regulatory roles. Besides, a functional modification of RNA polymerase by specific proteins or other molecules like ppGppp which controls transcription in bacteria , or differential regulation of two different RNA polymerases , might be involved (1).

In spite of numerous efforts, structure and site of synthesis of the chloroplast RNA polymerase are still controversial. Different purification procedures led to the hypothesis that chloroplasts of algae and higher plants may contain at least two different RNA polymerase activities which are distinguishable by their preference for specific genes and their biochemical properties (1). One type of RNA polymerase is tightly associated with the chloroplast DNA and synthesizes ribosomal rRNA . This type of enzyme may consist of a single polypeptide chain. Subsequently, other partially purified chloroplast RNA polymerase have been described to have more complex subunit compositions , and to be able to initiate transcription at tRNA- and mRNA-promoters (1). Some of the RNA polymerase subunits may be encoded by the chloroplast genome, since four ORFs (named *rpoA, rpoB, rpoC1* and *rpoC2*), have been identified and two of them, *rpoA* and *rpoB*, show a moderate degree of homology with the *E. coli* RNA polymerase a and b subunits (1). The exact contribution of these ORFs to the composition of the active enzyme is not understood. This finding, however, is in contrast to a previous report that all chloroplast RNA polymerase subunits are nuclear encoded (1).

Our finding that transcription of the *trnV* gene is developmentally regulated, allows some considerations of the problem of RNA polymerase. Since the presence of a functional and efficient protein synthesizing machinery is a prerequisite for the expression of chloroplast genes in all developmental stages, it is difficult to believe that (a) chloroplast encoded subunit(s) of RNA polymerase may be involved in starting the transcription of the *trnV* gene. It is still conceivable that a totally entirely nuclear encoded enzyme is responsible for this developmentally regulated transcription. This could provide a mechanism by which the nucleus controls organellar transcription by regulating the expression of a tRNA which is an essential part of the chloroplast protein synthesizing machinery.

ACKNOWLEDGEMENTS

The authors would like to express their gratitude to Profs. M. Terzi and V. Nuti-Ronchi for valuable suggestions. The authors thank Mr. G. De Simone for technical assistance. Work was supported by MAF, Italy, special grant "P. N. Biotecnologie MAF" and by CNR, Italy, special grants "P.F.- BBS, " and partially "P. F. - I. G.".

REFERENCES

1. Sugiura, M.: (1992) 'The chloroplast genome ', Plant Mol. Biol. 19, 149-168.
2. Klein , R.R., and Mullet, J.E. (1990) ' Light-induced transcription of chloroplast genes, psbA transcription is differntially enhanced in illuminated barley.', J.Mol.Chem. 265, 1895-1902
3. de Boer, H. A., Gilbert, S. F. and Nomura, M. (1979) ' DNA sequences of promoter regions for rRNA operons rrnE and rrnA in E.coli ' , Cell 17, 201-209.
4. Giuliano, G. D., Rosellini, D. and Terzi, M. (1983) ' A new method for the purification of the different stages of carrot embryoids ', Plant Cell Rep. 2, 216-218.

5. Del Giudice, L. (1981) ' Cloning of mitochondrial DNA from the petite-negative yeast Schz. pombe in the bacterial plasmid pBR322', Mol. Gen. Genet. 184, 465-470.
6. Del Giudice, L., Manna, F., Massardo, D.R., Wolf, K., Motto, M., (1987) ' Cryptic copies of Mu1 transposon in italian maize varieties ' , Maydica 32, 189-200
7. Sambrook, J., Fritsch, E. F. and Maniatis, T. (1989) 'Molecular cloning: a laboratory manual.
Cold Spring Harbor Laboratory Press.
8. Verwoerd, T. C., Dekker, B. M. M. and Hoekema, A. (1989) ' A small scale procedure for the rapid isolation of plant RNAs ', Nucleic Acids Res. 17, 2362.
9. Alifano, P., Rivellini, F., Limauro, D., Bruni, C.B. and Carlomagno M.S. (1991) 'A consensus motif common to all Rho-dependent prokaryotic transcriptuin terminators.'
Cell 64, 553-563.
10. Shinozaki, K., Ohme, M., Tanaka, M., Wakasugi, T.,Hayashida, N., Matsubayashi, T., Zaita, N., Chunwongse, J.,Obokata J., Yamagichi-Shinozaki, K., Ohto, C., Torazawa, K., Meng, B. Y., Sugita, M., Deno, H., Kamogashira, T.,Yamada, K., Kusuda, J., Takaiwa, F., Kato, A., Tohdoh, N., Shimida, H. and Sugiura, M. (1986) ' The complete nucleotide sequence of the tobacco chloroplst genome: its gene organization and expression.' EMBO J. 5 ,2043-2049
11. Delp, G., Igloi, G.L., Kossel, H. (1991) ' Identification of *in vivo* processing intermediates and a splice junctions of tRNAs from maize chloroplasts by amplification with the polymerase chain reaction. ' Nucl. Acids Res. 19, 713-716.

CRYOPRESERVATION OF SEVERAL TROPICAL PLANT SPECIES USING ENCAPSULATION/DEHYDRATION OF APICES

F. ENGELMANN[1,*], E.E. BENSON[1,2], N. CHABRILLANGE[1], M.T. GONZALEZ ARNAO[3], S. MARI[1], N. MICHAUX-FERRIERE[4], F. PAULET[5], J.C. GLASZMANN[5] and A. CHARRIER[1]

*: IPGRI, Via delle Sette Chiese 142, 00145 Rome, Italy
1: ORSTOM, BP 5045, 34032 Montpellier Cedex 1, France
2: DIT, Bell Street, Dundee DD1 1HG, UK
3: CNIC, Aptdo 6880, Cubanacan, La Habana, Cuba
4: CIRAD-BIOTROP, BP 5035, 34032 Montpellier Cedex 1, France
5: CIRAD-CA, BP 5035, 34032 Montpellier Cedex 1, France

ABSTRACT. Cryopreservation using encapsulation-dehydration of apices of *in vitro* plantlets was investigated with sugarcane (*Saccharum* sp.), cassava (*Manihot esculenta*) and two coffee species (*Coffea racemosa* and *C. sessiliflora*). Maximum survival rates of frozen apices were 91% with sugarcane, 60% with cassava and 38% with coffee. Growth recovery of cryopreserved sugarcane apices was rapid and direct. With cassava and coffee, callusing was often observed during recovery of frozen apices, thus reducing the number of plants regenerated. Restriction fragment length polymorphism analyses of sugarcane revealed several modifications in the band patterns of plants regenerated both from control and cryopreserved apices which were identical in both materials. These variations were therefore not due to freezing but preexisting among the *in vitro* plantlets used. These results demonstrate that the encapsulation-dehydration technique can be employed for cryopreservation of apices of some tropical species.

Introduction

In vitro collections of plantlets maintained under slow growth exist for sugarcane, coffee and cassava in various institutes. In Montpellier, France, the sugarcane *in vitro* collection of CIRAD-CA comprises 650 accessions (Paulet *et al.*, 1991). At ORSTOM, *in vitro* collections have been established for coffee (Bertrand-Desbrunais, 1991) and cassava (Brizard *et al.*, 1991) which include 486 and 76 accessions, respectively.

However, maintenance of large *in vitro* collections is time consuming. It allows short- or medium-term storage only, and risks of contamination and of somaclonal variation increase with time (Withers, 1987). Only cryopreservation (liquid nitrogen, -196°C) presently offers a safe long-term conservation option.

Cryopreservation has been applied presently to more than 80 plant species, among which around 40 from tropical origin (Engelmann, 1991). If cryopreservation can be considered as a routine technique for most cell and callus cultures, this is not the case for organized structures such as apices or embryos. However, the recent development of new techniques, among them encapsulation-dehydration, may change this situation.

The encapsulation-dehydration technique is based on the technology developed for the production of synthetic seeds. For cryopreservation, apices are embedded in alginate beads, precultured in liquid medium with high sucrose concentration and partially desiccated before

* Present address

M. Terzi et al. (eds.), Current Issues in Plant Molecular and Cellular Biology, 315-320.

freezing in liquid nitrogen. This technique has been developed with apices of several temperate species (Dereuddre, 1992).

In the present study, we summarize the results obtained with the application of encapsulation-dehydration of apices for the cryopreservation of three plants of tropical origin, sugarcane, cassava and coffee.

Materials and Methods

PLANT MATERIAL

The plant material consisted of *in vitro* plantlets originating from the collections of CIRAD-CA, France, and CNIC, Cuba, for sugarcane (*Saccharum* sp.), and of ORSTOM, France, for cassava (*Manihot esculenta*) and coffee (*Coffea sessiliflora, C. racemosa*). They were cultivated according to Paulet *et al.* (1991) for sugarcane, Bertrand-Desbrunais (1991) for coffee and Brizard *et al.* (1992) for cassava.

CRYOPRESERVATION PROCEDURES

The cryopreservation techniques used have been described in detail by Paulet et al. (1993) for sugarcane, Benson *et al.* (1992) for cassava and Mari *et al.* (1993) for coffee. In all cases, apices were placed after dissection for 24 hours on standard medium, encapsulated in the same medium with 3% alginate, desiccated in the air current of the laminar air flow cabinet for various periods, then frozen by direct immersion in liquid nitrogen. After slow thawing at room temperature, sugarcane apices were cultivated in the dark for one week on a medium supplemented with growth regulators, then transferred to standard conditions. Cassava and coffee apices were placed directly under light conditions, on standard medium (coffee) or on a medium modified by the addition of growth regulators (cassava).

Results

CRYOPRESERVATION OF SUGARCANE APICES

Preliminary experiments permitted determining the optimal pregrowth and desiccation conditions as a 24- or 48-hour culture period in a medium with 0.75 M sucrose, followed by 6 hours of desiccation (Paulet *et al.*, 1993; Gonzalez Arnao *et al.*, 1993a). During a large scale experiment including 14 different commercial varieties of sugarcane, survival of control apices ranged between 50 and 100%, with an average rate of 76% (Table 1). Survival of cryopreserved apices varied between 14 and 91%, with an average rate of 61%.

An histological study performed during the whole cryopreservation process revealed that cells were slightly harmed during pregrowth, desiccation and freezing (Gonzalez Arnao *et al.*, 1993b). Direct regrowth of cryopreserved apices occurred within three days after thawing and whole plants could be regenerated within 5 weeks.

Trueness to type of plants regenerated from cryopreserved apices of the variety Co6415 was first tested using two isozyme systems, without revealing any modification (Paulet *et al.*, 1994). Further investigation with Restriction Fragment Length Polymorphism (RFLP) involved a set of

17 nuclear probes covering all known linkage groups as well as 3 mitochondrial probes in combination with two restriction enzymes, representing more than 200 bands. Several molecular types were uncovered among plants regenerated both from control and cryopreserved apices which were identical in both materials. Variations were therefore not due to freezing but were instead pre-existing among the *in vitro* plants used in the experiments.

TABLE 1. Survival rates of control (-LN) and cryopreserved (+LN) apices of different sugarcane varieties.

		Survival (%)	
Origin	Variety	-LN	+LN
CIRAD	CO 6415	80	64
	B 69566	100	91
	CP 681026	100	64
	Q 90	100	82
	CO 740	80	38
	MY 5514	83	75
	JAC 5118	50	38
CNIC	C 26670	100	90
	C 8751	90	70
	B 34104	50	67
	B 4362	60	17
	POJ 2878	56	60
	CP 70-1133	50	80
	Ja 60-5	70	14
Average		76	61

CRYOPRESERVATION OF CASSAVA APICES

Apices of one genotype of cassava were submitted to various pregrowth durations in liquid media containing 0.5 to 0.75 M sucrose, followed by 2- to 6-hour desiccation periods and rapid immersion in liquid nitrogen. Survival after cryopreservation could be achieved in various conditions (Table 2). Optimal results (60% survival) were noted after 24 hours of pregrowth with 0.75 M sucrose followed by 6 hours of desiccation and 72 hours of pregrowth with 0.5 M sucrose followed by 6 hours of desiccation. These survival rates may be improved by increasing the desiccation period. However, callusing was noted on the recovery medium employed, thus reducing the final number of apices which could regenerate whole plantlets. Recent experiments (Chabrillange, unpublished results) indicated that transferring the cryopreserved apices on standard (hormone-free) medium 3 weeks after thawing significantly reduced callusing, thus increasing the number of apices developing directly into whole *in vitro* plantlets. Similar survival rates could be obtained with two additional cassava genotypes.

TABLE 2. Effect of pregrowth duration, sucrose concentration in the pregrowth medium and of desiccation duration on the survival (%) of cryopreserved cassava apices.

		Pregrowth duration (days)								
		1			3			5		
Desiccation (h)		2	4	6	2	4	6	2	4	6
Sucrose	0.5	0	7	0	0	13	60	0	0	20
(M)	0.75	0	33	60	0	7	27	0	7	27
	1	0	13	7	0	7	47	0	7	20

CRYOPRESERVATION OF COFFEE APICES

Increasing the pregrowth duration in medium with 0.75 M sucrose improved the desiccation tolerance of control apices of *C. sessiliflora* (Table 3). Survival of cryopreserved apices could be achieved after several pretreatment and desiccation durations.

TABLE 3: Effect of pregrowth duration in liquid medium with 0.75 M sucrose and of desiccation duration on the survival rate (%) of control (-LN) and cryopreserved (+LN) apices of *C. sessiliflora.*

	Pregrowth duration (days)							
	3		5		7		10	
Desiccation (h)	-LN	+LN	-LN	+LN	-LN	+LN	-LN	+LN
0	100	0	100	0	100	0	100	0
3	43	0	57	0	86	0	100	13
4.5	43	0	57	38	57	0	100	13
6	57	13	57	0	57	12	71	25

In the case of *C. racemosa*, survival of control apices decreased more rapidly if they had been pretreated directly for 3 days with 1 M sucrose (Table 4). However, survival of cryopreserved apices was noted only if pregrowth consisted of a daily increase in sucrose concentration from 0.5 to 1 M. The maximal survival rates obtained after cryopreservation were similar with both species (37-38 %).

An histo-cytological study performed during the cryopreservation process revealed that cells of some apices were severely damaged during desiccation and formed osmiophilic granules on the surface of the plasmalemma. In other apices, cells were less damaged and formed exocytosis vesicles in the periphery of the plasma membrane. Growth recovery of surviving cryopreserved apices originated from the most superficial cell layers of the meristematic dome only. It took place in the form of direct development of foliar primordia and/or callusing.

TABLE 4: Effect of pregrowth conditions (progressive increase of sucrose concentration (0.5 to 1 M) or direct pregrowth in 1 M sucrose) and of desiccation duration on the survival rate (%) of control (-LN) and cryopreserved (+LN) apices of *C. racemosa.*

		Progressive pregrowth		Direct pregrowth	
		-LN	+LN	-LN	+LN
Desiccation	0	100	0	60	0
(h)	3	43	0	29	0
	4.5	29	25	14	0
	6	29	27	0	0

Discussion

The results presented in this study demonstrate that encapsulation-dehydration can be employed for cryopreserving apices of some tropical species. The large-scale application of this technique can already be foreseen for sugarcane. Indeed, the average survival rate obtained with 14 different varieties was high (61%) and regeneration of plants from cryopreserved apices was direct and rapid. Moreover, the variations observed with this species during the assessment of genetic stability using RFLPs were not due to freezing but, instead, were preexisting among the *in vitro* plantlets which were used as starting material. They occurred either during bud culture and subsequent *in vitro* culture or were present among the plants which were initially used for bud culture. This latter hypothesis, which would imply that conventional propagation allows genetic drift, is being investigated.

Additional work is still needed with cassava and coffee. In the case of cassava, the survival rates obtained with three different genotypes were satisfactory. Improvements should mainly consist of modifying the recovery conditions in order to reduce callusing and stimulate direct regrowth of cryopreserved apices. With coffee, refinements of various steps of the cryopreservation process, including pregrowth, desiccation and recovery conditions, should give improved results.

Acknowledgements

M.T. Gonzalez Arnao gratefully acknowledges the support of FAO (project TCP/CUB/0056) and of IPGRI (special project 93/104).

References

Benson, E. E., Chabrillange, N. and Engelmann, F. (1992) "A comparison of cryopreservation methods for the long-term *in vitro* conservation of cassava", in Proc. SLTB Autumn Meeting, Stirling, U.K., 8-9/09/92.

Bertrand-Desbrunais, A. (1991) "La conservation *in vitro* des ressources génétiques des caféiers." PhD Thesis, Univ. Paris 6, 260 p.

Brizard, J.P., Noirot, M. and Engelmann, F. (1991) "Conservation d'une vitrothèque de manioc (*Manihot* sp.) en conditions de vie ralentie", Rapport de Fin d'Etude, ORSTOM, 21p.

Dereuddre, J. (1992) "Cryopreservation of *in vitro* cultures of plant cells and organs by vitrification and dehydration", in Dattée, Y., Dumas, C. and Gallais, A. (eds.), Reproductive Biology and Plant Breeding, Springer Verlag, Berlin, pp. 291-300.

Engelmann, F. (1991) "*In vitro* conservation of tropical plant germplasm - a review", Euphytica, 57, 227-243.

Gonzalez-Arnao, M. T., Engelmann, F., Urra, C. and Lynch P. (1993a) "Crioconservacion de meristemos apicales de plantas *in vitro* de caña de azucar mediante el metodo de encapsulacion/deshidratacion", Biotecnologia Aplicada, 10, 225-228.

Gonzalez-Arnao, M. T., Engelmann, F., Huet, C. and Urra, C. (1993b) "Cryopreservation of encapsulated apices of sugarcane: effect of freezing rate and histology", Cryo-Lett. 14, 303-308.

Mari, S., Engelmann, F., Chabrillange, N. and Charrier, A. (1993) "Cryopreservation of apices of *C. racemosa* using encapsulation/dehydration", in Proc. 12th Coll. ASIC, Montpellier, 6-11/06/93, pp. 674-676.

Paulet, F., Engelmann, F. and Glaszmann, J. C. (1993) "Cryopreservation of apices of *in vitro* plantlets of sugarcane (*Saccharum sp.*) hybrids using encapsulation/dehydration, Plant Cell. Rep., 12, 525-529.

Paulet, F., Acquaviva, C., Eksomtramage, T., Lu, Y.H., D'Hont, A. and Glaszmann, J. C. (1991) "La Vitrothèque ou conservation d'une collection de canne à sucre *in vitro*", in Rencontre Internationale en Langue Française sur la Canne à Sucre. 10-15 juin 1991, pp. 47-52.

Paulet, F., Eksomtramage, T., D'Hont, A., Guiderdoni, E., Glaszmann, J. C. and Engelmann, F. (1994) "Callus and apex cryopreservation for sugarcane germplasm management", in Proc. ISSCT Sugarcane Breeding Workshop "Germplasm Preservation - Accomplishments and Prospects", Montpellier, France, 7-11 March, 1994.

Withers, L. A. (1987) "Long-term preservation of plant cells, tissues and organs", Oxford Survey Plant Molec. Cell Biol., 4, 221-272.

APPLICATION OF BIOTECHNOLOGY TO *CARICA PAPAYA* AND RELATED SPECIES

R.A. DREW[1], J.N. VOGLER[1], P.M. MAGDALITA[2], R.E. MAHON[3] AND D.M. PERSLEY[4]
[1]Redlands Research Station, PO Box 327, Cleveland, Q 4163; [2]Department of Agriculture, University of Queensland, St Lucia, Q 4067 and Institute of Plant Breeding, University of the Philippines at Los Banos, Laguna, 4031; [3]Centre for Molecular Biotechnology, Queensland University of Technology, Brisbane, Q 4000; [4]Department of Primary Industries, 80 Meiers Road, Indooroopilly, Q 4068.

ABSTRACT. An efficient protocol has been developed for clonal multiplication of papaya based on rooted microcuttings from nodes of apically dominant plants *in vitro*. Other *Carica* species *cauliflora, parviflora, pubescens* and *goudotiana* have been micropropagated using this system. Four thousand tissue-cultured plants of one papaya clone have been established in the field with no apparent off-types. Plants established from adult tissue exhibited reduced juvenility compared with seedlings. Immature embryos of crosses between *C. papaya* and *C. cauliflora* have been rescued and grown *in vitro*. Three hundred and fifty interspecific hybrids have been established in pots and distinguished from parents by morphology and RAPD markers. Hybrids are being screened for resistance to papaya ringspot virus type P (PRSV-P) by sap inoculation followed by ELISA serological assay. Techniques are being developed for production of transgenic papaya plants resistant to PRSV-P. Transformation systems being investigated are microprojectile bombardment and the use of *Agrobacterium tumefaciens*. Plantlets are being regenerated via embryogenesis from immature embryos rescued from seeds, and from somatic embryos *in vitro*.

1. Introduction

The genus *Carica* originated in Central America and contains about 40 species (Purseglove (1968)), of which papaya is the only species of economic importance. *Papaya* is grown in tropical and sub-tropical regions as a fruit crop and for production of the enzyme, papain. Although hermaphrodite varieties are common in tropical countries, the papaya industry in Australia is based on seedlings of dioecious genotypes which exhibit considerable variation in agronomic traits and in disease susceptibility (Drew and Smith (1986)). The most serious problem facing the papaya industry worldwide is papaya ringspot virus (PRSV-P) (Litz (1984), Gonsalves (1994)), which continues to spread and decimate production in many countries. PRSV-P was first detected in papaya plantings in Australia in 1991 (Thomas and Dodman (1993)). Although the virus has a restricted distribution, its demonstrated ability to spread rapidly in other countries means that it poses a serious threat to the Australian industry.

This paper summarises our results on micropropagation (which should expedite the use of superior clones), field evaluation of tissue-cultured clones, and on procedures aimed at

M. Terzi et al. (eds.), Current Issues in Plant Molecular and Cellular Biology, 321-326.

developing PRSV resistance including interspecific hybridisation, embryo rescue, embryogenesis and transformation.

2. Micropropagation of *Carica papaya*

A number of problems have limited the application of *in vitro* techniques to commercial production. These include high levels of endogenous bacterial contamination in tissue cultures (Litz and Conover (1978), Drew (1988)), the difficulty in culturing mature tissue compared to juvenile tissue (Drew and Smith (1986), Rajeevan and Pandey (1986)), loss of viability when shoot cultures are repeatedly subcultured on proliferation media (Litz and Conover (1981), Rajeevan and Pandey (1986)), failure to consistently achieve high rooting percentages and high quality root systems (Drew and Miller (1989)) and problems associated with acclimatisation (Singh and Pandey (1988)).

Protocols have been developed for micropropagation of papaya and their efficacy demonstrated by clonal production in large scale field plantings on commercial plantations. Bacteria-free cultures were established *in vitro* from small bud explants following pretreatment of trees in the field as described by (Reuveni and Shlesinger (1990)) and then of cuttings in a glasshouse (Drew (1988)). Mature tissue from selected trees was established *in vitro* by alternately culturing bud explants on a roller drum in solution culture containing BAP and NAA, and on hormone-free agar medium (Drew (1988)). Apically dominant plantlets were produced when these shoots were rooted *in vitro*. Selected clones have been maintained *in vitro* for 10 years when multiplied by a system based on production of micro-cuttings from axillary buds of nodal segments from apically dominant plants (Drew (1992a)). A reliable procedure for production of high quality adventitious root systems on >90% of micro-cuttings was developed by minimising exposure to IBA (three days) either by subsequent transfer to hormone-free medium or by photo-oxidation of exogenous IBA by riboflavin (Drew *et al.* (1991), Drew *et al.* (1993)). One hundred percent of plantlets were established in soil either by controlled reduction of relative humidity in a natural light cabinet after deflasking (Drew (1988)) or by use of microporous membrane lids as described by (Kosai *et al.* (1990)) for two weeks prior to deflasking. The use of paclobutrazol as a foliage spray after deflasking also aided acclimatisation, however treated plants showed marked growth reduction in the field and many trees collapsed after five months when their reduced root systems could not support the weight of developing fruit.

A further advantage of this micropropagation system is that cultures of apically dominant rooted plants can be held for prolonged periods under normal incubation conditions at 25°C without transfer. Generally it is necessary to transfer papaya cultures every three to four weeks to maintain active growth (Drew 1988)). Cultures have been stored for four months on media containing 2% sucrose and for 12 months on media containing 1% fructose before sub-culture into nodal sections with viable axillary buds (Drew (1992a)). Fructose is rarely used in tissue culture media because it produces 5-hydroxymethyl-2-furaldehyde (HMF) when autoclaved. Although HMF is toxic to plant cultures, when 1% fructose was autoclaved at 121°C for 15 minutes <10 μM HMF was produced. HMF concentrations of 100-1000 μM were required to reduce growth of papaya shoots *in vitro*.

3. Micropropagation of Other *Carica* Species

The micropropagation system based on rooted micro-cuttings has been applied successfully to other *Carica* species: *cauliflora, parviflora, goudotiana* and *pubescens*. Of these *pubescens* has been the most recalcitrant *in vitro*. *Cauliflora* and *goudotiana* grow rapidly *in vitro* and micropropagated plants have been established in potting mix in a glasshouse.

4. Field Evaluation of Micropropagated Plants

When established in the field tissue-cultured plants derived from adult tissue displayed a reduced juvenile phase when compared to seedlings as evidenced by a reduced tree height, increased stem circumference, lower height of first flower, reduced time to harvest and higher fruit numbers per metre of stem (Drew and Vogler (1993)). These differences did not occur with clones established from juvenile tissue. In commercial plantings in sub-tropical regions, reduced internode length and thus reduced height to first flower is usually achieved with seedlings by planting them in autumn, so that the juvenile phase coincides with a period of reduced growth during cooler months. We have shown in these environments, that tissue-cultured clones from mature tissue can be planted in spring, reducing time to first harvest and thus production costs (Drew and Vogler (1993)). Further, the use of tissue-cultured plants in dioecious plantings eliminated the need for the commercial practice of overplanting to achieve the desired sex ratio. This also overcame the competitive effect of multiple plantings in terms of increased internode length and thus fruit height. Consequently, micropropagated plants have been characterised by stronger root systems and have established more quickly than seedlings (Drew and Vogler (1993)).

The first large scale planting of a tissue-cultured clone (500 plants) produced four dwarf off-types (Drew and Vogler (1993)). In that experiment, initial bud explants were numbered and used to distinguish sub-clones. All plants in one sub-clone were dwarf. Subsequently, 4 000 plants of one clone have been established on a grower's property on three planting dates over an 18 month period. The first 2 500 of these have now reached the fruiting stage, and no off-types are apparent.

5. Interspecific Hybridisation and Embryo Rescue

The principal aim of interspecific hybridisation in *Carica* has been to transfer PRSV resistance to *C. papaya* from related species. Early attempts failed because of embryo abortion within the seeds (Jimenez and Horovitz (1958), Mekako and Nakasone (1975)). In more recent attempts, embryo rescue has been employed (Manshardt and Wenslaff (1989), Rojkind *et al.* (1982)) however regeneration of plants is difficult, F_1 hybrids are generally infertile and no resistant cultivars have been produced.

In 1993, we attempted hybridisation between *C. cauliflora* and 15 *papaya* genotypes collected from six countries. Embryos occurred in only three local genotypes and were rescued 90-120 days after pollination. One genotype crossed more readily with *C. cauliflora* and has been micropropagated *in vitro* (Clone 2.001). Subsequently 2 099 (single and multiple) embryos were rescued when 43 736 seeds were dissected from 338 fruit of this cross. Single embryos were germinated and single plantlets produced. Plants were recovered from multiple embryos via callus and embryogenesis. Three hundred and fifty putative hybrids were established in pots in a glasshouse however survival rate was low due to hybrid breakdown, and 100 are surviving after six months.

Morphological characterisation of the putative hybrids at flowering showed leaf venation pattern, number of main leaf veins, shape of leaves, leaf tip serration and stem diameter to be intermediate between the two parents. General growth and vigour, and flowering ability were greatly inferior to the parents. Hybridity was also confirmed by randomly amplified polymorphic DNA technology (RAPD). Sixty-seven out of 100 hybrids tested were distinguished by intense double bands comprising an 850 base pairs fragment from *C. papaya* and an 800 base pairs fragment from *C. cauliflora*. A range of primers will be trialled to confirm the status of other putative hybrids.

Interspecific hybrids are being screened for resistance to PRSV-P by manual inoculation with Australian isolates of the virus. Inoculated plants are then assayed by ELISA serological tests, electron microscopy and back-inoculation of leaf samples to *Cucurbita pepo* which is highly susceptible to Australian PRSV. We also intend to screen plants in the field under conditions of high virus inoculum pressure. In work to date, PRSV has not been detected in *C. cauliflora* or in the interspecific hybrids. Inoculated standard cultivars have developed typical symptoms of PRSV in these tests.

In the current season, we are attempting further hybridisation between a range of *C. papaya* genotypes and *C. cauliflora, C. parviflora* and *C. goudotiana.* Embryos from three crosses (between the related species and *C. papaya* clone 2.001) have been rescued *in vitro*. *C. parviflora* x *C. papaya* crosses produced more embryos than other crosses and resultant embryos are growing more rapidly *in vitro* than those of other hybrids.

6. Transformation

A reliable procedure for transformation requires both high frequency of gene transfer and a highly efficient system for regeneration of plants from transformed cells. Transgenic papaya plants have been produced by both microprojectile bombardment (Fitch *et al.* (1990), Fitch *et al.* (1992)) and *Agrobacterium*-mediated transfer (Fitch *et al.* (1993)) via embryogenesis from transformed zygotic embryos. Plantlet production required culture for six to eight months prior to transformation (Fitch *et al.* (1993)) and six to nine months subsequently (Fitch *et al.* (1990), Fitch *et al.* (1993)). In sub-tropical regions this system is also dependent on seasonal availability of zygotic embryos.

In our laboratory, somatic embryos have been produced from immature zygotic embryos (90-120 days after pollination) after culture for two to three months in solution culture on an orbital shaker. Zygotic embryos were cultured initially on ½MS medium plus 2 μM BAP, 0.5 μM NAA and 400 μM adenine sulphate. Embryogenic callus was then transferred to solution culture containing ½MS medium plus 0.5 μM BAP and 0.05 μM NAA. Rapid proliferation of secondary embryos occurred in two to four weeks when single embryos from solution culture were transferred to agar medium containing ½MS medium plus 0.5 μM BAP and 0.05 μM NAA. These embryos can again be separated in three days on an orbital shaker and then remultiplied, or germinated on agar-based ½MS medium. This rapid system of secondary embryogenesis is currently showing potential for transformation experiments.

Techniques are being developed to produce transgenic papaya plants, resistant to the Australian PRSV-P strain. The transformation systems being tested are microprojectile bombardment and the use of *Agrobacterium tumefaciens*. Immature zygotic embryos (rescued 90-100 days after pollination) and somatic embryos from the secondary embryogenesis system are being used as target tissues for the microprojectile system. Somatic embryos are being used for transformation with *Agrobacterium tumefaciens*. Initial experiments are being carried out with the GUS reporter gene and the NPTII (Kanamycin resistance) gene. High levels of transient GUS expression have been observed in both tissue types.

Acknowledgments

The authors wish to acknowledge the contribution of Andrew Ballin of Queensland Department of Primary Industries for his work on embryogenesis. They also gratefully acknowledge financial support from the Queensland Papaw Industry via the Queensland Fruit and Vegetable Growers, the Horticultural Research and Development Corporation and the Australian Centre for International Agricultural Research.

References

Drew, R.A. (1988) 'Rapid clonal propagation of papaya *in vitro* from mature field grown trees', HortScience 23, 609-611.

Drew, R.A. (1992a) 'Improved techniques for *in vitro* propagation and germplasm storage of papaya', HortScience 27(10), 1122-1124.

Drew, R.A. (1992b) 'Determination of breakdown products from autoclaved fructose', in Tissue Culture and Field Evaluation of Papaya (*Carica papaya* L.), PhD Thesis, Murdoch Univ., Murdoch, Western Australia, pp 83-109.

Drew, R.A. and Smith, N.G. (1986) 'Growth of apical and lateral buds of papaw (*Carica papaya* L.) as affected by nutritional and hormonal factors', J. Hort. Sci. 61, 535-543.

Drew, R.A., McComb, J.A. and Considine, J.A. (1993) 'Rhizogenesis and root growth of *Carica papaya* L. *in vitro* in relation to auxin sensitive phases and use of riboflavin', Plant Cell, Tiss. Org. Cult. 33, 1-7.

Drew, R.A. and Miller, R.M. (1984) 'Nutritional and cultural factors affecting rooting of papaya (*Carica papaya* L.) *in vitro*', J. Hort. Sci. 64(6), 767-773.

Drew, R.A., Simpson, B.W. and Osborne, W.J. (1991) 'Degradation of exogenous indole-3-butyric acid and riboflavin and their influence on rooting response of papaya *in vitro*', Plant Cell, Tiss. Org. Cult. 26, 29-34.

Drew, R.A. and Smith, N.G. (1986) 'Growth of apical and lateral buds of papaw (*Carica papaya* L.) as affected by nutritional and hormonal factors', J. Hort. Sci. 61(4), 535-543.

Drew, R.A. and Vogler, J.N. (1993) 'Field evaluation of tissue-cultured papaw clones in Queensland', Aust. J. Exp. Ag. 33, 475-479.

Fitch, M.M., Manshardt, R.M., Gonsalves, D., Slightom, J.L. and Sanford, J.C. (1990) Stable transformation of papaya via microprojectile bombardment', Plant Cell Rep. 9, 189-194.

Fitch, M.M., Manshardt, R.M., Gonsalves, D., Slightom, J.L. and Sanford, J.C. (1992) 'Virus resistant papaya plants derived from tissues bombarded with the coat protein gene of papaya ringspot virus', Biotech. 10, 1466-1472.

Fitch, M.M., Manshardt, R.M., Gonsalves, D. and Slightom, J.L. (1993) 'Transgenic papaya plants from Agrobacterium-mediated transformation of somatic embryos', Plant Cell Rep. 12, 245-249.

Gonsalves, D. (1994) 'Papaya ringspot' in Compendium of Tropical Fruit Diseases, APS Press, St Paul, pp 67-68.

Jimenez, H. and Horovitz, S. (1958) 'Cruzabilidad entre especies de *Carica*', Agron. Trop. 7, 207-215.

Kozai, T., Lee, H. and Hayashi, M. (1990) 'Photoautotrophic micropropagation of *Rosa* plantlets under CO_2 enriched and high photosynthetic photon flux conditions', Abstracts of 7th Int. Cong. Plant Tiss. Cell Cult., Amsterdam, A3-112.

Litz, R.E. (1984) 'Papaya', in D.A. Evans, W.R. Sharp, P.V. Ammirato, Y. Yamada (eds), Handbook of Plant Cell Culture, Vol 2, Macmillan, New York, pp 349-368.

Litz, R.E. and Conover, R.A. (1977) 'Tissue culture propagation of papaya', Proc. Fla. State Hort. Soc. 90, 245-246.

Litz, R.E. and Conover, R.A. (1981) 'Effect of sex type, season, and other factors on *in vitro* establishment of *Carica papaya* L. explants', J. Am. Soc. Hort. Sci. 106, 792-794.

Manshardt, R.M. and Wenslaff, T.F. (1989) 'Zygotic polyembryony in interspecific hybrids of *Carica papaya* and *C. cauliflora*', J. Am. Soc. Hort. Sci. 114, 684-689.

Mekako, H.U. and Nakasone, H.Y. (1975) 'Interspecific hybridisation among 6 *Carica* species', J. Amer. Soc. Hort. Sci. 100, 237-242.

Purseglove, J.W. (1968) 'Caricaceae' in Tropical Crops, Dicotyledons, Vol 1, Wiley and Sons, New York, pp 45-51.

Rajeevan, M.S. and Pandey, R.M. (1986), 'Lateral bud culture of papaya (*Carica papaya* L.) for clonal propagation', Plant Cell, Tiss. Org. Cult. 61, 181-188.

Reuveni, O. and Shlesinger, D.R. (1990) 'Rapid vegetative propagation of papaya plants by cuttings', Acta Hort. 275, 301-306.

Rojkind, C., Quezada, N. and Gutierrez, G. (1982) 'Embryo culture of *Carica papaya, C. cauliflora* and its hybrids in vitro, in Plant Tissue Culture' (A. Fujiwara, ed.), 673-764, Jpn. Assoc. Plant Tissue Cult., Tokyo.

Singh, S.P. and Pandey, R.M. 'Note on a new device for harden-off of *in vitro* multiplied papaya plants', Indian J. Hort. 45, 271-273.

Thomas, J.E. and Dodman, R.L. (1993) 'The first record of papaya ringspot virus - type P from Australia', Australasian Plant Pathology 22, 2-7.

TOWARDS THE PRODUCTION OF TRANSGENIC TROPICAL MAIZE GERMPLASM WITH ENHANCED INSECT RESISTANCE

D.A. HOISINGTON and N.E. BOHOROVA

CIMMYT
Apdo. Postal 6-641
06600 Mexico, D.F.
MEXICO

ABSTRACT
It is estimated that over 50% of the area grown to maize in developing countries is adversely affected by insect pests. In an effort to develop enhanced insect resistance in maize germplasm for clients in developing countries, the Applied Biotechnology Unit at CIMMYT initiated a project in 1992 whose main research objectives are: (1) to identify *Bacillus thuringiensis* (Bt) strains that harbor d-endotoxin genes with effective insecticide action against major lepidopteran pests of maize; (2) to isolate the genes encoding specific Bt crystal proteins with high levels of toxicity for specific insect pests; (3) to develop gene constructs that allow the expression of these genes in transgenic maize plants; (4) to develop tissue culture technology for tropical maize; and (4) to introduce and express the Bt gene(s) in appropriate maize germplasm. Results to date have identified potential genes and Bt strains which are active against several of the targeted insect species. Screening of the regeneration potential of a wide range of CIMMYT maize inbreds has identified several inbreds which possess a high level of regeneration potential. Efforts are now underway to develop efficient transformation protocols for several of these inbred lines.

Introduction

Maize (*Zea mays* L.) is the most widely grown of the major crop species, being planted on over 80 million hectares in the developing world alone. This represents roughly 60% of the entire world's maize growing area and accounts for 40% of global maize production. Most of the maize grown in developing countries is in tropical or subtropical environments. The production challenge faced by farmers in these environements must be met while preserving and, if possible, improving the natural

M. Terzi et al. (eds.), Current Issues in Plant Molecular and Cellular Biology, 327-331.

resource base underpinning the world's food systems. Strategies for achieving this goal include as a key element reducing levels of chemical inputs.

CIMMYT data on the major maize production environments of the developing world indicate that in 29 countries with a maize area of 400,000 hectares or more, approximately 30 million, out of a total area of just over 55 million, hectares are seriously affected by insect pests. The effect of insects on production is extremely significant - it is estimated that field and storage insect pests typically cause losses of 10% or more. Lepidopteran insects are among the most important of these pests in both the developed and developing world. Annual losses caused by their infestations are estimated at over 4 million tons in Brasil and 1 million tons in Mexico, with a consequent cash loss for both countries of over US$600 million. Leprdopteran insects also cause extensive crop damage on an estimated 6.2 million hectares of maize in sub-Saharan Africa.

In recent years, the International Maize and Wheat Improvement Center (CIMMYT) has generated effective technology for the mass rearing of several important insect pests and for the artificial infestation of maize with those pests; both important aids in the development of insect resistant maize. Satisfying gains have been made at the Center in the work on multiple resistance to a pest complex that includes various species of borers and fall armyworm, and activities are under way on materials that are resistant to insect pests of stored grain.

Biotechnology provides the opportunity to enhanced host-plant resistance through the introduction of resistance genes from other organisms. If foreign resistance genes were added to resistant germplasm already available from CIMMYT's work, the resulting maize should possess a more powerful and stable insect resistance and would increase the genetic deversity of insect resistance in maize for the tropics and subtropics.

In an effort to develop enhanced insect resistance in tropical and sub-tropical maize for CIMMYT's clients in developing countries, the Applied Biotechnology Unit at CIMMYT initiated a project in 1992 whose main research objectives are: (1) to identify *Bacillus thuringiensis* (Bt) strains that harbor d-endotoxin genes with effective insecticide action against major lepidopteran pests of maize; (2) to isolate the genes encoding specific Bt crystal proteins with high levels of toxicity for specific insect pests; (3) to develop gene constructs that allow the expression of these genes in transgenic maize plants; (4) to develop tissue culture technology for tropical maize; and (4) to introduce and express the Bt gene(s) in appropriate maize germplasm.

To date, the principal activities have focused on screening toxins from isolated Cry genes and novel *Bacillus thuringiensis* (Bt) strains against tropical borers species and defining the culture conditions necessary for regenerating CIMMYT maize germplasm. Results from the tissue culture experiments will be presented in this paper.

Materials and Methods

Plants were grown at CIMMYT's experimental stations in El Batan and/or Tlaltizapan, Mexico. Selfed, whole ears were surface-sterilized with 70% ethanol for 1 min., followed by 20% Clorox containing 10 drops/liter of polyoxyethylene sorbitan monooleate (Tween 80) and rinsed three times with sterile de-ionized water. Immature embryos, 1.0-1.8 mm. in size, were aseptically removed from the kernels and placed flat-side down, scutellum up, on initiation medium. Fifty immature embryos collected from three to six plants of each genotype were placed in plastic Petri dishes (60 x 15 mm), five embryos in each. For callus initiation, the cultures were incubated in the dark at 26°C.

Four media were evaluated for callus formation. Modified N6 basal medium (Chu *et al.* 1975, Bohorova *et al.*, in preparation) was supplemented with 200 mg/l casein hydrolysate, 2.302 mg/l L-proline, 3% sucrose, 2 mg/l 3,6-dichloro-o-anisic acid (Dicamba) [=N6C1] and 15.3 mg/l silver nitrate [=N6C1SN]. For African and highland inbreds, N6C1 and N6C1SN were used, as well as N6 basal medium plus 2 mg/l 2,4-dichlorophenoxyacetic acid (2,4D) [=N6C3] and modified Murashige and Skoog (1962) medium supplemented with 150 mg/l L-asparagine, 6% sucrose, and 2.5 mg/l 2,4D [=MSE3]. The pH of all media was adjusted to 5.7 before sterilization with NaOH, and 0.8% agar (Bacto) was added.

The yellowish, irregularly shaped, compact tissue usually obtained on initiation media was transferred to the maintenance medium, which was the same as the initiation medium. Embryogenic tissue was subcultured every 21 days.

Plants were regenerated from embryogenic calli by transferring tissue to glass vials containing modified N6 hormone-free medium plus 2% sucrose (tropical and sub-tropical inbreds) or basal MS medium with 2% sucrose, 0.5 mg/l indol-3-acetic acid (IAA), 1 mg/l 6-bensylamino purine (6-BAP), and 0.8% agar (Bacto) (African and highland inbreds). Calli were incubated in a growth chamber at 26°C with a 16:8 light:dark photoperiod. After establishment of a good root system on modified MS medium containing 1 mg/l IAA, plantlets were transplanted to Jiffy pots, covered with glass cups and kept in a growth chamber for one week before being moved to the greenhouse.

Regeneration efficiency was determined after 15 days on regeneration media by dividing the number of embryos which had formed at least 3 plantlets from the embryogenic callus by the total number of embryos placed on callus initiation media.

Results

Embryogenic calli producing plant regeneration were obtained using at least one of the media tested from 50% of the tropical and sub-tropical, 35% of the African and 37% of the highland maize inbreds tested. Table 1 lists those inbred lines tested which gave the highest level of regeneration.

In general, we found that the developmental stage of the embryos is critical for successful callus induction. Immature embryos 1.5 mm in length were optimal for embryogenic callus formation. Vigorous callus formed as early as 3-4 days after culture initiation, and small calli were macroscopically visible by the seventh day. Most genotypes evaluated produced calli on at least one of the several media tested. However, differences were observed in the number of embryos producing calli and regenerating plantlets. Compact, nodular and embryogenic tissue was formed on N6C1SN medium, while friable, organized somatic embryos were formed on MSE3 medium. Of total of 23 African inbreds tested, 79% formed embryogenic calli and regenerated plantlets in N6C1SN medium, 75% in N6C3, 71% in N6C1 and 57% in MSE3.

Embryogenic calli showed a remarkable capacity to regenerate plantlets. Vigorous shoot formation was observed in calli previosly cultured in the presence of Dicamba and transferred to a medium containing IAA, 6-BAP, and low sucrose concentration. The highest rate of shoot differentiation and plantlet regeneration was

Table 1. Percent plant regeneration of maize inbreds showing highest level of regeneration potential.

	Culture Media			
Inbred	N6C1	N6C1SN	N6C3	MSE3
Tropical/Sub-tropical				
CML67g	56.0	75.0	n.t.	n.t.
CML67r	77.0	88.1	n.t.	n.t.
CML128	41.2	46.7	n.t.	n.t.
CML139	52.5	87.1	n.t.	n.t.
Ki14	45.8	64.6	n.t.	n.t.
African				
CML198	60.5	50.0	61.3	8.2
CML209	28.6	n.t.	100.0	0.0
CML213	51.8	72.1	58.8	31.0
CML214	85.0	96.7	40.0	95.6
CML215	100.0	100.0	85.7	n.t.
CML216	71.1	100.0	100.0	61.1
CML217	40.6	96.1	80.7	26.7
Highland				
CML241	59.7	72.7	46.1	65.7
CML242	19.4	43.8	15.6	51.0

observed during the first sub-culture period, but was maintained, at a lower rate, over nine subcultures.

The evaluation of the 44 tropical and subtropical lines resulted in 23 inbreds forming embryogenic callus and regenerating plants. The inbreds CML67, CML128, CML139, and CML131 are being used in current experiments for transformation. Most African inbreds produced calli on the four media tested and regenerating plantlets. Four inbreds CML213, CML214, CML215, CML216 showed highest rate of shoot differentiation and plants regeneration. Shoots and plantlets were obtained from compact primary calli on regeneration media from highland inbreds CML241 and CML242 with significantly high frequency.

Discussion

We have shown that under the appropriate culture condition immature embryos of several different maize genotypes can be successfully regenerated from morphogenetic cultures. It is determined that the scutelum region closest to the coleorhizal area from immature embryos contains cells capable of producing somatic embryos under defined experimental conditions. Factors influencing the expression of totipotency in cell cultures are the genotype, plant culture media composition, growth regulator and plant explant.

Callus initiation and maintainence by Dicamba is essential for the induction process. In our experiments, the use of higher levels of calcium chloride and manganese sulfate and lower levels of magnesium sulfate and ferrous sulfate in combination with Dicamba and silver nitrate in the initiation medium gave excellent results for embryogenic callus formation in most lines.

Most of the tropical and subtropical maize inbreds which possessed the highest regeneration potential originated from Antigua germplasm. In the tropical maize investigated by Prioli and Silva (1989), the genes controlling the response for plant regeneration under the described culture conditions were stated to be present in higher frequencies within the Cateto race than the Tuxpeño race. Our results suggest that Antigua germplasm may alse represent a source of gene(s) for regeneration potential, although further experiments would be necessary in order to compare the gene(s) from Antigua and Cateto sources.

This high regeneration potential demonstrated in tropical, sub-tropical, African and highland maize inbreds promises to allow the direct transformation of tropical germplasm. Efforts are now underway in the laboratory to develop transformation protocols for several of these inbredsusing the biolistics gun and BARGUS and anthocyanin gene constructs.

As transformation protocols are established, gene constructs containing Bt genes encoded toxins active against one or more of the targeted insect species will be incorporated into appropriate germplasm for evaluation as an additional source of resistance to these insect pests of tropical maize.

Partially purified spore/crystal complexes are being screened for toxicity against four corn borer species: southwestern corn borer (*Diatreae grandiosella),* sugacane borer *(Diatreae saccharalis),* fall armyworm *(Spodoptera frugiperda*) and corn earworm (*Heliothis zea*). Those toxins demostrating lathality are further tested to accurately determine the LC50.

The genes for toxins found to be extremely toxic to each of the four insects will be isolated, sequenced and made available in proper constructs for transformation and testing in transgenic maize varieties.

References

Bohorova N.E., B.L. Garrido, R.M. Brito, L.D. Huerta, and D.A. Hoisington. High regeneration potential of tropical, subtropical, African, and highland maize inbreds. *in preparation.*

Chu C.C., C.C. Wang, C.S. Sun, C. Hsu, K.C. Yin, C.V. Chu, C.Y. Bi. 1975. Establishment of an efficient medium for anther culture of rice though comparative experiments on the nitrogen sources. Sci. Sinica (Peking). 18:659-668.

Murashige T. and F.Skoog 1962. A revised medium for rapid growth and bioassays with tobacco tissue cultures. Physiol Plant 15:473-497.

Prioli L.M. and W.J. Silva. 1989. Somatic embryogenesis and plant regeneration capacity in tropical maize inbreds. Rev. Brasil. Genet. 12(3):553-566.

DNA-BASED CHROMOSOME MARKERS: BRASSICA CROPS AS CASE STUDY

C.F.QUIROS & J. HU
Department of Vegetable Crops
University of California
Davis, CA 95616

ABSTRACT. The genus *Brassica* includes three cultivated diploid species: *B. nigra* (2n=16, genome B), *B. oleracea* (2n=18, genome C) and *B. rapa* (2n=20, genome A), and three amphidiploid species: *B.carinata* (2n=4x=34, genomes BC), *B. juncea* (2n=4x=36, genomes AB) and *B. napus* (2n=4x=38, genomes AC). Marker development and chromosome mapping have disclosed important information on the structure of these genomes. They are complex genomes, highly duplicated with inter-genomic conservation of linkage blocks which permits intra- and inter-genomic homoeologous recombination. The chromosome maps are also being used for location of useful genes in vegetable and oil seed crops. Among the genes being mapped by various laboratories, disease resistance, cytoplasmic male sterility restorers, and oil seed quality are the most prominent ones. Eventually, linked markers to these genes will be used for marker based selection, gene isolation, molecular characterization and manipulation.

Introduction

Brassica encompass many diverse types of crops, grown as vegetables, fodder or sources of oils and condiments. The diploid species of Brassica range in genomic numbers from n=7 to n=12. The three diploid cultivated species are: 1) *B. nigra* (2n=2x=16, B genome), black mustard 2) *B. oleracea*, (2n=2x=18, C genome), cabbage group, and 3) *B. rapa* (syn. *B. campestris*) (2n=2x=20, A genome) turnips, rapeseed and oriental vegetables (Prakash and Hinata 1980).

Amphidiploidy and aneuploidy have played important roles during the differentiation and evolution of Brassica species (Prakash and Hinata 1980). The genomic relationships among cultivated diploid and derived amphidiploid species were elucidated by U (1935). The three basic diploid cultivated species mentioned above, have originated the three amphidiploids, *B. carinata* (2n=4x=34, genomes BC), *B. juncea* (2n=4x=36, genomes AB) and *B. napus* (2n=4x=n=38, genomes AC).

A common assumption has been that the n=8, 9 and 10 cultivated species have evolved in an ascending dysploid series from a common primitive genome, 'Urgenome' (Haga, 1938). Although there are no known Brassica species in nature with genomes of n=6, Sikka (1940) and Robbelen (1960) postulated that the ancestral genome consisted of six basic chromosomes, which originated by polysomy the n=8, 9 and 10 chromosome genomes. Thus, as a corollary of this hypothesis, the cultivated diploids are considered secondary polyploids (Prakash and Hinata, 1980). The lowest chromosome number observed in the tribe *Brassiceae* is n=7, thus it is regarded ancestral to higher genomic numbers.

M. Terzi et al. (eds.), Current Issues in Plant Molecular and Cellular Biology, 333-339.

Use of markers to study genome structure and evolution

Research from ours and other laboratories based on the development and utilization of molecular markers have demonstrated that indeed the Brassica genomes are highly duplicated (Kianian and Quiros 1992a, Song *et al.* 1988, 1990, Hoeneke and Chyi 1991). It has been estimated that 50% of RFLP loci are duplicated in the diploid species. Duplicated loci observed in the cultivated species is also observed in the n=7 wild species *Diplotaxis erucoides* and *Hirschfeldia incana* (*B. adpressa*), indicating that these duplications have occurred very early in the phylogeny of the Brassica species (Quiros *et al.* 1988). These findings support the hypothesis that Brassica diploids are actually secondary polyploids derived from ancestral genomes of fewer chromosomes.

We constructed F2 linkage maps for *B. oleracea* (Kianian and Quiros 1992a) and *B. nigra* (Truco and Quiros, submitted). The first one consists of 108 molecular markers covering 747 cM in 8 major and 3 minor linkage groups. The second one consists of 124 molecular markers covering 667 cM in 8 major and 3 minor linkage groups. Together with the *B. rapa* map we developed earlier (McGrath and Quiros 1991), we now have maps sharing common RFLP loci for the three cultivated genomes. This is serving us to start initial comparisons of chromosomal structure among genomes. In brief, we found extensive rearrangements for the markers, but it is also possible to detect conserved regions, which is in agreement with the previous observations of Slocum (1989) comparing the genomes A and C. The introduction for the first time the B genome into the comparison discloses similar level of marker arrangement conservation between the B genome with the A and C genomes. This is interesting because studies based on nuclear and chloroplast RFLP indicates that the B genome species are in a separate lineage from the A and C genome species (Warwick and Black 1991, Song *et al.* 1988). Thus this indicates that chromosomal arrangements and DNA sequences have evolved at different rates in the Brassica species.

In all three genomes we found extensive loci duplication, within the reported range of 50% duplicated loci for RFLP markers. However, we found that these duplications are highly rearranged and dispersed in the genomes arguing against polysomy . Thus the genomes of the Brassica species cannot simply be described by formulae including a few founder chromosomes being reiterated twice or trice as described by Sikka (1940) and Robbelen (1960). Similarly, the presence of 8, 9 or 10 chromosomes in the B, C and A genomes, respectively, cannot be explained by the duplication of one or two chromosomes. Furthermore, the DNA content on the three diploid species, regardless of genomic number, is similar (ranging from 470 to 660 Mbp/1C) (Arumuganathan and Earle 1991). Therefore, it is likely that they contain the same amount of genetic information but packed in different number of chromosomes.

We find that the highly duplicated nature of the genomes have important implications in structural changes of the chromosomes. These homoeologous regions under certain situations, such as those imposed by hybridization which results in aneuploidy and amphiploidy, will facilitate intra-genomic and inter-genomic recombination events. This provides extreme plasticity in the Brassica genome which are prone to frequent structural changes. These changes also include deletions which are frequently observed in derived aneuploids, such as the alien addition lines (Hu and Quiros 1991a).

Another important type of chromosomal rearrangements that are frequently found in the genus are reciprocal translocations. These have been reported in wild (Quiros *et al* 1987) as well as cultivated species of Brassica (Snogerup 1980, Kianian and Quiros 1992b). For example, we have found within the *B. oleracea* cytodeme, including cultivated *B. oleracea* and a series of n=9 wild species, at least three karyotypes differing by one or two reciprocal translocations (Kianian & Quiros 1992b).

Translocations often lead to aneuploidy and to chromosomal duplications (Khush and Rick 1967, Gottlieb 1983). Therefore, it is likely that this event has been predominantly involved, in the origin and evolution of the Brassica genomes.

Based on these preliminary findings, we are starting to put together the pieces of the Brassica genome evolution puzzle, which is summarized in Fig 1.

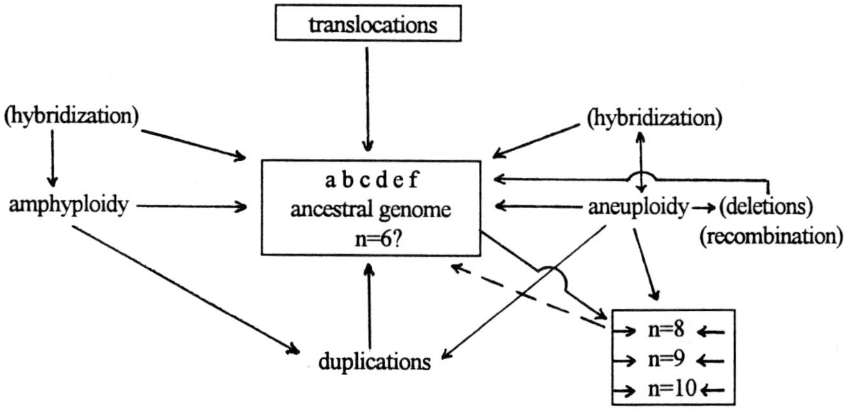

Figure 1. Events molding the brassica genomes.

Assuming that translocations have been one of the major forces in the evolution of the Brassica genomes, these could explain the appearance of initial duplications in the genome. Also, by originating aneuploids through uneven disjunction of tetravalents (3 to 1) further duplications will appear in the resulting hyperploids. Hybridization between these different "genomic types" will further generate additional aneuploids and duplications by homoeologous recombination generating different genomic numbers which could gain or lose chromosomes. Once the genomes have differentiated by isolation, when brought back together spontaneous amphiploidy by hybridization may have taken place, resulting in the existing allotetraploids. By hybridization of these back to their parental species, further aneuploidization will take place, resulting in additional opportunities from chromosomal changes through homoeologous recombination and deletions.

Mapping of useful genes:

A number of useful genes which may have economic impact have been mapped in Brassica crops (Table 1).

For this purpose two main approaches have been used, alien addition lines and F2 linkage analysis. For the latter, quantitative trait loci and bulk segregant analyses (BSA) (Michelmore *et al* 1991) have been added as refinements for assisting the mapping process.

Alien addition lines (aal) have served to locate genes on their respective chromosomes. For example, genes for blackleg resistance, glucosinolate, erucic acid, linolenic/linoleic acid and sinigrin content (Chevre *et al.* 1991, Struss *et al.* 1991). F2 isogenics have been used to construct extensive linkage maps in all three cultivated diploid species and in allotetraploid *B. napus*. These are serving to map useful genes as those listed in Table 1. Bulk segregant analysis have been used in F2 derived populations to tag Ogura cms restorer genes (Landry 1993). Quantitative trait loci analysis, also based on F2 segregating progenies, have defined chromosome segments carrying genes for complex traits such as vernalization response (Osborn 1994).

The application of molecular markers to plant breeding is practically in its beginning. As maps get more saturated, marker based selection approaches will become more feasible to use in the future. This is expected to expedite the development of improved varieties for most crops where this approach is feasible. Furthermore, gene mapping will support gene isolation, characterization and eventual transfer by molecular manipulation.

TABLE 1. Useful genes mapped in Brassica crops

Trait	Species	Method	Ref
Blackleg resistance	*B. napus/B. rapa*	F2 QTL	Ferreira
Blackleg resistance	*B. napus*	aal	Chevre/Struss
Blackleg resistance	*B. oleracea*	F2	Landry
Clubroot	*B. oleracea*	F2 QTL	Landry
Leaf Shape	*B. oleracea*	F2	Landry
Ogura cms restorers	*B. oleracea*	BSA	Landry
Glucosinolate	*B. napus*	aal, F2	Struss/Landry
Maturity date/	*B. oleracea*		
vernalization	*B. rapa*		
requirement	*B. napus*	F2, QTL	Osborn
Sinegrin	*B. napus*	aal	Struss
Self incompatibility	*B. oleracea*	F2	Kianian
Annual habit	*B. oleracea*	F2	Kianian
Glossy leaves	*B. oleracea*	F2	Kianian
Linolenic/	*B. napus*	F2	Hu/Chevre/
linolenic acid			Tanhuanpaa
Eruic acid	*B. napus*	aal	Chevre

References

Arumuganathan, K. and Earle, E.D. (1991) 'Nuclear DNA content of some important plant species', Plant Molec. Rpt. 9:208-219.

Chevre, A.M., This, P., Eber, F., Deschamps, M., Renard, M., Delseny, M.,and Quiros, C. (1991) 'Characterization of disomic addition lines *Brassica napus-nigra* by isozyme, fatty acid and RFLP markers', Theor. Appl. Genet. 81:43-49.

Delourme, R., Eber, F., and M. Renard, G. (1991) 'Radish cytoplasmic male sterility in rapeseed: breeding restorer lines with good female fertility', in Rapeseed in a Changing World, Proc. 8th International Rapeseed Cong. 5:1506-1510. Saskatoon, Saskatchewan, Canada.

Ferreira, M., Teutonico, R. and T., Osborn (1993) 'Mapping trait loci in oilseed', Brassica. VIII Crucifer Genetics Workshop. Saskatoon, Canada

Gottlieb, L.D. (1983) 'Isozyme number and phylogeny', in Proteins and Nucleic Acids in Plant Systematics. Jensen, U. and Fairbrothers, D.E. (eds.). Springer-Verlag, Berlin.

Haga, T. (1938) 'Relationship of genome to secondary pairing in Brassica (a preliminary note) ', Jpn. J. Genet. 13:277-284.

Hoenecke, M. and Chyi, Y.S. (1991) 'Comparison of *Brassica napus* and *B. rapa* genomes based on restriction fragment length polymorphism mapping', in Rapeseed in a Changing World, Proc. 8th International Rapeseed Cong. 4:1102-1107. Saskatoon, Saskatchewan, Canada.

Hosaka, K., Kianian, S.F., McGrath, J.M. and Quiros, C.F. (1990) 'Development and chromosomal localization of genome specific DNA markers of Brassica and the evolution of amphidiploid and n=9 diploid species', Genome 33:131-142.

Hu, J. and Quiros, C.F. (1991a) 'Molecular and cytological evidence of deletions in alien chromosomes for two monosomic addition lines of *Brassica campestris-oleracea*', Theor. Appl. Genet. 81:221-226.

Khush, G.S. and Rick, C.M. (1967) 'Tertiary trisomics: origin, identification, morphology and use in determining position of centromeres and arm location of markers',Can. J. Gen. Cytol. 9:610-631.

Kianian, S.F. and Quiros,C.F. (1992a) 'RFLP map of *B. oleracea* based on four crosses', Theor. Appl. Genet. 84:544-554.

Kianian, S.F. and Quiros, C.F. (1992b) 'Trait inheritance, fertility, and genomic relationships of some n=9 Brassica species', Genetic Resources and Crop Evol. 39:165-175.

Landry, B.S., Hubert, N., Etoh, T., Harada J.J. and Lincoln, S.E. (1991) 'A genetic map of *Brassica napus* based on restriction fragments length polymorphism detected with expressed DNA sequences', Theor. Appl. Genet. 34:543-552.

Landry, B.S., Hubert, N., Crete, R., Chang, M.S., Lincoln, S.E. and Etho, E. (1992) 'A genetic map of *Brassica oleracea* based on RFLP markers detected with expressed DNA sequences and mapping for resistance genes to race 2 of *Plasmodiophora brassicae*', Genome 35:409-420.

Landry, B.S. (1993) 'Towards map-based cloning in Brassica', VIII Crucifer Genetics Workshop, Saskatoon, Canada

McGrath, J.M. and Quiros, C.F. (1991) 'Inheritance of isozyme and RFLP markers in *Brassica campestris* and comparison with *B. oleracea*', Theor. Appl. Genet. 82:668-673.

Michelmore, R.W., Paran, I. and Kesseli, R.V. (1991) 'Identification of markers linked to disease resistance genes by bulk segregant analysis: a rapid method to detect markers in specific genomic regions by using segregating populations', Proc. Natl. Acad. Sci. 88:9828-9832

Prakash, S. and Hinata, K. (1980) 'Taxonomy, cytogenetics and origin of crop Brassicas, a review', Opera. Bot. 55:1-57.

Quiros, C.F., Ochoa, O., Kianian, S.F. and Douches, D.S. (1987) 'Analysis of *the Brassica oleracea* genome by the generation of *B. campestris-oleracea* chromosome addition lines: characterization by isozyme and rDNA genes', Theor. Appl. Genet. 74:758-766.

Quiros, C.F., Ochoa, O. and Douches, D.S. (1988) 'Brassica evolution: exploring the role of x=7 species in hybridization with *B. nigra* and *B. oleracea*', J. Hered. 79:351-358.

Robbelen, G. (1960) 'Beitrage zur Analyse des Brassica-Genoms', Chromosoma 11:205-228.

Sikka, S.M. (1940) 'Cytogenetics of Brassica hybrids and species', J. Genet 40:441-509.

Slocum, M.K. (1989) 'Analyzing the genomic structure of Brassica species using RFLP analysis,' in Helentjaris, T., Burr, B. (eds). Development and application of molecular markers to problems in plant genetics. Cold Spring Harbor Lab. Press, NY.

Snogerup, S. (1980) 'The wild forms of the *Brassica oleracea* group (2n=18) and their possible relations to the cultivated ones', in Brassica crops and wild allies: biology and breeding. Tsonuda, S., Hinata, K. and Gomez-Campo, C. (eds). Japan Sci. Soc. Press.

Song, K.M., Osborn, T.C. and Williams, P.H. (1988) 'Brassica taxonomy based on nuclear restriction fragment length polymorphisms (RFLPs). 1. Genome evolution of diploid and amphidiploid species', Theor. Appl. Genet. 75:784-794.

Song, K., Osborn, T.C. and Williams, P.H. (1990) 'Brassica taxonomy based on nuclear restriction fragment length polymorphisms (RFLPs): 3. Genome relationships in Brassica and related genera and the origin of *B. oleracea* and *B. rapa* (syn. *campestris*) ', Theor. Appl. Genet. 79: 497-506.

Struss, D., Quiros, C.F and Robbelen, G. (1991) 'Construction of different B genome addition lines in *Brassica napus*', in Rapeseed in a Changing World, Proc. 8th International Rapeseed Cong. 2:358-363. Saskatoon, Saskatchewan, Canada.

Osborn, T., Teutonico, R.A., Ferreira, M. and Camargo, L. (1994) 'Mapping and comparing genes for vernalization requirement in Brassica', Plant Genome II, Abs# P160, San Diego, CA

Tanhuanpaa, P.K., Vilkki, H.J., Vilkke, J.P. and Pulli, S.K. (1993) 'Segregation distortion of DNA markers in a microspore derived population, and identification of markers linked to a locus affecting linolenic acid content in oilseed rape (*Brassica napus* L) ', VIII Crucifer Genetics Workshop, Abs#23, Saskatoon, Canada

U.N. (1935) 'Genome-analysis in Brassica with special reference to the experimental formation of *B. napus* and peculiar mode of fertilization', Jpn. J. Genet. 7:389-452.

Warwick, S.I. and Black, L.D. (1991) 'Molecular systematics of Brassica and allied genera (Subtribe Brassicinae, Brassiceae), chloroplast genome and cytogene congruence', Theor. Appl. Genet. 82:81-92.

APPLICATION OF DNA AMPLIFICATION FINGERPRINTING IN THE BREEDING OF PHALAENOPSIS ORCHID

W.H.CHEN, Y.M.FU, R.M.HSIEH, W.T.TSAI, M.S.CHYOU, C.C.WU,
Y.S.LIN
Dept.of Horticulture,Taiwan Sugar Res.Inst.,
Tainan,Taiwan,R.O.C.

ABSTRACT. A method of DNA amplification fingerprinting (DAF) based on the DNA polymerase chain reaction (PCR) has been developed to identify different varieties of Phalaenopsis. This method used DNA extracted from trace amounts of leaf material. Distinguishable DAF patterns among the varieties with similar genetic background were obtainable if a suitable primer was used. These DAF patterns serve as molecular markers to protect patent rights to new Phalaenopsis varieties from Taiwan Sugar Corporation. Crosses were made between white and red floral varieties of Phalaenopsis equestris and Doritis pulcherrima. F_1 and parental plant were screened with 360 primers in PCR to analyze their DAF patterns. Results showed that 8 primers in Phalaenopsis equestris and 7 primers in Doritis pulcherrima produced distinct DNA polymorphism bands in red floral parent and F_1 progeny, but not in white floral parent. These molecular markers will be used for selection of progenies with red flowers at young seedling stage.

1. INTRODUCTION

The morphological characteristics, cytogenetic heredity and isozymic analysis are general used in breeding as genetic markers for detection and selection. However, these methods are limited by environmental effects and diagnosis resolution. Recently, DNA amplification fingerprinting (DAF), has been shown to be more effective in detecting polymorphism and will become a powerful method in plant breeding program (William et al. (1990), Paran et al. (1991), Welsh et al. (1992)).

Phalaenopsis, an indigenous orchid to Taiwan, is one of the most valued floriculture in the world because of its graceful colors and attractive appearance. Taiwan Sugar Research Institute (TSRI) has collected 29 wild species, including 183 clones and 873 superior varieties for the improvement of Phalaenopsis. Cytogentic studies and computer analysis of the parental genetic background have been carried out to understand sexual incompatibility and to design the breeding program. In this report, DAF technique has been used to protect the

M. Terzi et al. (eds.), Current Issues in Plant Molecular and Cellular Biology, 341-346.

patent rights of new varieties released from TSRI and to screen genetic markers related to red floral color at molecular level in the genus of Phalaenopsis and Doritis.

2. MATERIALS AND METHODS

Orchid materials as shown in Fig.1 included 5 genus, 5 species in the genus of Phalaenopsis and 5 clones in a species of Phal.equestris. All plants were cultivated in the greenhouse of Department of Horticulture, TSRI. Crosses were made between white and red floral varieties in Phal. equestris and Dor.pulcherrima in 1990 and F_1 progenies were maintained in TSRI greenhouse.

DNA of each plant was extracted from leaf by using a modified Gawel and Jarret (1991) method. The PCR reaction mixture contained 20 ng DNA, 0.5U Taq polymerase (Boehringer Mannheim), 20 μM dNTP, and 6 pmole primer in a volume of 10 μl. The primers were 10-mers purchased from Operon Technologies (Alameda, California). The amplification was performed in Air Thermo-Cycler (Idaho Technology), programmed for 94-38-72 ℃, 60-7-70 sec, 2 cycle, following 92-40-72 ℃, 2-7-70 sec, 55 cycle. Amplified products were run in a 2% agarose gel (Sigma, A-9539) and stained with ethidium bromide.

3. RESULTS AND DISCUSSION

3.1. DAF Patterns in the Orchids

Twenty random primers were tested to compare the DAF patterns among 5 genus,5 species in same genus and 5 clones in the same species. Polymorphisms were observed among them when a suitable primer was used in PCR reaction (Fig.1). It was also found that the more diversity of relationship there were, the more easy of suitable primer which distinguished the DAF patterns was selected.

Table 1 shows 9 and 8 primers produced considerable polymorphisms among genus of Phalaenopsis, Doritis, Cattleya, Dendrobium and Cymbidium, and among Phalaenopsis species of amabilis, amboinensis, mannii, violacea, and equestris, respectively. However, only 3 primers were able to detect genetic differences among the clones of Phal. equestris.

Distinguishable bands of DAF patterns among varieties with similar genetic background were obtained when a suitable primer was used. Similar methods have been successfully applied to detected polymorphisms in wheat (Devos and Gale (1992)) and banana (Kaemmer et al. (1992)). It is concluded that DAF is a powerful and useful method to find markers to protect patent rights of new varieties in Phalaenopsis.

3.2. Detection of Red Floral Markers

All F_1 plants from crosses between white flower and red flower plant had red flower. Primers were first screened by using a single plant of F_1 progeny. If specific DNA bands were found, further screening was conducted by using three F_1 plants.

Among 360 primers screened by DAF analysis, 41 primers in Phal. equestris and 75 primers in Dor.pulcherrima were found to produce polymorphisms which had special DNA bands in red floral plants of parent and one F_1 progeny to distinguish them from floral white parent. In addition, using three F_1 progenies, 8 primers in Phal.equestris included OPA-20(5'-GTT GCG ATC C-3'), OPC-7(5'-GTC CCG ACG A-3'), OPD-4 (5'-TCT GGT GAG G-3'), OPF-14(5'-TGC TGC AGG T-3'), OPJ-4(5'-CCG AAC ACG G-3'), OPK-18(5'-CCT AGT CGA G-3'), OPN-2(5'-ACC AGG GGCA-3'),OPQ-10(5'-TGT GCC CGA A-3'), and 7 primers in Dor.pulcherrima included OPA-11(5'-CAA TCG CCG T-3'), OPD-1(5'-ACC GCG AAG G-3'), OPJ-10(5'-AAG CCC GAG G-3'), OPJ-17(5'-ACG CCA GTT C-3'), OPM-11(5'-GTC CAC TGT G-3'), OPL-3(5'-CCA GCA GCT T-3'), OPX-12(5'-TCG CCA GCC A-3'), were found to amplify specific DNA bands in both red flower parent and F_1 offspring, but absent in white flower parent. Figs.2 and 3 show the DAF patterns in red and white parents and three F_1 plants in Phal.equestris and Dor. pulcherrima,respectively. F_2 and backcross plants are being raised and will be used as material for DAF analysis to confirm these molecular markers of floral gene in the following year. Furthermore,the breeding model of DAF markers in red flower will be extended to other important characteristics such as fragrance and disease resistance in orchids.

4. ACKNOWLEDGMENTS

This research was supported by NSC 82-0409-B058A-004.

5. REFERENCES

Devos,K.M. and Gale,M.D. (1992) 'The use random amplified polymorphic DNA markers in wheat', Theor.Appl.Genet. 84,567-572.

Gawel,N.J. and Jarret,R.L. (1991) 'A modified CTAB DNA extraction procedure for Musa and Ipomoea', Plant Molecular Biology Report 9, 262-266.

Kaemmer,D. Afza,R. Weising,K. Kahl,G and Novak,F.J. (1992) 'Oligonucleotide and amplification of wild species and cultivars of banana (Musa spp)', Bio/technology 10, 1030-1035.

Paran,I. Kesseli,R. and Michelmore,R. (1991) ' Identification of restriction fragment length polymorphism and random amplification polymorphic DNA markers linked to downy mildew resistance genes in lettuce using near-isogenic line', Genome 34,1021-1027.

Welsh,J. Honeycutt,R.J. McClell,M. and Sobral,B.W.S. (1991)'Parentage determination in maize hybrids using the arbitrarily primed polymerase chain reaction', Theor.Appl.Genet. 82, 437-476.

William,J.G.K. Kubelik,A.R. Livake,K.J. Rafalski,J.A. and Tingey,S. V. (1990) 'DNA polymorphisms amplified by arbitrary primers are useful as genetic markers', Nucleic Acids Research 18, 6532-6535.

Table 1. Primers available to distinguish DAF patterns among 5 genus, 5 species in Phalaenopsis and 5 clones in Phal.equestris listed in Figure 1.

primer	genus	species	clones	primer	genus	species	clones
OPF-1	-*	+**	+	OPF-11	-	-	-
OPF-2	+	+	-	OPF-12	+	-	-
OPF-3	-	-	-	OPF-13	-	-	-
OPF-4	-	+	+	OPF-14	-	-	-
OPF-5	+	-	-	OPF-15	+	+	-
OPF-6	+	-	-	OPF-16	-	+	-
OPF-7	+	-	-	OPF-17	-	+	-
OPF-8	+	-	-	OPF-18	-	+	-
OPF-9	+	-	-	OPF-19	-	-	-
OPF-10	+	+	-	OPF-20	-	-	+

* - :No difference of DAF patterns among the plants tested.
** +:Difference of DAF patterns among the plants tested.

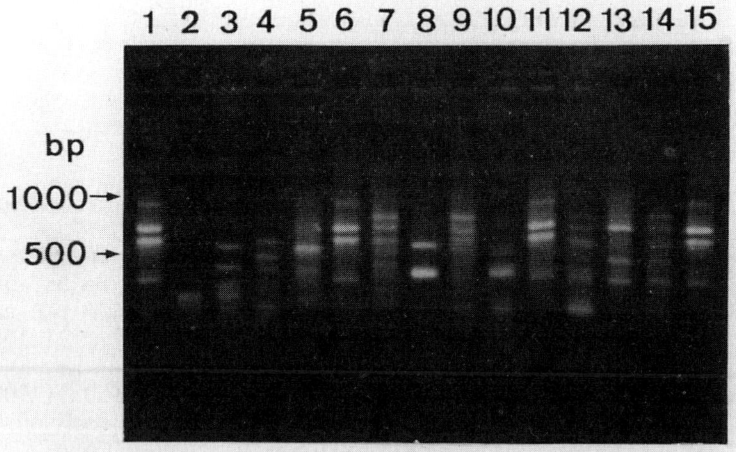

Figure 1. DAF patterns in genus of Phalaenopsis (1), Doritis (2), Cattleya (3), Dendrobium (4), Cymbidium (5); Phalaenopsis of amabilis (7), amboinensis (8), mannii (9), violacea (10); and Phal.equestris of 'W9-9' (1,6,11), 'W9-44' (12), 'W9-49' (13), 'W9-52' (14), 'W9-56' (15) using OPF-10 primer.

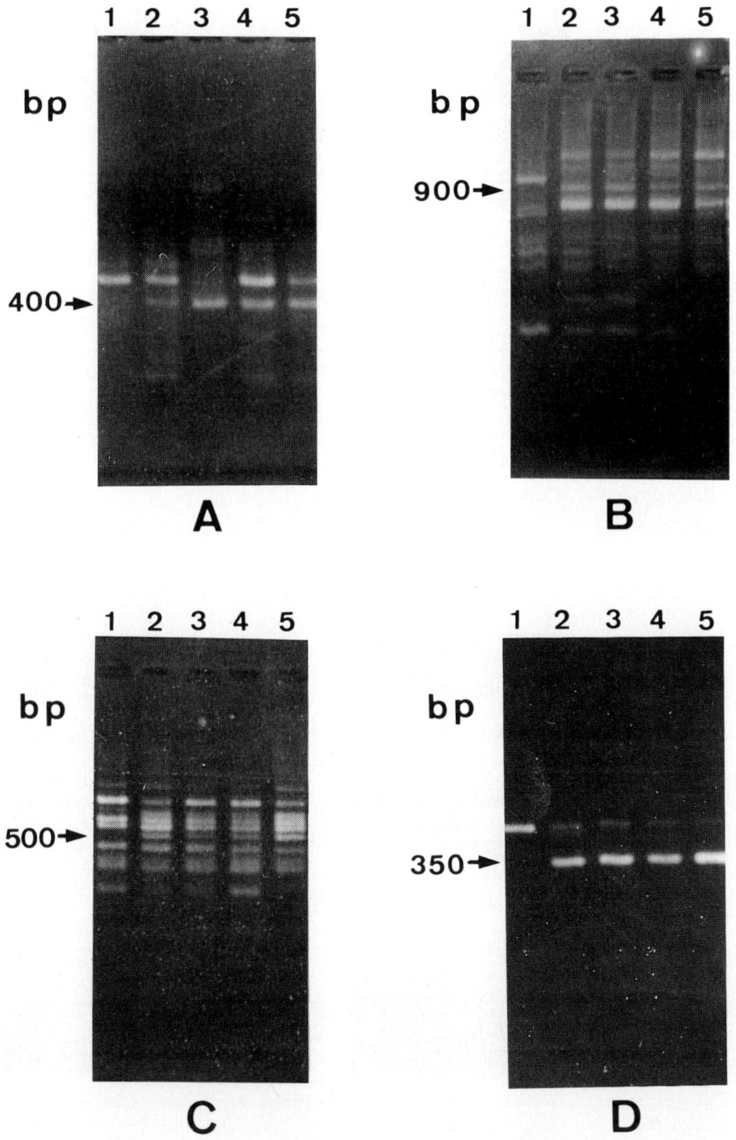

Figure 2. DAF patterns amplified by OPK-18 (A),OPC-7 (B), OPF-14 (C) and OPQ-10 (D) primers in Phal.equestris show that all F_1 plants(2-4) and red floral parent (5) have a 400bp band (A), 900bp band (B), 500 bp band (C), 350bp band(D), respectively, and white floral parent (1) does not have the band.

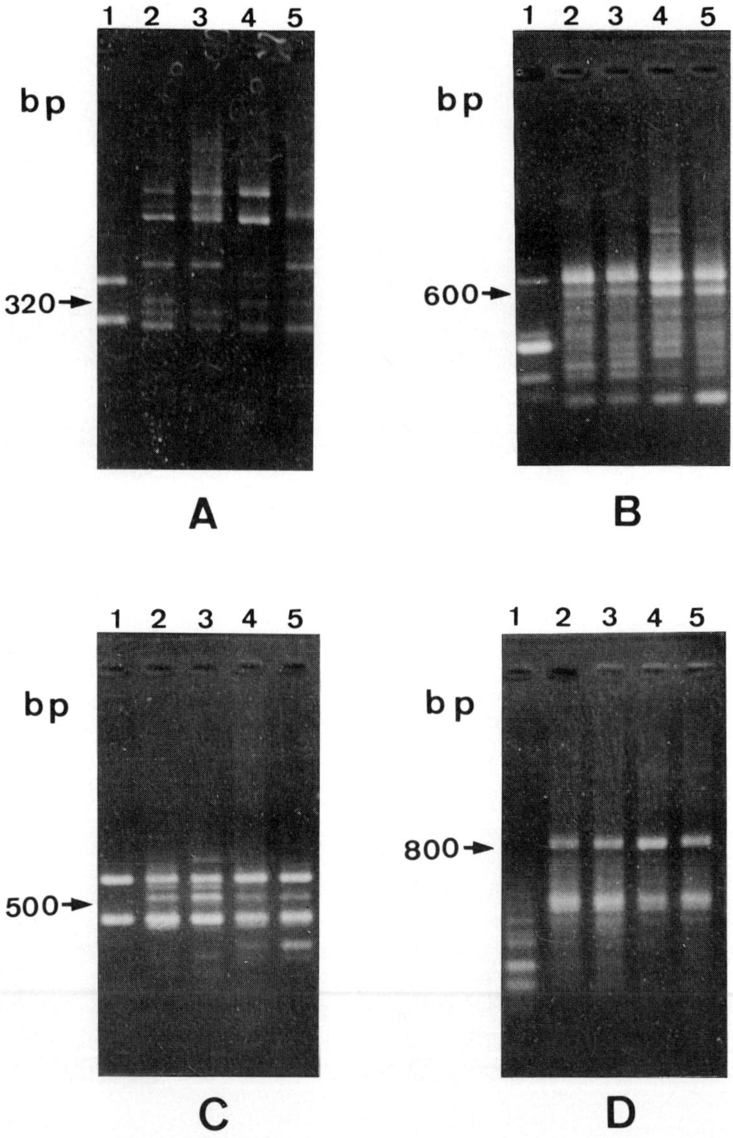

Figure 3. DAF patterns amplified by OPD-1 (A), OPJ-10 (B), OPJ-17 (C) and OPX-12 (D) primers in Dor.pulcherrima show that all F_1 plants (2-4) and red floral parent (5) have a 320bp band (A), 600bp band (B), 500bp band (C), 800bp band (D), respectively, and white floral parent (1) does not have the band.

MAINTENANCE OF AN IN VITRO COLLECTION OF ALLIUM IN THE GATERSLEBEN GENEBANK - PROBLEMS AND USE

E. R. J. KELLER, D.-E. LESEMANN[1], H. I. MAASS, A. MEISTER, H. LUX, I. SCHUBERT
Institute of Plant Genetics and Crop Plant Research, Corrensstraße 3, D-06466 Gatersleben, Germany;
[1] *Institute for Biochemistry and Plant Virology, Federal Biological Research Center for Agriculture and Forestry (BBA), Messeweg 11-12, D-38104 Braunschweig, Germany.*

ABSTRACT. In the Gatersleben Institute, about 3000 accessions of Allium are maintained for germplasm preservation, ethnobotanical and taxonomical research. Viroses accumulate in traditional field cultivation of vegetatively propagated accessions. Therefore, *in vitro* culture had been started to improve preservation. Meristem culture applied for virus elimination yielded a mean of 13.8 % of virus-free plantlets. Viability of the material stored under normal conditions and after cold or heat treatments was characterized by growth tests and ion efflux measurements. Stability of ploidy in haploid clones was checked by flow cytometric screenings. Culture of fertilized ovaries was successfully used for distant hybridization in 19 combinations of the genus *Allium*. Hybrids were verified by flow cytometry, karyotype, isoenzyme, and PCR analysis.

1. Introduction

The genus *Allium* is one of the main research targets of the Taxonomy and Genebank Departments within the Gatersleben Institute. In the last 20 years, a remarkable collection has been brought together comprising about 3000 accessions (Keller et al., 1994b). Respecting wild species (1350 determined, about 400 not yet determined accessions) the largest special *Allium* collection of the world exists in Gatersleben. One third of the accessions (830) has to be propagated vegetatively because of the total lack of seed production (garlic, top onion, haploids, distant hybrids etc.), problems of culture conditions (material with requirements which cannot be fulfilled completely under our conditions) or because the plants are outbreeders either prohibiting self fertilization or suffering of inbreeding depression (such material cannot be propagated identically by seeds). This was the reason for starting *in vitro* culture at the Gatersleben genebank in 1990. Initial material had been derived from investigations on gynogenesis (Keller 1990 a,b). Haploid lines have been considered to be useful as a model for testing genetic stability because of its lower chromosome number and simpler genetic structure.

Genetic stability is one of the critical points within an *in vitro* genebank. The spectrum of characters for assessing possible changes comprises morphological, cytogenetical and biochemical markers. The stability tests have been started with flow cytometrical screenings of the DNA content in haploid clones derived from gynogenesis in onion.

M. Terzi et al. (eds.), Current Issues in Plant Molecular and Cellular Biology, 347-352.

2. Material and Methods

In vitro culture: Primary explants are obtained from basal plates of the bulbs (cloves in garlic), flower heads or inflorescence bulbils. Plantlets are induced on medium BDS (Dunstan and Short 1977) with 1.25 mg/l benzyladenine (BA). When *in vitro* plantlets are established, they are maintained in a culture cycle consisting of two subcultures (1 month each) on medium BDS without hormones; increasing concentrations of sucrose (15 % in onion) for two months to induce bulb formation; cold storage in the dark, at + 3 °C at least for 7 months; 1 week at 10 °C; 1 month on hormone free medium at 25 °C; cutting phase on BDS + BA as above. Meristems with a diameter of approximately 0.3-0.5 mm were obtained from basal plates and inflorescence bulbils. Virus indexing was performed by ELISA and electron microscopy according to Barg et al. (1994) and Keller et al. (1994b).

Viability tests were performed according to the following scheme: thorough cleaning of bulblets or plantlets (washing and removing of dead plant parts) and removing the roots, 1 day later floating of the objects in distilled water for 2 h, conductivity measurements within the bath solution to characterize the ion leakage out of the plant tissue, weighing and measuring plant length, planting the material in vermiculite, cultivating plantlets for 14 days and again weighing and measuring plant length as well as measuring the root length.

Flow cytometric measurements were performed on a Becton Dickinson Flow Cytometer FacSTAR^{PLUS}. Plant material has been obtained during the cutting procedure of the *in vitro* plantlets when they have been transferred to their next subculture. Therefore, only the leaves were available. The material was cut directly in the dye solution (acridine orange or DAPI, 4',6-diamidino-2-phenylindole). Subsequently, it has been pressed through a sieve with meshes of 50 µm. DNA contents of 2000-3000 nuclei per sample were measured.

Distant crosses were performed by hand pollination of previously de-masculated flowers, followed by *in vitro* culture of the ovaries. For verification of the crosses, metaphase plates from root tips were used after staining with acetocarmine. Isoenzyme and RAPD- (random amplified polymorphic DNA-) patterns were used for cross verification as well.

3. Results

3.1. *IN VITRO* CULTURE FOR MAINTENANCE OF CLONAL MATERIAL IN *ALLIUM*

Establishment of *in vitro* clones is affected by factors which are especially important in long term preservation: contamination and vitrification. Explants from bulbs taken from the soil were twice as frequently infested as inflorescence bulbils. 11.5 % of the in vitro plantlets derived from meristems of 7 different clones vitrified up to 9 months after inoculation after having been transferred monthly on new medium BDS. Vitrification depended on the genotypes and varied between 4 and 19 % of the initial explants. The primary multiplication rate was higher for shallots (19.6 after 2 months) than for garlic (6.3 from cloves and 3.0 from aerial bulbils).

For safety maintenance two clonal lots are cultivated independently, each of them consisting of 18 plantlets (two per tube).

Storability of *in vitro* bulblets depends on several factors. After growing for two months on medium with high contents of sucrose, dormancy is broken in cold storage so that plantlets grow and consume the medium. Storability was improved after prolongation of the phase of warm culture on BDS with 15 % sucrose up to 6 months as well as after treatments with ethylene (Keller, 1993; Keller et al. 1994b).

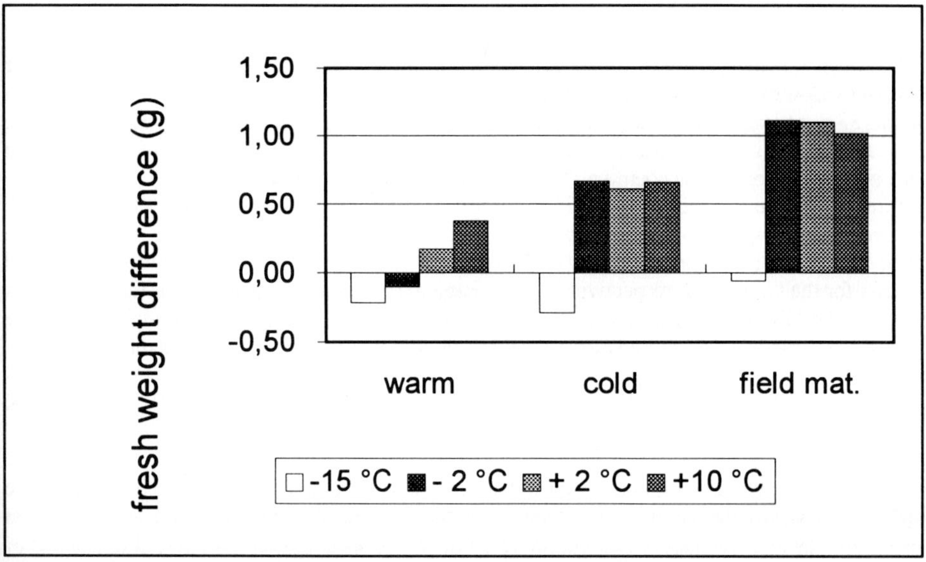

Fig. 1: Onion - cold adaptation of *in vitro* bulblets. Material stored at 25 °C is more sensitive to a 14-days cold treatment (-2 or +2 °C) than material stored at 4 °C. Right: field material for comparison; 20 bulblets per variant.

Viability tests were developed to improve the evaluation of the methods of cultivation and storage (Bonnier et al. 1992; Keller 1992). Clear positive correlations were found among the growth parameters (differences in fresh weight and plant length as well as root length) and negative correlations occurred between these parameters and the ion leakage. A sample of quantified description of viability is given in Fig. 1 for a case of cold adaptation in a clone of *Allium cepa*.

At present, the *in vitro* genebank comprises the following numbers of clones: garlic (72), shallot (12), material from haploid research in onion (129), hybrids from species crosses (77), other cultivated and wild species of *Allium* (178).

3.2. MERISTEM CULTURE FOR VIRUS ELIMINATION

Due to the small size of explants, not all of them survived in the first subculture. In garlic, 82.2 % of explants from inflorescence bulbils and only 75.1 % from cloves survived after 2 weeks. Lower survival rates were found in shallots (57.1 %).

The field material has been indexed to be virus infected to a high degree (Barg et al. 1994; Graichen et al., 1992; Graichen and Keller 1992). 6 different viruses have been detected: SLV (shallot latent virus), GCLV (garlic common latent v.), OYDV (onion yellow dwarf v.), LYSV (leek yellow stripe v.), MbFV (mite-borne filamentous v.) and a not yet identified potyvirus. After meristem culture, 13.8 % of the *in vitro* plantlets could be found to be virus-free. In 134 still infected plantlets 41.8% had only one and 25.4 % only two viruses, which represents a partial elimination success.

3.4. STABILITY ASSESSMENT OF PLOIDY IN HAPLOID CLONES

DNA contents should be one of the parameters indicating genetic (in)stability which can easily be recorded by flow cytometry (Walters 1992).

So far, 33 gynogenetic clones and 21 plants from distant crosses have been evaluated in 597 tests. In initial tests acridine orange proved to be not suitable because of high variation of the histogram peaks although chromosome counting showed clearly that all plants were haploid.

Therefore, for subsequent tests DAPI-staining was used. In 4 haploid lines, 108 records were obtained, 97 % of which gave clear histograms which are exactly one C-level lower than that of the diploid standard (Fig. 2). The two higher peaks represent one ploidy level (the left for the G0/G1 and the other for the G2 phases, respectively). In the haploid histogram, a small peak at the right side is caused by the G2 of diploid nuclei. It is likely due to nuclei near the vascular bundles which have higher ploidy levels also in diploid plants (see the logarithmic part of the standard's diagram).

3.5. *IN VITRO* CULTURE TO SUPPORT DISTANT CROSSES IN GENEBANK ACCESSIONS

In vitro culture is useful also for obtaining hybrids derived from distant crosses. Thus, it was possible to directly use the high degree of genetic diversity within the Gatersleben collection. Hand pollination was performed with subsequent ovary culture in a similar way as published for haploid production of unfertilized ovaries (Keller 1990 a,b). Metaphase plates clearly revealed both parental chromosome complements to be present in cross progenies of *A. cepa* with *A. globosum*, *A. hymenorrhizum*, *A. saxatile*, *A. rubens*, *A. senescens*, and *A. ledebourianum*. Isoenzymes were suitable for hybrid verification in hybrids from crosses of *A. cepa* with *A. carolinianum*, *A. globosum*, *A. hymenorrhizum*, *A. obliquum*, and *A. senescens*. RAPD patterns detected hybrids in crosses of *A. cepa* with *A. globosum*, *A. hymenorrhizum*, *A. obliquum*, and *A. albidum* (Keller et al. 1994a).

All cross progenies tested in the flow cytometrical screening have been diploid: crosses of *A. cepa* with *A. kurssanovii* (2), *A. saxatile* (2), *A. rubens* (2), *A. albidum* (2), *A. schoenoprasum* (1), *A. angulosum* (2), *A. jodanthum* (2), *A. chevsuricum* (2), *A. lineare* (2), *A. obliquum* (4).

4. Discussion

Maintaining an *in vitro* collection implies the application of various methods to a great diversity of material. On the one side, a compromise is to be found for maintaining the material with relatively low numbers of methodical variants. On the other is the great potential to use the various genotypes and culture methods for research and prebreeding as it is shown for the case of distant crosses.

The *in vitro* genebank will be developed into several directions. Storage methods have to be improved (reduction of manipulations necessary for subcultures), vegetatively propagated material has to be further freed of viruses and more samples have to be included, preservation of which will be important because of their special value for breeding, any unique characters or because they are especially endangered.

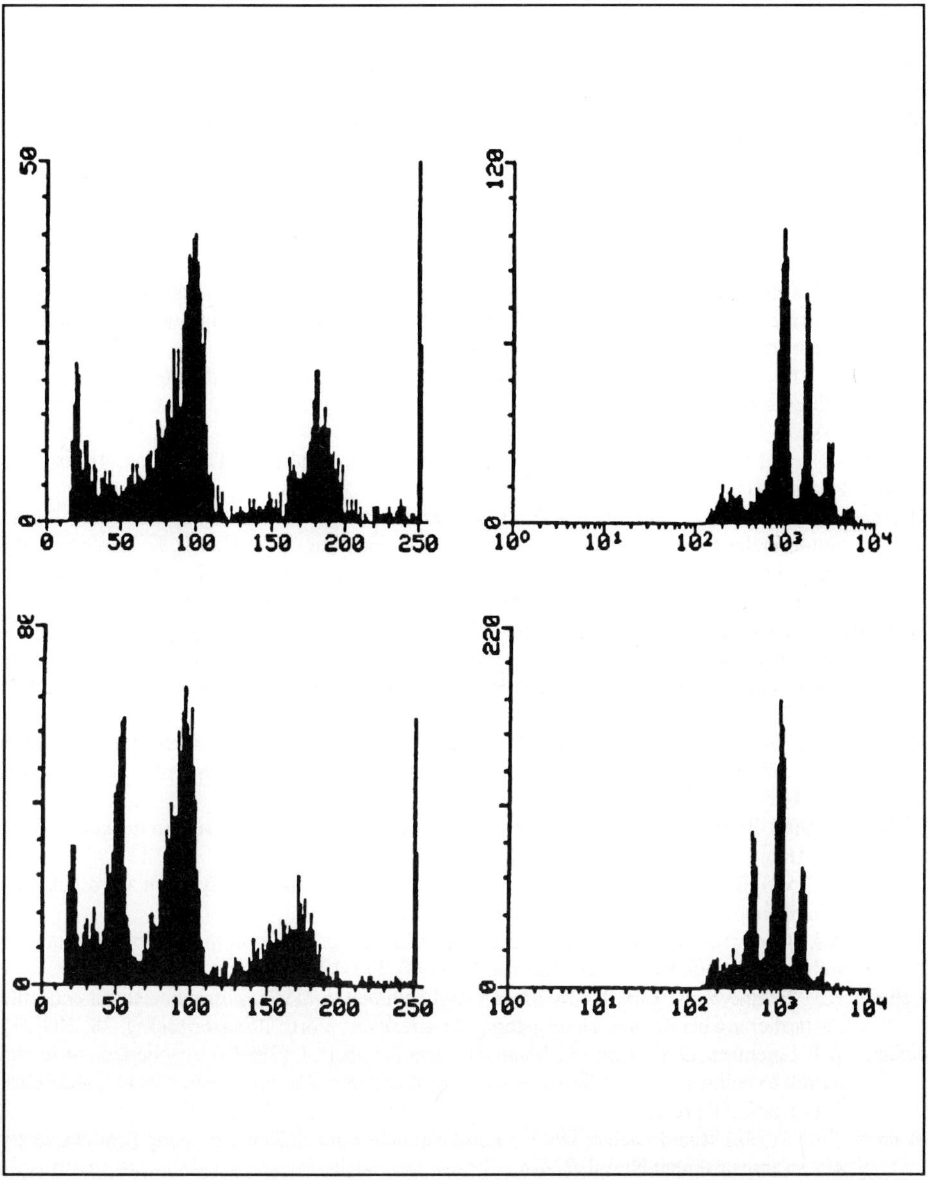

Fig. 2: Histograms derived from flow cytometry of a haploid regenerant of onion (below) and the diploid standard (above); left diagrams linear, right diagrams logarithmic; ordinates: numbers of stained particles (nuclei), abscissae: fluorescence intensity in arbitrary units; peaks result from the highest frequencies of stained particles representing the G0/G1 and G2 phases of the cell cycle at the different ploidy levels. Peaks of haploids are exactly one level lower than that of the diploid standard.

Acknowledgements

The authors thank for technical assistance in *in vitro* culture, measurements and cytological work: Mrs. D. Büchner, Miss I. Hoffmann, Mrs. D. Kriseleit, Mrs. D. Malur, Mrs. B. Hildebrandt, Mrs. Ch. Seidel, Mrs. T. Tölle, Miss K. Wendehake (IPK Gatersleben) and Mrs. Ch. Maaß (BBA Braunschweig).

References

Barg, E., Lesemann, D.-E., Green, S.K. and Vetten, M.J. (1994) 'Identification, partial characterization and distribution of viruses infecting *Allium* crops in South and South-East Asian countries', Acta Hortic. 358, 251-258.

Bonnier, F.J.M., Keller, J. and van Tuyl, J.M. (1992) 'Conductivity and potassium leakage as indicators for viability of vegetative material of lily, onion and tulip', Acta Hortic. 325, 643-648.

Dunstan, D.I. and Short K.C. (1977) 'Improved growth of tissue cultures of the onion, *Allium cepa*', Physiol. Plant. 41, 70-72.

Graichen, K., Hammer K. and Hanelt, P. (1992) 'Evidence of infections by aphid- and soil-borne viruses in the *Allium* collection of the Gatersleben institute', in P. Hanelt, K. Hammer and H. Knüpffer (eds.), The Genus *Allium* - Taxonomic Problems and Genetic Resources, Proc. Int. Symp. Gatersleben, June 11-13, 1991, Inst. Genet. Crop Plant Res. Gatersleben, pp. 89-92.

Graichen K. and Keller, J. (1992) 'Detection of viruses in *Allium* tissue culture - a prerequisite for virus elimination', in P. Hanelt, K. Hammer and H. Knüpffer (eds.), The Genus *Allium* - Taxonomic Problems and Genetic Resources, Proc. Int. Symp. Gatersleben, June 11-13, 1991, Inst. Genet. Crop Plant Res. Gatersleben, pp. 93-96.

Keller, J. (1990a) 'Culture of unpollinated ovules, ovaries, and flower buds in some species of the genus *Allium* and haploid induction via gynogenesis in onion (*Allium cepa* L.)', Euphytica 47, 241-247.

Keller, J. (1990b) 'Results of anther and ovule culture in some species and hybrids in the genus *Allium* L.', Arch. Züchtungsforsch. 20, 189-197.

Keller, J. (1992) 'Influence of different temperature treatments on viability of *in vitro* cultivated *Allium* shoots and bulblets', Acta Hortic. 31, 307-312.

Keller, E.R.J. (1993) 'Sucrose, cytokinin, and ethylene influence formation of *in vitro* bulblets in onion and leek', Genetic Resources and Crop Evol. 40, 113-120.

Keller, E.R.J., Eickmeyer, F., Lux, H., Maaß, H.I. and Schubert, I. (1994a) 'Ergebnisse zur entfernten Bastardierung bei *Allium*, Untergattung *Rhizirideum*', Vortr. Pflanzenzüchtg. 28, 309-311.

Keller, E.R.J., Lesemann, D.-E., Lux, H., Maaß, H.I. and Schubert, I. (1994b) 'Application of *in vitro* culture to onion and garlic for the management and use of genetic resources at Gatersleben', Acta Hortic. in press.

Walters, T.W. (1992) 'Rapid nuclear DNA content estimation for *Allium* ssp. using flow cytometry', *Allium* Improvement Newsl. 2, 4-6.

IN VITRO SELECTION OF A SALT-TOLERANT CELL LINE FROM PEA CALLI CULTURES

E. OLMOS, A. PIQUERAS and E. HELLIN
CEBAS-CSIC. Dpto. Fisiología y Nutrición Vegetal.
P.O.Box 4195.
30080 Murcia
Spain

ABSTRACT. Calli cultures of pea (*Pisum sativum* cv. Challis) were selected to NaCl resistance. The selected line grew well in 85.5 mM NaCl, whereas non-selected line was unable to grow in the same concentration of NaCl. However, growth of the selected line was reduced by 65% of dry weigth. Different parameters relative to salt tolerance have also been measured. We have observed an accumulation of different solutes as reducing sugars (glucose and fructose) as well as free proline. Several antioxidant enzymes have been studied (Superoxide dismutase, peroxidase, catalase and lipoxigenase). All of them, except catalase, were increased in their total activities from the cell line selected to 85.5 mM NaCl. Other enzymes related to salt stress have also been analized (ATPase, acid phosphathase and acid invertase), only ATPase not showed a significant change. Sodium and chloride levels were increased in salt selected line under salinity. However, calcium, phosphorus and magnesium levels were decreased. Studies are being carried out in order to stablish the correlation between salt stress and the physiological responses of the selected line.

1. Introduction

Salinity is known to affect in many ways to which a glicophyte must respond biochemically and physiologically to survive. Under saline conditions, the selective accumulation of inorganic and organic solutes have been observed as different mechanisms which cells perceive and adjust to stress conditions (Binzel et al., 1987). Different compatible osmolytes (nontoxic solutes) can be accumulate as protectant of proteins from the inhibitory effects of ions. Among these found in plants and callus cultures, proline and glycinebetaine are probably the most important (Daines & Gould, 1985, Rodhes, 1993). Other osmolytes are sugars (sucrose, reducing sugars, mannitol and sorbitol) (Muralitharan et al., 1993).

Superoxide free radicals ($O_2^{\cdot -}$) are generated during the metabolism of aerobic organisms by the univalent reduction of molecular oxygen (Fridovich, 1986). Superoxide dismutases catalyze the disproportionation of superoxide free radicals to H_2O_2 and O_2, and play an important role in protecting cells against the toxic effects of $O_2^{\cdot -}$ radicals (Fridovich, 1986). Hydrogen peroxide, produced by SOD and in the course of other enzymatic and nonenzymatic reactions, is removed by catalase and peroxidase (Fridovich, 1986).

H^+-ATPase plasma membrane is an enzyme which create a gradient of proton across the plasma membrane by extrusion of H^+ and contribute by this mechanism to osmotic adjusment under salinity (Watad et al., 1991). Changes in this activity can affect salt tolerance (Reuveni et al., 1993). Phosphorus deficiency has been observed under salt stress in plants and calli (Szabo-Nagy et al., 1992), parallel to this, it has also been observed an increment of the phosphatase activity (Szabo-Nagy et al., 1992).

M. Terzi et al. (eds.), Current Issues in Plant Molecular and Cellular Biology, 353-358.

2. Material and Methods

2.1 Cell culture

The pea calli used (*Pisum sativum* cv. Challis) were from a callus line selected to 85.5mM NaCl and a callus sensitive to that level of salt. The conditions of growth and the method for development of NaCl-tolerant lines have been decribed earlier (Olmos et al. 1994).

2.2 Enzyme extraction and assays

All operations were performed at 0-4°C. Pisum sativum calli were blended in 50 mM K-Phosphate buffer, pH 7.8, containing 0.1 mM EDTA, 5 mM Cysteine, 1% (w/v) polyvinil pyrrolidone and 0.2% (v/v) Triton X-100 (tissue/medium ratio 1:2; p/v) (Olmos et al. 1994). Total SOD (EC 1.15.1.1) activity was determined as described by McCord and Fridovich (1986). Catalase (EC 1.11.1.6) was assayed according to Aebi (1984). Peroxidase was determined as described by Bar-Akiba (1968). H^+-APTase and acid phosphatase activity was measured as the realise of Pi from ATP and pNP from pNPP (LeBel et al., 1978). Samples of the two cell line for H^+-ATPase were fractioned according to Hall (1992). The acid fosfatase was extracted in a 25mM Tris-MES buffer(pH 7.2) containing: 5mM EDTA and 250mM sucrose, determination was carried out according to Szabó-Nagy (1992). Lipoxigenase was extracted in 0.05M phosphate buffer (pH 7) and the assay was done using linoleate as substrate and estimated according to Sekhar & Reddy (1982). Invertase was extracted and assayed with the procedure described by Rather(1982). Protein determinations were carried out according to Bradford (1976) using crystalline BSA as standard.

2.3. Solute determinations

Sugars were extracted in water and analysis were carried out on an Interaction CHO-682 carbohydrate column, that was termostabilizated at 90 °C and Merk L-6200A HPLC pump; flow rate: 0.4mL/min; movil fase: H_2O; detection: Light Scattering Detector SEDEX 4X (S.E.D.E.R.E., France). Cholina and betaine were estimated according to Grieve and Grattan (1983). Free proline content was extracted in aqueous sulphosalicylic acid and estimated according to Bates et al. (1973). Four replicate cell samples were taken for each determination.

2.4. Ion analysis

Material of oven-dried (65°C, 24 h) callus samples were ashed at 490°C in muffle furnace for further analytical determinations. Sodium and potasium concentrations of both types of calli were estimated by flame photometer; magnesium and calcium by atomic absortion spectrophotometry (Hellín & Piqueras, 1990). Phosphorus was estimated colorimetrically (Kitson & Mellon, 1944). For chloride analysis, oven-dried calli (70°C, 24 hours) were extracted in destilled water for 1h and analyzed by ionic chromatography in a Dionex chromatograph.

3. Results

No growth was observed in calli exposed to concentrations higher than 85.5 mM of NaCl . At 85.5 mM of NaCl, calli were used to select tolerant cells. Callus in the initial subcultured was browning and compact, only small sectors with green color and good growth were used to next subcultures. The stability was obtained after four subcultured on the same medium (5% to 90%

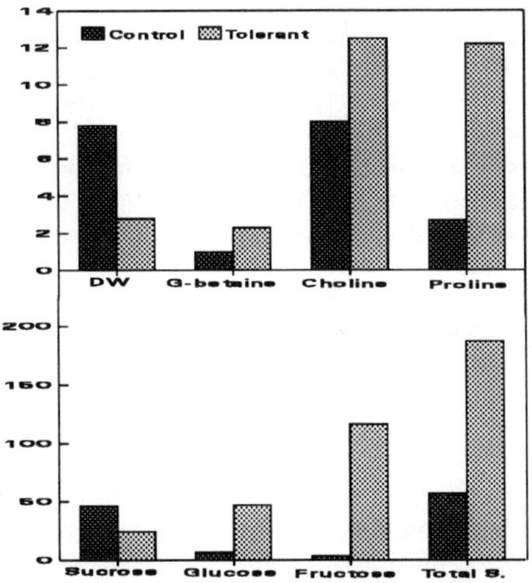

Fig 1. Solute concentrations in control and NaCl-tolerant callus. Concentration: Proline (μM/gFW), G-Betaine, Choline and Sugars (μM/gDW).

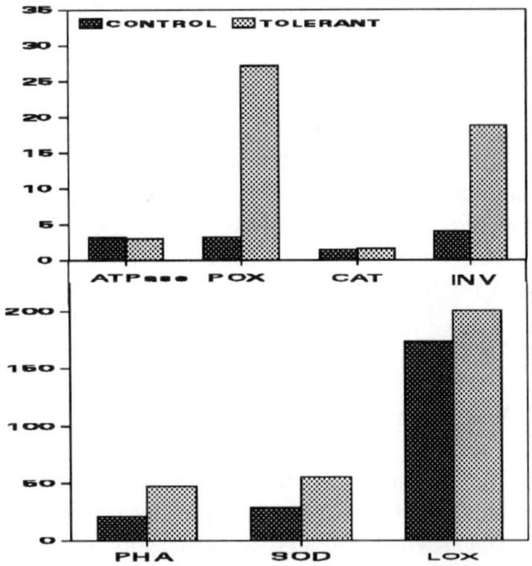

Fig. 2. Enzymatic activities in control and NaCl-tolerant callus. Units: ATPase (μM P_i/h x mg protein), Acid Phosphatase (nM pNP/min x mg protein), catalase (nM H_2O_2/min x mg protein), peroxidase, lipoxigenase and superoxide dismutase (U/mg protein), acid invertase (μM glucose/min x g FW)

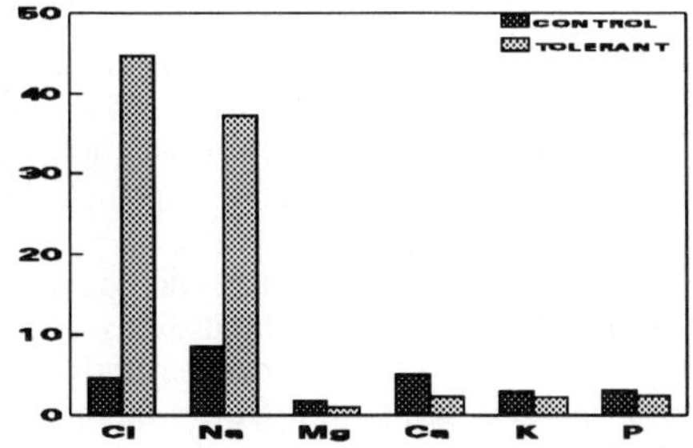

Fig. 3. Ion concentration (mg/gDW) in control and NaCl-tolerant callus

of survival). At 85.5 mM NaCl, callus dry weight was reduced by 65% compared to the control (Fig. 1).

The HPLC analysis of the crude aqueous extracts of sensitive and tolerant cell lines identified sucrose, glucose and fructose as main sugar by comparison with the external standards. Quantitative HPLC analysis showed that sucrose was the most abundant disaccharide present in sensitive pea calli. Selected line showed a marked decrease in the concentration of sucrose (Fig.1). Fructose and glucose levels in the sensitive line were much lower than those of sucrose, and were highly increased in the selected line (Fig. 1). Glicynebetaine and choline were detected in both sensitive and salt-tolerant pea calli. Although there was a small increase in the level of glicynebetaine in the salt-tolretant cell line (Fig.1), the concentration of this compound was considerably lower than the measured concentration of sugars and proline. Choline not showed a significant change in the selected line. Proline accumulated significantly more in the salt-tolerant line compared with the sensitive line (Fig 1).

Salt-tolerant calli showed a significant increase in total SOD especific activity in relation to Nacl-sensitive cells values (Fig.2). As for peroxidase activity, salt-tolerant cells shown a great increase in relation to control cells (Fig.2). However, catalase activity was not altered by salinity, although there was a slightly not significant increase in NaCl-tolerant calli (Fig.2). Changes in phosphatase activity in control and salt-tolerant calli are shown in Figure 2. The level of phosphatase activity in salt-resistant line was approximately 2-fold than in the control. This increase was parallel to the decrease in phosphorus content (Fig.3). The Mg-ATPase activity in cell wall fractions was slightly lower in salt-tolerant line (Fig.2). Lipoxygenase and invertase increased significantly in the selected line (Fig.2).

The results indicate that salt-tolerant cultures accumulate Na^+ and Cl^-, whereas the contents Mg^{+2}, K^+ and Ca^{+2} were lower in salt-tolerant line than in sensitive line (Fig.3).

4. Discussion

In plant resistant to different types of environmental stress conditions, increases levels of enzymes involved in superoxide detoxification have been found (del Río et al., 1991). NaCl-resistant pea cell showed a significant increase in specific SOD activity due to the induction of Cu,Zn-SODs activity, principally Cu,Zn-SOD I isozyme (Olmos et al., 1994). Catalase and peroxidase activities catalizes the breakdown of H_2O_2 produced at cellular level. In NaCl-adapted pea cells, catalase increased slighly as compared with control cells although not significantly. However, peroxidase activity was significantly increased in NaCl-tolerant cells in presence of 85.5 mM NaCl and was the most important activity in H_2O_2 elimination. In *Halimione portulacoides*, the high levels of SOD and peroxidase activities observed under salinity, as well as the increased catalase activity found at lower NaCl concentrations, have been reported to be essential for survival of this halophitic plant in natural salines (Kalir and Poljakoff-Mayber, 1981).

Reductions in sugar concentrations were higher on average in tolerant calli than in unselected calli. In pea selected calli, the low level of sucrose can be due to the high activity of the soluble acid invertase in this callus, transforming sucrose in glucose and fructose. However, glucose and fructose acccumulate and apparently were not metabolized further. These compounds can be used as osmolytes in the cells (Muralitharan et al., 1993).

Many plant species, particularly halophytes, genotypes in the Chenopodiaceae and Graminea accumulate large quantities of QACs as an osmotic solute (Rhodes, 1993). The inability of pea calli to accumulate glycinebetaine could be due to either the lack of choline oxidizing activity or betaine aldehyde dehydrogenase in pea calli (Rathinasabapathi et al., 1993). Levels of choline, however, were comparable to those found in other species (Storey & Jones, 1977). In pea selected calli, an effects such as NaCl stress on proline accumulation was observed. The differences in free proline content of sensitive calli and salt-tolerant calli could be due to different mechanisms. Since salt stress induces a secondary water stress effect that disturbs the metabolism and that elevated concentration of proline it might indicate the suceptibility of the callus tissue to water stress.

The increased level of phosphatase activity accompanied by a decrease of phosphorus in salt-stress have been reported in some varieties of wheat calli (Szabo-Nagy et al., 1992) and other whole plants. In our pea selected calli, we consider that phosphorus deficience can serve as a signal for the induction of acid phosphatase. The activity of plasma membrane H^+-ATPase has been reported to increase, decrease or not change (Reuveni 1993) after growth of plants or cells in different concentrations of salt. Pea selected calli showed a low decrease in the activity of the cell wall fraction. Reuveni (1993) observed a change in kinetic properties of the ATPase in tobacco cells adapted to salt as a posible adaptation mechanism to saline environments. Lipoxygenase activity was increased in salt-tolerant calli probably as a result of tissue injury, forming fatty acid hydroperoxides and these products can descompose into MDA and free radicals, which increased in *Pisum sativum* calli selected at 85.5 mM NaCl (Olmos et al., 1994). Lipoxygenase might be activated by increased levels of ABA in salt-stress (Eberhardt & Wegmann, 1989), according to a similar model proposed for wounding (Hildebrand, 1989).

The change in nutrients were important in the salt-tolerant calli, mainly the accumulation of Na and Cl. The low concentrations of Ca, K and Mg were similar to those described in callus of *Brassica campestris* and *Citrus limon* under NaCl stress and *B. napus* and *B. carinata* in seawater(Paek et al., 1988; He & Cramer, 1993; Hellín & Piqueras, 1990). Under salt stress cells can response by ion inclusion or exclusion (Daines & Gould, 1985). In pea selected calli not showed a Na exclusion mechanism at cellular level ,but is not clear if cells acumulate Na in vacuoles or cytosol by a Na compartmentation. Similar results have also been found in Brassica species (He & Cramer, 1993).

5. Acknowledgements

This work has been suppoted by grant of CICYT n°BIO90-800

6. References

Aebi, M. (1984) 'Catalase in vitro', Methods in Enzymology 105, 121-126.

Bar-Akiba, A. (1968) 'Induced formation of enzymes as a possible measure on micronutrient requirement of Citrus trees', in CEBAC-CSIC (eds.) Control de la fertilizacion de las plantas cultivadas. II Coloquio Europeo y Mediterráneo. Sevilla, pp. 573-75.

Bates, L.S., Waldren, R.P. and Teare, I.D. (1973) 'Rapid determination of free proline for water stress studies', Plant Soil 39, 205-207.

Binzel, M.L., Hasegawa, P.M., Rhodes, D., Handa, S., Handa, A.K. and Bressan, R.A. (1987) 'Solute accumulation in tobacco cells adapted to NaCl', Plant Physiol. 84, 1408-1415.

Bradford, M.M. (1976) 'A rapid and sensitive method for the quantification of microgram quantities of protein utilizing the principle of protein-dye binding', Anal. Biochem. 72, 248-254.

Daines, R.J. and Gould, A.R. (1985) 'The cellular basis of salt tolerance studied with tissue cultures of the halophytic grass *Discthilis spicata*', J. Plant Physiol. 119, 269-280.

del Río, L.A. Sevilla, F., Sandalio, L.M. and Palma, J.M. (1991) 'Nutritional effect and expression of superoxide dismutases: induction and gene expression, diagnostics, prospective protection against oxygen toxicity', Free Rad. Res. Commun. 12-13, 819-828.

Eberhardt, H.-J, and Wegmann. (1989) 'Effects of abscisic acid and proline on adaptation of tobacco callus cultures to salinity and osmotic shock', Physiol. Plant. 76, 283-288.

Fridovich, I. (1986) 'Biological effects of the superoxide radical', Arch. Biochem. Biophys. 247, 1-11.

Grieve, C.M. and Grattan, S. R. (1983) 'Rapid assay for determination of water soluble quaternary ammonium compounds', Plant Soil 70, 303-307.

Hall, J.L. and Nelson, S.J. (1990) 'Propierties of APTase activity associated with cell wall fractions from higher plants', J. Plant Physiol. 137, 241-243.

He, T. and Cramer, R. (1993) 'Cellular responses of two rapid-cycling Brassica

species, *B. napus* and B. *carinata*, to seawater salinity', Physiol. Plant. 87, 54-60.
Hellín, E. and Piqueras, A. (1990) 'Relación entre tolerancia a la sal y concentracion de iones de lineas celulares de *Citrus limon*', in UIB (eds) Nutrición mineral bajo condiciones de stress. III Simposio Nacional de Nutricion Mineral de las Plantas. Palma de Mallorca. pp. 319-324.
Hildebrand, D.F. (1989) 'Lipoxygenases', Physiol. Plant. 76, 249-253.
Kalir, A. and Poljkoff-Mayber, A. (1981) 'Changes in activity of malate dehydrogenase, catalase, peroxidase and superoxide dismutase in leaves of *Halimione portulacoides* (L.) Aellen exposed to high sodium chloride concentrations'. Ann. Bot. 47, 75-85.
Kitson, R.E. and Mellon, M.G. (1944) 'Colorimetric determination of phosphorus as molybdovanado phosphoric acid', Ind. Eng. Chem. Anal. Ed. 16, 379-383.
LeBel, D., Poirier, G.G. and Beaudoin, A.R. (1978) 'A convenient method for the ATPase assay'. Anal. Biochem. 85, 86-89.
Muralitharan, M.S., Chandler, S.F. and Steveninck, R.F.M. (1993), 'Physiological adaptation to high ion concentrations or water deficit by callus cultures of *Highbush Blueberry*, Vaccinium corymbosum', Aust. J. Plant Physiol. 20, 159-172.
Murashige, T. and Skoog, F. (1962) 'A revised medium for rapid growth and bioassays with tobacco tissue cultures', Physiol. Plant. 15, 473-497.
Olmos, E., Hernandez, J.A., Sevilla, F. and Hellín E. (1994) 'Induction of several antioxidant enzymes in the selection of a salt-tolerant cell line of Pisum sativum'. J. Plant Physiol. (in press).
Paek, K.Y., Chandler, S.F. and Thorpe, T.A. (1988) 'Physiological effects of Na_2SO_4 and NaCl on callus cultures of *Brassica campestris* (Chinese cabbage). Physiol. Plant, 72, 160-166.
Rathert, G. (1982) 'Influence of extreme K:Na ratios and high substrate salinity on plant metabolism of crops differing in salt tolerance', J. Plant Nutrition. 5, 97-109.
Rathinasabapathi, B., Gage, D.A., Mackill, D.J. and Hanson, A.D. (1993) 'Cultivated and wild rices do not accumulate glycinebetaine due to deficiencies in two biosynthetic steps', Crop Sci. 33, 534-538.
Reuveni, M., Bressan, R.A. and Hasegawa, P.M. (1993) 'Modification of proton transport kinetics of the plasma membrane H^+-ATPase after adaptation of tobacco cells to NaCl', J. Plant Physiol. 142, 312-318.
Rhodes, D. (1993) 'Quaternary ammonium and tertiary sulfonium compounds in higher plants', Annu. Rev. Plant Physiol. Plant Mol. Biol. 44, 357-384.
Sekhar, B.P.S and Reddy, G.M. (1982) 'Studies on lipoxygenase from rice (*Oryza sativa* L.)', J. Sci. Food Agric. 33, 1160-1163.
Storey, R. and Wyn Jones, R.G. (1977) 'Quaternary ammonium compounds in plants in relation to salt resistant', Phytochemistry. 16, 447-453.
Szabó-Nagy, A. Galiba, G. and Erdei, L. (1992). 'Induction of soluble phosphatases under ionic and nonionic osmotic stresses in wheat', J. Plant Physiol. 140, 629-633.
Watad, A.A., Reuveni, M., Bressan, R.A. and Hasegawa, P.M. (1991) 'Enhanced net K^+ uptake capacity of NaCl adapted cells', Plant Physiol. 95, 1265-1269.

THE ROLE OF SECRETED PROTEINS IN CARROT SOMATIC EMBRYOGENESIS

Theo Hendriks and Sacco C. De Vries
Department of Molecular Biology
Agricultural University of Wageningen
Dreijenlaan 3
6708 HA Wageningen
The Netherlands

1. Introduction

Plant cells are enveloped by a cell wall, the plant extracellular matrix (Roberts, 1989), and there is a growing awareness that besides providing support and protection, the cell wall has an important function in plant development and differentiation. Initially, newly formed cells deposit a thin, semi-rigid primary cell wall to accommodate elongation. Later, when growth has ceased and cells acquire specialized functions, a more rigid secondary cell wall is produced, either by thickening the primary cell wall or by depositing new wall components with a different composition.

Plant cell walls harbour a complex array of secreted proteins, some of which have a structural function whereas others may shape and re-shape cell walls during development. Furthermore, secreted proteins may function in cell-cell communication, by functioning as part of a signal transduction chain. Following laser microsurgery of two-celled embryos it was established that prolonged contact of a *Fucus* thallus cell protoplast with the wall of the ablated rhizoid cell resulted in the formation of cells with the characteristics of the rhizoid cell. This experiment provides evidence for the presence of stable non-diffusible wall components able to change the fate of algal cells (Berger et al., 1994). Despite the growing interest in the plant cell and its role in plant development (for recent reviews: Carpita and Gibeaut, 1993; Fry, 1990; Iiyama et al., 1994; Kieliszewski and Lamport, 1994; McCane and Roberts, 1989; Showalter, 1993), studies that directly demonstrate the role of an individual secreted protein in a specific developmental process are still rare.

In this contribution we present an overview of the individual secreted proteins present in the conditioned medium of carrot suspension cultures that have sofar been identified in our laboratory (Table 1). In addition, in view of the suggested general role of arabinogalactan proteins in growth and development (Knox, 1993), and recent findings with respect to the effect and presence of these highly glycosylated proteins in carrot somatic embryogenesis (Kreuger and Van Holst, 1993; Pennell et al., 1992), the possible role of AGPs in somatic embryogenesis will also be discussed shortly.

Carrot somatic embryo formation is accompanied by the secretion of a limited number of proteins into the culture medium that are either absent from or present in reduced amounts in cultures that do not form somatic embryos (De Vries et al., 1988a). This suggests that several secreted proteins are correlated with somatic embryo development. That secreted proteins may

M. Terzi et al. (eds.), Current Issues in Plant Molecular and Cellular Biology, 359-368.

TABLE 1. Secreted proteins in carrot cell suspension cultures identified by the sequence of their respective cDNA clones or on the basis of their biological actvity.

Protein	Mr (kD)	Origin	Biological activity
EP1 (S-locus like) clone/α serum/protein	52-54	Wall non-embryogenic cells	unknown
EP2 (nsLTP) clone/α serum/protein	10	Wall embryogenic cell clusters	Cutin monomer export ?
EP3 (endochitinase) clone/α serum/protein	32	unknown	Rescue ts11 somatic embryogenesis
EP4 clone/α serum/protein	47	Wall non-embryogenic dividing cells	unknown
EP5 (peroxidase) protein	38	unknown	Restores tunicamycin- impaired embryogenesis

actually play a role in somatic embryogenesis, was indicated by three independent observations: (1) the speed with which carrot cells released from hypocotyl tissue acquire the ability to produce somatic embryos was increased considerably by the addition of unfractionated heat-labile, high-molecular mass components present in medium of an established embryogenic suspension culture (De Vries et al., 1988b); (2) the defect in somatic embryogenesis in several non- or low-embryogenic carrot lines could be partially restored by the addition of secreted components from wild-type embryogenic cell lines (De Vries et al., 1988a; LoSchiavo et al., 1990); and (3) carrot somatic embryogenesis inhibited by tunicamycin, an inhibitor of the N-glycosilation, could be restored by the addition of secreted glycoproteins present in embryo and suspension cultures (De Vries et al., 1988a).

On the basis of these observations, in subsequent studies two strategies were employed to further resolve the role of secreted proteins in carrot somatic embryogenesis.The first was to obtain specific probes corresponding to several secreted proteins. To this end, antisera were raised against all proteins secreted by an embryogenic suspension culture and used to screen a cDNA expression library. Individual proteins corresponding to the cDNA clones obtained were then identified by Western blotting using specific antisera raised against the E.coli-produced lacZ-cDNA fusion protein. This way, cDNA clones corresponding to individual secreted proteins as well as specific antisera against these proteins were obtained, allowing a much more detailed analysis. One important finding was that individual cell types in the heterogenous cell populations produced different spectra of proteins, suggesting that the total profile of secreted proteins is a direct reflection of the cellular diversity in the cultures (Van Engelen et al., 1991, 1992; Sterk et al., 1991).

The second strategy was based on the observed promoting effects of secreted proteins present in the conditioned medium on somatic embryogenesis inhibited by tunicamycin, or impaired by a temperature sensitive mutation (De Vries et al., 1988a; LoSchiavo et al., 1990). Fractionation of the secreted proteins present in embryo or embryogenic suspension cultures and testing the individual fractions obtained for their embryo-rescue activity, allowed the identification of the individual causative secreted proteins (Cordewener et al., 1991; De Jong et al., 1992).

2. Proteins identified by the cloning strategy

2.1. EP1: A GLYCOPROTEIN HOMOLOGOUS TO *BRASSICA* S-LOCUS GLYCOPROTEINS

The EP1 (extracellular protein 1) mRNA was found to code for a small number of glycoproteins with different molecular weights that are secreted by carrot suspension cells. The observed heterogeneity appears to arise from different post-translational modifications, including processing of the four N-linked oligosaccharides present (Van Engelen et al., 1991; Sturm, 1991). Using specific antibodies against the lacZ-EP1 fusion protein it was shown that the EP1 proteins were secreted by expanding, non-embryogenic cells only. This finding and the higher levels of EP1 proteins in basal parts of seedling hypocotyls and roots as compared to apical parts, suggested a role of EP1 in cell elongation (Van Engelen et al., 1991; Van Engelen and De Vries, 1992).

These initial results indicated that the EP1 proteins are not likely to be involved in somatic embryogenesis, and in fact suggested quite the opposite. To obtain additional information that might provide clues concerning the function of the EP1 proteins, the EP1 cDNA was sequenced and the expression of the gene in seeds and developing seedlings analysed by in situ hybridization (Van Engelen et al., 1993). These studies revealed that the EP1 cDNA contained a region sharing homology with the *Brassica* S-locus glycoproteins and that the EP1 gene is preferentialy expressed in the epidermis of seedling root, hypocotyl and cotelydon, the root cap, a patch of cells at the surface of the apical meristem, and in the outer cell layers of the integument and fruit wall of the developing seed. Taken together, these results indicated that the EP1 proteins are not likely to be involved in cell elongation, nor in cell adhesion, but perhaps may have a function similar to other *Brassica* S-locus-like glycoproteins found in other plant species (Nasrallah and Nasrallah, 1993). This function would obviously be of a more general nature then the specialized cell-cell interactions involved in the self-incompatibility system.

2.2. EP2: A SECRETED LIPID TRANSFER PROTEIN

The cDNA derived amino acid sequence of EP2, a 10 kD protein present only in the medium of carrot embryo and embryogenic cultures, showed homology to lipid transfer proteins (LTPs) from several other plant species. Western blotting revealed that the EP2 protein was recognized by antibodies raised against LTPs from spinach, maize, and Arabidopsis, confirming the homology. Employing in situ mRNA localization, the expression of the EP2 gene was shown to be limited to the protoderm of both zygotic and somatic embryos, and in epidermal cells of leaf primordia, flower organs and maturing seeds. The expression of the gene in (pro)epidermal tissues and the extracellular location of the encoded protein, led to the proposition that the EP2 protein may be involved in the transfer of cutin monomers trough the cell wall to sites of cutin polymerization (Sterk et al., 1991).

The EP2 protein was purified from the medium of carrot embryo cultures, with the aid of specific antibodies against the lacZ-EP2 fusion protein, or against heterologous LTPs (Meijer et al., 1993). The identity of the purified EP2 protein was confirmed by showing that the amino acid composition of the purified protein closely matched the one predicted by the cDNA-derived amino acid sequence of the mature protein (Hendriks et al., 1993). The purified protein was shown to be able to enhance the transfer of fluorescent phospholipid analogs between artificial membranes, and it was the only secreted protein with this property. In addition, by employing a simple gel permeation assay it was shown that the EP2 protein was capable of binding palmitic acid, oleic acid and oleoyl-CoA in a near equimolar ratio (Meijer et al., 1993; Hendriks et al., 1993). Since all of these molecules may be considered as precursors of cutin monomers (Kolattukudy, 1980), these results suggest that the EP2 protein has at least the biochemical properties to transfer precursors of cutin monomers or cutin monomers themselves.

The results obtained sofar suggest that the EP2 protein could be involved in the deposition of a cuticle that surrounds carrot embryos (cf. Rubos, 1984). Its function may be two-fold: 1) the targetting of cutin monomers towards the site of cutin synthesis and thus preventing them from being trapped in membranes of endoplasmic reticulum, Golgi or plasmalemma or in the cell wall; and 2) protection of these molecules from enzymatic and/or non-enzymatic loss of the CoA-ester linkage, which is essential for the incorporation in the cutin polymer (Kolattukudy, 1980). Calculations indicated that when the EP2 protein is discarded after delivery of a cutin monomer, the amount of EP2 protein needed to form a cuticle surrounding a globular somatic embryo exceeds by far the amount of EP2 protein produced. This may indicate that either EP2 may be involved in the transfer of a particular subset of cutin monomers, or, alternatively, that the EP2 is reused and functions as a shuttle in the transfer of cutin monomers from their site of synthesis to the site of cutin polymerization. This possibility is more easy to envisage in the intact plant, where the EP2 protein is likely to be localized in the cell wall, as shown for LTP in *Arabidopsis* (Thoma et al., 1993). In somatic embryo cultures the shuttle will function inefficiently, since a major portion of the EP2 protein will be lost in the medium. The expression pattern of the *Arabidopsis* LTP1 gene was analyzed by in situ mRNA localization studies as well as by promoter-reporter studies. This revealed an expression pattern similar to and at least as complex as in carrot (Thoma et al. 1994).

What purpose would a cuticle serve an embryo? In view of the aqueous environment surrounding embryos this might be different from cuticles that cover the epidermis of areal parts of plants, which mainly provide a barrier that prevents the loss of water. In embryos the function of a cuticle might be the opposite, namely the prevention of precocious turgor-driven water uptake that might disturb embryo development. In addition, a cuticle might serve to protect the embryo from activities of hydrolytic enzymes, either in the endosperm, or in the case of somatic embryos, in the surrounding medium.

3. Proteins identified by their biological activity

3.1. EP3: A 32 kD ACIDIC ENDOCHITINASE

The ts11 somatic embryo variant cell line, obtained after EMS mutagenesis, was selected on the basis of the arrest in its embryo development at an elevated temperature (Giuliano et al., 1984). The temperature sensitivity of ts11 appeared to be restricted to a relative short period around the transition from globular to heart stage embryo. In the same time interval ts11 embryo development at the permissive temperature was sensitive to replacement of the medium by fresh medium. Both temperature- and fresh medium-induced arrest could be overcome by supplementing the medium with secreted proteins from wild type carrot cells (LoSchiavo et al., 1990).

In search for the causative factor in the rescue of the temperature-impaired somatic embryogenesis in ts11, a 32 kD acidic protein was purified from wild type competent medium and identified as an endochitinase (De Jong et al., 1992). Though this protein was initially thought to be glycosylated, further detailed analysis revealed that this was not the case. In addition, sequence data of putative EP3 cDNA clones indicated that there are no N-glycosylation consensus amino acid sequences present in the protein (Kragh et al., in preparation). The sequence data also revealed that the 32 kD endochitinase is a class IV chitinase (cf. Collinge et al., 1993).

In the absence of known chitinase substrates in the plant cell wall, the possible function of the 32 kD endochitinase in the rescue of impaired ts11 somatic embryogenesis was approached by investigating the effect of putative chitinase products on ts11 somatic embryogenesis at the nonpermissive temperature. Testing several N-acetylglucosamine containing compounds revealed that a *Rhizobium*-produced lipo-oligosaccharide or nodulation factor , NodRlv-V(Ac, C18:4), was able to promote ts11 embryo formation with a similar efficiency as the 32 kD endochitinase (De Jong et al., 1993). Based on this result it was proposed that the function of the 32 kD endochitinase is the release of a Nod-factor-like signal molecule from a larger plant-produced precursor that might be present in the cell wall.

The main morphological effect of the 32 kD endochitinase on the ts11 globular embryos appeared to be a restoration of the the embryo protoderm, suggesting that the arrest in ts11 embryo formation is caused by the inability to form a proper protoderm at the nonpermisive temperature (De Jong et al., 1992). In addition to this, the 32 kD endochitinase, as well as the Nod-factor, are able to promote the formation of proembryogenic masses and globular embryos (De Jong et al., 1993a).

Recently, it was found that a 32 kD endochitinase is present in ts11 cultures that was indistinguishable from the purified wild-type form with respect to its biochemical properties and rescue activity (De Jong et al., submitted). This indicated that the rescue activity of the 32 kD endochitinase cannot be explained by the total absence of this protein or the presence of an abberant, inactive form. At the non-permissive temperature, however, the concentration of the 32 kD in the medium of ts11 embryo cultures appeared to be transiently decreased during the temperature-sensitive period. This observations suggested that a transient decrease in the concentration of the 32 kD endochitinase in the medium of ts11 embryo cultures grown at the non-permissive temperature causes the arrest in ts11 embryo development. The rescue of ts11 embryo development by both the 32 kD endochitinase and the Nod-factor was also restricted to a relatively short period.

More recently, it was shown that in the medium of 10-line wild-type carrot cultures there are at least five 32 kD endochitinase present in the medium that are recognized by anti-32 kD antibodies on western blots obtained after native polyacrylamide gel electrophoresis (Hendriks et al., in preparation). In accordance with this was the cloning of different mRNAs encoding highly homologous endochitinases (Kragh et al., in preparation). All five isoforms were active chitinases, while the two must abundant ones were purified to homogeniety. Upon testing the effect on ts11 embryo development at the nonpermissive temperature it was found that both promote the number of embryos formed, whereas only one was capable of promoting the globular-heart stage embryo transition (Hendriks et al., in preparation). These results indicate that subtle differences may exist in the activity of different chitinase isoforms of the same family. Chitin fragments have also been reported to elicite specific biological effects (Felix et al., 1993), suggesting that plant cells may have more receptors that recognize and respond to GlcNac-oligomer-containing compounds not previously identified. Whether plant chitinases are involved in the generation of Nod-factor like plant analogues remains to be established (Schmidt et al. 1994).

3.2. EP4: A 47 kD GLYCOPROTEIN

A 47 kD glycoprotein was obtained in the same fractionation of secreted proteins which yielded the 32 kD endochitinase, and initial experiments suggested that this protein could increase the number of embryos formed in ts11 about tenfold (Fiorella LoSchiavo and Anke de Jong, unpublished results). It was for this reason that this protein, designated EP4, was purified and further characterized (Van Engelen et al., 1993). A specific antiserum against the purified protein was obtained and used to determine its origin. EP4 was found present in media from both embryogenic and non-embryogenic cell lines, and appeared to be secreted by a specific subset of non-embryogenic cells that are different from the non-embryogenic cells producing EP1. EP4 was not secreted by embryogenic or embryo cells.

The antiserum raised against EP4 recognized a second protein of 45 kD that was ionically bound to the cell walls separating adjacent cells in clusters of non-embryogenic cells. Upon closer inspection these cell walls in the clusters appeared to be cross walls formed by cell division in elongated cells, suggesting that EP4 may have a function in this type of cell division. In seedlings, the presence of EP4 was found to be restricted to the roots.

Two EP4 cDNA clones were obtained upon screening a cDNA expression library with a oligonucleotide probe designed on the bases of the amino acid sequence of an EP4 peptide obtained after tryptic digestion (Van Engelen et al., submitted). The clones were sequenced and one was found to encode the 47 kD protein, whereas the other clone encoded a highly homologous, but different protein, presumably the 45 kD protein. The sequences shared significant homology with the early nodule-specific gene ENOD8 of alfalfa, a gene proposed to be involved in root nodule formation in resonse to *Rhizobium* infection (Dickstein et al., 1993)

3.3. EP5: A 38 kD CATIONIC PEROXIDASE

Tunicamycin inhibits carrot somatic embryogenesis without effecting the proliferation of other cells in the culture (LoSchiavo et al., 1984), and it was shown that this inhibition could be overcome by the addition of secreted (glyco)proteins present in embryo and embryogenic suspension cultures (De Vries et al., 1988a). Upon fractionation of the secreted proteins and testing the embryo-rescue activity of the fractions obtained, a 38 kD cationic peroxidase was purified that was able to overcome the tunicamycin-inhibited embryogenesis. This 38 kD cationic peroxidase fraction contained four isoforms of which only one, or possibly two, appeared responsible for the observed rescue activity (Cordewener et al., 1991).

The endogenous carrot peroxidase isoenzyme(s) that appeared to be missing from the medium of tunicamycin-treated cultures, could also be effectively replaced by commercial cationic horseradish peroxidase isoenzymes. The horseradish peroxidase isoenzymes lost this property however upon the removal of the heme, indicating that it was the peroxidase activity that was responsible for the rescue activity (Cordewener et al., 1991).

Tunicamycin blocked somatic embryogenesis before the globular stage and appeared to promote the expansion of surface cells of proembryogenic masses, followed by the complete disrupture of the entire structure. Addition of purified 38 kD peroxidases prevented this disruptive process, but only to a certain extent, since torpedo stage embryos formed under these conditions still had irregular surfaces when compared to untreated torpedo stage embryos. From these observations it was proposed that the 38 kD cationic peroxidase(s) may have a function in somatic embryogenesis by reducing cell wall extensibility (Cordewener et al., 1991), possibly by invoking the cross-linking of cell wall polymers via the oxidative coupling of phenolic side chains (Iiyama et al., 1994; Kato et al., 1994).

4. Arabinogalactan proteins (AGPs)

Arabinogalactan proteins are proteoglycans with poly- and oligosaccharide units consisting of 1,3-ß-D-galactopyranosyl backbones with side chains of (1,3-ß- or 1,6-ß-) D-galactopyranosyl and L-arabinofuranosyl residues (Fincher et al., 1983). AGPs are large molecules with a molecular mass of 100,000 or more, whereas the protein core usually constitutes less than 10% of that mass. In plants AGPs are very heterogenous and each tissue seems to contain a specific set of AGPs. This tissue-specificity of AGPs can be determined by cross-electrophoresis or the use of monoclonal antibodies directed to AGP-specific epitopes (Van Holst and Clarke, 1986; Knox, 1993). AGPs have been detected in the plasma membrane, in the space between plasma membrane and cell wall and in the cell wall (Knox, 1993).

A carbohydrate epitope in cell walls of certain single cells in embryogenic, but not in non-embryogenic, suspension cultures of carrot, recognized by the monoclonal antibody JIM8, was reported by Pennell et al. (1992). The JIM8 monoclonal antibody was previously shown to recognize an AGP that was present in the plasmalemma of carrot suspension cells (Pennell et al., 1991), and therefore the cell wall epitope was possibly another molecule bearing the same epitope. The expression of the JIM8 cell wall epitope was found to be regulated during the initiation, proliferation , and prolonged growth of suspension cultures, while changes in the abundance of JIM8-reactive cells preceded changes in embryogenic potential. It was shown that the JIM8-reactive cell wall epitope was present in only a few different cell types that seemed developmentally related. The switch by one of these cell types, amongst which cells designated type 1 cells by Nomura and Komamine (1985), to somatic embryogenesis was accompanied by the dissipation of the JIM8-reactive epitope. These results suggested that the JIM8-reactive cell wall carbohydrate epitope represents a marker for a very early transitional cell state in the developmental pathway to carrot somatic embryogenesis (Pennell et al., 1992).

Recently, direct evidence for a role of AGPs in somatic embryogenesis was reported by Kreuger and Van Holst (1993). They demonstrated that by the addition of AGPs isolated from carrot seeds to a two-year-old, non-embryogenic cell line embryogenic potential was re-induced. Addition of AGPs from a non-embryogenic cell line to cultures newly initiated from hypocotyl explants resulted in non-embryogenic suspension cultures, whereas in untreated controls or upon the addition of seed AGPs embryogenic cultures were obtained. These results indicate that the developmental fate of cells in suspension cultures can be influenced by the addition of specific AGPs at very low concentrations (10-100 nM). It was proposed that AGPs play a role in cell-cell interactions during differentiation (Kreuger and Van Holst, 1993).

In view of our interest in the early events that determines the transition of somatic cells into embryogenic cells (De Jong et al., 1993b), a cell tracking system was established that allowed the determination of the capability of individual single carrot suspension cells to develop into somatic embryos (Toonen et al., 1994). Using this system it proved however impossible to distinguish cells capable of forming somatic embryos on the basis of their morphology (Toonen et al., 1994).

5. Conclusions

Of the secreted proteins present in carrot suspension cultures that have been identified in our laboratory sofar (Table 1), three seem to play a role in somatic embryogenesis. Of these, a lipid transfer protein (EP2) is a marker for the presence of embryogenic cell clusters, while two others, a chitinase (EP3), and a peroxidase (EP5), appear to have a more direct role.

Interestingly, each is a representative of a class of proteins that has been implicated in plant defence against biotic and abiotic stresses. Two of these proteins are apparently either associated with (EP2) or affect the formation of (EP3) the embryo protoderm. It is obvious that in order to provide more evidence concerning the function of secreted proteins in plant somatic embryogenesis their substrates and products need to be indentified. Of equal importance is it to develop better and more diverse complementation systems. Irrespective of their true functions, the proteins identified and their encoding mRNAs are clearly markers of the diverse cell types present in carrot suspension cultures and thus provide important tools in the study of plant cells that are in the process of reacquiring embryogenic potential (cf. Sterk and De Vries, 1992). The possibility to monitor the behaviour of large numbers of individual plant cells should also create new insights into the series of highly complex events that take place in the initiation and establishment of plant tissue and cell cultures.

6. Acknowledgements

We thank our colleagues for communicating unpublished results. Primary research in our laboratory is supported by the EC-BRIDGE programme, the Foundation of Life Sciences and the Technology Foundation, both subsidized by the Netherlands Organization for Scientific Research, and the EC-Biotech-PTP programme.

7. References

Berger, F., Taylor, A., Brownlee, C. (1994) Cell fate determination by the cell wall in early *Fucus* development. Science 263, 1421-1423.

Carpita, N.C., Gibeaut, D.M. (1993) Structural models of primary cell walls in flowering plants: consistency of molecular structure with the physical properties of the walls during growth. Plant J. 3, 1-30.

Collinge, D.B., Kragh, K., Mikkelsen, J.D., Nielsen, K.K., Rasmussen, H., Vad, K. (1993) Plant chitinases. Plant J. 3, 31-40.

Cordewener, J.H.G., Booij, H., van der Zandt, G., van Engelen, F., van Kammen, A., de Vries, S.C. (1991) Tunicamycin-inhibited carrot somatic embryogenesis can be restored by secreted cationic peroxidase isoenzymes. Planta 184, 478-486.

De Jong, A.J., Cordewener, J.H.G., LoSchiavo, F., Terzi, M., Vanderkerckhove, J., van Kammen, A., de Vries, S.C. (1992) A carrot somatic embryo mutant is rescued by chitinase. Plant Cell 4, 425-433.

De Jong, A.J., Heidstra, R., Spaink, H.P., Hartog, M.V., Meijer, E.A., Hendriks, T., LoSchiavo, F., Terzi, M., van Kammen, A., de Vries, S.C. (1993a) Rhizobium lipooligosaccharides rescue a carrot somatic embryo mutant. Plant Cell 5, 615-620.

De Jong, A.J., Schmidt, E.D.L., De Vries, S.C. (1993b) Early events in higher-plant embryogenesis. Plant Mol. Biol. 22, 367-377.

Dénarié, J., Cullimore, J. (1993) Lipo-oligosaccharide nodulation factors: A minireview. New class of signalling molecules mediating recognition and morphogenesis. Cell 74, 951-954.

De Vries, S.C., Booij, H., Janssens, R., Vogels, R., Saris, L., Lo Schiavo F., Terzi, M., van Kammen, A. (1988a) Carrot somatic embryogenesis depends on the phytohormone-controlled expression of correctly glycosylated proteins. Genes Develop. 2, 462-476.

De Vries, S.C., Booij, H., Meyerink, P., Huisman, G., Wilde, H.D., Thomas, T.L., van Kammen, A. (1988b) Acquisition of embryogenic potential in carrot cell suspension cultures. Planta 176, 205-211.

Dickstein, R., Prusty, R., Peng, T., Ngo, W., Smith, M.E. (1993) ENOD8 a novel early nodule-specific gene, is expressed in empty alfalfa nodules. Mol. Plant-Microbe Interact. 6, 715-721.

Felix, G., Regenass, M., Boller, T. (1993) Specific perception of subnanomolar concentrations of chitin fragments by tomato cells: Induction of extracellular alkalinization, changes in protein phosphorylation, and establishment of a refractory state. Plant J. 4, 307-316.

Fincher, G.B., Stone, B.A., Clarke, A.E. (1983) Arabinogalactan proteins; structure, biosynthesis, and function. Annu. Rev. Plant Physiol. 34, 47-70.

Fry, S.C. (1990) Roles of the primary cell wall in morphogenesis. In: Progress in plant cellular and molecular biology, HJJ Nijkamp, LHW van der Plas, J. van Aartrijk, eds. Kluwer Academic Publishers, Dordrecht, pp. 504-513.

Giuliano, G., LoSchiavo, F., Terzi, M. (1984) Isolation and developmental characterization of temperature-sensitive carrot cell variants. Theor. Appl. Genet. 67, 179-183.

Hendriks, T., Meijer, E.A., Thoma, S., Kader, J-C., De Vries, S.C. (1993) The carrot extracellular lipid transfer protein EP2: Quantitative aspects with respect to its putative role in cutin synthesis. In: Molecular-genetic analysis of plant metabolism and development, NATO-ASI Seminars, G. Coruzzi, P. Puidgomenech, eds. Plenum Press, New York, in press.

Iiyama, K., Lam, T.T-B., Stone, B.A. (1994) Covalent cross-links in the cell wall. Plant Physiol. 104, 315-320.

Kato, Y., Yamanouchi, H., Hinata, K., Ohsumi, C., Hayashi, T. (1994) Involvement of phenolic esters in cell aggregation of suspension-cultured rice cells. Plant Physiol. 104, 147-152.

Kieliszewski, M.J., Lamport, D.T.A. (1994) Extensin: repetitive motifs, functional sites, post-translational codes, and phylogeny. Plant J. 5, 157-172.

Knox, J.P. (1993) The role of cell surface glycoproteins in differentiation and morphogenesis. In: Post-translational modifications in plants, N.H. Battey, H.G. Dickinson, A.M. Hetherington, eds, Cambridge University Press, pp. 267-283.

Kreuger, M., van Holst, G-J. (1993) Arabinogalactan proteins are essential in somatic embryogenesis of *Daucus carota* L. Planta 189, 243-248.

LoSchiavo, F., Giuliano, G., Pitto, L., de Vries, S.C., Bollini, R., Genga, A.M., Nuti-Ronchi, V., Cozzani, F., Terzi, M. (1990) A carrot variant temperature-sensitive for somatic embryogenesis reveals a defect in glycosylation of extracellular proteins. Mol. Gen. Genet. 223, 385-393.

LoSchiavo, F., Quesada-Allue, L.A., Sung, Z.R. (1986) Tunicamycin affects somatic embryogenesis but not cell proliferation in carrot. Plant Sci. 44, 65-71.

Meijer, E.A., De Vries, S.C., Sterk, P., Gadella Jr., D.W.J., Wirtz, K.W.A., Hendriks, T. (1993) Characterization of the non-specific lipid transfer protein EP2 from carrot (*Daucus carota* L.). Mol. Cell. Biochem. 123, 159-166.

McCann, M.C., Roberts, K. (1991) Architecture of the primary cell wall. In: The Cytosceletal Basis of Plant Growth and Form, C.W. Lloyd, ed., Academic Press, London, pp. 109-128.

Nasrallah, J.B., Nasrallah, M.E. (1993) Pollen-stigma signalling in the sporophytic self-incompatibility response. Plant Cell 5, 1325-1335.

Nomura, K., Komamine, A. (1985) Identification and isolation of of single cells that produce somatic embryos at a high frequency in carrot suspension cultures. Plant Physiol. 79, 988-991.

Pennell, R.I., Janniche, L., Scofield, G.N., Booij, H., De Vries, S.C., Roberts, K. (1992) Identification of a transitional cell state in the developmental pathway to carrot somatic embryogenesis. J. Cell Biol. 119, 1371-1380.

Roberts, K. (1989) The plant extracellular matrix. Curr. Opin. Cell Biol. 1, 1020-1027.

Rubos, A.C. (1985) Isolation and culture of single cells from carrot embryos. Plant Sci. 38, 107-110.

Schmidt, E.D.L., De Jong, A.J., De Vries, S.C. (1994) Signal molecules involved in plant embryogenesis. Plant Mol. Biol., in press.

Showalter, A.M. (1993) Structure and function of plant cell wall proteins. Plant Cell 5, 9-23.

Spaink, H.P., Wijfjes, A.H.M., Van Vliet, T.B., Kijne, J.W., Lugtenberg, B.J.J. (1993) Rhizobial Lipo-oligosaccharide signals and their role in plant morphogenesis; are analogous lipophilic chitin derivatives produced by the plant? Aust. J. Plant Physiol. 20, 381-392.

Sterk, P., Booij, H., Schellekens, G.A., Van Kammen, A., De Vries, S.C. (1991) Cell-specific expression of the carrot EP2 lipid transfer protein gene. Plant Cell 3, 907-921.

Sterk, P., De Vries, S.C. (1992) Molecular markers for plant embryos. In: Synseeds: Applications of synthetic seeds to crop improvement, K. Redenbauch, ed, CRC Press, London, pp. 115-132.

Thoma, S.L., Kaneko, Y., Somerville, C. (1993) The non-specific lipid transfer protein from *Arabidopsis* is a cell wall protein. Plant J. 3, 427-437.

Thoma, S.L., Hecht, U., Kippers, A., Botella, J., S.C. De Vries, Somerville, C. (1994) Tissue-specific expression of a gene encoding a cell wall-localized lipid transfer protein from *Arabidopsis*. Plant Physiol. 105, 35-45.

Toonen, M.A.J., Hendriks, T., Schmidt, E.D.L., Verhoeven, H.A., Van Kammen, A., De Vries, S.C. (1994) Description of somatic embryo-forming single cells in carrot suspension cultures employing video cell tracking. Planta, in press.

Van Engelen, F.A., Sterk, P., Booij, H., Cordewener, J.H.G.,Rook, W., Van Kammen, A., De Vries, S.C. (1991) Heterogeity and cell-type specific localization of a cell wall glycoprotein from carrot suspension cells. Plant Physiol. 96, 705-712.

Van Engelen, F.A., Hartog, M.V., Thomas, T.L., Taylor, B., Sturm, A., Van Kammen, A., De Vries, S.C. (1993) The carrot secreted glycoprotein gene *EP1* is expressed in the epidermis and has sequence homology to *Brassica* S-locus glycoproteins. Plant J. 4, 855-862.

Van Engelen, F.A., De Vries, S.C. (1992) Extracellular proteins in plant embryogenesis. Trends in Genet. 8, 66-70.

Van Engelen, F.A., De Vries, S.C. (1993) Secreted proteins in plant cell cultures. In: Morphogenesis in Plants, K.A.Roubelakis-Anelakis, K. Tran Thanh Van, eds, Plenum Press, New York, pp. 181-200.

Van Holst, G-J., Clarke, A.E. (1986) Organ specific arabinogalactan-proteins of *Lycopersicon peruvianum* (Mill.) demonstrated by crossed electrophoresis. Plant Physiol. 80, 786-789.

EXPRESSION OF FLORAL SPECIFIC GENES IN TOMATO HYPOCOTYLS IN LIQUID CULTURE

L.PITTO, L.GIORGETTI, C.MIARELLI, G.LUCCARINI, C.COLELLA AND V.NUTI RONCHI
Institute of Mutagenesis and Differentiation, C.N.R.
via Svezia 2/10 56100 Pisa
Italy

ABSTRACT. Cytological and histological analysis of hypocotyl explants of many species revealed the occurrence of meiotic-like events (somatic meiosis and prophase reduction) located into structures differentiated around the vascular strands. These structures could be assimilated to primitive reproductive organs (pistil- and anther-like) containing pollen-like and embryo-sac like cells. Previous work demonstrated that segregating events in carrot are a prerequisite for the acquisition of totipotency. Here the floral nature of the structures developed from hypocotyls in culture was demonstrated, at the molecular level, in explants of tomato,Lycopersicon being a specie where floral specific genes are available. In situ hybridization experiments showed, in tomato hypocotyls cultured in vitro,the induction of two tomato floral specific genes: the TM8, a gene of the MADS family which in vivo is highly expressed specifically in floral meristems, and the MON 9612, a floral specific gene very tightly spatially and temporally regulated in tomato pistils. A study of the temporal pattern expression of these genes was performed by RT-PCR.

1. Introduction

Chromosome segregation events, named somatic meiosis and prophase reduction, have been demonstrated to occur in highly embriogenic carrot cell lines and in carrot hypocotyl explants from which new embryogenic cell cultures are generated (Nuti Ronchi et al.1992 a,b). In hypocotyls, meiotic-like events were detected in different zones of the vascular strands, during the 20 days of culture, forming a population of segregated cells (Nuti Ronchi et al. 1990). More recent results showed, at the cytological and at the molecular level, that the segregational events are related to the acquisition of embryogenic competence as shown by the segregation of molecular markers in somatic embryos (Giorgetti et al.submitted). Moreover, the reductional events were localized on structures similar to primitive reproductive organs, developed along the vascular cylinder, as rough sporangium- and ovary-like forms or rudimentary inflorescence, suggesting that a more general genetic reprogramming event could be the main response of carrot hypocotyls to stress of culture (Pitto et al. 1992, Nuti Ronchi et al. 1992).
The results obtained utilizing the same culture protocol on hypocotyl explants from other species (Lycopersicon esculentum, Helianthus annuus, Prunus, Vicia faba, etc) confirmed the occurrence of segregational events also among different plant species during the first days of culture *in vitro*. In Lycopersicon a wave of reductional divisions is detectable already at the 7th hour of culture, whereas the reproductive structures were well developed after 10 days.

M. Terzi et al. (eds.), Current Issues in Plant Molecular and Cellular Biology, 369-375.

This work has been carried out with the aim to confirm, at the molecular level, the real identity of the homeotic structures that so strikingly resemble floral organs. The strategy chosed to obtain this objective has been offered by the possibility to use specific floral cloned genes that, however, are available only for some plant.Whereas it has been possible to obtain a genetic proof of the segregation events in carrot, thanks to the single cell origin of carrot somatic embryos, no floral genes have been cloned in this specie. Therefore the expression of floral genes have been studied on tomato which is, among the species studied up to now showing segregational events, the one where more cloned floral specific genes are available.

2. Materials and Methods

2.1. HYPOCOTYL CULTURE

Lycopersicon esculentum (cv.Roma) hypocotyls were explanted as described by de Vries et al. (1988) for carrot hypocotyls.

2.2. CYTOLOGICAL ANALYSIS

Cytological and istological analysis were performed as described by Nuti et al. (1990). The callose specific staining is described in Nuti (1981).

2.3. RNA EXTRACTION

Total RNAs were prepared essentially as described by Giuliano et al. (1993) except for the following alterations: 1-5 gr of tissue was ground to a fine powder using a mortar with pestel under liquid nitrogen. Powdered cells were homogenized in buffer A (4M guanidine thiocianate,1% sarcosyl, 0.1M sodium acetate ph 5.2, 0.7% 2-mercaptoethanol), 5ml/1gr of tissue. The final RNA precipitation was performed in Na acetate ph 4,2, instead of lithium chloride.

2.4. RT-PCR

The following ologonucleotides, selected utilizing the OLIGO program, were used in the amplifications: TM8, TCTCCTTTTCTCTCCTTCTG (upstream) and TGCTTCAATCTTTCATACCA(downstream);MON9612, ACGGCAACATTAGGGACTCA (upstream) and CCTCTTTGCGGTATTTTTCA (downstream).

RT-PCR was performed using the RNA PCR kit (Perkin Elmer Cetus) according to the manifacturer's instructions. The amounts of plant RNA used ranged between 100ng to 1ug .In each samples a control transcript with relative oligonucleotides (as described in Giuliano et al. 1993) was included .The reverse transcriptase reactions were performed at 42°C for 1hr followed by 5 min. at 99°C to inactivate the reverse transcriptase. Retrotranscribed samples were amplified in a Omnigene temperature cycler (Hybaid), with the following protocol: 95°C (2min), then 35 cycles at 93°C (1min),53°C (1 min) for the MON 9612 amplification, 93°C(1 min),51°C (1 min)for the TM8. Taq polymerase was generally added separately to each samples after the first minute at 95°C.

2.5. IN SITU HYBRYDIZATION

For production of labeled RNA probes, the cDNA TM clones were isolated from the pTZ 18 vector in which they were inserted and cloned into the p GEM 7Zf(+) (Promega). The pMon cDNA clones received were already subcloned into Bluescript Ks^- (Gasser et al. 1989). In all the *in situ* experiments probes were made for each strand to control that only one of the two probes hybridized to sections of the same hypocotyls. *In situ* hybridization experiments were performed as described in Smith et al. (1987).

3. Results and Discussion

3.1. CYTOLOGICAL AND HISTOLOGICAL ANALYSIS

Cytological analysis of the tomato hypocotyls explants cultured in liquid medium in presence of 2,4-D showed the same chromosomal reducing events already described in carrot, namely somatic meiosis and prophase reduction (Nuti et al. 1992 a, b). Meiotic activity starts after only 3 hours of culture and is localized in specific cells along the vascular strands. The proliferation rate is very high in this kind of cells remaining constant from the 7th to the 72th hours of culture. As reported in table 1, the cytological anlysis reveals a high percentage of reductional events with a maximum of 83% at the 23 hrs

TABLE 1. Cytological analysis of tomato hypocotyl explants cultured in liquid B5 medium additioned with 2,4 D.

time (h)	14	20	23	48	72
reductional events (%)	58.33	61	83	38.3	30

Frequently , in tomato, these divisions are of the type prophase reduction, where the prophase nucleus with a large nucleolus, persistent through all the phases of the division, is split to the final formation of two nuclei clearly composed each of 12 pairs of chromosomes and one nucleolus.

The consequence of these reductional events is, as it occurs *in vivo*, the formation of gametic like cells: embryo sac-like cells may be identified in the very large and vacuolized cells showing nuclei which, for number and distribution, remind the *in vivo* features, whereas smaller cells were similar to immature mononucleated pollen grains.A specific staining for the callose wall (see methods) a typical marker for male gametophyte *in vivo*, revealed the presence of this kind of wall also in these pollen-like cells originated *in vitro*. Gametic-like cells expressing other markers of the "gametic" conditions were also identified in hypocotyl explants of other species (Giorgetti et al.1993). In the carrot hypocotyl cultures these reproductive-like cells are released into the medium and are able, having acquired the embryogenic competence, to form Proembryogenic Masses (PEMs) and somatic embryos when transferred to the proper medium (Giorgetti et al. submitted). Tomato gametic-like cells , instead, are released into the medium only late in culture and undergo a degradative process that make them unable to divide further and therefore to perform embryogenesis..

The aspect of tomato hypocotyl explants changes drammatically after only few days in culture: the cortical layers appear swollen at one end, assuming a resemblance to carpels, in most cases also with some lengthwise thickenings as sutures imitating the fusions zones of the tomato carpels.

Contemporaneously, around the vascular strand, a placenta-like is formed into or over which structures similar to ovules develop.Several variants of these features may be present in different hypocotyls of the same culture. In explants of 6BaP treated seedlings the pistil-like appears as closed in a calix-like compressed structure, and the ovules soon degenerate, their cells being differentiated in short tracheids. Scattered in the placenta, particularly in proximity of the ovules, strand of tissue with cytological features of the stylar transmitting tissue are found.

Another-like structures are also present, and others resembling a very rough inflorescence, where a proper corolla is never formed, but the primordia, (always lacking a well formed epidermis), develop, as in tomato *in vivo*, as a forked racemose cyme.

3.2. MOLECULAR ANALYSIS

In a previous study RNAs poly A+ were isolated from tomato flowers at different stages of development, from tomato leaves and from hypocotyls mantained in liquid culture for different periods of time. Northern blot experiments were performed with these samples utilizing , as probes, homologous and heterologous floral specific clones isolated in other laboratories (Pitto et al. 1992).Whereas clear hybridization signals were obtained with flower RNAs,with hypocotyl RNAs, most of the probes tested showed a very weak signal often irreproducible from different RNA preparations. These results clearly indicate that the method used was not sensitive enough to monitor possible induction of floral specific genes during the development of hypocotyl structures.A very low signals can be justified considering the large variability of developmental stages of the hypocotyl floral structures growing in the flasks. Moreover, many of the clones which we tested represent genes whose expression is tightly regulated in time and position.Therefore we could expect that only a small portion of our *in vitro* developed structures express these functions. The *in situ* hybridization approach, giving us the possibility to choose the best formed structures, was considered the most suitable technique to solve the problem.

Most of the floral genes cloned up to now in tomato are expressed only or mainly in mature floral organs (Gasser et al. 1989; Twell et al.1990; Pnueli et al. 1991). Exceptions were represented by the TM 4 and TM 8 genes. These genes are part of the MAD family.They were isolated from a libraries generated from floral meristems of the tomato mutants *anantha* utilizing as probe the *deficiens* cDNA clone (Pnueli et al.1991). *Anantha* is a recessive mutation which blocks floral promordia is before organogenesis. Accordingly to their origin these two genes showed, in Northern blot experiments, the maximum of expression in floral meristems, a reduced expression at the level of mature flower and no expression in vegetative organs (Pnueli et al. 1991). Genes involved in the early organization of floral development may have, hypothetically, a better chance to be expressed in our primitive immature structures. Therefore we tested the expression of TM4 and TM8 on tomato flowers and hypocotyls by *in situ* hybridization experiments.

As expected, both genes were expressed in tomato flowers; in hypocotyls, whereas the TM4 did not give any signal, TM8 expression was limited to the early days of culture, namely from the 2nd to the 10th day. The pattern of expression was exclusively related to the developing structures around the vascular strands particularly intense at both ends, where inflorescence primordia and ovaries are formed.

A more accurate histological analysis of the structures resembling the tomato ovary, revealed the presence of a particular tissue, cytologically similar to the glandular stigmatic tissue. This tissue, that is called stigmatoid tissue (Esau 1965) or transmitting tissue (Arber 1937), is present in the stigma but continues along the stylo to the placenta, normally in the form of strands of considerably elongated cells staining deeply with cytoplasmic stains. It is known that stigmatoid tissue occurs on the placenta within the ovary and in some species in the funiculus of the ovule as well. Due to the easy cytological recognition of this tissue in the developing tomato ovary-like forms, it seemed convenient to test the expression of pistil specific clones, even if the activity of these genes was mainly restricted to mature flower The more specific clones available was the MON 9612 (Gasser et al. 1989; Budelier et al.1990), that is expressed on the upper third of the transmitting stylar and stigmatic tissue of the mature tomato flower. *In situ* hybridization experiments have confirmed this localization in mature flowers; besides, a strong specific signal was evident also around the ovules. This fact, not reported by Gasser et al.(1989), may be due to the different cv. of tomato used.

A strong signal was detected in the tomato hypocotyl structures in strands of tissue strongly resembling, cytologically, the pistil stigmatoid tissue. Even in these ancestral primitive structures, the expression of this gene appears regulated spatially and temporally. The signal is evident from the 6th day, when, due to the lack, in the ovary-like forms, of a proper style and stigma, the expression was located, in strands, along the structure, outside the vascular cylinder or at the margin, underneath the hypocotyl epidermis. These swollen cortical layers assume the aspect of carpels enclosing the placenta and ovules. At later times (15 days), the expression is more strong and mainly located around the ovules, in the placenta that at this time may often grow from the vascular strands to fill the swollen carpels-like structure

The hypocotyl sections of 0 time of culture did not show any expression of either probes. As a controll sections of the same structures used for the "antisense" experiments, were submitted to "sense" hybrydization. No expression was detected in all specimens..

To have a more general view of the temporal pattern expression of the TM8 and MON 9612 genes on structures developing from hypocotyls in culture, we decided to use a technique that, for the high level of sensitivity has been used to study rare transcripts: the reverse transcriptase-polymerase chain reaction (RT-PCR) (Chelly et al. 1988).

The two pairs of primers designed for these experiments resulted in a 0,45 kb and a 0,22 kb bands of amplification when used, as control, in a PCR experiments with DNA respectively from the MON9612 and TM8 plasmids .Total RNAs were isolated from different tomato organs and from hypocotyls cultured in liquid medium for different periods of time as described in Materials and Methods. All the RNA preparations were blotted and hybrydized to a ^{32}P labelled soybean actin gene. Actin gene transcripts were found at a nearly equivalent levels in all RNA preparations tested. Omitting the reverse transcriptase step, no PCR amplifications were obtained from any of the RNAs samples, indicating the absence of any genomic DNA contaminations.Using RT-PCR we were able to detect expression of both genes in the floral tomato organs and, with variable intensity, in hypocotyl explants. For both genes , the signal was already quantifiable in RNAs extracted from hypocotyls mantained in culture for 5 days, although the level of expression was higher for the MON9612 gene.The expression of the TM8 gene seemed to decrease in hypocotyls cultured for longer period of time, and it is not influenced by 6BaP treatment. Last treatment seemed to extend the expression of the MON9612 gene for longer periods of time.

These data, besides confirming the cytological and histological observations about the floral nature of the *in vitro* cultured tomato hypocotyls, show that, though extremely primitive and immature, the structures may express genes very tightly regulated both spatially and temporally in the flower. Therefore the structures we have studied may be defined homeotic, accepting the definition of homeosis as the "assumption by one part of an organism of likeness to another part" (Sattler, 1988).

The conclusion would therefore follow that a determinate program leading to floral induction and including a gametophytic phase is initiated any time the system is challenged by an alterated enviroment, as an attempt of the plant to afford the unsuitable enviroment trying to propagate the species. In this respect the concept of plant cell totipotency might be reconsidered.

4. Acknowledgements

We thank Prof. Gasser and the Monsanto company who provided the pMON clones and Prof . Lifschtz who provided the TM clones. We thank Giovanni Giuliano for the helpful suggestions about the RT-PCR and for providing the oligonucleotides and transcripts for control amplifications.This work was supported partly by MAF (Ministry of Agricolture and Foresty)-"Sviluppo tecnologie innovative"program and partly by National Research Council of Italy, special project RAISA, sub-project n.2, Paper n.1543.

5. References

Arber,A. (1937) "The interpretation of the flower: a study of some aspects of morphological thought" Cambridge Phil.Soc.Biol.Rev. 12, 157-184.

Budelier,K.A., Smith,A.G. and Gasser,C.S. (1990) "Regulation of a stylar transmitting tissue Mol.Gen.Genet. 224, 183-192.

Cherlly,J., Kaplan,J.C., Gautron,S. and Khan,A. (1988) "Transcription of the dystrophin gene in human muscle and non muscle tissue" Nature vol.333, 858-860.

De Vries,S.C., Booij,H., Meyerink,P., Huisman,G., Wilde,H.D.Thomas,T.L. and Van Kammen,A.B. (1988)" Acquisition of embryogenic potential in carrot cell-suspension cultures " Planta 176, 196-204.

Esau,K. (1965) Plant anatomy John Wiley & Sons publishers, New York.

Gasser,C.S., Budelier,K.A., Smith, A.G., Shah,D.M. andFraley,R.T. (1989) "Isolation of tissue specific cDNAs from tomato pistils" The Plant Cell vol.1, 15-24.

Giorgetti,L., Pitto,L., Miarelli,C., Luccarini,G., Colella,C., Scarano,M.T. and Nuti Ronchi,V. (1993) "Auxin induced reproductive organs as primaary response to culture of different plant species" In vitro 29A,75A.

Giuliano,G., Bartley,G.E. and Scolnik,P.A. (1993) "Regulation of carotenoid biosynthesis during tomato development" The Plant Cell 5, 379-387.

Nuti Ronchi,V.(1981) "Histological studies of organogenesis *in vitro* from callus cultures of two Nicotiana species" Can. J. Bot. 59,1969-1977.

Nuti Ronchi,V.,Giorgetti,L. and Tonelli,M. (1990) "The commitment to embryogenesis, a cytological approach" in H.J.J. Nijkamp, L.H.W. Van Der Plas, J. Van Aartrijk (eds.) Progress in Plant Cellular and Molecular Biology Kluwer Academic Publishers Dordrecht pp437-442.

Nuti Ronchi,V.,Giorgetti,L., Geri,C., Pitto,L., Vergara,R. and Martini,G.(1992) "Cytological, anatomical and morphological aspects of somatic embryogenesis" in M.Griga and E.Teklova Regulation of plant somatic embryogenesis, Research Institute of Technical Crops and Legumes, Sumperk publishers pp.67-72.

Nuti Ronchi,V.,Giorgetti,L., Tonelli,M. and Martini,G. (1992) "Ploidy reduction and genome segregation in cultured carrot cell lines.I. Prophase chromosome reduction" Plant Cell Tissue and Organ Culture, 30, 107-114.

Nuti Ronchi,V.,Giorgetti,L., Tonelli,M. and Martini,G. (1992) "Ploidy reduction and Genome segregation in cultured carrot cell lines.II Somatic meiosis". Plant Cell Tissue and Organ Culture, 30, 115-119.

Pitto,L., Miarelli,C.,Giorgetti,L., Colella,C., Luccarini,G. and Nuti Ronchi,V. (1992) "Occurrence of floral homeotic structures in carrot, tomato and Helianthus hypocotyls in liquid cultures "Proceedings of theXXXVIII congress of the Italian Genetic Association 38, 107-108.

Pnueli,L., Abu-Abeid,M., Zamir,D., Nacken,W., Schwarz-Sommer,Sh.and Lifschitz,E. (1991) "The MADS box gene family in tomato: temporal expression during floral development, conserved secondarystructures and homology with homeotic genes from Anthirrinum and Arabidopsis" The Plant Journal 1,255-266.

Sattler,R. (1988) "Homeosis in plant" Am. J. Bot. 75, 1606-1607.

Smith,A.G., Hinchee,M.A. and Horsch,R.B. (1987) "Cell and tissue specific expression localized by in situ RNA hybridization in floral tissue" Plant Molec.Biol.Rep 5, 237-241.

Twell,D., Wing,R., Yamaguchi,J. and McCormick,S. (1989) "Isolation and expression of an another specific gene from tomato" Mol. Gen. Genet. 217, 240-245.

REGENERATION OF PLANTS FROM PROTOPLASTS OF *ARABIDOPSIS THALIANA* CV. COLUMBIA L. (C24), VIA DIRECT EMBRYOGENESIS

C. M. O'NEILL AND R. J. MATHIAS
Brassica and Oilseeds Research Department
John Innes Centre, Colney,
Norwich NR4 7UJ,
United Kingdom.

Abstract

This report describes the reproducible regeneration of fertile plants from mesophyll protoplasts of *Arabidopsis thaliana* Columbia (C24), Landsberg erecta, Estland and Columbia genotypes. The protocol is novel in that a large proportion of protoplasts isolated from *in vitro* grown plants regenerate via direct somatic embryogenesis. In Columbia C24, the most responsive genotype, 0.8-5.3% of protoplasts gave rise to macroscopic colonies, of which 70-80% were regular embryo-like structures. Four weeks after transfer to solid medium 68-88% of these structures had produced well developed shoots. Plants could be established in the glasshouse and the regenerated plants were fertile. The potential for exploiting direct embryogenesis from protoplasts in a well characterised model plant is discussed.

Introduction

Embryogenesis is a key developmental process of fundamental biological importance in the plant life cycle. It is also of agronomic importance as embryo and seed development are closely coupled and the embryo can have a major role in determining storage product accumulation. This is especially significant in crop species where the principal seed storage tissues are the cotyledons, as in the major European temperate oilseed species, sunflower (*Helianthus annuus*) and oilseed rape (*Brassica napus*). The developmental regulation of starch and oil synthesis, and the switch from one to the other, has led to speculation that manipulation of developmental genes could have a long term role in increasing oil yields. Embryo and seed development is a temporally and spatially regulated complex of interacting genetic, environmental and biochemical steps. While a lot of descriptive, and some correlative, data on embryo development is available, comparatively little is known of its biological determination and genetic control.

Somatic embryogenesis has been suggested as a tool in the analysis of the environmental, biochemical and genetic control of embryogenesis, although the precision with which somatic embryogenesis mimics zygotic embryo development varies a great deal according to the species and culture systems involved. Somatic embryogenesis has been reported from many systems but perhaps the best characterised system is carrot (*Daucus carota*). The dedifferentiation of somatic

M. Terzi et al. (eds.), Current Issues in Plant Molecular and Cellular Biology, 377-382.

cells, into pro-embryogenic and embryogenic cells, in culture, is itself an interesting developmental phenomenon. The determination of embryogenic versus organogenic development *in vitro* is an area in which a number of contributory factors, including osmotic stress [1,2], genotype [3,4], and growth regulator regime [1,2,5,6] among others, have been recognised but no mechanism has been identified.

Arabidopsis thaliana, a Cruciferous weed distantly related to the *Brassica* crop group, has been used extensively in classical genetic studies and a comprehensive library of mutant lines exists [7]. It is argued that *Arabidopsis*' small size, short life-cycle and large bank of genetic stocks make it an ideal subject for genetic analysis [8]. Recent molecular and genetic mapping of the *Arabidopsis* genome has increased its potential role in molecular and whole plant studies. In seeking to exploit protoplast fusion between Arabidopsis and Brassica spp. for genome analysis we discovered that cell and tissue culture techniques in *Arabidopsis* are poorly developed. Plant regeneration from protoplasts of mesophyll [9-11] and suspension culture cells, [12], has been described, but various workers have found the responses in these culture systems to be erratic. We were unable to regenerate plants from mesophyll protoplasts using either of two published methods [9,10]. However, the replacement of 2,4-dichlorophenoxyacetic acid (2,4-D) in the culture medium with 3,6-dichloro-2-methoxybenzoic acid (dicamba) enabled us to regenerate plants [13]. Protoplasts cultured in the presence of dicamba underwent a distinctive pattern of early divisions and subsequently many of the dividing protoplasts directly formed structures resembling somatic embryos. These produced the majority of the shoots that were regenerated in culture.

Materials and Methods

The plant materials used in this study were from inbred populations of the *Arabidopsis thaliana*, races Columbia C24, Columbia, Estland and Landsberg Erecta, maintained at the John Innes Centre, Norwich. We have previously described in detail the preparation of plant material, isolation and culture of protoplasts, regeneration of plants and the method used for sectionning tissues for microscopic analysis [13].

Results

Using published protocols [9,10] first divisions were observed in 9 to 13% of protoplasts 6-7 days after initial plating. Some protoplasts continued to develop normally for up to 2 weeks but we were unable to routinely culture cells beyond the 2-3 week stage. (In only one experiment were microclonies produced). However, there was a dramatic improvement in culture response when dicamba replaced 2,4-D in the culture medium. The first cell divisions were seen after 3-5 days, the rate of division was more rapid than in 2,4-D and did not decrease after 10-14 days. Columbia C24 was the most responsive genotype with 6.7-22.3% of protoplasts entering first division and 0.8-5.3% producing macroscopic structures (>0.1mm in diameter) (table 1). The majority of early cell divisions in the protoplasts treated with dicamba followed a very distinctive pattern. Initial divisions were not accompanied by significant cell expansion, consequently the protoplasts were sub-divided into small, densely cytoplasmic cells, resulting in extremely compact micro-colonies. Seventy to 80% of the colonies were oval, with a smooth outline and made up of extremely compact and highly cytoplasmic cells (Fig. 1a). The remainder (20-30%) were typical friable, "organogenic" calli, consisting of loose cell aggregates without a clearly defined shape or surface. Over 3-5 weeks in culture the smooth structures developed into globular and, in some cases, heart-shaped structures, 0.1-1.0mm long. On transfer to the light all the macroscopic structures quickly became green and when transferred to shoot regeneration medium 68-88% produced shoots (table 1). The smooth, compact structures produced shoots and roots without an intervening callus phase. Once the shoots were ≥ 1 cm long they could be rooted and then transferred to compost where they eventually flowered.

The densely cytoplasmic cells and distinctive early divisions of dicamba treated protoplasts which led to distinct oval structures that "germinated" suggested that this was direct somatic embryogenesis from protoplasts. Semi-thin sections through representative callus and "embryo" structures revealed that while the former consisted of disorganised and loose aggregates of vacuolated cells (Fig. 1b) the internal structure of the latter was highly organised and contained compact, densely cytoplasmic cells (Fig. 1c). In sections of older structures the anatomy was similar to that of an early zygotic embryo; including a distinct epidermal layer and meristemmatic regions.

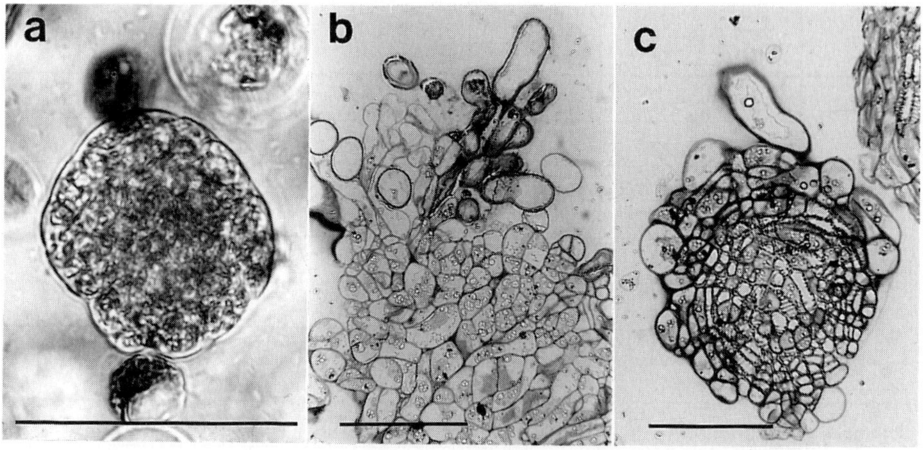

Figure 1. Stages in the regeneration of protoplasts of A. thaliana Columbia C24 in cultures containing dicamba. a: Early compact cell colony of densely cytoplasmic cells. b: Semi-thin section of "organogenic" type cell colony. c: Semi-thin section of "embryogenic" type cell colony. Bar = 100μm.

Initially dicamba was used, as a direct substitute for 2,4-D, at a concentration of 1mg/l and this was effective in promoting Columbia C24 protoplast regeneration via embryogenic structures. A range of dicamba concentrations, and combinations of dicamba with other growth regulators, were tested on Columbia C24 [13]. The most effective treatment for promoting protoplast division and regeneration of macroscopic colonies was 8mg/l dicamba + 0.1 mg/l 2,4-D and 10mg/l dicamba. The other genotypes were not exhaustively tested on the full range of dicamba and other growth regulator regimes used with Columbia C24. They were less responsive on the three regimes that were tested and the beneficial effects of dicamba were not so marked as in Columbia C24. In Columbia and Estland the 8mg/l of dicamba + 0.1mg/l 2,4-D treatment resulted in higher first division frequencies than the 1mg/l dicamba treatment. In contrast, in all three genotypes shoot regeneration frequencies appear to be better in 1mg/l of dicamba than 8mg/l dicamba + 0.1mg/l 2,4-D. However, the regeneration figures and differences between them are small and have not been tested statistically so variation between treatments should be interpreted with caution. In all cases the substitution of dicamba for 2,4-D enabled the recovery of colonies from protoplasts and the regeneration of plants (table 1). The protoplast development in these genotypes, (early cell divisions with minimal cell expansion producing compact, organised macroscopic colonies of highly cytoplasmic cells) resembles the process initially observed in Columbia C24.

Table 1. Protoplast regeneration responses of four *A. thaliana* genotypes on three growth regulator regimes ([a] range of response and [b] mean from 2-5 replicate experiments).

Genotype	Hormone regime	1st division frequency (%)		Plating efficiency (%)		Shoot regeneration (%)	
Columbia C24	1mg/l 2,4-D	[a]9.0-13.0	[b]11.0	0		0	
	1mg/l dicamba	6.7-13.0	9.0	[a]0.8-1.4	[b]1.0	[a]53.0-82.0	[b]60.0
	8mg/l dicamba + 0.1mg/l 2,4-D	13.0-18.7	16.6	3.2-5.3	4.3	68.0-88.0	78.0
Landsberg erecta	1mg/l 2,4-D	3.0-4.0	3.5	0		0	
	1mg/l dicamba	2.0-10.0	6.6	0.2-1.2	0.7	0-8.5	3.4
	8mg/l dicamba + 0.1mg/l 2,4-D	5.0-11.0	7.4	0.1-1.0	0.6	0-3.0	0.5
Estland	1mg/l 2,4-D	n/a		n/a		n/a	
	1mg/l dicamba	4.0-7.0	5.3	0.1-1.0	0.6	0-6.0	2.7
	8mg/l dicamba + 0.1mg/l 2,4-D	8.0-18.0	12.0	0.7-1.3	1.0	1.2-2.5	1.9
Columbia	1mg/l 2,4-D	10.0-11.0	10.5	0-0.6	0.3	0	
	1mg/l dicamba	2.0-13.0	7.7	0.29-1.3	0.5	0.1-0.6	0.3
	8mg/l dicamba + 0.1mg/l 2,4-D	8.0-18.0	13.0	0-0.9	0.5	0-0.1	0.1

Discussion

Attempts to regenerate plants from protoplasts of four genotypes of *A. thaliana* using published protocols [9,10] were unsuccessful. However, the substitution of dicamba, for 2,4-D, in the culture medium enabled plants to be recovered from protoplasts. The most responsive genotype was Columbia C24 were first division frequencies (6.7-18.7%), plating efficiencies (0.8-5.3%) and shoot regeneration (68-88%) were comparable with reported figures [9-11]. Among the other races that were tested the overall efficiency of the regeneration system was much lower, principally because in these genotypes the frequency of shoot regeneration from macroscopic colonies was less than 5%.

The majority of protoplasts (70-80%) cultured in dicamba, alone or in combination with 2,4-D, underwent a series of distinctive divisions to produce very compact micro-colonies which develop into distinct, compact structures that produce shoots and roots without an intervening callus phase and thus appear to be somatic embryos. The remaining 20-30% of colony forming protoplasts develop into friable calli.

Protoplasts embedded in a semi-solid matrix often produce rather compact cell colonies and colonies of this type predominate in 2,4-D treated cultures but occur at low frequencies (20-30%) in dicamba treatments. The majority of the colonies in dicamba treated cultures are distinct in appearance and anatomy, being very compact, globular colonies of densely cytoplasmic cells. The occurrence, in *Arabidopsis* cultures, of globular colonies that may be equivalent to those we observed has been described. However, in one case shoot formation from regenerated calli was described as organogenic [9]. In the other, colonies labelled "type B", which were described as compact and smooth, and may have been embryogenic structures, did not grow beyond 50-100 cells [11].

Direct embryogenesis from mesophyll protoplasts and embryogenic development in cell cultures has been reported among the *Brassiceae* [14-18] and a reduction in the auxin content of the medium has been suggested as the trigger that switches cells from organogenic to embryogenic development [5,6]. Dicamba induces the development of embryogenic calli in other species [2,19,20] but the mechanism by which it does so is unknown. It is known that catabolism of dicamba and 2,4-D differs in whole plants [21] so variation in the rate and mechanism of detoxification of growth regulator pools may account for the different responses they induce *in vitro*. In this culture system dicamba exhibits none of the toxicity commonly associated with high concentrations of 2,4-D and consequently is effective up to 10 mg/l. The synergistic effect of low concentrations of 2,4-D combined with high concentrations of dicamba is also of interest with regard to growth regulator metabolism and developmental control.

Recently monoclonal antibodies raised to developmentally regulated arabinogalactan proteins (AGPs) have been used to immunochemically map the development of plants and cells *in vivo* and *in vitro*. AGP antibodies have been used as probes for key biochemical changes in cells that mark developmental changes or presage entry into particular developmental pathways [22,23]. In carrot suspension cultures, cell divisions that lead to embryogenic and non-embryogenic division pathways can be discriminated on the basis of unequal cell divisions. These divisions are biochemically unequal as the expression of the AGPs on the plasmamembranes of daughter cells is different. They may also be visibly unequal, producing daughter cells of unequal size [24]. Initial experiments with the *A. thaliana* protoplast system indicate that similar morphological and biochemical differentiation accompanies early divisions in dicamba-treated cultures.

In common with other protoplast cultures this system enables isolated, differentiated, somatic cells to de-differentiate and re-enter the cell division cycle. Unusually the entry of *A. thaliana* protoplasts into "organogenic" and "embryogenic" developmental pathways, from the earliest cell divisions, can be manipulated by the application of a chemical "switch". The availablility of such a model in a species that is well characterised at the level of conventional and molecular genetics provides unique opportunities to study a key developmental process. This culture system will facilitate studies of the interaction of environmental and chemical signals with wild-type, embryo-development and hormone mutants [25,26] and potentially contribute to the unravelling of the genetic control of embryo development.

References

1. Brown, C., Brooks, F., Pearson, D., Mathias, R.J. (1989) J.Plant Physiol. **133**: 727-733
2. Close, K.R., Ludeman, L.A. (1987) Plant Sci. **52**: 81-98
3. Brown, D.C.W., Atanassov, A. (1985) Plt.Cell Tiss.Org.Cult. **4**: 111-122
4. Tomes, D.T., Smith, O.S. (1985) Theor.Appl.Genet. **70**: 505-509
5. Kirti, P.B., Chopra, V.L. (1989) Plant Breeding **102**: 73-78
6. Gupta, V., Agnihotri, A., Jagannathan, V.. (1990) Plant Cell Rep. **9**: 427-430
7. Koornneef, M. (1990) In: Genetic maps. Ed. O'Brien, S.J. Cold Spring Harbour Laboratory Press, Cold Spring Harbour, N.Y.. pp 95-99
8. Meyerowitz, E.M. (1987) Ann.Rev.Genet. **21**: 93-111
9. Damm, B., Willmitzer, L. (1988) M.G.G. **213**: 15-20
10. Karesch, H., Bilang, R., Potrykus, I. (1991) Plant Cell Rep. **9**: 575-578
11. Masson, J., Paszkowski, J. (1992) The Plant Journal **2**: 829-833
12. Ford, K.G. (1990) Plant Cell Rep. **8**: 534-537
13. O'Neill, C.M., Mathias, R.J. (1993) J.Exp.Bot. **44**: 1579-1585
14. Li, L.C., Kohlenbach, H.W. (1982) Plant Cell Rep. **1**: 209-211
15. Maheswaran, G., Williams, E.G.. (1986) J Plant Physiol. **124**: 455-463
16. Kranz, E. (1988) Plt.Cell Tiss.Org.Cult. **12**: 141-146
17. Eapen, S., Abraham, V., Gerdemann, M., Scheider, O. (1989) Ann.Bot. **63**: 369-372
18. Kirti, P.B., Chopra, V.L. (1990) Plt.Cell Tiss.Org.Cult. **20**: 65-67
19. Hunsinger, H., Schauz, K. (1987) Plant Breeding **98**: 119-123
20. Papenfuss, J.M., Carman, J.G. (1987) Crop Sci. **27**: 588-593
21. Broadhurst, N.A., Montgomery, M.L., Freed, W.H. (1966) J.Agric.Food Chem. **14**: 585-588
22. Pennell, R.I., Roberts, K. (1990) Nature **344**: 547-549
23. Pennell, R.I., Janniche, L., Kjellbom, P., Scofield, G.N., Peart, J.M., Roberts, K. (1991) The Plant Cell **3**: 1317-1326
24. Pennell, R.I., Janniche, L., Scofield, G.N., Booij, H., de Vries, S.C., Roberts, K. (1992) J.Cell Biol. **119**: 1371-1380
25. Mayer, U., Torres-Ruiz, R.A., Berleth, T., Misera, S., Jurgens, G. (1991) Nature **353**: 402-407
26. Wilson, A.K., Pickett, F.B., Turner, J.C., Estelle, M. (1990) M.G.G. **222**: 377-383

CONTROLLING FACTORS AND MARKERS FOR EMBRYOGENIC POTENTIAL AND REGENERATION CAPACITY IN BARLEY (*HORDEUM VULGARE* L.) CELL CULTURES

A. P. MORDHORST [1], S. STIRN, T. DRESSELHAUS, H. LÖRZ
University of Hamburg
Institute for General Botany
Centre for Applied Plant Molecular Biology, AMP II
Ohnhorststr. 18
22609 Hamburg
Germany

[1]Present adress: *Wageningen Agricultural University*
Department of Molecular Biology
Dreijenlaan 3
6703 HA Wageningen
The Netherlands

Abstract

Factors affecting *in vitro* embryogenesis as well as markers for this developmental process have been characterized in barley cell cultures. Through modifications of the nitrogen composition in culture media of microspore cultures it was possible to manipulate plating efficiency as well as embryogenesis and plant regeneration independently. A certain combination of nitrate, ammonium, and glutamine (N24 A3 G3 medium) e.g. led to the formation of germinating androgenetic embryos in a high frequency, while glutamine as sole nitrogen source (G30 medium) led to the development of embryogenic, but non-regenerating aggregates. This microspore system was used in Northern experiments to analyse the expression of two barley cDNA-clones (B15C, pG22-69) expressed during zygotic embryogenesis. The analysis indicated that these clones are early markers for *in vitro* embryogenesis and that this development is also induced in non-regenerating aggregates, but is stopped before differentiation in scutellum, shoot and root primordia. These markers could also distinguish embryogenic (regenerable and non-regenerable) from non-embryogenic barley cell suspensions. In these cultures we identified a cellular 85 kDa polypeptide which accumulated during the loss of regenerative capacity. Following the pattern of proteins secreted to the medium polypeptides were characterized correlating with embryogenic capacity (46 kDa), or the loss of regenerative potential (17.4 and 40.5 kDa), respectively. The data indicated that induction of division, prolonged proliferation, induction of embryogenesis and embryo development, respectively, are independently regulated processes. The markers can be used to characterize embryogenic cultures as well as to elucidate the molecular mechanisms of cell differentiation.

Introduction

Although all economically important cereal crops, including wheat and barley have been transformed recently (Becker et al. 1994, Jähne et al. 1994), these techniques are of low efficiency and limited up to now to a few laboratories and a low number of genotypes. Efficient transformation methods as tools for plant breeding and research require the transfer and the integration of foreign DNA into transformation as well as regeneration competent cells (Potrykus

M. Terzi et al. (eds.), Current Issues in Plant Molecular and Cellular Biology, 383-388.

1990). The low and unpredictable loss of embryogenic capacity during prolonged time of culture is the most limiting factor for the routine production of transgenic cereals. Factors influencing *in vitro* embryogenesis and regeneration are still poorly understood and need therefore to be investigated in more detail. Barley microspore cultures are an ideal system for these investigations, since regeneration occurs via germination of directly formed androgenetic embryos derived from single cells (without previous callus formation). As shown e. g. for carrot (Wetherell and Dougall 1976) and rice (Grimes and Hodges 1990) the nitrogen composition is one of these factors and was therefore studied intensively in this system.

In barley, cDNA-clones have been identified which are expressed in zygotic embryos (B15C, dormancy related, Aalen et al. 1994, and pG22-69, linked to dessication tolerance, Bartels et al. 1991). We report here the expression of these cDNA-clones during *in vitro* embryo development of microspore cultures and their use as markers for embryogenic potential in cell suspensions. Furthermore patterns of cytosolic proteins and proteins secreted to the medium were compared in order to develop molecular markers to be able to describe the actual state of the embryogenic/ regenerative potential of individual cell cultures.

Material and Methods

SUSPENSION AND MICROSPORE CULTURES

Barley cell suspensions were initiated from anther cultures (cv. Igri) and subcultured according to Jähne et al. (1991). The regenerative capacity was tested in regular intervals by plating suspension aggregates on solid L3DB1 medium (Jähne et al. 1991). Suspensions able to regenerate plants are referred as embryogenic/regenerable, whereas embryogenic suspensions can only produce embryogenic, nodular callus on regeneration medium but subsequent regeneration is failed. Non-embryogenic cell cultures do not develop embryogenic callus and have lost their regenerative capacity.

Barley microspores were pretreated, isolated and cultured as decribed previously (Mordhorst and Lörz 1993). The nitrogen composition in the media was composed of a mixture of maximal three nitrogen compounds: nitrate (KNO_3), ammonium (NH_4Cl) and L-glutamine. Media were defined through the total nitrogen content (5 - 50 mM), the NO_3^-:NH_4^+-ratio (100:0 - 0:100) and the ratio of inorganic:organic nitrogen (N_{inorg}:N_{org}-ratio; 100:0 - 0:100). One of these parameters was varied while the others were kept constant (Mordhorst and Lörz 1993).

RNA EXTRACTION, NORTHERN BLOTS AND PROTEIN ANALYSIS

Total RNA was extracted from suspension culture aggregates of different age (and therefore different embryogenic and regeneration potential) as well as from microspore cultures of different stages during development according to Stirn et al. (1994). Northern blots (Sambrook et al. 1989) were hybridized either with the DIG-labeld cDNA-clone B15C (Aalen et al. 1994) and signals were visualized according to Düring (1991) or with the α^{32}dCTP-labeled cDNA-clone pG22-69 (Bartels et al. 1991). The concentration of proteins from the culture supernatant and the extraction of cellular proteins from barley cell suspensions as well as protein gel electrophoresis will be described elsewere (Stirn et al. 1994).

Results

INFLUENCE OF NITROGEN COMPOSITION

All tested parameters of the nitrogen supply reacted in the same way: they did not effect the frequency of initial divisions (Fig. 1), while plating efficiency (percent of microspores able to form aggregates) was influenced moderately. Dramatic effects could be seen concerning embryogenesis and plant regeneration, where very specific optima were determined, which interestingly do not correspond to the optima for plating efficiency. Highest numbers of plants were obtained with a total nitrogen content between 20 and 35 mM and a NO_3^-:NH_4^+-ratio of

90:10 (data not shown). Analyzing different N_{inorg}:N_{org}-ratios the highest plating efficiency was obtained with a ratio of 0:100 (only glutamine as sole nitrogen source : G30 medium), which led to the formation of embryogenic aggregates but nevertheless inhibited plant regeneration (Fig. 1). With a N_{inorg}:N_{org}-ratio of 90:10 (N24 A3 G3 medium with 24.3 mM nitrate, 2.7 mM ammonium and 3 mM nitrogen as glutamine) the highest number of plants was scored (Fig. 1). High frequencies of induction of division did not necessarily led to good developing cultures as shown for media containing only NO_3^- (N30) or NH_4^+ (A30) as sole nitrogen source (Fig. 1).

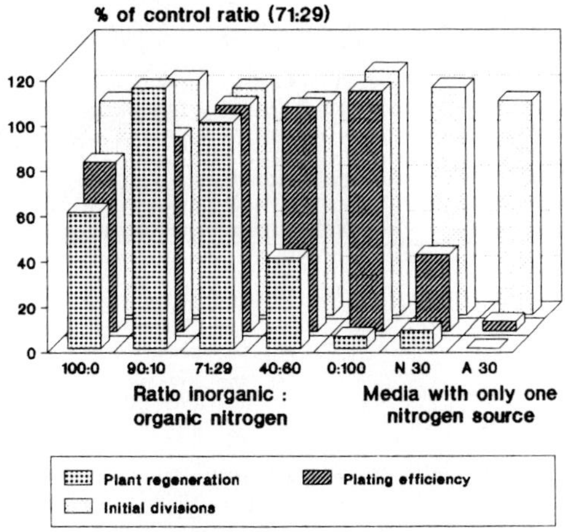

Fig. 1: Frequency of initial divisions, plating efficiency and regenerated plants of Igri microspores in response of different ratios of inorganic: organic nitrogen as well as nitrate and ammonium as sole nitrogen source. Values are given relative to the ratio of 71:29 which was set as control to 100 %.

EXPRESSION ANALYSIS OF EMBRYO SPECIFIC CLONES

The expression pattern of the embryo specific clone B15C was analyzed in different developmental stages during microspore derived embryo formation (N24 A3 G3 medium) and compared to similar stages from microspore derived cultures in G30 medium which inhibit embryo formation and plant regeneration. In both media the corresponding transcript of B15C could first be detected in two week-old cultures and increased in a similar way in both media in three week-old cultures (Fig. 2 a,b; lanes 2 and 3). During this time the cultures consisted of globular aggregates of increasing size. The expression decreased again in developing embryos on N24 A3 G3 medium (Fig. 2a, lanes 4-6) and furthermore in germinating embryos (Fig. 2a, lane 7). In contrast to that the expression increased in developing non-regenerating aggregates on the G30 medium (Fig. 2b, lanes 4-6). No transcripts could be detected in microspores (Fig. 2a) or any other plant material tested, except embryos (data not shown). This cDNA-clone was able to distinguish between embryogenic/regenerable and non-embryogenic cell suspensions. Hybridization signals of lower intensity were also detected in embryogenic suspensions which lost the ability to regenerate plants (Fig. 2c). Similar hybridization patterns were obtained by hybridisation of filters with the embryospecific cDNA-clone pG22-69 (data not shown).

PROTEIN ANALYSIS

The protein patterns of embryogenic/regenerable and non regenerable cell suspensions are very similar. Nethertheless, in one-dimensional gels of proteins secreted to the culture medium a 17.4 kDa protein (glycosilated; data not shown) could be identified in cultures lost their regenerative potential (Fig. 3). After two dimensional gelelectrophoresis (2D) we detected a polypeptide of 46 kDa with an isoelectric point of 6.1 (Fig. 4a), which was exclusively found in embryogenic culture supernatants and inversely correlated to the 17.4 kDa polypeptide (Tab. 1). A 40.5 kDa polypeptide was further identified exhibiting the same expression pattern as the 17.4 kDa negative marker (Fig. 4b). Comparing the 2D polypetide pattern of cytosolic proteins a 85 kDa polypeptide (pI 5.8) was identified which accumulated in correlation with the loss of regenerative capacity (data not shown).

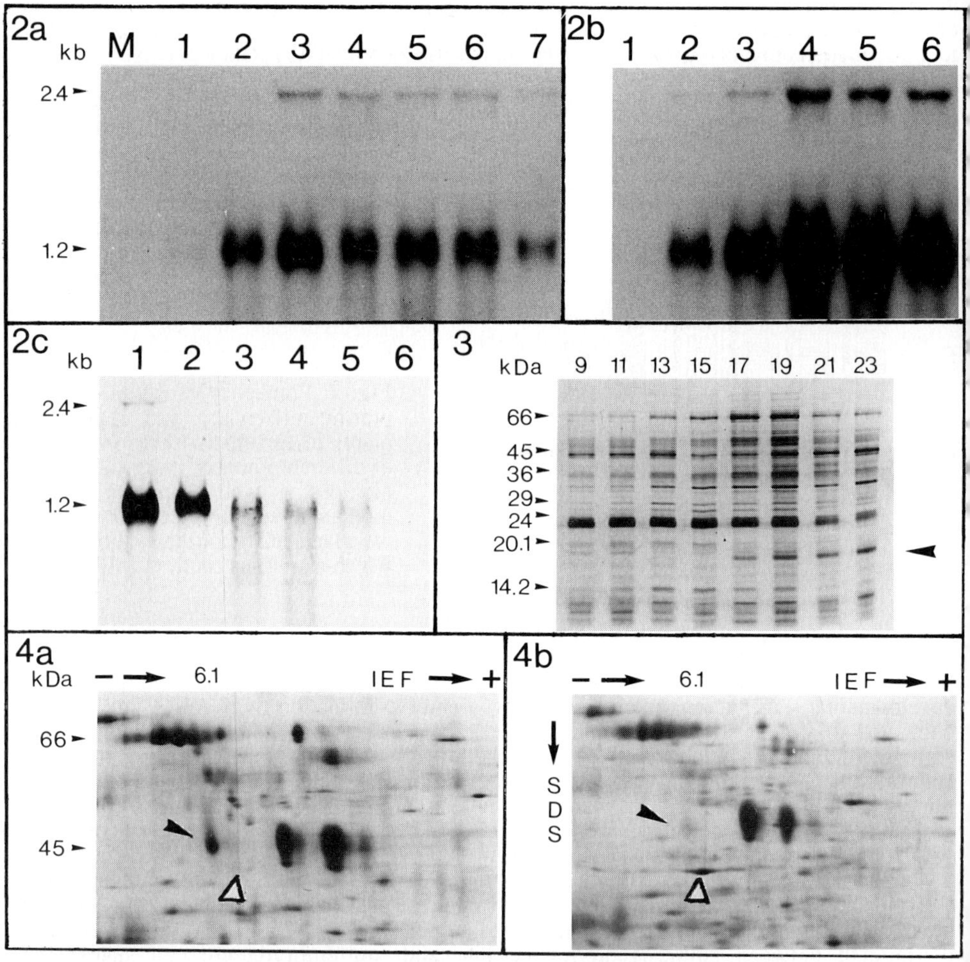

Fig. 2: Hybridization with the cDNA-clone B15C to RNA gel blots (15 μg RNA per line). **a**,**b**: development of microspore cultures in N24 A3 G3 and G30 media, respectively; **a**: M freshly isolated microspores **1**, **2**, **3**, 1, 2, and 3 weeks of culture; **4**, **5**, **6**, androgenetic embryos 0.5-0.8, 0.8-1.2, and 1.2-2.0 mm in diameter, respectively, **7**, germinating androgenetic embryos. **b**: **1**, **2**, **3**, 1, 2, and 3 weeks of culture, **4**, **5**, **6**, developing aggregates 0.5-0.8, 0.8-1.2, and 1.2-2.0 mm in diameter, respectively. **c**: barley suspension cultures initiated from anther cultures: **1**, **2**, embryogenic/regenerative suspensions (3 and 5 months old, respectively), **3**, **4**, **5**, suspensions 3, 4, and 6 weeks after the last plant regeneration had occured, **6**, non-embryogenic suspension, 2 years old. **Fig. 3**: Comparison of proteins secreted to the culture medium during the course of *in vitro* culture. Numbers refere to weeks in suspension culture. The 17.4 kDa polypeptide is marked with an arrow head. **Fig. 4**: Comparison of proteins secreted to the culture medium after 2 D-gel electrophoresis. **a**: embryogenic/regenerable suspension (9 weeks old): **b**: non-regenerable suspension (23 weeks old). The 46 kDa polypeptide is marked with an arrow head, the 40.5 polypeptide with an open arrow head. Details of 2 D-gels in the range of ph 7.7 to pH 5 and 29 kDa to 66 kDa are shown.

Tab. 1: Correlation between regenerative capacity and ocurrence of marker polypeptides in the culture supernant during prologned time of culture. [a] percentage of plated suspension aggregates regenerating plants; - not present; + present; (+) decreasing; (-) weak spot only

in vitro culture (weeks)	Plant regene-ration [a]	Occurence of 17.4 kDa marker	Occurence of 46 kDa marker	*in vitro* culture (weeks)	Plant regene-ration [a]	Occurence of 17.4 kDa marker	Occurence of 46 kDa marker
9	20	-	+	17	0	+	(+)
11	27	-	+	19	0	+	(-)
13	20	-	+	21	0	+	-
15	10	-	+	23	0	+	-

Discussion

The total nitrogen content, the NO_3^-:NH_4^+-ratio as well as the N_{inorg}:N_{org}-ratio play crucial roles in androgensis in barley microspore cultures. The more advanced the development of cultures the more visible is this influence. While frequency of initial divisions could not be altered through different nitrogen sources, plating efficiency (parameter refering to prologned proliferation) was moderately influenced and plant regeneration (via androgentic embryos) strongly regulated. Of special interest is the fact that the optima differ from each other (Fig. 1).

The possibility of directing the development of subpopulations from the same microspore preparation in different directions (e. g. germinating embryos in medium N24 A3 G3 and non-regenerating aggregates in G30 medium, repectively) was used in Northern experiments. The development of *in vitro* embryos is discussed to be analogous to zygotic embryogenisis: macroscopical and microscopical analysis revealed similar structures formed during the course of embryo development (Magnusson and Bornman 1985, Michaux-Ferrier and Schwendiman 1992). Similarity could also be shown at the molecular level. Both analysed cDNA-clones (dormancy and dessication related) found in zygotic embryos could also be detected in *in vitro* embryos and embryogenic cultures which undergo obviously neither a period of dormancy nor of dessication. Compared to the zygotic situation they are both expressed in much earlier developmental stages: signals were abtained alredy in two-week old globular aggregates in contrast to zygotic embryos were expression is limited to the differentiation in scutellum and embryo axis (Aalen et al. 1994, Bartels et al. 1991).

The high expression of embryo specific clones in non-regenerating aggregates (G30 medium) assumes that embryogenesis is also induced there. The further differentiation to distinct embryos (N24 A3 G3 medium) or the multiplication of early stages (G30 medium) was dependent of the nitrogen composition in culture media. The early expression of gene products related to relatively late events of embryogenesis in a high level might be involved in suppression of further embryo development. A high expression of these clones seemed not to be the result of feeding with only glutamine containing medium (Fig. 2b) since embryogenic suspensions (with low or no regeneration capacity) cultured in medium with a mixture of nitrate ammonium and organic nitrogen compounds show also high hybridization signals (Fig. 2c). The putative involvement of aspects of dormancy and dessication in non-regeneration cultures might be helpful to overcome the regeneration problem in monocots, because they are under hormonal control (ABA, GA).

In barley, proteins secreted to the medium have alredy been described by Nielsen and Hansen (1992). No homologies could be detected after comparing their protein patterns with the one we obtained in the course of these investigations. Up to now, there are no reports on proteins correlated with the loss of regenerative ability in barley suspension cultures, but there are investigations on proteins secreted to the medium of non-embryogenic carrot cultures (Van Engelen et al. 1991). Although the function of the identified proteins in barley suspensions remained to be analyzed further, they are of help to predict the embryogenic and regeneration capacity of barley cell cultures.

This communication shows that induction of division, further prologned proliferation, induction of embryogenesis as well as the embryo development itself are processes which are regulated independently and could therefore e. g. manipulated through the nitrogen composition of culture media. This resulted in different expression patterns of embryospecific cDNA-clones. The RNA- as well as the protein markers are uefull to characterize the embryogenic state of cell cultures and may be of interest to elucidate the mechanisms of differentiation.

Acknowledgements

We would like to tank Dr. R. Aalen (Division of General Genetics, Oslo, Norway) and Dr. D. Bartels (Max Planck Institute, Cologne, Germany) for the plasmids. We are also very thankful to Elisabeth Roßa and Thomas Winkler for their excellent technical assistance. This work was supported by grant BEO 0319336A from the Ministry of Science and Technology (BMFT, Bonn).

References

Aalen, R. B., Opsahl Ferstad, H.-G., Linnestad, C., and Olsen, O.-A. (1994) 'Transcripts encoding an oleosin and dormancy related protein are present both in the aleuron layer and in the embryo of developing barley (*Hordeum vulgare* L.) seeds', Plant J. 5, 385-396 .

Bartels, D., Engelhardt, K., Roncarati, R., Scheider, K., Rotter, M., and Salamini, F. (1991) 'An ABA and GA modulated gene expressed in the barley embryo encodes an aldose recutase related protein', EMBO J. 10, 1037-1043.

Becker, D., Brettschneider, R., and Lörz, H. (1994) 'Fertile transgenic wheat from microprojectile bombardment of scutellar tissue', Plant J. 5, 299-307.

Düring. K. (1991) 'Ultrasensitive chemiluminescent and colorigenic detection of DNA, RNA and proteins in plant molecular biology', Anal. Biochem. 196, 433-438.

Grimes, H. D., and Hodges, T. K. (1990) 'The inorganic NO_3^-:NH_4^+ ratio influences plant regeneration and auxin sensitivity in primary callus derived from immature embryos of indica rice (*Oryza sativa* L.)', J. Plant Physiol. 136, 362-367.

Jähne, A., Lazzeri, P. A., Jäger-Gussen, M., and Lörz, H. (1991) 'Plant regeneration from embryogenic cell suspensions derived from anther cultures of barley (*Hordeum vulgare* L.)', Theor. Appl. Genet. 82, 74-80.

Jähne, A., Becker, D., and Lörz, H. (1994) 'Regeneration of transgenic, microspore-derived, fertile barley', Theor. Appl. Genet. (in press).

Magnusson. I., and Bornman, C. H. (1985) 'Anatomical observations on somatic embryogenesis from scutellar tissues of immature zygotic embryos of *Triticum aestivum*', Physiol. Plant. 63, 137-145.

Michaux-Ferrier, N., and Schwendiman, J. (1992) 'Histology of somatic embryogenesis', in Y. Dattee, C. Dumas, and A. Gallais (eds.), 'Reproductive biology and plant breeding', Springer, Berlin, Heidelberg, New York.

Mordhorst, A. P., and Lörz, H. (1993) 'Embryogenesis and development of isolated barley (*Hordeum vulgare* L.) microspores are influenced by the content and composition of nitrogen source in culture media', J. Plant Physiol. 142, 485-492.

Nielsen, K., A., and Hansen, I. B. (1992) 'Appearence of extracellular proteins associated with somatic embryogenesis in suspension cultures of barley (*Hordeum vulgare* L.)', J. Plant Physiol. 139, 489-497.

Potrykus, I. (1990) 'Gene transfer to cereals: an assessment', Bio/Technol. 8, 535-542.

Sambrook, J., Fritsch, E. F., Maniatis, T. (1989) 'Molecular cloning: a laboratory manual', Cold Spring Habor Laboratory Press, USA.

Stirn, S., Mordhorst, A. P., Fuchs, S., and Lörz, H. (1994) 'Molecular and biochemical markers for embryogenic potential and regenerative capacity in barley (*Hordeum vulgare* L.) cell cultures', (submitted).

Van Engelen, F. A., Sterk, P., Booij, H., Cordewener, J. H. G., Rook, W., Van Kammen, A., and De Vries, S. C. (1991) 'Heterogeneity and cell type specific localization of a cell wall glycoprotein from carrot suspension cells', Plant Physiol. 96, 705-712.

Wetherell, D. F., and Dougall, D. K. (1976) 'Sources of nitrogen supporting growth and embryogenesis in cultured wild carrot tissue', Physiol. Plant. 37, 97-103.

IMPORTANCE OF EXTRACELLULAR PROTEINS FOR SOMATIC EMBRYOGENESIS IN PICEA ABIES

S. VON ARNOLD, U. EGERTSDOTTER and L.H. MO
The Swedish University of Agricultural Sciences
Uppsala Genetic Center
Department of Forest Genetics
P.O. Box 7027
S-750 07 Uppsala
Sweden.

ABSTRACT. Embryogenic cell lines of *Picea abies* have been divided into two groups. Group A contains somatic embryos having a large and dense embryonic region composed of small and closely adhering cells. Group B contains somatic embryos with loosely aggregated cells in their embryonic region. Somatic embryos belonging to group A can mature after being treated with abscisic acid, which group B embryos will usually not do. When considering embryogenic cell lines of *Picea abies* belonging to group A and B as developmentally different we assume that group B embryos are blocked in their development. The blockage would indicate that cells in the embryonic region expand and separate from each other early during development, instead of developing large and dense embryonic regions. When group B embryos are treated with extracellular proteins from group A embryos, they are stimulated to develop group A-morphology. The pattern of extracellular proteins reflects the embryo morphology. Extracellular proteins which vary between group A and group B embryos, and therefore have putative regulatory effects on embryo development include, arabinogalactan proteins, zeamatin-like proteins, chitinases and peroxidases.

1. Introduction

The process of somatic embryogenesis can be divided into different steps: (1) initiation of somatic embryos from the primary explant, (2) proliferation of embryogenic cultures, (3) maturation of somatic embryos and (4) plant regeneration. The process offers unique opportunities to study embryology. It also has practical applications in forest tree improvement and reforestation. The whole procedure is described by von Arnold et al. (1994).

When the primary explants are cultured on medium containing both auxin (2,4-dichlorophenoxyacetic acid; 2,4-D) and cytokinin (N^6-benzyladenine; BA) three different kinds of tissues are formed (von Arnold and Hakman, 1988): (1) a green nodulated tissue with meristemoids which under suitable conditions develop further into adventitious buds, (2) a non-embryogenic callus composed of small rounded cells and (3) an embryogenic tissue which is translucent, mucilaginous and composed of many small somatic embryos. The first sign of differentiating embryogenic structures is observed on the explant within 4 weeks of culture (Mo and von Arnold, 1991). Embryogenic structures can differentiate from the epicotyl, the cotyledons and the hypocotyl of embryos and seedlings of *Picea abies*. The primary event during the initiation of somatic embryos is that epidermal, subepidermal and/or cortical cells become meristematic and form nodules (Mo and von Arnold, 1991). The nodules develop into spherical structures consisting of densely packed cells, which are separated from the surrounding cells. Embryogenic structures differentiate from

M. Terzi et al. (eds.), Current Issues in Plant Molecular and Cellular Biology, 389-392.

the nodules. Our impression is that the separation of the nodules from the surrounding cells is critical. At present we do not know how this process is regulated, although we think that extracellular proteins might be involved.

The embryogenic cell lines of *Picea abies* have been divided into two main groups based on morphology and growth characteristics of the somatic embryos (Jalonen and von Arnold, 1991; Bellarosa et al., 1992; Egertsdotter and von Arnold, 1993). The first group of cell lines, group A, consist of somatic embryos with an embryonic region with small, densely packed cells from which the vacuolated suspensor cells extend. Cell lines belonging to this group have the ability to go through the maturation process when cultured on maturation medium containing abscisic acid (ABA). The second group of cell lines, group B, consist of somatic embryos comprised of only a few loosely aggregated cells in their embryonic region. Frequently the cells in the embryonic region are intermingled with vacuolated cells. Normally, B-type somatic embryos do not form mature somatic embryos after an ABA treatment. Cell lines classified as belonging to group A or B do not change in morphology or growth habit over time as long as the cultures are maintained on medium containing both auxin and cytokinin. Furthermore they retain their characteristics after cryopreservation (Nörgaard et al., 1993).

The somatic embryos are stimulated to go through a maturation process, when treated with ABA, in which the embryos stop proliferating, increase in size and start to accumulate storage material, including carbohydrates, proteins and lipids. Mature somatic embryos can develop further when cultured individually on medium lacking growth regulators. Desiccation treatment of mature somatic embryos of *Picea abies,* according to the method developed for *Picea glauca engelmannii* complex (Roberts et al., 1990), stimulates the development of plants. Small plants with cotyledons, a root and an epicotyl can be transferred to unsterile conditions in growth chambers where they grow fast. Most of the regenerated plants have a normal appearance and survive when transferred to the field.

Since only somatic embryos belonging to group A can undergo maturation, a critical step for plant regeneration, it is important to learn how to stimulate group B-embryos to develop into group A-embryos. Two central questions in this context are how cell lines belonging to group A and group B differ at the cellular and the molecular level, and how their development into different groups is controlled. Our hypothesis is that somatic embryos belonging to group A are maintained in culture at a more developed stage than group B-embryos. Taking this into consideration the question was raised as to whether there are differences between cell lines belonging to group A and group B in extracellular proteins in accordance with what has been found in embryogenic cultures of *Daucus carota* (for review see van Engelen and de Vries, 1992).

2. Results

To initially study the possible influence of extracellular proteins on the development of somatic embryos of Norway spruce a bioassay was set up in which concentrates of proteins from group A- and B-cell lines were added to group A-and B-embryos (Egertsdotter et al., 1993). The tests showed that extracellular proteins from group A-cell lines influence the morphology of embryos from group B-cell lines so that they become more similar to group A-embryos in appearance.

Further studies on the actual composition of extracellular proteins showed that approximately 20 protein bands could be detected in the culture medium of embryogenic suspension cultures of Norway spruce by SDS PAGE and *in vivo* labelling experiments (Egertsdotter et al., 1993). The major part of these protein bands were present in all of the 10 cell lines studied. Interestingly, there were three protein bands (28, 66 and 85 kD) exclusively found in the medium from group A-cell lines and never in group B-cell lines. In order to make sure that the pattern of extracellular proteins are reflecting the embryo morphology and not just differs among genotypes, representative cell lines from group A and B were analyzed more in detail. It was found that in medium containing only cytokinin the B-type embryos developed further, i.e. the cells in the embryonic region remained attached so that distinct and dense embryonic regions developed (Bellarosa et al., 1993; Mo et al., 1994). Simultaneously as the B-type embryos developed a large and dense embryonic region they started to secrete proteins specific for A-type embryos. Therefore, we concluded that the 28, 66 and 85 kD protein bands represent proteins which are specifically secreted by somatic embryos having a large and dense embryonic region. One of these proteins with a molecular weight of 28 kD showed around 55% N-terminal identity to a newly discovered class of antifungal proteins which up to now only have been found in monocotyledonous species (Vigers et al., 1991). Further studies on the composition of the culture medium showed differences in peroxidases and chitinases between group A-and group B- cell lines (Mo at al., 1994). More detailed studies of the protein pattern of extracellular chitinases and zeamatin-like proteins, as well as the extracellular peroxidase activity, revealed a close correlation between the presence of specific chitinases and embryo morphology.

In embryogenic cultures of carrot it has been shown that the embryogenic potential can be regulated by arabinogalactan proteins (AGPs) (Kreuger and van Holst, 1993). Furthermore, isolated AGPs from seed extracts of carrot had a positive effect on the development of embryogenic cultures, while AGPs from a non-embryogenic culture had a negative effect. In accordance with this we found AGPs in the culture medium of non-embryogenic cell lines, group A-and group B-cell lines and in seed extracts from Norway spruce (Egertsdotter and von Arnold, 1994). The amount and composition of AGPs, as detected by ß-glycosyl Yariv reagent and monoclonal antibodies, varied among the different types of tissues and cell lines. Group A-cell lines contained five times more extracellular AGPs than group B-cell lines. The precipitation profile for AGPs was specific for A-and B-cell lines. The content of AGPs recognized by the antibody JIM 13 was higher in B-cell lines than in A-cell lines. Seed extracts as well as isolated AGPs from seed extracts and from concentrated conditioned medium influenced morphology of somatic embryos as revealed in bioassays. Isolated AGPs from seed extracts promoted the aggregation of somatic embryos and the formation of enlarged embryonic regions. Total seed extracts stimulated B-type embryos to stably develop into A-type embryos characterized by a large embryonic region with densely packed cells, a distinct suspensor region and a high degree of aggregation of embryos. Simultaneously, the extracellular AGP profile changed from B-to A-type. The stable conversion was confirmed by maturation of the somatic embryos of group B origin.

3. Conclusions

Based on our knowledge today of extracellular proteins in embryogenic cultures of *Picea abies* the following conclusions have been drawn. 1. The conifers seem to have a similar system as angiosperms in regulation of somatic embryogenesis, with extracellular proteins. 2. The morphology of the somatic embryos varies among different cell lines, but it can also be regulated by changing the addition of growth regulators to the culture medium. The presence of specific extracellular proteins can be correlated to the morphology of the somatic embryos. Specific proteins are only secreted by somatic embryos which have reached a certain size and/or developmental stage. 3. Concentrated extracellular proteins from a well developed cell line (group A) can influence the morphology of less developed cell lines (group B) towards a more developed stage. Extracellular proteins which varies between group A-cell lines and group B-cell lines and therefore have putative regulatory effects on embryo development include arabinogalactan proteins, zeamatin-like proteins, chitinases and peroxidases. 4. Seed extracts stimulate a stable transition of less developed embryos into well developed embryos.

4. References

Bellarosa, R., L.H. Mo and S. von Arnold (1992) 'The influence of auxin and cytokinin on proliferation and morphology of somatic embryos of *Picea abies* (L.) Karst', Ann. Bot. 70, 199–206.

Egertsdotter, U. and S. von Arnold (1993) 'Classification of embryogenic cell lines of *Picea abies* as regards protoplast isolation and culture', J. Plant Physiol. 141, 222–229.

Egertsdotter, U., L.H. Mo and S. von Arnold (1993) 'Extracellular proteins in embryogenic suspension cultures of Norway spruce (*Picea abies*)', Physiol. Plant. 88, 315–321.

Egertsdotter, U. and S. von Arnold (1994) 'Importance of arabinogalactan proteins for the development of somatic embryos of *Picea abies*', (manuscript).

Jalonen, P. and S. von Arnold (1991) 'Characterization of embryogenic cell lines of *Picea abies* in relation to their competence for maturation', Plant Cell Rep. 10, 384–387.

Kreuger, M. and G.J. van Holst (1993) 'Arabinogalactan proteins are essential in somatic embryogenesis of *Daucus carota*', Planta 189,243–248.

Mo, L.H. and S. von Arnold (1991) 'Origin and development of embryogenic cultures from seedlings of Norway spruce (*Picea abies*)', J. Plant Physiol. 138, 223–230.

Mo, L.H., U. Egertsdotter and S. von Arnold (1994) 'Secretion of specific extracellular proteins by somatic embryos of *Picea abies* is dependent on embryo morphology' , Plant Sci. (submitted).

Nörgaard, J.V., V. Duran, Ö. Johnsen, P. Krogstrup, S. Baldursson and S. von Arnold (1993) 'Variations in cryotolerance of embryogenic *Picea abies* cell lines and the association to genetic, morphological and physiological factors', Can. J. For. Res. 23, 2560–2567.

Roberts, D.R., B.C.S. Sutton and B.S. Flinn (1990) 'Synchronous and high frequency germination of interior spruce somatic embryos following partial drying at high relative humidity', Can. J. Bot. 68, 1086–1090.

van Engelen, F.A. and S.C. de Vries (1992) 'Extracellular proteins in plant embryogenesis', Trends Genet. 8, 66–70.

Vigers, A.J., W.K. Roberts and C.P. Selitrennikoff (1991) 'A new family of plant antifungal proteins', Mol. Plant-Microbe Interactions 4, 315–323.

von Arnold, S. and I. Hakman (1988) 'Plantlet regeneration *in vitro* via adventitious buds and somatic embryos in Norway spruce (*Picea abies*)' in: J.W. Hanover and D.E. Keathley (eds). Genetic Manipulation of Woody Plants. Basic Life Sciences 44, 199–215. Plenum Press, New York and London.

von Arnold, S., D. Clapham, U. Egertsdotter, I. Ekberg, L.H. Mo, H. Yibrah (1994) 'Somatic embryogenesis in *Picea abies*' in: Y.P.S. Bajaj (ed.). Biotechnology in Agriculture and Forestry, (in press).

NUTRIENT ABSORPTION AND THE DEVELOPMENT AND GENETIC STABILITY OF CULTURED MERISTEMS

J. G. CARMAN
Plants, Soils and Biometeorology Department
Utah State University
Logan, UT 84322-4820, USA

ABSTRACT. Meristem development *in vitro* often is abnormal and perturbed by developmental anomalies. Some of these anomalies may be ameliorated by providing more complete nutrient regimes that mimic those found *in situ*. At the whole plant level, water, nutrients and hormones pass through xylem and phloem, both of which differentiate precociously in root and shoot apices, and finally to individual meristem cells, primarily via the symplast. Xylem solute composition near meristems largely depends on the mineral absorption properties of roots, which vary extensively among species. The organic and inorganic content of phloem also varies extensively among species. Such differences beg the questions, "can the development and genetic stability of tissue-cultured meristems and other small explants be improved by providing more complete species-specific combinations of organic and inorganic nutrients, and can such combinations be predicted from physiological principles of nutrient absorption coupled with reasonably thorough chemical characterizations of appropriate phloem, xylem, and cell sap solutions?" Using this approach we have improved the development and genetic stability of wheat embryogenic tissue cultures. Principles important to species-specific media refinement are discussed.

1 Introduction

Shoot meristem culture involves removing and culturing the shoot apical dome (and frequently one or more leaf primordia) and has been used for over 40 years to eradicate viruses (Quak 1977, Kartha 1986, Van Zaayan et al. 1992). Today, meristem culture is receiving renewed interest as a tool in plant transformation (McCabe et al. 1988, Christou et al. 1989, Bidney et al. 1992, Pérez-Vincente et al. 1993, Iglesias et al 1994, Smith et al. 1994). Success in these applications is a direct result of the work of many scientists who have developed and tested media and techniques that support meristem growth and development *in vitro* (reviewed by Quak 1977).

The major physiological functions of meristems are to synthesize protoplasm and produce cells. In accomplishing these functions, rapidly growing meristems synthesize large quantities of cell-wall components, plasma membranes, endomembranes, cytoskeleton, ribosomes, mitochondria, plastids and nucleoplasm. Appropriate quantities of various biomolecules are in constant demand, including *i*) macromolecular components, such as mono and disaccharides, amino acids, nucleotides, fatty acids, and glycerol, *ii*) metabolic intermediates, such as pyruvate, citrate, malate, succinate and glyceraldehyde 3-phosphate, and *iii*) vitamins and coenzymes, such as thiamin, nicotinic acid, riboflavin, biotin, folic acid and pantothenic acid. Shoot apical meristems can synthesize these molecules from a simple, auxin-containing, chemically-defined nutrient medium, such as MS (Murashige and Skoog 1962), and differentiate into plantlets *in vitro* (Smith and Murashige 1970). However, an examination of the meristem culture literature reveals a preponderance of temporal and spacial developmental anomalies in such cultures.

M. Terzi et al. (eds.), Current Issues in Plant Molecular and Cellular Biology, 393-403.

Of the tobacco meristems cultured by Smith and Murashige (1970), none continued normal growth and only 30% eventually formed whole plants. Four types of developmental anomalies were noted: *i*) root formation at cut surfaces followed by shoot growth and plantlet formation, *ii*) root and leaflet formation followed by cessation of growth and death, *iii*) root formation only, and *iv*) cellular distention, especially at the cut surface, followed by necrosis. These types of anomalies are typical for other species as well. Cassava shoot tips less than 0.2 mm long formed only roots (Kartha et al. 1974). Percentage plantlet formation from cultured meristems of annual ryegrass between 0.2 and 0.5 mm long was only 45%, but increased to 59% for shoot tips between 0.5 and 1.5 mm long (Dale 1975). With orchard grass, percentage plantlet formation from meristems between 0.3 and 0.5 mm long was only 27%, but increased to 44% for shoot tips between 0.5 and 0.9 mm long (Dale 1979). Only 4% of poplar meristems survived culture, and these required 12 weeks to produce shoots (Rutledge and Douglas 1988). It is apparent from these and other studies (Quak 1977, Stone 1982, Stein 1991, Lim et al. 1993) that low frequency regeneration is the norm for cultured meristems, as are lengthy regeneration periods, which often extend from several weeks to many months (Quak 1977, Stone 1982, Rutledge and Douglas 1988, Griffiths et al. 1990).

Sluggish and abnormal development of cultured meristems may be caused by a variety of physiological stresses, including *i*) wounding, *ii*) desiccation during explantation, *iii*) osmotic stresses experienced during culture, *iv*) insufficient chemical energy for synthesizing macromolecules and for maintaining turgor and tolerable ionic gradients, *v*) nutrient supplies less complex than those experienced *in situ* and causing diversion of energy supplies from growth and development to precursor biosynthesis, and *vi*) slow or variable rates of nutrient absorption across plasmalemmas. Studies relevant to these factors are reviewed in this chapter and are grouped so as to address the following four questions: *i*) by what transport and absorption mechanisms do nutrients enter meristems *in situ*? *ii*) what molecules nourish meristems *in situ*? *iii*) what nutrient transport and absorption mechanisms are available to meristems *in vitro*? *iv*) would media and procedures that simulate nutrient availability *in situ* represent an improvement over current procedures? Disparities between current media formulations and *in situ* nutrient supplies are identified, and research leading to improved media formulations is described.

2 Meristems and Source-sink Transport

2.1 ANATOMICAL CONSIDERATIONS

Several anatomical features limit nutrient transfer in meristems, including the distance from the meristem apex to mature phloem and xylem. These distances vary with rates of root growth; longer distances occur in fast growing roots while shorter distances occur in slower growing roots (reviewed by Demchenko 1994). These relationships may represent a nutritional control on root growth rate.

Phloem generally develops prior to xylem. In general, 15 to 25 cell files span the distance between meristem tips and functional phloem, and about 25 to 35 cells span the distance between meristem tips and functional xylem. Nutrients in the phloem are either *i*) unloaded into the apoplast at the distal regions of phloem maturation, where they pass osmotically to the apex, or *ii*) retained in the symplast where they travel via pressure flow to meristem apex cells through plasmodesmata. Two anatomical features suggest that the latter mode of transfer is more prevalent. First, meristem cells grow close together; they do not have well developed secondary walls (Esau 1977), the absence of which increases the resistance to intercellular water movement (Nonami and Boyer 1987) and severely limits the "wicking action" characteristic of the apoplast. Second, meristem cells have many more plasmodesmata than most other types of plant cells, with from 1,000 to 10,000 per cell being common (Robards 1976, Meiners et al. 1988). Furthermore, plasmodesmata in young wheat roots are sensitive to oxygen supply; their apertures constrict during periods of adequate oxygen supply and expand during periods of poor aeration. Fully expanded plasmodesmata allow for increased rates of oxygen diffusion through the symplast and permit larger molecules (five to ten-fold increase in molecular weight, includes ATP) to pass more freely (Cleland et al. 1994).

2.2 PHYSIOLOGICAL CONSIDERATIONS

Colligative properties also favor movement of nutrients to meristems via the symplast. Movement of nutrients and water in the xylem occurs down a water potential (Ψ_w) gradient driven primarily by evapotranspiration, but secondarily by the water requirements of either photosynthesizing or elongating cells. Shoot meristems are enclosed by young leaf tissues, and evapotranspiration-induced Ψ_w gradients from the xylem to the atmosphere are not strong. Hence, apoplast water near meristems moves down Ψ_w gradients into dividing and elongating cells (Nonami and Boyer 1987). In contrast, solute movement in the phloem occurs down pressure potential (Ψ_p) and solute concentration gradients, both of which are driven by *i*) the loading of osmotically-active solutes into the phloem at sources, *ii*) the incorporation of these solutes into osmotically-inactive macromolecules at sinks, *iii*) isolation of solutes in vacuoles via active transport across tonoplasts, and *iv*) the yielding of cell membranes and walls (cell volume expansion) during cell division and differentiation (Boyer 1985, Nonami and Boyer 1987).

Phloem solutions often flow symplastically from the phloem into sink cells (Oparka 1990) down a Ψ_p gradient but against an osmotic (solute) potential (Ψ_s) gradient. In general, the Ψ_w of cytosolic, apoplastic, and phloem solutions range from about -3.7 to -0.2 MPa and will be near equilibrium among these three solutions at any given location along the various transport pathways of the plant. The apoplast Ψ_w equals its solute potential, which ranges from about -0.2 to -0.03 MPa (Nonami and Boyer 1987), plus the xylem tension (Ψ_p), which generally ranges from about -3.5 to -0.2 MPa. The Ψ_w of the cytosol equals the Ψ_s (about -4.0 to -0.5 MPa) plus the Ψ_p (about 0 to 1.0 MPa). Sieve element Ψ_w equal their Ψ_s (about -4.0 to -2.0 MPa) plus their Ψ_p (about 0.5 to 2.5 MPa). Potentials beyond these ranges commonly occur in some taxa but occur only under stress conditions in most other taxa (Salisbury and Ross 1992).

Meristem excision disrupts symplastic pressures and apoplastic tensions, which *in situ* are critical to the maintenance of a physiological cytosolic Ψ_w. For example, a healthy meristem with a cytosolic Ψ_s of -0.8 MPa and a cytosolic Ψ_p of 0.5 MPa (i.e., total Ψ_w = -0.3 MPa), which is typical for growing tissues (Grignon and Sentenac 1991), would have an adjoining apoplast Ψ_w also equal to -0.3 MPa, but contributed by a Ψ_s of perhaps -0.05 MPa and a Ψ_p (tension) of perhaps -0.25 MPa. The excision and culture of such a meristem (under conditions of 100% RH) would result in an immediate and substantial increase to a less negative value in the extracellular Ψ_p (apoplastic tension) and a substantial decrease to a less positive value in the cytosolic Ψ_p (symplastic pressure). Disequilibrium would then occur: The cytosol Ψ_w would become more negative, the apoplast Ψ_w would become less negative, and free water would diffuse from the apoplast into the excised cells. Mechanisms of osmotic adjustment, involving transport of ions from the cytosol to the apoplast, might also become operative (Grignon and Sentenac 1991). Restoration of the pre-excision cytosolic Ψ_w would occur if the explants were immediately transferred to a medium with a Ψ_w equal to or slightly higher (less negative, to account for some apoplastic tension) than the pre-existent apoplast Ψ_w, which in this example was a -0.3 MPa.

The above considerations eliminate, on the basis of Ψ_w alone, phloem exudate as an ideal medium for meristem culture. This is because the Ψ_s of phloem exudate is generally < -2.0 MPa whereas the cytosol Ψ_s is generally > -1.0 MPa (Salisbury and Ross 1992). Cultured meristems will remain turgid as long as the medium Ψ_w remains higher than the cytosolic Ψ_s. Nevertheless, there are probably physiological advantages to matching as closely as possible the meristem culture medium Ψ_w to that of the pre-existent *in situ* apoplast Ψ_w at or near the meristem. Such Ψ_w can best be measured by isopiestic thermocouple psychrometry using intact plants (Boyer et al. 1985, Nonami and Boyer 1987).

2.3 CELLULAR CONSIDERATIONS

Excising meristems terminates symplastic delivery of nutrients from phloem to meristems. Thus, cultured meristems are forced to rely on passive and active absorption mechanisms for nutrient uptake. That active absorption is fully functional in growing tissues is evidenced by the fact that apoplast fluids from such tissues contain extremely low levels of solutes (Nonami and Boyer 1987). Hence, active absorption mechanisms must exist for cytosolic solutes such as sugars, amino acids, organic acids, sugar phosphates, inorganic constituents, vitamins, hormones, and other compounds normally transported in the symplast. Such mechanisms adequately prevent the accumulation of these compounds in the apoplast via simple diffusion down their concentration gradients. Elucidation of such mechanisms in plants has lagged behind that for animals (Darnell et al. 1990). Nevertheless, three different types of facilitated membrane transport mechanisms in plants have now been characterized: ion ATPase pumps, membrane permeases (channel proteins), and cotransport proteins.

Plant ATPase pumps in plasma membranes and tonoplasts have been described for Ca^{2+}, Na^+-K^+, and H^+ (Darnell et al. 1990, Salisbury and Ross 1992). These enzymes use energy from ATP to transfer cations (against concentration gradients) from the cytosol into either vacuoles or the apoplast. The transfer of cations across the plasma membrane produces steep *p*H and electropotential gradients. Gradients of from -100 to -150 mV are common as are H^+ gradients that exceed 100-fold (cytosol *p*H ≈ 7, apoplast *p*H ≈ 5). Electrochemical gradients facilitate passive uptake of cations through membranes or membrane permeases while H^+ gradients power cotransport of numerous compounds either into or out of the cytosol (ΔpH-dependent absorption).

Plasma membranes are semi-permeable; nevertheless, they readily absorb only a few compounds, such as water, CO_2, N_2, O_2, urea, ethanol and other small uncharged polar molecules. Ions, sugars, amino acids, and other charged polar molecules such as ATP generally are poorly absorbed. Absorption of many of these entities is facilitated by a variety of permeases that span the width of the plasma membrane and permit passive but rapid one-way absorption down electrochemical potential gradients. Permeases specific to sugars, amino acids, K^+, NH_4^+, Mg^{2+}, Na^+, Cl^- and Ca^{2+} have been described, and rates of transfer across plant membranes as high as 10^8 ions per second have been detected (reviewed by Darnell et al. 1990, Salisbury and Ross 1992).

Cotransport proteins rely on proton-motive forces generated by ATP-facilitated pumping of H^+ from the cytosol to the apoplast. Energy released as these H^+ move back into the cytosol, down their electrochemical potential gradients, is captured by cotransport proteins and used to transfer solutes either into the cytosol (symport) or out of the cytosol (antiport), usually against concentration gradients. Specific cotransporters have been identified for neutral, acidic, and basic amino acids, sugars, anions (NO_3^-, Cl^-, $H_2PO_4^-$, SO_4^{2-}), and cations ($Na^+$and K^+). The ATPases required to maintain H^+ gradients for these transporters may consume 20 to 40% of a cell's ATP (reviewed by Darnell et al. 1990, Lie and Bush 1990, Salisbury and Ross 1992). Hence, cultured meristems should benefit from medium nutrients and a medium *p*H that minimize unnecessary ATP consumption. Thus, unmodified phloem exudate is inappropriate as a tissue culture medium not only because of its low Ψ_w but also because of its high *p*H (7 to 7.5).

3 Phloem Exudate Composition: A Guide to Media Development?

3.1 RATIONALE

The composition of biomass in meristems provides insights into the forms of nutrients they use. In roots of both maize (Silk and Erickson 1980) and pea (Brown and Broadbent 1950), maximal cell division and protein accumulation per millimeter of root tissue occurs between 0.5 and 2.0 mm from the root tip (Fig. 1). Hence, large quantities of amino acids are required for protein synthesis in mitotically-active cells. Protein accumulation rates on a per cell basis continue to increase with distance from the root tip until they reach a maximum in the zone of elongation at about 5 mm from the root tip (Brown and Broadbent 1950). Due to cell expansion, protein accumulation rates on a per mm basis drop to near zero beyond about 3 to 4 mm from the tips of both pea (Brown and Broadbent 1950) and maize (Silk and Erickson 1980) roots. In contrast, the accumulation of total dry mass (DM) per millimeter peaks between 2 and 3 mm from pea (Brown and Broadbent 1950) and maize (Fig. 1) root tips, which coincides with the accumulation of solutes and primary-wall components during cell elongation. On a per cell basis, rates of DM accumulation increase with distance from the root tip through the zone of differentiation and secondary-wall formation, which extends at least 9.0 mm from the root tip (Brown and Broadbent 1950).

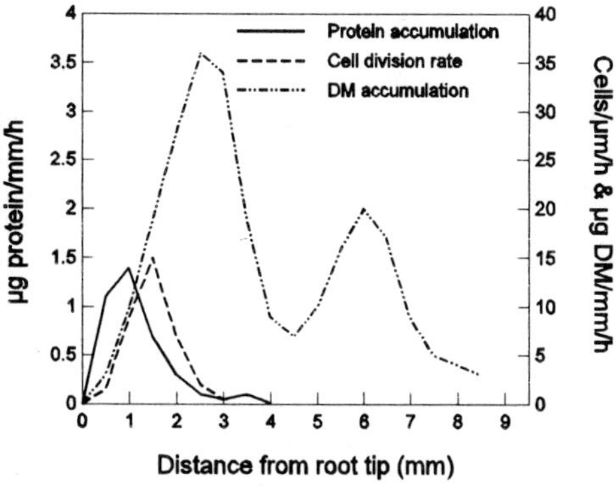

Figure. 1. Protein, cell number, and DM accumulation as a function of distance from a maize root tip. Data from Gandar (1983) and Silk and Erickson (1980). Redrawn from Silk (1984).

These data emphasize the particularly high demand meristematic cells have for amino acids and expanding and differentiating cells have for both amino acids and carbohydrates. They suggest that appropriate amino-acid supplements may be beneficial to small explant cultures, as has been documented for protoplasts and microspores (Thompson et al. 1986, Olsen 1987, Torrizo and Zapata 1992, Holm et al. 1994). Many of the proteins synthesized by meristematic cells are membrane proteins, which may account for 20 to more than 40% of the DM of plasmalemmas (Darnell et al. 1990, Salisbury and Ross 1992).

Intact meristems receive most of their nutrition symplastically from phloem solutions, whereas cultured meristems rely primarily on ATPase-facilitated nutrient absorption via the apoplast. It is reasonable to assume that phloem solutions near a meristem are well suited nutritionally for the *in situ* growth and development of that meristem. Therefore, it may be desirable to pattern the composition of meristem culture media after the nutrient composition of such phloem solutions. However, as noted above, the Ψ_w of phloem solutions are far too low to support meristem development *in vitro*. Raising the Ψ_w of simulated-phloem solutions (Ψ_s = -4 to -2 MPa) by simple dilution to an appropriate level (≈ -0.3 to -0.6 MPa) is a reasonable alternative. However, rates of ΔpH-dependent absorption into the cytosol differ for different nutrients (such as for different amino acids, Lie and Bush 1990), such that their absorption into cells is not proportional to their rate of flow through the cell-to-cell symplast via plasmodesmata. In fact, available but fragmented data from Michaelis-Menten kinetic studies indicate that ΔpH-dependent nutrient absorption rates vary among molecules, species, and among tissues within species (Ruiter et al. 1984, Lie and Bush 1990, Opaskornkul et al. 1994). Furthermore, some studies indicate that preferential nutrient passage may also occur through plasmodesmata (see Gunning and Robards 1976, Meiners et al. 1988).

Regardless of their shortcomings, dilute simulated phloem solutions, which have a low *p*H (5 to 5.5), should be good starting points for small explant culture, from which modifications based on Michaelis-Menten kinetic studies of the uptake of specific nutrients, as well as other physiological studies, can be made. This approach provides small explants with approximately the same broad range of nutrients as occurs *in situ*, which should reduce, for example, the unnecessary use of ATP for ΔpH-dependent absorption of ammonium and nitrate, and subsequent synthesis of amino acids. We have increased the Ψ_w of our media by reducing the concentration of only the major nutrients (major salts, sugars, and/or amino acids) while leaving the concentrations of the minor components intact (minor mineral salts, vitamins and hormones). The *p*H of simulated-phloem medium is adjusted downward, from ≈7 to ≈5, to simulate the apoplast (mainly to facilitate ΔpH-dependent nutrient transport). Simulated-phloem solutions are well buffered, as are real phloem solutions, with an overall pK_a close to seven. To deter *p*H shifts during culture, the medium pK_a may be lowered by titration trials accompanied by minor adjustments of the major amino and organic-acid concentrations.

3.2 COMPOSITION

A broad range of organic and inorganic substances move by mass flow through the phloem of higher plants. These include (with general ranges of concentration in parentheses) water, carbohydrates (400 to 1000 mM), amino acids (30 to 200 mM), inorganic substances (2.5% DM of phloem solution), lipids (approximately 0.3% DM of phloem solution), organic acids (1 to 20 mM), sugar phosphates (5 to 10 mM), nucleic acids (0.4 mM, mostly ATP), vitamins (<20 μM), and growth regulators (usually <0.1 μM). Sucrose is the most abundant sugar in the phloem of most taxa, often accounting for from 40 to 80% of the solution DM. However, all sugars combined account for only about 1% of the DM of Cucurbitaceae phloem solutions, with the majority of DM being composed of amino acids (Ziegler 1975, Smith and Milburn 1980). Amino acid levels are also high in phloem solutions of castor bean, with levels reaching 230 mM (Schobert and Komor 1989). High levels of organic acids (20 mM) have been found in embryo-sac fluids that are supplied by the phloem (Murray 1988), and in phloem exudate from castor bean (Smith and Milburn 1980). In general, there are distinct qualitative and quantitative differences among plant taxa in the types and concentrations of substances translocated, and large seasonal variations occur within plants (Douglas 1993, Kelner et al. 1993). Intraspecies variation suggests that meristems can tolerate rather broad concentration ranges for the specific nutrients translocated through the phloem.

Variations exist among plants in the types of sugars they translocate. As mentioned above, most plants translocate primarily sucrose, including ferns, pines, lilies, legumes, maples, cashews, asters, and oaks. Plants that translocate sucrose plus raffinose-based oligosaccharides include catalpas, various euonymus species, almonds, mint, eucalyptus, olives, some ash, privet, fuchias, evening-primroses, verbena, cashews, jojoba, citrus, lindon, and elm. Plants that translocate sucrose plus raffinose-based oligosaccharides plus sugar alcohols include some ash, lilacs, apples, prunes, pears, amelanchiars, various crataegus species, cotoneasters, plums, cherries, apricots, and peaches (Ziegler 1975, Salisbury and Ross 1992). It may be of value when tissue culturing these species to simulate their respective carbon sources.

Other major nutrients found in phloem solutions occur in forms not present in most conventional plant tissue culture media. Nitrogen is translocated in plants primarily in the form of amino acids. Ammonium may be found, but it generally contributes less than 10% of the total N; NO_3 is essentially absent (Ziegler 1975, Salisbury and Ross 1992). Table 1 summarizes the major forms of translocated nitrogen (primarily amino acids). Phosphorus is translocated in the phloem as $H_2PO_4^-$ and as sugar phosphates (Ziegler 1975, Salisbury and Ross 1992). Sulfur is most efficiently reduced in chloroplasts (reviewed by Marschner 1986), which may not be abundant in small explants, and is translocated in reduced forms, such as glutathione, methionine and cysteine (Rennenberg et al. 1979). Other components of phloem, such as organic acids, are only rarely used in plant tissue culture media.

The construction of tissue culture media using nutrient forms reflective of phloem solutions and modified to reflect the physiological constraints of excised tissues is not well represented in the tissue culture literature (George et al. 1987, 1988; Murray 1988). Formulation of such media and of the attending, sometimes novel, culture procedures required to simulate *in situ* physicochemical environments may prove fruitful, especially with the culture of small explants (Carman 1988, 1989, 1990; Hess and Carman 1993).

4 Simulating the Wheat Ovule

We have characterized the mineral, free amino acid, water soluble carbohydrate, hormone (IAA, ABA and six cytokinins), and dissolved oxygen levels in wheat ovules from anthesis through early embryo maturation (Hess 1992, Hess and Carman 1993, Carman et al. 1993, and other manuscripts in preparation). From these data, a wheat embryo culture (WEC) medium was designed (Table 2). Procedures for preparing one version of this medium (nutrient levels averaged over the first 18 d of zygotic embryo formation) are listed below. These procedures should be of value as a pattern for developing a variety of simulated phloem-solution media for a variety of species and small explants.

Table 1. Amino acids found in phloem.

	Species				
Amino acid	Wheat[1]	Barley[2]	Tobacco[3]	Spinich[4]	Mean ± SD of 23 species[5]
	mM				
Aspartate	4.9	26	1.5	24	-
Asparagine	1.8	2.4	2.0	3	4.6±6.0
Glutamate	0.8	89	15.3	65	-
Glutamine	36.5	8	28.8	18	7.3±9.0
Serine	8.2	25	5.6	-	13.7±8.9
Glycine	2.4	5.8	T	7	-
Threonine	4.0	7.3	1.7	4	2.3±2.3
Alanine	3.4	11.3	1.2	13	-
Valine	2.6	3.7	-	9	3.2±2.8
Isoleucine	1.4	1.1	2.4	-	2.6±2.0
Leucine	1.2	1.1	-	-	2.4±2.5
Lysine	1.7	1.7	1.4	-	-
Tyrosine	0.6	.5	0.5	-	-
Phenylalanine	1.0	-	3.7	-	1.5±1.5
Proline	0.1	-	14.4	-	-
Arginine	2	-	-	-	-
NH_4	-	-	-	-	3.5±4.2

[1]Fisher and Macnicol (1986); [2]Winter et al. (1992); [3]Hocking (1980); [4]Riens et al. (1991); [5]Ziegler (1975)

Table 2. Summary composition of double strength MS (dMS) and 1/2 strength WEC media.

Metabolite	dMS	½WEC
	mM	
Sucrose equ.	58	≈100
Nitrogen	120 (NO_3 + NH_4)	≈55 (amino acids + NH_4)
Potassium	40	25
Phosphorus	2.5 ($H_2PO_4^-$)	8 (sugar phosphates)
Calcium	6	3
Sulfur	3 (Sulfate)	5 (sulfate + S-amino acids)
Magnesium	3	3
Organic acids	0	≈2
Ψ_s (MPa)	≈-0.56	≈-0.61
*p*H	≈5.2	6.5

The amino acid stock for WEC medium is prepared by adding, in order, the following chemicals (in g) to 450 mL of water and bringing the solution to a volume of 500 mL: malic acid, 7.00; succinic acid, 3.50; Asp, 2.31; Asn, 2.31; Cys-HCl, 0.53; Trp, 0.20; Ile, 1.97; Leu, 2.62; His, 0.78; Phe, 1.24; Thr, 0.48; Pro, 3.45; Ala, 14.24; Gly, 3.00; Val, 2.93; Lys, 2.92; Arg, 2.61; and Met, 1.49. The major mineral stock is prepared by adding the following chemicals (in g) to 225 mL of water and bringing the solution to a volume of 250 mL: $MgSO_4•7H_2O$, 21.58; KNO_3, 7.58; $MgCl_2•6H_2O$, 12.70; $NaNO_3$, 2.13; and NH_4NO_3, 6.00. The minor mineral stock is prepared by adding the following chemicals (in mg) to 225 mL of water and bringing the solution to a volume of 250 mL: H_3BO_3, 772.5; $ZnSO_4•7H_2O$, 1796.9; $MnSO_4•H_2O$, 380.3; KI, 20.8; $CuSO_4•5H_2O$, 6.2; $NaMoO_4•2H_2O$, 6.0; $CoCl_2•6H_2O$, 0.6.

Basal WEC medium is prepared by adding in order the following stocks and chemicals to 600 mL of H_2O: amino acid stock, 50.0 mL; malic acid, 420 mg; Glu, 150 mg; Gln, 150 mg; Ser, 2.63 g; major mineral stock, 10.0 mL; minor mineral stock, 5.0 mL; 1.5 X MS iron stock (Murashige and Skoog, 1962); B_5 vitamin stock (Gamborg et al., 1968); D-fructose 1,6-diphosphate dicalcium salt (Sigma, practical grade), 4.54 g; fructose, 20.00 g; glucose, 14.4 g; sucrose, 13.7 g; and myoinositol, 1.8 g. A solution of tyrosine is then prepared by mixing 136 mg in 42 mL of 1.0 M KOH. This is added slowly while the nutrient medium is vigorously stirred. The volume is brought to 1000 mL and the *p*H adjusted to 6.5, which is common for solutions that bath plant embryos *in situ* (Murray 1988), with dilute KOH or malic acid, and the medium is filter sterilized. Half strength semi-solid WEC medium is prepared by adding half the regular amount of medium components (except vitamins and 2,4-D, which are added at regular concentrations) to half of the total amount of water. This is then filter sterilized. Phytogel is added at 1.8 g L^{-1} (final concentration) to the remaining portion of water, which is autoclaved for 15 min. The two components are combined, prior to gelling, and immediately dispensed to sterile plastic Petri dishes.

Comparisons of WEC medium with other media have been made. These involved the rescue of immature wheat embryos (8 to 10 d past flowering) and the induction of somatic embryogenesis from immature embryos taken at 12 d past flowering. WEC medium out-performed other media in both cases. In the latter case, WEC and two dilutions of WEC (1/2 and 3/4 strength) were compared with each other and with double-strength MS (dMS). The frequency and quality of embryogenic callus after 30 days was superior in the WEC treatments (Fig. 2). The WEC medium closest to dMS in Ψ_w, 1/2WEC (Table 2) showed the greatest improvement over dMS (Fig. 2), which suggests that nutrient composition, rather than Ψ_w was the major contributing factor. Minor variations of this medium have performed very well in our lab for the induction of embryogenic maize and rice tissue cultures and for the culture of potato meristems (unpublished). Though not modified for cotton, this medium shows promise for the culture of young cotton embryos (Norma Trolinder, personal communication).

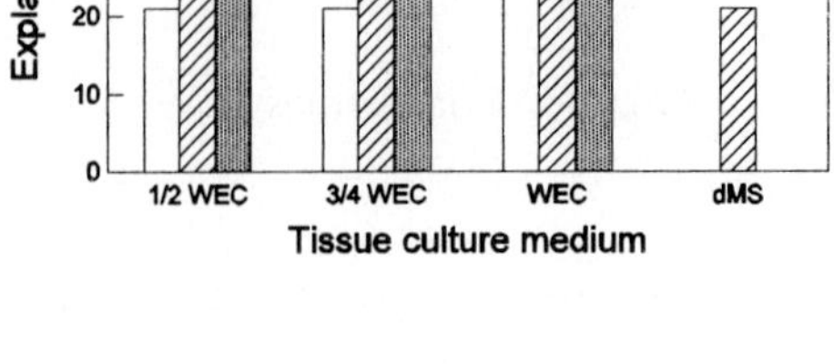

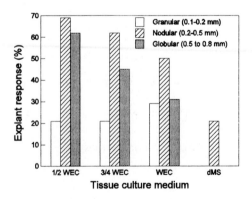

Figure 2. Media effects on the frequency of production of embryogenic callus of various textures from immature wheat embryo explants after 30 d in culture. A spring wheat line, PCYT10, was used and each medium was ammended with 5 μM 2,4-D.

5 Conclusions

Large disparities between conventional media formulations and *in situ* nutrient supplies exist. Differences include *i*) the predominance of amino acid nitrogen in phloem and its relative absence in conventional tissue culture media, *ii*) the presence of sugar phosphates, reduced sulfur, and organic acids in phloem and their relative absence in conventional media, and *iii*) unique combinations of sugars in the phloem of many taxa, which may not have been adequately tested in tissue cultures of the respective taxa. These disparities justify attempts at improving performance of cultured meristems and other small explants by using the nutrient composition of the phloem as a baseline for media development.

Phloem solutions are ideal for nurishing meristems via pressure-driven symplastic delivery. However, their low Ψ_w and high *p*H prohibit them from being used directly for nurishing meristems apoplastically. The Ψ_w of such solutions must be increased by as much as 10-fold for *in vitro* culture, and the H^+ concentration must be increased 100-fold (*p*H lowered from 7 to 5). The Ψ_w may be increased by reducing the concentrations of sugars and amino acids, and levels of these major nutrients may be fine-tuned by taking into consideration differential rates of ΔpH-dependent absorption across plasmalemmas.

Our results with simulated phloem media are promising. Nevertheless, this area of tissue culture research (*in vitro* cell nutrition) is in its infancy, and additional knowledge is required before we can expect to supply meristems, zygotes, protoplasts, microspores, and other small explants with levels of nutrients and hormones commensurate with their genetic potentials. As such knowledge increases, practitioners will undoubtedly be supplied with brilliant new palettes from which to conduct the art of plant tissue culture.

6. References

Bidney, D., C. Scelonge, J. Martich, M. Burrus, L. Sims and G. Huffman. 1992. Microprojectile bombardment of plant tissues increases transformation frequency by *Agrobacteium tumefaciens*. Plant Mol Biol 18:301-313.

Boyer, J. S. 1985. Water transport. Ann Rev Plant Physiol 36:473-516.

Brown, R. and D. Broadbent. 1950. The development of cells in the growing zones of the root. J Exp Bot 1:249-263.

Carman, J. G. 1988. Improved somatic embryogenesis in wheat by partial simulation of the in-ovulo oxygen, growth-regulator and desiccation environments. Planta 175:417-424.

Carman, J. G. 1989. The *in ovulo* environment and its relevance to cloning wheat via somatic embryogenesis. In Vitro 25:1155-1162.

Carman, J. G. 1990. Embryogenic cells in plant tissue cultures: occurrence and behavior. In Vitro 26:746-753

Carman, J. G., J. R. Hess, D. L. Bishop, and D. J. Hole. 1993. *In ovulo* environments and embryo dormancy in wheat. - *In* Pre-Harvest Sprouting in Cereals 1992 (M. K. Walker-Simmons and J. L. Ried, eds), pp. 163-170. Am Assoc Cereal Chem, Inc, St Paul, MN. ISBN 0-913250-81-3.

Christou, P., W. F. Swain, N. S. Yang, D. E. McCabe. 1989. Inheritance and expression of foreign genes in transgenic soybean plants. Proc Nat Acad Sci 86:7500-7504.

Cleland, R. E., T. Fujiwara, and W. J. Lucus. 1994. Plasmodesmal-mediated cell-to-cell transport in wheat roots is modulated by anaerobic stress. Protoplasma 178:81-85.

Dale, P. J. 1975. Meristem tip culture in *Lolium multiflorum*. J Exp Bot 26:731-736.

Dale, P. J. 1979. The elimination of cocksfoot mottle virus from *Dactylis glomerata* by shoot tip and tiller bud culture. Ann Appl Biol 9:285-288.

Darnell, J., J. Lodish, and D. Baltimore. 1990. Molecular cell biology. Scientific American Books, W. H. Freeman and Company, New York.

De Ruiter, H., J. Schuurmans, and C. Kollöffel. 1984. Amino acid leakage from cotyledons of developing and germinating pea seeds. J Plant Physiol 116:47-57.

Demchenko, N. P. 1994. Temporal aspects of protophloem development in roots of *Triticum aestivum* L. Environ Exp Bot 34:95-106.

Douglas, A. E. 1993. The nutritional quality of phloem sap utilized by natural aphid populations. Ecol Entom 18:31-38.

Esau, K. 1977. Anatomy of seed plants. John Wiley and Sons, Inc.

Fisher, D. B. and P. K. Macnicol. 1986. Amino acid composition along the transport pathway during grain filling in wheat. Plant Physiol 82-1019-1023.

Gamborg, O. L., R. A. Miller, and K. Ojima. 1968. Nutrient requirements of suspension cultures of soybean root cells. Exp Cell Res 50:151-158.

Gandar, P. W. 1983. The analysis of growth and cell production in root apices. Bot Gaz 141:131-138.

George, E. F., D. J. M. Puttock, and H. J. George. 1987. Plant Culture Media, Vol. 1. Formulations and uses. Exegetics, Limited, Edington.

George, E. F., D. J. M. Puttock, and H. J. George. 1988. Plant Culture Media, Vol. 2. Commentary and analysis. Exegetics, Limited, Edington.

Griffiths, H. M., S. A. Slack, and J. H. Dodds. 1990. Effect of chemical and heat therapy on virus concentration in *in vitro* potato plantlets. Can J Bot 68:1515-1521.

Grignon, C. and H. Sentenac. 1991. *p*H and ionic conditions in the apoplast. Ann Rev Plant Physiol Plant Mol Biol 42:103-128.

Gunning, B. E. S. and A. W. Robards. 1976. Intercellular communication in plants: studies on plasmodesmata. Springer, Berlin.

Hess, J.R. 1992. The ovular hormonal environment during embryo development in wheat and its relevance to embryogenesis. Ph.D. Dissertation. Utah State University, Logan, Utah.

Hess, J. R. and J. G. Carman. 1993. Normalizing development of cultured *Triticum aestivum* L. embryos. I. Low oxygen tensions and exogenous ABA. J Exp Bot 44:1067-1073.

Hocking, P. J. 1980. The composition of phloem exudate and xylum sap from tree tobacco (*Nicotiana glauca* Grah.). Ann Bot 45:633-643.

Holm, P. B., S. Knudsen, P. Mouritzen, D. Negri, F. L. Olsen, and C. Roué. 1994. Regeneration of fertile barley plants from mechanically isolated protoplasts of the fertilized egg cell. Plant Cell 6:531-543.

Iglesias, V. A., A. Gisel, I. Potrykus, and C. Sautter. 1994. *In vitro* germination of wheat proembryos to fertile plants. Plant Cell Rep 13:377-380.

Kartha, K. K. 1986. Production and indexing of disease-free plants. In: L. A. Withers and P. G. Alderson (eds.) Plant tissue culture and its agricultural applications. pp 219-238. Butterworths, London.

Kartha, K. K., O. L. Gamborg, F. Constabel, and J. P. Shyluk. 1974. Regeneration of cassava plant from apical meristems. Plant Sci Lett 2:107-113.

Kelner, J.-J., S. Lachaud, and J.-L. Bonnemain. 1993. Seasonal variations of the tissue distribution of [^{3}H]ABA and [^{3}H]nutrients in apical buds of beech. Plant Physiol Biochem 31:531-539.

Li, Z.-C. and D. R. Bush. 1990. ΔpH-dependent amino acid transport into plasma membrane vesicles from sugar beet leaves. I. Evidence for carrier-mediated, electrogenic flux through multiple transport systems. Plant Physiol 94:268-277.

Lim, S. T., S. M. Wong, and C. J. Goh. 1993. Elimination of cymbidium mosaic virus and odontoglossum ringspot virus from orchids by meristem culture and thin section culture with chemotherapy. Ann Appl Biol 122:289-297.

Marschner, H. 1986. Mineral nutrition of higher plants. Academic Press, London.

McCabe, E. E., W. F. Swain, B. J. Martinell, and P. Christou. 1988. Stable transformation of soybean (*Glycine max*) by particle acceleration. Bio/Technology 6:923-926.

Meiners, S., O. Baron-Epel, and M. Schindler. 1988. Intercellular communication - filling in the gaps. Plant Physiol 88:791-793.

Murashige, T. and F. Skoog. 1962. A revised medium for rapid growth and bioassays with tobacco tissue cultures. Physiol Plant 15:473-497.

Murray, D. R. 1988. Nutrition of the angiosperm embryo. John Wiley & Sons, New York.

Nonami, H. and J. S. Boyer. 1987. Origin of growth-induced water potential. Solute concentration is low in apoplast of enlarging tissues. Plant Physiol 83:596-601.

Olsen, F. L. 1991. Induction of microspore embryogenesis in cultured anthers of *Hordeum vulgare*. The effects of ammonium nitrate, glutamine and asparagine as nitrogen sources. Carlsberg Res Commun 52:393-404.

Oparka, K. J. 1990. What is phloem unloading? Plant Physiol 94:393-396.

Opaskornkul, C., M. Greger, and J.-E. Tillberg. 1994. Effects of apoplastic sucrose on carbohydrate pools and sucrose efflux of mesophyll protoplasts of pea (*Pisum sativum*). Physiol Plant 90:685-691.

Pérez-Vincente, R., X. D. Wen, Z. Y. Wang, N. Leduc, C. Sautter, E. Wehrli, I. Potrykus, and G. Spangenberg. 1993. Culture of vegetative and floral meristems in ryegrasses: potential targets for microballistic transformation. J Plant Physiol 142:610-617.

Quak, F. 1977. Meristem culture and virus-free plants. In: J. Reinert and Y. P. S. Bajaj (eds.) Plant cell, tissue, and organ cutlure. Springer, Berlin.

Rennenberg, H., K. Schmitz, and L. Bergmann. 1979. Long-distance transport of sulfur in *Nicotiana tabacum*. Planta 147:57-62.

Riens, B., G. Lohaus, D. Heineke, and H. W. Heldt. 1991. Amino acid and sucrose content determined in the cytosolic, chloroplastic, and vacuolar compartments and in the phloem sap of spinach leaves. Plant Physiol 97:227-233.

Robards, A. W. 1976. Plasmodesmata in higher plants. In: B. E. S. Gunning and A. W. Robards (eds.) Intercellular communication in plants: studies on plasmodesmata. Springer, Berlin.

Rutledge, C. B. and G. C. Douglas. 1988. Culture of meristem tips and micropropagation of 12 commercial clones of poplar *in vitro*. Physiol Plant 72:367-373.

Salisbury, F. B. and C. W. Ross. 1992. Plant physiology. Wadsworth Publishing Company, Belmont.

Schobert, C. and E. Komor. 1989. The differential transport of amino acids into the phloem of *Ricinus communis* L. seedlings as shown by the analysis of sieve-tube sap. Planta 177:342-349.

Silk, W. K. 1984. Quantitative descriptions of development. Ann Rev Plant Physiol 35:479-518.

Silk, W. K. and Erickson, R. O. 1980. Local biosynthesis rates of cytoplasmic constituents in growing tissues. J Theor Biol 83:701-703.

Smith, J. A. C., and J. A. Milburn. 1980. Osmoregulation and the control of phloem-sap composition in *Ricinus communis* L. Planta 148:28-34.

Smith, R. H. and T. Murashige. 1970. *In vitro* development of the isolated shoot apical meristem of angiosperms. Amer J Bot 57:562-568.

Smith, R. H., T. S. Ko and S. H. Park. 1994. Transfer and expression of T-DNA into rice and sorghum via *Agrobacterium*. Congress on Cell and Tissue Culture, Tissue Culture Association, June 4-7, 1994, Research Triangle Park, NC.

Stein, A., S. Spiegel, G. Faingersh, and S. Levy. 1991. Responses of micropropagated peach cultivars to thermotherapy for the elimination of prunus necrotic rignspot virus. Ann Appl Biol 119:265-271.

Stone, O. M. 1982. The elimination of four viruses from *Ullucus tuberosa* by meristem-tip culture and chemotherapy. Ann Appl Biol 101:79-83.

Thompson, J. A., R. Abdullah, and E. C. Cocking. 1986. Protoplast culture of rice (*Oryza sativa* L.) using media solidified with agarose. Plant Sci 47:123-133.

Torrizo, L. B. and F. J. Zapata. 1992. High efficiency plant regeneration from protoplasts of the indica cultivar IR58. J Exp Bot 43:1633-1637.

Van Zaayen, A., C. Van Eijk, and J. M. A. Versluijs. 1992. Production of high quality ornamental crops through meristem culture. Acta Bot Neerl 41:425-433.

Winter, H., G. Lohaus, and H. W. Heldt. 1992. Phloem transport of amino acids in relation to their cytosolic levels in barley leaves. Plant Physiol 99:996-1004.

Ziegler, H. 1975. Nature of translocated substances. In: M. H. Zimmermann and J. A. Milburn (eds.) Encyclopedia of plant physiology, new series, vol 1. Transport in plants I, phloem transport, pp 59-100. Springer, Berlin.

MULTIPLE FLORAL BUD FORMATION *IN VITRO* ON THE GIANT DOME OF *ELEUSINE CORACANA* GAERTN.

T. WAKIZUKA[1] and T. YAMAGUCHI[2]
1 Science Education Institute of Osaka Prefecture,
Karita 4, Sumiyoshi-ku, Osaka 558, Japan
2 College of Agriculture, University of Osaka prefecture,
Gakuen-cho 1, Sakai, Osaka 593, Japan

ABSTRACT. The shoot apex is an autonomous structure that forms leaf primordia and buds, and becomes committed to floral buds that are complex. Our aim was to establish an *in vitro* culture system that maintains autonomous morphogenic competence as the shoot apical meristem does. The giant dome (an enlarged apical dome produced *in vitro*) of *Eleusine coracana* Gaertn., which propagates itself without any leaf primordia, developing multiple shoots, could be useful for this purpose. We investigated the possibility of such domes on entering the prefloral state, and of regenerating floral buds. Light microscopy and scanning electron microscopy showed that many floral buds developed directly from the surface of the dome, without vegetative shoot formation. Hence the giant dome is comparable with the shoot apical meristem in its function as well as in the tunica-corpus arrangement of cells. Multiple floral buds consisted of a mean of 36 floral buds per square millimeter and the florets that developed were gramineous in form.

INTRODUCTION

The shoot apex has been used in studies of the histogenesis and morphogenesis of vascular plants for a long time (Nougarede (1967), Gifford and Corson (1971), Huala and Sussex (1993)). *In vitro* cultures of the shoot apex are adaptable to examinations of the development mechanisms, floral determination (McDaniel et al. (1991)), the degree of autonomy of the floral apex (Hicks and Sussex (1970)), and genetic transformation (Simmonds et al. (1992), Perez-Vicente et al. (1993)).

One reason for the attention paid to the shoot apex is its autonomy. There are different patterns of morphogenesis in the apex, which gives rise to leaf primordia, axillary buds, and internodes, while still growing itself. The shoot apex changes from vegetative to floral development in response to the surroundings. Methods for the study of the shoot tip culture (Styer and Chin (1983)), and floral bud cultures (Rastogi and Sawhney (1989)) have been reviewed. In culture of the shoot apical meristem, complex growth occurs concurrently or

M. Terzi et al. (eds.), Current Issues in Plant Molecular and Cellular Biology, 405-410.

consecutively.

We have established a system for *in vitro* culture of finger millet, in which the development of the apical meristematic dome can be studied separately from the initiation of leaf primordia; with this system, the induction and multiplication of enlarged apical domes are possible (Wakizuka and Yamaguchi (1987)). The cultures resemble shoot apical domes histologically except for their enormous size. They have a tunica-corpus structure, with an outer layer with anticlinal dividing cells and inner tissue with mainly periclinal dividing cells. The enlarged domes proliferate, because they remain in a vegetative state for years although the plant is an annual. The domes are unusual in keeping a capacity for multiple shoot formation for an indefinitely long time during subculture. We call the cultures "giant domes." The only morphogenetic pattern that has been reported similar to that of giant domes is from the *in vitro* culture of *Musa* spp., in which expanded apical domes with leaf primordia form several new apices (Banerjee et al. (1986)).

The culture of excised plant parts has shown that the reception of environmental flowering cues such as photoperiod and temperature does not depend upon an intact plant system or even leaves (Scorza (1982)). Still, few *in vitro* studies have been done of the culture of shoot apical domes without any subjacent organs. Because (1) the structure of giant domes resembles that of shoot apex domes, (2) giant domes form multiple shoots more readily that shoot apices do, and (3) giant domes proliferate without leaf primordia for an indefinitely long period, giant domes may enter the reproductive state via a prefloral state from the vegetative state.

This paper reports the production of multiple floral buds directly from giant domes of finger millet. Giant domes and multiple floral buds provide a new *in vitro* culture system.

MATERIALS AND METHODS

The procedures of induction and culture of giant domes of *Eleusine coracana* Gaertn. (finger millet) were as described previously (Wakizuka and Yamaguchi (1987)). Giant domes subcultured for 5 years or more were used.

For the induction of floral buds, three induction media were prepared. One medium consisted of KNO_3 at 2 mmol l^{-1} and $CaCl_2 \cdot 2H_2O$, $MgSO_4 \cdot 7H_2O$, and KH_2PO_4, all at 1.5 mmol l^{-1}, with the inorganic and organic micro-elements for MS medium (Murashige and Skoog (1962)), except for the concentration of thiamine-HCl (1 mg l^{-1}). The concentration of KNO_3 in another medium was 20 mmol l^{-1}, which was close to the N-level of most culture media. MS basal medium was used as well. The media were supplemented with zeatin at 4.6×10^{-6} mol l^{-1} or not. Sucrose (30 g l^{-1}) and Gelrite (2 g l^{-1}) were added. The pH was adjusted to 5.7 before mixtures were autoclaved at 0.118 MPa for 15 min. Disposable polystyrene dishes (60 x 15 mm) were filled with 10 ml of a medium.

Pieces of a giant dome (about 1 x 1 mm) were cultured first on an induction medium for one week. The cultures were transferred to MS basal medium with or without zeatin. All cultures were kept at 25 °C under fluorescent light of 2,000 lx and with a 10-h daily photoperiod.

RESULTS AND DISCUSSION

Giant domes cultured first on the medium with 20 mmol l^{-1} KNO_3 did not produce floral buds whatever the later culture conditions, but they gave rise to multiple shoots vigorously, as reported before (Wakizuka and Yamaguchi (1987)). MS basal medium gave the same results.

Floral differentiation was identified when an apical dome elongated (Fig. 1). Such elongation is the first morphological indication of spike formation in intact plants (Gifford and Corson (1971)). Floral buds were detected one week after transfer from the medium with 2 mmol l^{-1} KNO_3 and with zeatin. Transfer to the MS basal medium was essential for development of floral buds. Numerous floral initials were visible through a binocular. They tended to be synchronous in development (Figs. 2-4). During initial culture, zeatin was needed for the giant dome to move from a prefloral state from a vegetative state, and the concentration of KNO_3 had to be low.

Zeatin had no effect on inflorescence differentiation of the multiple floral buds, but it was needed in the initial culture if floral initials were to form during later culture. These results are consistent with those reported from studies of *in vitro* flowering of gramineous plants. Benzyl-aminopurine induces precocious flower development of bamboo *in vitro* (Nadgauda et al. (1990)). Kinetin does not cause inflorescence differentiation of excised shoot apices of *Lolium temulentum* (McDaniel et al. (1991)). However, cytokinins are required for the development of excised flower buds of many dicotyledonous species (Scorza (1982)).

The morphogenetic potential for reproductive organogenesis in *in vitro* culture is dependent on explants being conditioned before excision. Tran Thanh Van (1973) showed that superficial tissues of the flower stem of tobacco give rise to floral buds *in vitro*, and George and Eapen (1990) reported that inflorescence development occurs directly from callus cultures derived from inflorescence segments of finger millet.

In our study, floral organogenesis seemed to be controlled between the time when the giant dome enters the prefloral state and the time it becomes committed to form floral buds. Accordingly, the giant domes in the prefloral state gave rise to multiple floral buds autonomously in ordinary culture conditions. This result implies that floral meristems of giant domes are stably determined. In similar situations, prefloral meristems of day-neutral tobacco complete floral development in culture *in vitro* (Hicks and Sussex (1970)), and floral determination of maize apical meristems isolated after vegetative development is completed is stable (Irish and Nelson (1991)).

The cultures grew to a diameter of 5-6 mm by 2 weeks after being transferred to the MS basal medium. The floral buds were accompanied with glume primordia (Figs. 1-3). A tiny protuberance appeared at the

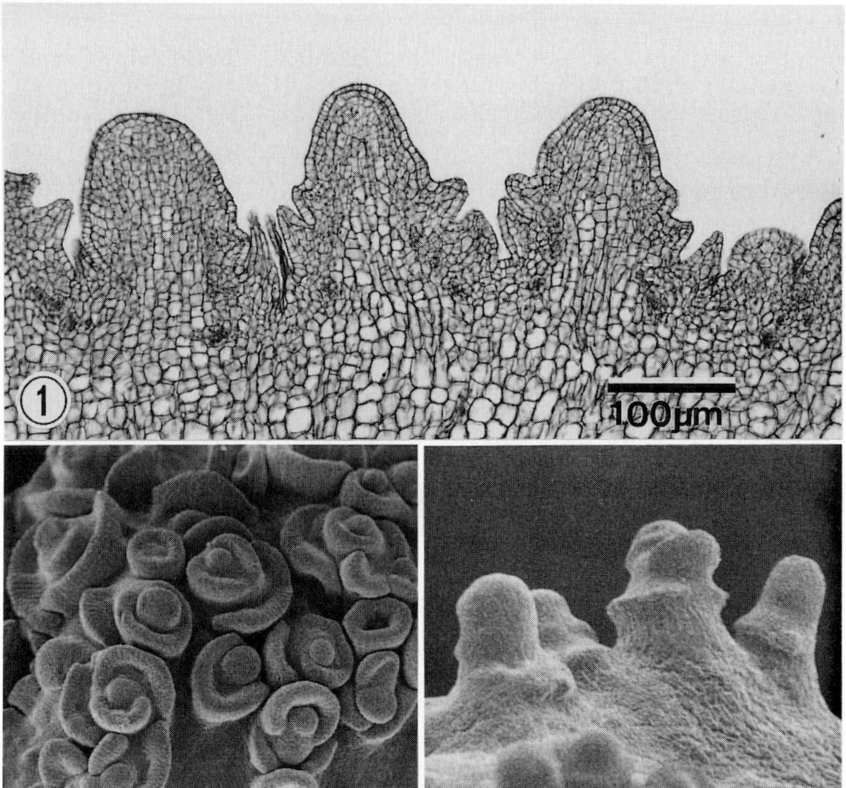
1
100µm
2
100µm
3
100µm
4
1 mm
5
1 mm

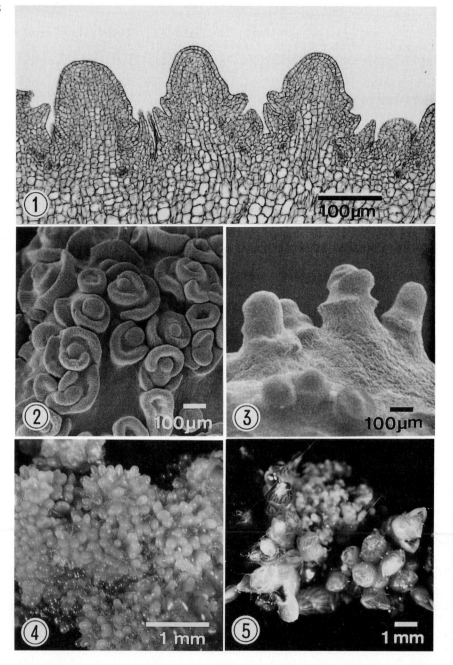

adaxial basal site of the most advanced glume primordium. Many floral buds stood close together on the surface of the culture (Fig. 4). There was no leaf structure. The number of floral buds initiated on giant domes was between 16 and 41 per square millimeter of the surface of the dome; the mean was 36.

After the next 2 weeks (ending 5 weeks after the start of culture for floral induction), the floral buds had given rise to sets of spikelets consisting of florets in which the lemma and palea were visible. Some of the florets had well-developed floral organs (pistil, stamens, lemma, and palea). In all, 1-10 florets per culture developed well (Fig. 5).

Multiple floral buds are probably highly stable genetically as long as the giant dome produces them directly without dedifferentiation to callus formation. The cuticle layer at the surface of the dome was further evidence of this stability (results of transmission electron microscopy not shown). A surface topography is investigated on the dedifferentiation and redifferentiation of rice callus (Maeda (1990)).

The formation of multiple floral buds increases the multiprogrammable morphogenetic potential of giant domes, and this increase makes approaches to gene tranfer more numerous.

In conclusion, multiple floral buds were induced directly from giant dome cultures of *Eleusine coracana*, showing that giant domes are similar to shoot apical meristems in function as well as structure. Giant domes, which do not occur other than experimentally, seem to be the result of metamorphosis of shoot apical meristems. Giant dome cultures provide new methods for plant tissue culture.

ACKNOWLEDGEMENTS

We thank Prof. E. Maeda, Nagoya University, for discussions, and Dr. T. Esaka, Science Education Institute of Osaka Prefecture, for assistance with the electron microscopy. This work was supported by Grant-in-Aid No. 01304010 from the Ministry of Education, Science, and Culture, Japan.

REFERENCES

Banerjee, N., Vuylsteke, D., and De Langhe, E. A. L. (1986) 'Meristem tip culture of *Musa*: Histomorphological studies of shoot bud

---

Figures 1–5. Multiple floral buds derived from giant domes of *Eleusine coracana* Gaertn. **1:** Longitudinal section of multiple floral buds after 3 weeks of culture (including 1 week of culture on the 2 mmol l^{-1} KNO_3 medium with zeatin described in the text). **2** and **3:** Scanning electron micrographs taken after 2 weeks (**2**) or 3 weeks (**3**) of culture. **4:** Multiple floral buds of a dome after 3 weeks of culture. **5:** Florets that developed in 8 weeks of culture. Note the lemma with clear nerves, palea, stamens, and pistil.

proliferation', in L. A. Withers and D. G. Alderson (eds.), Plant Tissue Culture and its Agricultural Applications, Butterworths, London, pp. 139-147.

George, L. and Eapen, S. (1990) 'High frequency plant-regeneration through direct shoot development and somatic embryogenesis from immature inflorescence cultures of finger millet (*Eleusine coracana* Gaertn)', Euphytica 48, 267-274.

Gifford, E. M., Jr., and Corson, G. E., Jr. (1971) 'The shoot apex in seed plants', Bot. Rev. 37, 143-229.

Hicks, G. S. and Sussex, I. M. (1970) 'Development in vitro of excised flower primordia of *Nicotiana tabacum*', Can. J. Bot. 48, 133-139.

Huala, E. and Sussex, I. M. (1993) 'Determination and cell interactions in reproductive meristems', Plant Cell 5, 1157-1165.

Irish, E. E. and Nelson, T. M. (1991) 'Identification of multiple stages in the conversion of maize meristems from vegetative to floral development', Development 112, 891-898.

Maeda, E. (1990) 'Surface topography and regeneration from rice callus', 7th Int. Cong. Plant Tissue Cell Culture, Amsterdam.

McDaniel, C. N., King, R. W., and Evans, L. T. (1991) 'Floral determination and in-vitro floral differentiation in isolated shoot apices of *Lolium temulentum* L.', Planta 185, 9-16.

Murashige, T. and Skoog, F. (1962) 'A revised medium for rapid growth and bio-assays with tobacco tissue cultures', Physiol. Plant. 15, 473-497.

Nadgauda, R. S., Parasharami, V. A., and Mascarenhas, A. F. (1990) 'Precocious flowering and seedling behaviour in tissue-cultured bamboos', Nature 344, 335-336.

Nougarede, A. (1967) 'Experimental cytology of the shoot apical cells during vegetative growth and flowering', Int. Rev. Cytol. 21, 203-351.

Perez-Vicente, R., Wen, X. D., Wang, Z. Y., Leduc, N., Sautter, C., Wehrli, E., Potrykus, I., and Spangenberg, G. (1993) 'Culture of vegetative and floral meristems in ryegrasses: Potential targets for microballistic tranformation', J. Plant Physiol. 142, 610-617.

Rastogi, R. and Sawhney, V. K. (1989) 'In vitro development of angiosperm floral buds and organs', Plant Cell Tissue Organ Cult. 16, 145-174.

Scorza, R. (1982) '*In vitro* flowering', Hortic. Rev. 4, 106-127.

Simmonds, J., Stewart, P., and Simmonds, D. (1992) 'Regeneration of *Triticum aestivum* apical explants after microinjection of germ line progenitor cells with DNA', Physiol. Plant. 85, 197-206.

Styer, D. J. and Chin, C. K. (1983) 'Meristem and shoot-tip culture for propagation, pathogen elimination, and germplasm preservation' Hortic. Rev. 5, 221-277.

Tran Thanh Van, M. (1973) 'Direct flower neoformation from superficial tissue of small explants of *Nicotiana tabacum* L.', Planta (Berl.) 115, 87-92.

Wakizuka, T. and Yamaguchi, T. (1987) 'The induction of enlarged apical domes *in vitro* and multi-shoot formation from finger millet (*Eleusine coracana*)', Ann. Bot. 60, 331-336.

EFFECT OF GIBBERELLIC ACID (GA_3) AND MICROPROPAGATION ON AXILLARY SHOOT INDUCTION IN MONOSTEM GENOTYPE (*TO*-2) OF TOMATO (*L. ESCULENTUM* MILL)

P. BIMA, F. MENSURATI, G. P. SORESSI
Dept. Agrobiology and Agrochemistry, University of Tuscia, 01100 Viterbo - Italy.

ABSTRACT. The *torosa*- 2 tomato (*to*-2) is characterized by inhibition of axillary shoot formation. The effect of gibberellic acid (GA_3) appears essential for the induction of lateral shoots, when applied at an early development stage of *torosa* plants. The effect of micropropagation *per se* on the *in vitro* formation of axillary shoots indicates that probably there is some inhibition in hormone transport. The highest efficiency in axillary shoot formation was observed when germination and micropropagation occurred on media supplemented with GA_3. The lateral branches developed in the basal portion of the main stem of *torosa* -2 plants from apex and node microcuttings grown in open field prove that root system plays an essential role in the hormone unbalance which is assumed to be responsible for the *torosa* monostem phenotype.

Introduction

The nature of development process is closely coordinated and there is a relationship among the interacting plant structures where hormones play a fundamental role. Many works demonstrated that lateral outgrowth is controlled by a balance of apically produced auxins, cytokinins, and abscissic acid (5; 6; 10). On the other hand roots, may generate or modify several kinds of hormonal messages (1).

Mutants have been a classical tool of biochemists and in the recent years mutants of higher plants have come into widespread use (King, 1991). The side-shootless *torosa* -2 (*to* -2) mutant, with a monostem phenotype of potential value for tomato crop improvement (Soressi, 1979), represents an excellent material for studying the hormonal mechanism leading to the meristem formation responsible of the plant habit. Analysis of endogenous level of growth substances gives evidences that the control of axillary bud differentiation and branching in the *to* -2 genotype is dependent on a complex mechanism involving gibberellins, cytokinins and ABA (3).

Aseptic culture techniques have figured prominently in the study of plant growth and development. The identification of hormones and the role of growth regulators came about a large measure as a result of this studies. The culture of plantlet or excised shoot apex *in vitro* may offer insights into branching process and about the developmental program of the shoot meristem.

The aim of this work is to define the effect of gibberellic acid (GA_3) and micropropagation on the axillary shoot induction in two different breeding lines carrying *to-2* mutation.

Materials and Methods

A first assay (figure 1) concerned the propagation of two monostem tomato lines, STU 32228 and STU 34859, both homozygous for the *to- 2* gene , the latter also homozygous for the genes *d* and *pat-2*. The seeds were sterilized in 2% sodium hypochlorite for 15 minutes, rinsed three times in

M. Terzi et al. (eds.), Current Issues in Plant Molecular and Cellular Biology, 411-416.

sterile water and then germinated in Magenta GA-7 vessels on three different culture media, each one containing MS (Murashige and Skoog, 1962), 3% sucrose, 0.6% agar (Oxoid), adjusted to pH 5.8 prior to autoclaving (20' at 120°C), and as plant hormones, 0.0, 0.5 and 1 mg/l of GA_3. Culture conditions were 25°C and a photoperiod of 16 h with cool white light. After 10 days germination, the apices, isolated from seedlings at two-leaves stage were transferred to fresh medium. Apices from each germination group were further distributed onto three different treatments (0.0; 0.5 and 1 mg/l GA_3) correspondent to the germination ones. The plantlets from these apices were then micropropagated once, separating the shoot tip from the micronode, which consisted of a portion of stem with a node. A number of 20 to 30 cuttings obtained from 4-5 plantlets were cultured on the same medium as the apex. The number and position on stem of the micronodes with axillary shoots were recorded.

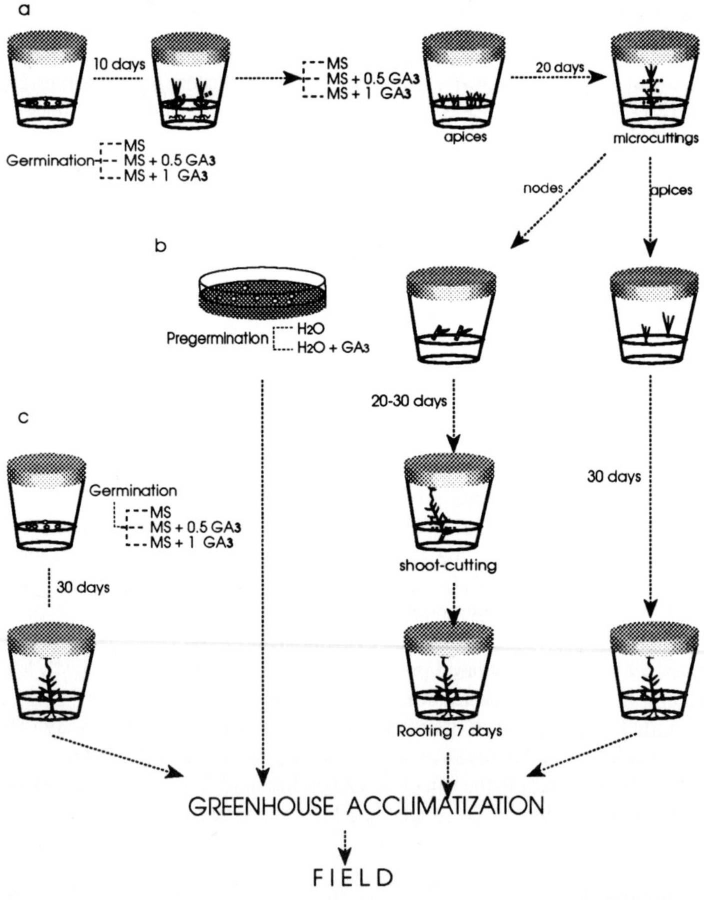

Figure 1: Scheme of experimental design used with the phases from germination to field growth: **a**: germination and micropropagation *in vitro* on MS medium with and without GA_3; **b**:- pre-germination in Petri dish with direct transferring to greenhouse and field; **c**: germination and seedling growth *in vitro* .
All seedlings and plantlets obtained were transplanted to field after greenhouse acclimatization.

After 30 days these shoots were transferred onto hormone-free MS medium for rooting. Apical nodes could form roots directly. After acclimatization in a greenhouse, all the plantlets were transferred to open field. Localization and number of lateral branches was recorded at fruit ripening.

In a second assay (figure 1) seeds were germinated in Petri dishes in presence of water solution with 0.0, 0.5 and 1 mg/l of GA_3. Seedlings were transferred to field. At the same time, other seeds were sterilized and sown on solid MS with 3% sucrose and 0.0, 0.5, and 1 mg/l GA_3. After 20 days, plantlets with 4-5 leaves were transferred to pots for the acclimatization and then to open field. Lateral branches have been counted on fruiting plants by distinguishing three main stem portions: basal, central, top.

Results

As far as development of shoots from micronodes *in vitro* is concerned, line STU-34859 showed a better response. The highest concentrations of GA_3 caused phytotoxic reactions during micropropagation in line STU-32228, as plants turned chlorotic and the formation of buds was even lower than that of control (MS/MS); the best result was obtained by sowing on MS and micropropagating in presence of 0.5 mg/l GA_3 (62%). Applied at sowing, 1 mg/l GA_3 shows some effect on the production of axillary shoots when these are propagated on MS; anyway, this effect is weaker (47 %) than that of the above-mentioned treatment (62 %). The other treatments gave a production of axillary buds equal to or lower than the control.

On hormone-free MS as well, the production of shoots *in vitro* of the line STU-34859 was higher than STU-32228. If one considers the germination medium, by sowing on hormone-free MS, the percentage of shoots increases when micropropagation is carried out with 0.5 mg/l GA_3. Sowing on 0.5 mg/l GA_3, the percentage of shoots that develop after micropropagation on MS or on 0.5 mg/l GA_3 is very similar.

Table 1: Percentage of microcuttings with axillary shoot on monostem tomato lines STU 34859 and STU 32228 according to GA3 concentration in the culture media.

culture media		STU - 34859	STU - 32228
sowing	micropropagation.	%	%
	MS	47	32
MS	0.5 GA3	73	62
	1 GA3	(40)	18
	MS	70	33
0.5 GA3	0.5 GA3	69	25
	1 GA3	58	38
	MS	63	47
1 GA3	0.5 GA3	73	38
	1 GA3	100	23

( ) : data from 10 microcuttings from 2 plantlets

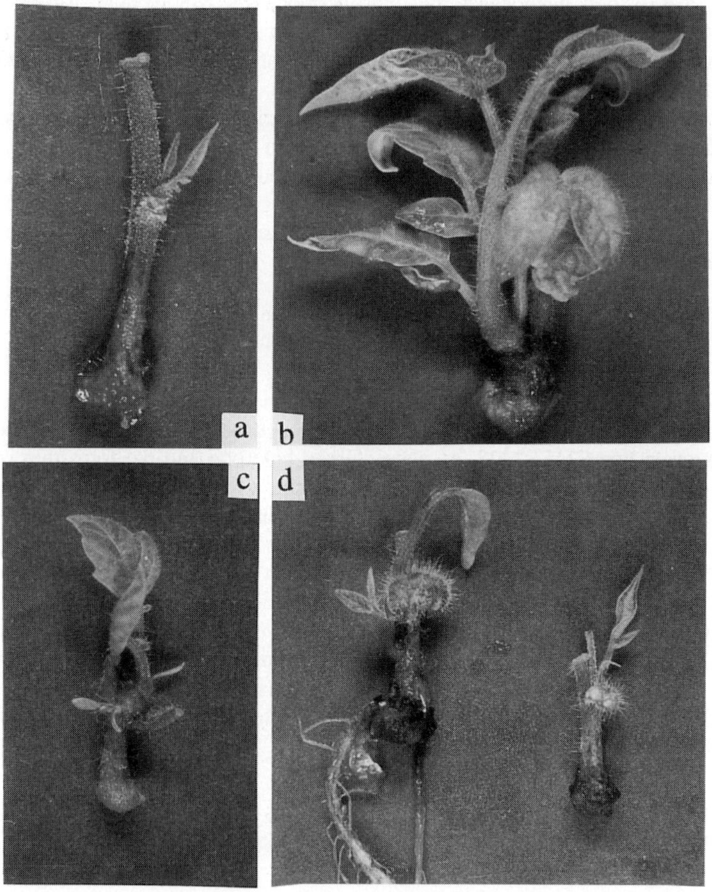

Fugure 2. Microcuttings with developing axillary shoots of the line STU-34859: **a:** normal axillary shoot formation; **b:** elongated axillary shoot; **c:** micronode with little bulbous formation on wich regenerates a shoot; **d:** microcuttings with callus like structures in correspondence of the developing axillary shoots.

When germination occurred on 1 mg/l GA_3, a rise in the number of shoots has been observed in STU-34859 parallel to the GA_3 increasing dose in apex and micropropagation media culture (63%, 73%, 100% respectively). STU-32228 behaved in opposite way (47%, 38%, 23% respectively) (Table 1).

With regard to the position of the node from which shoots developed *in vitro*, line STU-32228 shows a considerable increase from the 3rd node onwards in individuals that germinated in presence of GA_3 and were micropropagated on hormone-free MS; in line STU-34859 the percentage is always greater , with a higher rate for 1st, 2nd and 3rd nodes .

Formation of small tumours or bulblet-like structures was occasionally observed at the axil of microcuttings in both lines (figure 2).

With regard to the behaviour of plants in open field, only STU-34859 gave a reliable response. The plants that regenerated from apices whose seeds germinated with 1 mg/l GA_3 and were then transferred to medium with 1 mg/l GA_3 or hormone-free MS, produced in open field up to 4 and 6 lateral branches on their 2/3 basal portion (figure 3). Of the plants coming from micronodes, only those that had been sown on MS and micropropagated with 1 mg/l GA_3 gave 2 and 3 basal branches. The response to the other treatments did not differ from normal. No development of lateral shoots was observed in field on plants originated from seeds pre-germinated with GA_3 or from plantlets grown up *in vitro* in presence of GA_3 without micropropagation.

Figure 3. Plant with fruits of the monostem line (*d, to-2, pat-2*) STU- 34859, just defoliated to clearly show the 4 lateral branches (a1; a2; a3; a4) formed at the basis on the main stem. GA_3 treatment: 1 mg/l sowing/ 1 mg/l micropropagation

Discussion

In vitro experiments aimed to induce axillary shoot formation in monostem tomato genotypes, prove that micropropagation *per se* is at some degree effective even if a better response was obtained when MS cultured microcuttings were excised from seedling germinated on GA_3. The axillary shoot formation increased further when apex microcuttings from GA_3 germinated seedlings were cultured on media containing GA_3. Thus it can be assumed that gibberellins have to be active at a very early development stage, on embryo- apical meristem during germination and a few weeks after, as observed in maize (7). Therefore the action of GA seems to be very important in the apex, a zone of active cellular division where the axillary bud primordia are formed.
To this regard, the influences of roots and root- sinthesized hormones appear to play a fundamental role by allowing differentiation and development of shoots. In fact roots may modificate hormonal messages (1) and Mapelli's hypothesis (3;9) claims the lower endogenous gibberellins activity could inhibit the catabolism and translocation of ABA so that cytokinins are not traslocated to the shoot meristem.

The importance of the root system is furtherly stressed by the feature of the field grown plants originated from apices cultured in presence of GA_3. These plants exhibit lateral branches only at the base- stem portion, whereas the control plants are practically sideshootless. Likely as soon as the root system has been reestablished, the induction and differentiation of new shoots has been hampered.

The two monostem genotypes reacted differently to GA_3 treatments: the better performance of STU- 34859 is probably due to the dwarf (*d*) gene involving a GA endogenous concentration, lower than the non *d* STU- 32228. Indeed, GA deficient shoots are much more responsive to a very small increase in GA (1).

The distribution of the axillary shoots on the main stem is similar to that reported by Mapelli and Kinet (1992) with plants grown in hydroponic solution with gibberellins or cytokinins proving that gibberellins unblock the transport of cytokinins. Moreover, a synergistic interaction between the two growth hormones when simultaneously presents can not be excluded. In agreement with Mapelli (3) gibberellins result efficient in inducing axillary shoot formation in *torosa* -2 genotype but our data emphasize the importance of GA_3 action at an early stage of apex ontogeny. These results must be complemented with hystological and biochemical investigations to find out where and when shoot meristems are formed.

The monostem *to*-2 tomato genotype appears an excellent material to study, also through *in vitro* techniques, the process of meristem differentiation, the control of plant architecture, and to approach the hormones- stress relationships.

References

1. Jackson, M. B. (1993) Are plant hormones involved in root to shoot communication? Advances in Botanical Research, 19:104 - 189
2. King, J. (1991) The genetic basis of plant phisiology processes. Oxford Univ. Press, NY. ISBN 0- 19- 50 48 57- 1.
3. Mapelli, S.and Kinet, J. M. (1992) Plant growth regulator and graft control of axillary bud formation and development in the *to*-2 mutant tomato. Plant Growth Regulation, 11: 385 - 390
4. Murashige, T. and Skoog, F. (1962) A revised medium for rapid growth and bioassays with tobacco tissue cultures Physiol. Plant., 15:473 - 497
5. Prochazca, S. and Jacobs, W. P. (1984) Transport of benzyladenine and gibberellic acid from roots in relations to the dominance between axillary buds of pea (*Pisum sativum* L)cotyledons. Pl. Physiol.,1990) Phytohormone Mutants in plant research. J. of Plant Growth Regul. 9: 97 - 111.
7. Sheridan, W. F. (1988) Maize developmental Genetics: genes of morphogenesis. Annu. Rev. Genet., 22: 353 - 385.
8. Soressi, G. P., Mapelli, S. (1992) Genetical, physiological and biochemical aspects of a monostem tomato phenotype suitable for mechanization and processing. Acta Horticulturae, 301: 229 - 236.
9. Soressi, G. P. (1979) Potential of mutation in tomato. In: Israeli - Italian Joint Meeting on Genetics and Breeding of Crop Plants, 245 - 258. Roma: Istituto Sperimentale Cerealicoltura.
10. Sossountzov, L, Maldiney, R., Sotta, B., Sabbagh, I., Habricot, Y., Bonnet, M. and Miginiac, E. (1988) Immunocytochemical localization of cytokinins in Craigella tomato and a sideshootless mutant. Planta, 175: 291 - 304.

THE PLANT ONCOGENE *ROLB* ENHANCES MERISTEM FORMATION IN TOBACCO THIN CELL LAYERS

M. M. ALTAMURA*, F. CAPITANI*, M. TOMASSI*, I. CAPONE** and P. COSTANTINO**

** Department of Plant Biology; ** Department of Genetics and Molecular Biology, University "La Sapienza"*
P.le A. Moro, 5
00185 Rome
Italy

ABSTRACT

The expression and the effects of *Agrobacterium rhizogenes rolB* gene during *in vitro* flower, root and vegetative organogenesis were investigated. Thin cell layers were excised from floral pedicels, inflorescence rachis and stems of tobacco plants containing either *rolB* and its promoter fused to the GUS reporter gene or only this latter construct. Organogenesis and GUS activity were monitored in pedicel and stem explants cultured under conditions for floral, root, and vegetative neoformation. Flowering, rhizogenesis and caulogenesis were enhanced in *rolB*-transformed explants as compared to untransformed controls: the neoformed meristems were much more numerous and appeared sooner. *rolB* promoter was specifically active in the initial cells of all the meristem types. Rachis explants from *rolB*-plants produced, upon culture under flowering conditions, also an exceeding number of vegetative buds. Thus, *rolB* seems to stimulate the formation of meristems regardless of their subsequent differentiation and of the hormone controlling the morphogenic event.

INTRODUCTION

The affinities in the structural patterns of plant apical meristems - the presence of zones showing differences in frequency of cell division and the occurrence of different planes of cell division in surface and deep layers - suggest the existence of common factors (gene?) controlling the early organization of all types of primary meristems regardless of the type of organ which is subsequently formed.

In vitro tecniques allow to study the factors governing meristem formation. In particular, with the thin cell layer system it is possible to produce different organ types depending on the hormonal concentrations and balance in the culture medium [1]. Other parameters, mainly the excision site and the stage of development of the donor plant, must also be taken into account in order to optimize the production of a specific organ [2,3]. The flowering commitment sharply decreases - while the vegetative commitment increases - in thin cell layers from branches of different order of the same inflorescence. As a consequence, the rachis is a multicompetent site: when cultured on hormonal conditions suitable for flowering, its explants produce not only flowers, but also vegetative buds and even some roots [4]. A gene that might be involved in meristem formation, is the plant oncogene *rolB* from the Ri-T-DNA of *Agrobacterium rhizogenes*. Transgenic plants containing *rolB* show developmental aberrations such as flower heterostyly [6] and adventitious rooting [7]. The traits of *rolB*-transformed tissues suggest a role for the gene in altering either auxin sensitivity [8] or auxin concentration in plant cells [9]. As far as the regulaton of *rolB* in the plant host, an extensive cis-analysis of its promoter has shown that this oncogene is developmentally regulated and its expression is tissue-specific and controlled by auxin in mature embryos and adult plants [10,11]. Several

M. Terzi et al. (eds.), Current Issues in Plant Molecular and Cellular Biology, 417-422.

regulatory domains have been identified, different combinations of which allow selective gene expression in different populations of cells in the plant meristems [12]. Thus *rolB*, a bacterial oncogene, is finely regulated in plant cells by plant regulatory factors.

It is well established that the induction of root meristems is under auxin control and a key role of auxin has been shown also in flower neoformation [13]. In contrast, it is widely accepted that caulogenesis is a differentiative programme mainly controlled by cytokinins [14].

The aim of the present work is to understand - by means of thin cell layers excised from pedicel, rachis, and stem of *rolB* transformed tobacco plants - whether *rolB* is a "meristem inducing gene" - which stimulates the formation of meristems of different types regardless of the hormones controlling them - or else its effects are restricted to organogenesis controlled by auxin.

MATERIALS and METHODS

The constructs harbouring *rolB* and its 5' non-coding region and the gene fusion of the GUS reporter gene with the 5' non-coding region of *rolB* have been described elsewhere [5].

Transgenic plants containing both constructs (BpB-GUS) or only the latter (pB-GUS) were obtained from leaf disc infections of *Nicotiana tabacum* SR1, with appropriate recombinant *Agrobacterium* strains [5]. F1 transgenic plants were grown up to the stage of five expanded leaves, or to infructescence with flowers and fruits. Explants were excised at these stages for the caulogenic, and the flowering and rooting programmes, respectively. Thin cell layers of BpB-GUS and pB-GUS plants and of untransformed SR1 were excised from pedicels and rachises (flowering programme; hormonal conditions: 1μM indole-3-acetic acid, 1μM kinetin), and from stem internodes between the 5th and 8th node (rhizogenic programme; hormonal conditions: 10μM indolebutyric acid, 0.1 μM kinetin). Thin cell layers were also excised from stem internodes of plants with five expanded leaves (caulogenic programme; hormonal conditions: 10μM benzyladenine, 1μM indole-3-acetic acid). Sterilization and culture conditions were according to Altamura *et al.* [4] and Torrigiani *et al.* [3]. Histochemical staining and embedding of samples were carried out as previously described [11]. At the culture end (30 d), stocks of 100 explants per plant type, excision site and culture conditions were scored with the stereomicroscope and the results expressed as percentage of explants with organogenic response, and as mean number of organs (±SE) per explant. Significance of differences between means was evaluated by the Student's t-test and between percentages by the χ^2 test.

RESULTS

The constructs harboured in pB-GUS and BpB-GUS plants are shown in Fig. 1.

Thin cell layers from pedicels and rachises produce flowers. While pedicel explants form *de novo* flowers only (pure flowering programme), a number of rachis explants exhibits also flowers associated with vegetative buds (mixed programme).

In the case of the pure flowering response of pedicel explants, the presence of *rolB* gene increases both the percentage of explants showing flowers ($P < 0.01$), and the mean number of flowers per explant, as compared to untransformed controls (Tab. 1 and Fig. 2).

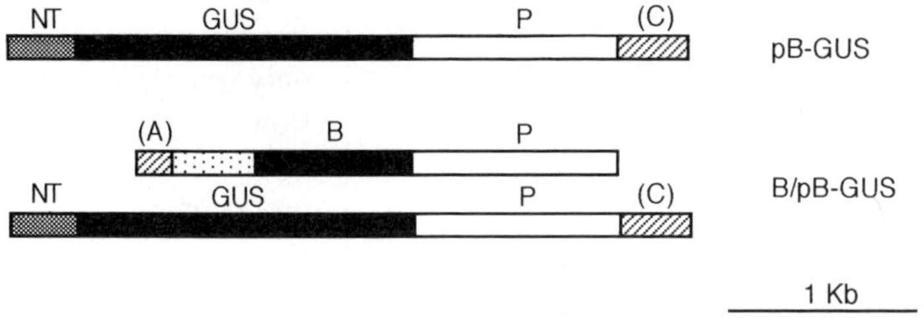

Figure 1. GUS and *rolB* constructs. GUS, ß-Glucuronidase; NT, transcription termination of the nopaline synthase gene; (A), (C), truncated coding region of *rolA* and, respectively, *rolC*; P, *rolB* promoter.

For the mixed programme, significant differences can be observed between BpB-GUS and control rachis explants. The number of transgenic explants showing flowers + vegetative buds is about three times that of the controls (Fig. 2), and the mean number of vegetative buds on such explants is twice that of the controls (Tab.1). Furthermore, some explants exhibit even the pure vegetative programme (6% explants, with a mean number of 6.2 vegetative buds per explant). Thus, in BpB-GUS rachis explants, both the total flowering response (30% of explants with the pure flowering programme + 30% of explants exhibiting the mixed programme), and the total vegetative response (30% of explants with the mixed programme + 6% of explants with the pure vegetative programme) are highly enhanced as compared to the untransformed controls (40% total flowering and 10% total vegetative budding).

TABLE 1. Mean number (± SE) of roots, flowers, and vegetative buds (pure and mixed programmes) per BpB-GUS and control (SR1) explant.

	Roots (pure progr.)	Flowers (pure progr.)	Veg.buds (pure progr.)	Flowers (mixed progr.)	Veg.buds (mixed progr.)
BpB	9.6±0.5	5.1±0.4	22.6±2.8	3.4±0.3	6.3±1.2
SR1	4.1±0.2	3.7±0.5	10.9±1.9	4.8±1.0	3.0±1.1

As for the pure vegetative response, *rolB* stimulates the explants to produce buds. Both the percentage of budding explants and the mean number of buds per explant are in fact much higher than in the controls (Tab. 1 and Fig. 2).

Analogous observations, though even more pronounced, hold for the rooting programme: significant ($P<0.01$) increases are recorded in both the percentage of rooting explants (Fig.2) and in the mean number of roots per explant (Tab. 1).

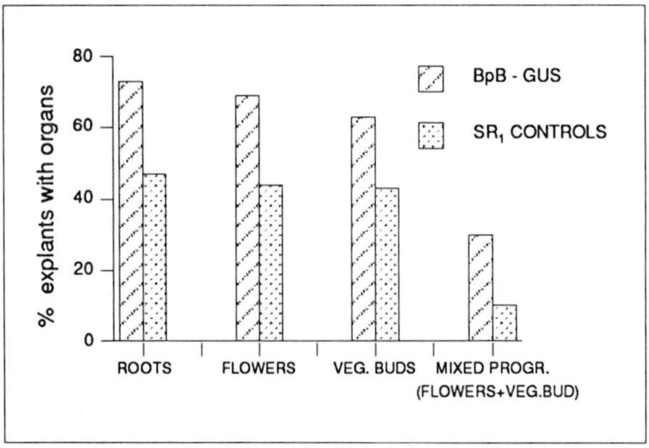

Figure 2. Percentage of BpB-GUS and control explants with roots only (pure programme), flowers only (pure programme), vegetative buds only (pure programme), and with flowers plus vegetative buds on the same sample (mixed programme).

Histological analysis of *rolB* expression during the culture period was also performed. No GUS-specific staining was ever observed in control explants. In all organogenic programmes, callus is macroscopically evident, sooner on BpB-GUS explants than on controls. Regardless of the programme, starting from day 4, the proliferated areas of BpB-GUS explants are filled with groups of intensely stained meristematic cells (Fig. 3). The first meristems appear on the pB-GUS and control explants at least 4 days later. The massive production of meristems in BpB-GUS explants continues up to the culture end, regardless of the organogenic programme.

On BpB-GUS explants only, meristem production is never associated with intense callus proliferation.

The general pattern of GUS activity during flower, vegetative and root formation does not differ in BpB-GUS and pB-GUS explants. In each type of *de novo* formed organ, the *rolB* promoter is specifically active in the initial cells of the different meristems, as exemplified in Fig. 4.

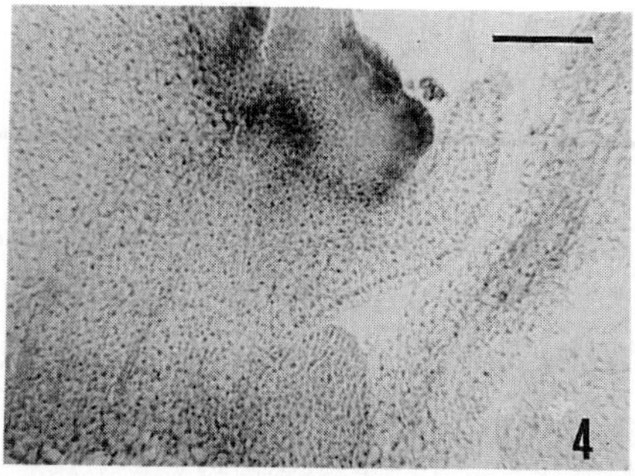

Figures 3 and 4. Thin cell layers from BpB-GUS plants. Fig.3: the meristems appear on callus as spots of intense GUS activity (dark zones). Fig.4: pattern of GUS activity (dark zones) in flowers . Bars= 0.5 mm and 100 μm.

DISCUSSION

We show here that *rolB* activity in meristematic cells does not require that these derive from embryonic initials. The meristems formed *in vitro* by the differentiated cells of thin cell layers show, in fact, the same pattern of *rolB* expression as the embryo-derived meristems *in planta* [11].

rolB -containing thin cell layers show a very high rhizogenic response (Fig.2, Tab.1), a result which is not surprising in view of the well established strong root-inducing capability of this gene *in planta* [7]. However, we show here that the effects of *rolB* on organogenesis are more general, as this gene enhances also the formation of meristems of organs other than roots, such as flowers and vegetative buds (Fig.2, Tab.1). It remains to be assessed whether this meristem-promoting action of *rolB* is a consequence of the increased auxin sensitivity of *rolB*-transformed cells. Whereas flowering and rooting *in vitro* are auxin-controlled programmes, vegetative budding is apparently not. In addition our results show that *rolB* stimulates both flowering and vegetative budding on rachis explants cultured under the same hormonal conditions (flowering medium). Thus *rolB* seems capable of enhancing meristem formation regardless of the exogenous hormonal input and of the type of organ which is produced.

The question is now whether *rolB* acts by enhancing a specific endogenous commitment of the cells or whether it is capable *per se* to induce meristems. A culture system where the commitment of the explant has been nullified is currently being utilized in order to better characterize the effects and role of *rolB* in meristem formation. Through this study it will be possible to identify plant genes specifically involved in the molecular events that preside meristematic condition.

ACKNOWLEDGEMENTS

This work was supported by funds from Ministero dell'Università e della Ricerca Scientifica e Tecnologica (40%), Italy.

REFERENCES

1. Tran Thanh Van, M., Thi Dien, N. and Chlyah, A. (1974) ' Regulation of organogenesis in small explants of superficial tissue of *Nicotiana tabacum* L.', Planta 119, 149-159.
2. van den Ende, G., Croes, A.F., Kemp, A. and Barendse, G.W.M. (1984) 'Development of flower buds in thin-layer cultures of floral stalk tissue from tobacco: role of hormones in different stages', Physiologia Plantarum 61, 114-118.
3. Torrigiani, P., Altamura, M.M., Capitani, F., Serafini-Fracassini, D. and Bagni, N. (1989) '*De novo* root formation in thin cell layers of tobacco: changes in free and bound polyamines ', Physiologia Plantarum 77, 294-301.

4. Altamura M.M., Monacelli, B. and Pasqua, G. (1989) 'The effect of photoperiod on flower formation *in vitro* in a quantitative short-day cultivar of *Nicotiana tabacum*', Physiologia Plantarum 76, 233-239.
5. Capone, I., Spanò, L., Cardarelli, M., Bellincampi, D., Petit, A. and Costantino, P. (1989) 'Induction and growth properties of carrot roots with different complements of *Agrobacterium rhizogenes* T-DNA', Plant Molecular Biology 13, 43-52.
6. Schmülling, T., Schell, J. and Spena, A. (1988) 'Single genes from *Agrobacterium rhizogenes* influence plant development', EMBO J. 7, 2621-2629.
7. Spena, A., Schmülling, T., Koncz, C. and Schell, J. (1987) 'Independent and synergistic activity of *rolA, B* and *C* in stimulating abnormal growth in plants', EMBO J. 6, 3891-3899.
8. Maurel, C., Barbier-Brygoo, H., Spena, A., Tempé, J. and Guern, J. (1991) 'Single *rol* genes from the *Agrobacterium rhizogenes* TL-DNA alter some of the cellular responses to auxin in *Nicotiana tabacum*', Plant Physiology 97, 212-216.
9. Estruch, J.J., Schell, J. and Spena, A. (1991) 'The protein encoded by the *rolB* plant oncogene hydrolyses indole glycosides', EMBO J. 10, 3123-3128.
10. Capone, I., Cardarelli, M., Mariotti, D., Pomponi, M., De Paolis, A. and Costantino, P. (1991) 'Different promoter regions control level and tissue specificity of expression of *Agrobacterium rhizogens rolB* gene in plants', Plant Molecular Biology 16, 427-436.
11. Altamura, M.M., Archilletti, T., Capone, I., and Costantino, P. (1991) 'Histological analysis of the expression of *Agrobacterium rhizogenes rolB*-GUS gene fusions in transgenic tobacco', New Phytologist 118, 69-78.
12. Capone, I., Frugis, G., Costantino, P. and Cardarelli, M. (1994) 'Expression in different populations of cells in the root meristem is controlled by different domains of the *rolB* promoter', Plant Molecular Biology, in press.
13. Smulders, M.J.M., Visser, E.J.W., Croes, A.F. and Wullems, G.J. (1990) 'The dose of 1-naphthaleneacetic acid determines flower-bud regeneration in tobacco explants at a large range of concentrations', Planta 180, 410-415.
14. Brock, T.G. and Kaufman, P.B. (1991) 'Growth regulators: an account of hormones and growth regulation', in F.C. Steward (ed.), Plant Physiology. A treatise. Volume X: Growth and development, Academic Press, San Diego, pp. 277-340.

RESULTS, SHOWING POSSIBILITIES OF MERISTEM METHOD FOR IMPROVING SOME CHARACTERISTICS OF POTATO VARIETIES

V.ROSENBERG
The Estonian Plant Biotechnical Research Centre EVIKA
Saku, Harjumaa, EE3400 ESTONIA

ABSTRACT. We have studied the differences of meristemic clones during 15 years. Up to the present time we have compared in field trials 175 meristemic clones of 35 varieties. With some cultivars a more profound research has been conducted. We have been interested in the possibilities of using meristems for producing variations, which preserve the morphological characteristics and positive traits of the variety, but have a higher disease resistance, higher yielding ability, higher dry matter content, better tuber shape, etc. We have also tried to find out how persistent the variation traits are and how they are transmitted to the meristemic clone of the next generation. The majority of the studied meristemic clones were regenerated with our virus eradication technique. According to this technique, meristems are derived from buds of specially prepared green plants. Both the parent and plants of meristemic clones have undergone selection on several levels. As the purpose of our system of virus eradication from seed potatoes is to obtain positive results, we did not study all the clones but only the vigorous ones. Weakly developed regenerants and those of vague shape or colour were sorted out at the 1st stage. In the majority of cases, no morphological variation from variety characteristics was observed on field grown plants of meristemic clone. In some cases the plants differed in the intensity of flowering, in the number of stalks, as well as in the height of plants and in leaves. No significant differences in the shape and colour of tubers were observed. In case of the cultivar "Vigri", some meristemic clones differed in the colour of flowers and leaves. As to the yielding ability, disease resistance and to the starch content of tubers, some varieties displayed great variability. Our results indicate that with the help of meristems it is possible to improve the characteristics of cultivars, and it is far simpler, cheaper and more resultful than with protoplast culture.

INTRODUCTION

Potato has a lot of diseases and therefore potato growers use yearly a vast amount of chemicals, which is both expensive and harmful to nature. In the future we cannot rely only on chemicals, because we have to protect soil as well as the entire environment. There is also tendency that pathogens form strains which are resistant to new chemicals making the pest control thus ineffective. The occurrence of late blight has shown it very clearly in the recent years. E.g. in 1994, potato fields in European countries were sprayed against late blight 10 times or even more. A smaller number of treatments did not give results.

The best solution would be disease resistant varieties. Therefore, plant breeders try to find new possibilities for creating resistant cultivars. In the past century approximately 1000 potato varieties have been bred in Europe, which all have one or another valuable characteristic. Although it seems that we have enough varieties already, the characteristics of some varieties should be improved. E.g. the widely grown variety "Bintje" which has many good characteristics, but is very susceptible to late blight. The same could be said about the very early and high-yielding variety "Premiere". In addition to

M. Terzi et al. (eds.), Current Issues in Plant Molecular and Cellular Biology, 423-426.

disease resistance, there are also other problems; for example the dry matter content of some varieties is too low, the tubers are too big , too small or poorly shaped.

We have studied potato's meristem culture since 1966. In the beginning, the primary aim was to remove virus diseases from the initial material of seed production. Thereafter we investigated the factors affecting the regeneration of plants from meristems and the efficiency of virus eradication. Later, when the meristem method was more widely used in seed production it was suspected that meristem culture might changed varietal characteristics. Then we started with field trials, in which we compared seed potatoes improved by traditional clone selection with the meristemic material from which viruses had been eradicated in our laboratory. During 12 trial years we compared several varieties. We did not find variation from varietal characteristics, but the plants grown from sees potatoes which were improved by meristem method, were more vigorous and had higher yields, depending on the variety and material 4 - 21 t/ha. It could be seen that plants regenerated from meristemic seed material were more resistant to late blight.

Since the very beginning, we have kept separately the subsequent generation, meristemic clone of each meristeme, and now for almost 15 years we have studied the characteristics of meristemic clones, trying to find an answer to the question of if it is possible to improve one or another trait of the variety with the help of meristem culture. Our opinion is that with meristem it would be far cheaper, simpler and more resultful than with usual crossing, protoplast, or some other tissue culture. In or research we have observed several positive variations. At the present time we cannot say, what are the interactions between meristem's characteristics and variation traits, how to evoke them preserving at the same time the positive characteristics and morphological traits of the variety. At the moment we still cannot knowingly direct the results and therefore, we have started studying the interactions between meristem and variations, as well as some specific moments of our virus eradication system. Up to now we have used no chemical substances to produce variants.

MATERIAL AND METHODS

Meristemic clones are derived from the buds of green plants and cultivated on growth medium, which is our modification of the Murashige cake. Of growth promoting hormones the medium contains kinetin and gibberellin. Before operating the meristem, we grow the plants 6 - 8 weeks in thermotherapy at the temperature 38 ± 1° C in light 16 hours and in darkness 8 hours, t^{o} 20 - 22°C. 2 - 3 months later the regenerated meristemic plants are multiplied by microcutting in vitro. Thereafter the part of plants is preserved in test tubes, a part, usually 3 of each meristem, is planted into greenhouse, where they are controlled for virus infection.

In further studies we use only virus-free meristemic clones. The plants selected out for the trial are multiplied in greenhouse in plastic rolls, planted into field and the 1st generation tubers are grown. These tubers are used in following years in field comparison trials. The field trial has 4 replications, 90 plants in every plot. During growth period we observe the morphological characteristics, at harvest we fix the amount of tubers, the weight, the yield (t/ha) and determine the dry matter and starch contents, the nitrate content, the infection of tubers. In some cases the susceptibility of tubers to late blight on the tubers has been determined in laboratory, as well as their resistance to mechanical injuries.

No plant protection products are used in trial fields to control diseases and pests.

RESULTS

We have studied in field trials meristemic clones of many varieties. It seems that there are great differences between varieties, some varieties are more stable and the meristemic clones of the differed in our trial less than those of other varieties. Especially interesting for us have been the cultivars "Eba", "Vigri", "Premiere", "Bintje", "Kondor". The results discussed below are about the cultivars "Eba" and "Vigri", which have been studied more profoundly.

TABLE 1.Comparison results of meristemic clones of the variety "Eba" on the average of 3 years

Number of meristemic clone	Size of meristemic cut and year	Number of tubers per plant	Yield t/ha	Some morphological differences
295	0,3 mm (1981)	13,6	28,5	leaves narrow stalks low typical to "Eba
3373	0,2 mm (1983)	18,2	47,5	leaves broader earlier emergence
3471	0,3 mm (1983)	15,4	44,0	typical but leaves broader

Thereafter we were interested in the persistency of characteristics of the meristemic clone 3373, and how these characteristics are transmitted to the meristems of new generation. Therefore we established the following trials. We chose from the field, which was the subsequent generation of 3373, a plant with especially broad leaves and high yield. The plant had 117 tubers. We chose one tuber, eradicated viruses with our technique and selected out 3 meristemic clones which are shown in the scheme.

THE SCHEME OF DIFFERENT MERISTEMIC CLONES OF "EBA"

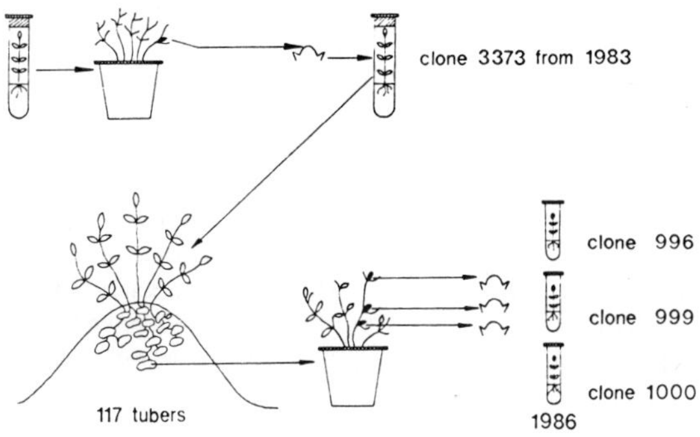

Comparison results are given in Table 2.

TABLE 2.Results of meristemic clones of the variety "Eba" on the average of 5 years

Meristemic clone	Tubers per plant	Tubers' weight g	Yield (t/ha)	Starch content %
996	7,9	58	26	16,5
999	7,5	59	23	17,6
1000	7,3	60	24	16,9
3373	9,3	71	37	15,8

It became evident that in this way obtained progeny of meristemic clone resembled more the clone 295 both in morphological characters and yielding ability. The yielding potential of 3373 has not displayed any decrease, the average of 5 years in Table 2 is lower than the great differences in the yielding level of those years. We continue the research on 3373 characteristics and their transmission. In the field trials of 1994 we have 11 meristemic clones which have been taken from 3373 plants kept in vitro.

The meristemic clones of the variety "Vigri" were compared parallely in EVIKA's trial field in Saku and in the trial field of the Estonian Agricultural University (EPMÜ) in the vicinity of Tartu. "Vigri" is susceptible to late blight, and as we did not use any plant protection products and also the fertiliser rate was low, 500 kg of complex fertilizer per ha in spring, the yields in those trials were not high. Table 3 shows that in both trial fields clone number 260 was the most high-yielding and No 1272 had the lowest yields, which confirms the results.

TABLE 3. Yielding ability of meristemic clones of the variety "Vigri" in EVIKA's and EPMÜ's trial fields

Clone number	EVIKA 2-year average	EPMÜ 4-year average	EVIKA & EPMÜ average
260	29,0	26,2	27,6
918	24,7	23,7	24,2
264	24,9	22,4	23,6
1225	22,8	24,1	23,5
1221	23,3	21,3	22,3
242	21,4	22,4	21,9
284	22,5	21,4	21,9
298	21,6	21,5	21,6
1275	23,3	19,8	21,6
1219	21,7	16,4	19,0
1272	19,5	16,5	18,0

Meristemic clones of "Vigri" have not been derived at the same time and from the same parent. There are several combinations. Some clones have been preserved in vitro for a longer period (12 years), some are younger. Meristemic clones of "Vigri" have differences in the susceptibility to late blight, to mechanical Injuries and in the dry matter content of tubers. Clone 284 had the highest dry matter content. The most resistant to mechanical injuries was 260, the most susceptible 1219.

Interesting results have been obtained with the cultivar "Premiere". At the present time we have trials with 43 meristemic clones and we hope to produce the variant, which would have preserved the earliness, good flavour and high yielding ability of the variety but which would be more resistant to late blight. The preliminary results are promising. We have selected out 11 meristemic clones of "Premiere" with better characteristics than others. We also study 4 meristemic clones of the cultivar "Bintje". Two of them have higher tops, but the other morphological characteristics are similar. In our field trials there are meristemic clones of other countries as well: meristemic plants of "Kondor" and "Sante" brought from Netherlands, "Nevski", "Verba", "Uralski Ranni" from Russia. The comparison of these with our meristemic clones has also given interesting results.

On the basis of present results we can conclude that with the help of meristemic culture it is possible to improve some characteristics of potato varieties very easily and quickly. To direct the process knowingly, the research work should be continued and extended.

SIEVE ELEMENTS ISOLATED FROM CALLUS CULTURES CAN BE USED TO RAISE PHLOEM-SPECIFIC MONOCLONAL ANTIBODIES

RICHARD D. SJÖLUND
Department of Biological Sciences, Rm 312 CB
University of Iowa
Iowa City, Iowa 52242, USA
e-mail: rsjolund@vaxa.weeg.uiowa.edu

ABSTRACT. Phloem sieve elements function in phloem loading *in vitro* and can be separated from callus cultures of *Streptanthus tortuosus* and *Arabidopsis thaliana* by cell wall digestion, filtration and density gradient centrifugation. Isolated sieve elements have been used to immunize mice and to raise monoclonal antibodies that recognize phloem cells in the callus cultures and in organs of the parent plants. The monoclonal antibodies also identify unknown proteins on immunoblots, providing a technique for the identification of novel antigens from cell cultures based on cell specificity and immuno-microscopy.

1. Introduction

1.1 DOES PHLOEM FORMED IN CALLUS FUNCTION?

Plant tissue cultures offer unique advantages for the study of phloem because sieve tubes formed in callus cultures are usually short and discontinuous, often not more than 5-7 sieve elements in length. Although this discontinuous nature of callus phloem limits their role in long-distance transport (Hanson & Edelman, 1970), these "islands" of phloem allow for investigations that are more difficult to conduct with intact plants where the sieve elements form a continuous tube throughout the plant body. Studies of callus phloem based on plasmolysis experiments demonstrate that these sieve elements do develop high solute levels (Lackney & Sjolund, 1991) and high hydrostatic pressures that result in the plugging of sieve pores by P-protein when they are damaged (Sjolund, Shih, & Jensen, 1983). These data suggest that callus phloem is

M. Terzi et al. (eds.), Current Issues in Plant Molecular and Cellular Biology, 427-432.

"active" and that the sieve elements and companion cells formed *in vitro* have properties similar to those of the intact plant, and that studies of phloem biochemistry, molecular biology, and development based on callus phloem may provide clues to phloem function in the intact plant.

1.2 ISLANDS OF CALLUS PHLOEM CAN BE SEPARATED FROM PARENCHYMA CELLS AND USED TO RAISE ANTIBODIES

Because sieve elements rely on high hydrostatic pressure to move loaded molecules, it is not surprising that their cell walls are thickened and chemically modified (Lucas & Franceschi, 1982). In callus cultures, that feature has allowed us to differentially digest the walls of callus parenchyma cells and the walls of sieve elements. Parenchyma cells are converted into true, wall-free and round protoplasts after digestion with cellulase and pectinase, while the sieve elements retain some wall material and remain attached by their joining sieve plates (Sjolund, 1990) allowing them to be separated from the callus parenchyma cells by filtration and density gradient centrifugation. We have used these partially-purified sieve elements as complex antigens to immunize mice and to develop hybridomas that secrete phloem-specific monoclonal antibodies (Köhler & Milstein, 1975) which have been screened using immunofluorescence microscopy This technique may be useful for identifying other tissue-specific antigens following the partial isolation of differentiated cells, tissues or organizing regions from tissue cultures.

2. Methods and Materials

2.1. CALLUS CULTURES

Callus cultures of two members of the Brassicaceae, *Streptanthus tortuosus* and *Arabidopsis thaliana* were initiated from cotyledons using a Murashige and Skoog (MS) medium (Murashige & Skoog, 1962) containing 1 mg/liter kinetin and 5 mg/liter 2,4-dichlorophenoxyacetic acid (2,4-D) (our medium 205). After 2 transfers at 2-week intervals, cultures to be used for phloem studies were transferred to a similar MS medium (our medium 103) containing 1 mg/liter kinetin and 0.1 mg/liter napthalene acetic acid (NAA). The phloem-differentiating cultures (phloem[+] cultures) were maintained on the 103 medium.

Other cultures were kept on 205 medium to provide a cell line that does not differentiate phloem cells (phloem[-] cultures).

2.2. ANTIBODY PRODUCTION

The isolation of sieve elements and the production of hybridomas from Balb/C mice using callus sieve elements as immunogens is as previously described (Tóth, Wang, & Sjölund, In Press).

3. Results

The differences in cellular differentiation of the *Streptanthus* callus tissue when grown on a medium (103) that induces phloem differentiation (phloem[+] culture) and on a medium (205) that does not initiate phloem development (phloem[-] culture) is shown in figures 1 and 2. Sieve elements (SE) and companion cells (CC) are recognizable in the phloem [+] cultures and resemble those of intact plants. Xylem vessels (X) are also differentiated on the 103 medium. Phloem[-] cultures do not contain sieve elements or xylem vessels. When phloem[+] cultures of *Streptanthus* or *Arabidopsis* are digested with 2% pectinase and 1% cellulase (Worthington Biochem, New Jersey), The islands of phloem and xylem (Fig. 3, X/P) resist digestion and can be retained on a 74µm nylon filter. The phloem sieve elements can be further separated from parenchyma cells and xylem vessels by centrifugation in a Percoll® (Pharmacia, Uppsala) gradient (in 0,25M sucrose); phloem is collected at a density of 1.050 as labeled by a Percoll® marker bead.

An isolated sieve element from *Arabidopsis* is shown in figure 4. Note that a sieve plate (SP) joins two sieve elements, and that some wall material (CW) still surrounds the cells.

These isolated sieve elements have been used to produce phloem-specific monoclonal antibodies that are able to recognize phloem proteins on immunoblots and to label sieve elements in free-hand sections of *Streptanthus* and *Arabidopsis* plants. An example of the labeling of phloem-specific proteins is shown in figure 5. In lane 1, total proteins from a phloem[-] culture that lacks sieve elements fail to bind a P-protein-specific monoclonal antibody (RS22), while proteins from a phloem[+] culture (lane 2) or from isolated sieve elements (lane 3) demonstrate the binding of the antibody to an 89kD protein identified as filamentous PP-1 P-protein. Figure 6 shows an immunofluorescent microscope image of the labeling of a free-hand section of *Streptanthus* stem by RS5, a monoclonal that labels a phloem-specific ß-*amylase*. Stem sections such as this were used to screen hybridomas for the producion of phloem-specific monoclonal antibodies.

4. Discussion

Although the lack of long-distance transport has, in the past, been interpreted as a reason *not* to study phloem in callus cultures (Spanner, 1978), it now appears that the ability to regulate the differentiation of these cells *in vitro* can be used to facilitate the study of sieve element cell and molecular biology. The discontinuous nature of these islands of phloem, although a shortcoming for the study of translocation, can be used to advantage for the isolation and characterization of the cells. Further immunological and molecular approaches to the study of the phloem, including subtractive hybridization (Herfort & Garber, 1991)of phloem[+] and phloem[-] cDNAs may also be aided greatly by the ability to regulate phloem development using plant tissue culture techniques.

5. Acknowledgement

This work was supported by a grant from NSF, MCB-9018880

6. References

Hanson, A. D., & Edelman, J. (1970). Phloem in carrot calluses. Planta, 93, 171-174.
Herfort, M. R., & Garber, A. T. (1991). Simple and efficient subtractive hybridization screening. Biotechniques, 11, 602-605.
Köhler, G., & Milstein, C. (1975). Continuous cultures of fused cells secreting antibody of predefined specificity. Nature (London), 256, 495-497.
Lackney, V. K., & Sjolund, R. D. (1991). Solute concentrations of the phloem and parenchyma cells present in squash callus. Plant Cell Envir. 14, 213-220.
Lucas, W. J., & Franceschi, V. R. (1982). Organization of the Sieve-Element Walls of Leaf Minor Veins. Journal of Ultrastruct Research, 81, 209-221.

Murashige, T., & Skoog, F. (1962). A Revised Medium for Rapid Growth and Bio Assays with Tobacco Tissue Cultures. Physiologia Plantarum, 15, 473-497.
Sjolund, R. (1990). Sieve elements in plant tissue cultures: Development, freeze-fracture, and isolation. In: Sieve Elements. Comparative structure, induction and development. H.-D. Behnke and R. D. Sjolund, eds. Springer-Verlag, New York. pp. 179-198. .
Sjolund, R. D., Shih, C. Y., & Jensen, K. G. (1983). Freeze fracture analysis of phloem structure in plant tissue cultures 3. P-protein, sieve area pores, and wounding. J Ultrastruct Res, 82, 198-211.
Spanner, D. (1978). Sieve-plate pores, open or occluded? A critical review. Plant, Cell and Envir, 1, 7-20.
Tóth, K., Wang, Q., & Sjölund, R. (In Press). Monoclonal antibodies against phloem P-protein from plant tissue cultures I. Microscopy and biochemical analysis. American J. of Botany.

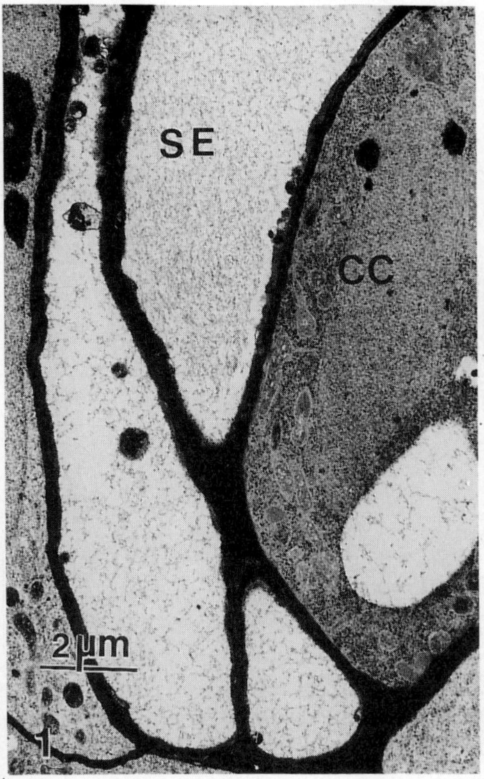

Fig. 1. Phloem (+) cells on 103 medium.

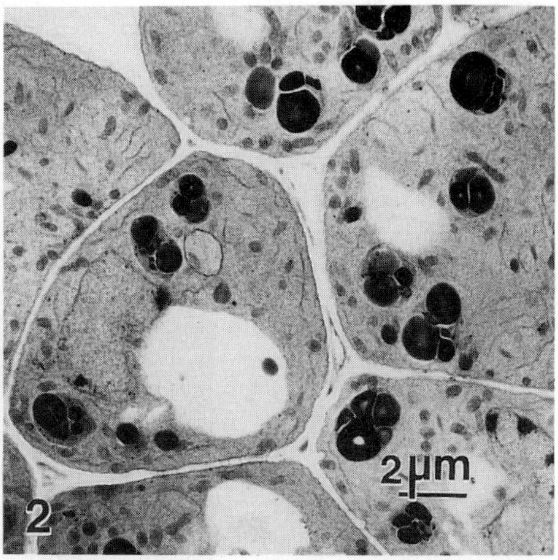

Fig. 2 Phloem(-) cells

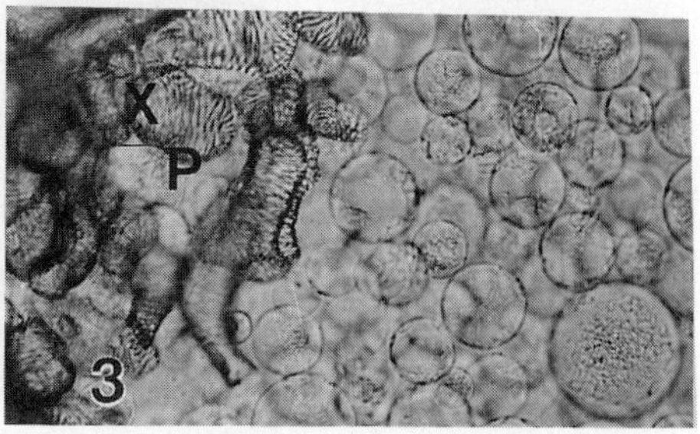

Fig. 3 Digested phloem(+)

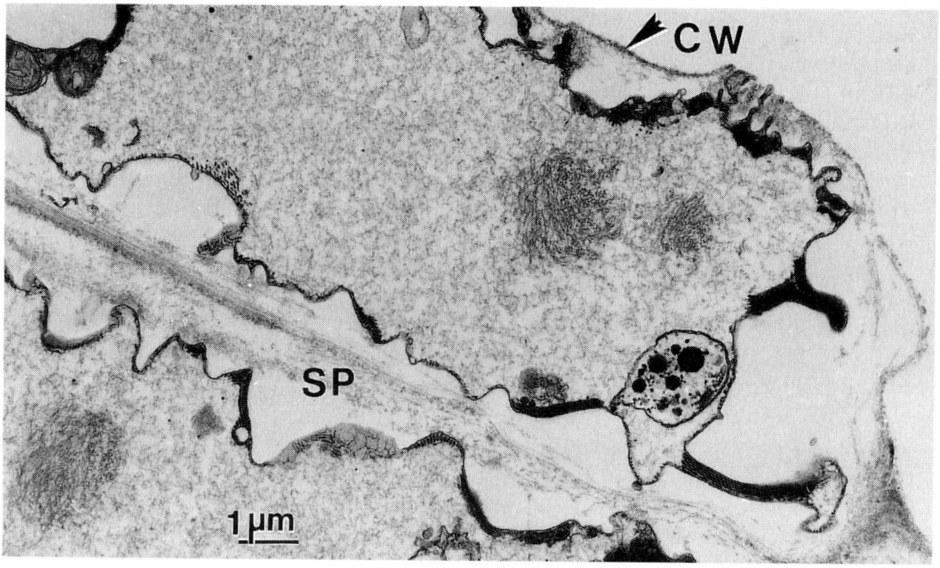

Fig. 4. An *Arabidopsis* sieve element isolated from a phloem(+ culture). Note the persistance of the sieve plate (SP) and remnants of the cell wall (CW)

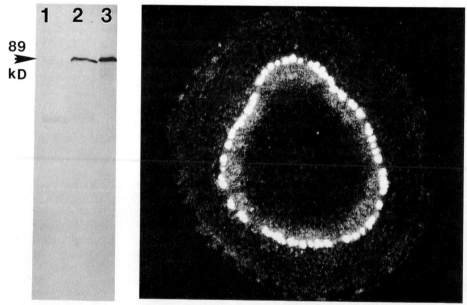

Fig. 5. Immunoblot of proteins of (1) phloem (-), (2) phloem (+), and (3) sieve elements.

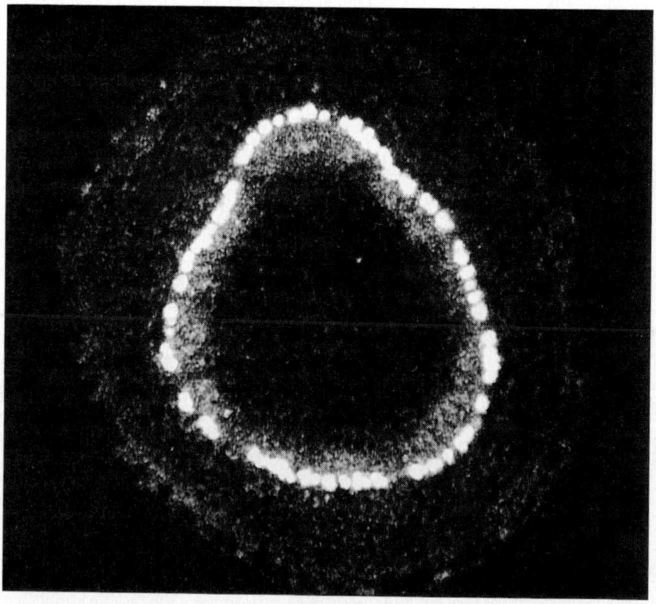

Fig. 6. A stem of *Streptanthus* labeled with monoclonal RS5.

OLIGOSACCHARIDES FROM XYLOGLUCAN AFFECT THE DEVELOPMENT OF *RUBUS FRUTICOSUS* CELL SUSPENSION CULTURE

J.P. JOSELEAU, G. CHAMBAT, A.L. CORTELAZZO, A. FAIK, B. PRIEM AND K. RUEL
Centre de Recherches sur les Macromolécules Végétales (CERMAV-CNRS) B.P. 53 38041 Grenoble-cedex 9 (France)

ABSTRACT. Changes in the activity of several wall-bound glycohydrolases and glycanohydrolases have been monitored during a culture cycle on *Rubus fruticosus* suspension. All the activities did not have their maximum at the same stage of the growth.

Among all the enzyme activities that hydrolyse xyloglucans, α-L-fucosidase was high during the exponential growth phase, suggesting that the substrate polysaccharide may undergo *in muro* structural rearrangements of its side chain substituents. The influence of oligosaccharides from the xyloglucan series on the activity of wall-associated enzymes was investigated. The same oligosaccharide did not affect all the enzymes in the same manner, some being activated when the others are unaffected or even inhibited.

1. Introduction

The biochemical and biophysical processes by which plant cell walls develop in a growing cell are complex and still not well understood, particularly as far as the factors that regulate and controle the extension of cell walls are concerned.The plant cell is surrounded by a wall which is a complex composite organized around cellulose microfibrils, and which owes its cohesion to various types of bondings and interactions that occur between its polymer constituents. This explains that extension and expansion of primary plant cell walls which accompany the cell growth can take place only at the expense of molecular rearrangements that occur at the level of these bondings and interactions (Taiz, 1984 ; Talbot and Ray, 1992).

The primary plant cell wall is essentially made of heteropolysaccharides whose average structures are relatively well described (Carpita and Gibeault, 1993), although less is known about their precise sequence. The actual models of vascular plant cell walls emphasize the underlying roles of the cellulose microfibril weft onto which a certain type of xyloglucan is linked *via* hydrogen bonds, of the pectic polymers which form a gel network ionically stabilized by calcium bridges and of a certain number of structural glycoproteins covalently cross-linked (Carpita and Gibeault 1993 ; Talbot and Ray, 1992 ; Cosgrove, 1993). Despite the great number of bonds and of chemical and physical interactions that maintain together the wall polymers, the growing plant cell wall remains a dynamic structure. However the biochemical processes that allow modifications in the polymer network, or/and of the constitutive polymers during plant cell development are not fully understood and remain largely speculative. Among the factors that may participate and control cell wall extension and cell wall polymer rearrangements are the wall-associated enzymes which act on matrix polymers. Cell wall glycanases have received considerable attention, and although the concept of their participation in the so-called wall loosening has been well documented (Taiz, 1984) it is still debated (Rayle and Cleland, 1992), and their role during primary cell wall development is not yet clarified.

The enzymes associated to the wall of growing cells are numerous and diverse. Particularly, the enzymes which are active on oligo and polysaccharides may participate in the wall rearrangements occuring during growth in modifying the constitutive polysaccharides (Labavitch, 1981). Polysaccharide hydrolases may thus act by *endo*

M. Terzi et al. (eds.), Current Issues in Plant Molecular and Cellular Biology, 433-443.

cleavage at random in the main chain, and by an *exo* attack in the side chains. Both modes of action would alter the structure of the polysaccharides, therefore, their mutual interactions and, finally, would affect the wall internal cohesion (Inouhe and Nevins, 1991). Another consequence of the hydrolytic action of these enzymes is the generation of polysaccharide fragments which were shown to have regulation activity in growth, development and defense response in plants (Darvill et al., 1992).

In the wall of cell grown in suspension and in the culture medium several enzymes with glycosidase and polysaccharide hydrolase have been identified (Asamizu et al., 1981 ; Konno et al., 1986 ; Masuda et al., 1988). Variations of the activities during growth show that the enzymatic modifications of the wall polysaccharides are regulated during cell wall development (Simola and Sopanen, 1970 ; Joseleau and Chambat, 1980 ; Asamizu et al., 1981 ; Amino and Komamine, 1989). The most studied factors that can affect enzymes activity during growth are the auxins. It has been demonstrated that auxin induces the synthesis of *endo* -1,4-β-glucanase in growing peas (Verma et al., 1974). The products of some endo-β-glucanases acting on the wall xyloglucans is a series of oligosaccharides, some of which were shown to have antiauxin (York et al., 1984 ; McDougall and Fry, 1988) or auxin mimicking activities (McDougall and Fry, 1990). In addition of cell-wall polysaccharides as precursors of biologically active oligosaccharides, some glycoproteins are likely to be other potential precursors. Indeed, glucosamine-containing oligosaccharides, also called 'unconjugated *N*-glycans', have been found in several plant tissues and possess biological properties (Priem et al., 1993). One of them was found to influence flax hypocotyl development , and both auxin-synergistic and anti-auxin effect were reported, depending of the auxin amount used for treatment (Priem et al., 1990). Such oligosaccharides could be produced from the partial degradation of extracellular *N*-glycosylated proteins. It thus appears that a relationship may exist between wall-associated enzymes, auxin and oligosaccharides in the controle of plant cell wall development.

In the present study, we report on the changes in several glycanohydrolase and glycohydrolase activities during a cell cycle in the suspension culture of *Rubus fruticosus*. These changes are indicative of the main polysaccharide rearrangements that occur during cell wall metabolism. The influence of oligosaccharides from the xyloglucan series on the principal enzyme activities is also reported. It seems that a same oligosaccharide does not affect all the enzymes in the same manner, some being activated when the others are unaffected or even inhibited.

2. Materials and methods

Cell culture. Rubus fruticosus cells were subcultured in the medium of Heller supplemented with vitamin B1 (1 mg litre^{-1}) and glucose (20 g litre^{-1}).

At different stages of growth between 4 and 24 days of culture, the cells were harvested by filtration on a scintered glass funnel (40 to 90 µm) and washed with citrate buffer (50 mM ; pH 6.2).

Cell wall preparation. Washed cells were resuspended in Tris-HCl buffer (50 mM ; pH 7.2) at 4°C and disrupted seven times for 2 min with a sonicator (at 20-25 KHz in a Branson Sonic Power Company Sonified model B-12). The efficacy of the treatment was monitored by light microscopy examination of the preparation. The cell walls were centrifuged at 12000 g for 15 min at 4°C. After a second washing in Tris-HCl buffer the crude cell wall preparation was ultrafiltrated on an Amicon membrane YM 10. The cell walls were stored after lyophilization. This preparation was then used for *in situ* enzyme activities measurements.

Extraction of enzymes. The crude cell wall preparation from the second washing in Tris-HCl buffer was treated with 4M LiCl in the same buffer containing 0.5 mM of phenyl methyl sulfonyl fluoride and sodium azide. A series of successive extractions was performed until there was no more absorption at 280 nm. The supernatants were pooled and dialyzed on the YM10 Amicon membrane to give the "LiCl extract".

Protein estimation. Proteins in cell walls and in LiCl-extract were estimated according to the method of Potty (1969) which allows protein estimation in the presence of pectic polymers. Sample (3-5mg) was dissolved in distilled water (1-2 ml) and the reagents allowed to react for 10 min at 20°C. The absorbance was read at 500 nm. For cell wall preparations the suspension was centrifuged *in fine*.

Enzyme assays. Glycosidase activities were measured using the corresponding methyl glycosides as substrates plus α-L-fucosyl lactose in the case of α-L-fucosidase activity. For estimation of total activities on polysaccharides, xyloglucan from *R. fruticosus* (Joseleau et al., 1992), carboxymethylcellulose (7H35XF Hercules - France) and polygalacturonic acid (NBC) were used as 0.3% w/v solution in acetate buffer (100 mM, pH 5.0). Activity was measured by the method of Somogyi (1952). For the LiCl-extract, 0.1 ml substrate solution was incubated with 0.1 ml enzyme mixture at 37°C for 16 h. The reaction was terminated by the addition of 0.2 ml of Somogyi reagent. After keeping the reaction mixture at 100°C for 20 min, and cooling, 0.2 ml of Nelson reagent was added and adjusted to 2.5 ml with distilled water. The mixture was allowed to stabilize in the dark for 30 min and the absorbance was measured at 660 nm or 500 nm.

For measurement of the *in situ* activities, 0.35 ml of the substrate in acetate buffer was added to 5mg of the cell wall preparation in 0,35 ml water. After incubation at 37°C for 16 h the reaction was terminated with 0.7 ml of Somogyi reagent and 0.7 ml of Nelson reagent was added as before. The final volume was adjusted to 8.75 ml with distilled water. Before reading the absorbance the suspension was centrifuged at 12000 g for 5 min. The results were expressed in reducing equivalents of glucose released per 100 mg dry weight of cell wall.

Peroxidase activity. This activity was assayed spectrophotometrically by the oxidation of guaiacol. The reaction mixture consisted of 40μM guaiacol and 20μM H_2O_2 in phosphate buffer (40 mM, pH 6.7) in a total volume of 2.4 ml. Soluble enzyme extracts (0.1 ml) were used for the assay. The increase in absorbance at 470 nm was followed (Evans, 1990). Activity was expressed a ΔA_{470}/g.dry weight x min^{-1}. Enzyme extracts were prepared from sonicated cell wall preparations successively extracted with 50 mM phosphate buffer then 1M NaCl. The supernatant from the 1M NaCl extraction was used for ionically bound peroxidase assays.

Oligosaccharide activation assays. 10 Day-old cultures were used for the assays. Cells (10 g, fresh weight) were transferred into 250 ml Erlenmeyer flasks containing 100 ml of medium. The oligosaccharides were added at the desired final concentration. For enzymatic activity measurements the cells were prepared as above. Controls lacking the oligosaccharides were run during each experiment.

3. Results and discussion

In order to get a picture as complete as possible of the modifications that the constitutive polysaccharides may undergo from wall-associated enzymes, the enzymes were not separated nor purified, but assayed as a crude extract or still associated to the cell walls. When considering the possibilities of action of wall-bound hydrolases on the constitutive polysaccharides, it was important to verify the respective activities of the enzymes whether extracted by a 4M LiCl solution or kept *in situ* in the microenvironment of the cell wall. Figure 1 shows that on pure substrates, carboxymethylcellulose (CMC) xyloglucan (XG) and polygalacturonic acid (PG A), respectively, the enzymes activities could be easily measured both in the LiCl extract and *in situ.* The measured activities expressed in reducing-end equivalent relative to the dry weight of cell wall were of the same order of intensity except for the enzymes active on xyloglucan which were much higher *in situ*. This shows that even when they are within the cell wall network, the enzymes exhibit activities which do not seem to be affected by the influence of the environing polymers, nor by the fact that the catalytic action of the *in situ* immobilised enzymes was heterogeneous as compared to the catalytic action of the soluble enzyme of the LiCl extract on soluble substrates. Similar results were obtained by Nagahashi et al. (1987) with a glucosaminidase which showed comparable activity in the free state or wall-bound. However, it appears (Figure 1) that all the substrate polysaccharides do not undergo the same extent of hydrolysis whether the enzymes are extracted or wall-bound. Xyloglucans underwent a stronger attack from the enzymes in the cell walls than from the free enzymes. This was also the case when CMC was used as a substrate. This may be the indication that LiCl did not remove all of the enzymes active on these substrates, or that the microenvironment of the cell wall influenced differently the various activities (Noat et al., 1980 ; Ricard and Noat, 1986). In any case, from the above results it can be inferred that the use of the whole cell wall preparation for measuring the wall-bound enzyme activities provides significant representation of their *in vivo* activity.

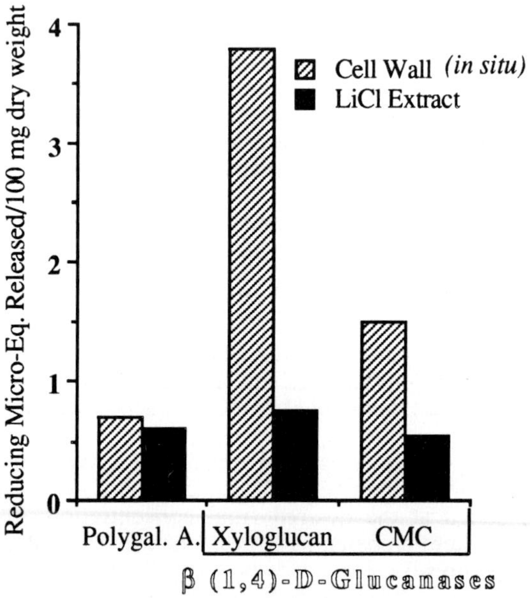

Figure 1. Comparison of the activity of enzymes measured *in situ* in the cell wall, and in LiCl extract from 10-day-old cells.

The measurement of the activities based on the released reducing sugars corresponds to the sum of all the hydrolytic cleavages undergone by a polysaccharide and therefore reflects all the activities directed against this polymer that are present in the cell wall. Previous results showed that beside enzymes acting on polygalacturonic acid, xyloglucan and CMC, the cell walls of *R. fruticosus* also contains activities directed against chitin, 1,3-β-D-glucanase (Joseleau et al. 1992) and xylan and galactan (data not shown). On the other hand glycosidases are also present as shown by their action on various methylglycosides (Table 1).

TABLE 1. Enzymatic activities measured in LiCl extract and *in situ* in the walls of *R. fruticosus* cells.

Substrate	LiCl (*) Extract	*In situ* (*)
Me-α-D-xyloside	0.7	0.6
Me-β-D-galactoside	0.4	0.5
Me-β-D-glucoside	1.4	1.8
Me-α-L-fucoside	trace	0.25
α-L-fucosyl-lactose	0.8	1.4

(*) Enzymatic activities are expressed in micro-equivalents of reducing sugars for 100 mg of dry cell walls.

Here again a good correspondence was found between the LiCl-extracted and the *in situ* enzymes. It is noteworthy that only a weak α-L-fucosidase activity was observed on methyl-α-L-fucoside whereas this activity becomes much stronger when the trisaccharide α-L-fucosyl-lactose (Fuc-Gal-Glc) was used as a substrate.

3.1. CHANGES OF WALL-BOUND ENZYME ACTIVITIES DURING A CULTURE CYCLE

Variation in the activity of the wall-associated enzymes could be the result of protein synthesis or of activation or inhibition of the existing enzymes. Figure 2 shows that the protein content in the walls of *R. fruticosus* cell culture varies almost in the same manner as the cell wall dry weight during a cell culture cycle. The maximal protein proportion appeared at around d.8 whereas the maximal dry weight was observed at d.10, that is during most active cell division periode in the exponential growth phase. These nearly parallel variations indicated that the relative proportion of proteins in the wall does not change significantly. Therefore, a direct comparison of the activities measured at various stages of growth can be performed and will reflect the behaviour of the wall-bound enzymes on a given substrate.

The changes of activities on polygalacturonic acid, xyloglucan and carboxymethyl cellulose showed different trends (Figure 3), although for the three substrates the activity was maximal at the begining of the growth curve, soon after the latence phase (d.4). All the activities tended to cancel out at the end of the stationnary phase (d.24). It can be seen (Figure 3) that the activities on polygalacturonic acid decreased rapidly to reach a low value as soon as d.10. Whereas the 1,4-β-D-glucanase activity, as observed on CMC, decreased progressively and regularly all along the culture cycle, it is interesting to note that the enzymes that act on xyloglucan varied differently, remaining at a high activity up to day 10 of the culture and then weakened drastically during the stationnary phase of the growth.

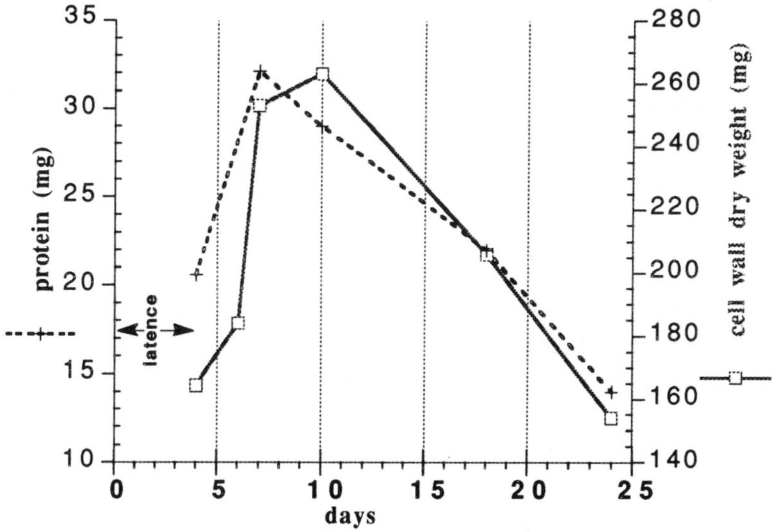

Figure 2. Variation of dry weight and proteins in cell walls during a cell culture cycle. (mg protein for 10 g of fresh weight cells).

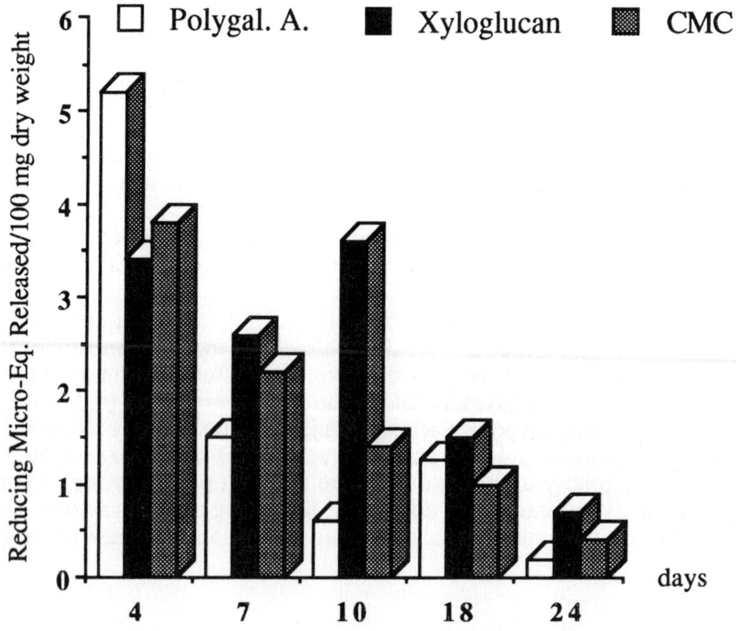

Figure 3. Changes in the activities of wall-associated enzymes acting on Polygalacturonic acid, CMC and xyloglucan.

Among the different activities that can participate to xyloglucan breakdown or structural modification, α-L-fucosidase is of particular importance since terminal α-L-fucosyl substituents are characteristic of primary wall hemicellulose components, and since this terminal residue seems to be directly involved in the binding of xyloglucan to cellulose, as shown by molecular conformation analysis (Levy et al., 1991). The change of activity of α-L-fucosidase on the substrate fucosyl-lactose followed exactly the same trend as that observed for the total activity on the polysaccharide XG, with the highest activity during the exponential growth and a maximum at d-10. This confirms the importance of the metabolism of fucosyl side chains in xyloglucan *via* an α-fucosidase (Augur et al. 1993).

3.2. EFFECTS OF XYLOGLUCAN OLIGOSACCHARIDES ON WALL-BOUND ENZYME ACTIVITIES

When the biologically active XG nonasaccharide (XXFG) (McDougall and Fry, 1990) was added in the culture of *R. fruticosus* , a rapid response ensured that was characterized by a reorientation of the cellulose microfibrils and a loosening of the wall surface (Ruel and Joseleau, 1993). Another consequence of the presence of added oligosaccharide in the culture medium was the activation of the release of xyloglucans and other wall polymers in the medium (Joseleau et al., 1992). Such reorganisations within the walls of growing cells may be regarded as a manifestation of the action of wall-associated enzymes. It is therefore of interest to try to correlate the elicitation activity of oligosaccharides and their effects on the cell wall enzymes.

The cells in active growth phase (d.10) were incubated with purified xyloglucan oligosaccharides, the heptasaccharide XXXG and the nonasaccharide XXFG, (Joseleau et al. 1992 ; Joseleau et al. 1994) at different concentrations and for one hour. The activities measured on CMC, xyloglucan and polygalacturonic acid showed a definite dependance towards the concentration of the incubated oligosaccharide (Figure 4) and towards the nature of the oligosaccharide. However for both oligosaccharides an activation was observed on the hydrolysis of the three polysaccharides assayed and for both also the maximal activation was provided by a 10^{-8} molar concentration of the oligomer. This agrees with other reports (York et al. 1984 ; McDougall and Fry, 1988) showing the dose-dependent activity of these oligosaccharides.

In a similar manner the effect of XXXG and XXFG on the α-L-fucosidase activity showed a clear dose-dependence and structure-dependence (Table 2). The heptasaccharide promoted the fucosidase activity at all concentrations, the maximum being at 10^{-8} M with a factor of enhancement of more than 2. On the other hand the nonasaccharide caused an inhibition of the fucosidase at 10^{-7} M and 10^{-8} M. This inhibition was cancelled at 10^{-7} M at which concentration an activity superior to the normal was regained.

The effect of the nonasaccharide on wall associated peroxidases was also measured. Factors of enhancement of only 1.3 to 1.5 could be observed. The maximal enhancement was measured after 3 hours of culture in the oligosaccharide-added medium. The effect disappeared at a longer times (24 to 72 h). No significant differences were noted with the concentration of the oligosaccharide between 10^{-8} M and 10^{-6} M. However stronger enhancement of the peroxidases by factors of 2 to 2.3 were induced by chitin and chitosan.

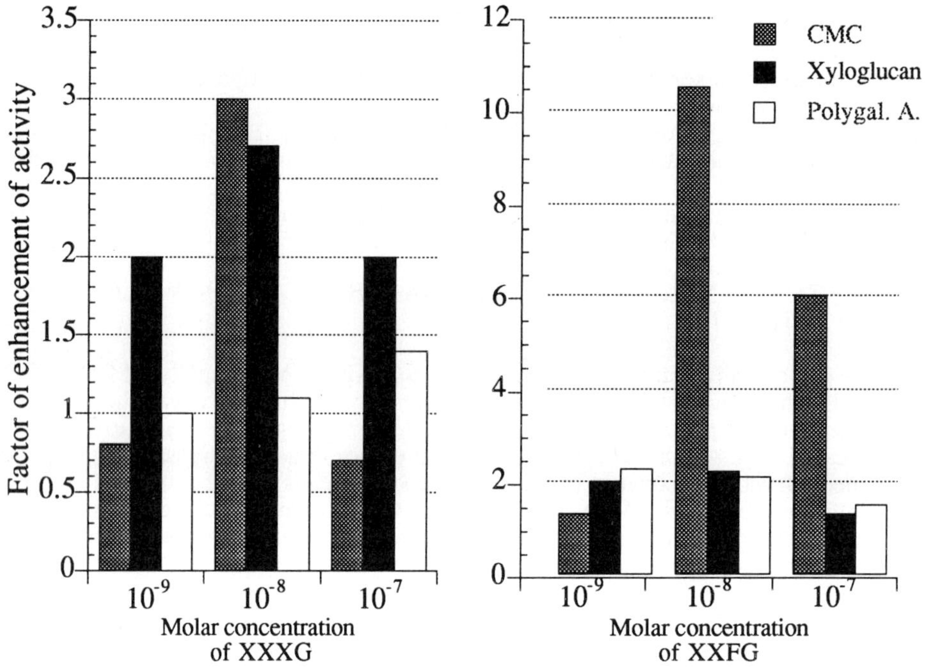

Figure 4. Effect of the concentration of heptasaccharide XXXG and nonasaccharide XXFG on the induction of wall-bound enzymes active on CMC, xyloglucan and polygalacturonic acid.

TABLE 2 : Effect of oligosaccharide structure and concentration on the wall bound α-L-fucosidase activity (Factor of enhancement) as estimated on fucosyl-lactose as a substrat.

Oligosaccharide	Concentration		
	10^{-9}M	10^{-8}M	10^{-7}M
Heptasaccharide XXXG	1.6	2.2	1.2
Nonasaccharide XXFG	0.2	0.3	1.4

4. Conclusion

In growing cells the polymers of the wall matrix undergo *in muro* rearrangements which accompany cells extension. Polysaccharides are modified by several wall-associated enzymes with endo- and exo-glycohydrolase activities. During a culture cycle, over a period of 24 days, all the wall-bond enzymes do not exhibit their maximal activity at the same stage of growth, suggesting that there is a permanent and selective control of the action of these enzymes on the wall polysaccharide structures. Among all the activities that can have xyloglucans as a substrate, α-L-fucosidase was highest during the exponential growth phase. This activity which seems to be very specific for the terminal fucosyl residue of xyloglucan side chains, may play a key role in both the modification of the interaction between xyloglucans and cellulose microfibrils (Levy et al. 1990) and the modification of xyloglucan oligosaccharins role, in affecting their structure (O'Neill et al., 1988). The involvement of α-L-fucosidase in plant growth regulation is suggested by the variation of activity observed during the cell culture cycle and by its susceptibility to the oligosaccharins. Evidence that this enzyme is developmentally regulated and that it can interact with xyloglucan oligosaccharin was shown in etiolated pea stems (Farkas et al. 1991 ; Augur et al. 1993). Several wall-bound enzymes have been shown to be developmentally regulated (Kishor et al 1992 ; Nakamura and Hayashi, 1993) but the various factors that can control their regulation are still largely unknown.

References

Amino, S. and Komamine, A. (1989) "Changes in the wall-bound glycosidase activities during cell cycle in a synchronous culture of *Catharanthus roseus*', Z. Naturforsch. 44c, 754-756.

Asamizu, T., Inoue, Y. and Nichi, A. (1981) 'Glycosidases in carrot cells in suspension culture, localization and activity change during growth', Plant Cell Physiol. 22, 469-478.

Augur, C., Benhamou, N., Darvill, A. and Albersheim, P. (1993) 'Purification, characterization and cell wall localization of an α-fucosidase that inactivates a xyloglucan oligosaccharin', The Plant Journal 3, 415-426.

Carpita, N.C. and Gibeault, D.M. (1993) 'Structural models of primary cell walls in flowering plants: consistency of molecular structure with the physical properties of the walls during growth', The Plant Journal 3, 1-30.

Cosgrove, D.J. (1993) 'How do plant cell walls extend?', Plant Physiol. 101, 1-6.

Darvill, A., Augur, C., Bergmann, C., Carlson, R.W., Cheong, J.J., Eberhard, S., Hahn M.G., Lo', V.M., Marfa', V., Meyer, B., Mohnen, D., O'Neill, M.A., Spiro, M.D., van Halbeek, H., York, W.S. and Albersheim, P. (1992) 'Oligosaccharins-oligosaccharides that regulate growth, development and defence responses in plants', Glycobiology 2, 181-198.

Evans, J.J. (1990) 'Cell wall bound and soluble peroxidases in normal and dwarf tomato', J. Agric. Food Chem. 38, 948-951.

Farkas, V., Hanna, R. and Maclachlan, G. (1991) 'Xyloglucan oligosaccharide α-L-fucosidase activity from growing pea stems and germinating nasturtium seeds', Phytochemistry 30, 3203-3207.

Inouhe, M. and Nevins, D.J. (1991) 'Inhibition of auxin-induced cell elongation of maize, coleoptiles by antibodies specific for cell wall glycanases', Plant Physiol. 96, 426-431.

Joseleau, J.P. and Chambat, G. (1980) 'Variation des activités de type cellulase et polygalacturonase dans les cellules de *Rosa glauca* cultivées en milieu liquide', Physiol. Vég. 18, 443-451.

Joseleau, J.P., Cartier, N., Chambat, G., Faik, A. and Ruel, K. (1992) 'Structural features and biological activity of xyloglucans from suspension cultured plant cells', Biochimie 74, 81-88.

Joseleau, J.P., Chambat, G., Cortelazzo, A., Faïk, A. and Ruel, K. (1994) 'Putative biological action of oligosaccharides on enzymes involved in cell-wall development' Biochem. Soc. Trans. 22, 404-409.

Kishor, P.B.K., Rao, J.D. and Reddy, G.M. (1991) 'Activity of wall-bound enzymes in callus cultures of *Gossipium hirsutum* L. during growth', Annals Bot. 69, 145-149.

Konno, H., Yamasaki, Y. and Katoh, K. (1986) 'Characteristic of β-galactosidase purified from cell suspension cultures of carrot', Physiol. Plant 68, 46-52.

Labavitch, J.M. (1981) 'Cell wall turnover in plant development', Annu. Rev. Plant Physiol. 32, 385-406.

Levy, S., York, W.S., Stuike-Prill, R., Meyer, B. and Staehelin, L.A. (1991) 'Simulations of static and dynamic molecular conformations of xyloglucans. The role of the fucosylated sidechain in surface-specific sidechain folding' The Plant J. 1, 195-215.

Masuda, H., Komiyama, S. and Sugawara, S. (1988) 'Extraction of enzymes from cell walls of sugar belt cells grown in suspension culture', Plant Cell Physiol. 29, 623-627.

McDougall, G.J. and Fry, S.C. (1988) 'Inhibition of auxin-stimulated growth of pea stem segments by a specific nonasaccharide of xyloglucan', Planta 175, 412-416.

McDougall, G.J. and Fry, S.C. (1990) 'Xyloglucan oligosaccharides promote growth and activate cellulase: evidence for a role of cellulase in cell expansion', Plant Physiol. 93, 1042-1048.

Nagashi, G., Tu, S.I. and Barnett, M. (1987) 'Properties of cell wall-associated enzymes in the bound and free state in physiology of cell expansion during plant growth', Cosgrove, D.J. and Knievel, D.P. eds., the American Society of Plant Physiologists, 265-266.

Nakamura, S. and Hayashi, T. (1993) 'Occurence of endo-1,4-β-glucanase activities in suspension-cultured poplar cells during growth', Mokuzai Gakkaishi 39, 1056-1061.

Noat, G., Crasnier, M. and Ricard, J. (1980) 'Ionic control of acid phosphatase activity in plant cell walls', Plant Cell Environ. 3, 225-232.

O'Neill, R.A., White, A.R., York, W.S., Darvill, A.G., and Albersheim, P. (1988) 'A gas chromatographic-mass spectrometric assay for glycosylases', phytochemistry 27, 329-333.

Potty, V.H. (1969) 'Determination of proteins in the presence of phenols and pectins', Anal. Biochem. 29, 535-539.

Priem, B., Gitti, R., Bush, A.C. and Gross, K.C. (1993) 'Structure of ten free N-glycans in ripening tomato fruit', Plant Physiol. 102, 445-458.

Priem, B., Morvan, H., Hafez, A.M.A. and Morvan, C. (1990) 'Influence of a plant glycan of the oligomannoside type on the growth of flax plantlets', CR Acad. Sci. Paris 311, 411-416.

Rayle, D.L. and Cleland, R.E. (1992) 'The acid growth theory of auxin-induced cell elongation is alive and well', Plant Physiol. 99, 1271-1274.

Ricard, J. and Noat, G. (1986) 'Electrostatic effects and the dynamics of enzyme reactions at the surface of plant cells. I - A theory of the ionic control of a complex multi-enzyme system', Eur. J. Biochem. 155, 183-190.

Ruel, K. and Joseleau, J.P. (1993) 'Influence of xyloglucan oligosaccharides on the micromorphology of the walls of suspension-cultured *Rubus fruticosus* cells', Acta Bot. Neerl. 42, 363-378.

Simola, L.K. and Sopanen, T. (1970) 'Changes in the activity of certain enzymes of *Acer pseudoplatanus* cells at four stages of growth in suspensions cultures', Physiol. Plant. 23, 1212-1222.

Somogyi, M. (1952) 'Note on sugar determination', J. Biol. Chem. 195, 19-23.
Taiz, L. (1984) 'Plant cell expansion: regulation of cell wall mechanical properties, Ann. Rev. Plant Physiol. 35, 585-657.
Talbot, L.D. and Ray, P.M. (1992) 'Changes in molecular size of previously deposited and newly synthesized pea cell wall matrix polysaccharides', Plant Physiol. 98, 369-379.
Verma, D.P.S., Maclachlan, G.A., Byrne, M. and Ewings, D. (1975) 'Regulation and *in vitro* translation of messenger ribonucleic acid for cellulase from auxin-treated pea epicotyls' J. Biol. Chem. 250, 1019-1026.
York, W.S., Darvill, A.G. and Albersheim, P. (1984) 'Inhibition of 2,4-dichlorophenoxyacetic acid-stimulated elongation of pea stem segments by xyloglucan oligosaccharide', Plant Physiol. 75, 295-297.

RAPID WALL SURFACE REARRANGEMENTS INDUCED BY OLIGO-SACCHARIDES IN SUSPENSION-CULTURED CELLS

K. RUEL, A.L. CORTELAZZO, G. CHAMBAT, A. FAÏK, M.F. MARAIS AND J.P. JOSELEAU
Centre de Recherches sur les Macromolécules Végétales (CERMAV-CNRS) B.P. 53 38041 Grenoble-cedex 9 (France)

ABSTRACT. When *Rubus fruticosus* cells cultured in suspension were transferred in a medium to which oligosaccharides had been added, morphological modifications could be observed in transmission electron microscopy. Typically, medium containing 10^{-8}M xyloglucan hepta or nonasaccharide induced ultrastructural reorganizations localized at the external surface of the walls. Another consequence of the wall modification was an acceleration of cell separation in the cell aggregates. All these responses could be observed within short times of exposition to the oligosaccharides, 10 to 15 minutes. When the culture was maintained on the oligosaccharide added medium for a longer time, excretion of cell wall material resulted. This material contained all types of wall polymers including crystalline cellulose microfibril fragments. Ultrastructural modifications could also be induced by other saccharides like arabinogalactan fragments, chitin and chitosan.

1. Introduction

As a result of continuous cell division, the natural tendency of plant cells grown in liquid cultures is to form aggregates that could comprise hundred of cells. However, it is a well known observation that at the end of a subculture-period in which the culture is kept for some time at the stationnary phase, the number of aggregates decreases and the clusters are reduced to fewer cells (Street et al., 1972). This change in the morphological aspect of the culture is not well understood for the reason that nothing is well established about the exact nature of the material or about the linkages that keep cells aggregated (Hayashi and Yoshida, 1988). Several factors may affect the aggregation and de-aggregation of the cells along the culture period. The mechanical agitation which is applied to the culture to maintain the cells in suspension may participate in the separation of the aggregates. However, the walls of cells cultured in suspension undergo chemical and biochemical changes during growth (Asamizu et al. 1983 ; Blaschek and Franz, 1983 ; Bacic et al., 1988 ; Joseleau et al. 1992) which are reflected in the ultrastructural organization of the growing cells (Roberts, 1990). Rearrangements at the wall surface could be observed when the cells dissociate (Ruel and Joseleau, 1993). These local changes in the cell wall micromorphology were the consequence, or the origin, of the disruption of aggregates. This phenomenon which takes place normally when a suspension culture ages was accelerated when a biologically active oligosaccharide from the xyloglucan series was added to the medium (Ruel and Joseleau 1993). The biochemical effects induced by xyloglucan-derived oligosaccharides are numerous and may change according to their structure and to the concentration at which they are applied (York et al, 1984 ; McDougall and Fry, 1989 ; Fry, 1992 ; Joseleau et al. 1994). Influence of active xyloglucan fragments on the development of cells in liquid cultures has to be expected since these oligosaccharides have been shown to be naturally excreted *in vivo* by the breakdown of the parent polysaccharide (Mc Dougall and Fry, 1991).

The induction of cell separation by oligosaccharides or polysaccharides had already been observed by Hayashi et al. (1988) who showed that a purified 18-KDa galacturonan added to soybean callus tissue cultured in a medium containing colchicine caused an immediate enhancement of cell separation. The authors could confirm this action of galacturonan on

M. Terzi et al. (eds.), Current Issues in Plant Molecular and Cellular Biology, 445-455.

other dicotyledons, such as carrot, tobacco and hibiscus suspension cells (Hayashi et al. 1990).

In the present work, since oligosaccharides of the xyloglucan series have been identified in the medium of the suspension of *Rubus fruticosus* (Joseleau et al. 1992), we have investigated their effects on the ultrastructure of the developping cell walls. Depending on the time during which the culture was kept in the presence of the oligosaccharide the effect on the ultrastructure of the cell wall showed different aspects. The observed micromorphological changes may be interpreted as the consequence of the action of the oligosaccharides on wall-associated enzyme activities.

2. Materials and Methods

PLANT MATERIAL. - *Rubus fruticosus* cells were cultured as described in Joseleau et al., (this issue)

OLIGOSACCHARIDE ACTIVATION ASSAYS. - Two 100 ml Erlenmeyer flasks containing 10 day-old cultures were respectively submitted to XXFG nonasaccharide and a mixture rich in arabinogalactan 12 < DP < 25, the final concentration of which was 10^{-8}M. Two other flasks were for controls. The fourth cultures originate from the same subculture growing as described in Joseleau et al., (this issue). The cells were harvested from control and experimental flasks after 10 or 15 min, 30 min, 1 h, 3 h, ... of incubation. They were immediately filtered through a scintered glass funnel (40 μm) and fixed for electron microscopy.

ELECTRON MICROSCOPY. - Preparation of the samples for thin sectioning: normal cells or elicited ones were filtered through nylon cloth before being fixed in a freshly prepared mixture of 0.1% glutaraldehyde, 2% paraformaldehyde (or 2% glutaraldehyde, 0.5% paraformaldehyde) in 0.05 M phosphate buffer as described in Ruel and Joseleau, (1993). Samples dehydrated up to ethanol 70% were embedded in LR White resine (hard mixture) and polymerized 24 h at 50°C .

ANTIBODY PROBES. - Three antibodies were used. A polyclonal antixyloglucan antiserum (A-XXFG noted A-XG) described in Ruel et al. (1990) and Ruel and Joseleau (1993). JIM5 and JIM7 are two monoclonal antibodies respectively directed against low-esterified polygalacturonans (< 30%) and highly methyl esterified (35-90%) polygalacturonans (Vanden Bosch et al. 1989, Knox et al. 1990). They were a gift from J.P. Knox and K. Roberts.

IMMUNOCYTOCHEMICAL LABELLINGS. - They were performed on thin sections floating in plastic rings. For the polyclonal A-XG, the protocol was as described in Joseleau et al. (1992), with a few modifications: BSA was replaced by 5% of non-fat dried milk both in TBS_{500} (0.01 M Trisphosphate buffer, pH 7.4, containing 0.5 M NaCl) and Tris buffer (TB) (0.01 M Trisphosphate buffer, pH 7.4). The antiserum was diluted 1/10 to 1/40 in TBS_{500} depending on the type of fixator used. Protein A (PA 10 or 5) (Janssen Pharmaceuticals) was 1/20 in Tris buffer containing 0.5% gel fish plus 0.02% PEG (polyethylene glycol) or TB milk 5%. There were no differences in the quality of the secondary labelling using either one of the two buffers. For the monoclonal antibodies JIM5 and JIM7: incubation with the primary antibodies diluted 1/10 (JIM5) 1/20 (JIM7) were two in TBS_{500} plus Normal Goat Serum (NGS) 10%. After rinsing in TBS_{500}, TB, the secondary marker was GARaG5 (Janssen Pharmaceuticals) diluted 1/20 or 1/30 in

Tris buffer, gel fish, PEG or in Tris buffer-milk as described for the anti-XG antibody. Post staining was 2 min in 2% aqueous uranyl acetate.
Controls: They included incubation with secondary antibody alone, preimmune antisera and antibodies preabsorbed with purified antigens.

ENZYME GOLD PROBE. - The purified β-1,4 endoglucanase I (EGI) from the fungus *Humicola insolens* was a gift from M. Schülein (Novo Nordisk A/S, Denmark). The complex EGI-colloïdal gold was prepared as described by Ruel and Joseleau (1984) with some adaptations according to the proper characteristics of the enzyme (M = 50 KDa, PI = 5.5, optimal pH = 7.5).
Briefly, the pH of a colloïdal gold solution (Au 8.5) prepared according to De Mey (1986) was adjusted to 5.4 with a 0.1 M solution of potassium carbonate K_2CO_3. Water suspension (1.3 ml) of the enzyme (1 mg of protein/ml as measured with the coomassie protein assay reagent kit for proteins) diluted 1/40 was added to 6.5 ml of the gold suspension. After 5 min of contact at room temperature 156 µl of polyethylene glycol (PEG) 1% were added. The mixture was centrifuged at 34.000 t/min using a Beckman Rotor Ti 50 for 75 min. at 4°C. The mobile pellet was collected and further diluted in 0.01 M citrate buffer pH 6.0 containing 0.02% PEG +0.5% gel fish.

ENDOGLUCANASE GOLD COMPLEX LABELLING. - Ultrathin sections floating in plastic rings were treated for 30 min. with the mobile pellet diluted 1/20 in citrate buffer, PEG, gel fish. After rinsing on citrate buffer, a short fixation (3 min.) with a 2% glutaraldehyde solution was performed before uranyl acetate post staining.

3. Results

An indirect demonstration that the addition of xyloglucan nonasaccharide in the culture medium of the suspension of *R. fruticosus* cells triggered a body of effects that resulted in cell wall desorganization was provided by the higher yield of extraction of wall-associated proteins from the cells of the treated culture. Table 1 shows that the easiness with which the proteins became extractable with LiCl solutions was dependent of the oligosaccharide concentration applied.

TABLE 1. Effect of xyloglucan nonasaccharide upon wall-associated protein extraction.

Protein Yield (a)	Oligosaccharide concentration 0	10^{-9}M	10^{-8}M	10^{-7}M
LiCl extractable	270	500	480	470
Residual	750	630	610	620

(a) In microgram protein per gram fresh weight.

3.1. MICROMORPHOLOGICAL ASPECTS OF CELL WALL REORGANIZATION INDUCED BY OLIGOSACCHARIDES.

When the nonasaccharide XXFG (Fry et al. 1993) was added at a 10^{-8}M concentration in the medium of *R. fruticosus* suspension (10 day-old culture) several effects could be observed on the morphology of the cells.

3.1.1. *Surface Modifications.* A rapid modification induced by the addition of the oligosaccharide in the medium could be seen on those cells from clusters and individualized cells, the wall of which was directly in contact with the culture medium. The modification consisted in the partial release of a narrow fringe of microfibrils which became oriented perpendicular to the surface and progressively individualized (Figure 1a). Associated to the surface modifications, an enhanced reactivity to uranyl acetate staining was clearly visible. The microfibrils appeared desentangled from other amorphous material. Interestingly the first morphological response to the elicitor was observed early after the introduction of the oligosaccharide, that is after only 10 to 15 min of contact with the oligosaccharide. In clusters where the cells share adjacent surfaces, forming a three-way junction, a loosening appeared in the filled corner and the desorganization progressed along the "middle lamella" to eventually lead to cell separation (Figure 1b).

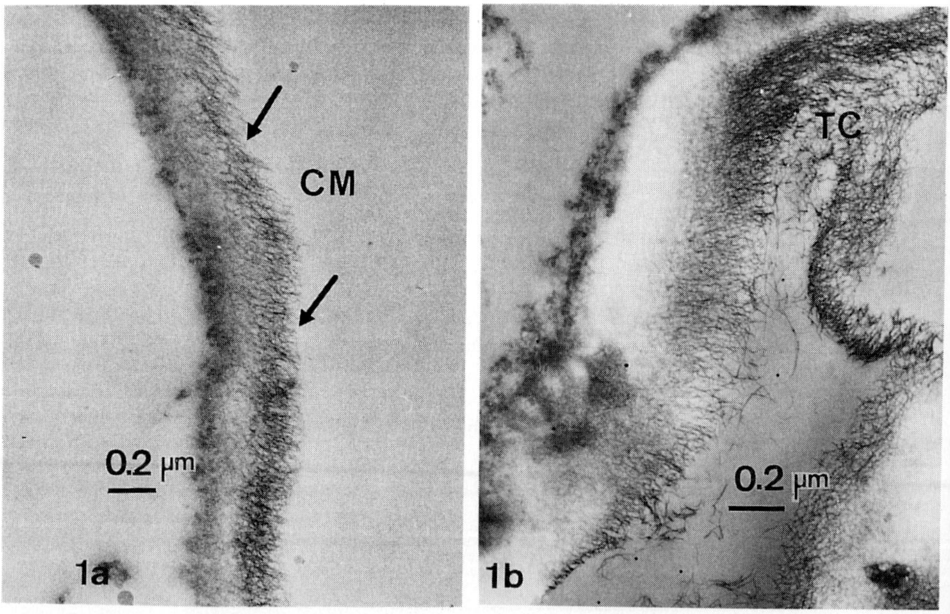

Figure 1a,1b. - Morphological aspect of *R. fruticosus* cells after 10 min. of contact with the nonasaccharide XXFG. Uranyle acetate post staining a : a narrow fringe with a reoriented microfibrillar material individualized on the outer part of the wall (arrow); b: loosening of the tricellular junction. A "middle lamella" (arrow-head) differentiates and the cells tend to separate (CM: culture medium, TC: tricellular junction).

3.1.2. *Material Released into the Medium.* A consequence of the superficial defibrillation induced by the oligosaccharides on the outermost part of the cells was that the cell wall material first extending outwards detached into the medium (Figure 2). The material released into the medium showed a fibrillar criss-crossed aspect and a particular fragility under the electron beam.

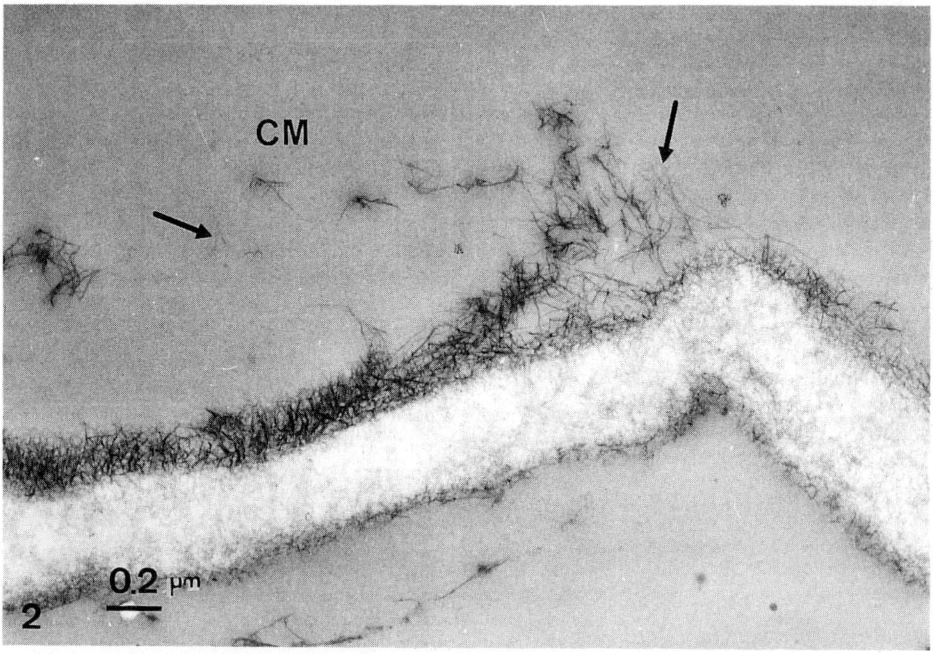

Figure 2.- *R. fruticosus*. Cell in contact with the culture medium (10 min. of contact with XXFG). The outer-most part of the wall extends outwards giving rise to a thin network of microfibrils (Uranyl-acetate post-staining). Note the strong reactivity of the outer-part of the wall (CM: culture medium).

3.2. SPECIFIC LABELLING OF THE POLYSACCHARIDE CONSTITUENTS

Three antisera and one enzyme-gold complex were used to identify the material. The results concern observations performed after 10 to 15 minutes of culture in the presence of the added XG nonasaccharide.

3.2.1. *Xyloglucans*. The polyclonal anti-XG gave a positive labelling when the external hairy surface began to differentiate (Figure 3a).Then, when the culture was maintained for a longer time (30 to 60 minutes) in the oligosaccharide-containing medium, the labelling weakened and became virtually negative as the surface microfibrils became individualized (Figure 3b).

3.2.2. *Pectic Polysaccharides*. The monoclonal antibody JIM5 directed against weakly esterified polygalacturonans gave a positive labelling at the beginning of the appearance of the hairy coat (Figure 4a,b). This reactivity was maintained when the network of individualizing microfibrils formed, and finally became negative when the microfibrils became more separated (Figure 4c). JIM7 directed against heavily methylated polygalacturonans was very weakly recognized by the elicited cells (see positive labelling with control). (Figure 5a,b).

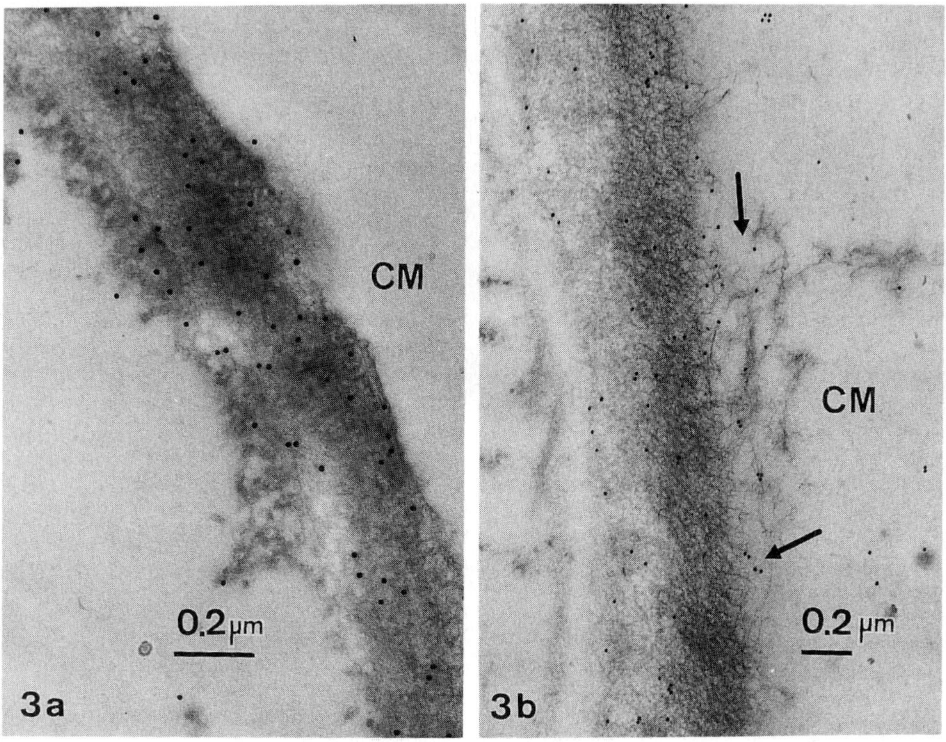

Figure 3a,b. - *R. fruticosus*, anti-XG gold labelling. a: control: gold particles (10 nm in diam.) are regularly distributed on the wall. b: after 10 min. with XXFG, the gold distribution is modified ; XG molecules tend to leach outside, accompanying the fibrillar material (arrows) (CM: culture medium).

3.2.3. *Characterisation of Cellulose*. The enzyme gold complex was prepared with a 1,4-β-D-glucanase I (EGI) from the fungus *Humicola insolens*. This enzyme, specifically hydrolyses amorphous cellulose and is devoid of cellulose-binding domain (Henrissat, personal communication). A comparison of normal 10d-old cells with the elicited ones shows that the gold labelling moved outwards, associated to the criss-crossed reticulated microfibrillar material (Fig. 6).

3.3. *Effect of Other Saccharides*. Similar addition of oligo and polysaccharides were performed using different series of polysaccharides. Thus, arabinogalactan fragments isolated from *R. fruticosus* culture and consisting of a mixture of oligosaccharides of degree of polymerization superior to 20 were added at a concentration of about 10^{-5} mg per milliliter. Chitin and chitosan which are known as powerful elicitors (Hadwiger and Beckman, 1980) were also used at similar concentrations. In each case a typical pattern of ultrastructural modification could be observed which corresponded to a more or less intense loosening of the cell walls (data not illustrated).

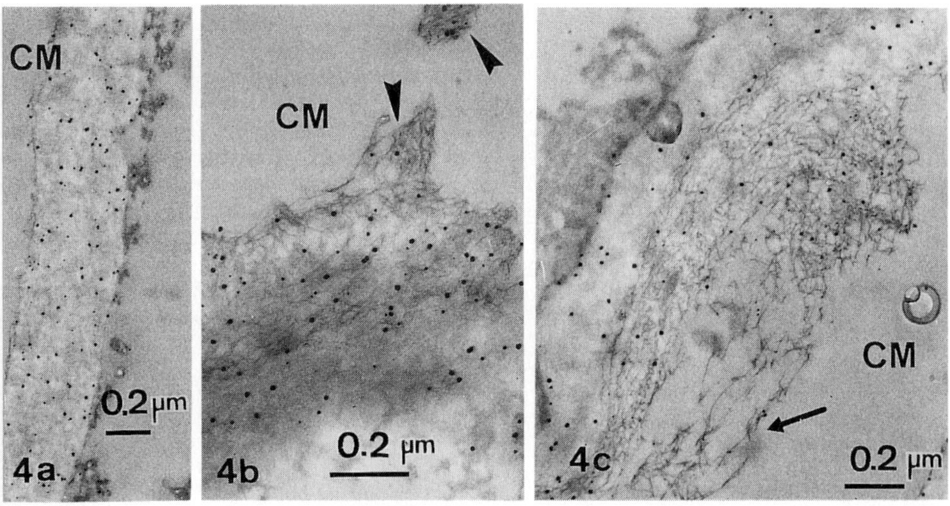

Figure 4 a,b,c: JIM5. a: homogeneous distribution of gold particles on the control. b: after 10 min. of incubation with XXFG, JIM5 labelling extends to the reoriented fibrillar material in contact with the culture medium (CM). c: when the fibrillar material individualizes, gold particles disappear (arrow).

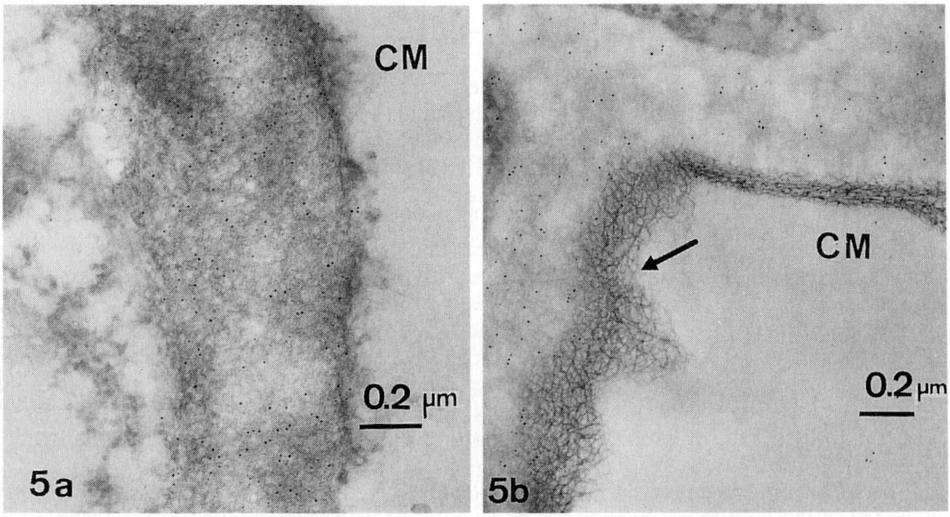

Figure 5a,b: JIM7. a: control. An abundant gold deposit in homogeneously distributed. b: after 10 min. of XXFG elicitation, gold particles become rare and scattered on the wall and do not attach to the hairy layer (arrow).

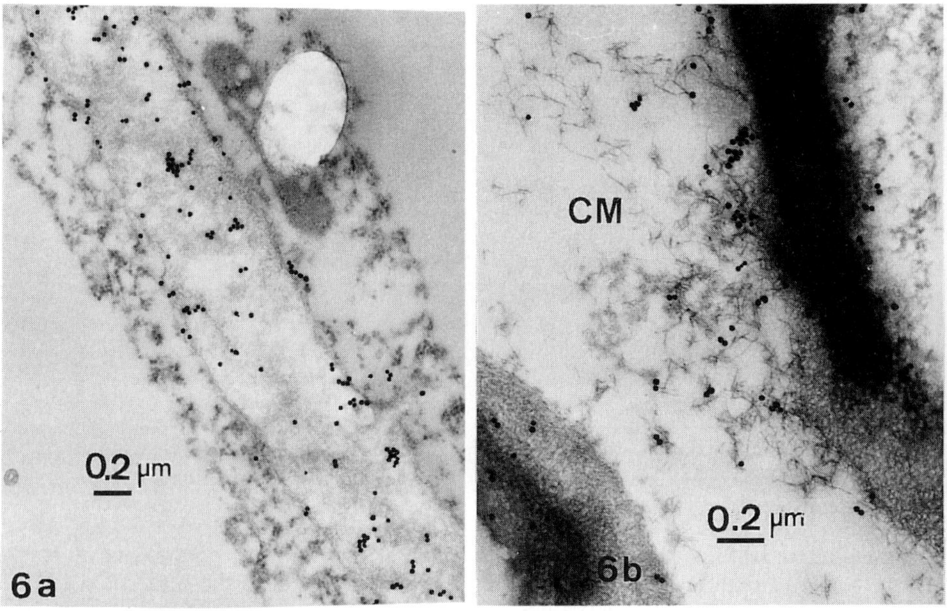

Figure 6 a,b. Au8.5 - EGI gold complex. a : control, b: *R. fruticosus* elicited by XXFG. The enzyme attaches to the microfibrillar network and seems to loose its affinity for the wall.

4. Discussion

The easier extraction of the wall-associated proteins from the oligosaccharide-treated culture reflects the disorganization of the polymer network which retained a part of the wall proteins. This could be the result of the cleavage of covalent linkages in the chain of polysaccharides by activation of wall-bound polysaccharidases by the active oligosaccharins (Joseleau et al. 1994 ; Joseleau et al. preceding article). It thus ensues internal loosening in the polymer network. The easier release of proteins may also be due to some changes in the ionic interactions between the cell wall polymers and the proteins by modification of the charges of the pectins. This possibility correlates with our observation that the cell walls from the treated culture showed a decrease of esterified polygalacturonans, suggesting that a demethylation was induced by the action of the oligosaccharides. The increase in the yield of LiCl-extractable proteins essentially corresponds to enzyme extraction as indicated by the higher activities found in the 4M LiCl extracts of the oligosaccharide-treated cells (data not shown), and not to an extraction of structural glycoproteins which are probably covalently bound to the wall polysaccharides (Lamport, 1986).

The surface modifications observed in the cells which have a part of their wall directly exposed to the culture medium could be considered as a consequence of the leaching of non cellulosic polymers (McDougall and Fry 1991). This phenomenon is progressive and gives rise to a fibrillar network which finally breaks into fibrillar elements. These observations suggest that both the hemicelluloses and the pectic polymers were progressively removed and solubilized. A result is the stripping of the cellulose microfibrils as indicated by the characteristic fragility of this residual material under the electron beam. . It is interesting to note that the uranyl acetate staining was stronger on the hairy surface material. This could indicate that acid polysaccharides are subjected to chemical modifications displacing acid groups outwards. The localized dissolution of the junction material that connects the cells together and of a portion of the wall of the adjacent cells is part of the process of cell separation. Cell separation is a precisely controlled process in which the enzymes that mediate the dissolution of the junction material have not yet been identified (Roberts, 1990). In this process it seems that the oligosaccharides added in the culture have the capacity to activate enzymes involved in the dissolution of polysaccharides at the external face of the cell walls.

Immunolabellings were used in order to try to identify the true nature of the polysaccharides, the excretion of which was stimulated by the XXFG nonasaccharide. Clearly, xyloglucans are rapidly solubilized and leached out from the surface of the wall. Once in the culture medium, they seem to be quickly hydrolyzed since their labelling becomes negative.

Besides internal cleavages and side chain modifications in the wall polymers catalyzed by the activated hydrolases it seems that the process also involves demethylation of pectic polysaccharides as indicated both by the enhancement of the uranyl acetate staining in the outer part of the wall and by the immunolabelling given by JIM5 and JIM7 probes. The fading of the wall labelling by JIM7 and the displacement of the gold grains on the hairy surface seem to indicate that a demethylation of the pectic polysaccharides is also induced very rapidly when the cells are in contact with the nonasaccharide. This would result in a multiplication of the negative charges and create repulsing forces that could participate into the separation. All these mechanisms lead to the accumulation in the medium of all types of wall polymers, including cellulose, and of their degradation products. This has already been observed with suspension cells cultured in a medium that induces cell separation (Hayashi et al. 1994). The β-1,4-endoglucanase I-gold complex was prepared in view of characterizing the modifications undergone by the cellulose of the walls on the basis of EGI specificity for amorphous cellulose. Thus, an important labelling was obtained on control sections of non-elicited cells. After 10 minutes elicitation, the lower labelling of the internal part of the wall showed that cellulose underwent an endo cleavage in its amorphous regions. This agrees with the induction of β-1,4-glucanase activities by the oligosaccharide (Joseleau et al., this issue). On the other hand, amorphous cellulose remains temporarily in the crosslinked network at the surface of the wall as indicated by the stronger labelling of this part of the material. This positive reaction with EGI will eventually disappear on prolonged time of culture (Figure 6), leaving a microfibrillar cellulose mostly in its crystalline form. Such distinction could be made because EGI from *Humicola insolens* has been demonstrated as devoid of cellulose binding domain (Henrissat, personnal communication) and therefore can complex its substrate only *via* its catalytic site on the amorphous glucan chain.

Different types of oligosaccharides and polysaccharides seemed to be able to induce similar effects on the cell wall micromorphology. The present results showed that comparable cell separation activation and surface modifications were observed when xyloglucan hepta and nonasaccharides were added to the culture, as well as when a mixture of higher oligosaccharides from *R. fruticosus* such as arabinogalactan, or chitin, were used. These results are in agreement with previous observation (Hayashi and Yoshida, 1988) that cell separation was enhanced by the addition of extracellular galacturonan to the medium of soybean suspension containing colchicine. It it thus clear that the polysaccharides and the oligosaccharides originating from the walls of the cells grown in liquid medium exert a control on the development of the suspension. It appears from our results that in addition to the well demonstrated effect of the concentration of the saccharides (Fry, 1992) the duration of their interaction with the cells takes also an important part in the extent of the modifications that they induce in the cell wall. It is important for the mode of action of these extracellular activators that an effect could be observed after a time as short as ten minutes after the introduction of the oligosaccharides. The study of the action of the nonasaccharides from 10 minutes to 24 hours revealed that on prolonged contact, the part of the cell wall undergoing the autolysis remained limited to a superficial external layer, and that the residual wall had lost the "hairy" aspect that characterized the early removal of polysaccharides and the reorientation of the cellulose microfibrils. Even when the culture was maintained for a longer time in the presence of the oligosaccharides, all the cell aggregates did not totally separate into single cells. This should correspond to a different organization and may be to a different nature of the junction zones between cells whether they concern older cells deeply buried in the aggregates or younger cells at the exterior.

The restricted localization of the modification to the surface of the cell walls induced by the oligosaccharides raise different questions. Is the limitation to the surface due to a limitation of penetration of the oligosaccharide into the cell wall ? This would suggest that the XG oligosaccharides that were used in this study acted as biochemical activator of the wall-associated enzymes involved in the wall polymer hydrolysis and chemical modification. This would also suggest a particular localization of these enzymes in the external part of the wall. Another possible interpretation could be an heterogeneous ultrastructural organization of the wall polysaccharides, with the external part differently organized than the inner part, and therefore more susceptible to ultrastructural modifications. This is one of the interest of biochemical studies of the plant cell walls connected to ultrastructural visualization to bring elements to solve such interrogations.

Acknowledgements

Thanks are expressed to Drs K. Roberts and J.P. Knox for the gift of JIM5 and JIM7 antibodies. We are grateful to Dr. M. Schülein (Novo) for giving us a sample of EGI and to Dr. B.Henrissat for helpful discussion.

References

Asamizu, T., Nakano, N. and Nishi, A. (1983) 'Changes in non-cellulosic cell wall polysaccharides during the growth of carrot cells in suspension cultures'. Planta, 158, 166-174.

Bacic, A., Harris, P.J. and Stone, B.A. (1988) 'Structure and function of plant cell walls' in Preiss, J. (ed.) The Biochemistry of Plants. A Comprehensive Treatise, vol 14 : Carbohydrates, Academic Press, New-York, London, pp.. 297-371.

Blaschek, W. and Franz, G. (1983) 'Influence of growth conditions on the composition of cell wall polysaccharides from cultured tobacco cells'. Plant Cell Reports, 2, 257-260.

De Mey, J. (1986) 'Colloïdal gold probes', in Polak J. and Van Noorden, S. (eds) Immunocytochemistry, Modern Methods and Applications, Wright-PSG, Bristol, pp. 82-106.

Fry, S.C. (1992). 'Xyloglucan : a metabolically dynamic polysaccharide' Trends in Glycosci. and Glycotechnol., 4, 279-289.

Fry, S.C., York, W.S., Albersheim, P., Darvill, A., Hayashi, T., Joseleau, J.P., Kato, Y., Lorences E.P., Maclachlan, G.A. McNeil, M. Mort, A.J., Reid, J.S.G., Seitz, H.U., Selvendran, R.R. Voragen, A.G.J. and White, A.R. (1993). An unambiguous nomenclature for xyloglucan-derived oligosaccharides. Physiol. Plant. 89 : 1-3.

Hadwiger, L. and Beckman J.M. (1980) 'Chitosan as a component of pea *Fusarium solani* interaction'. Plant Physiol. 67, 170-173.

Hayashi, T. and Yoshida, K. (1988) 'Cell expansion and single-cell separation induced by colchicine in suspension-cultured soybean cells'. Proc. Natl. Acad. Sci. USA, 85, 2618-2622.
Hayashi, T., Yoshida, K. and Ohsumi, C. (1990) 'Regeneration of colchicine-induced single carrot cells' Agric. Biol. Chem. 54, 1567-1568.
Hayashi, T., Oshumi, C., Kato, Y., Yamanouchi, H., Toriyama, K., and Hinata, K. (1994). 'Effects of amino acid medium on cell aggregation in suspension-cultured rice cells'. Biosci. Biotech. Biochem. 58, 256-260.
Joseleau, J.P., Cartier, N., Chambat, G., Faïk, A. and Ruel, K. (1992) 'Structural features and biochemical activity of xyloglucans from suspension-cultured plant cells'. Biochimie, 74, 81-88.
Joseleau, J.P., Chambat, G., Cortelazzo, A., Faïk, A. and Ruel, K. (1994) 'Putative biological action of oligosaccharides on enzymes involved in cell-wall development' Biochem. Soc. Trans. 22, 404-409.
Knox, J.P., Linstead, P.J., King, J. , Cooper, C. and Roberts, K. (1990) 'Pectin esterification is spatially regulated both within cell walls and between developing tissues of root apices'. Planta, 181, 512-521.
Lamport, D.T.A. (1986) 'The primary cell wall : a new model' in Young, R.A. and Rowell, R.M. (eds), Cellulose, Structure, Modification and Hydrolysis'. John Wiley & Sons, New-York, pp. 77-90.
McDougall, G.J. and Fry, S.C. (1989) 'Anti-auxin activity of xyloglucan oligosaccharides: the role of groups other than the terminal α-L-fucose residue'. J. Exp. Bot. 40, 233-239.
Mc. Dougall, G.J. and Fry (1991) 'Xyloglucan nonasaccharide, a naturally occuring oligosaccharin, arises *in vitro* by polysaccharide breakdown'. J. Plant Physiol. 137, 332-336.
Roberts, K. (1990) 'Structures at the plant cell surface', Current Opinion in Cell Biology, 2, 920-928.
Ruel, K. and Joseleau, J.P. (1984) 'Use of enzyme-gold complexes for the ultrastructural localization of hemicelluloses in the plant cell wall', Histochem. 81, 573-580.
Ruel, K. and Joseleau, J.P. (1993) 'Influence of xyloglucan oligosaccharides on the micromorphology of the walls of suspension-cultured *Rubus fruticosus* cells'. Acta Bot. Neerl. 42, 363-378.
Ruel, K., Joseleau, J.P. and Franz, G. (1990) 'Aspects cytologiques de la formation des xyloglucanes dans les cotylédons des graines de *Tropaeolum majus* L.' C.R. Acad. Sci. Paris, t. 310, SER III, 89-95.
Street, E.M., Davey, M.R. and Sutton-Jones, B. (1972) 'Ultrastructure of plant cells growing in suspension culture'. Symp. Biol. Hung., 14, 145-159.
Vanden Bosch, K.A., Bradley, D.J., Knox, J.P., Perotto, S. Butcher, G.W. and Brewin, N.J. (1989). 'Common components of the infection thread matrix and the intercellular space identified by immunocytochemical analysis of pea nodules and uninfected roots'. EMBO J. 8, 335-342.
York, W.S., Darvill, A.G. and Albersheim, P. (1984) 'Inhibitor of 2,4-dichlorophenoxyacetic acid-stimulated elongation of pea stem segments by a xyloglucan oligosaccharide'. Plant Physiol. 75, 295-297.

CELL WALL BREAK DOWN AND ENHANCEMENT OF CELLULASE ACTIVITY DURING THE EMERGENCE OF CALLUS FROM RICE ROOTS IN THE PRESENCE OF 2,4-D

K. YOSHIDA & K. KOMAE [1]
Biotechnology Research Center, Taisei Corporation
Akanehama 3-6-2, Narashino, Chiba, 275 Japan
1. National Agriculture Research Center
Kannondai 3-1-1, Tsukuba, 305 Japan

ABSTRACT. Roles of cell wall break down and cell wall hydrolases have been studied in the callus formation. We have done microscopic observations, the determination of cell wall compliances and assays for 6 glycanases and 8 glycosidases during the formation of callus in root explants of rice (*Oryza sativa* L. cv. Sasanisiki). The separation of cortical cells from root explants of rice was stimulated by 2,4-D and both the plastic and the elastic compliances of cell walls increased during the formation of callus. These events, in particular, the separation of cortical cells, may be important for the generation of clumps of callus cells that are initiated at the interior of root tissues, specifically around the vessels. Buffer-soluble cellulase activity was significantly enhanced by treatment with 2,4-D (6- to 10-fold) at the early stage of formation of root callus (1-2 days after the start of treatment with 2,4-D). The optimum concentration of 2,4-D and time course studies indicated that this enhancement was correlated with and preceded both the separation of cortical cells and loosening of the cell wall. The enhancement of cellulase activity by 2,4-D in rice is the such first finding in a monocot.

1. Introduction

Auxin-induced callus formation is the first stage and a prerequisite for the culture of plant cells *in vitro*. Callus formation is considered to be the result of a combination of wound healing and extensive cell division (Yeoman, 1970). An understanding of the complete mechanism should answer to two important questions in plant sciences. The first is what is callus ? The answer to it may provide new ideas on how to use callus cells in plant biotechnology. The second is how is the division of plant cells regulated in differentiated and organized tissues ? This latter question cannot be answered by the results of studies done with protoplasts or suspension-cultured cells because these types of cells have already been liberated from tissues. Even though the study of callus formation is considered classical, it is still an informative system for plant science studies.

2. Tissue disorganization is essential for the callus emergence

Callus formation involves not only the division of cells but also the disorganization of tissues (Davidson et al. 1976). We observed the separation of epidermal and cortical cells (i.e. disorganization of the tissue) in response to 2,4-D during the formation of callus in root explants of rice and in hypocotyl explants of carrot and *Nicotiana* (Fig. 1). This separation of cells has also been reported in the extensive auxin-induced division

M. Terzi et al. (eds.), Current Issues in Plant Molecular and Cellular Biology, 457-462.

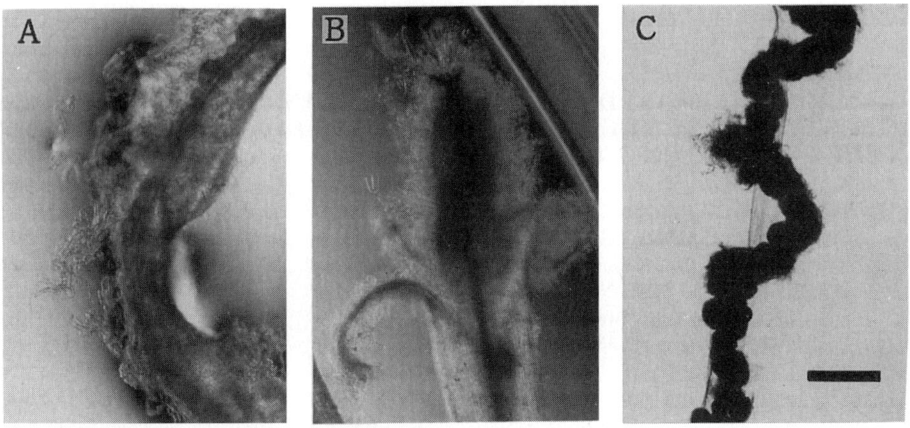

Figure 1. Cortical dissociation in the callus formation. **A.** Root explants of rice cultured in AA liquid medium (Toriyama and Hinata, 1985) that contained 58.4 mM sucrose and 13.5 µM 2,4-D. **B.** Hypocotyl explants of carrot (cv. Kurodagosun) cultured in B5 liquid medium that contained 58.4 mM sucrose and 4.5 µM 2,4-D. **C.** Hypocotyl explants of *Nicotiana* spp. cultured in MS liquid medium that contained 58.4 mM sucrose and 4.5 µM 2,4-D. Bar = 0.5 mm. Tissue explants were cultured in plastic cell wells (6 well plates) at 28°C in the dark, for 7 days with shaking at 80 rpm

of cells in bean petioles (Linkins et al. 1973). The outermost layer or two of cells autolyses and collapses without cell division during the early development of callus on cylindrical explants of Jerusalem artichoke tubers (Davidson et al. 1976).
The separation of cells in the cortex or the outermost layer may be essential for the release (emergence) of callus cells, whose primordia are induced in the interior of tissues or around the vessels, that is, pericycle region (Nishimura and Maeda 1982). This disorganization of tissue allows us to isolate callus cells from the rigid tissues of plants without vigorous mechanical procedures, such as homogenization (Yoshida and Komae 1993).

3. Cell wall break down is a basis of the tissue disorganization

We can consider the disorganization as being a disruption of the normal positions of individual cells in the plant tissue and this process may involve the loss of cell - adhesive properties, as occurs during metastasis in animal systems. Separated cortical cells of rice roots remained fluorescein diacetate (FDA) - positive, an indication that the cortical cells remained viable during the dissociation process (Fig. 2). This confirms that the cortical dissociation is not based on the disruption of cells but on the separation of cells or on the dissociation of cell walls between cells. We investigated the mechanical properties of root tissues of rice during the callus formation (Yoshida and Komae 1993). Elastic (DE) and plastic(DP) compliances were determined according to Cleland' s method (Cleland 1967). Both DE and DP of rice roots increased in proportion to the concentration of 2,4-D (Fig. 3 a). These increases in both DE and DP caused by 2,4-D started after 2 days (after the initiation of callus primordia) and continued with the enlargement of callus primordia (Fig. 3 b), reflecting the cortical

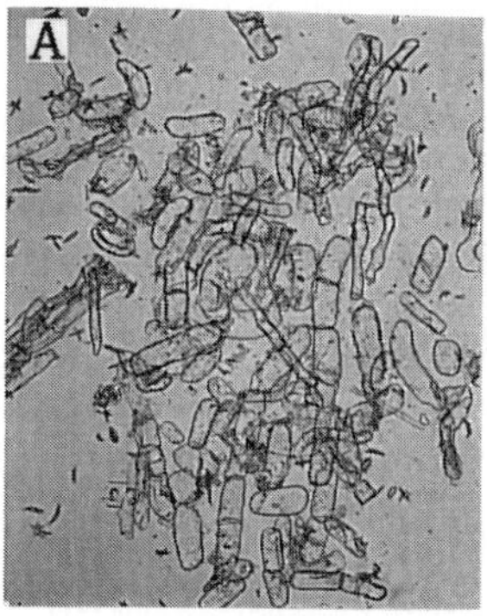

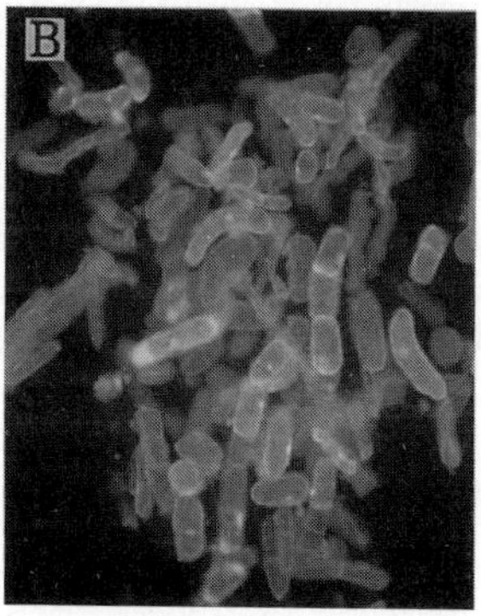

Figure 2. Cortical cells separated from root tissues of rice during the callus formation. **A.** Light micrograph. **B.** Fluorescence micrograph of cortical cells after the fluorescein diacetate (FDA)-staining.

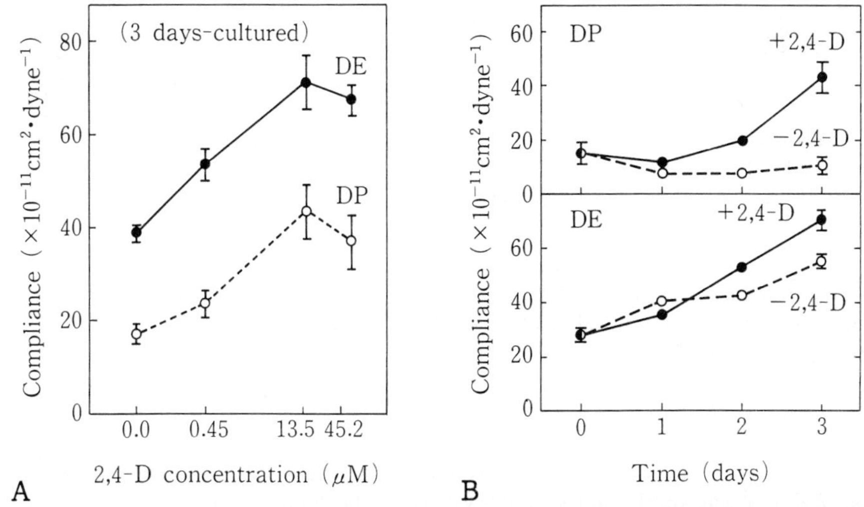

Figure 3 Effects of 2,4-D o the elastic (DE) and plastic (DP) compliances of rice root explants. **A.** Optimum concentration of 2,4-D. **B.** Time course. Error bars: SE (n = 15-20). Rice root explants were cultured with or without 13.5 μM 2,4-D.

dissociation observed under microscope. These increases in DE and DP could be an evidence of the extent of cell separation and/or of an increase in the extensibility of each cell. DE and DP increase during auxin-induced elongation in the explants of the stems or the coleoptiles (Cleland 1967, Masuda 1969), even though the elongation of rice root explants was not affected by 2,4-D (Yoshida and Komae 1993). At the very least, the analysis of mechanical properties suggests that some components of cell wall in rice roots are broken down as a result of treatment with 2,4-D.

4. Enhancement of cellulase activity occurs before the cell wall break down

Extracellular matrix components (collagen, fibronectin and laminin) of animal cells are degraded by metalloproteases and serine proteases. This process is believed to be involved in the cell migration in response to injury and in the metastasis (Chen 1992). Cell wall components of rice cells can be degraded by glycanases. Monocot cell wall consists of arabinoxylan, cellulose, (1,3;1,4)-ß-glucan, xyloglucan and pectin (Bacic et al., 1988). We investigated the effects of 2,4-D on 8 glycosidases and 6 glycanases

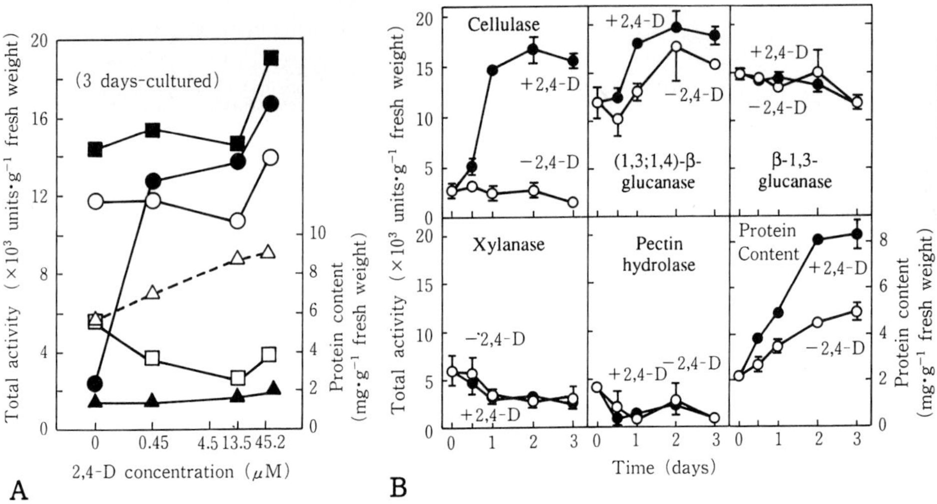

Figure 4. Effects of 2,4-D on the activity of buffer-soluble glycanases in rice root explants, as measured viscometrically. **A.** Optimum concentration, Total activities of cellulase (—●—), ß-1,3-glucanase (—○—), (1,3;1,4)-ß-glucanase (—■—), pectin hydrolase (—▲—), xylanase (—□—) and protein content (- -△- -) after 3 days. **B.** Time course, Root explants were cultured with or without 13.5 μM 2,4-D. Activities were determined in duplicate reactions. Error bars: SEs for the triplicate cultures (n = 3).

with the photometric assays. This assay has shown that cellulase activity in buffer-soluble fraction was significantly enhanced by 2,4-D treatment (See Tables 1 and 2 in Yoshida and Komae 1993). The viscometric assay has also shown that the cellulase activity specifically increased in proportion to 2,4-D concentration (Fig. 4a) and that the maximum activity (6-10 times over the control) occurred after treatment with around 13.5 μM 2,4-D 1 day after the start of the culture (Fig. 4 b). The cellulase seems to function in the tissue disorganization of rice roots like proteolytic enzymes in animal system.

5. Conclusion

The enhancement of cellulase activity by 2,4-D preceded both the separation of cortical cells and the increase in cell wall compliances in rice roots (Fig. 5). The cellulase activity was roughly correlated with the cell wall compliances (Figs 3a and 4). These

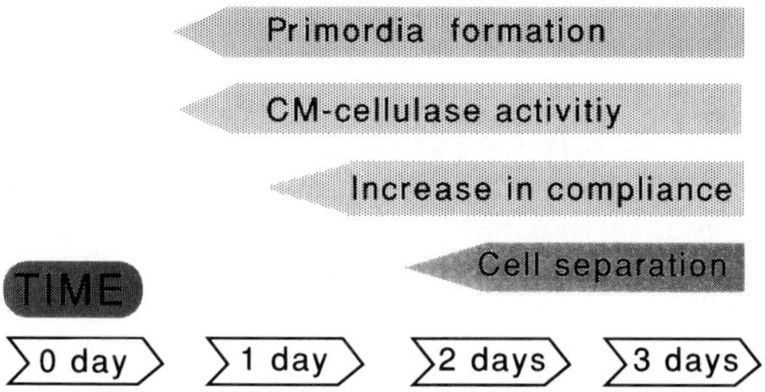

Figure 5 Primordia formation, cellulase activity, cell wall break down and cortical dissociation in rice root explants.

results suggest that, in rice, the cellulase activity might be involved in the separation of cortical cells (Linkins et al. 1973), in the loosening of cell wall (Fry 1989), in the cell division (Kemmerer and Tucker 1994, Linkins et al. 1973, Fan and Maclachlan 1966) and/or in the cell growth (Nakamura and Hayashi 1994, O' Neil and Scott 1987). This involvement might occur via the degradation of cellulose, xyloglucan (Hayashi et al. 1984) or (1,3;1,4)-ß-glucan (Hatfield and Nevins 1987) in the walls of rice cells in vivo, since all of these components has been shown to be endogenous substrates of plant cellulases (Fry 1989). To confirm the involvement of rice cellulase in the dissociation of cell walls and in the formation of callus, we are purifying rice cellulase to investigate its substrate specificity and to raise antibodies for immunocytochemical studies in root tissues of rice.

Acknowledgments

We thank Dr. Takahisa Hayashi of Kyoto University for his technical advice and critical discussions. We also thank 6 students of Chiba University and a student of Toho University for their technical assistance.

References

Bacic, A., Harris, P. J. and Stone, B. A. (1988) Structure and function of plant cell walls. In Biochemistry of Plants: A Comprehensive Treatise (J. Preiss, ed.), Vol. 14: Carbohydrates, pp. 298-371. San Diego, CA: Academic Press.

Chen, W.-T. (1992) Membrane Proteases: roles in tissue remodeling and tumor invasion. Curr. Opin. Cell Biol. 4: 802-809.

Cleland, R. (1967) Auxin and the mechanical properties of the cell wall. Ann. N.Y. Acad. Sci. 144: 3-18.

Davidson, A. W., Aitchison, P. A. and Yeoman, M. M. (1976) Disorganized systems. In Cell Division in Higher Plants. Edited by Yeoman, M. M. pp. 407-431. Academic Press, London.

Fan, D. F. and Maclachlan, G. A. (1966) Control of cellulase activity by indoleacetic acid. Can. J. Bot. 44: 1025-1034.

Fry, S. C. (1989) Cellulases, hemicelluloses and auxin-stimulated growth: a possible relationship. Physiol. Plant. 75: 532-536.

Hatfield, R. D. and Nevins, D. J. (1987) Hydrolytic activity and substrate specificity of an endoglucanase from *Zea mays* seedling cell walls. Plant Physiol. 83: 203-207.

Hayashi, T., Wong, Y. S. and Maclachlan, G. (1984) Pea xyloglucan and cellulose II. Hydrolysis by pea endo-1,4-ß-glucanase. Plant Physiol. 75: 605-610.

Kemmerer, E. C. and Tucker, M. L. (1994) Comparative study of cellulases associated with adventitious root initiation, apical buds, and leaf, flower, and pod abscission zones in soybean. Plant Physiol. 104: 557-562.

Linkins, A. E., Lewis, L. N. and Palmer, R. L. (1973) Hormonally induced changes in the stem and petiole anatomy and cellulase enzyme patterns in *Phaseolus vulgaris* L. Plant Physiol. 52: 554-560.

Masuda, Y. (1969) Auxin-induced cell expansion in relation to cell wall extensibility. Plant Cell Physiol. 10: 1-9.

Nakamura, S. and Hayashi, T. (1993) Purification and properties of an extracellular endo-1,4-ß-glucanase from suspension-cultured poplar cells. Plant Cell Physiol. 34: 1009-1013.

Nishimura, S. and Maeda, E. (1982) Cytological studies on differentiation and dedifferentiation in pericycle cells of excised rice roots. Japan J. Crop Sci. 51: 553-560.

O'Neill, R. A. and Scott, T. K. (1987) Rapid effects of IAA on cell surface proteins from intact carrot suspension culture cells. Plant Physiol. 84: 443-446.

Toriyama, K. and Hinata, K. (1985) Cell suspension and protoplast culture in rice. Plant Sci. 41: 179-183.

Yeoman, M. M. (1970) Early Development in Callus Cultures. Int. Rev. Cytol. 29: 383-409

Yoshida, K. and Komae, K. (1993) Dissociation of cortical cell walls and enhancement of cellulase activity during the emergence of callus from rice roots in the presence of 2,4-D. Plant Cell Physiol. 34: 507-514.

AUXIN PERCEPTION AT THE PLASMA MEMBRANE OF PLANT CELLS : RECENT DEVELOPMENTS AND LARGE UNKNOWNS

H. BARBIER-BRYGOO, C. MAUREL, J.M. PRADIER, A. DELBARRE, V. IMHOFF & J. GUERN
Institut des Sciences Végétales, CNRS, Avenue de la Terrasse, 91198, Gif sur Yvette, France

ABSTRACT. Studies of membrane responses of tobacco protoplasts to auxin have demonstrated the existence of elementary response chains to auxins at the plasma membrane. In the presence of extracellular auxin molecules or auxin agonists, plasmamembrane auxin responsive proteins modulate the activity of a variety of electrogenic units, ultimately leading to changes in membrane potential. The plasma membrane auxin responsive proteins are immunologically related to ZmER-abp1, the major auxin-binding protein from maize. Evidence suggests a complex organization for the functional auxin perception units at the plasma membrane, comprising a soluble excreted auxin-binding protein associated to a transmembrane protein. Major unknowns concern the consequence(s) of the activation of the plasma membrane elementary chains of auxin perception on more integrated auxin responses. Strategies to answer the important question of whether the auxin effects result from the activation of a unique ubiquitous receptor or involve a multitargetted activation of independent receptors are discussed.

Introduction

The chemical factors used by plants as growth and differentiation signals elicit a remarkable variety of responses at the level of plants, excised organs or individual cells. Auxins modify ionic exchanges at the plasma membrane, selective transcription of genes, elongation growth, cell division and vascular tissue differentiation. These responses are likely elicited after binding of the hormone to specific receptors. Despite considerable progress made in the last few years towards the understanding of auxin action, only a few pieces of the puzzle have been identified yet and much concerning the molecular mechanisms underlying the action of these hormones remain to be elucidated (Palme, 1992; Barbier-Brygoo, 1994; Jones, 1994).
The most important recent breakthroughs concerned the isolation and molecular characterization of a variety of auxin-binding proteins (ABPs) which may constitute auxin receptors (Venis and Napier, 1994 and references herein). The major question is now to identify those which really function as hormonal receptors (Napier and Venis, 1991; Barbier-Brygoo, 1994; Jones, 1994). The present review contributes to this question with a special focuss on plasma membrane receptors.

1. Auxin binding proteins and auxin perception units at the plasma membrane

The use of azido auxins has resulted in the photoaffinity labeling of plasma membrane proteins of 40 and 42 kDa in zucchini (Hicks *et al.*, 1989, 1993), 23, 24, 58 and 60 kDa in maize (Feldwisch *et al.*, 1992), and 23 and 24 kDa in Arabidopsis (Zettl *et al.*, 1994). Little is known about the function of these proteins except for the 24 kDa protein from *Arabidopsis* identified

M. Terzi et al. (eds.), Current Issues in Plant Molecular and Cellular Biology, 463-472.

as a GST (Zettl *et al.*, 1994). On the basis of the pattern of competition of photolabeling by auxin analogs, the 40-42 kDa proteins from zucchini and tomato have been suggested as components of the auxin influx carrier. Since the 23 and 24 kDa proteins from maize could also be photolabeled as well by an azido-NPA derivative it was concluded that these proteins could be components of the efflux carrier (Zettl *et al.*, 1992).

The identification of the function of auxin binding proteins and more precisely the search for auxin receptors strictly depend on the availability of a suitable functional assay. This is the reason why we have been interested since several years in monitoring early auxin effects on the properties of the plasma membrane. It is now well established that the application of exogenous auxin rapidly modifies the ion transport properties of this membrane, as revealed by auxin-induced changes in the electrical transmembrane potential difference measured by classical intracellular microelectrodes in a variety of material (Barbier-Brygoo, 1994 and references therein).

The most extensive study concerning the auxin-induced modifications of the electrical properties of the plasma membrane has been performed on tobacco mesophyll protoplasts. When treated with auxins these protoplasts are hyperpolarized within 1-2 minutes (Ephritikhine *et al.*, 1987; Barbier-Brygoo *et al.*, 1989). The protoplast response is specific for active auxins (Barbier-Brygoo *et al.*, 1991) with a specificity pattern closely resembling that of the auxin-controlled proliferation of cells derived from mesophyll protoplast suspensions.

This protoplast response to auxins has been criticized on the ground that the membrane potential of the protoplasts is strongly depolarized, likely due to the introduction of a shunt resistance upon penetration of the microelectrode (Van Duijn *et al.*, 1988). However, all the results obtained so far demonstrate that this electrical response corresponds to a specific action of auxin at the plasma membrane. Furthermore, as discussed later, most of the main conclusions drawn from this assay have been confirmed by independent approaches (Rück *et al.* 1993), strengthening its interest as one of the rare convenient cellular assay for auxin activity.

An important step further has been to demonstrate that the membrane response of mesophyll protoplasts could be triggered by protein conjugates of 5-substituted NAA analogs expected to be membrane impermeant (Venis *et al.*, 1990). The fact that these conjugates acted as rapidly as free NAA suggested that both types of molecules were recognized at the outer face of the plasma membrane. This conclusion has been further strengthened by the discovery of the activity of agonist antibodies as discussed later in section 2.1. Independent sets of experimental data show that one of the mechanisms of the auxin-induced hyperpolarization of the plasma membrane corresponds to an activation of the H^+-pump ATPase (Barbier-Brygoo *et al.*, 1989; Santoni *et al.*, 1990; 1991; Lohse and Hedrich, 1992; Rück *et al.*, 1993). Recent data has shown that, aside the ATPase, auxin modulates the activity of anion (Marten *et al.*, 1991; Zimmerman *et al.*, in preparation) and cation channels (Blatt and Thiel, 1994). These results show that the electrical responses induced by auxins at the plasma membrane involve several types of electrogenic units. This complexity has now to be taken into account in interpreting the variety of electrical responses to auxins reported in the literature in terms of kinetics and intensity, occurence or not of an initial depolarization followed by an hyperpolarization and bimodal character of the dose-response curves.

The presence of elementary perception chains operating at the plasma membrane independently of cytoplasmic factors is demonstrated by the auxin stimulated H^+-translocation activity of the ATPase of plasma membrane vesicules from tobacco leaves (Santoni *et al.*, 1990; 1991) and by the auxin effects on the ion channel GCAC1 of *V. faba* guard cell protoplasts, observed at the level of single channels in isolated membrane patches (Marten *et al.*, 1991). The critical question now is to understand how are organized these plasma membrane elementary perception chains connecting auxin receptors to the electrogenic proteins.

2. Homologies and differences between the proteins involved in the perception of auxins at the plasma membrane and the ER-ABPs.

2.1. SIMILARITIES AND DIFFERENCES

Experimental evidence strongly suggest that the tobacco auxin responsive proteins (tARPs), located at the plasma membrane and involved in the electrical responses to auxin, are related to the endoplasmic reticulum auxin binding protein, ZmER-abp1, initially identified in maize. Different antibodies to the major auxin-binding protein from maize coleoptile, ZmER-abp1 (Venis and Napier, 1994), corresponding to epitopes located around the glycosylation site of this protein, were shown to inhibit the auxin-induced hyperpolarization of tobacco mesophyll protoplasts, without interfering with the fusicoccin-induced response (Barbier-Brygoo *et al.*, 1989). At the opposite, polyclonal antisera D16 and D51, produced against an oligopeptide embracing a region of abp1 putatively involved in auxin binding mimick auxin effects on protoplasts (Venis *et al.*, 1992). Similar agonist activity is exhibited by monovalent Fab fragments from D16 IgGs. The antagonist activity of PAB124 (a polyclonal antibody to the ZmER-abp1) and the agonist activity of D16, first demonstrated by us in an heterologous system (tobacco protoplasts, anti-maize antibodies) were fully confirmed recently (Rück *et al.*, 1993) in homologous conditions (maize coleoptile protoplasts, anti-maize antibodies), providing in this system as well, evidence for the presence of abp1-related perception sites for auxin at the outer face of the plasma membrane.
This set of data further confirms that auxin responsive proteins involved in the electrical responses are located at the outer face of the plasma membrane. The plasma membrane auxin responsive proteins of tobacco mesophyll as well as those of maize coleoptile are immunologically related to maize ER abp1, at least at the level of the epitopes included in the putative auxin-binding domain and of those mapped around the glycosylation site. The essential difference concerns the plasma membrane versus the ER locations of these proteins but, as it will be discussed later, a set of evidence suggests that the functional abp-related protein located at the protoplast surface may be a relocalized fraction of the endoplasmic reticulum protein.

2.2. A WORKING MODEL FOR THE ORGANIZATION OF THE PERCEPTION UNITS

The idea that a fraction of the ER ABP could be secreted came from experiments consisting in incubating tobacco mesophyll protoplasts with minute amounts of purified maize abp1 (Fig. 1). This "complementation" resulted in a marked enhancement of the auxin sensitivity of the protoplasts (Barbier-Brygoo *et al.*, 1991), suggesting that this non-integral membrane protein was able to bind to the protoplast surface and play an active role in auxin signalling by interacting with a transmembrane protein (TP) of tobacco plasma membrane essential for the transmission of the auxin signal (see also Klämbt, 1990). To account for these results, we proposed a model for the organization of auxin perception units at the plasma membrane (Barbier-Brygoo *et al.*, 1991) where tobacco auxin-responsive proteins (tARPs), secreted from an intracellular pool, form functional auxin perception units at the extracellular surface by associating with TP. Heterologous functional units can result from the interaction between maize abp1 and the tARP-binding site of TP.
Recent results showing that a synthetic peptide reproducing the C-terminus of abp1 was able to influence the activity of potassium channels at the plasma membrane of *Vicia faba* guard cells (Thiel *et al.*, 1993) further consolidate our discovery of the complementation effect of exogenous ABP described above. These results point to the C-terminus of abp1 as an important region for abp coupling to intracellular transduction cascades and provide tools for isolating the putative TP.

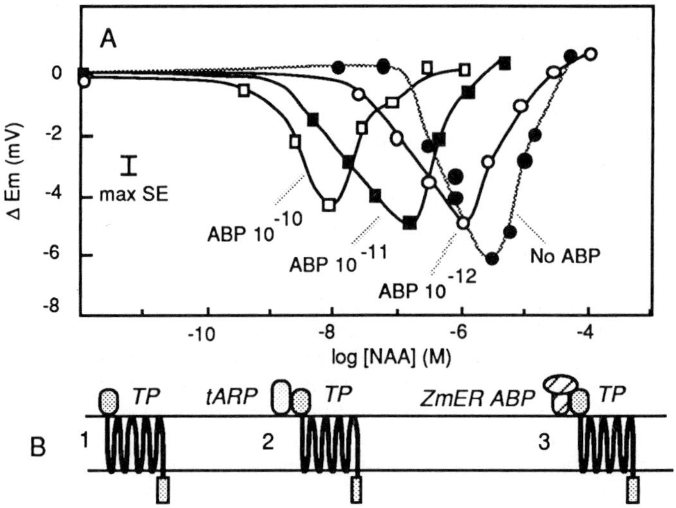

Figure 1. Increase in the auxin sensitivity of tobacco mesophyll protoplasts incubated with maize ER ABP at different concentrations. A : Protoplasts were prepared and treated with different concentrations of maize ABP as described in Barbier-Brygoo *et al.* (1991) and then assayed for their electrical response to a range of NAA concentrations. B : Working model of the organization of auxin perception units at the plasma membrane with homologous units composed of extracellular tobacco auxin responsive proteins tARP bound to transmembrane proteins TP (1, 2) or heterologous units associating the exogenous ZmER ABP and TP (3).

3. From the auxin perception at the plasma membrane to integrated hormonal responses

3.1. AUXIN PERCEPTION AT THE CELL SURFACE VERSUS MULTIPLE PERCEPTION PATHWAYS WITH DIFFERENT CELLULAR LOCALIZATIONS.

Major unknowns concern the consequence(s) of the activation of the elementary chains of auxin perception located at the plasma membrane on more integrated auxin responses.

The identity of the specificity pattern of the electrical response of mesophyll protoplast to auxin with that of the auxin-controlled proliferation of cells derived from these protoplasts associated with the fact that in protoplasts from an auxin-resistant mutant, the sensitivity of both responses towards a set of auxin analogues was similarly decreased (Ephritikhine *et al.*, 1987) suggested that the auxin perception units located at the plasma membrane could be involved in the regulation of cell division. On the contrary, as shown in Fig. 2, in tobacco mesophyll protoplasts and protoplast-derived cells expressing the *rolB* gene from *A. rhizogenes*, the electrical responses displayed a very high sensitivity to auxin (Fig. 2A) whereas the induction of cell division (Fig. 2B, D) as well as the induction of the expression of the *rolB* gene itself (Fig. 2C) require much higher auxin concentrations (Maurel *et al.*, 1991). These results apparently suggest that independent pathways for auxin perception could coexist, one corresponding to cell surface responses as evidenced by changes in membrane potential, others corresponding to cell division and to the regulation of specific genes.

Further work is obviously needed to resolve these apparent contradictions in order to investigate the important question of whether the auxin effects results from the activation of a unique ubiquitous receptor or involve a multitargetted activation of independent receptors. In this context, promising approaches consisting in specifically blocking or activating surface receptors with antibodies or impermeant auxin agonists to follow the consequences of such modulations of membrane responses on other responses, like gene expression, are currently used in different laboratories.

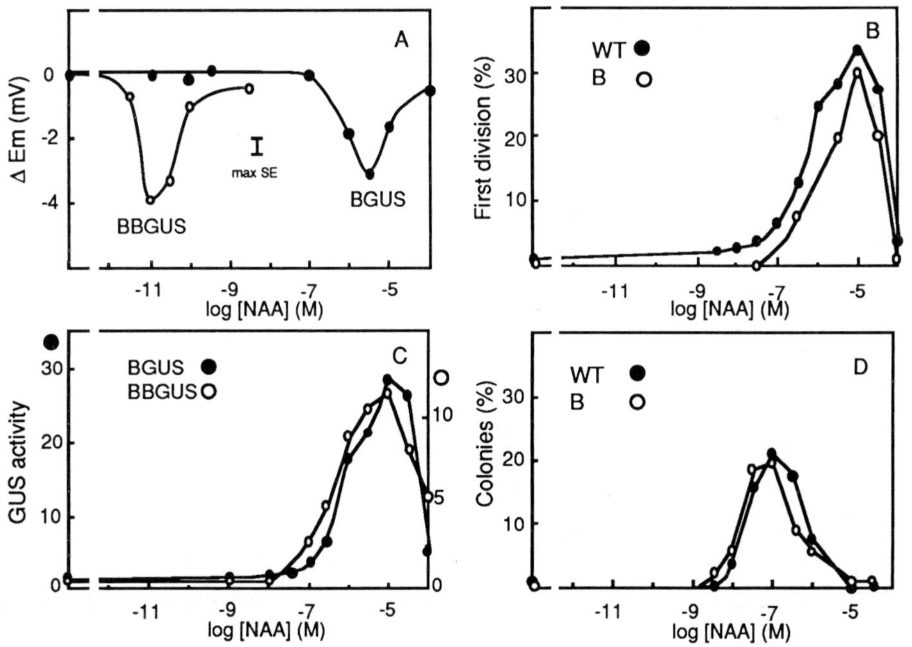

Figure 2. Comparison of the dose-response curves of different auxin effects on tobacco mesophyll protoplasts or protoplasts derived cells. Control protoplasts were isolated from wild type plants (WT) or from plants containing a GUS gene under the control of the *rolB* gene promoter (BGUS). Protoplasts expressing the *rolB* gene were isolated from plants containing a *rolB* gene alone (B) or a GUS gene associated with a functional *rolB* gene (BBGUS) as described in Maurel *et al.* (1994). The electrical response illustrated in A was measured as described in Barbier-Brygoo *et al.* (1991). The GUS activity reported in C has been assayed according to Maurel *et al.* (1990). The occurence of the first division in protoplast-derived cells was estimated after 5 days of culture of protoplast suspensions plated at 50,000 protoplasts/mL (B). The influence of auxin on cell division of protoplast-derived cells plated at the low density of 100 cells/mL was monitored from the number of developed colonies after 4 weeks (D) as described in Maurel *et al.* (1991).

In our group, the exploration of the perception pathway leading to the auxin-induced expression of the *rolB* gene in transgenic tobacco protoplasts has shown that neither the expression of this gene can be triggered by the impermeant auxin-agonist antibody D16 nor can it be blocked by polyclonal antibodies to the ZmER-abp1 which inhibit the plasma membrane responses (Fig.3). These results which suggest that the expression of the *rolB* gene is controlled by auxins through a pathway distinct from that involved in the electrical responses favour the idea of the coexistence of independent perception pathways responsible for the variety of auxin responses. This conclusion has however to be considered with some caution. First, the stability of the antibodies over a 24 h period has not been checked. Second, the data only concern the *rolB* gene expression and the same approach is currently used in our group to investigate the dependence or independence of the expression of different auxin-regulated genes towards modulations of the activity of surface receptors. Third, as discussed in the next chapter, the comparison of short term and long term auxin responses can be biaised by several factors of complexity which can mask their relationships.

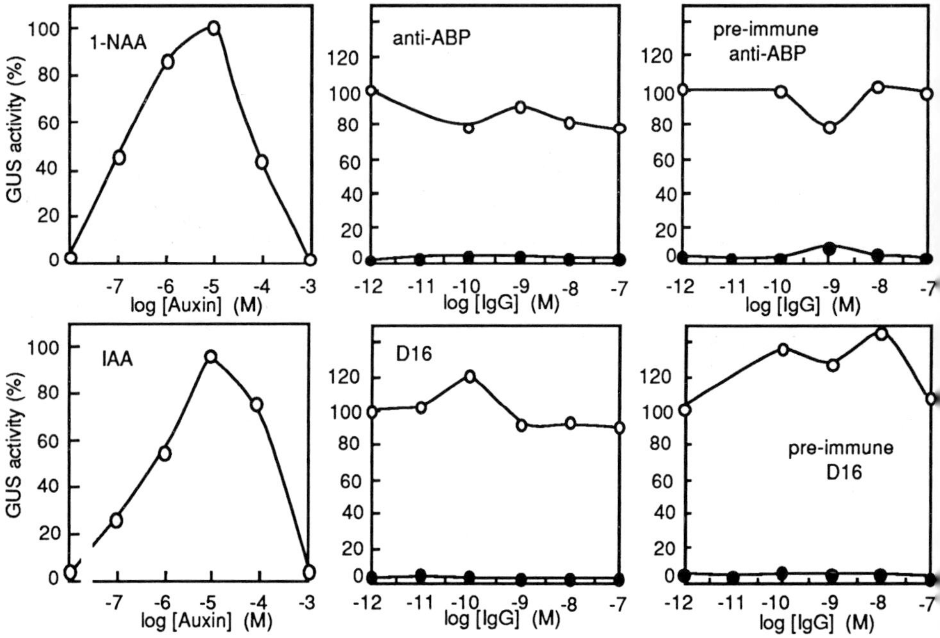

Figure 3. Influence of the permeant auxins 1-NAA and IAA, of the auxin agonist antibody D16, of antibodies directed against the ER ABP from maize and of the corresponding preimmune sera on the expression of the *rolB* gene in tobacco protoplasts prepared from plants containing a GUS gene under the control of the *rolB* gene promoter (BGUS). The effect of the different antibodies on rolB expression has been assayed in the absence (filled circles) or presence of 1µM NAA (open circles). Protoplasts were incubated for 24 h at 26°C, in immunotitration plates at a density of 100,000 cells per well in 300 µL of T00 medium. Gus activity was determined by a fluorometric assay adapted from Maurel *et al* (1990).

3.2. FACTORS OF COMPLEXITY THAT SHOULD BE TAKEN INTO ACCOUNT

3.2.1. *Auxin metabolism and auxin responses.* Delbarre *et al.* (1994a, 1994b) have shown that in the conditions used for studying cell responses to auxin, such as gene expression in protoplasts or cell division of protoplasts or cell suspensions, the exogenous auxins used to trigger these responses are extensively metabolized. The intensity of this metabolization is such that the extracellular and intracellular auxin concentrations rapidly and strongly decrease with time. This problem, of negligible importance when monitoring short-term responses, can be of high relevance for long term ones, the auxin signal experimentally imposed being eventually transient in terms of intensity. The dose-response curve of an auxin effect developing over a long period will be shifted towards concentrations higher than those really necessary to trigger the primary responses and will bring biased information on the affinity of the receptors involved. This is examplified by the fact that the optimal NAA concentration

necessary for cell division of protoplasts and protoplast-derived cells is decreased by 1-2 orders of magnitude for cell populations at low density of inoculation (Fig. 2B and D).
The operational conclusion is that the study of any auxin response triggered by the application of exogenous auxin should be associated with measurements of extracellular and intracellular levels of free auxin and metabolites all along the period of response monitoring. We have performed such a study in the case of tobacco protoplasts expressing the *rolB* gene, in order to check the claim from Spena that the morphogenetic effect of the RolB protein resulted from its glucosidase activity releasing free active auxin from auxin glycosides (Estruch *et al.*, 1991). Delbarre *et al.* (1994b) have shown that this hypothesis is not valid : neither increases in free auxin concentration nor enhanced hydrolysis of auxin glucose ester could be demonstrated in transgenic tobacco mesophyll protoplasts precisely monitored in conditions were they expressed the *rolB* gene and displayed their typical enhanced membrane sensitivity to auxin (Maurel *et al.*, 1994). Similar conclusions have been reached from parallel approaches in other biological systems (Nielsson *et al.*, 1993; Schmulling *et al.*, 1993). The *rolB* function is still unknown but a positive consequence of the debate concerning this protein has been to raise a strong renewal of the interest concerning auxin metabolism and more generally hormone metabolism.

3.2.2. *Evidence and speculations concerning the dynamics of plasma membrane receptors.* The idea that auxin could regulate the exocytic flux has been proposed long ago as coherent with the well known effect of auxin on cell elongation and deposition of cell wall material. More specifically, it has been recently shown that the stimulatory effect of auxin on exocytosis affects the traffic of the plasma membrane ATPase, in a manner apparently critical for the regulation of proton excretion and cell elongation (Hager *et al.*, 1991). An even more exciting version of the idea that auxin could regulate the endomembrane traffic was raised by Cross (1991) who suggested that the binding of auxin to the ER ABP in the endoplasmic reticulum could be involved in the traffic of the plasma membrane auxin receptors themselves, i.e. in the exocytosis of the small fraction of ER ABP excreted through this pathway. This hypothesis is largely speculative but it fits with a few independent pieces of data. The first evidence has been provided by the discovery that auxin binding to abp1 induces a conformational change of its C terminal end revealed by a reduction the binding of a monoclonal antibody directed to epitopes of this region (Napier and Venis, 1990). This could explain how the KDEL signal for ER retention can be by-passed through the masking of the KDEL sequence following from auxin binding to abp1. Second, the fact mentioned by Rück *et al.* (1993) that a fraction of Zm-abps extracted from maize coleoptiles has an aminoacid sequence identical to that of Zm-abp1 but contains sugar residues added only in the Golgi apparatus is consistent with the export of this fraction to the plasma membrane. Third, the recent work from Jones and Herman (1993) on maize cell suspensions has provided experimental evidence for the secretion of abp molecules or immunochemically related proteins within the cell wall space and the external medium. Further work is however still necessary to undoubtedly identify these secreted proteins as abp1 and really demonstrate the export of abp1 to the cell surface.
While the proposal of J. Cross (1991) was that the main functional role of auxin binding to the ER ABP would be to facilitate the secretion of cell wall precursors, we provide here an alternative (or additional) view focusing on the consequences that the dynamic organization of the traffic of plasmamembrane auxin receptors would have on the activation of the plasma membrane auxin receptors and related responses. The first prediction is that, as ABP is unstable in conditions of acidic pH (i.e. in conditions prevailing in the cell wall), a continuous excretion would be needed in order to insure the renewal of the degraded extracellular ABP. If this export of ABP molecules to the plasma membrane requires the intracellular binding of auxin to the ER ABP, the second prediction is that impermeant auxin analogs or auxin agonists would trigger the activation of the external ABP molecules but would not insure the activation

of the secretion pathway. As a consequence, only short term auxin responses should be induced by impermeant auxins. The third prediction is based on the fact that the affinity of auxin binding to ABP is drastically decreased at pH about 7 in the ER lumen. Consequently, one should expect that any hormonal responses lasting for longer than a few times the half life of the external ABP would require rather high auxin concentrations to insure the export of receptors to the plasma membrane, concentrations much higher than those necessary to saturate the external receptors. These predictions have been evaluated by a quantitative model as illustrated by Figure 4. It shows that auxin responses expressed rapidly enough not to be affected by the ABP turn over rate would be fully triggered by external impermeant auxin agonists. On the contrary, longer term responses which obligatory need a renewal of external receptors, would not be triggered to a significant extent in the absence of permeant auxin agonists and would exhibit a dose response curve to permeant auxins mostly reflecting the shifted affinity of ABP in the ER environment. More complex hypotheses are however necessary to account for the lack of effect of the anti-ABP antibodies.

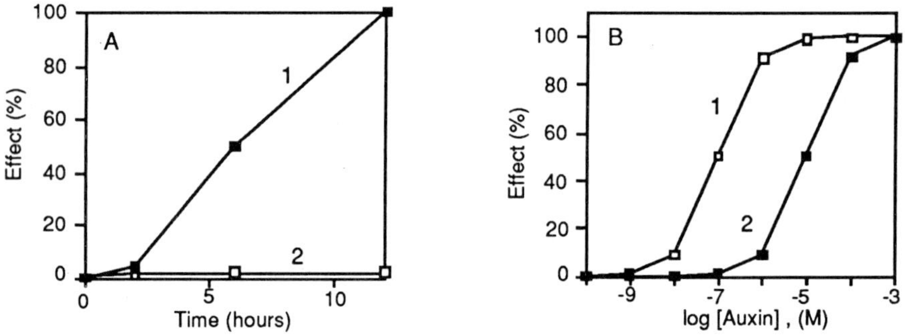

Figure 4. Modelisation of the influence of an auxin-dependent renewal of the plasma membrane receptor on the short term and long term auxin responses triggered by the activation of this receptor. The short term (5 min) response and the long term response (12 h) are assumed to be triggered by a plasma membrane receptor with an affinity of 0.1 μM and a half life of 30 min. The renewal of the receptor at the plasma membrane is assumed to correspond to a secretion of ER-ABP triggered by the binding of intracellular auxin to ABP molecules with an affinity of 10 μM. A : Evolution with time of the long term response triggered by the activation of the plasma membrane receptor by a permeant auxin analog (1) or an impermeant one, unable to induce the renewal of the surface receptor (2). B : Dose-response curves of the short term (1) and long term responses (2) induced by permeant auxins.

4. Conclusions

From the experimental evidence gathered up to now concerning the variety of auxin binding proteins already discovered, the operation of a complex set of different receptor types is envisaged (Jones, 1994), including multiple isoforms of receptor types, differing in their affinities, specificities, cell and tissue localizations and functions. However, the recognition of those proteins acting as auxin receptors is still in infancy. The presence of elementary response chains to auxin located at the plasma membrane is the most documented area of this domain but the detailed organization of these chains awaits further characterization as well as their connection with more integrated auxin responses. The model discussed here of an auxin-regulated traffic of plasma membrane receptors gives a coherent interpretation of various experimental data and shows that the strategies aiming at filling the gap between short-term and long-term responses have to take into account several factors of complexity which can mask the real organization of the perception pathways for auxins.

5- Literature

Barbier-Brygoo, H., Ephritikhine, G., Klämbt, D., Ghislain, M., and Guern, J. (1989) 'Functional evidence for an auxin receptor at the plasmalemma of tobacco mesophyll protoplasts', Proc. Natl. Acad. Sci. USA 86, 891-895.

Barbier-Brygoo, H., Ephritikhine, G., Klämbt, D., Maurel, C., Palme, K., Schell, J., and Guern, J. (1991) 'Perception of the auxin signal at the plasma membrane of tobacco mesophyll protoplasts', Plant J. 1, 83-93.

Barbier-Brygoo, H. (1994) 'Tracking auxin receptors using functional approaches', Critical Reviews in Plant Sciences, in press.

Blatt, M.R., and Thiel, G. (1994) 'K^+ channels of stomatal guard cells: bimodal control of the K inward-rectifier evoked by auxin', Plant J. 5, 55-88.

Cross, J.W. (1991) 'Cycling of auxin-binding protein through the plant cell: pathways in auxin signal transduction', New Biologist 3, 813-819.

Delbarre, A., Muller, P., Imhoff, V., Morgat, J.L., and Barbier-Brygoo, H. (1994a). Uptake, accumulation and metabolism of auxins in tobacco leaf protoplasts', Planta, in press.

Delbarre, A., Muller, P., Imhoff, V., Barbier-Brygoo, H., Maurel, C., Leblanc, N., Perrot-Rechenmann, C., and Guern, J. (1994b) 'The *rolB* gene of *Agrobacterium rhizogenes* does not increase the auxin sensitivity of tobacco protoplasts by modifying the intracellular auxin concentration', Plant Physiol. 105, in press.

Ephritikhine, G., Barbier-Brygoo, H., Muller, J.-F., and Guern, J. (1987) 'Auxin effect on the transmembrane potential difference of wild-type and mutant tobacco protoplasts exhibiting a differential sensitivity to auxin', Plant Physiol. 83, 801-804.

Estruch, J.J., Schell, J., and Spena, A. (1991) 'The protein encoded by the *rolB* plant oncogene hydrolyses indole glucosides', EMBO J. 10, 3125-3128.

Feldwisch, J., Zettl, R., Hesse, F., Schell, J., and Palme, K. (1992) 'An auxin-binding protein is localized to the plasma membrane of the maize coleoptile cells , identification by photoaffinity labeling and purification of a 23-kDa polypeptide', Proc. Nat. Acad. USA 89, 475-479.

Hager, A., Debus, G., Edel, H.-G., Stransky, H., and Serrano, R. (1991) 'Auxin induces exocytosis and the rapid synthesis of a high turnover pool of plasma-membrane H^+-ATPase', Planta 185, 527-537.

Hicks, G.R., Rayle, D.L., Jones, A.M., and Lomax, T.L. (1989) 'Specific photoaffinity labeling of two plasma membrane polypeptides with an azido auxin', Proc. Natl Acad. Sci. USA 86, 4948-4952.

Hicks, G.R., Rice, M.L., and Lomax, T.L. (1993) 'Characterization of auxin-binding proteins from zucchini plasma membrane', Planta 189, 83-90.

Jones, A.M., and Herman, E.M. (1993) 'KDEL-containing auxin-binding protein is secreted to the plasma membrane and cell wall', Plant Physiol. 101, 595-606.

Jones, A.M. (1994) 'Auxin-binding proteins', Annu. Rev. Plant Physiol. Plant Mol. Biol., in press.

Klämbt, D. (1990) 'A view about the function of auxin-binding proteins at plasma membrane', Plant Mol. Biol. 14, 1045-1050.

Lohse, G., and Hedrich, R. (1992) 'Characterization of the plasma membrane H^+-ATPase from *Vicia faba* cells. Modulation by extracellular factors and seasonal changes', Planta 188, 206-214.

Marten, I., Lohse, G., and Hedrich, R. (1991) 'Plant growth hormones control voltage-dependent activity of anion channels in plasma membrane of guard cells', Nature 353, 758-762.

Maurel, C., Barbier-Brygoo, H., Spena, A., Tempé, J., and Guern, J. (1991) 'Single *rol* genes from the *Agrobacterium rhizogenes* T_L-DNA alter some of the cellular responses to auxin in *Nicotiana tabacum'*, Plant Physiol. 97, 212-216.

Maurel, C., Brevet, J., Barbier-Brygoo, H., Guern, J., and Tempé, J. (1990) 'Auxin regulates the promoter of the root-inducing *rolB* gene of *Agrobacterium rhizogenes* in transgenic tobacco', Mol. Gen. Genet. 223, 58-64.

Maurel, C., Leblanc, N., Barbier-Brygoo, H., Perrot-Rechenmann, C., Bouvier-Durand, M., and Guern, J. (1994) 'Alterations of auxin-perception in *rolB*-transformed tobacco protoplasts, time course of *rolB* mRNA expression and increase in auxin sensitivity reveal multiple control by auxin', Plant Physiol., in press.

Napier, R.M., and Venis, M.A. (1990) 'Monoclonal antibodies detect an auxin-induced conformational change in the maize auxin-binding protein' Planta 182, 313-318.

Napier, R.M., and Venis, M.A. (1991) 'From auxin-binding protein to plant hormone receptor ?' Trends Biochem. Sci. 16, 72-75.

Nilsson, O., Crozier, A., Schmülling, T., Sandberg, G., and Olsson, O. (1993) 'Indole-3-acetic acid homeostasis in transgenic tobacco plants expressing the *Agrobacterium rhizogenes rolB* gene', Plant J. 3, 681-689.

Palme, K. (1992) 'Molecular analysis of plant signaling elements : relevance of eukaryotic signal transduction models', Int. Rev. Cytol. 132, 223-283.

Rück, A., Palme, K., Venis, M.A., Napier, R.M., and Felle, H. (1993) 'Patch-clamp analysis establishes a role for an auxin-binding protein in the auxin stimulation of plasma membrane current in *Zea mays* protoplasts', Plant J. 4, 41-46.

Santoni, V., Vansuyt, G., and Rossignol, M. (1990) 'Differential auxin sensitivity of proton translocation by plasma membrane H^+-ATPase from tobacco leaves', Plant Sci. 54, 177-184.

Santoni, V., Vansuyt, G., and Rossignol, M. (1991) 'The changing sensitivity to auxin of the plasma membrane H^+-ATPase: Relationship between plant development and ATPase content of membranes', Planta 185, 227-232.

Schmülling, T., Fladung M., Grossmann, K., and Schell, J. (1993) 'Hormonal content and sensitivity of transgenic tobacco and potato plants expressing single *rol* genes of *Agrobacterium rhizogenes*', Plant J. 3, 371-382.

Thiel, G., Blatt, M.R., Fricker, M.D., White, I.R., and Millner, P. (1993) 'Modulation of K^+ channels in *Vicia* stomatal guard cells by peptide homologs to the auxin-binding protein C-terminus', Proc. Natl Acad. Sci. USA 90, 11493-11497.

Van Duijn, B., Ypey, D.L., and Van der Molen, L.G. (1988) 'Electrophysiological properties of *Dictyostelium* derived from membrane potential measurements with microelectrodes', J. Membrane Biol. 106, 123-134.

Venis, M.A., Thomas, E.W., Barbier-Brygoo, H., Ephritikhine, G., and Guern, J. (1990) Impermeant auxin analogues have auxin activity', Planta 182, 232-235.

Venis, M.A., and Napier, R.M. (1994) 'Auxin receptors and auxin-binding proteins', Crit. Rev. Plant Sci. , in press.

Zettl, R., Feldwisch, J., Boland, W., Schell, J., and Palme, K. (1992) '5'-azido-[3,6-H]-1-naphthylphtalamic acid, a photoactivatable probe for naphthylphtalamic acid receptor proteins from higher plants, identification of a 23 kDa protein from maize coleoptile plasma membrane', Proc. Nat. Acad. USA 89, 480-484.

Zettl, R., Schell, J., and Palme, K. (1994) 'Photoaffinity labeling of *Arabidopsis thaliana* plasma membrane vesicles by azido-[7-^{3}H]indole-3-acetic acid, Identification of a glutathione S-transferase', Proc. Nat. Acad. USA 91, 689-693.

PERCEPTION OF THE AUXIN SIGNAL IN CARROT CELL MEMBRANES: ABPs, HORMONE-CONJUGATE PROCESSING AND AUXIN-REGULATED ENZYME ACTIVITIES

F. Filippini, C. Laveder, F. Lo Schiavo and M. Terzi
Department of Biology and C.R.I.B.I., University of Padova
via Trieste 75, 35121 Padova, Italy

ABSTRACT: Most of the auxin-binding capacity of microsomal preparations from carrot and tobacco cell suspensions is lost if the microsomes are washed with KCl. The material removed by salts (MaRS) seems to have a relevant role in auxin perception, since it contains auxin-binding sites, responsible for ABP modulation and immunologically related to the product of the plant oncogene *rolB* (Filippini F. et al., 1994). Preliminary findings indicate the presence, in the MaRS, of glycosidase and IA-aminoacid hydrolase activities. Moreover, MaRS shows a high phosphatase activity which could represent part of the cell membrane phosphorylation-dephosphorylation system.

Introduction

In the sixties the first reports on auxin-binding in plants from Hertel and co-workers (Hertel R. and Leopold A. C., 1963, Hertel R. and Flory R., 1968) opened a new era in the study of the role and the mechanism of action of auxin. Since then, searching for auxin receptors became mainly a search for auxin-binding proteins (ABPs), and in a few cases it led to the identification and characterization of ABPs with a possible receptor function.

Often the conclusions about the physiological role of a novel ABP were controversial, in most cases owing to an incomplete purification. Even the physiological function of ABP1, the well-characterized 22 kDa ABP identified in *Z. mays* (Löbler M. and Klämbt D., 1985) was not completely clarified, nor a precise mechanism of action in the auxin perception-transduction pathway was precisely defined. The data on other ABPs or auxin-binding fractions are still limited: protein structure, cell and tissue localization, biosynthesis, turnover and physiological function are still largely undefined.

One point however has become very clear: both experimental strategy and theoretical approach to plant hormone biology could not be simply traslated from the animal system, as it was intended up to the early '80s (Rubery P. H., 1981).

Since auxins elicit a wider range of cellular responses than animal hormones, it is conceivable that not only their mode of action but also their receptors will have distinctive features. Recent studies on ABPs in maize plasma membranes suggest the existence of several distinct ABPs sharing one or more epitopes (Feldwisch J. et al., 1992). Although it is possible that all auxin responses are mediated by the same primary biochemical events, the studies conducted over the last two decades, with the development of molecular and genetic approaches to the study of hormone action (Klee H. J. and Estelle M., 1991), are more consistent with multiple modes of auxin action (Estelle M., 1992).

M. Terzi et al. (eds.), Current Issues in Plant Molecular and Cellular Biology, 473-479.

Hence, although it is a widely held view that presumed auxin receptors could play a critical role in the integration of various auxin responses (Trewavas A. J., 1982), several groups today look at ABPs not necessarily as to putative auxin receptors but, more generally (and perhaps more correctly), as to proteins involved in auxin perception. Therefore, there is an increasing interest in studying genes and related protein products possibly involved in the regulation of auxin sensitivity, such as the *iaa* and *rol* genes from *Agrobacterium* (Romano C. P. et al., 1991, Spanò L. et al., 1988). These genes alter the morphogenetic programme of the transformed plants by altering the sensitivity to the hormone and, as such, they are likely to provide invaluable tools for studying the mechanisms underlying growth and development in plants (Palme K. and Schell J., 1993).

Recently, a cytokinin-glycosidase activity was reported for the *rolC* protein (Estruch J. J. et al., 1991a), and it was postulated to regulate cytokinin sensitivity of plant cells by increasing intracellular level of "free" cytokinin through a speeded-up cytokinin-glycoside processing. An auxin-glycosidase activity was attributed to *rolB* protein, because of its capability to hydrolyze indoxyl-ß-D-glucoside (plant indican), and it was hypotesized that *rolB* exerts its physiological and developmental alteration by increasing intracellular concentrations of IAA released from its glucoside conjugates (Estruch J. J. et al., 1991b).

Such a hypotesis turned out to be invalid when the product of *rolB* showed to be incapable to hydrolyze the 'true conjugate' IAA-glucose nor to increase the level of free auxin in the transformed plants (Nilsson O. et al., 1993). Therefore, the *rolB* gene effects cannot depend on a stable and non-specific increase in endogenous IAA levels. Moreover, although it was shown that the *rolB* gene activity dramatically increases the sensitivity of tobacco protoplasts to exogenous auxins, as seen by changes in the electrophysiological properties of their plasma membranes (Maurel C. et al., 1991), the auxin sensitivity of germinating P_{35S}-rolB transgenic seedlings, as well as auxin requirement in cells of dividing calli, is similar to that of non-transformed tissues (Schmülling T. et al., 1993).

In recent years some auxin-binding proteins have been identified which possesse enzyme activities such as glycosidase or glutathione-transferase (Brzobohaty B. et al., 1993, Bilang J. et al., 1993, Droog F. N. J. et al., 1993). Particularly, p60 protein of *Z. mays* has been identified by photoaffinity labeling (Campos N. et al., 1992) and is capable to hydrolyze cytokinin-ß-D-glucosides; it is interesting that *rolC* protein has a similar enzyme activity, except for some difference in substrate specificity. In the roots of maize seedlings, p60 was localized to the meristematic cells and may function *in vivo* to supply the developing maize embryo with active cytokinin.

The role of 'bound auxins' and, more generally, of hormone conjugates has been debated (Cohen J. D. and Bandurski R. S., 1982, Palme K. and Schell J., 1993) and more and more evidence has led to the idea that hormone-conjugate turnover can play a crucial role in hormone regulation of plant growth and development. For example, it has been demonstrated that cytokinin controls floral development through the activation of homeotic genes (Estruch J. J. et al., 1993); several genes are auxin-regulated (van der Zaal F. J. et al., 1987, Boot K. J. M. et al., 1993); so it is reasonable to say that activities releasing 'free' hormone from inactive conjugates represent an additional step in the control system.

In a preceding paper on carrot ABPs, we showed that the level of 2,4-D in the medium determines the level of auxin-binding proteins on the membranes of carrot cells grown in suspension, and that the ABP modulation response varies in different physiological conditions or developmental stage (Lo Schiavo F. et al., 1991). We used somatic embryogenesis as a morphogenetic model system to study ABP modulation (Lo Schiavo F. and Filippini F., 1992) and by using developmental mutants we demonstrated that the embryogenicity of a cell line is related to its capability to modulate auxin-binding capacity (Filippini F. et al., 1992).

We recently demonstrated that at least two classes of auxin-binding sites can be distinguished in carrot and tobacco cell membranes, the first one being immunologically related to ABP1 and the second one to RolB (Filippini F. et al., 1994).

The two classes of auxin-binding sites show a different response to auxin induction (Filippini F. et al., in preparation) and can be separated by treating the cell membranes with salts. Our interest was focused on the salt-removable membrane protein fraction, due to its prominent auxin-binding capacity and to its putative role in the control of auxin sensitivity.

Materials and Methods

CELL LINES AND CULTURE CONDITIONS

Carrot wild-type cell line, A^+t_3 was highly embryogenic and derived via subsequential rejuvenation from a dedifferentiated pool of seed-germinated plantlets of *D. carota* L., cultivar S. Valery. Carrot *rolB* -transformed cell line, cl6 was obtained by infecting a carrot disc with *A. rhizogenes* harbouring a plasmid containing the *rolB* gene and its 5' promoter region. Carrot cells were grown in suspension in Gamborg's B5 medium supplemented with 2% sucrose and 0.5 mg/L 2,4-D, 0.25 mg/L 6-BAP, at 24°C under continuous shaking (70 rpm) and sub-cultivation time was 8-10 days.

PREPARATION OF HOMOGENATES, MEMBRANES AND MaRS

Cells were collected at different times from subcultivation and washed twice with sterile distilled water; homogenate and membrane preparation was performed according to Lo Schiavo F. et al., 1991. Protein concentrations were determined by the BioRad assay, following manifacturer's instructions. The preparation of washed membranes and MaRS (*Ma* terial *R* emoved by *S* alts) was performed according to Filippini et al., 1994.

IAA-BINDING ASSAYS

IAA-binding assays were performed according to Lo Schiavo F. et al., 1991.

ENZYMATIC ASSAYS

Membrane or MaRS enzymatic activities were evaluated by incubating samples at 30°C for 2 hrs in 100 mM Na_3-citrate, 5mM $MgCl_2$, 4 mM substrate, in a total volume of 0.4 ml.

The pH of buffers was 5.0 for glycosidase and 6.0 for phosphatase assay. The reactions were stopped by adding 0.6 ml of 2M Na_2-CO_3. The specific activity was evaluated as the amount of *p* -nitrophenol released (1 O.D. at 405 nm = 1.22 mM PNP) per hour per mg protein.

THIN-LAYER CHROMATOGRAPHY (TLC)

The activity on indoxyl-derivatives or IA-aminoacids was detected through TLC; it was performed on Silica-gel 60 glass plates (Merck) with or without fluorescent indicator.

The solvent phase was chloroform/ethyl acetate/ethanol/formic acid 4:4:1:1; the spots were detected by UV-transillumination.

Results

WASHING MEMBRANES WITH SALTS REVEALS TWO CLASSES OF AUXIN-BINDING SITES

If a carrot microsomal preparation is washed with increasing concentrations of KCl, the auxin-binding capacity of the membrane is progressively reduced (Fig 1). Auxin-binding capacity of washed membranes reaches a plateau at approximately 0.15 M KCl.

When carrot washed membranes are treated with anti-ABP1 antibodies, the auxin-binding is almost completely inhibited, whereas the ABC of the protein fraction contained in the material removed by salts (MaRS) is partially inhibited by anti-RolB antibodies, both in the wild-type and in the *rolB* -transformed cell lines (not shown). The thermal stability and pH-optimum of the IAA-binding to untreated or washed membranes are different, and Scatchard analysis reveals differences in affinity and number of sites for the two classes of auxin-binding sites (not shown). Membranes with a different auxin-binding capacity, due to a different auxin induction, when washed show a constant residual auxin-binding capacity (Fig. 2).

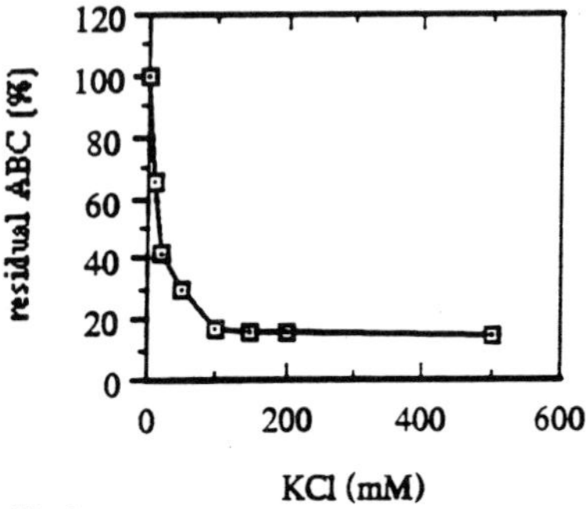

Fig.1
Auxin-binding capacity (ABC) of carrot cell membranes after washing with increasing concentrations of KCl.

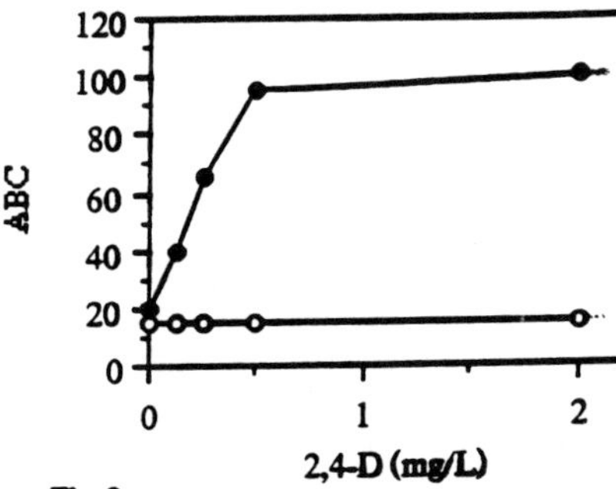

Fig.2
Auxin dependent modulation of auxin-binding capacity (ABC, arbitrary units) of carrot untreated membranes (closed symbols) and washed membranes (open symbols).

THE SALT-REMOVABLE MEMBRANE PROTEIN FRACTION SHOWS ß-D-GLYCOSIDASE, PHOSPHATASE AND IA-AMINOACID HYDROLASE ACTIVITIES

The glycosidase activity of MaRS protein fraction obtained from carrot cells cultivated in the presence or the absence of 2,4-D was tested and compared with the activity shown by washed membranes and cytosolic fraction. Glycosidase specificity was stated by using different substrates: *p* -nitrophenyl-D-glucoside (α and ß), *p* -nitrophenyl-ß-D-galactoside, indoxyl-ß-D-glucoside (plant indican), indoxyl-ß-D-glucuronide, 5-Br,4-Cl-indoxyl-ß-D-glucoside (X-glc), 5-Br,4-Cl-indoxyl-ß-D-galactoside (X-gal). MaRS revealed to hydrolyze only *p* -nitrophenyl- or indoxyl- derivatives with a ß-D-glucose or a ß-D-galactose moiety. Some difference was shown by B5- and B5+ MaRS with respect to the specific activities of glucosidase and galactosidase. The effect of KCl, LiCl, NaCl, $CaCl_2$ and $MgCl_2$ was also tested. The salts were added to the incubation mixture in the range 1-50 mM, and no significant effect was observed except for 1-5 mM $MgCl_2$ that produced a slight increase in MaRS glucosidase activity. The pH-optimum for the glucosidase assay was around pH 5, as it was stated both from spectrophotometric and TLC analysis. MaRS showed a ten-fold increase in the specific ß-D-glucosidase activity with respect to microsomal fraction, and a fifty-fold increase with respect to cytosol.

MaRS proteins derived from membranes of carrot cells grown both in the absence or presence of auxin revealed to be able to hydrolyze plant indican, but their glucosidase activities were slower than those shown by tobacco counterparts. Both in carrot and in tobacco the activity of MaRS on plant indican was increased when UTP and magnesium were added to the incubation mixture. Preliminary data indicate that IAA interferes *in vitro* with the ß-glucosidase activities present in the extrinsic membrane protein fraction, which seem to be inhibited by cytokinin (not shown).

Indoxyl-phosphate and *p* -nitrophenyl-phosphate are processed by carrot and tobacco membranes; in both cases the specific activity is increased in the MaRS fraction. MaRS phosphatases are differentially expressed in cells cultivated in the presence or absence of auxin, as demonstrated by differences in the temperature dependance of the assay and by the use of inhibitors; IAA interferes also *in vitro* with MaRS phosphatases (not shown).

Seven commercially available IA-aminoacids (IAA conjugated with Ala, Asp, Gly, Ile, Leu, Phe and Val) were used to search for MaRS IA-aminoacid hydrolases. Preliminary evidence of minor activities on IA-Ala and IA-Phe awaits confirmation; on the contrary MaRS from cells cultivated in the absence of auxin clearly processes IA-Asp, since free IAA is released, as shown from TLC analysis (not shown).

Discussion

In carrot, the capability of a cell line to undergo somatic embryogenesys is in some way related to its capability to modulate ABP level in response to the hormone signal; moreover, developmental switches determine a change in the sensitivity to auxin, in terms of ABP-modulation response (Lo Schiavo et al., 1991, Filippini et al., 1992). We recently reported that *rolB* -transformed tobacco cell membranes show an increased auxin-binding capacity and that binding sites immunologically related to the *rolB* protein are also present in the extrinsic membrane protein fraction of wild-type cells; the salt resistant auxin-binding sites instead, were inhibited by anti-ABP1 antibodies (Filippini et al., 1994).

Here we report that also in carrot, two classes of binding sites can be identified in cell membranes by washing with salts, and that these sites show properties similar to their tobacco counterparts. It seems that a site immunologically related to maize ABP is tightly associated to membranes both in carrot and in tobacco, and that the auxin-binding capacity of this site (or class of sites) is not modulated by auxin concentration. The salt-removable sites are conversely auxin-modulated, and IAA-binding to such sites is completely (tobacco) or partially (carrot) inhibited by anti-RolB antibodies.

Several unpublished results from reconstitution experiments and displacement analysis allowed us to have an idea of the heterogeneity of the 'modulatable ABPs'; contemporarily our study on *rolB* (in collaboration with the group of P. Costantino in Rome) led us to 'link' the characterization of 'morphogenetic ABPs' with that of 'morphogenetic oncoproteins', and the bridge was constituted by MaRS. We chose this membrane protein fraction as a good starting material for searching activities possibly involved in hormone regulation of developmental processes; particularly, we focused our attention to the characterization of MaRS hormone-regulated enzymes or for activities involved in the processing of auxin and cytokinin conjugates.

Since an indoxyl-glucosidase activity was reported for RolB, the existance in MaRS of binding sites immunologically related to RolB suggested us to search for ß-D-glucosidase activities in MaRS. In fact MaRS showed a several-fold increase in the glucosidase activity with respect to other cell fractions. Some parameters of the assay, such as pH optimum and the Mg^{++}, UTP-dependant increase of the activity, resemble those shown by RolB; conversely auxin-binding or the effect of auxin and cytokinin could be related to p60-like and RolC-like activities. MaRS phosphatase activities seems also IAA-regulated *in vivo* and *in vitro* ; work is in progress toward the characterization of the auxin regulation of the microsomal phosphorylation-dephosphorylation system.

Phytohormone conjugates are abundant in plant tissues (Cohen J. D. and Bandurski R. S., 1982), but their normal biological function remains largely obscure. The apparent physiological activity of any particular conjugate correlates with its rate of hydrolysis in plant tissue. IAA conjugates are slow-release sources of auxin (Hangarter R. P. and Good N. E., 1981); particularly, IA-Asp is the most represented aminoacidic conjugate of IAA. The identification in MaRS, obtained from membranes of cells cultivated in the absence of auxin, of an enzyme capable of releasing free IAA from IA-Asp matches with the idea that several activities involved in the hormone regulation can be identified among a protein fraction reversibly associated to cell membranes.

The characterization of the activities present in the MaRS of membranes obtained from carrot cell cultures, will possibly contribute to shed more light on the characterization of the elements involved in the hormone regulation of plant cell differentiation and plant development.

Acknowledgements

We are indebted to Prof. P. Costantino for *rolB* transformation and to Dr. M. Trovato for anti-RolB antibodies, to Dr. M.Venis for anti-ABP1 antibodies, to Prof. L. Quesada-Allué for help in defining TLC protocols. This work was supported by Biotechnology plan of MRAAF and by MURST 40%.

References

Bilang J., Macdonald H., King P. J. and Sturm A. (1993) 'A soluble auxin-binding protein from *Hyosciamus muticus* is a glutathione S-transferase'. Plant Physiol. 102, 29-34

Boot K. J. M., van der Zaal B. J., Velterop J, Quint A., Mennes A. M., Hooykaas P. J. J. and Libbenga K. R. (1993). 'Further characterization of expression of auxin-induced genes in tobacco (*N. tabacum*) cell-suspension cultures'. Plant Physiol. 102, 513-520

Brzobohaty B., Moore I., Kristoffersen P., Bako L., Campos N., Schell J. and Palme K. (1993). 'Release of active cytokinin by a ß-glucosidase localized to the maize root meristem'. Science 262, 1051-1054

Campos N., Bako L., Feldwisch J., Schell J. and Palme K. (1992) 'A protein from maize labeled with azido-IAA has novel ß-glucosidase activity'. Plant J. 2(5), 675-684

Cohen J. D. and Bandurski R. S. (1982). 'Chemistry and physiology of the bound auxins'. Annu. Rev. Plant Physiol. 33, 403-430

Droog F. N. J., Hooykaas P. J. J., Libbenga K. R. and van der Zaal E. J. (1993). 'Proteins encoded by an auxin-regulated gene family of tobacco share limited but significant homology with glutathione S-transferases and one member indeed shows *in vitro* GST activity'. Plant Mol. Biol. 21, 965-972

Estelle M. (1992) 'The Plant Hormone Auxin: Insight In Sight', BioEssays 14(7), 439-444

Estruch J. J., Chriqui D., Grossmann K., Schell J. and Spena A. (1991a). 'The plant oncogene *rolC* is responsible for the release of cytokinins from glucoside conjugates'. EMBO J. 10, 2889-2895

Estruch J. J., Schell J. and Spena A. (1991b). 'The protein encoded by the *rolB* plant oncogene hydrolyses indole glucosides'. EMBO J. 10, 3125-3128

Estruch J. J., Granell A., Hansen G., Prinsen E., Redig P., Van Onckelen H., Schwarz-Sommer Z. and Spena A. (1993). 'Floral development and expression of floral homeotic genes are influenced by cytokinins'. Plant J. 4(2), 379-384

Feldwisch J., Zettl R., Hesse F., Schell J. and Palme K. (1992). 'An auxin-binding protein is localized to the plasma membrane of maize coleoptile cells: Identification by photoaffinity labeling and purification of a 23-kDa polypeptide'. Proc. Natl. Acad. Sci. USA 89, 475-479

Filippini F., Terzi M., Cozzani F., Vallone D. and Lo Schiavo F. (1992). 'Modulation of auxin-binding proteins in cell suspensions. II. Isolation and initial characterization of carrot cell variants impaired in somatic embryogenesis'. Theor. Appl. Genet. 84, 430-434

Filippini F., Lo Schiavo F., Terzi M., Costantino P. and Trovato M. 'The plant oncogene rolB alters binding of auxin to plant cell membranes' Plant Cell Physiol. (*in press*)

Hangarter R. P. and Good N. E. (1981). 'Evidence that IAA conjugates are slow-release sources of free IAA in plant tissues'. Plant Physiol. 68, 1424-1427

Hertel R. and Leopold A. C. (1963). 'Versuche zur analyse des auxintransports in der koleoptile von *Zea mays* L.'. Planta (Berl.) 59, 535-562

Hertel R. and Flory R. (1968). 'Auxin movement in corn coleoptiles.' Planta (Berl.) 82, 123-144

Klee H. J. and Estelle M. (1991). 'Molecular genetic approaches to plant hormone biology'. Annu. Rev. Plant Physiol. Plant Mol. Biol. 42, 529-551

Löbler M. and Klämbt D. (1985). 'Auxin-binding protein from coleoptiles membranes of corn (Z. *mays* L.). Purification by immunological methods and characterization'. J. Biol. Chem. 260(17), 9848-9853

Lo Schiavo F., Filippini F., Cozzani F., Vallone D. and Terzi M. (1991) 'Modulation of auxin-binding proteins in cell suspensions. I. Differential responses of carrot embryo cultures'. Plant Physiol. 97, 60-64

Lo Schiavo F. and Filippini F. (1992). 'Somatic embryogenesys as developmental system to study modulation of auxin-binding capacity'. in Progress in Plant Growth Regulation, Karssen, van Loon, Vreughdenhil eds., Kluwer Acad. Publ., pp. 202-205

Maurel C., Barbier-Brygoo H., Spena A., Tempé J. and Guern J. (1991). 'Single *rol* genes from the *Agrobacterium rhizogenes* T_L-DNA alter some of the cellular responses to auxin in *Nicotiana tabacum*'. Plant Physiol. 97, 212-216

Nilsson O., Crozier A., Schmülling T., Sandberg G. and Olsson O. (1993). 'Indole-3-acetic acid homeostasis in transgenic tobacco plants expressing the *Agrobacterium rhizogenes rolB* gene'. Plant J. 3(5), 681-689

Palme K. and Schell. J. (1993). 'On plant growth regulators and their metabolites: a changing perspective'. Sem. Cell Biol. 4, 87-92

Romano C. P., Hein M. B. and Klee H. J. (1991). 'Inactivation of auxin in tobacco transformed with the indoleacetic acid-lysine syntetase gene of *Pseudomonas savastanoi* '. Gen. & Dev. 5, 438-446

Rubery P. H. (1981). 'Auxin receptors'. Annu. Rev. Plant Physiol. 32, 569-596

Schmülling T., Fladung M., Grossmann K. and J. Schell (1993). 'Hormonal content and sensitivity of transgenic tobacco and potato plants expressing single *rol* genes of *Agrobacterium rhizogenes* T-DNA'. Plant J. 3(3), 371-382

Spanò L., Mariotti D., Cardarelli M., Branca C. and Costantino P. (1988). 'Morphogenesis and auxin sensitivity of transgenic tobacco with different complements of Ri T-DNA'. Plant Physiol. 87, 479-483

Trewavas A. J. (1982). 'Growth substance sensitivity: the limiting factor in plant development'. Physiol. Plant. 55, 60-72

van der Zaal F. J., Memelink J., Mennes A. M., Quint A. and Libbenga K. R. (1987). 'Auxin-induced mRNA species in tobacco cell cultures'. Plant Mol. Biol. 10, 145-157

ENDOGENOUS LEVELS OF FOUR PLANT HORMONES MAY AFFECT THE CULTURE CONDITIONS OF POPLAR PROTOPLASTS TO REGENERATE PLANTS

H. SASAMOTO[1], Y. HOSOI[1] & M. KOSHIOKA[2]

1. For. & Forest Products Res. Inst., POB 16, Tsukuba, Ibaraki, 305 Japan;
2. Nat. Res. Inst. Veg. Orn. Tea, Ano, Mie, 514-23 Japan.

ABSTRACT. Capability of protoplasts isolated from three kinds of cells of poplar (Populus alba) ,leaf cells of 10-year-old shoot culture; suspension cells of two stages derived from a seed, to regenerate plants was the same as the capability of original cells for plant regeneration after the culture with auxin (2,4-D, NAA) and cytokinin (BAP, 4-PU). Endogenous levels of four plant hormones; IAA, cytokinins, gibberellins(GA) and abscisic acid(ABA), were measured in the protoplasts. After extraction and purification steps, ELISA method was used for IAA, cytokinins (anti-ZR, anti-DHZR, anti-IPA), and ABA. GA were assayed with modified micro-drop method (dwarf rice seedlings, Tanginbozu). Little difference between capable and incapable cells was found in the levels of auxin and cytokinins. However, a marked(ten times) difference was found in the content of GA and ABA among them. Exogenous application of uniconazole-P and ABA had the reverse effect on the cell division activity of capable and incapable protoplasts. Thus, endogenous levels of hormones may have very important role for the establishment of the culture system of protoplasts to regenerate plants.

1. INTRODUCTION

Establishment of protoplast to plant regeneration system is prerequisite to the genetic engineering of woody plant, through somatic cell fusion or transformation by electroporation or microinjection. We have recently succeeded in formation of callus and regeneration of plants from protoplast cultures of several species, by combinations of optimal hormonal concentrations of auxins and cytokinins (1-4). In addition, cell fusion(2) and microinjection(3) methods were applied to Populus species. However, improvement of the regeneration capacity of recalcitrant protoplasts, still remains to be investigated. In this report, we investigated the relationship between exogenous hormonal condition and endogenous levels of hormones in protoplasts for regeneration of Populus species.

M. Terzi et al. (eds.), Current Issues in Plant Molecular and Cellular Biology, 481-486.

2. MATERIALS AND METHODS

2.1. PROTOPLAST CULTURE OF LEAF CELLS AND REGENERATION OF PLANTS

Protoplasts were isolated from leaves of shoot culture of Populus alba, subcultured for more than 10 years, which was derived from callus culture induced from mature tree, using 1% Cellulase RS and 0.25% Pectolyase Y-23 in 0.6M mannitol solution for 1-2 hrs of static incubation of cut leaves at room temperature as described earlier(4). After filtration through a 40µ nylon mesh, protoplasts were washed by centrifugation at 100g with 0.6M mannitol solution. 10-30 µl of concentrated protoplast suspension (80-90% of viability) was put into a 0.3 ml of medium in a 24-well or 96-well culture flask to make cell density of $2\text{-}7\times10^4$/ml. They were cultured at 27°C in a humidified CO_2-incubator without supply of CO_2. Liquid ammonium nitrate-free MS medium containing 0-100µM of 2,4-D and 0-10µM of BAP or N-(2-chloro-4-pyridyl)-N'-phenylurea(4-PU)(5) was used.

Proliferated colonies were transferred to MS agar medium containing 1µM of 2,4-D and 0.1µM of BAP without mannitol. Then, proliferated callus was transferred to MS agar medium with or without 1µM of NAA and/or 0.1, 1, 10µM each of cytokinins; 4-PU, Thidiazuron(TDA), Zeatin, BAP and 2-iP.

Differentiated shoots were excised and planted in the MS medium containing 4µM of IBA for rooting.

2.2. INDUCTION AND SUBCULTURE OF SUSPENSION CELLS FROM A SEED AND REGENERATION OF PLANTS

Seeds were stored at -30°C under desiccation with silica gel until use. They were sterilized with 0.5% NaClO solution for 10min. Each seed was put into the 0.5ml of liquid medium in a 10ml flat-bottomed tube, and cultured in the dark on a shaker at 100 rpm. MS medium containing 1µM of 2,4-D and 0.1 µM of BAP was used.

Proliferated cells were transferred to the same medium as above in a 100 ml Erlenmeyer flask, and subcultured at an interval of 2 weeks.

Suspension cells were washed with hormone-free MS medium and transferred to the medium described in 2.1, or to medium with four combinations of hormones such as 0.1 and 1 µM of TDA or 4-PU with and without 1 µM of NAA.

2.3. PROTOPLAST CULTURE OF SEED-DERIVED SUSPENSION CELLS

Suspension cells were washed with 0.6M mannitol solution, and put into the enzyme solution containing 1% each of Cellulase RS and Driselase, and 0.25% of Pectolyase Y-23. After 5-6 hrs incubation, isolated protoplasts were treated with the same way as described above in the case of leaf protoplasts, except for MS instead of ammonium nitrate-free MS as the basic medium.

Proliferated colonies cultured in the medium containing 10µM of 2,4-D and 1 µM of BAP were transferred to the liquid medium of same composition but not including mannitol.

Cells were transferred to the differentiation medium, the same composition as for original suspension cells. Differentiated shoots were cut and planted in the MS medium containing 4μM of IBA.

2.4. MEASUREMENT OF ENDOGENOUS LEVELS OF PLANT HORMONES

2.4.1. Extraction and purification of hormones. Protoplast pellet after centrifugation in a 1.4 ml glass tube was quickly frozen with liquid nitrogen and stored until use in a -70°C freezer.

IAA was extracted from protoplasts with iso-Propanol-imidazole buffer solution for overnight extraction. After addition of acidic H_2O, partitioning against Methylene chloride was done (T. Hayashi, personal communication). The acidic $MeCl_2$ fraction was concentrated in vacuo using a centrifugal evaporator and used for TLC. Cytokinin and ABA were extracted with 80% EtOH, and after partitioning, acidic $MeCl_2$ fr. was used for TLC of ABA, and aq. phase was further partitioned against water-sat.-n-BuOH, which was used for TLC of cytokinins. Gibberellins were extracted with 80% MeOH, and after partitioning, acidic ethylacetate fr. was used for TLC. Otherwise acidic ethylacetate fr. was further purified by partitioning against phosphate buffer, 0.1M, pH 7.0 or 0.5M, pH8.0.

2.4.2. Thin layer chromatography. Avicel or Funacel(Cellulose) TLC with a solvent system (iso-Propanol/28%ammonia/water=10:1:1) was done for IAA and gibberellins and ABA. After cutting out the each plate according to Rf values of hormones, they were eluted with 80% MeOH or EtOH. Silica gel TLC was used for cytokinins, with a solvent system of Chloroform:MeOH(4:1).

2.4.3. Measurement of hormones. Using the commercial ELISA assay kits, contents of cytokinins; Zeatin(-riboside); Dihydrozeatin(-riboside); IPA, and ABA were measured at least three different dilution samples.

IAA content was measured by ELISA method using a monoclonal antibody to IAA (6).

Gibberellin activity, eluted with 50% acetone, was measured by a modified microdrop bioassay method using dwarf rice (Tan-ginbozu) seedlings(7).

3. RESULTS AND DISCUSSIONS

3.1. CELL DIVISIONS OF PROTOPLASTS WITH EXOGENOUS AUXINS AND CYTOKININS

The combination of 1μM of 2,4-D and 0.1μM of BAP was most effective for cell divisions of leaf protoplasts. 10μM of 2,4-D was also effective. 4-PU was also effective instead of BAP, and lower concentrations such as 0.001 and 0.01μM. Optimal cell density was 2-5 x 10^4/ml.

In the case of protoplasts of suspension cells subcultured within 6 months (GSN). The best combination was 10 μM of 2,4-D and 1 μM of BAP. 10 to 30 μM of 2,4-D was also effective, but not 1μM. 0.1 μM of BAP was sometimes effective for colonies formation. Optimal cell density was very low compared with the cell density for leaf protoplasts. Sometimes, similar cell density to of leaf protoplast was effective for the protoplasts from 2 months subcultured suspension cells (GSN-2), which differentiated in broad range of hormonal conditions (3.2.).

However, protoplasts of suspension cells subcultured more than 1 year (GSO) had no colony in any hormonal condition. These cells enlarged at the presence of 10-30 µM 2,4-D, however, no cell division occurred at any hormonal condition tested.

Table 1. Hormonal condition and cell density for colony formation in protoplast culture

	2,4-D(µM)	BAP(µM)	cell density
leaf	1-10	0.1	$1\text{-}7\times10^4$/ml
GSN-2*	10-30	0.1-1	$5\times10^3\text{-}4\times10^4$/ml
GSN-3	10-30	0.1-10	$2\times10^3\text{-}10^4$/ml
GSN-4	10	1	7×10^3/ml
GSN-6	10-30	1	10^4/ml
GSO-15	no	no	
GSO-17	no	no	

Suspension culture was induced also from petiole of 3-year old tree, which was derived from a leaf protoplast culture from shoot culture. These suspension cells and their protoplasts behaved in the same way as the seed derived suspension cells.

3.2. DIFFERENTIATION WITH AUXIN AND CYTOKININ

Leaf protoplast- derived callus, and suspension cells subcultured within 6 months(GSN) differentiated shoots on the medium containing 1µM of NAA and 0.1 to 1µM of 4-PU. And suspension cells(GSN)-protoplasts-derived callus also differentiated shoots.

Table 2. Differentiation of shoots from original suspension cells, and their protoplasts derived cells in the medium containing 1µM of NAA and 0.1-1µM of 4-PU

	original cells	protoplast derived callus
Leaf	/	+
GSN-2	+	/
GSN-3	+	+
GSN-4	+	+
GSO-15	-	/
GS0-17	-	/

+:shoots, -:no shoot /:no experiment

TDA worked in the same way as 4-PU. However, the hormonal combination of 1µM of NAA plus 10µM of zeatin, which was effective for the differentiation of leaf protoplast-derived callus, was not effective for suspension cell-protoplast derived callus. Similar reduction of differentiation acitivity was also observed when leaf protoplasts were treated by electric cell fusion or electroporation. All kind of shoots differentiated roots in the medium containing IBA.

However, no differentiation occurred with suspension cells subcultured for more than 1 year(GSO) in any condition tested.

Capability of protoplasts for cell divisions and capability of original suspension cells and protoplast-derived callus for differentiation of shoots was almost the same each other.

3.3. ENDOGENOUS LEVELS OF HORMONES

Recovery of each authentic hormone ranged from 50%(IAA, t-Zeatin), 80%(ABA) to 100%(GA_3) at most.

Table 3. endogenous levels of hormones in protoplasts (mean value pmoles eq. / $6x10^6$cells)

	IAA	Cytokinin(IPA)	Gibberellins(GA_3eq.)	ABA
Leaf	21	3.0	8.4	0.89
GSN	35	4.7	12	24
GSO	16	2.1	85	4.9

Levels of IAA or cytokinin did not differ much between three kind of protoplast. Within three kind of cytokinins tested, higher content of IPA was measured compared to that of other cytokinins which reacts to other antibodies.

Content of gibberellins differed much between leaf protoplast and suspension cell-protoplast. And it was higher in old suspension cell(GSO)-protoplast than capable suspension cell(GSN)-protoplast. Activities at 3 different Rf value were found, however, no drastic difference was found between three kind of materials.

ABA content was higher in suspension cell-protoplast than leaf protoplast, and capable suspension cell-protoplast is higher than in old incapable suspension cell-protoplasts.

Such data suggested the possibility of the improvement of cell division activity in incapable protoplasts using the growth regulators other than auxin or cytokinins.

3.4. EFFECTS OF EXOGENOUS ABSCISIC ACID AND UNICONAZOLE-P ON CELL ENLARGEMENT AND CELL DIVISIONS

A little increase of colony formation of leaf protoplasts was obtained by 0.1 and 1 µM of uniconazole-P. However, cell enlargement was inhibited in suspension cell(GSO)-protoplast by it as shown in Fig 1. On the other hand, GA_3 inhibited the cell division of leaf protoplast but not the initial cell enlargement. ABA affected inhibitory on the colony formation or cell enlargement of leaf protoplast. On the contrary, ABA increased the cell enlargement of suspension cell-protoplast. There is a possibility of colony formation from suspension cell-protoplast, which was subcultured more than 1 year and incapable of colony formation in the survey of conditions using auxins and cytokinins.

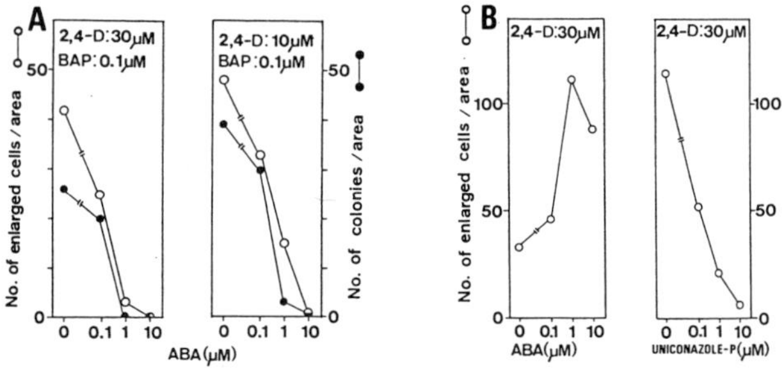

Figure 1. Effects of ABA and uniconazole-P on cell enlargement and colony formation of leaf protoplasts(A) and suspension cells-protoplasts(B). Cell densities were $2x10^4$/ml(A) and 10^5/ml(B).

4. REFERENCES

(1). Sasamoto, H., Kondo, A., Hosoi, Y., Maki, H. and Odani, K.(1992) 'Callus regeneration from cotyledon protoplasts of Chamaecyparis obtusa (Hinoki cypress), In Vitro Cell Div. Biol. 28P, 132-136.

(2). Hosoi, Y. and Sasamoto, H. (1993) 'Plantlet regeneration from electro-fused poplar protoplasts selected by a micromanipulator', Abst. 15th Int. Bot. Congr., p546.

(3). Hosoi, Y. and Sasamoto, H. (in press) 'Direct gene transfer mediated by microinjection in poplar single cell', Abst. 8th Intl. Congr. Plant Tissue & Cell Culture,

(4). Sasamoto, H., Hosoi, Y., Ishii, K., Sato, T., and Saito, A.(1989) 'Factors affecting the formation of callus from leaf protoplasts of Populus alba', J. Jpn. For. Soc. 71,449-455.

(5). Takahashi, S., Sudo, K., Okamoto, T., Yamada, K., and Isogai, Y.(1978)'Cytokinin activity of N-phenyl-N'-(4-pyridyl)urea derivatives', Phytochemistry, 17, 1201-1207.

(6). Matsumoto, R., Yamamoto, M., and Okudai, N. (1989) 'Monoclonal antibodies to indole-3-acetic acid against IAA conjugated to BSA on the indolic rings of IAA', J. Jpn. Soc. Hort. Sci.,89(2),122-123.

(7). Nishijima, T., Koshioka, M., and Yamazaki, H. (1994) 'Use of several gibberellin biosynthesis inhibitors in sensitized rice seedling bioassays', Biosci. Biotech. Biochem., 58, 572-573.

AUXIN AND ETHYLENE IN HAIRY ROOT CULTURES OF *HYOSCYAMUS MUTICUS*

S. BIONDI, M. MENGOLI and N. BAGNI
Department of Biology
University of Bologna
via Irnerio, 42
40126 Bologna
Italy

Treatment of normal and *Agrobacterium rhizogenes*-transformed root cultures of *Hyoscyamus muticus* with three different auxins (IAA, IBA, NAA) revealed that the response varied considerably between auxins, between transformed and normal roots and depending on the parameter examined (length of primary root, number and length of lateral roots). While IBA reduced primary root elongation only at the highest concentration tested, the other two auxins were strongly inhibitory at all concentrations in both root lines. All three auxins promoted branching in normal roots (except 2.5 μM IAA) but not in transformed roots where they were generally without effect (IBA) or inhibitory. In normal roots, which produced less ethylene than transformed roots, the auxin stimulation of ethylene formation was more pronounced than in transformed roots. The ethylene precursor ACC and its conjugated form MACC were undetectable in both root lines but accumulated considerably when NAA was supplied. The ethylene synthesis inhibitor AVG had, at high concentrations, a drastic effect on root growth and development while silver thiosulphate, an ethylene antagonist, did not. The latter compound, however, resulted in no or very limited root hair formation in transformed and normal roots, respectively.

1. INTRODUCTION

Agrobacterium rhizogenes is a soil bacterium which induces, mostly in dicotyledonous plants, the formation of adventitious roots (known as hairy roots) at the site of infection. Root formation is due to the transfer and expression of genes located in a DNA fragment (TL-DNA) of the Ri plasmid delivered by the bacteria into the host cells during infection. Hairy roots, which are composed of genetically transformed cells (Chilton et al., 1982), can be isolated and grown in axenic culture in the absence of exogenous hormones where they display rapid growth, prolific branching and abundant root hairs. Among the transferred genes, the products of the open reading frames 10, 11 and 12, denoted *rol*A, *rol*B and *rol*C respectively (White et al., 1985), were shown to induce morphogenic events: the formation of the hairy roots themselves and morphological alterations in the transgenic plants regenerated from them. These loci act independently and synergistically (Spena et al., 1987). The *rol*B gene appears to play a key role in that it is able to induce root formation even when expressed singly. It has been reported that the *rol*B product is an indoxyl-β-glucosidase (Estruch et al., 1991a); perhaps this enzyme may also hydrolyze inactive IAA-β-glucoside conjugates to give free IAA. As for *rol*C, the gene product is a cytokinin-β-

M. Terzi et al. (eds.), Current Issues in Plant Molecular and Cellular Biology, 487-492.

glucosidase (Estruch et al., 1991b) which alters endogenous levels of the cytokinin iPA in transgenic plants (Nilsson et al., 1993b). Alternatively, the biological effects of *rol* gene activity might be due to an altered hormonal sensitivity of transgenic cells/tissues (Spanò et al., 1988; Maurel et al., 1991).

In addition to the alterations in hormonal content and sensitivity reported to date, other direct or indirect effects of the *rol* gene products might influence the physiology and morphology of transformed tissues or plants. For instance, different endogenous polyamine levels have been observed in normal and transformed tobacco plants (Mengoli et al., 1992) and roots of *Hyoscyamus muticus* (Biondi et al., 1993). In an attempt to further characterize the hormonal status and response of *H. muticus* hairy root cultures, we investigated ethylene production, the effects of inhibitors of ethylene synthesis or action and the response to several auxins supplied in a wide range of concentrations in both normal and transformed root cultures.

2. MATERIALS AND METHODS

2.1 ESTABLISHMENT OF ROOT CULTURES

Transformed hairy root clones of *Hyoscyamus muticus* L., initiated by inoculating 3 week-old seedlings with *Agrobacterium rhizogenes* (strain A4), were kindly supplied by Prof. Hector Flores. Untransformed root cultures were established from excised root tips of sterile seedlings grown *in vitro* on basal White's (White, 1938) medium. Both transformed (T) and normal (N) root cultures were maintained in Petri dishes, kept in darkness at 23°C, on the same hormone-free B5 (Gamborg et al., 1968) solid medium and subcultured every 2-3 weeks.

2.2 EFFECTS OF CHEMICAL TREATMENTS ON GROWTH

The compounds to be tested (the ethylene inhibitors silver thiosulphate, STS and aminoethoxyvinylglycine, AVG; the auxins indoleacetic acid, IAA, indolebutyric acid, IBA and naphthaleneacetic acid, NAA) were filter-sterilized and added to previously autoclaved medium. Growth was monitored, approximately 10 days after subculture, by measuring primary root length and number and length of lateral roots.

2.3 INCORPORATION OF LABELLED METHIONINE

Measurement of ethylene synthesis by incorporation of 3,4-[^{14}C]-methionine was based on the procedure previously described by Biondi et al. (1990). Root explants (approx. 1 g) were transferred to 100-ml conical flasks containing 5 ml liquid B5 medium. Where indicated 100 μM NAA and/or 100 μM $CoCl_2$ or 20 μM salicylic acid were also supplied. Flasks were left standing overnight and then 37 kBq of 3,4-[^{14}C]-methionine (Dositek, France, 2.18 TBq mol^{-1}) were added to the medium. An ethylene trap (0.25 M mercuric perchlorate in a small well) and a CO_2 trap (a filter paper disc imbibed with 1 N KOH) were set in place and the flasks sealed with rubber stoppers. After 18 h of incubation at room temperature, the radioactive ethylene captured by the mercury was measured in a scintillation counter.

2.4 ETHYLENE DETERMINATION BY GAS CHROMATOGRAPHY

After aerating the Petri dishes for ca. 30 min, approximately 1 g of roots were transferred to 11-ml test tubes containing 250 μl distilled water. After sealing the tubes with rubber caps, 1 ml gas samples were withdrawn at 1-h intervals with a hypodermic syringe and injected into a gas chromatograph equipped with an activated alumina column and a flame ionization detector.

2.5 DETERMINATION OF ACC AND MACC

Root samples were frozen in liquid nitrogen and ground in a mortar on ice. Extraction was carried out with 80% (v/v) ethanol. After centrifugation for 15 min at 1500 g, the supernatant was concentrated under vacuum at 45°C and then resuspended with 2 ml water and 0.5 ml chloroform. After a brief centrifugation, an aliquot of the aqueous phase was used to determine ACC using the method of Lizada and Yang (1979). MalonylACC (MACC) was assayed in the same way after hydrolyzing an aliquot of the aqueous phase with 7.2 N HCl at 100°C for 3 h.

3. RESULTS

3.1 AUXIN EFFECTS ON GROWTH

In general terms the following effects were observed. In N roots, which generally display rapid elongation growth but no or sporadic branching, auxins, as expected, led to abundant lateral root (LR) formation and growth. From this point of view, IBA 0.5 and 2 μM were the most effective treatments, while 5 μM IBA produced an anomalous morphology (many short LRs with abundant root hairs and some very long and thin ones lacking hairs; some callusing). The latter morphology resulted from treatment with IAA at all 3 concentrations. NAA, from 0.01 to 1 μM also induced many LRs which grew normally, the response being proportional to the amount given. On the other hand, 10 μM was inhibitory. In T roots already at 0.5 μM both IAA and IBA resulted in stunted growth and some callusing, while NAA was inhibitory at 1 μM and above.

In another experiment where 0.5, 1 or 2.5 μM IAA, IBA or NAA were supplied to N and T roots, the following results were obtained:

a) In both N and T roots elongation growth of the primary root was strongly inhibited by IAA and NAA at all 3 concentrations. IBA reduced growth much less and indeed in N roots 0.5 μM stimulated and 1 μM had no effect on growth.

b) As mentioned above, numerous LRs appeared in auxin-treated N roots. IBA gave the highest number of LRs at the highest concentration tested, while NAA and especially IAA (which blocked LR formation totally) were inhibitory at this concentration. In T roots only IAA was inhibitory at 1 μM while NAA and, to a lesser extent IBA, significantly reduced branching only at the highest concentration.

c) The average length of LRs displayed a somewhat different trend. In N roots treated with IAA, LR growth exhibited a distinct optimum at 0.5 μM. The effect of NAA and IBA, on the other hand, differed only slightly between auxins and between the various concentrations tested. In T roots LR length became progressively shorter with increasing hormone concentration without major differences between auxins.

3.2 ETHYLENE, ACC AND MACC LEVELS

As revealed by gas chromatography, T roots produced more ethylene than N roots (Fig. 1). This ethylene production was enhanced in response to exogenously supplied ACC, the immediate precursor of ethylene, suggesting that this compound is limiting and that ACC oxidase, the enzyme which converts ACC to ethylene, is present and active.

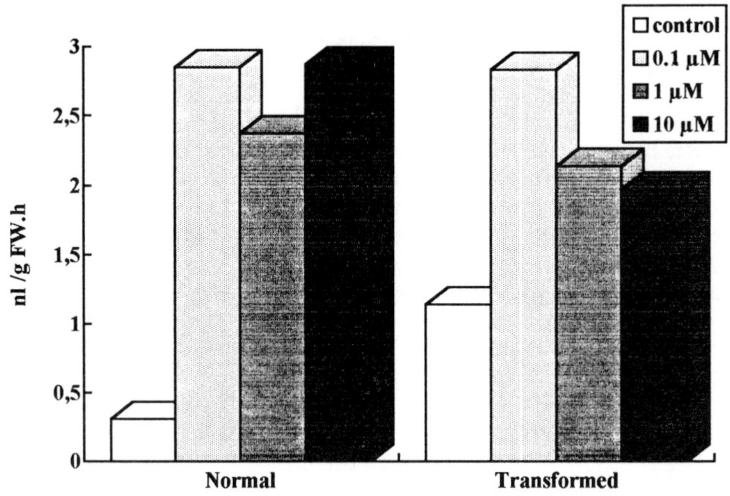

Fig. 1. Ethylene production in normal and transformed root cultures of *H. muticus* in the presence or absence of NAA.

It is well known that auxins stimulate ethylene production. Indeed, 0.1 µM NAA enhanced ethylene formation in both N and T roots by 9.5- and 2.5-fold, respectively (Fig. 1). This stimulation remained more or less constant up to 10 µM NAA in N roots but tended to decrease slightly in T roots at the higher concentrations. The stimulating effect of auxin on ethylene production in T roots was confirmed when ethylene synthesis was measured by labelled methionine incorporation.

That ACC is present in limited amounts was confirmed when endogenous levels of this compound were determined in T roots. In fact, both ACC and its conjugate MACC were not detectable in control roots. In the presence of NAA, ACC and especially MACC accumulated significantly. A similar effect, however, was not observed when either cobalt or salicylic acid, both inhibitors of ACC oxidase, were supplied unless NAA was also present. It is noteworthy that the MACC to ACC ratio was rather high (4:1).

3.3 EFFECT OF ETHYLENE INHIBITORS ON GROWTH

The ethylene synthesis inhibitor AVG and the ethylene antagonist STS both affected growth and development of hairy roots. At high concentrations (100 µM) AVG drastically inhibited the growth of both primary and lateral roots and formation of the latter. At the same concentration STS did not significantly alter any of these parameters but did provoke a very conspicuous absence of root hairs. When examined under the microscope, the absence of root hairs was confirmed in T roots, whereas N roots in fact exhibited a large number of very short root hairs.

Silver salts, in particular silver nitrate but to some extent also STS, have in some cases been shown to stimulate ethylene synthesis (Liu et al., 1990). We checked this possibility and found no differences between control and STS-treated roots.

4. CONCLUSIONS

Schmulling and coworkers (1993) reported that the expression of single *rol* genes of the TL-DNA of *A. rhizogenes* strain A4 alters the endogenous concentrations of, and the sensitivity to, several plant hormones in transgenic plants. In particular, the overexpression of *rol*C led to an altered response to all the main hormones (auxins, cytokinins, abscisic acid, GA_3) as well as to ACC. The overexpression of *rol*B in transformed tobacco calli led to necrosis at lower auxin concentrations than in the wild-type. However, other parameters of auxin action remained unchanged. This suggests that this sort of analysis must take into consideration not only each class of hormones but also different parameters of growth and development which respond to hormonal action.

This is also evident from the data presented here: primary and lateral root elongation, branching and root hair formation can all be influenced in different ways by hormone treatment. A further complication arises due to the different responses obtained using different auxins. The effects observed were most pronounced with IAA followed by NAA and IBA in that order. What emerges clearly, however, is that T roots are, on the whole, negatively affected by the presence of externally supplied auxins. This probably reflects their greater sensitivity to, rather than an overproduction of, this hormone as the latter phenomenon was excluded by the findings of Nilsson et al. (1993a) in tobacco plants expressing *rol*B.

Ethylene production under standard conditions and its stimulation by exogenous auxin are two other parameters which distinguish T roots from their normal counterparts. Again, the fact that it is higher in T roots and that the auxin stimulation begins to decline at lower concentrations than for N roots seem to support the contention that T roots are more sensitive to (endogenous and exogenous) auxin.

It is noteworthy that the only evident morphological response to the ethylene antagonist in T roots is the absence of hairs. The fact that the absence of root hairs in STS-treated T roots is complete whereas in N roots they are present but blocked at their intial stages of growth is another indication of their different sensitivity.

Acknowledgements - This research was supported by funds from the Italian Ministero della Pubblica Istruzione (to N.B.).

REFERENCES

Biondi, S., Diaz, T., Iglesias I., Gamberini G. and Bagni, N. (1990) 'Polyamines and ethylene in relation to adventitious root formation in *Prunus avium* shoot cultures', Physiol. Plant. 78, 475-483.

Biondi, S., Mengoli, M., Mott, D. and Bagni, N. (1993) 'Hairy root cultures of *Hyoscyamus muticus*: effect of polyamine biosynthesis inhibitors', Plant Physiol. Biochem. 31, 51-58.

Chilton, M.-D., Tepfer, D.A., Petit, A., David, C., Casse-Delbart, F. and Tempé, J. (1982) '*Agrobacterium rhizogenes* inserts a T-DNA into the genomes of the host plant root cells', Nature 295, 432-434.

Estruch, J.J., Chriqui, D., Grossmann, K., Schell, J. and Spena, A. (1991) 'The plant oncogene *rol*C is responsible for the release of cytokinins from glucoside conjugates', EMBO J. 10, 2889-2895.

Estruch, J.J., Schell, J. and Spena, A. (1991) 'The protein encoded by the *rol*B plant oncogene hydrolyses indole glucosides', EMBO J. 10, 3125-3128.

Gamborg, O.L., Miller, R.A. and Ojima, K. (1968) 'Nutrient requirements of suspension cultures of soybean root cells', Exp. Cell Res. 50, 151-158.

Liu, J., Mukherjee, I. and Reid D.M. (1990) 'Adventitious rooting in hypocotyls of sunflower (*Helianthus annuus*) seedlings. III. The role of ethylene', Physiol. Plant. 78, 268-276.

Maurel, C., Barbier-Brygoo, H., Spena, A., Tempé, J. and Guern, J. (1991) 'Single *rol* genes from the *Agrobacterium rhizogenes* TL-DNA alter some of the cellular responses to auxin in *Nicotiana tabacum*', Plant Physiol. 97, 212-216.

Mengoli, M., Chriqui, D. and Bagni, N. (1992) 'Protein, free amino acid and polyamine contents during development of hairy root *Nicotiana tabacum* plants', J. Plant Physiol. 139, 697-702.

Nilsson, O., Crozier, A., Schmulling, T., Sandberg, G. and Olsson, O. (1993a) 'Indole-3-acetic acid homeostasis in transgenic tobacco plants expressing the *Agrobacterium rhizogenes rol*B gene', Plant J. 3, 681-689.

Nilsson, O., Moritz, T., Imbault, N., Sandberg, G. and Olsson, O. (1993b) 'Hormonal characterization of transgenic tobacco plants expressing the *rol*C gene of *Agrobacterium rhizogenes* TL-DNA', Plant Physiol. 102, 363-371.

Schmulling, T., Fladung, M., Grossmann, K. and Schell, J. (1993) 'Hormonal content and sensitivity of transgenic tobacco and potato plants expressing single *rol* genes of *Agrobacterium rhizogenes* T-DNA', Plant J. 3, 371-382.

Spanò, L., Mariotti, D., Cardarelli, M., Branca, C. and Costantino, P. (1988) 'Morphogenesis and auxin sensitivity of transgenic tobacco with different complements of Ri T-DNA', Plant Physiol. 87, 479-483.

Spena, A., Schmulling, T., Koncz, C. and Schell, J. (1987) 'Independent and synergistic activity of *rol*A, B and C loci in stimulating abnormal growth in plants ', EMBO J. 6, 3891-3899.

White, F.F., Taylor, B.H., Huffman, G.A., Gordon M.P. and Nester, E.W. (1985) 'Molecular and genetic analysis of the transferred DNA regions of the root-inducing plasmid of *Agrobacterium rhizogenes*', J. Bacteriol. 164, 33-44.

White, P.R. (1938) 'Cultivation of excised roots of dicotyledonous plants', Am. J. Bot. 25, 348-356.

EFFECTS OF PHYTOHORMONES ON LATERAL ROOT DIFFERENTIATION IN *EUCALYPTUS GLOBULUS* AND TRANSGENIC *NICOTIANA TABACUM* SEEDLINGS

A. Pelosi[1] E.K.F. Chow[1], M.C.S. Lee[1], S.F. Chandler[2] and J.D. Hamill[1]
[1]*Department of Genetics and Developmental Biology*
Monash University
Clayton, Melbourne, Victoria 3168 Australia
[2]*Calgene Pacific Pty Ltd*
16 Gipps Street
Collingwood, Melbourne, Victoria 3066 Australia

ABSTRACT. Auxin induced differentiation of Lateral Root Primordia (LRP) was studied in roots of *Eucalyptus globulus* and transgenic *Nicotiana tabacum* seedlings. In *E. globulus*, IBA was the most effective auxin for LRP induction. IBA at 10^{-5}M increased LRP density 5-10 fold compared to non-auxin treated controls, with LRP differentiation being completed within 48hr when explants were cultured at 20^0C. Transfer of roots from auxin medium to hormone-free medium resulted in conversion of the majority of LRP to lateral roots (LR) after a further 7 days of culture. Pretreatment of tissues with IBA for 16 hr followed by transfer to medium containing BAP showed that conversion of LRP to LR was inhibited by concentrations of 10^{-5}M or greater, but stimulated at low concentrations of BAP ($10^{-9}10^{-11}$M). IBA induction of LRP was inhibited when actinomycin-D or cycloheximide (both at 10^{-3}M) were supplied prior to or within the first 48 hr of IBA treatment suggesting that *de novo* transcription and translation of essential gene(s) are required for LRP formation following exposure to IBA. LRP induction was also studied in roots of *N. tabacum* seedlings transformed with individual *rol* genes from *Agrobacterium rhizogenes*. After 72 hr exposure to IBA at different concentrations, roots of *rol*A seedlings showed a slightly reduced sensitivity to IBA with respect to LRP induction compared to control seedlings. On the other hand seedlings containing the *rol*C or *rol*B genes showed an increase in sensitivity to IBA with respect to LRP induction.

1. Introduction

The formation of adventitious root primordia and their subsequent development into actively growing lateral roots is essential for the successful clonal propagation of elite genotypes of many species, including those in the genus *Eucalyptus*.

Auxins are known to have a major role in the formation of lateral roots (Thimann 1936; Torrey 1950, 1956; Wightman *et al* 1980). Cytokinins, when supplied with auxin, will inhibit root formation at high concentrations (~10^{-5}M) whereas lower levels of cytokinin (10^{-8}M) can show a slight stimulatory effect on LR formation (Wightman *et al* 1980; Biddington and Dearman 1982; MacIsaac *et al* 1989).

It has previously been shown that differentiation of LR following auxin induction involves two stages in lettuce seedlings (MacIsaac *et al* 1989). The first stage requires a period of

M. Terzi et al. (eds.), Current Issues in Plant Molecular and Cellular Biology, 493-499.

less than 24 hours in contact with high auxin levels to induce cell division in the pericycle. The second stage occurs independently of auxin, with continued cell division to form LRP.

An increase in the concentration of specific soluble proteins has been observed during root formation following auxin induction (Kantharaj *et al* 1985; Dhindsa *et al* 1987; MacIsaac and Sawhney 1990). MacIsaac and Sawhney (1990) showed that the synthesis of specific polypeptides in roots of auxin-treated lettuce seedlings could be inhibited by either kinetin or the translation inhibitor cycloheximide. The transcription inhibitor actinomycin-D inhibited LRP formation in hypocotyls of *Phaseolus vulgaris* (Kantharaj *et al*, 1985) if applied within the first 40 hours after IBA induction.

We have investigated the interacting effects of auxin and cytokinin on LRP and LR formation in *Eucalyptus globulus* seedlings. These results should facilitate further studies aimed at identifying molecular events associated with LR formation in this species.

We have also undertaken studies into the auxin sensitivity of roots of *N. tabacum* containing *rol* genes from *Agrobacterium rhizogenes*. The transformation of plant tissue with *A. rhizogenes* Ri T-DNA genes has been shown to increase the efficiency of root differentiation in a number of species with three T_L-DNA genes, *rol*A, *rol*B and *rol*C being particularly important (Spena *et al* 1987; Schmülling *et al* 1988). Although the biochemical function of the genes remains unclear, there is evidence that expression of these *rol* genes affects the hormone balance of, or sensitivity to phytohormones in transformed tissues (reviewed in Hamill 1993). We have compared the sensitivity of *N. tabacum* seedlings transformed with individual *rol* genes to non-transformed seedlings with respect to the concentrations of auxin which induce LRP induction.

2. Material and Methods

2.1 PLANT MATERIAL

Eight day old seedlings of *E. globulus* germinated *in vitro* were selected and root explants excised. Five explants were transferred to Petri dishes containing modified MS medium solidified with 0.8% agar in the presence or absence of growth substances as described in the text. Petri dishes were sealed and placed at a 90° angle to orientate the explants vertically. *N. tabacum* var SR1 seedlings, transformed with either the *rol*A, *rol*B or *rol*C gene (under control of endogenous promoters and kindly supplied by Dr A. Spena, Köln), were grown in a vertical orientation on modified MS medium solidified with 0.8% agar and containing 50mg/ml kanamycin for selection of transformed seedlings. Control tissues contained the GUS gene (under control of Domain A of the CaMV35S promoter in pBin19 and termed CaMVAGus), and were also selected on kanamycin-containing medium. Seedlings were 11 days old when they were used.

2.2 TREATMENTS

2.2.1 *E. globulus* explants or *N. tabacum* seedlings were cultured on the surface of medium containing a growth substance and/or inhibitor for the time periods described. All explants were rinsed in sterile double distilled water between transfers to fresh medium. Ten explants were usually sampled per treatment. Growth substances used were the cytokinins BAP, kinetin, zeatin or 2ip, and the auxins IBA, IAA, NAA or 2,4-D. The RNA synthesis inhibitor actinomycin-D (act-D) and the protein synthesis cycloheximide (CH) were also used as indicated.

2.3 DETECTION OF LRP AND LR IN EXPLANTS

Roots were cleared in 2% chromium trioxide for 16 hours (*E. globulus*) or 12 hours (*N. tabacum*) at room temperature after all final treatments. Roots were then rinsed several times in distilled water before examination with a stereo microscope placed on top of a light box. Roots were scored for the number of LRP and LR per centimetre of root length.

3. **Results**

3.1 Hormonal Control of Lateral root Differentiation in *E. globulus*.

3.1.1 *Auxin Induction of Root Primorida.* In *E. globulus* root explants, IBA was found to be slightly more effective than NAA for LRP induction which was increased 5-10 fold compared to the non-hormone treated controls. Both IAA and 2,4-D were much less effective at stimulating LRP production. After 72 hours contact with different concentrations of IBA, 10^{-5}M was found to be the most effective concentration, producing about 14 LRP/cm of primary root tissue. When treated with 10^{-5}M IBA, the number of LRP began to increase significantly 24 hours after initial contact with the auxin. Maximal densities of LRP were achieved after a further 18-24 hrs(Figure 1). Transfer of IBA treated root explants to hormone free medium for a further seven days of culture resulted in the conversion of the majority of LRP into LR. However, when roots were maintained on 10^{-5}M IBA, conversion of LRP to LR was inhibited. When explants were stimulated by IBA, and then cultured on hormone free medium, maximum LR formation was observed after treatment with 10^{-5}M IBA for 16-18 hours.

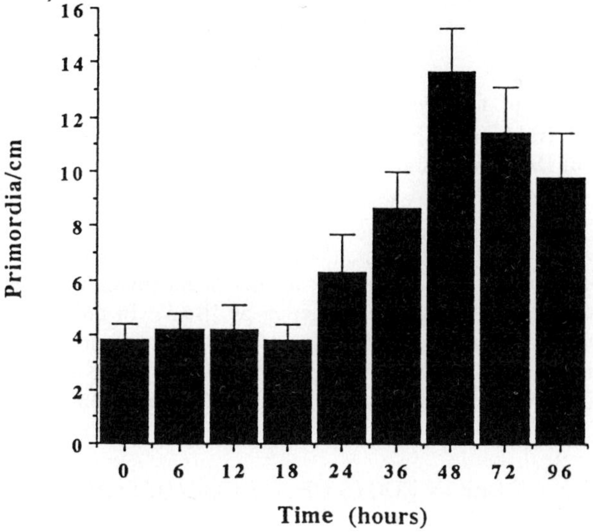

Figure 1. Time course of LRP induction in root explants (mean of 10 replicates ±SE) of *E. globulus* seedlings treated with 10^{-5}M IBA.

3.1.2 *Effects of Cytokinin Pretreatment.* A six hour pretreatment of root explants with BAP, kinetin, zeatin or 2ip was inhibitory to subsequent LRP induction by 10^{-5}M IBA. All cytokinins were effective inhibitors of LRP induction at concentrations between 10^{-3}M-10^{-5}M, with BAP and 2ip being also effective inhibitors at 10^{-6}M. Pretreatment of root explants with IBA for 16 hours, followed by transfer to BAP, showed that the conversion of LRP to LR was inhibited by concentrations of 10^{-5}M or greater and stimulated at low concentrations (10^{-9}-10^{-11}M) (Figure 2).

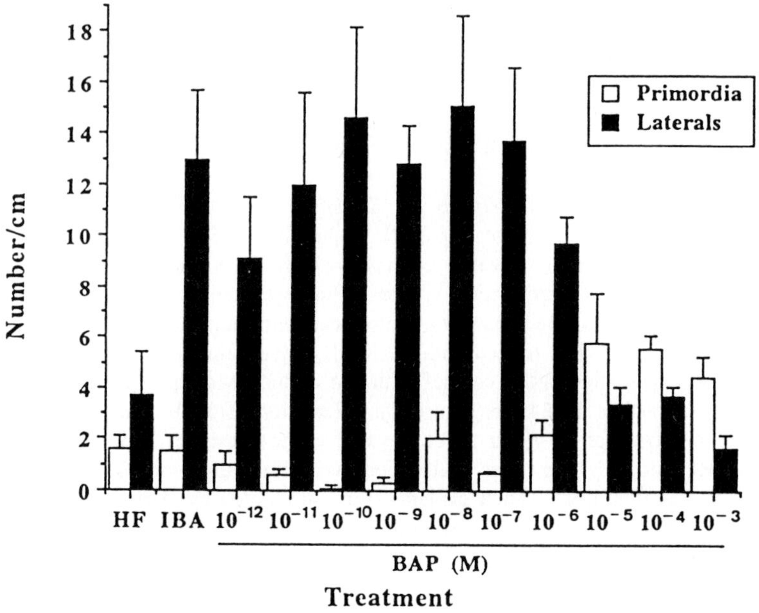

Figure 2. Conversion of LRP to LR in root explants (mean of 10 replicates ±SE) after induction by 10^{-5}M IBA for 16 hrs followed by cultivation on auxin free medium which contained a range of BAP concentrations.

3.1.3 *Effects of Actinomycin-D and Cycloheximide on LRP Induction.* Pretreatment of root explants with act-D or CH for six hours prior to IBA treatment showed that both inhibitors at 10^{-3}M were effective in preventing LRP induction by IBA. . In order to determine the stage at which act-D and CH inhibited IBA induction of LRP, root explants were exposed to 10^{-5}M IBA for various time periods before tissue was supplied with either act-D or CH. Both act-D and CH were inhibitory to LRP induction when applied up to 42-48 hrs after initial exposure of the explants to IBA.

3.2 INDUCTION OF LRP IN ROOTS OF *N. TABACUM* CONTAINING *ROL* GENES

Control seedlings of *N. tabacum* (CaMVAGus) showed maximum rates of LRP induction when treated with 10^{-4}M IBA for 48-72 hrs with ~24 LRP being found per cm of root tissue.

There was a slight decrease in the sensitivity of roots of seedlings containing the *rol*A gene to the LRP stimulating effects of IBA (figure 3). On the other hand, seedlings containing *rol*B or *rol*C genes showed a measurable increase in sensitivity to IBA with respect to LRP induction (figure 3). Roots of *rol*C seedlings showed maximal numbers of LRP (~27±1 per cm) when treated with 10^{-5}M IBA.

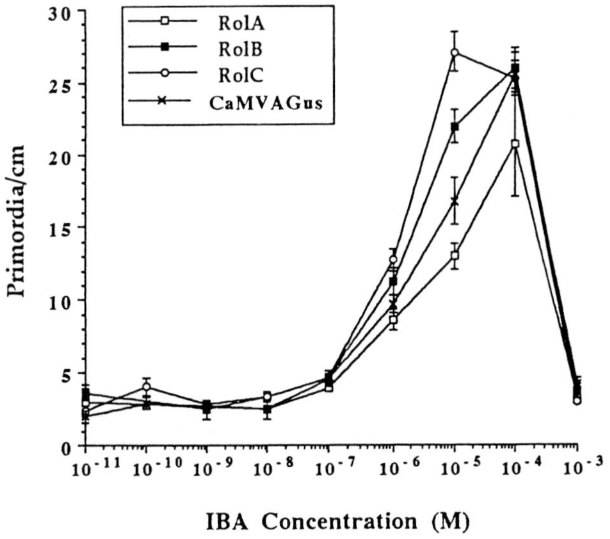

Figure 3. LRP formation in IBA treated roots of *N. tabacum* var SRI (CaMVAGus) or seedlings containing an individual *rol* gene. Data represent mean of 10 replicates ±SE.

4. Discussion

LRP initiation in seedlings of *E. globulus* was influenced by exogenous application of auxin and cytokinin. While very high levels of auxin (10^{-3}M, 10^{-4}M) were inhibitory to LRP induction in seedling roots, lower levels (10^{-5}M) were stimulatory, increasing the density of primordia to about 14 per cm of root tissue. Continual exposure to auxin prevented the emergence of LRP into LR. When auxin was removed, however, the process of conversion of LRP to LR took place efficiently with essentially all LRP converting to LR within a few days.

Differentiation of LRP in *E. globulus* root explants was completed within 48 hours of initial exposure to IBA. Exposure to 10^{-5}M IBA for 16-18 hours was sufficient to induce maximal numbers of LR when tissues were transferred to hormone free medium for 7 days after IBA stimulation. These results are comparable to those reported for lettuce seedlings roots by

MacIsaac *et al* (1989) who found that 20 hour exposure to 10^{-5}M NAA was sufficient to induce optimal rates of LRP formation even though a further 1-2 days of culture were required to visualize all the LRP.

These results suggest that there are two stages involved in the formation of lateral roots in *E. globulus* as has been reported in both pea (Wightman *et al* 1980) and lettuce (MacIsaac *et al* 1989). The first stage occurs in the presence of high levels of exogenous auxin enabling the initiation of LRP to take place, while the second stage occurs independently of exogenous auxin allowing LRP to develop into growing lateral roots.

Pretreatment of *E. globulus* root tissue with a range of cytokinins at concentrations of 10^{-5}M or higher completely abolished the stimulatory effects of 10^{-5}M IBA on LRP induction. This is in general agreement with the hypothesis that a high auxin:cytokinin balance is required for LRP induction (Wightman *et al* 1980; MacIsaac *et al* 1989). Unlike the situation in lettuce seedlings, application of cytokinin during the first 20 hours after auxin was applied did not prevent subsequent LRP induction in *E. globulus* although a high cytokinin level (10^{-3}M) prevented LRP conversion into LR. However, the presence of low levels of cytokinin (10^{-9-11}M) tended to increase the frequency of conversion of LRP to LR.

Application of a transcription inihibitor (actinomycin-D) or translation (cycloheximide) inhibitor, either before or during the first 42-48 hours of auxin treatment, prevented the formation of LRP in *E. globulus* roots suggesting that transcription and translation of essential genes are required for LRP differentiation. Previous studies have correlated the appearance of specific proteins with LRP induction following auxin treatment of tissues (Kantharaj *et al* 1985; Dhindsa *et al* 1987; MacIsaac and Sawhney 1990). MacIsaac and Sawhney (1990) also showed that treatment of lettuce roots with cycloheximide inhibited LRP induction following exposure to NAA. The results of the current study involving *E. globulus* are therefore, in broad agreement with results obtained from seedlings of more distantly related species. In addition, they are extended by the observations that exposure to a transcription or translation inhibitor prevents auxin induced LRP formation at all stages of the process.

The slightly reduced sensitivity of *rol*A transgenic seedlings of *N. tabacum* to IBA, with respect to LRP induction, and increased sensitivity of *rol*B and *rol*C seedlings is also of interest. Transgenic plants containing either the *rol*C gene alone or *rol* genes in combination have been shown to have increased rates of root branching (Schmülling *et al* 1988, Kurioka *et al* 1992, Lee *et al* 1994). It seems likely this is due to increased rates of LRP induction due to increased sensitivity to auxin and/or to increased levels of auxin in cells giving rise to lateral root primordia. Currently we are examining the sensitivity of seedlings containing various *rol* genes in combination to ascertain whether there are any further changes in the optimal concentrations of auxin which promote LRP induction.

ACKNOWLEDGMENTS

The financial support of DITARD (Genetic Technology Grant Number 14040), APM Forests, ANM Forest Management and North Forest Product is gratefully acknowledged. We also thank Dr Angelo Spena for supplying seeds of *N. tabacum* containing *rol* genes.

References

Biddington, N.L. and Dearman, A.S. (1982). 'The involvement of the root apex and cytokinins in the control of lateral root emergence in lettuce seedlings'. Plant Growth Reg. 1, 183-193.

Dhindsa, R.S., Dong, G. and Lalonde, L. (1987). 'Altered gene expression during auxin-induced root development in hypocotyls from excised mung bean seedlings'. Plant Physiol. 84, 1148-1153.

Hamill, J.D. (1993). 'Alterations in auxin and cytokinin metabolism of higher plants due to expression of specific genes from pathogenic bacteria - a review'. Aust. J. Plant Physiol. 20, 405-423.

Kantheraj, G.R., Mahaderan, S. and Padmenabhan, G. (1985). 'Tubulin synthesis and auxin-induced lateral root initiation in *Phaseolus vulgaris*. Phytochem. 24, 23-28.

Lee, M.C.S., Pelosi, A., Chow, E.K.F. and Hamill, J.D. (1994). Transformation of *Arabidopsis thaliana* with T-DNA from *Agrobacterium rhizogenes*: Effects on the root branching pattern and root:shoot biomass of transgenic plants grown *in vitro*. Transgenic Res. (submitted).

MacIsaac, S.A. and Sawhney, V.K. (1990). 'Protein changes associated with auxin-induced stimulation and kinetin-induced inhibition of lateral root initiation in lettuce (*Lactuca sativa*) roots'. J. Exp. Bot. 41, 1039-1044.

MacIsaac, S.A., Sawhney, V.K. and Pohorecky, Y. (1989). 'Regulation of lateral root formation in lettuce (*Lactuca sativa*) seedling roots: Interacting effects of α-naphthaleneacetic acid and kinetin'. Physiol. Plant. 77, 287-293.

Schmülling, T., Schell, J. and Spena, A. (1988). 'Single genes from *Agrobacterium rhizogenes* influence plant development'. EMBO J. 7, 2621-2629.

Spena, A., Schmülling, T., Koncz, C. and Schell, A. (1987). 'Independent and synergistic activity of *rol*A, B and C loci in stimulating abnormal growth in plants'. EMBO J. 6, 3891-3899.

Thimann, K.V. (1936). 'Auxins and the growth of roots'. J. Bot. 23, 561-567.

Torrey, J.G. (1950). 'The induction of lateral roots by indoleacetic acid and root decapitation'. Amer. J. Bot. 37, 257-264.

Torrey, J.G. (1956). 'Chemical factors limiting lateral root formation in isolated pea roots'. Physiol. Plant. 9, 370-388.

Wightman, F., Schneider, E.A. and Thimann, K.V. (1980). 'Hormonal factors controlling the initiation and development of lateral roots. II Effects of exogenous growth factors on lateral root formation in pea roots'. Physiol. Plant. 49, 304-314.

Theoretical and Experimental Definition of Minimal Photoresponsive Elements in *cab* and *rbcS* genes

Argüello-Astorga, G.R. and Herrera-Estrella, L.R.
Centro de Investigaciones y Estudios Avanzados del I.P.N. Unidad Irapuato. Departamento de Ingeniería Genética en Plantas. Ap 629 36500 Irapuato, Gto. MEXICO. Fax (462) 5-0759

Abstract. Light responsive elements have been identified in a number of genes whose transcription is modulated by light. However, the presence of multiple *cis*-acting sequences in these elements has made difficult to determine whether single DNA motifs or more complex arrays of cis-acting sequences are responsible for the light-responsive properties of thesepromoters. Here we present a theoretical and experimental analysis that allowed us to identify a minimal light-responsive unit which is evolutively conserved in *rbcS* promoters.

Introduction

Light constitutes the primary energy source for plant growth and provides much of the environmental signals regulating their development. Thus, plants have evolved complex biochemical systems and diverse molecular mechanisms to detect and differentially respond to radiation of different wavelengths. Higher plants have three classes of photoreceptors: blue/UV-A light photoreceptor (cryptochrome), red light-absorbing phytochromes, and a UV-B photoreceptor (31). The transcription of some plant genes is activated as a response to light absorbed by the three classes of photoreceptors (e.g. *cab*, *rbcS* and *chs*), whereas others seem to be regulated by a single type of receptor (31).
The perceived light signals are transduced to photoresponsive genes via largely unknown signal pathways, but which seem to involve the participation of GTP-binding proteins and calcium/calmodulin-dependent and independent pathways (23, 24). At the end of light-signal transduction chains, regulatory proteins which interact with specific cis-acting elements determine the activation or repression of photoresponsive genes.
Despite the common dependence on defined light-signal reception/transduction systems, photoresponsive genes display marked differences in their response to light, in terms of the intensity and spectral quality required for their activation, time course of induction and phytochrome escape kinetics (17, 31). For instance, it has recently been shown that the promoters of eight nuclear genes encoding plastid proteins from spinach respond to the same intensity and quality of light in remarkably different ways (17). Moreover, even closely related genes in the same organism exhibit different responses to light. For example, pea *cab* genes have been classified according to their response to red light, some having a rapid and strong response (e.g. AB96), and others a slow and weak response (e.g. AB80; 33). Additionally, it has been found that in many cases the ability of the same gene to respond to light of a given spectral quality changes during plant development (13, 31).
The transcriptional properties of photoresponsive genes, partially correlate with apparent differences in the architecture of their promoters, in which elements regulating their expression either quantitatively or qualitatively, seem to be located in different regions (6). Such differences are evident even between members of the same gene

M. Terzi et al. (eds.), Current Issues in Plant Molecular and Cellular Biology, 501-511.

family (18,19). Of the several *trans*-acting factor binding-sites identified in *cab* and *rbcS* genes, two classes of *cis*-acting sequences have been proposed as essential components of their molecular light switches: 1) GT-1 binding sites, GPu(T/A)AA(T/A) (6, 16), and the so-called "G-box", (CACGTGGCA) (7). These motifs are present in the 5' flanking sequence of many other light-regulated, and seem to be critical components of experimentally defined **L**ight **R**esponsive **E**lements (LRE's). Since neither single or multiple copies of these elements are able to confer light-responsiveness to heterologous minimal promoters (16, 17), its function in native LREs must be dependent on specific interactions with other *cis*- and *trans*-acting elements. This observations led to the idea that *cab* and *rbcS* LREs are in fact composite elements.

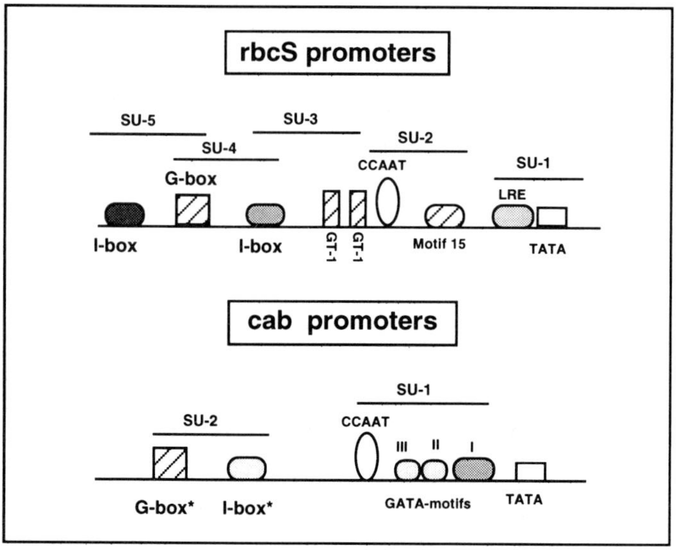

Figure 1. Simplified scheme of *cis*-acting elements present in *rbcS* and *cab* promoters. Structural units (SU) conserved in phylogenetically related plant species are shown. *rbcS* SU-1 and SU-2 are conserved in Solanaceae, SU-3 in dicots and SU-4 and -5 are conserved from gymnosperms to monocots. In *cab* promoters SU-1 is conserved from gymnosperms to monocots, and SU-2 in dicots.

Despite the apparent diversity and linear arrangement of *cis*-acting elements in the regulatory region of photoresponsive genes, the possibility that LRE's could have common characteristics remains open. The identification of such common features (if they exist) would be a major advance in understanding the molecular mechanisms involved in light-regulation of plant genes. With this aim, we decided to systematically analyze the *cab* and *rbc*S regulatory regions in which the presence of at least one LRE has unequivocally been experimentally determined. For this analysis we made use of a phylogenetic-structural approach, which has proven useful to unravel the organization of complex regulatory regions of geminiviruses (1). As a result of this analysis, structurally conserved units in the 5' flanking sequences of *cab* and *rbcS* genes were identified. Here, we present such analysis and experimental data showing that at least one of these evolutively conserved structural units functions as a minimal light responsive unit.

RESULTS

Phylogenetic and Structural Analysis of *cab* and *rbcS* genes

Studies on *cab* and *rbcS* genes from diverse monocot and dicot plant species, indicate the presence of LREs in at least two regions of their 5' flanking sequences: 1) the region immediately adjacent to the transcription initiation site (~100 bp) (14) and 2) the region comprised between positions -400 and -100, that in some cases (e.g. pea *rbcS* 3A gene) includes at least 2 independent LREs (13). The phylogenetic-structural analysis of the specific *cab* and *rbcS* segments experimentally defined as LREs, as well as of the corresponding regions of a number of members of these gene families that have not been experimentally examined, led us to the following observations:
a) In all of these regions there are stretches of DNA structurally conserved in at least some members of each gene family, and within more or less broad lineages of plants. By "structural conservation" we understand the consistent association of two or more conserved sequence motifs, whose arrangement and/or spacing has remained relatively invariant in the course of evolution (figures 1-3). Some of these "structural units" (SU) are conserved from Gymnosperms to Monocots, and in one case (i.e. *Cab*-SU1), even the spacing between the two most conserved motifs apparently has remained constant since the divergence of the lineage that gave rise to modern gymnosperms and angiosperms (fig.3).

DICOTS

Arabidopsis *rbcS* 3A
CGATAAGG - 12 - TCCACGTGGC A - 19 - CGATAAGG

Sunflower rbcS 1
AGATAAGA - 21 - GACACGTGGC T -20 - TGATAAGG

MONOCOTS

Maize *Zm*3
CGATAAGG - 16 - GACATGTGGTG --------------------

Maize *Zm* 1
------------------TCCATGTGGC G - 8 - CGATAAGG

GYMNOSPERMS

***Larix laricina rbcS* 1**
CGATAAGG - 12 - GACACGTAGT A - 27 AGATATGG

I box	G box	I box

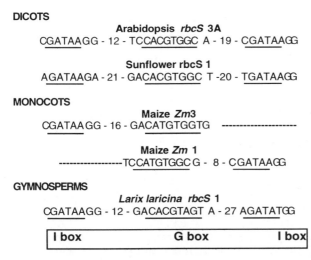

Figure 2. Phylogenetically conserved SU-4 and SU-5 associated with light-responsive elements in *rbcS* promoters.

b) Elements related to the G-box and GT-1-binding sites are found in only a subset of these phylogenetically conserved LRE-associated SU's.
c) The most relevant structural feature of the LRE-associated structural units, is the conspicuous presence in all of them of sequence elements highly homologous to the so-called I-box (AGATAAGPu) (7). These are, besides the canonical I-box, the GATA-I motif of *cab* promoters (see figure 3), the motif 15 of Solanaceae *rbcS* genes (AGATGAGG, 20), and the CCTTATCAT (inverse: ATGATAAGG) motif associated to the TATA box in rbcS genes (9). From a structural viewpoint, this is the only

common feature in the *cab* and *rbcS* sequences with properties of LREs that could be established in our study. The relevance of this finding will be discussed below.

DICOTS

Petunia 22L

CCAATGAAATTGTAGATAGAGATATCATAAGATAAGA..19..TATA

Pea AB80

CCAACTAGCCATAGCTTTATGATAACACACGATAA GA..10...TATA

MONOCOTS

***Lemna* AB30**

CCAATGGCGTGCGGCCAGTAGATATCGGTGGATAATG...31...TATA

Maize Cab1

CCAATGGCAACTCGTCTTAAGATTCCACGAGATAAGG..18...TATA

GYMNOSPERMS

***Pinus thunbergii* Cab6**

CCAATAAGAAGGTAAGAAGGCTGGCGTTCGGATAAGG..18...TATA

Figure 3. Phylogenetically conserved structural unit associated with light-responsive elements in the 5'-flanking sequence of *cab* genes

Functional Analysis of a conserved *rbcS* structural unity.

Although the evolutive conservation of this type of structural units in LRE's is rather suggestive, it remains to be determined whether some of them indeed constitute Minimal Photoresponsive Units (MPU's) or whether they are only part of more complex light-responsive elements. To answer this question we are analyzing whether synthetic oligonucleotides comprising these structural units can confer light-responsive properties to heterologous promoters.

A synthetic oligonucleotide was designed that represents a structural unit conserved in *rbcS* genes from Solanaceae (52 bp long), and which is composed of an I- and a G-boxes (I-G oligonucleotide, figure 4). This oligonucleotide was cloned in one or two copies upstream of three different heterologous promoters fused to the GUS reporter gene (11): the CaMV 35S minimal promoter (-46 to +8; mp35S), the truncated CaMV 35S (-90 to +1; tp35S), and a truncated pea *cab*(AB-80) promoter (-100 to +55; tpAB80). With the use of three different promoters, the transcriptional properties of I-G box element, in different regulatory contexts, can be examined, namely: a simple context in which the only other active *cis*-acting element is the TATA box (mp35S); in the presence of a *cis*-acting element that directs root specific expression (i.e. ASF-1, which interacts with the AS-1 element present in tp35S; 15); and in association with another LRE-associated SU (i.e. *Cab*-SU1, contained in tpAB80). Transgenic tobacco plants harboring these gene constructs were produced using the *Agrobacterium*-mediated transformation system, and the level and pattern of GUS expression examined in the T1 progeny of the primary transformants. Surprisingly, it was found that even a single copy of the I-G oligonucleotide was able to increase dramatically the strength of the three truncated promoters in photosynthetic tissue (20 to 50% of the activity of the 35S complete promoter, data not shown).

(I-G) oligonucleotide functions as Light Responsive Element.

In order to determine whether the *cis*-acting elements contained in the (I-G) oligonucleotide function as an LRE, T1 plants harboring a single copy of this oligonucleotide fused to mp35S [(I-G)-mp35S-GUS construction] were subjected to different light treatments and then analyzed for GUS expression. It was found that when (I-G)-mp35S-GUS plants were germinated and grown in dark (etiolated) for 10, 15 or 20 day periods showed no detectable GUS activity in any tissue, with the exception of a weak expression in the upper part of the hypocotyl in several, but not all, independent transgenic lines. Subsequent exposure to light of the etiolated plantlets promoted the expression of the reporter gene in their cotyledons, and a seemingly decreased activity in the hypocotyl. When (I-G)-mp35S-GUS plantlets grown in continuous white light for 10-15 days and then transferred to darkness for 4 days, it was observed that GUS expression significantly decreases when compared to plant grown in the light. (fig.5). It was also found that a 1 minute pulse of white or red light is sufficient to induce the expression of the (I-G)-mp35S-GUS gene construct. Taken as a whole, these results clearly demonstrate that the 52-bp oligonucleotide containing the I and G boxes, functions as a LRE.

Light induction of (I-G)-mp35S-GUS is influenced by plastid development.

The analysis of the tissue-specific pattern of expression of the reporter gene in (I-G)-mp35S-GUS transgenic plants, showed that the chimaeric promoter is active mainly in green tissues: leaves (mesophyll cells), outer region of petioles and stem, sepals, flower carpel and stigma, but not in vascular bundles of petioles and stem, petals, mature stamens, and roots (data not shown). It is well established that expression of *cab* and *rbcS* genes is strongly influenced by the developmental stage of plastids (30). In order to determine whether the transcriptional activity of (I-G)-mp35S-GUS shows the same dependence on plastid development, transgenic plants were grown with or without the herbicide norfluorazon (10 μM) for 18-20 days. In plants grown in the presence of this herbicide, GUS activity could only be observed in cotyledonary leaves, but not in true leaves, whereas non-treated control plants displayed an elevated expression in all leaves. Transgenic control plants harboring a 35S-GUS construction (pBI121; 11) showed the same pattern of expression with or without norfluorazon. These results demonstrate that the activity of the I-G element is strongly influenced by the presence or absence of functional chloroplasts, but this dependence is not absolute, as indicated by the activity of the chimaeric promoter in cotyledonary leaves.

Is the I-G oligonucleotide a Minimal Photoresponsive Unit?

Since the I-G oligonucleotide contains binding sites for, at least, two different protein factors (GBF and GAF-I or IBF; 2, 6, 7), the relative contribution of each regulatory protein to the functional properties of the entire 52-bp sequence remains to be explored. In order to establish the role of the I- and G- boxes in the light-responsive properties of this sequence, oligonucleotides were synthesized in which either the I-(oligo Im-G) or the G-boxes (oligo I-Gm) were mutated, and cloned upstream of the mp35S-GUS chimaeric gene (figure 4). Preliminary experiments of transient expression in tobacco and pea leaves using the biolistic method, indicate that both mutant oligonucleotides are unable to enhance the transcriptional activity of mp35S, in contrast with the high activity observed for (I-G)mp35S-GUS in the same assay (data no shown). If these results are corroborated in more quantitative assays (i.e. protoplast) or in transgenic plants, this would demonstrate that the presence of both conserved sequence motifs is necessary for the integral functioning of the I-G structural unit, and that this sequence is indeed a minimal photoresponsive unit.

Oligonucleotide	I- BOX / G-BOX
I-G	TAGGATG AGATAAGA TTACTGAGGTGCT TACACGTGGCA CCTCCATTGTGGT
Im-G	TAGatct atgt ggtcc TACTGAGGTGCTTACACGTGGCACCTCCATTGTGGT
I-Gm	TAGGATGAGATAAGATTACTGAGGTGCT agctgcat at gCCTCCATTGTGGT

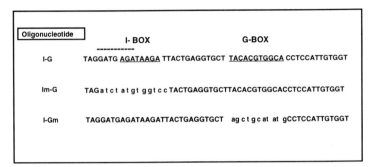

Figure 4. Oligonucleotides used in this study. I-G represents a native 52 bp conserved region present in the regulatory region of *rbcS* genes. Im and Gm are variants mutated in the I and G boxes, respectively.

DISCUSSION

Using a phylogenetic-structural approach, we have identified in the 5'-flanking sequence of *cab* and *rbcS* genes, short stretches of DNA that have been structurally conserved throughout large periods of plant evolution, and which are contained in diverse experimentally-defined LREs. It was determined that one of these evolutively conserved structural units (SU-5 in figure 1), functions as a LRE. Both the I- and G-boxes contained within this structural unit seem to be essential for its function and, therefore, it can be considered as a minimal functional unit of transcriptional photoresponse. A previous study on the *Arabidopsis rbcS*-1A promoter already suggested the existence of a composite light-responsive element. Indeed, it had been demonstrated that the activity of this *rbcS* promoter decreased >90% by mutating the G-box, and that a similar decrease result of the simultaneous mutation of the two I-boxes which flanking the G-box (4).

The 52 nt I-G oligonucleotide, as far as we know, is the shortest "native" *rbcS* sequence defined as LRE. A somewhat similar sequence was previously identified as a LRE in a parsley chalcone synthase gene (32). There are several similarities between both photoresponsive units: 1) they confer photoresponsivity to the 35S minimal promoter in a single copy ; 2) both units are composed by two conserved sequence motifs which bind different protein factors *in vitro*, and which are necessary for the functional integrity of the unit; 3) they have a common sequence-motif, CACGTGGC, called G-box in *rbcS* genes, and Box II in *chs* promoter (7, 29), which is bound in vitro by proteins belonging to the basic-leucine zipper class of DNA-binding factors (28, 32). This last feature could suggest that a G-box binding factor (GBF) is the link between the early steps in the reception/transduction of light-signals, and the responsive genes (10, 28). However, several observations are not easy to reconcile with the assumption that a GBF is indeed the light-activating component of photoresponsive units in *rbcS*, *cab* and *chs* genes. G-box like sequences are ubiquitous elements that have been proposed to be involved in the response to a variety of regulatory pathways, besides those related with light signals. For example, in genes regulated by abscisic acid (21), methyl-jasmonate (12), ethylene (22) and anaerobic stress (5). Additionally, G-box binding activities are contained in nuclear protein extracts from leaves, fruits, and roots.

The relative concentration of GBFs in leaf nuclei is not altered by light-dark conditions (7, 27). More important, it has been shown that multiple copies of the G-box fused to tp35S-GUS, directs a high GUS expression in roots and cotyledons of etiolated plantlets, which is inhibited by light of different wavelengths (17, 25).

Are the I-box-like elements the critical components of *cab*- and *rbcS*-LREs?

A growing body of experimental and formal evidence indicates that I-box-like elements could be critical components of molecular light switches in several plant genes: **1)** they are present in virtually all *rbcS* and *cab* genes, and apparently in many other phytochrome-regulated genes (6, 9); **2)** the most evolutively-conserved LRE-associated SU's identified in this work have only one feature in common: the presence of an element with the core sequence GATPuAGpu (the I-box core); **3)** only few of the many nuclear factors which interact with regulatory sequences of *rbcS* and *cab* genes, are found in significantly higher concentrations in light-grown as compared to dark-adapted plants. Most of these light responsive factors bind I-box-like elements (2, 3, 6); **4)** the only DNase-I hypersensitive site found induced by light in the pea *rbcS*-3.6 and -3A genes is centered around the I-box (8); **5)** *In vivo* methylation interference experiments demonstrate that the methylation patterns around motif 15 (another I-box related sequence) of tomato *rbcS* genes differ between dark-grown and light-grown tomato cotyledons (19); **6)** the pea *rbcS* 3A promoter sequence from -50 to +15 mediates light responsiveness (14); this short region includes an I-box like element as component, which is protected from DNase I digestion by pea nuclear extracts (6). **7)** a 9 bp.deletion, which eliminate specifically a I-box in a maize *rbcS* gene promoter, reduced 50% the expression in light of the reporter gene (26).

It should be pointed out however, that the nucleotidic sequence of these I-box-like elements is not identical, and that in some cases nucleotides flanking them are also conserved, thus defining a more broad consensus sequence. Differences among I-boxes from the same gene (e.g. in rbcS-SU4 and -SU5), or from different photoregulated genes are clearly observed. Additionally, the arrangement of I- and G-boxes varies among different genes. Thus, it is possible that differences in arrangement or specific sequence of these *cis*-acting elements play an important role determining the particular characteristics of each LRE.

CONCLUSION

Our theoretical and experimental data, as well as much of the experimental evidence regarding transcriptional regulation of photoresponsive genes, suggest that their LREs are made up of complex units, which include two or more factor-binding sites in a defined arrangement. Plausibly, at least one of these sequence motifs is bound by a transcription factor that is a direct target of the signal transduction pathway that activates transcription as a response to a light stimulus. Some of these light-responsive factors have as cognate site an I-box-like element. The combined action of the factors constituting a bipartite or multipartite protein complex associated to the LRE, determines the activation or down regulation of the gene when light is present. This organization of LREs allows to couple the light signal to other endogenous signals, such as tissue or developmental stage specific-factors, or metabolic and hormonal signals. In this way, it is altogether possible to produce a more plastic and differentiated transcription regulation of photo-responsive genes.

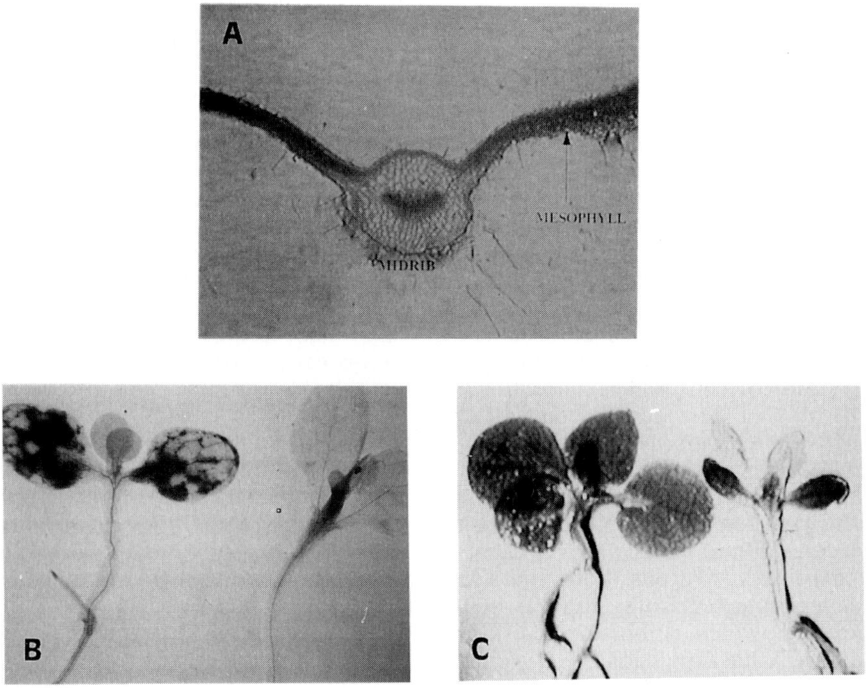

Figure 5. Expression of (I-G)-mp35S-GUS in transgenic tobacco plants. A) Histochemical analysis of a leaf transversal section of a (I-G)-mp35S-GUS transgenic tobacco plant. Expression is restricted to mesophyll cells. B) GUS expression in light (left) and dark (right) grown I-G-mp35S-GUS tobacco plantlets. C) Effect of norfluorazon on the expression of I-G-mp35S-GUS transgenic plants. Expression in every leave is observed in control plants (left), whereas norfluorazon treated plants show expression only in the cotyledons (right).

References.

1.Argüello-Astorga, G.R., Guevara-Gonzalez, R.G., Herrera-Estrella, L.R., Rivera-Bustamante, R.F. 1994. Geminivirus replication origin have a group-specific organization of iterative elements. A model for replication. Virology (in press).

2.Borello, U., Ceccarelli, E., and Giuliano, G. 1993. Constitutive, light-responsive and circadian clock-responsive factors compete for the different I box elements in plant light-regulated promoters. The Plant Journal 4: 611-619.

3. Buzby, J. S., Yamada, T., and Tobin, E. M. 1990. A Light-Regulated DNA-Binding Activity Interacts with a Conserved Region of a *Lemna gibba rbcS* Promoter. The Plant Cell 2: 805-814.

4. Donald, R. G. K., and Cashmore, A. R. 1990. Mutation of either G box or I sequences profoundly affects expression from the Arabidopsis rbcS-1A promoter. The EMBO Journal 9: 1717-1726.

5. Ferl, R. J., and Laughner, B. H. 1989. *In vivo* detection of regulatory factor binding sites of *Arabidopsis thaliana Adh*. Plant Mol. Biol. 12: 357-366.

6 Gilmartin, P. M., Sarokin, L., Memelink, J., and Chua, N.-H. 1990. Molecular Light Switches for Plant Genes. The Plant Cell 2: 369-378.

7.Giuliano, G., Pichersky, E. Malick, V. S., Timko, M. P., Scolik, P. A., and Cashmore, A. R. 1988. An evolutionarily conserved protein binding sequence upstream of a plant light-regulated gene. Proc. Natl. Acad. Sci. 85: 7089-7093.

8. Görz , A., Schäefer, W., and Kahl, G. 1988. Constitutive and light-induced DNAseI hypersensitive sites in the rbcS genes of pea (*Pisum sativum*). Plant Mol. Biol. 11: 561-573.

9. Grob, U., and Stüber, K. 1987. Discrimination of Phytochrome dependent light inducible from non-light inducible plant genes. Prediction of a common light responsive element (LRE) in phytochrome dependent light inducible plant genes. Nucleic Acids. Res. 15: 9957-9972.

10. Harter, K., Kircher, S., Frohnmeyer, H., Krenz, M., Nagy, F., and Shäfner, E. 1994. Light-Regulated Modification and Nuclear Translocation of Cytosolic G-Box Binding Factors in Parsley. The Plant Cell 6: 545-559.

11. Jefferson RA, Kavanagh TA, Bevan MW: GUS fusions: ß-glucuronidase as a sensitive and versatile gene fusion marker in higher plants. EMBO J. 6, 3901-3907 (1987).

12. Kim, S. R., Choi, J. L., Costa, M. A., and An, G. 1992. Identification of G-Box Sequence as an Essential Element for Methyl Jasmonate Response of Potato Proteinase Inhibitor II Promoter. Plant Physiol. 99: 627-631.

13. Kuhlemeier, C., Cuozzo, M., Green, P. J., Goyvaerts, E., Ward, K., and Chua, N.-H. 1988. Localization and conditional redundancy of regulatory elements in rbcS-3A, a pea gene encoding the small subunit of ribulose- bisphosphate carboxylase. Proc. Natl. Acad. Sci. 85: 4662-4666.

14. Kuhlmeier, C., Strittmetter, G., Ward, K., and Chua, N.-H. 1989. The Pea rbcS-3A Promoter Mediates Light Responsiveness but not Organ Specificity. The Plant Cell 1: 471-478.

15. Lam, E., Benfey, P.N., Gilmartin,P., Fang, R.-X., and Chua N.-H. 1989. Site-specific mutations alter in vitro factor binding and change promoter expression in transgenic plants. Proc. Natl. Acad. Sci. USA. 86, 7890-7894.

16. Lam, E., and Chua, N.-H. 1990. GT-1 Binding Site Confers Light Responsive Expression in Transgenic Tobacco. Science 248: 471-474.

17 Lübberstedt, T., Bolle, C. E. H., Sopory, S., Flieger, K., Herrmann, R. G., and Oelmüler, R. 1994. Promoters from Genes for Plastid Proteins Possess Regions with Different Sensitivities toward Red and Blue Light. Plant Physiol. 104: 997-1006.

18. Manzara, T., Carrasco, P., and Gruissem, W. 1993. Developmental and organ-specific changes in DNA- Protein interactions in the tomato rbcS1, rbcS2 and rbcS3A promoter regions. Plant Mol. Biol. 21: 69-88.

19. Manzara, T., Carrasco, P., and Gruissem, W. 1991. Developmental and Organ-Specific Changes in Promoter DNA-Protein Interactions in the Tomato rbcS Gene Family. The Plant Cell 3: 1305-1316.

20. Manzara, T., and Gruissem, W. 1988. Organization and expression of the genes encoding ribulose-1,5-biphosphate carboxylase in higher plants. Molecular Biology of Photosyntesis 621-644.

21. Marcotte Jr., W. R., Russell, S. H., and Quatrano, R. S. 1989. Abscisic Acid-Responsive Sequences from the Em Gene of Wheat. The Plant Cell 1: 969-976.

22. Meller, Y., Sessa, G., Eyal, Y., and Flhur, R. 1993. DNA protein interactions on a cis-DNA element essential for ethylene regulation. Plant Mol. Biol. 23: 453-463.

23. Neuhaus, G., Bowler, C., Kern, R., and Chua, N.-H. 1993. Calcium/Calmodulin-Dependent and -Independent Phytochrome Signal Transduction Pathways. Cell 73: 937-952.

24. Romero, L. C., and Lam, E. 1993. Guanine nucleotide binding protein involvement in early steps of phytochrome- regulated gene expression. Proc. Natl. Acad. Sci. 90: 1465-1469.

25. Salinas, J., Oeda, K., and Chua, N. H. 1992. Two G-Box-Related Sequences Confer Different Expression Patterns in Transgenic Tobacco. The Plant Cell 4: 1485- 1493.

26. Schäffner, R. A., and Sheen, J. 1991. Maize rbcS Promoter Activity Depends on Sequence Elements Not Found in Dicot rbcS Promoters. The Plant Cell 3: 997-1012.

27. Schindler, U., and Cashmore, A. R. 1990. Photoregulated gene expression may involve ubiquitous DNA binding proteins. The EMBO Journal 9: 3415-3427.

28. Schindler, U., Menkens, A. E., Beckmenn, H., Ecker, J. R., and Cashmore, A. R. 1992. Heterodimerization between light-regulated and ubiquitously expressed Arabidopsis GBF bZIP proteins. The EMBO Journal 11: 1261-1273.

29. Schulze-Lefert, P., Dangl, J. L., Becker-André, M., Hahlbrock, K., and Schulz, W. 1989. Inducible *in vivo* DNA footprints define sequences necessary for UV light activation of the parsley chalcone synthase gene. The EMBO Journal 8: 651-656.

30. Simpson, J., Van Montagu, M., and Herrera-Estrella, L.1986. Photosynthesis-associated gene families: differences in enviromental and tissue specific responses. Science 233, 34-38 .

31. Thompson, W. F., and White, M. J. 1991. Physiological and molecular studies of

light-regulated nuclear genes in higher plants. Annu. Rev. Plant Physiol. Plant Mol. Biol. 42: 423- 466.

32. Weisshaar, B., Armstrong, G. A., Block, A., de Costa e Silva, O., and Hahlbrock, K. 1991. Light-inducible and constitutively expressed DNA-binding proteins recognizing a plant promoter element with funcional relevance in light responsiveness. The EMBO Journal 10: 1777-1786.

33. White, M. J., Fristensky, B. W., Falconet, D., Childs, L. C., Watson, J. C., Alexander, L., Roe, B. A., and Thompson, W. F. 1992. Expression of the chlorophyll-a/b-protein multigene family in pea (*Pisum sativum* L.). Planta 188: 190-198.

T-DNA TAGGING OF A GENE INDUCING DESICCATION TOLERANCE IN *CRATEROSTIGMA PLANTAGINEUM*

Antonella Furini, Francesco Salamini and Dorothea Bartels
Max-Plank-Institut für Züchtungsforschung
Carl-von-Linné-Weg 10 50829 Köln (Germany)

ABSTRACT

Drought tolerance in the resurrection plant *Craterostigma plantagineum* is restricted to fully developed plants. In callus this phenomenon can be induced by treatment with abscisic acid. We have used a recently established protocol for *Craterostigma* transformation by *Agrobacterium tumefaciens* (Furini *et al.,* 1994) for the stable integration of a T-DNA carrying a transcriptional enhancer element. By this method a gene was tagged whose induced expression resulted in the formation of a desiccation-tolerant callus in the absence of exogenously applied abscisic acid. *Craterostigma* genes normally induced by desiccation are constitutively induced in the transgenic callus. The tagged gene was cloned from a genomic library and used for the isolation of the corresponding cDNA clone.

INTRODUCTION

A representative of the resurrection plants, *Craterostigma plantagineum* Hochst. (*Scrophulariaceae*) has been adopted as a model system to study the molecular biology of desiccation tolerance (Bartels *et al.,* 1993). In this species the drought responses can be studied in fully developed plants as well as in callus cultures in which a pretreatment with abscisic acid (ABA) is necessary to induce desiccation tolerance (Bartels *et al.*, 1990, 1991). Genes that are induced in leaves by desiccation, or by ABA-treatment and/or ABA-treated callus have been isolated from *Craterostigma* tissues (Bartels *et al.*, 1990).

Changes in gene expression in response to drought have been described (Bray, 1988, 1991; Mundy and Chua, 1988), and most of the genes that have been isolated to date are also induced by ABA (Skriver and Mundy, 1990; Bray, 1991). This indicates that because of its elevated concentrations during water deficit, ABA is a good candidate to mediate drought responsive gene expression (Bray, 1991). However, the isolation of genes induced by drought, but not by ABA (Guerrero *et al.,* 1990; Yamaguchi-Shinozaki *et al.,* 1992) suggests the existence of both ABA-dependent as well as ABA-independent mechanisms of signal transduction leading from water stress to the expression of particular genes (Yamaguchi-Shinozaki and Shinozaki, 1993).

For a molecular analysis of desiccation tolerance, the genes involved have to be identified and cloned. A mutational approach for the clarification of a biological phenomenon may be advantageous if the genetic changes can be easily identified. T-DNA tagging offers the possibility to identify genes in higher plants (Feldmann, 1991; Walden *et al.* 1991). The transfer of T-DNA from *Agrobacterium tumefaciens* into the plant genome

M. Terzi et al. (eds.), Current Issues in Plant Molecular and Cellular Biology, 513-518.

can be considered as insertional mutagenesis, when the inserted element affects the transcriptional activity of genes located in the vicinity of the insertion: the gain or the loss of function of a tagged gene might therefore result in a screenable phenotype (Walden *et al.*, 1994). Moreover, the T-DNA is preferentially integrated into transcribed regions of the plant genome (Koncz *et al.*, 1989), and this fact increases its potential use for gene tagging.

Recently a T-DNA tagging vector has been constructed (Hayashi *et al.*, 1992) that produces dominant mutations and allows selection from populations of primary transformants. This vector contains multiple transcriptional enhancers derived from the cauliflower mosaic virus (CaMV) 35S RNA promoter positioned near the right border sequence of the T-DNA. The insertion of such a T-DNA into the plant genome will result in a constitutive overexpression of genes flanking the T-DNA.

In order to apply this approach to the tagging of genes in the drought-related signal transduction pathway in *Craterostigma*, a reproducible transformation protocol for this resurrection plant was developed (Furini *et al.*, 1994). Upon transformation T-DNA-directed overexpression of desiccation tolerance genes in transgenic tissues was expected to result in calli resistant to dehydration and capable to "resurrect" in the absence of exogenously supplied ABA.

MATERIALS AND METHODS

Plant material. Plants of *Craterostigma plantagineum* used in this study were propagated under sterile conditions on MS basal medium (Murashige and Skoog, 1962) in a growth chamber at 24 ±1°C with 16 h/day fluorescent light providing 200μE/ m^2/ s.

Agrobacterium and tagging vector. The *Agrobacterium tumefaciens* strain GV3101 rifampicin resistant and carrying the helper plasmid pMP90RK conferring kanamycin and gentamycin resistance (Koncz and Schell, 1986) was used as the host strain. The T-DNA tagging vector pPCVICEn4HPT, kindly provided by R. Walden from Max-Planck-Institut (Hayashi *et al.,* 1992), was transferred to *Agrobacterium* by transformation (Ebert *et al.*, 1987). It has 4 components:

1. The plasmid backbone based on the vector pPCV002 (Koncz and Schell, 1986).
2. A plant-specific gene conferring hygromycin resistance.
3. An origin of replication in *E. coli* as well as a gene conferring ampicillin and carbenicillin resistance for selection in *E. coli* and *Agrobacterium.*
4. The CaMV 35S RNA transcriptional enhancer (-90 to -427; Odell *et al.,* 1985) cloned as a tetramer close to the right T-DNA border sequence.

Bacterial culture. Agrobacterium tumefaciens harboring the T-DNA tagging vector was grown for 36 h at 28 °C on a gyratory shaker at 150 rpm in 50 ml of YEB medium containing kanamycin (100 mg/l) and carbenicillin (100 mg/l). The cells were pelleted by centrifugation at 4 °C and 3000 rpm and resuspended for leaf explant infection in MSAR medium (Koncz *et al.*, 1990) supplemented with 0.25 g/l of antioxidant mixture (150 mg/l ascorbic acid and 100 mg/l citric acid) to a final density of OD_{550} = 0.5.

Transformation procedure and Tissue culture media and conditions. Leaf explant transformation procedure, media used for callus induction and maintainance as well as for plant regeneration were as previously described (Furini *et al.*, 1994).

Identification of desiccation-tolerant calli. About 25,000 calli were selected on hygromycin-containing medium and maintained by subculturing at 3 week-intervals for about 2 - 3 months. Calli were then transferred to sterile filter paper and dried in a constant air stream of a ventilating hood for 20 h at 20 °C. During this time calli lost about 94 - 96%

of their initial fresh weight. Untransformed calli (negative control) and ABA-treated calli (positive control) were also placed in the hood at each drying treatment. Dried calli were then transferred to the callus medium for rehydration and after 2 days screened for survivors using a stereoscope.

ABA content. Abscisic acid was extracted from a callus that was viable after the drying treatment and a non-desiccation tolerant callus. The amount of ABA was quantified using a competitive ABA-ELISA (enzyme-linked immunosorbent assay) as described by Weiler (1986).

Extraction of RNA and Northern hybridization. Poly(A)+ RNA was prepared from transgenic as well as untransformed tissue as described by Bartels and Thompson (1983). Total RNA was extracted according to De Vries *et al.* (1986). Northern analysis was carried out according to Bartels *et al.* (1990).

Southern blot analysis. Genomic DNA was isolated (Doyle and Doyle, 1990) and digested with selected restriction enzymes. Hybridizations were performed according to Maniatis *et al.* (1982). The hygromycin gene and the 35S transcriptional enhancer were isolated from the pPCVICEn4HPT vector and used as probes. DNA fragments were radiolabelled as described by Feinberg and Vogelstein (1983).

Contruction and screening of a genomic library. Genomic DNA was isolated from leaflets of transformed plants regenerated from the desiccation-tolerant callus according to standard procedures (Schwarz-Sommer *et al.*, 1984). Fragments of 14 to 23 kb produced by partial digestion with *Mbo*I and isolated after agarose gel separation were cloned into the *Bam*HI-cut EMBL4 lambda vector, packaged *in vitro* and plated on K803 cells to yield from 800,000 to 1 million recombinants. The 35S transcriptional enhancer (337) was used as a probe for screening of the library.

Northern analysis of the genomic clones. The isolated genomic clones were restricted by *Eco*RI and fractionated in a 0.8% agarose gel. Each *Eco*RI fragment of two genomic clones of approximately 14 Kb, selected as representative, was used in a Northern hybridization analysis. For this experiment poly (A)+ RNA was extracted from transgenic callus and fresh and dried leaves from transgenic plants as well as wild-type callus, ABA treated callus and fresh and dried leaves of the untransformed plants. A fragment of 5.8 Kb that hybridized to the transgenic tissues and to ABA-treated callus and dried leves was used for screening of the cDNA library.

Screening of a cDNA library. A cDNA library from mRNA isolated from ABA-treated callus was previously constructed. A fragment of 5.8 Kb of the genomic clone was used as a probe for the isolation of the corresponding cDNA clone.

RESULTS AND DISCUSSION

In *Craterostigma plantagineum* desiccation tolerance is expressed in the whole plant and detached leaves. Although in callus a drying treatment increases the endogenous ABA concentration (Bartels *et al.,* 1990), this does not result in the induction of specific transcripts and hence of the desiccation tolerance. Rather exogenous ABA treatment is necessary for the accumulation of drought-induced mRNAs and proteins. Also in fully developed plants an ABA treatment mimics the desiccation process for the induction of desiccation-related genes. Thus, in *Craterostigma* ABA is apparently an important component of the pathway from water stress to gene activation.

The *Agrobacterium*-mediated transformation using the T-DNA tagging vector pPCVICEn4HPT and selection is summarized as follows:

Leaf explants transformation
|
Calli culture
|
Calli desiccation
|
Rehydration
|
Screening for survivors

A transgenic *Craterostigma* callus was selected for its ability to withstand desiccation without an exogenous application of ABA. Under the same conditions only ABA-treated wild-type calli were tolerant to desiccation, while untreated calli failed to "resurrect". This transgenic callus was more friable than normal *Craterostigma* callus and it had a reddish appearence similar to that of ABA-treated wild-type calli. When giberellic acid, an antagonist of ABA action, was added to the callus medium, the reddish callus turned green and its tolerance to desiccation was suppressed.

The insertion of the T-DNA into the plant genome was confirmed by Southern hybridization analysis. Northern analysis indicated that genes normally induced by desiccation or by ABA treatment were induced in the transgenis callus. Plants regenerated from transgenic callus did not display any particular phenotype, and calli newly produced from these plants were still desiccation-tolerant. The endogenous ABA level was determined in transgenic and in wild-type callus. The ABA-ELISA assay showed a similar ABA content in the samples tested.

DNA extracted from leaves of regenerated callus was used to construct a genomic library and 15 positive clones were selected by hybridization to the 35S transcriptional enhancer. DNA from purified genomic clones was restricted by *Eco*RI digestion and each isolated *Eco*RI fragment from two representative clones was used as a probe in a mRNA blot analysis. A fragment of 5.8 Kb containing a part of the transcriptional enhancer hybridized to mRNA: the tagged gene contained in this 5.8 Kb *Eco*RI fragment was constitutively expressed in the transgenic callus and leaves of plants regenerated from this callus, while in wild-type *Craterostigma* it was expressed only in dried leaves and ABA-treated callus. A mRNA of approximtely 1.0 Kb was recognized, and the corresponding cDNA clone was isolated from a cDNA library constructed from ABA-treated wild-type callus.

By T-DNA insertional mutagenesis an ABA-regulated gene has been tagged. Its induction is necessary in the pathway leading from water stress to desiccation tolerance. To confirm that the tagged gene is able to confer desiccation tolerance, its cDNA will be subcloned under the expression of the constitutive 35S promoter and used for transformation of *Craterostigma* leaf explants. Transgenic calli from this experiment should withstand desiccation without an exogenous ABA treatment, and prove that the isolated gene is essential in the ABA-regulated signal transduction pathway leading to desiccation tolerance.

ACKNOWLEDGEMENTS We thank R. Walden and C. Koncz for suggestions and support to the project. H. Sommer for constructing the genomic library and C. Sänger for technical assistance. A. F. is recepient of an EC fellowship within the Bridge programme.

REFERENCES

Bartels D., Thompson R.D. (1983) The characterization of cDNA clones coding for wheat storage proteins. Nucleic Acids Res. 11: 2961-2978.

Bartels D., Schneider K., Terstappen G., Piatkowski D., Salamini F. (1990) Molecular cloning of abscisic acid-modulated genes which are induced during desiccation of the resurrection plant *Craterostigma plantagineum.* Planta 181: 27-34.

Bartels D., Schneider K., Piatkowski D., Elster R., Iturriaga G., Terstappen G., Le Tran Binh, Salamini F. (1991) Molecular analysis of desiccation tolerance in the resurrection plant *Craterostigma plantagineum.* VI Nato Advances Study. Institute Plant Molecular Biology (eds.) R.G. Herrma, B. Larkins pp 663-671.

Bray E.A. (1988) Drought- and ABA-induced changes in polypeptide and mRNA accumulation in tomato leaves. Plant Physiol. 88: 1210-1214.

Bartels D., Velasco R., Schneider K., Forlani F., Furini A., Salamini F. (1993) Resurrection plants as model systems to study desiccation tolerance in higher plants. In Mabry T.J., Nguyen H.T., Dixon R.A., Bonness M.S. IC^2 Institute Austin, Texas, pp 47-58.

Bray E.A. (1991) Regulation of gene expression by endogenous ABA during drought stress. In Davies W.J., Jones H.G. (eds) Abscisic acid: physiology and biochemestry. Bios Scientific Publishers, Oxford, pp 81-98.

De Vries S., Hoge H., Bisseling T. (1986) In: Plant molecular biology manual B6: 1-13 (eds.)Kluwer Academic Publishers, Belgium.

Doyle J.J., Doyle J.L. (1990) Isolation of plant DNA from fresh tissue. Focus 12: 13-15.

Dure III L., Crouch M., Harada J., Ho T.-H.D., Mundy J., Quatrano R., Thomas T., Sung Z.R. (1989) Common amino acid sequence domains among the LEA proteins of higher plants. Plant Mol. Biol. 12: 475-486

Ebert P.R., Ha S.B., An G. (1987) Identification of an essential upstream element in the nopaline synthase promoter by stable and transient assays. Proc. Natl. Acad. Sci. USA 84: 5745-5749.

Feinberg A.P., Vogelstein B. (1983) A technique for radiolabelling DNA-restriction endonuclease fragments to high specific activity. Anal. Biochem. 132: 6-13.

Feldmann K.A. (1991) T-DNA insertion mutagenesis in Arabidopsis: mutational sprectrum. Plant J. 1: 71-82.

Furini A., Koncz C., Salamini F., Bartels D. (1994) *Agrobacterium*-mediated transformation of the desiccation tolerant plant *Craterostigma plantagineum* Plant Cell Reports (in press).

Guerrero F.D., Jones J.T., Mullet J.E. (1990) Turgor-responsive gene transcriptional and RNA levels increase rapidly when pea shoots are wilted. Sequence and expression of three inducible genes. Plant Biol. 15: 11-26.

Hayashi H., Czaja I., Schell J., Walden R. (1992) Activation of a plant gene by T-DNA tagging: Auxin independent growth in vitro. Science 258: 1350-1353.

Koncz C., Schell J. (1986) The promoter of the T_L-DNA gene 5 controls the tissue-specific expression of chimeric genes carried by a novel type of Agrobacterium binary vector. Mol. Gen. Genet. 204: 383-396.

Koncz C., Martini N., Mayerhofer R., Koncz-Kalman Zn., Körber H., Redei G.P., Schell J. (1989) High-frequency T-DNA-mediated gene tagging in plants. Proc. Natl. Acad. Sci. USA, 86:8467-8471.

Koncz C., Mayerhofer R., Koncz-Kalman Zn., Nawrath C., Reiss B., Re'dei G.P., Schell J. (1990) Isolation of a gene encoding a novel chloroplast protein by T_DNA tagging in *Arabidopsis thaliana.* EMBO J. 9: 1334-1346.

Maniatis T., Fritsch E.F., Sambrook J. (1982) Molecular cloning: a laboratory manual. Cold Spring Harbor Laboratory press, Cold Spring Harbor, New York, pp 187-198.

Mundy J., Chua N.H. (1988) Abscisic acid and water-stress induce the expression of a novel rice gene. EMBO J. 2279-2286.

Murashige T., Skoog F. (1962) A revised medium for rapid growth and bioassays with tobacco tissue cultures. Physiol. Plant. 15: 473-479.

Odell J.T., Nagy F., Chua N.H. (1985) Identification of DNA sequences required for the activity of the cauliflower mosaic virus 35S promoter. Nature 313: 810-812.

Schwarz-Sommer Zs, Gierl A, Klösgen R.B., Wienand U., Peterson P.A. Saedler H. (1984) The Spm (En) transposable element controls the excision of a 2-Kb DNA insert at the WX^{m-8} allele of *Zea mays*. EMBO J. 3: 1021-1028.

Skriver K., Mundy J. (1990) Gene expression in response to abscisic acid and osmotic stress. Plant Cell 2: 503-512.

Walden R., Koncz C., Schell J. (1990) The use of gene vectors in plant molecular biology. Meth. Mol. Cel. Biol. 1: 175-194.

Walden R., Hayashi H., Schell J. (1991) T-DNA as a gene tag. Plant J. 1: 281-288.

Weiler E.W. (1986) Plant hormone immunoassays based on monoclonal and polyclonal antibodies. In: Modern methods of plant analysis. Immunology in plant sciences. N.S., vol. 4, pp 1-17. Linskens H.F., Jackson J.F. (eds.) Springer, Berlin Heidelberg New York.

Yamaguchi-Shinozaki K., Koizumi M., Urao S., Shinozaki K. (1992) Molecular cloning and characterization of 9 cDNAs for genes that are responsive to desiccation in *Arabidopsis thaliana:* sequence analysis of one cDNA clone that encodes a putative transmembrane channel protein. Plant Cell Physiol. 33: 217-224.

Yamaguchi-Shinozaki K., Shinozaki K. (1993) The plant hormone abscisic acid mediated the drought-induced expression but not the seed-specific expression of rd22, a gene responsive to dehydration stress in *Arabidopsis thaliana.* Mol Gen Genet 236: 331-340.

ARE ARRESTIN-LIKE PROTEINS INVOLVED IN PLANT SIGNAL TRANSDUCTION PATHWAYS ?

A. NATO[1], A. MIRSHAHI[1a], J.M. CAVALCANTE ALVES[1b], D. LAVERGNE[1], G. DUCREUX[1], M. MIRSHAHI[2], J.-P. FAURE[2], J. DE BUYSER [3], G. TICHTINSKY[3], M. KREIS[3] and Y. HENRY[3]

1 : Laboratoire de Morphogenèse Expérimentale Végétale, Bt 360, Université Paris XI, 91405 Orsay, France. 2 : INSERM U86, Centre de Recherches Biomédicales des Cordeliers, 15 rue de l'école de médecine, 75270 Paris, France 3 : Institut de Biotechnologie des Plantes, URA CNRS 1128, Université Paris XI, 91405 Orsay, France. Felloships from a : AKF-Genève and b : CNPq-Brazil

ABSTRACT. Arrestin molecules represent a class of proteins involved in controlling different transduction pathways mediated by G-protein related receptors. Visual arrestin (or S-antigen) is a 48 kDa soluble protein of retinal photoreceptor cells where it participates in the regulation of phototransduction through its binding to activated and phosphorylated rhodopsin. The aim of the present study is to identify in plant cells arrestin-related proteins which could play a similar role in transduction processes involved in somatic embryogenesis. Immunological approaches, using two monoclonal antibodies (S8D8 and S9E2) raised against bovine arrestin were performed. We demonstrated the general distribution of a soluble protein of 38 kDa recognized by S8D8 in leaf and root extracts of dark- or light-grown wheat seedlings. Another 28 kDa polypeptide was recognized by S9E2 in soluble and insoluble fractions only in the green leaves. This second protein was also highly accumulated in light-grown wheat somatic embryos. Our results using aneuploid wheat lines lacking different chromosome arms showed that the 3DL arm which controls somatic embryogenesis was also involved in the expression of the gene coding for the 28 kDa protein. This study also suggests that the 28 kDa protein can be a regulatory element of the phytochrome signal transduction pathway. In addition its presence in membrane fractions of green leaves and light-grown somatic embryos is probably in relation with chloroplast differentiation and implicated in photosynthetic energy transduction mechanisms.

1. Introduction

In contrast to other eucaryotes, the differentiation program in plants is flexible. Since Stewards et al. (1958) using carrot cell suspension cultures, it has been evidenced that plant somatic cells were susceptible to develop into embryos when cultured *in vitro*. Somatic embryogenesis is the process by which single somatic cells from a wide range of plant tissues develop into differentiated whole plants through characteristic embryological stages, during *in vitro* culture. Auxin stimuli are likely to induce somatic embryogenesis from competent explants (Carman, 1990). The whole process involved stages requiring auxin and stages where auxin could be suppressed, and sometimes light stimuli can take more importance. The ability of cultured wheat somatic cells to produce embryos without a sexual process suggests that signals for embryogenesis were supplied *in vitro* to a wide range of cells that have the potential to express an egg-cell gene expression program. Aneuploid genetic stocks of the wheat variety Chinese Spring have been used to identify chromosome arms involved in the different steps of short- and long-term wheat somatic embryogenesis (Henry et al., in press). In a previous paper we determined the genetic basis of wheat somatic embryogenesis (De Buyser et al., 1992).

M. Terzi et al. (eds.), Current Issues in Plant Molecular and Cellular Biology, 519-524.

Up to now, little is known about the regulatory proteins involved in signal transduction pathways acting during plant embryogenesis. Light is used not only for photosynthesis, but also as a modulator of complex developmental and regulatory mechanisms. The aim of our work was to analyze analogies between retinal phototransduction regulation and plant photomorphogenesis through the existence of immunoreactive arrestin-like molecules as a new class of regulating proteins in plant transduction processes.

Visual transduction is a complex transmembrane signalling system that converts light stimuli into intracellular changes leading to neuronal impulse. A photon activates the visual receptor rhodopsin that catalyzes the exchange of nucleotides on the G-protein transducin. Transducin in its GTP-bound form activates a cyclic GMP phosphodiesterase (PDE). The quenching of metarhodopsin II is thought to arise from its phosphorylation by a rhodopsin kinase followed by the binding of a regulatory protein, arrestin (also called S-antigen). This process desensitizes the rhodopsin receptor and prevents continuous transducin activation. Arrestin thus plays a key role in arresting the phototransduction process (reviewed by Wilson and Applebury, 1993). Arrestins is the name given to a class of soluble proteins conserved among animal species. Homologs of visual arrestin have been characterized in various cells (Mirshahi et al., 1989, 1992b) and genes encoding β–arrestins regulating β-adrenergic transduction have been isolated (reviewed by Wilson and Applebury, 1993). We have previously identified in plants the presence of proteins immunoreactive with monoclonal antibodies raised against arrestin (Mirshahi et al., 1991, 1992a).

In this work, we show that proteins recognized by antibodies to arrestin are present in different parts of wheat seedlings. One of these proteins was also related to the *in vitro* wheat somatic embryogenesis capacity, suggesting that it could act as a signal regulator in transduction systems involved in plant embryogenesis.

2. Materials and Methods

2.1. PLANT MATERIAL

The spring wheat (*Triticum aestivum* L.) Chinese Spring (CS) used as a reference has a high somatic embryogenesis capacity (De Buyser et al., 1992). The Chinese Spring ditelosomic (DT) and nullisomic-tetrasomic (NT) lines were provided by S.M. Reader (AFRC, Cambridge Laboratory, UK) and G. Kimber (University of Missouri, USA). Each DT line lacks one pair of homologous chromosome arms, allowing the effect of the deficient genes to be measured. Each NT line lacks a given pair of homologous chromosomes and compensates with an extra pair of one of their homoeologous. Fourty of the 42 arms were analyzed.

2.2. TISSUE CULTURE

Fourteen days after anthesis the CS, CS DT and CS NT immature seeds were harvested and surface sterilized in 2% Ca hypochlorite for 1-2 min. The embryos were excised and placed, 10 per 9 cm Petri dish, with the scutellum exposed. The tissue culture conditions were described in Henry et al. (in press). The capacity for somatic embryogenesis was maintained by selecting the organized callus pieces at subculture. Nonembryogenic (NE) callus lines were developed by selecting unorganized callus pieces. Seedlings were germinated three days on sterile moistened paper in Petri dishes, under 16 h photoperiode or maintained in darkness.

2.3. TISSUE EXTRACTS

Young seedlings and callus tissues were ground in a prechilled mortar and then homogeneized in extraction buffer containing 50 mM Tris-HCl pH 7.6, 10 mM $MgCl_2$, 10 mM DTT and 0.2 mM PMSF. The homogenates were centrifuged at 9000 *g* for 15 min. The supernatant was taken as a soluble protein extract. The pellet was washed three times by resuspension in the extraction buffer containing 0.7 M NaCl (1 h at 4°C) and centrifugation at 9000 *g* for 20 min. The final pellet was resuspended in the extraction buffer and solubilized in 1% Triton X-100 (1 h at 4°C). Protein contents of the extracts were determined as described by Sedmak and Grossberg (1977).

2.4. POLYACRYLAMIDE GEL ELECTROPHORESIS AND WESTERN BLOTTING

SDS-PAGE was perfomed according to Laemmli (1970). For each sample 20 µg of denatured protein extract were electrophoresed for 2 h in a 15% polyacrylamide gel. Separated proteins were electrotransferred to PVDF membranes for 1 h using a semi-dry apparatus. The membranes were incubated with monoclonal antibodies (mAbs) against bovine retinal arrestin (Faure et al., 1984). We used mAbs S9E2 and S8D8 that respectively recognize a C-terminal epitope in the sequence 361-368 (EDPDTAKE) and a N-terminal epitope in the sequence 40-50 (PVDGVVLVDPE) of bovine arrestin. Antigen-antibody complexes were detected with sheep anti-mouse biotinylated antibody, followed by streptavidin-biotinylated horseradish peroxidase complex (Amersham). 4-Chloro-1-naphthol-H_2O_2 was used as substrate reaction mixture.

3. Results

3.1. PROTEINS CROSSREACTIVE WITH ARRESTIN IN WHEAT SEEDLINGS

The mAb S8D8 detected a polypeptide of 38 kDa on Western blots of soluble fractions of leaves and roots from green or etiolated seedlings. The mAb S9E2 labelled a polypeptide of 28 kDa in soluble and insoluble fractions from green leaves (Fig. 1). The absence of this protein in roots and etiolated leaves suggests that its synthesis is probably in relation with chloroplast differentiation.

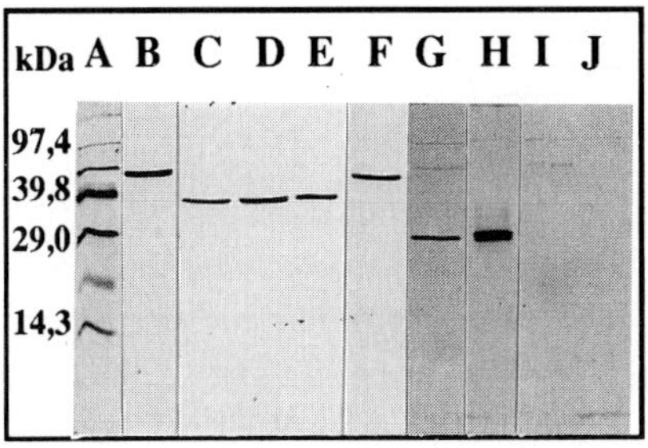

Figure 1. Western blot analysis of wheat seedlings.
Lane A : standard molecular weight markers, lanes B and F : bovine arrestin. Soluble fractions from green leaves (lane C), roots (lane D) and etiolated leaves (lane E) with the mAb S8D8. Soluble (lane G) and insoluble (lane H) fractions from green leaves, and soluble fractions from root (lane I) and etiolated leaves (lane J) with the mAb S9E2.

3.2. SOLUBLE PROTEINS AS A MARKER OF WHEAT EMBRYOGENIC POTENTIAL

Morphological observations reveal differences between embryogenic and nonembryogenic wheat cultures grown under light. Nonembryogenic calli are friable, white and translucent. The compact embryogenic calli are pale-yellow with green areas. The total soluble protein content is used as a simple marker of the embryogenic potential of wheat somatic tissue cultures. The embryogenic cultures were discriminated from nonembryogenic by a higher content of soluble proteins (Table 1) whereas there were no difference in the SDS-PAGE profile of the extract (Fig. 2).

TABLE 1. Soluble protein content of *in vitro* somatic wheat tissue culture expressed in $mg.g^{-1}$ fresh weight

GENOTYPE	CS E	CS DT3DL	CS DT3DS	CS NE
Somatic embryogenesis	Yes	Yes	No	No
Protein content	8.0	8.1	5.8	3.4

3.3. PROTEINS CROSSREACTIVE WITH ARRESTIN IN WHEAT SOMATIC TISSUE CULTURES

Western blot analysis with mAb S8D8 revealed positive signals in soluble proteins from both embryogenic (E) and nonembryogenic (NE) tissues. A polypeptide of 38 kDa was always detected by this mAb in E and NE extracts under dark or light culture conditions. S9E2 reacted with a 28 kDa polypeptide only in embryogenic cultures. This band was present in the soluble extract, whereas a doublet of 26-28 kDa was revealed in the insoluble fraction (Fig. 2). This polypeptide appears to be specific of the wheat somatic embryogenesis process under light or dark conditions, but its concentration was the highest in embryogenic cultures grown in the light.

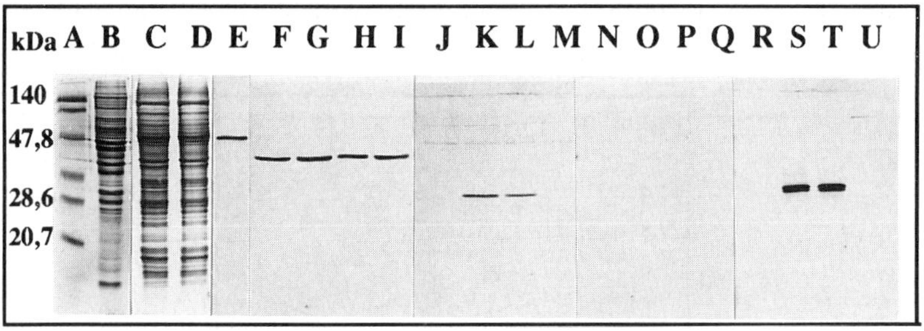

Figure 2. Western blot analysis of wheat somatic tissue cultures.
Lane A : standard molecular weight markers, lane E : bovine arrestin.
Soluble fractions from bovine retina (lane B), wheat embryogenic (lane C) and nonembryogenic (lane D) tissue culture, separated by SDS-PAGE and stained by Coomassie brillant blue G.
Western blot of soluble (lanes F to M) and insoluble (lanes N to U) fractions from somatic tissue cultures with mAbs S8D8 (lanes F to I and N to Q) or S9E2 (lanes J to M and R to U).
Embryogenic cultures are from Chinese Spring (lanes C, H, L, P, T) and DT 3DL (lanes G, K, O, S). Non embryogenic cultures are from Chinese Spring (lanes D, I, M, Q, U) and DT 3DS (lanes F, J, N, R).

3.4. DITELOSOMIC ANALYSIS OF PLANT ARRESTIN-CROSSREACTIVE PROTEINS

In order to confirm that a protein immunoreactive with S9E2 is a marker of the embryogenic potential of wheat tissue cultures, we used the 36 ditelosomic and 7 nullisomic-tetrasomic Chinese Spring lines previously tested for their short- and long-term somatic embryogenesis ability, including the nonembryogenic DT 3DS (Henry et al., in press). Western blot analysis of the total soluble protein extract from the embryogenic cultivar Chinese Spring was used as a reference. When compared to Chinese Spring and for example to the embryogenic DT 3DL, only the extract from the non embryogenic DT 3DS cultures did not show any immunoreactivity with mAb S9E2 (Fig . 2). This result confirms that the accumulation of a 28 kDa protein in somatic tissue cultures is correlated with embryogenic competence. Our results also suggest that a gene located on the wheat chromosome arm 3DL controls the expression of this protein in the callus.

4. Discussion

Little is known in plants about signal transduction pathways involving G-proteins. During the past few years, many efforts were done to characterize receptors and associated functions in plants (Barbier-Brygoo *et al.*,1989; Guern *et al.*, 1990; Weiss *et al.*, 1993). It is now well admitted that there is a potential coupling of various receptors to G-proteins (Hasunuma *et al.*, 1987; Guern *et al.*, 1990; Neuhauss et al., 1993; Weiss *et al.*, 1993). Studies on signal transduction in plant systems used the concepts from the animal kingdom. In our work, we refer to the well known regulation of phototransduction in retinal cells, where arrestin plays a central role, and we consider the possibility that similar molecules could participate in transduction pathway regulation in plants.

MAbs S8D8 and S9E2 raised against N- and C-terminal epitopes of bovine arrestin detected two types of immunoreactive proteins in wheat seedlings. The first type (38 kDa) recognized by S8D8 was detected in different parts of the wheat seedlings (roots, leaves) grown in the dark or under light. The second type (28 kDa) recognized by S9E2 was only observed in wheat green leaves from seedlings grown under light and was present in both soluble and membranous fractions. This suggests that the expression of the second protein is under the control of at least one photoregulated gene, indicating the implication of a light-regulated receptor. Phytochrome is a well characterized plant photoreceptor influencing processes such as seed germination, stem elongation, leaf expansion, chloroplast biogenesis and the induction of flowering (Neuhaus et al., 1993). Our working hypothesis is that the 28 kDa protein, light dependent, is an intermediate which participates in a signal transduction mechanism involving phytochrome.

The implication of the 28 kDa protein in the regulation of signal transduction pathways was searched through the study of the wheat somatic embryogenesis process. The most striking feature of the present work is the characterization of the enhancement of the protein recognized by mAb S9E2 in embryogenic wheat tissue cultures grown under light. If we consider that embryo induction and development (i.e. polarity acquisition, cell division and elongation, nutrient accumulation and transport, cell differentiation) are under intrinsic hormonal influences or regulations (Zimmerman, 1993), we shall suppose that the multiple signal transduction pathways being activated would mobilize arrestin-like molecules for the regulation of receptor-mediated responses. This could explain the important accumulation of the 28 kDa protein. The light stimulation of this protein crossreactive with arrestin in embryogenic wheat tissues suggests its implication as a regulatory element in a photomorphogenetic signal transduction pathway via phytochrome. The light responses under phytochrome control may involve activity of one or more G-protein as already evidenced by Bossen et al. (1990). Recently, high level of the α subunit of G-protein was observed in developing embryos of *Arabidopsis* (Weiss *et al.*, 1993). We postulate that desensitization of receptors (i.e. the loss of responsiveness of receptors that are continuously stimulated) is a widespread process also present in the plant kingdom. Our experiments using aneuploid lines indicate that the soluble 28 kDa protein is encoded by a nuclear gene. This protein correlates with events in wheat somatic embryogenesis. The protein recognized by the mAb S9E2 is a new protein marker for somatic embryogenesis. The involvement of this molecule in somatic embryogenesis gene regulation is a stimulatory working hypothesis. Such a protein could play a role in the transduction of signals produced both by a hormonal factor and a light stimulus, and mediated by G-protein related receptors during plant development.

To gain more insight into the localization and function of these proteins during plant growth and development, we are now conducting more extensive gel blotting and immunolocalization studies. Future researches will address the question of whether the plant arrestin-like molecules are part of novel signal transduction pathways.

5. References

Barbier-Brygoo, H., Ephritikine, G., Klämbdt, D., Ghislain, M. and Guern, J. (1989) 'Functional evidence for an auxin receptor at the plasmalemma of tobacco mesophyll protoplasts', Proc. Natl. Acad. Sci.,USA 86, 891-895.

Bossen, M.E., Kendrick, R.E. and Vredenberg, W.J. (1990) 'The involvement of a G-protein in phytochrome-regulated, calcium dependent swelling of etiolated wheat protoplasts', Physiol. Plant. 80, 50-62.

Carman, J.G. (1990) 'Embryogenic cells in plant tissue cultures : occurence and behaviour'. In vitro Cell. Dev. Biol. 26, 746-753.

De Buyser, J., Marcotte, J.L. and Henry, Y. (1992) 'Genetic analysis of *in vitro* wheat somatic embryogenesis', Euphytica 63, 265-270.

Faure, J.-P., Mirshahi, M., Dorey, C., Thillaye, B., De Kozak, Y. and Boucheix, C. (1984) 'Production and specificity of monoclonal antibodies to retinal S-antigen', Curr. Eye Res. 3, 867-872.

Guern, J., Ephritikhine, G., Imhoff, V. and Pradier, J.M. (1990) 'Signal transduction at the membrane bound level of plant cells', in H.J.J. Nijkamp, L.H.W.Van Der Plas and J. Van Aartrijk (eds), Progress in Plant Cellular and Molecular Biology, Kluwer Acad. Pub., Dordrecht/Boston, pp. 466-479.

Hasunuma, K., Furukawa, K., Tomita, K., Mukai, C. and Nakamura, T. (1987) 'GTP-binding proteins in etiolated epicotyls of *Pisum sativum* (Alaska) seedlings', Biochem. Biophys. Res. Comm. 148, 133-139.

Henry, Y., Marcotte, J.L. and De Buyser, J. 'Chromosomal location of genes controlling short-term and long-term somatic embryogenesis in wheat revealed by immature embryo culture of aneuploid lines', Theor. Appl. Genet. (in press).

Laemmli, U.K (1970) 'Cleavage of structural proteins during the accessibility of the head of bacteriophage T4', Nature 227, 680-685.

Mirshahi, M., Borgese, F., Razaghi, A., Scheuring, U., Garcia-Romeu, F., Faure, J.-P. and Motais, R. (1989) 'Immunological detection of arrestin, a phototransduction regulatory protein, in the cytosol of nucleated erythocytes', FEBS Lett. 258, 240-243.

Mirshahi, A., Nato, A., Razaghi, A., Mirshahi, M. and Faure, J.P. (1991) 'Présence de protéines apparentées à l'arrestine (antigène-S) dans les cellules végétales', C. R. Acad. Sci., Paris 312, 441-448.

Mirshahi, A., Nato, A., Lavergne, D., Ducreux, G., Faure, J.P. and Mirshahi, M. (1992a) 'Chloroplastic membrane-bound arrestin-like immunoreactive proteins in tobacco and *Chlamydomonas* cells, in J.L. Rigaud (ed)., Structure and Functions of Retinal Proteins, John Libbey Eurotext, Paris, pp. 351-354.

Mirshahi, M., Razaghi, A., Vandewalle, A., Cluzeaux, F., Tarraf, M. and Faure, J.-P. (1992b) 'Immunodetection and localization of protein(s) related to retinal S-antigen (arrestin) in kidney', Biol. Cell 76, 175-184.

Neuhaus, G., Bowler, C., Kern, R. and Chua, N.-H. (1993) 'Calcium/calmodulin-dependent and independent phytochrome signal transduction pathways', Cell 73, 937-952.

Sedmak, J.J. and Grossberg, E. (1977) 'A rapid, sensitive and versatile assay for protein using Coomassie brilliant blue G 250', Anal. Biochem. 79, 544-552.

Stewards, F.C., Mapes, M.O. and Mears, K. (1958) 'Growth and organized development of cultured cells. II. Organization in cultures grown from freely suspended cells', Am. J. Bot. 45, 705-708.

Weiss, C.A., Huang, H. and Ma, H. (1993) 'Immunolocalization of the G-protein α subunit encoded by the *GPA1* gene in *Arabidopsis.* ', The Plant Cell 5, 1513-1528.

Wilson, C.J. and Applebury, M.L. (1993) 'Arresting G-protein coupled receptor activity', Current Biology 3, 683-686.

Zimmerman, J.L. (1993) 'Somatic embryogenesis : a model for early development in higher plants', The Plant Cell 5, 1411-1423.

AN ALTERNATIVE APPROACH TOWARDS UNDERSTANDING MONOCOT ZYGOTIC EMBRYOGENESIS

CH. FISCHER & G. NEUHAUS
Institute of Plant Sciences (Prof. I. Potrykus), ETH - Zürich
Universitätstrasse 2
CH - 8092 Zürich

Abstract. We are focusing our attention on the mechanisms that determine the change from radial to bilateral symmetry during early monocot embryo development. The model plant chosen for this purpose is wheat (*Triticum aestivum* L.). An *in vitro* culture system for early zygotic embryos (globular stage) has been established. As a first approach, the influence of auxins in early embryo pattern formation will be studied. For this purpose, antiauxins (2,4,6-trichlorophenoxyacetic acid) as well as auxin polar transport blockers (2,3,5-triiodobenzoic acid) will be used.

Introduction.

Our aim is to gain information on the signals influencing early plant embryogenesis. In particular, we have focused our attention on the mechanisms underlying the change from the globular proembryo to the polar organisation (shoot - root axis) of monocot embryos. This shift in the embryonic symmetry represents the initial delineation of the two major embryonic organ systems: the scutellum and the axis. Little is known concerning the mechanisms underlying this change of symmetry. Nevertheless, recently, it has been shown that auxins influence the establishment of bilateral symmetry during early dicot embryogenesis (*Brassica juncea)* (Liu *et al.* (1993b)). By contrast related experiments in monocots are missing. Therefore, while there are many putative candidates (signal transduction intermediates, homeobox genes, and hormones) that may influence this shift of symmetry, we will, as a first approach, study the role of auxins during this important step. For this purpose, the main prerequisite is to establish an *in vitro* culture system for young zygotic embryos. Unfortunately, until now, there have been only a few publications that describe the normal development of a proembryo to a mature embryo *in vitro* (Monnier (1976); Liu *et al.* (1993); Mol *et al.* (1993)). In many cases, related studies have been more descriptive and have not discussed the quality of the embryogenesis, i.e. the percentage of embryos that display multiple shooting, etc. (Kost *et al.* (1992); Kranz and Lörz (1993)). Often, the starting material comprised older embryos (i.e Iglesias *et al.* (1994)). All these difficulties encounted are mainly due to the lack of *in vitro* culture conditions which would allow normal direct embryogenesis to proceed, particulary in the case of young embryos. Embryogenesis studies have also been hindered by the relative inacessibility of the plant embryo at early stages. Several requirements have to be fulfilled in order to establish an *in vitro* culture system for zygotic embryos. The development *in vitro* has to be very close to *in vivo*

M. Terzi et al. (eds.), Current Issues in Plant Molecular and Cellular Biology, 525-530.

development. The time needed to achieve complete development has to be in the same range as for *in vivo* development. Finally, the system should be simple to handle and be accessible for manipulation. The model plant chosen for these studies is wheat (*Triticum aestivum*).

Material and methods

Wheat plants (*Triticum aestivum* L. cv Sonora, TDS seeds, Switzerland) were grown in a greenhouse under 17 h light / 7hr dark at 18 ºC day / 15 ºC night. The daylight was supplemented with 400 W fluorescent lamps (HPL-N 400W mercury lamp, Philips), if a minimum of 5000 lux was not reached during the day. The minimum humidity was 35 % during the day and 50 % during the night. The ears, collected 3 days after anthesis, were surface sterilized with 70 % ethanol. Immature embryos, having an embryo proper diameter of 100 to 160 µm, were excised aseptically under a dissecting microscope, keeping the tissues in the isolation N6P medium (table 1) droplet. The excised embryos were then plated on the surface of a double layer N6P medium system (table 1) in a 3.5 cm diameter petri dish and cultured in the light. As soon as the first leaf primordia covered the meristem, the embryos were transferred to the N6P bottom medium of the double layer system. For germination, the embryos were transferred to MS medium (Murashige and Skoog (1962)) without hormones containing 2% sucrose.

Table 1: Composition of culture media.

Reference	Composition
Isolation N6P medium	N6 inorganic salts (Chu *et al.* (1975)) N6 vitamins (Chu *et al.*(1975)) 2 mg/l glycine 100 mg/l Casein hydrolysate 16 % sucrose 0.6 mg/l 6-Benzyl -aminopurine (BAP) 0.6 mg/l MES pH 5.6
N6P double layer system	1 ml N6P top medium 2 ml N6P bottom medium
N6P top medium	Isolation N6P medium solidified with 8 g/l Sea Plaque Agarose (Sea Plaque, FMC Bioproducts, Rockland, USA)
N6P bottom medium	N6P top medium without hormones containing only 5 % sucrose

Results

IN PLANTA EARLY DEVELOPMENT OF WHEAT IMMATURE EMBRYO

The first cell division of the zygote gives the terminal cell which develops into the hemispherical embryo proper, and the basal cell which gives rise to the cylindrical suspensor (fig. 1, A). The development of the proembryo into a well differentiated embryo is marked by the initiation and rapid enlargement of the scutellum and the concomittant exogeneous formation of the shoot meristem. The first step of this process is the transition phase characterized by a shift from the radial symmetry of the proembryo to the bilateral symmetry of the embryo. This transition phase starts with the initiation of the scutellum. The scutellum grows in both the axial and the lateral directions resulting in a prominence on one side of the embryo proper. Rapidly following the scutellum initiation, the swelling on the other face of the embryo proper indicates

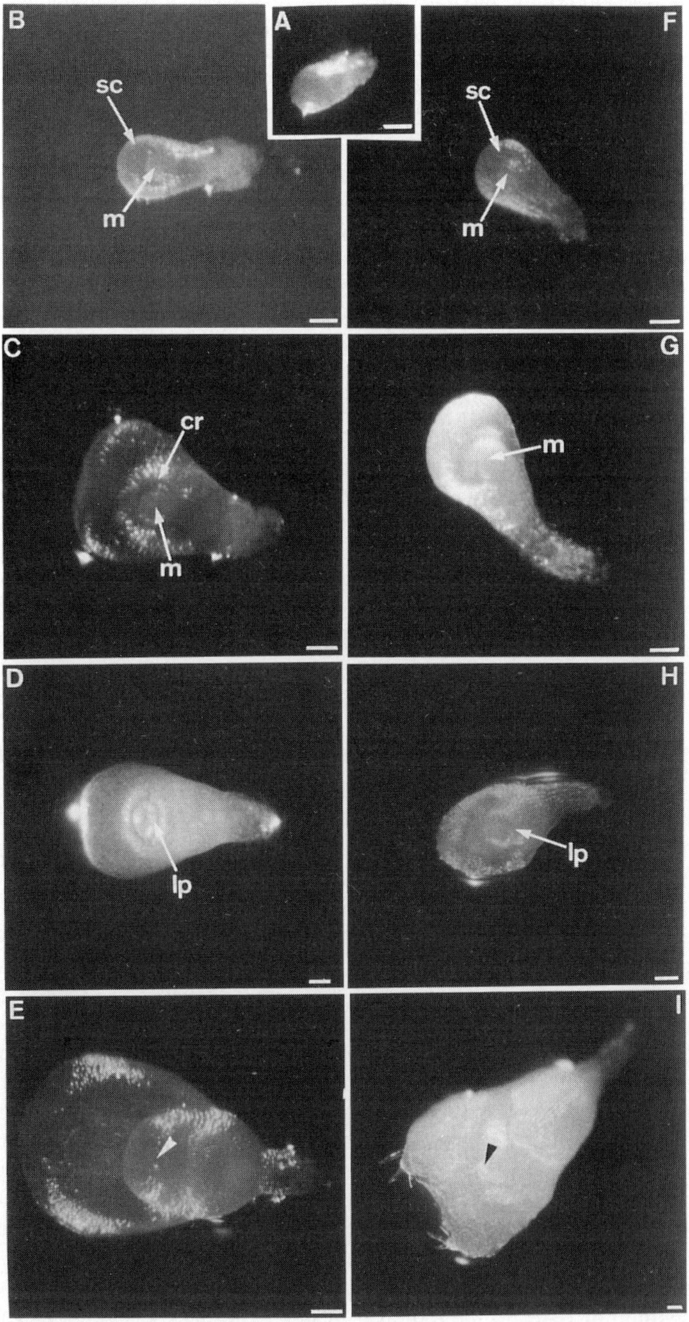

Figure 1: Comparison between *in planta* and *in vitro* development of wheat globular zygotic embryos. A: globular proembryo; B, C, D, E: *in planta* development; F, G, H, I: *in vitro* development; bars represent 100 μm. (sc: scutellum, m: shoot apical meristem, cr: coleoptilar ring, lp: first leaf primordium, arrowhead: pore not enclosed by the coleoptilar ring)

the initiation of the shoot meristem. The final stage of transition is represented by the embryo in figure 1, B. During subsequent development, the shoot meristem will be surrounded by the coleoptilar ring (fig. 1, C). Then, in between the apical meristem and the coleoptilar ring, the first leaf primordium appears at the base of the meristem (fig. 1, D). This leaf primordium progressively covers the shoot meristem. The result of this process is that the meristem of the well developed immature embryo is entirely covered by the first leaf primordium which is itself, except for a small pore, enclosed in the coleoptilar ring (fig. 1, E). The scutellum develops, in parallel to the apical meristem, into an oval shield-like organ (fig. 1, E). It is attached to one side of the axis structure and makes up most of the bulk of the embryo. By contrast to the shoot meristem, the root meristem is not located at the exterior of the embryo and hence can not be detected morphologically.

IN VITRO DEVELOPMENT OF WHEAT GLOBULAR EMBRYOS

There are several differences between *in planta* and *in vitro* development. *In vitro*, the scutellum has variable shapes in contrast to the well defined shield-like structure observed during *in planta* development. The bulk of the embryo is not protuding but is instead surrounded by the scutellum (compare fig. 1, E and H). This is probably due to the fact that the coleoptilar "ring" is not, in the great majority of the cases (fig. 1, I represents an exception), enclosing the shoot meristem and the leaf primordia but remains largely open (fig. 1, H). The growth of the first leaf primordium is also different. Instead of covering the meristem, it forms a sheath surrounding the meristem (fig 1, H). However, even with these differences, 20 % of the embryos are able to undergo direct embryogenesis which results in the germination of normal plantlets.

IMPORTANT FACTORS FOR "NORMAL" DIRECT EMBRYOGENESIS IN VITRO

The embryos excised from the kernels were mainly in the globular to late globular stage (fig. 1, A). In these stages, the embryo proper had a diameter of 100 to 150 µm. In some cases, young transition embryos were also isolated (160 µm). Certain factors were found to have a prevailing influence on the culture of young wheat embryos. Among them, the osmotic pressure of the medium, adjusted by the addition of sucrose, was very important. We observed that the younger the excised embryos were, the higher the medium osmolarity had to be. The sucrose concentration necessary to allow the growth and the development of freshly excised young globular embryos was 16 %. Below this concentration, the embryo was either not able to growth or it developed into an abnormal structure. Likewise, in order to avoid osmotic shock, the embryo isolation had to be carried out in the culture medium. We observed that embryos cultured in liquid medium died soon after culture or did not develop normally. Therefore, embryos were placed on the top of an agarose medium. Based on the consideration that the embryo *in vivo* is subjected to a higher osmotic pressure at early globular stages, which rapidly decreases during development (Smith, 1973), a double layer system was developed. In this system, the top layer medium contains 16 % of sucrose and the bottom layer 5 %. This system allowed up to 20 % of the embryos in culture to initiate a scutellum which subsequently enlarged and to form a shoot meristem. Finally, these embryos differentiate leaf primordia (table 2). Unfortunately, these embryos were blocked at this stage and could not germinate afterwards. The cytokinins, BAP, also has a positive effect on embryo development *in vitro*. The addition of BAP, in the top layer medium enhances leaf primordia formation (table 2) which reflects an improvement in the quality of the shoot meristem differentiation. The optimal concentration of BAP was 0.6 mg/l. Nevertheless, even under these conditions, the embryos were not able to germinate, probably due to a lack or excess of another component. Without addition of BAP to the medium, a shoot meristem can be formed but is, in the majority of the cases, blocked early before leaf primordia formation. The medium used, based on N6 medium, (Chu *et al.* (1975)) initially contained a high concentration of proline (25 mM). This high concentration as well as lower concentrations had inhibitory effect on embryo germination. In fact, in order to obtain qualitatively and quantitatively an optimal germination rate, it was necessary to remove proline from the medium. If the proline was not totally removed but added at lower concentrations, germination of the embryos would occur but with lower efficiencies, depending on the concentrations used. Nevertheless, germination only occurred if the embryo had developed to the point where the first leaf primordium covered the meristem. Otherwise in order to avoid vitrification and the resulting arrest of plantlet growth, the sucrose concentration had to be reduced progressively to 2 %. Subsequently, the optimal efficiency of normal plantlet formation obtained was 18 % (table2) .

Table 2: Quality of embryo development obtained with different culture parameters

Parameter	Highest differentiation stage achieved	Efficiency*
Double layer system	Leaf primordia	20
BAP (0.6 mg/l)	Leaf primordia	18
Absence of proline	Plantlet	18

* % reported to the total amount of embryos in culture.

THE STUDY OF THE EFFECT OF ANTIAUXINS ON CULTURED YOUNG EMBRYOS

We are presently manipulating the established cultured system in order to study the influence of auxins on early embryo pattern formation. In the initial experiments, we will focus on the effect of auxins on the onset and initial development of the embryo axis and the scutellum. For this purpose, antiauxins will be added directly to the culture medium. Two classes of antiauxins will be used: auxin polar transport inhibitors (2,3,5-triiodobenzoic acid) and molecules that are assumed to have an antagonistic effect (2,4,6-trichlorophenoxyacetic acid). The plant material selected for this approach comprises very young embryos, namely globular embryos as well young transition embryos (embryo propers having a diameter of 100 to 160 μm). These studies are currently being performed.

Discussion

An *in vitro* culture system has been established allowing the development through direct embryogenesis of very young wheat embryos. The pattern of development of these embryos mimics the one observed *in planta.*

Some of the important factors required for the culture of these embryos have been highlighted by our experiments. Our observation that the younger the excised embryo is, the higher the medium osmolarity should be, coincides with the conclusion obtained previously by Rietsema *et al.* (1952). A double layer culture system has been set up so that the osmotic pressure of the medium decreases progressively during embryo culture without having to transfer the embryos. This system mimics to a certain extent the osmotic decrease that occurs in the endosperm of the developping kernel (Smith (1973); Iglesias *et al.* (1994)). Monnier (1976) was the first to recognize the importance of this kind of system and this system has been used with success in several other studies (Liu *et al.* (1993a); Iglesias *et al.* (1994)). Hormones, in particular, BAP, have a benefitial effect on shoot meristem differentiation in young globular embryos, in contrast to their effect on older wheat embryos which already have a clearly differentiated shoot meristem (Iglesias *et al.* (1994),. Another important improvement of the culture system was the removal of proline which has an inhibity effect on the germination of these embryos. In previous studies (Armstrong and Green (1985)), proline has been shown suitable for the generation of highly embryogenic maize calli.

We now possess an *in vitro* culture system that allows the normal growth of proembryos to mature embryos. Taking advantage of this culture system, the proembryos can now be manipulated *in vitro* in order to study the changes from radial to bilateral symmetry. This change has to be considered on several levels i.e. cell division patterns, modification of the cytoskeleton...etc. In monocots, there is a lack of information concerning the mechanisms involved. By contrast, in dicots, Liu and coworkers showed in a recent report that polar auxin transport is essential for the establishment of bilateral symmetry of *Brassica juncea* embryos (Liu *et al.* (1993b)). When auxin polar transport inhibitors were added to the culture medium of globular indian mustard embryos, instead of two cotyledons forming on opposite sites of each embryo other as in the wild untreated embryo, the embryos each formed a ring of cotyledon tissus. Therefore, we will first focus our attention on the influence of auxins on this important step by using auxin polar transport inhibitors as well as auxin "antagonists".

Acknowledgements: C. F. is supported by an EMBO fellowship (ALTF 287-1993)

References

Armstrong, C.L., and Green, C.E. (1985) 'Establishment and maintenance of friable, embryogenic maize callus and the involvement of L-proline', Planta 164, 207 - 214.

Chu, C.C., Wang, C.C., Sun, C.S., Hsu, C., Yin, K.C., and Chu, C.Y., and Bi, F.Y. (1975) 'Establishment of an efficient medium for anther culture of rice, through comparative experiments on the nitrogen sources', Scientia Sinic. 18, 659 - 668.

Iglesias, V.A., Gisel, A., Potrykus, I., and Sautter, C. (1994) '*In vitro* germination of wheat proembryos to fertile plants', Plant Cell Reports 13, 377 - 380.

Kost, B., Potrykus, I., and Neuhaus G. (1992) 'Regeneration of fertile plants from excised immature zygotic embryos of *Arabidopsis thaliana*', Plant Cell Reports 12, 50 - 54.

Kranz, E., and Lörz, H. (1993) '*In vitro* fertilization with isolated, single gametes results in zygotic embryogenesis and fertile maize plants', The Plant Cell 5, 739 - 746.

Liu, C. M., Xu, Z. H., and Chua, N. H. (1993, a) 'Proembryo culture: *in vitro* development of early globular-stage zygotic embryos from *Brassica juncea*', The Plant Journal 3, 291 - 300.

Liu, C. M., Xu, Z. H., and Chua, N. H. (1993, b) 'Auxin polar transport is essential for the establishment of bilateral symmetry during early plant embryogenesis', The Plant Cell 5, 621 - 630.

Mol, R., Matthys-Rochon, E., and Dumas, C. (1993) '*In vitro* culture of fertilized embryo sacs of maize: Zygotes and two-celled proembryos can develop into plants', Planta 189, 213 - 217.

Monnier, M. (1976) 'Culture *in vitro* de l'embryon immature de *Capsella Bursa-pastoris* Moench', Rev. Cyt. Biol. végét. 39, 121 - 138.

Murashige, T., and Skoog, F. (1962) 'A revised medium for rapid growth and bio-assays with tobacco tissue cultures.', Physiol. Plant. 15, 473 - 497.

Rietsema, J., Satina, S., and Blackeslee, A.F. (1952) 'The effect of sucrose on the growth of *Datura stramonium* embryos *in vitro*', Am. J. Bot. 40, 538 - 545.

Smith, J.G. (1973) ' Embryo development in *Phaseolus vulgaris* ii. Analysis of selected inorganic ions, ammonia, rganic acids, aminoacids and sugars in the endosperm liquid', Plant Physiol. 51, 454 - 458.

MUTATIONS CONFERRING EARLY FLOWERING ON A LATE-FLOWERING ECOTYPE OF *ARABIDOPSIS THALIANA*

D.P. Wilson and A.D. Neale
Department of Genetics and Developmental Biology,
Monash University, Clayton, Victoria, Australia, 3168.

The late-flowering *Arabidopsis thaliana* ecotype Pitztal requires approximately 90 days under standard growth conditions before the onset of flowering occurs. Early-flowering ecotypes of *A. thaliana* grown under the same conditions will flower in around 20-30 days. The flowering time for many late ecotypes can be significantly reduced by vernalization of the germinating seed. Using both chemical mutagenesis and ionising radiation mutagenesis we have obtained several independent mutants of the Pitztal ecotype which have similar flowering characteristics to the early ecotypes. These mutants are being characterized and strategies to clone the mutated genes are underway.

Introduction

During the development of angiosperms leaves, stems and flowers arise from apical meristematic tissue. The apical meristem is a reservoir of cells that sequentially gives rise to embryonic, juvenile and adult leaves and ultimately forms reproductive floral structures containing the plant gametes. In *A. thaliana* the onset of flowering involves reorganization of the vegetative shoot meristem into an inflorescence meristem from which the different parts of the inflorescence are derived. The inflorescence meristem then generates floral meristems which produce floral organ primordia. Wild-type *Arabidopsis* has an indeterminate inflorescence with the inflorescence meristem retaining its identity throughout the remainder of the life of the plant (Weigel *et al.*, 1992).

In a number of plant species the transition of the apical meristematic tissue from the formation of vegetative to floral structures requires the perception by the plant of the appropriate environmental signals. In *Arabidopsis* the timing of the transition can be affected by changes in daylength (photoperiodism) or exposure to cold (vernalization) (Napp-Zinn, 1979). The flowering time for many late-flowering ecotypes can be significantly reduced by vernalization.

Arabidopsis contains both early and late flowering ecotypes. Recent reports on the control of flowering in *Arabidopsis* have largely focussed on flowering mutants of early-flowering ecotypes. It is expected that genetic and physiological analysis of these mutants will provide insight into the fundamental processes involved in floral induction. Similarly the study of flowering mutants in late-flowering ecotypes should also lead to an understanding of floral induction with the possibility of isolating a gene encoding an inhibitor of floral induction and with the further possibility of isolating genes which are regulated by the inhibitor. To this end we have

M. Terzi et al. (eds.), Current Issues in Plant Molecular and Cellular Biology, 531-538.

undertaken a mutagenesis and screening program on the late flowering *Arabidopsis* ecotype Pitztal to isolate early-flowering mutants.

Materials and Methods

Seed of the ecotype Pitztal originated from the Langridge and Brock *Arabidopsis* seed collection. Seeds of Pitztal, Columbia or Landsberg *erecta* were sown in punnets containing a 1:1 (w/w) mixture of Debco Seed Raising mix and Perlite. Trays of punnets were covered with a layer of polyethane to prevent moisture loss. Seed not undergoing vernalisation were placed in the Greenhouse. Vernalisation treatment involved germinating the seed at 4°C for the appropriate length of time and subsequent transfer to the greenhouse. Plants were grown under constant fluorescent light and watered with standard nutrient solution. Flowering times were recorded daily following appearance of flower primordia and bolting among plants undergoing various treatments.

For EMS mutagenesis 10,000 seeds of the Pitztal ecotype were placed in a 40 mM solution of Ethyl Methanosulfonate (EMS). The mixture was stirred for eight hours at room temperature. Seeds were washed eight times over a period of two hours with distilled water. After the final wash seeds were resuspended in a 0·15% agar solution. Seeds were sown by pipetting 5ml of this mixture into each of 100 punnets. Seeds were vernalised for 14 days.

For gamma-irradiation mutagenesis 10,000 seeds of the Pitztal ecotype contained in a small seed envelope were placed in the chamber of the Gamma cell 1000 Gamma irradiator. Seeds were exposed to the Caesium-137 source for the length of time required to achieve a 100krad dose (6hrs 22min). Again seeds were suspended in a 0·15% solution of Agar and sown as for the EMS treated seeds. Seeds were vernalised for 14 days.

For *A. thaliana* crosses female recipients were early flowering mutants of the Pitztal ecotype. Male donors were wild-type Columbia ecotype. Using fine forceps cleaned in 100% ethanol, and a binocular microscope, flowers of the female recipient were hand emasculated by removing all six stamens. Any pollinated flowers within the inflorescence were removed before proceeding with the crosses. Anthers of mature flowers of the Columbia donor were removed and the pollen applied to the stigmatic surface of the female recipients.

Results

FLOWERING TIMES OF *ARABIDOPSIS* ECOTYPES.

Early-flowering ecotypes of *A. thaliana*, under standard conditions, will flower in around 20-30 days following germination. The late flowering *A. thaliana* ecotype Pitztal requires approximately 90 days under the same conditions before the onset of flowering occurs. The lateness of this ecotype is dominant when crossed with early flowering ecotypes (Burn *et al.*, 1993).

The flowering times of wild type ecotypes was determined under the conditions to be utilized for mutant screening. The mean flowering time for non-vernalized Pitztal ecotype was 110.8 days. Vernalization treatment of the Pitztal ecotype reduced the time to flower considerably. 14 days vernalization produced a mean flowering time of 86.4 days and a 19 day treatment gave a mean flowering time of 58.8 days. A similar decrease in Pitztal flowering time due to vernalization has been reported by Burn *et al.*, (1993) where a longer vernalization period of 21 days resulted in a mean flowering time decreased from 74.4 days to 22.9 days. The results of Burn *et al.* (1993) suggest that full vernalization is achieved by a cold period longer that 19 days. The flowering time recorded for fully vernalized Pitztal is similar to the mean flowering time of the non-vernalized early ecotypes Landsberg *erecta*, Niederzenz and Columbia which is 21.8, 21.1 and 21.9 days respectively under the growth conditions utilized here. Vernalization decreases the distribution of flowering times as well as the average flowering time (Figure 1.).

Figure 1. Flowering times of non-vernalized and vernalized Pitztal ecotype.

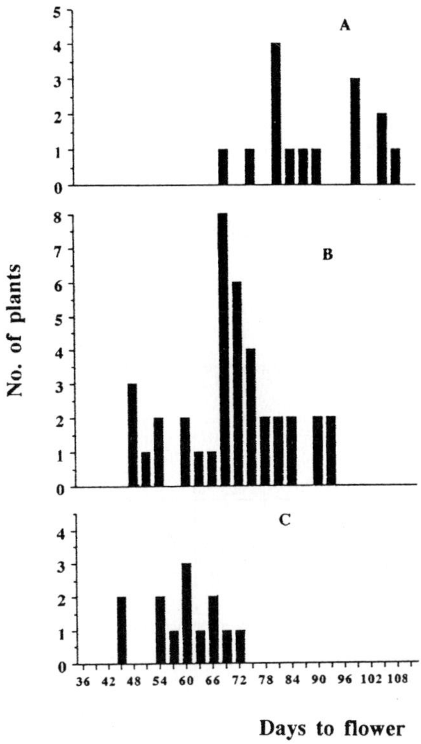

MUTAGENESIS OF PITZTAL ECOTYPE.

10,000 Pitztal seeds were mutagenized with EMS and another 10,000 seeds irradiated using a Cs^{137} source. The percent germination for mutagenized seed was significantly lower than untreated seed with chemically treated seed having the lowest percent germination value (see Table 1). A portion of the mutagenized germinating plants were slow growing and unhealthy with the most severe effects evident in plant derived from irradiated seeds. Death at the cotyledon stage was 5.3%. Very few deaths were observed at a later stage of development. Seed yield of M1 plants was also lower than normally found with untreated plants

Table 1. Percent germination and survival of M1 plants.

Type of plant	Percent germination	Percentage of deaths at the cotyledon stage
EMS treated Pitztal	18·6	2·0
γ-irradiated Pitztal	23·5	5·3
Pitztal control	59·3	---
Landsberg.*erecta* control	60·5	---

The flowering time for M1 plants following mutagenesis and propagation of M1 seed was within or longer than the range expected for partially vernalized wild-type Pitztal as indicated in Table 2.

Table 2. Mean flowering times of wild-type and treated Pitztal plants.

Type of plant	Mean flowering time (days)
Wild-type Pitztal	110·8
M1 Pitztal plants (vernalised 14 days)	
EMS	77·3
Gamma-irradiation	101·7
Early-flowering M2 Pitztal plants (Non-vernalised)	
EMS	25·9
Gamma-irradiation	27·5

SCREENING FOR FLOWERING-TIME MUTANTS.

Non-vernalized progeny from M1 plants were screened for early flowering. Plants flowering in less than 35 days were selected. A flowering time of 35 days was less than the earliest flowering (45 days) observed for partially vernalized W/T Pitztal. A total of 32 mutants (see Table 2) were obtained from the M2 population with the earliest flowering in 22 days and the latest selected at 33 days. EMS treatment resulted in 19 mutants being selected and 13 mutants were selected from the irradiation experiment. Figure 2 shows an early flowering mutant and a wild-type plant of the same age.

The mean flowering time was 25.9 days for EMS derived mutants and 27.5 days for the radiation induced mutants (see Table 2).

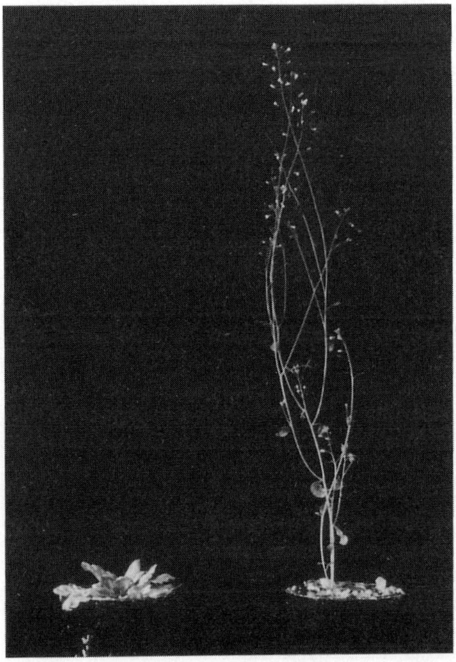

Figure 2. Non-vernalized early-flowering M2 Pitztal plant (right) and a wild-type Pitztal plant (left) 30 days following germination.

M3 lines were established and flowering times determined (see Table 3).

The mean flowering times for both EMS and radiation induced mutant lines showed a similar range. The mean flowering time of EMS M3 plants combined being 22.9 days compared with the value of 23.4 days for radiation induced M3 plants.

MUTANT CHARACTERIZATION

The M3 lines were crossed with early flowering Columbia ecotype and flowering times examined (see Table 4). When wild-type Pitztal is crossed with the early-flowering Columbia ecotype the progeny are late flowering (Burn *et al.*, 1993). The progeny from the crosses of the Pitztal early-flowering mutants with Columbia were all early flowering indicating that the mutation/s cannot be overcome by the Columbia genotype.

Table 3. Flowering times of radiation induced early-flowering mutant lines.

Mutant	number of plants	Mean flowering time
1	39	22·9 ± 3·7
2	61	22·2 ± 1·7
3	80	19·6 ± 5·4
4	24	24·9 ± 8·1
5	06	39·3 ± 4·2
6	40	21·9 ± 1·8
7	16	20·4 ± 2·9
8	53	18·9 ± 2·3
9	36	25·2 ± 3·9
10	28	27·9 ± 6·6
11	48	17·8 ± 1·9
12	47	19·6 ± 3·2

Table 4. Mean flowering times of the progeny resulting from crosses of irradiation induced mutants and Columbia ecotype.

Type of mutant crossed	Mean flowering time
2	20·8 ± 2·2
3	20·8 ± 4·0
4	26·1 ± 5·7
5	21·9 ± 3·3
6	29·3 ± 2·9
7	22·0 ± 1·5
8	26·8 ± 3·9
9	28·0 ± 0·0
10	23·5 ± 2·1
11	29·3 ± 6·6
12	19·8 ± 1·2

Discussion

Using both chemical mutagenesis and ionising radiation mutagenesis we have obtained several independent mutants of the late-flowering Pitztal ecotype which flower in less than 35 days without vernalization. The flowering time (appearance of flower primordia) of the mutants was characterized. Selected floral-induction mutants were crossed with early ecotypes to eliminate lines possibly containing non-specific stress-related mutations. No non-specific mutations were found.

Studies with a number of plant species have provided strong evidence for the existence of both flower-inducing and flower-inhibiting pathways (Evans, 1975; Lang, 1977; Murfet, 1989). In at least some plants both the long day response and the vernalization response may be under the control of factors which block flowering in the absence of the appropriate environmental signals.

Mutageneses in *Arabidopsis* have identified up to 30 loci which are required for normal flower development. Accumulating results from these studies suggest that flower development is under the control of a genetic hierarchy in which earlier-acting genes control the expression of later-acting ones. It is thought that the factors mediating floral induction establish the inflorescence meristem by activating genes which control the identity of inflorescence and floral meristems (Weigel and Meyerowitz, 1993). It is further proposed that these activated meristem identity genes act in a synergistic manner to promote floral over inflorescence meristem identity (Weigel *et al.* , 1992; Bowman *et al.* , 1992).

From genetic studies of various early- and late-flowering ecotypes of *A. thaliana* it is apparent that the number and interactions of genes controlling onset of flowering differs between ecotypes (see Karlovska, 1974). Up to 15 loci affect the floral transition in early-flowering ecotypes. Mutagenesis of these genes results in the isolation of mutants which have delayed flowering time and are known as *late flowering genes* (Kornneef *et al.*, 1991). These loci appear to act through three independent pathways and can be divided into groups according to their daylength and vernalization responses and being either recessive or semidominant are likely to act positively in floral induction (Kornneef *et al.*, 1991; Lee *et al.*, 1993; Burn *et al.*, 1993). Several mutants which flower earlier than wild-type have been reported (Sung *et al.*, 1992; Zagotta *et al.*, 1992; Goto *et al.*, 1991) however the mode of action in accelerating flowering is not yet clear.

From a number of studies utilizing early and late ecotypes (Harer, 1950; Napp-Zinn, 1957) it is thought that one major and a few minor factors influence the degree of lateness in most late ecotypes and that none of the mutations in the late-flowering early ecotype mutants would be allelic to any of the major factors controlling late-flowering in the late ecotypes (Redei, 1962). The late flowering ecotype Pitztal appears to contain a single major dominant gene which delays flowering and the location of this locus has been mapped to within 4.3 cM (approximately 700kB) of an RFLP marker (Burns *et al.*, 1993). It is likely that this locus corresponds to the *FLA/FRI* gene found to be responsible for late flowering in several late ecotypes (Napp-Zinn 1979; Clarke and Dean 1994; Lee *et al.*, 1993).

In this study radiation conditions were optimised to allow deletion production favourable for gene cloning. The generation of radiation-induced deletion mutants along with genomic subtraction has allowed the isolation of a number of genes from *Arabidopsis* (Sun *et al.*, 1992; Wilkinson and Crawford, 1991). As the early flowering mutants were selected in M2 lines, gross chromosomal aberrations should have been reduced due to their low transmissibility through pollen (Koornneef *et al.*, 1983).

Characterization of these mutants including segregation analysis, allelism tests, linkage analysis and mapping will be undertaken. This should provide some indication of the numbers and locations of genes, along with the dominance relationships of the various alleles that are associated with the control of late flowering in the Pitztal ecotype.

References

Bowman, J.L, Alvarez, J., Weigel, D., Meyerowitz, E.M and Smyth D.R (1993) 'Control of flower development in *Arabidopsis thaliana* by *APETALA1* and interacting genes', Development 119, 721- 743.

Burn, J.E., Smyth, D.R. Peacock, W.J. and Dennis, E.S. (1993) 'Genes conferring late flowering in *Arabidopsis thaliana*', Genetica 90, 147-155.

Clark, J.H and Dean, C. (1994) 'Mapping *FRI* a locus controlling flowering time and vernalisation response in *Arabidopsis thaliana.*', Mol. Gen. Genet 242, 81-89.

Evans, L.T. (1975) 'Daylength and the Flowering of Plants', in Benjamin Modular Program in Biology, W.A. Benjamin, Inc. California.

Goto, N., Kumagai, T., and Koorneef, M. (1991) 'Flowering responses to light breaks in photomorphogenic mutants of *Arabidopsis thaliana*, a long-day plant', Physiol. Plant 83, 209-215.

Härer, L. (1950) 'Die Verebung des Blühalters früher und später sommereinjähriger Rassen von *Arabidopsis thaliana* (L.) Heynh', Beitr. Biol. Pflanz. 28, 1-35.

Koornneef, M., Han Hart, C.J. and van der Veen, J.H. (1991) 'A genetic and physiological analysis of late flowering mutants in *Arabidopsis thaliana*', Mol. Gen. Genet. 229, 57-66.

Lang, A., Chailakhyan, M.Kh., and Frolova, I.A. (1977) 'Promotion and inhibition of flower formation in a day neutral plants in grafts with a short-day plant and a long-day plant', Proc. Natl. Acad. Sci. 74, 2412-2416.

Lee, I.,Bleeker, A. and Amasino, R. (1993) 'Analysis of naturally occuring late flowering in *Arabidopsis thaliana*', Mol. Gen. Genet. 237, 171-176.

Murfet, I.C. (1989) 'Flowering genes in *Pisium*,in E.Lord and G.Bernier (eds), Plant Reproduction: From Floral Induction to Pollination, Amer. Soc. Plant Physiologists, Rockville, pp 10-18.

Napp-Zinn, K. (1957) 'Untersuchungen zur genetik des kaltebedurfnisses bei *Arabidopsis thaliana*' Ind. Abst. Vererb. 88, 253-285.

Napp-Zinn, K. (1979) 'On the genetic basis of vernalisation requirement in *Arabidopsis thaliana* (L.) Heyne', in P.Champagnat and R.Jacques(eds.), La Physiologie de la Floraison. Coll. Int. CNRS Paris, 217-220.

Redei, G.P. (1962) 'Supervital mutants of *Arabidopsis*', Genetics 47, 443-460.

Sun, T.P., Goodman, H.M. and Ausubel, F.M. (1992) 'Cloning the *Arabidopsis* GA1 locus by genomic subtraction.', Plant Cell 4, 119-128.

Sung, Z.R., Belachew, A., Bai Shunong and Bertrand-Garcia, R. (1992) '*EMF*, an *Arabidopsis* gene required for vegetative shoot development', Science 258, 1645-1647.

Weigel, D., Alvarez, A., Smyth, D.R. Yanofsky, M.F. and Meyerowitz, E.M. (1992) '*Leafy* controls floral meristem identity in *Arabidopsis*', Cell 69, 843-859.

Weigel, D. and Meyerowitz, E.M (1993) 'Genetic hierarchy controlling flower development', in M. Bernfield (ed.), Molecular basis of morphogenesis, Wiley-Liss, N.Y. pp 91-105.

Wilkinson,J.Q. and Crawford,N.M. (1991) 'Identification of the *Arabidopsis CHL3* gene as the nitrate reductase structural gene *NIA2.*', Plant Cell 3, 461-471.

Zagotta, M.T., Shannon, S., Jacobs, C. and Meeks-Wagner, D.Ry. (1992) 'Early-flowering mutants of *Arabidopsis thaliana*', Aust. J. Plant Physiol 19, 411-418.

STRATEGIES FOR INCREASING TOLERANCE AGAINST LOW TEMPERATURE STRESS BY GENETIC ENGINEERING OF MEMBRANE LIPIDS

A. KUCH, D.C. WARNECKE, M. FRITZ, F.P. WOLTER & E. HEINZ

University of Hamburg, Institut für Allgemeine Botanik, Ohnhorststr. 18, 22609 Hamburg, Germany

Abstract

Function and survial of plants at widely varying temperatures require, among others, a temperature-compatible fluidity of organellar membranes. Some plants have developed adaptation mechanisms by changing membrane components. This guarantees the appropriate homeoviscous/homeophasic fluidity as required in particular temperature regimes. Due to differences in organellar membrane lipid profiles, different lipids and different membranes are involved in changes at different temperatures. A continuous adaptation from higher to lower temperatures can be separated into discrete replacements of critical membrane lipids which block the next step into subsequently lower ranges. We have cloned several of the genes controlling the biosynthesis of lipid components in different compartments relevant for adjustment to progressively lower temperature regimes (acyltransferase, glucosyltransferase). Transformation experiments towards sense/antisense expression are in progress to prove this concept and to enhance cold tolerance of agriculturally interesting crops.

Introduction

Upper and lower limits of growth temperature are important in agriculture, and extensive research efforts have been directed towards an understanding of primary lesions occuring at these limits. In general, such investigations include exposure or adaptation of plants to their growth limits followed by analyses of changes in composition and function. As expected for complex organisms of different geographic origin, many potentially primary targets have been identified. Just to mention a few, temperature-induced changes in concentration of intracellular calcium, association of tubulin, structure of cytoskeleton or membrane functions are considered as critical factors (Wang 1990). They contribute to the immediate temperature stress, which in sensitive plants gives rise to a complex set of symptoms. These vary widely between plants and sometimes require extended periods of time for manifestation. Therefore, it is often difficult to recognize or separate cause and consequence. The situation is further complicated by the fact that some plants or some of their organs have inducible abilities to acclimate to lower temperatures, provided an

M. Terzi et al. (eds.), Current Issues in Plant Molecular and Cellular Biology, 539-544.

appropriate phase of adaptation is given, which again may widely differ in duration or intensity of stress signals. The compositional changes occuring during acclimation processes are particularly interesting, since they may provide clues for identifying the factors required for winterization of plants. But again, caution is necessary to differentiate a change introduced for protective means from a consequence of this change without relevance for protection.

As mentioned above, membranes may be among the primary sites, where lower temperatures can interfere with metabolism in many ways. Subcellular compartments are limited by different membranes, which are characterized by both specific sets of proteins and lipids. The different membranes guarantee functionality by providing unique lipid matrices of appropriate and optimal, but not necessarily identical fluidity, which is due to different and organelle-specific lipid profiles. Within a limited temperature span these lipid mixtures can maintain the required fluidity, whereas a shift to extreme ranges may result in unsuitable matrix characteristics (Lyons 1973). The critical temperature ranges, their resulting changes and the functional consequences may be different in the various membrane types. Accordingly, acclimation often involves a homeoviscous/homeophasic adaptation of sensitive membrane matrices by modulation of critical lipid components. Depending on the plant and the temperature to be reached, such changes may be limited to a single membrane type and its most critical component or extend to other membranes and accordingly to other lipid components or lipid portions.

Detailed and complete lipid analyses of individual membrane types before and after acclimation resulted in the identification of components which may be of relevance for low temperature compatibility of a particular membrane. In this context, analyses of whole plants or organs are of limited value, since subtle variations on a background of unchanged lipids from dominating membranes may not be recognized. As a result of research carried out by several groups in this field, the following lipids have been recognized as limiting the functions of particular membranes at temperatures lowered to different degrees. Phosphatidylglycerol with two nonfluid acyl residues may interfere with the proper functioning of thylakoid and envelope membranes of plastids, when temperatures fall into chilling ranges between 15-0°C (Murata 1983, Roughan 1985). Plants are either chilling sensitive due to the synthesis of this particular lipid species or they are not, since adaptative mechanisms due to removal of this component are unknown. In plasma membranes, cerebrosides and steryl glucosides may become critical at subzero temperatures (frost or freezing tolerance, Steponkus 1984, Wojciechowski 1991). The same components are considered to be involved in water stress resistance, since both phenomena are ascribed to membrane changes initiated by hydration problems of lipid headgroups. As the tonoplast reflects the plasma membrane in lipid composition, its temperature characteristics may also contribute to frost tolerance. The role of cerebrosides and steryl glucosides were deduced from adaptative changes observed in winter rye, which can acclimate to lower subzero temperatures than other cereals (Lynch et al. 1987, Webb et al. 1994). In addition, the degree of fatty acid unsaturation was increased in plasma membrane phospholipids as observed for lipids from several frost resistant perennial plants.

These correlations are the basis for our approach of studying low temperature resistance and of eventually contributing to its improvement by genetic engineering of membrane lipids. According to this concept critical groups of components have to be modified for survival and successful metabolism at lower temperature. Recently cloned relevant genes from lipid metabolism will allow such changes and enable new experiments not possible before. In a first series of investigations we are engaged in a constitutive manipulation of critical components by sense or antisense engineering of single enzymes. The concept for a stepwise increase of low temperature tolerance by altering membrane lipids and the corresponding critical enzymes are depicted in figure 1. Our first targets are nonfluid phosphatidylglycerol in plastids and steryl glucosides from plasma membranes and tonoplasts, which either can be increased or decreased. Cerebrosides and polyenoic fatty acids will be manipulated subsequently. The performance of accordingly transformed plants at lower temperature will show, whether the concept outlined above is correct and particularly, to which extent the replacement of specific lipids can improve the cold tolerance of formerly sensitive plants. In this context it will be interesting to see, whether these critical lipid components are required at all, i.e. whether they can be reduced permanently or whether their manipulation has to mimic changes occurring in adaptable plants during acclimation. This would require expression under control of a promoter displaying corresponding sensitivity. In terms of an analogous picture referring to human practice, membranes of different organelles may be compared with socks, gloves, caps and underwear, which can be put on and off in appropriate quality according to temperature demand, and our transformation experiments could be compared with the exchange of various pieces of clothes. An open question is whether transgenic organisms would care to be dressed in winter wear all the time. In the following a short outline of our first projects and their present status is given.

Nonfluid Phosphatidylglycerol

The biosynthesis of these lipid molecules in plastids is controlled by a soluble acyltransferase in the chloroplast stroma. Chilling sensitive plants have an unselective enzyme, whereas chilling resistant species contain a highly selective enzyme which prevents

Step by Step
Changing of Membrane Lipids
by Acclimation or Genetic Engineering

lipid change	relevant enzymes	effect
reduce nonfluid phosphatidyl-glycerol from plastids	acyl-ACP:glycerol-3-phosphate acyltransferase	reduction of chilling injury
reduce steryl glucoside, acyl-steryl glucoside and cerebroside from plasma membrane and tonoplast	UDP-glucose: sterol glucosyltransferase and UDP-glucose: ceramide glucosyltransferase	reduction of freezing injury
convert fatty acid 18:1 to 18:2 and 18:3	phospholipid desaturases	reduction of freezing injury

Figure 1: Cold acclimation is accompanied by adaptation of membrane properties through alterations of critical lipid components. Such changes should be mimicked by genetic manipulation of relevant enzymes in attempts to increase the cold tolerance of crops.

the incorporation of palmitic acid into the C1-position of plastid lipids (Murata 1983). To examine this hypothesis, we expressed the unselective acyltransferase from *E. coli* in *Arabidopsis* chloroplasts, which as a consequence accumulated appreciable proportions of a nonfluid phospatidylglycerol (Wolter et al. 1992). Chilling treatment of transgenic plants resulted in chilling symptoms, but detailed studies of photosynthetic parameters failed to detect significant differences between wild type and transgenic plants and chloroplasts. A reverse experiment was carried out with chilling sensitive *Nicotiana tabacum* by expression of the selective acyltransferase from *Arabidopsis* (Murata et al. 1992). In this case the transformed plants showed increased chilling resistance.

At present we are extending our investigations to *Cucumis sativus* and *Phaseolus vulgaris*, both of which are chilling sensitive crops. We have isolated the acyltransferase cDNA from *Phaseolus* to be replaced by the selective enzyme from pea (Weber et al. 1991) or spinach, which we have also cloned. The cDNA for the *Cucumis* acyltransferase has been isolated before (Johnson et al. 1992), and we are presently establishing transformation protocols for this species, which we can regenerate with high frequency. The regeneration and transformation of *Phaseolus* is carried out in a cooperating laboratory (H. J. Jacobsen).

Steryl glucosides and free sterols

The adaptation experiments with winter rye mentioned above have demonstrated a reduction of steryl glucosides and a concomitant increase of free sterols in plasma membranes. From their effects in bilayers, the reciprocal changes of both components can be rationalized as contributing to the hardening effects. Therefore, we purified a key enzyme of sterol lipid interconversion, the UDP-glucose:sterol glucosyltransferase, raised antibodies against the isolated protein and determined the N-terminal amino acid sequence (Warnecke & Heinz 1994). With these tools we have started experiments on cloning of a corresponding cDNA which in terms of our picture codes for another piece of clothes to be exchanged. The cDNA will be used in transformation experiments similar to those described above, and investigations with the transgenic plants will enable a new approach towards the elucidation of steryl lipid functions in plasma membranes and tonoplasts and their contribution to hydration stress as exemplified in frost tolerance.

References

Johnson, T. C., Schneider, J. C., Sommerville, C. (1992), "Nucleotide sequence of acyl-acyl carrier protein: glycerol-3-phosphate acyltransferase from cucumber", Plant Physiol. 99, 771-772.

Lynch, D. V., Steponkus, P. L. (1987), "Plasma membrane lipid alterations associated with cold acclimation of winter rye seedlings (*Secale cereale* L. cv Puma)", Plant Physiol. 83, 761-767.

Lyons, J. M. (1973), "Chilling injury in plants", Annu. Rev. Plant Physiol. 24, 445-466.

Murata, N. (1983) "Molecular species composition of phosphatidylglycerols from chilling-sensitive and chilling-resistant plants", Plant & Cell Physiol. 24, 81-86.

Murata, N., Ishizaki-Nishizawa, Higashi, S., Hayashi, H., Tasaka, Y., Nishida, I. (1992), "Genetically engineered alteration in the chilling sensitivity of plants", Nature 356, 710-713.

Roughan, G. P. (1985), "Phosphatidylglycerol and chilling sensitivity in plants ", Plant Physiol. 77, 740-746.

Steponkus, P. L. (1984), "Role of the plasma membrane in freezing injury and cold acclimation", Annu. Rev. Plant Physiol. 35, 543-584.

Wang, C. Y. (1990), Chilling injury of horticultural crops, CRP, Boca Raton.

Warnecke, D.C., Heinz, E. (1994), "Purification of a membrane-bound UDP-glucose:sterol ß-D-glucosyltransferase based on its solubility in diethyl ether", Plant Physiol., in press.

Webb, M. S., Uemura, M., Steponkus, P. L. (1994), "A comparison of freezing injury in oat and rye: two cereals at the extremes of freezing tolerance", Plant Physiol. 104, 467-478.

Weber, S., Wolter, F. P., Buck, F., Frentzen, M., Heinz. E. (1991), "Purification and cDNA sequencing of an oleate-selective acyl-ACP:*sn*-glycerol-3-phosphate acyltransferase from pea chloroplasts", Plant Mol. Biol. 17, 1067-1076.

Wojciechowski, Z. A. (1991), "Biochemistry of phytosterol conjugates", in G. W. Patterson, W. D. Nes (eds.), Physiology and Biochemistry of Sterols, American Oil Chemists' Society, Champaign, Illinois, pp. 361-395.

Wolter, F. P., Schmidt, R., Heinz, E. (1992), "Chilling sensitivity of *Arabidopsis thaliana* with genetically engineered membrane lipids", EMBO Journal 11, 4685-4692.

ALUMINUM EFFECTS ON IN VITRO TISSUE CULTURES OF *PHASEOLUS VULGARIS*

F.J. ESPINO, M.T. GONZALEZ-JAEN, J. IBAÑEZ, A.M. SENDINO, A.M.VAZQUEZ.
Dpto. Genética, Facultad de Biología
Universidad Complutense
28040 Madrid. Spain

ABSTRACT. The toxicity conditions for an aluminum salt (Al) were simulated in vitro using calli of *Phaseolus vulgaris* obtained from cataphyls of 10 day old seedlings. To simulate these conditions it was necessary to modify the culture medium. The culture medium used was composed of the salts of MS and the vitamins of B5. The modifications introduced were the following: reduction of the phosphates and Ca, using Fe without chelating agent and adjusting the medium at pH 4.0. The lethal dose was 500 mg/l ($AlCl_3$. $6H_2O$) at which the callus growth completely stopped. To study how gene expression is modified by Al presence (500 mg/l) in the culture medium, calli were cultured for 4 weeks in the presence of Al. Total RNA were isolated from these calli, and translated in vitro using wheat germ system. The translated products, resolved by 2D-PAGE, have demonstrated the existence of polypeptides related to the stress situation, although the majority seemed to be also induced by the acidification of the culture medium. The presence of Al inside the calli was demonstrated by histological stains.

Introduction

Al is one of the most abundant elements on the earth's surface, and its toxic effects on the growth and development of plants in acid soils have been documented, particularly the inhibition of root elongation (Foy et al. 1993, Ikeda and Tadano, 1993). However, there is little information on the molecular level in relation to Al toxicity.

Plant tissue culture has been used by different authors to identify genotypes tolerant to Al (Conner and Meredith, 1985; Parrot and Bouton, 1990; Kamp-Glass et al., 1993). In these works, it was necessary to modify the culture media to simulate Al toxicity because this is enhanced in soils with low calcium saturation and low pH. These works have suggested that whole-plant tolerance to Al is expressed at the cellular level.

We have used plant tissue culture in order to study Al toxicity at the cellular level analyzing gene expression in Al-treated and untreated calli of *Phaseolus vulgaris*.

Material and Methods

PLANT MATERIAL

Calli of *Phaseolus vulgaris* (cv. Palmeña) were obtained from cataphyls of 10 day old seedlings.

M. Terzi et al. (eds.), Current Issues in Plant Molecular and Cellular Biology, 545-549.

A cellular line, derived from only one seedling, was established and maintained for several years.

CULTURE MEDIA

The medium to induce callus formation was called L3. It was composed of MS salts, B5 vitamins, 2,4-D (2 mg/l), sucrose (30 g/l) and bactoagar (10 g/l). The pH was adjusted to 5.7. Calli were maintained on the same medium and were transferred to a fresh one at 30 day intervals. To simulate Al toxicity the L3 medium was modified according to Conner and Meredith (1985): the concentration of phosphate ions was reduced to 10 μM and that of calcium to 100 μM, the pH was lowered to 4.0, unchelated iron was used, and the medium was solidified with 10 g/l of vitroagar. Two different concentrations of Al were used: 200 and 500 mg/l of $AlCl_3$. 6 H_2O. The $AlCl_3$ was autoclaved separately and added to the rest of the medium before pouring. These two media were called Al200 and Al500. Two other media, L3m and L3m4, were used as controls. Both of them were similar in composition to Al media but without $AlCl_3$. L3m was adjusted to pH 5.7 and L3m4 to pH 4.0. All cultures were maintained at 25 ± 1 °C and 16 h of diffuse light.

CALLUS GROWTH

Two grams of calli belonging to the same cellular line were transferred to Petri dishes containing the 5 media (L3, L3m, L3m4, Al200 and Al500). For each treatment the samples of 10 Petri dishes were taken at 1, 2, 3, and 4 weeks, and their increase in fresh weight measured.

AL DETECTION

Calli growing on L3, L3m, L3m4, Al200 and Al500 for 2, 4, 6 and 8 days were stained with haematoxilin according to Rincon and Gonzales (1992). The presence of Al in calli was observed by an intense blue colour while calli without it were brown.

RNA ISOLATION AND IN VITRO TRANSLATION

Total RNA was isolated from calli cultured 4 weeks on L3m, L3m4 and Al500. The RNA extraction was performed according to de phenol-SDS method combined with selective precipitation using LiCl. RNA concentration was estimated by absorption at 260 nm. Cell-free protein synthesis was accomplished using a wheat germ system (Promega) and ^{35}S-Methionine (ICN, Flow, 1000 Ci/mmol at 10 mCi/ml). 6 μg of total RNA were used for each reaction.

2-D ELECTROPHORESIS

Isoelectric focussing was accomplished according to O'Farrel and O'Farrel (1977), and the second dimension following Laemmli (1970). After fixation, gels were treated with En^{3}hance (Dupont) and dried. Gels were autoradiographed using X-ray film (Kodak X-OMAT AR). Exposure was made at -80°C for 8 days. Each experiment was repeated twice at least. Only changes observed in all replicate experiments were taken into account.

Results and Discussion

IN VITRO AL TOXICITY

The modified medium employed reproduced in vitro the Al toxicity, as has been reported by other authors using similar modifications (Conner and Meredith, 1985; Parrot and Bouton, 1990; Kamp-Glass et al., 1993). Table 1 summarizes the data concerning callus growth using two Al concentrations. At 500 mg/l, callus growth was completely stopped during the first days of culture, but at 200 mg/l, calli showed a small increase in weight. Lower concentrations, as 90 mg/l, had no noticeable effect on callus growth, however this dose was enough to inhibit root development during seed germination (data not shown).

TABLE 1. Callus growth in the presence of aluminum

	1 week	2 weeks	3 weeks	4 weeks
L3	2.15±0.250	2.65±0.250	4.28±0.410	5.67±0.600
L3m	2.11±0.234	1.98±0.214	4.22±0.860	5.45±1.472
3m4	2.08±0.282	2.02±0.500	4.16±0.742	5.10±0.827
Al200	1.95±0.281	1.93±0.350	2.40±0.228	2.75±0.210
Al500	1.96±0.210	1.95±0.190	1.96±0.125	2.12±0.157

Measurements are given in grams

The stain procedure developed by Rincon and Gonzales (1992) detects Al in the plant tissues. Through this technique, Al was detected in the calli after few days of culture. So, after two days calli were completely stained with blue colour, indicating that in the modified media (Al200 and Al500) Al easily penetrated the calli during culture.

GENE EXPRESSION RELATED TO AL STRESS

In a first attempt to study the modification promoted by Al in the gene expression 2-D electrophoresis with total proteins isolated from L3m, L3m4 and Al500 calli were performed. When applying this technique, no changes in the protein patterns were detected but a general decreasing in the protein amount from Al500 calli was noticed (data not shown).

To obtain more precise data concerning the gene expression a most powerful method was applied and in vitro translations were carried out. Patterns of in vitro translations, obtained from calli cultured for a long time (4 weeks) on Al500 are shown in Figure 1. Comparing the differential polypeptides, they can be grouped into four classes:

1. polypeptides that only appear in L3m4 calli: spots numbered 1 and 2.
2. polypeptides detected in L3m4 and Al500 calli: spots 3, 4, 5, 6, 7 and 8.
3. polypeptides that appear in Al500 calli: only spot 9.
4. polypeptides that disappear in L3m4 and Al500 calli, only present in L3m calli: spots 10 and 11.

We found some modifications which could be due to two different causes, that is to say the media acidity, as the pH had been lowered, and the toxic presence of Al. As can be observed, some of the spots are common in both media L3m4 and Al500 so they must be due to the first modification already mentioned, the low pH. However, spot number 9 is only found in the Al medium so it would seem to be due to its presence.

In conclusion, in our system the Al seems to decrease the rate of protein synthesis in a general manner, and only one transcript seems to be differentially expressed in its presence. However, the low pHs modify the mRNA population in a most noticeable way.

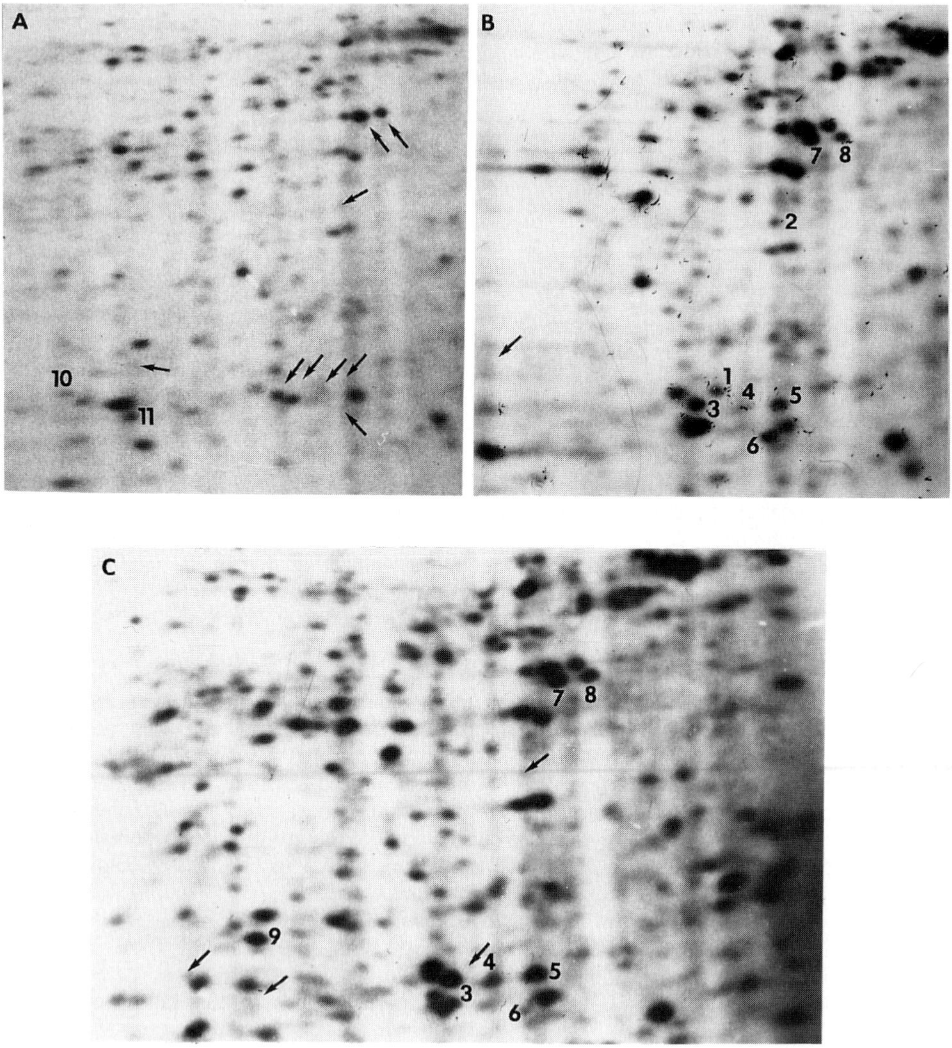

Figure 1. In vitro translation of mRNA derived from L3m, L3m4 and Al500 calli (Respectively A, B and C). Arrows indicate absence of protein spot.

Acknowledgements

This work was supported by a grant of CICYT (BIO 90-1034-C0201).

References

Conner, A.J. and Meredith, C.P. (1985) 'Simulating the mineral environment of aluminum toxic soils in plant cell culture'. J. Exp. Bot. 366, 870-880.

Foy, C.D., Carter JR., T.E., Duke, J.A. and Devine, T.E. (1993) 'Correlation of shoot and root growth and its role in selecting for aluminum tolerance in soybean'. Journal of Plant Nutrition 16, 305-325.

Ikeda, H. and Tadano, T. (1992) 'Ultrastructural changes of the root tip cells in barley induced by a comparatively low concentration of aluminum'. Soil Sci. Plant Nutr. 39, 109-117.

Kamp-Glass, M., Powell, D., Reddy, G.B., Baligar, V.C. and Wrigth, R.J. (1993) 'Biotechniques for improving acid aluminum tolerance in alfalfa'. Plant Cell Reports 12, 590-592.

Laemmli. U.K. (1970) 'Cleavage of structural proteins during the assembly of the head of bacteriophage T4'. Nature 227, 680-685.

O'Farrel, P.H. and O'Farrel, P.Z. (1977) 'Two-dimensional polyacrilamide gel electrophoretic fractionation'. Methods Cell Biol. XVI, 405-420.

Parrot, W.A. and Bouton, J.H. (1990) 'Aluminum tolerance in alfalfa as expressed in tissue culture'. Crop Sci. 30, 387-389.

Rincón, M. and Gonzales, R.A. (1992) 'Aluminum partitioning in intact roots of aluminum-tolerant and aluminum-sensitive wheat (*Triticum aestivum* L.) cultivars'. Plant Physiol. 99, 1021-1028.

EFFECT OF HYPEROSMOTIC 3-O-METHYL-D-GLUCOSE IN THE MEDIUM ON METABOLIC PARAMETERS IN *Actinidia deliciosa* CALLUS

G. A. SACCHI[1], A. ABRUZZESE[1], C. ALISI[1], S. MORGUTTI[1], L. ESPEN[1], N. NEGRINI[1], M. COCUCCI[1], S. COCUCCI[1], R. MULEO[2] & A. R. LEVA[3].

1 University of Milan, Dept. Fisiologia delle Piante Coltivate e Chimica Agraria-via Celoria 2, 20133 Milano-Italy. 2 University of Pisa, Dept. Coltivazione e Difesa delle Specie Legnose-via Borghetto 80, 56100 Pisa-Italy. 3 C.N.R.-Istituto Propagazione delle Specie Legnose-via Ponte di Formicola 76, 50018 Scandicci-Italy.

ABSTRACT. Calli of *Actinidia deliciosa* were cultured on medium with 30 mM sucrose and 0.2 M 3-O-methyl-D-glucose (3-OMG) and their metabolic adaptation to the hyperosmotic condition was studied. 3-OMG concentration in the tissue reached that of the medium. Experiments with labelled 3-O-methyl-D-[U-^{14}C]glucose (3-OM[U-^{14}C]G) showed that it was partially metabolised but not oxidised to CO_2. *In vivo* ^{31}P-NMR showed that in 3-OMG-treated calli a phosphorylated compound, with a chemical shift similar to that of monophosphoesters, was accumulated. The 3-OMG-treated calli showed more negative water (ψ_w) and osmotic (ψ_π) potentials than the controls. In the following of subcultures an increase in organic and inorganic osmolytes, together with increase in O_2 uptake and decrease in ATP/ADP ratio occurred. The data show that, even if 3-OMG can not be considered a classic osmotic agent due to its permeability, it induced metabolic changes similar to those described for osmotic adaptation, suggesting that the adaptations might be due to a physical or metabolic effect of 3-OMG not linked to a difference between ψ_π and $\psi_{w\ agar}$.

1. Introduction

The most widely used experimental approach for the selection of cell lines adapted to osmotic conditions is to culture cells or callus tissue in a hyperosmotic medium obtained with high concentrations of non-penetrating and not metabolizable compounds. The experimental conditions used are different with respect to the osmotic stress naturally experienced by plant tissues: in fact, this generally occurs in the presence of permeable organic or inorganic solutes. Moreover, recent findings showing that mannitol directly interferes with the metabolism of cultured cells (Thompson et al., 1986), that polyethylene glycol, even if slightly, is absorbed by sorghum calli (Newton et al., 1990) and that it could affect membrane structure and functionality in maize seedlings (Izzo et al., 1989), raise some question on the mechanisms involved in the effects of these widely used compounds.

Taking into account these considerations and findings, we have hypothesised to use as osmotic agent, in *Actinidia deliciosa* callus cultures, a permeant, non-trophic (Gogarten and Bentrup, 1983) carbohydrate: 3-O-methyl-D-glucose (3-OMG). We have thus verified the permeability of 3-OMG and its effect, when present at high concentration in the culture medium, on growth, tissue water parameters, respiratory activity, levels of K^+ and of some metabolites indicative of the metabolic state of tissue or involved in osmotic adjustment. Three subsequent 45 day-subcultures (S I, S II, S III) of *A. deliciosa* callus were conducted and the above cited parameters were analysed at the 35th day of each subculture, when metabolic suffering, probably linked to medium exhaustion, is not yet present (Abruzzese et al., 1994).

M. Terzi et al. (eds.), Current Issues in Plant Molecular and Cellular Biology, 551-556.

2. Materials and Methods

Cambium tissue from adult female woody vine of *A. deliciosa* cv. Hayward was grown on Gamborg B5 medium supplemented with 18 μM 2,4–D to induce callusing. The subsequent subcultures were conducted on the same medium (pH 5.9) supplemented with 60 mM sucrose, 23 μM zeatin and 230 nM NAA, solidified with 0.4% (w/v) Bacto–Agar. After two 45 day–subcultures at 23±1°C (under 16 h of 40 μE m^{-2} s^{-1} daily irradiation), calli were again subcultured for three 45 day–periods on the B5 medium supplemented with 60 mM sucrose (controls) or 30 mM sucrose plus 0.2 M 3–OMG (3–OMG calli). At the 35th day of each subculture, growth (FW increase) of calli was monitored and physiological and metabolic parameters determined on a fraction of the total material available. For this purpose, the whole callus was cut into small (2 mm^2, 2 mm thickness) slices that were rapidly washed, blotted with paper towel and either immediately used (*in vivo* experiments) or divided into 150 mg–samples, which were frozen in liquid N_2 and stored at –80°C (*in vitro* determinations).

Water (ψ_w) and osmotic (ψ_π) potentials (comprising also, in our conditions, the value of callus matric potential) of callus and ψ_w of the agarized medium were measured (Cocucci et al., 1990) with a thermocouple psychrometer sample changer connected to a nanovoltmeter–thermometer.

3–OMG and glucose uptakes were measured at 26°C on samples (300 mg FW) in an incubation medium (henceforth called IM) consisting in a non–agarized B5 medium without sucrose plus 1 mM 3–OM[U–^{14}C]G (S.A.= 3.7 GBq mol^{-1}) or 1 mM D–[U–^{14}C]glucose (S.A.= 3.7 GBq mol^{-1}). At the end of incubation, samples were washed twice to remove apoplastic radioactivity, homogenised in 0.1 M HNO_3, heated to 100°C for 10 min and radioactivity determined.

In order to assess whether 3–OMG reached equilibrium between tissue and medium, samples (600 mg FW) were incubated in 3 ml IM supplemented with 0.2 M 3–[U–^{14}C]G (S.A.= 11.1 GBq mol^{-1}). At different times, radioactivity was determined on aliquots of the medium. The water volume of the tissue was measured by equilibration in IM with 0.2 M 3–OMG and ^3H_2O (S.A.= 259 kBq mol^{-1}). The ^{3}H fixed into cell material was evaluated by measuring radioactivity in the lyophilised tissue.

O_2 uptake was measured at 26°C in a differential respirometer on samples (250 mg FW) in IM plus 30 mM sucrose and 0.2 M 3–OMG (only for the treated tissue).

In order to assess whether 3–OMG was metabolised, samples (250 mg FW) were incubated in respirometer vessels in IM solution plus 1 mM 3–OM[U–^{14}C]G (S.A.= 11.1 GBq mol^{-1}). After six hours, the radioactivity trapped by the 2 M NaOH in the central well of the vessels was measured and the tissue was homogenised in 0.5 M PCA as described below. The neutralised PCA–soluble fraction was analysed according to Neal and Beevers (1960). The radioactivity present in 0.5 cm paper segments was then measured.

In vivo ^{31}P–NMR spectra were obtained with a Bruker AMX 600 spectrometer by packing the callus slices in a Wilmad tube (10 mm in diameter) equipped with a perfusion system in which the aerated IM added with 0.2 M 3–OMG was flowing. Spectra were recorded at 242.9 MHz. Resonance identification was obtained according to Roberts et al., 1980.

For *in vitro* determinations, samples (450 mg FW) were homogenised in a mortar with 0.5 M ice–cold PCA. K_2CO_3 was then added to the collected 13 000 *g* supernatants to remove PCA; aliquots were used for the assays of malic acid (pH 10.0), G–6–P (pH 7.6), sucrose, amino acids and adenosine phosphates (pH 7.0). Malic acid and G–6–P were determined enzymatically. Sucrose was determined by the colorimetric Nelson's method. Total free amino acids were measured by the ninhydrin method. ATP was assayed by bioluminescence emission using the LKB 1243–200 ATP monitoring reagent in a LKB–Wallac 1250 luminometer. ADP was determined after enzymatic conversion to ATP.

Levels of K^+ were determined by atomic absorption spectroscopy on samples (150 mg FW) mineralised at 100°C in HNO_3:H_2SO_4:$HClO_4$ (5:1:1 v:v:v) and diluted with distilled water.

3. Results

3–OMG affected callus growth. Growth (percent increase in fresh weight) in the 3–OMG–treated calli was, at the 35th day of the three subcultures, about 30% of the control (data not shown).

Table 1 shows the values of callus water parameters at the 35th day of the first subculture (S I). The values of ψ_π of the callus tissue were more negative than the $\psi_{w\ agar}$. The value of ψ_w of the 3–OMG–treated callus was lower than that of the control, as a consequence of the strong decrease in its ψ_π value. The pressure potential ($\psi_p = \psi_w - \psi_\pi$) was higher in the 3–OMG–treated calli. These general features did not change in the subsequent subcultures (data not shown).

TABLE 1. Values of water parameters of *A. deliciosa* calli grown in the absence or in the presence of 0.2 M 3–OMG and of the agarized medium.

	Water parameter, MPa			
	ψ_w	ψ_π	ψ_p	$\psi_{w\ agar}$
Control	–0.44	–0.61	+0.17	–0.17
3–OMG	–1.11	–1.40	+0.29	–0.87

The data are the means of five measurements. S.E. did not exceed ±6%.

Figure 1 shows that *A. deliciosa* calli absorbed 3–OMG. The rate of uptake of 1 mM 3–OM[U–^{14}C]G in the control callus, at the beginning of S I, was linear for at least two hours and only slightly lesser than the rate of uptake of 1 mM D–[U–^{14}C]glucose. When, at the 35th day of each

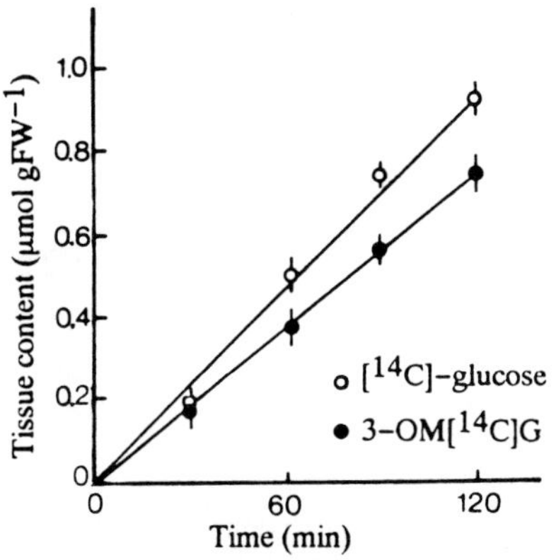

Figure 1. Uptake of 3–OM[U–^{14}C]G and D–[U–^{14}C]glucose by *A. deliciosa* callus. Vertical bars represent S.E. (n=9).

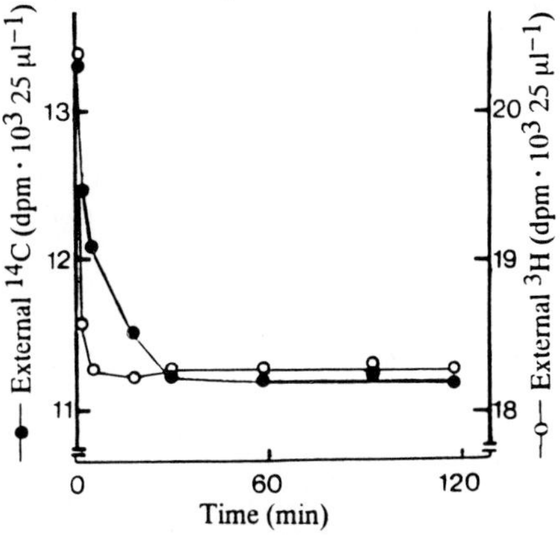

Figure 2. Equilibration of 0.2 M 3–OM[U–^{14}C]G and ^3H_2O in *A. deliciosa* callus. Data from one typical experiment.

subculture, slices of the 3–OMG–treated calli were incubated in the presence of 0.2 M 3–OM[U–^{14}C]G, the radioactivity (dpm ml^{-1}) in the incubation solution quickly decreased reaching a steady state equilibrium (Fig. 2). By taking into account the dilution due to the tissue volume (evaluated by means of ^3H_2O equilibration, Fig. 2), it was calculated that, at steady state, the concentration of 3–OMG in the tissue was about the same than that the medium, *i.e.* ca. 0.16 M.

Calli grown for 35 days on 3–OMG showed a rate of O_2 consumption lower than the control calli (–41%). At the 35th day of S III, this activity was almost completely recovered reaching a value similar to that of the control (practically constant in all subcultures, Tab. 2). Though the adenylate pool was practically the same in the two culture conditions and constant throughout S I–S III (data not shown), the ATP/ADP ratio in the 3–OMG–treated calli showed a tendency to decrease in the subsequent subcultures, while in the control it was practically constant (Tab. 2). The levels of G–6–P significantly decreased in S I on 3–OMG, but, just starting from S II, they rose to levels similar to those of the control (again, practically constant throughout the subsequent subcultures).

TABLE 2. O_2 uptake, ATP/ADP ratio and G–6–P levels in *A. deliciosa* calli grown for three subsequent subcultures (S I–S III) in the absence or in the presence of 0.2 M 3–OMG.

	O_2 uptake $\mu l\ h^{-1}\ gFW^{-1}$		ATP/ADP		G–6–P nmol gFW^{-1}	
	C	3–OMG	C	3–OMG	C	3–OMG
S I	100 ± 3	59 ± 2	3.5	3.5	129 ± 7	96 ± 6
S II	97 ± 5	71 ± 3	3.2	3.2	119 ± 5	114 ± 3
S III	96 ± 4	88 ± 3	3.3	2.6	124 ± 5	110 ± 4

The data are the means of three measurements run in triplicate ± S.E. (n=9)

The 3–OMG–treated calli showed, at the 35th day of S I, lower levels of some organic (sucrose, malic acid) and inorganic (K^+) osmolytes (Tab. 3). In the passage from S I to S II and S III on 3–OMG, the sucrose and amino acid levels progressively increased; K^+ increased from S II to S III, while the levels of malic acid remained practically unchanged.

TABLE 3. Levels of sucrose, free amino acids, malic acid and K^+ in *A. deliciosa* calli grown in the absence or in the presence of 0.2 M 3–OMG.

	Sucrose $\mu mol\ gFW^{-1}$		Amino acids $\mu mol\ gFW^{-1}$		Malic acid $\mu mol\ gFW^{-1}$		K^+ $\mu mol\ gFW^{-1}$	
	C	3–OMG	C	3–OMG	C	3–OMG	C	3–OMG
S I	3.2±0.2	1.4±0.1	4.5±0.2	4.1±0.3	3.2±0.2	2.3±0.1	99±3	85±5
S II	3.8±0.3	7.5±0.4	4.9±0.3	10.0±0.6	2.0±0.1	1.9±0.1	87±3	85±5
S III	3.4±0.2	11.2±0.6	5.2±0.3	16.1±0.9	2.2±0.1	2.4±0.1	85±4	118±7

The data are the means of three measurements run in triplicate ± S.E. (n=9).

When, at the 35th day of each subculture, slices of calli from either the control or the 3–OMG–condition were incubated in 3–OM[U–^{14}C]G for six hours, virtually no radioactivity was found as [^{14}C]–CO_2, indicating that no complete oxidation of 3–OMG occurred. A fraction (ca. 60% of total in the 3–OMG–treated calli and a smaller fraction in the control) of the absorbed radioactivity was associated with one or more, non–charged (retained by neither the anionic nor the cationic exchange resins) molecules different than 3–OMG (Fig. 3).

In vivo ^{31}P–NMR spectra of 3–OMG–treated calli (Fig. 4) showed, at the 35th day of each subculture, a large increase in the peak of vacuolar inorganic phosphate (Pi_{vac}, 3) and in a peak (1) with a chemical shift (δ) in the range of monophosphoesters. The latter was particularly evident as compared with the peaks of cytoplasmic inorganic phosphate (Pi_{cyt}, 2) and phosphoadenylate compounds (4, 5 and 6), which were similar in both the 3–OMG–treated– and the control calli. The assignment of the peak 1 was difficult because the δ of none commercially available monophosphoester corresponded exactly to it. Phosphorylated 3–OMG obtained *in vitro* by the hexokinase reaction showed a δ very similar to that of the investigated peak (datum not shown).

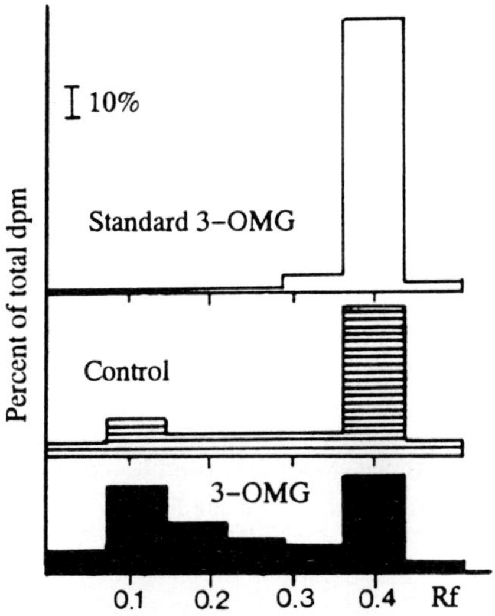

Figure 3. Radiotracer scans of the PCA–soluble fraction from *A. deliciosa* calli (S II) incubated with 3–OM[U–^{14}C]G.

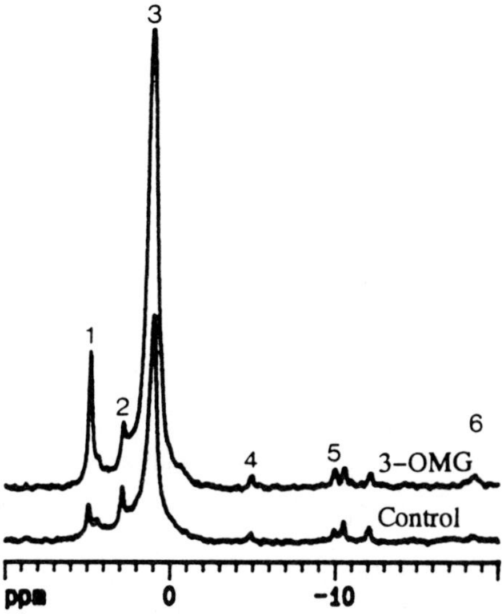

Figure 4. *In vivo* ^{31}P–NMR spectra of *A. deliciosa* calli grown in the absence or in the presence of 3–OMG (S III).

4. Discussion

In cultured *A. deliciosa* calli, a high concentration (0.2 M) of 3–OMG induced physiological and metabolic changes consistent with the onset of tissue adaptation to hyperosmotic conditions. In particular, in the subsequent subcultures on 3–OMG, the callus cells showed increasing levels of organic and inorganic osmolytes (Tab. 3). Nevertheless, the extent of these increases did not justify the low value of callus ψ_π (Tab. 1) observed. This, together with the finding that 3–OMG permeates the tissue (Fig. 1), supports the results of the equilibration experiments (Fig. 2) indicating that 3–OMG reached in the cultured callus a concentration close to that of the medium. The selection is thus conducted in the presence of both imposed low ψ_π in the cells and high ψ_p (Tab. 1) probably due to a non–adequate cell–wall plasticity.

Even if callus cells did not oxidise 3–OMG to CO_2, 3–OMG was partially metabolised to a more polar compound (Fig. 3) according to the findings of Gogarten and Bentrup (1989). This might be due to an involvement of 3–OMG in some step of hexose metabolism, and/or to the induction of detoxification mechanisms. We can speculate that induction of this latter activity might represent an aspect of selection of cell lines characterised by enhanced oxidase activities; this hypothesis is also suggested by the tendency to decrease in the ATP/ADP ratio in the subsequent subcultures on 3–OMG, when O_2 consumption became higher (Tab. 2), even if it cannot be excluded that this be due to an enhanced request of ~P in the stress condition. Moreover, preliminary results suggest that, in the 3–OMG–treated calli, higher extra–mitochondrial O_2 consumption activities are operating. The perspective on the possibility of selecting cell lines able to detoxify also the active forms of oxygen eventually originating from 3–OMG degradation appears particularly interesting, since several indications positively relate these activities to the stress tolerance characteristics of plant materials (Polle and Rennenberg, 1993).

The phosphorylated compound detected by ^{31}P–NMR that accumulated to a greater extent in the 3–OMG callus (Fig. 4) is probably phosphorylated 3–OMG. Absence of detectable radioactivity associated to charged (phosphorylated) compounds from calli fed with 3–OM[U-^{14}C]G might be due to the low specific activity for the high internal level of non–radioactive 3–OMG. A sugar phosphate derived from 3–OMG was found in long–term experiments also by Gogarten and Bentrup (1989). Our data indicate that, in our conditions, 3–OMG might be involved in the hexokinase reaction and might thus interfere with the regulation of carbon metabolism. The processes leading to osmotic adaptation in our material (which appear only partially dependent on the sensing of sudden and large changes in cell ψ_p, conceivably not occurring for the permeability characteristics of 3–OMG) could be triggered by this action of 3–OMG. Differently from what observed with non–permeant osmotica, the adaptive mechanism/s evoked by 3–OMG might thus depend not on the drop in cell ψ_p, but to either the sharp decrease in cell ψ_π or the effects of 3–OMG itself on cell metabolism. It is noteworthy that, when sucrose or glucose are utilised as osmotic agents, they also decrease cell ψ_π, but they practically suppress the morphogenetic potentiality, differently than 3–OMG (Muleo et al., 1994), probably for their action on the trophic mechanisms. For this reason, 3–OMG could represent a compound suitable for conducting molecular and physiological studies on the aspects involved in the response to osmotic stress.

Acknowledgement: Research supported by Italian Ministry of Agricultural, Food and Forestry Resources in the framework of the Project "Resistenze genetiche delle piante agrarie a stress biotici ed abiotici".

5. References

Abruzzese, A., Alisi, C., Espen, L., Morgutti, S., Negrini, N., Sacchi, G.A., Cocucci, M., Cocucci, S. and Leva, A.R. (1994) 'Metabolic suffering in *Actinidia deliciosa* callus growing on solid medium' VIII International Congress IAPTC, Poster No S17/20.

Cocucci, S., Morgutti, S., Abruzzese, A. and Alisi, C. (1990) 'Response to osmotic medium and fusicoccin by seeds of radish (*Raphanus sativus*) in the early phase of germination', Physiol. Plant. 80, 294–300.

Gogarten, J.P. and Bentrup, F.W. (1983) 'Fluxes and compartmentation of 3–O–methyl–D–glucose in *Riccia fluitans*', Planta 159, 423–431.

Gogarten, J.P. and Bentrup, F.W. (1989) 'Substrate specifity of the hexose carrier in the plasmalemma of *Chenopodium* suspension cells probed by transmembrane exchange diffusion', Planta 178, 52–60.

Izzo, R., Navari–Izzo, F. and Quartacci, M.F. (1989) 'Growth and mineral content of roots and shoots of maize seedlings in response to increasing water deficits induced by PEG solutions', J. Plant Nutrition 12, 1175–1193.

Muleo, R., Leva, A. R., Bartolini, G. and Sacchi, G. A. (1994) 'Role of sucrose and glucose on morphogenetic induction in kiwi (*Actinidia deliciosa*) somatic tissue: osmotic and trophic aspects', VIII International Congress IAPTC, Poster No. S1/168.

Neal, G.E. and Beevers, H. (1960) 'Pyruvate utilization in castor–bean endosperm and other tissues', Biochem. J. 74, 409–416.

Newton, R.J., Puryear, J.D., Bhaskaran, S. and Smith, R.H. (1990) 'Polyethylene glycol content of osmotically stressed callus cultures', J. Plant Physiol. 135, 646–652.

Polle, A. and Rennenberg, H. (1993) 'Significance of antioxidants in plant adaptation to environmental stress', in L. Fowden, T. Mansfield and J. Stoddart (eds.), Plant Adaptation to Environmental Stress, Chapman and Hall, London, pp. 263–273.

Roberts, J.K.M., Ray, P.M., Wade–Jardetzky, N. and Jardetzky, O. (1980) 'Estimation of cytoplasmic and vacuolar pH in higher plant cells by ^{31}P NMR', Nature 283, 870–872.

Thompson, M.R., Douglas, T.J., Obata–Sasamoto, H. and Thorpe, T.A. (1986) 'Mannitol metabolism in cultured plant cells', Physiol. Plant. 67, 365–369.

GENE EXPRESSION IN PHOTOSYNTHETICALLY ACTIVE TOMATO CELL CULTURES IS INFLUENCED BY POTATO SPINDLE TUBER VIROID INFECTION

H.-P. MÜHLBACH, S. STÖCKER, R. WERNER, F. HARTUNG, U. GITSCHEL and M.-C. GUITTON
Department of Genetics, Institut für Allgemeine Botanik, University of Hamburg, Ohnhorststrasse 18, D-22609 Hamburg, Germany

ABSTRACT. Long-term cultures of photosynthetically active callus and cell suspensions established from potato spindle tuber viroid (PSTVd) infected tomato plants (*Lycopersicon esculentum* Mill.) show remarkable cytopathic effects. To study the influence of PSTVd infection on gene expression in these cells, RNA gel blot hybridization analyses using a series of tomato cDNA probes were performed. As an example, accumulation of a ß-glucosidase mRNA was reduced in PSTVd infected cell cultures as well as in infected tomato leaves, whereas the mRNA encoding the cathepsin D inhibitor protein was accumulated upon PSTVd infection in cell cultures and in tomato leaves. Thus, in photosynthetically active tomato cell cultures the expression of particular genes in response to PSTVd infection resembled that of the entire plant. To identify further "viroid-responsive" genes, cDNA from either cell culture was enriched via repeated cycles of amplification by PCR following in vitro subtractive hybridization. The enriched cDNAs were cloned and sequenced. With this approach, we have isolated 7 cDNA clones specific for PSTVd infected cell cultures and 5 clones specific for uninfected ones.

1. Introduction

Plant cell and tissue cultures proved to be helpful tools in studies on the interactions of viruses and viroids with their host cells (Mühlbach 1982). Heterotrophically grown cell suspension cultures of *Solanum demissum* allowed the detection of PSTVd replication intermediates (Mühlbach et al. 1983). Using such cell cultures it could be shown that PSTVd replicates in the cell nucleus (Spiesmacher et al. 1985) with the involvement of DNA dependent RNA polymerase II (Schindler and Mühlbach 1992). Cytopathic responses to viroid infection, however, could hardly be detected in heterotrophically grown cell cultures. On the other hand, photosynthetically active callus and cell suspension cultures established from PSTVd infected and from uninfected tomato plants, exhibited typical morphological alterations, visible at the light and electron microscopical level, and dramatic changes in their carbohydrate metabolism (Stöcker et al. 1993). We have therefore analysed the accumulation of mRNA in photosynthetically active tomato cell cultures using as probes cDNA clones isolated from PSTVd infected tomato plants. To detect further "viroid-responsive" genes, we established cDNA libraries from uninfected and from PSTVd infected tomato cell cultures. By the highly sensitive procedure of in vitro subtractive hybridization and PCR amplification we could isolate several additional cDNAs, which indicate the specific induction or repression of particular genes in response to PSTVd infection.

M. Terzi et al. (eds.), Current Issues in Plant Molecular and Cellular Biology, 557-562.

2. Material and Methods

2.1. CALLUS AND CELL SUSPENSION CULTURES

Photosynthetically active callus and cell suspension cultures were established from either PSTVd infected or uninfected plants of the PSTVd-sensitive tomato cultivar "Rutgers" exactly as described by Stöcker et al. (1993). Cell suspensions were cultured at low sucrose levels (photomixotrophic) or without any sucrose (photoautotrophic) in a CO_2-enriched atmosphere with white light at 117-120 $\mu E\ m^{-2}\ s^{-1}$ according to Stöcker et al. (1993).

2.2. PHYSIOLOGICAL CHARACTERIZATION OF SUSPENSION CULTURE CELLS

The relative photosystem II activity was determined by pulse amplified modulated fluorescence (PAMF) according to Schreiber et al. (1986). Incorporation of $^{14}CO_2$ and starch accumulation were measured according to Herzbeck (1985).

2.3. RNA EXTRACTION AND GEL BLOT HYBRIDIZATION

For PSTVd detection, RNA extracted from cultured cells was separated on 5% (w/v) polyacrylamide gels, blotted onto Hybond N (Amersham) according to Mühlbach et al. (1992), and hybridized to digoxigenin (Boehringer, Mannheim) labelled antisense PSTVd RNA transcripts as described by Stöcker et al. (1993). The accumulation of particular mRNA species was analysed in RNA extracts prepared according to MacDonald et al. (1987) after separation in 2% agarose gels containing 6% (v/v) formaldehyde. Digoxigenin (Boehringer, Mannheim) labelled RNA probes were generated by T7 transcription of the corresponding cDNA clones following the manufactors instruction and used for hybridization as described above. Digoxigenin labelled DNA probes were prepared by PCR amplification of the cDNA inserts of vector pT3T7BM (Boehringer, Mannheim), using T3 and T7 specific primers.

2.4. IN VITRO SUBTRACTIVE HYBRIDIZATION AND PCR AMPLIFICATION

cDNA libraries were established from polyA^{+} RNA preparations of either PSTVd infected or uninfected tomato cell cultures of the same physiological stage, using the Promega cDNA synthesis kit. Preparation of cDNA and PCR amplification were performed according to Wang and Brown (1989). In vitro subtraction was done in two rounds with a driver DNA to target DNA ratio of 20:1 in the first and 10:1 in the second round. The target DNA remaining in solution was amplified, EcoRI restricted and cloned into the vector pT3T7BM (Boehringer, Mannheim).

3. Results

3.1. CHARACTERISTICS OF PHOTOSYNTHETICALLY ACTIVE SUSPENSION CULTURES OF UNINFECTED AND PSTVd INFECTED TOMATO CELLS

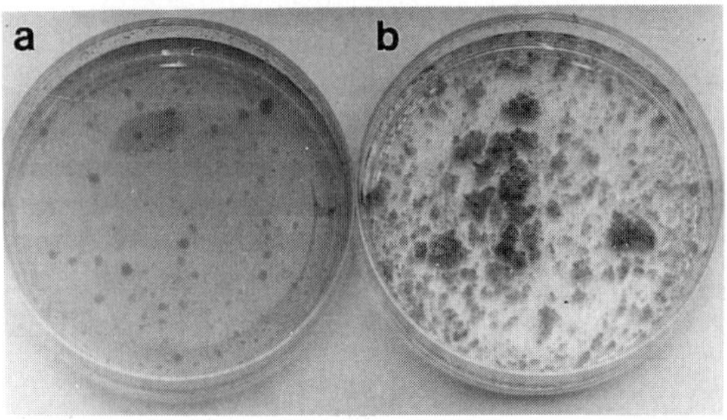

Fig. 1: Morphological differences of uninfected (**a**) and PSTVd-infected (**b**) cells of photosynthetically active suspension cultures of *Lycopersicon esculentum* Mill.

Striking morphological differences were observed, when cells from uninfected and PSTVd-infected plants were cultivated under conditions that promote growth of photosynthetically active cells. Uninfected cells formed a fine homogeneous suspension, as shown in Fig. 1a, whereas PSTVd infected cells grew as tight aggregates (Fig. 1b), which always spontaneously emerged after selection of single cells or small colonies by filtering through 1 mm sieves. A detailed investigation of morphological, ultrastructural and physiological characteristics revealed a series of alterations in PSTVd-infected cells, which are summarized in table 1.

TABLE 1. Morphological and physiological properties of photosynthetically active suspension cultures of PSTVd-infected and uninfected tomato cells

	PSTVd-infected	Uninfected
Cell aggregation	high	low
Cell shape[1]	round	longitudinal
Chloroplast envelope[1]	deformed	regular
Plasmalemmasome structure[1]	irregular	regular
Relative cell mass increase[2]	52 %	230 %
Relative photosynthetic capacity (F_{max}/F_o)[3] during photomixotrophic growth[4]	4.0	2.8
Relative photosynthetic capacity (F_{max}/F_o)[3] during photoautotrophic growth[5]	-	4.4
$^{14}CO_2$ incorporation (μmol $^{14}CO_2$/mg Chl·h)	0.9	1.5
Starch accumulation (mg/g dry weight)	56.6	32.7

[1] Data from Stöcker et al. (1993)
[2] Over 10 d culture
[3] F_{max}/F_o was determined by pulse amplified modulated fluorescence according to Schreiber et al. (1986)
[4] Photomixotrophic growth with 1% sucrose in culture medium
[5] Photoautotrophic cultures (without any sugar in culture medium) could only be established with uninfected cells

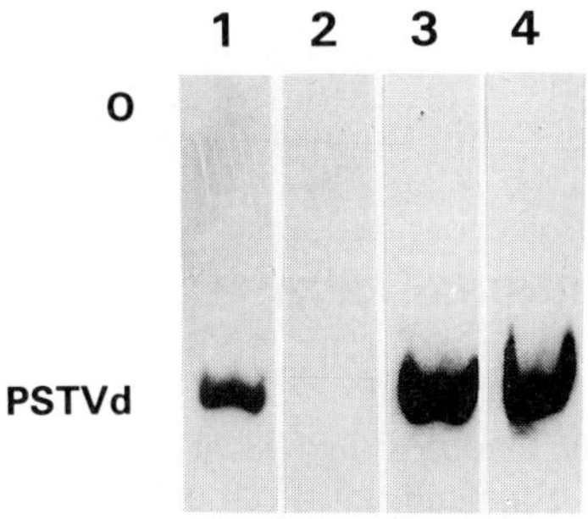

Fig. 2: Detection of PSTVd in photosynthetically active cell cultures of *Lycopersicon esculentum* Mill. by RNA gel blot hybridization. **1**: PSTVd infected plants; **2**: Uninfected callus; **3**: PSTVd infected callus; **4**: PSTVd infected cell suspension.

PSTVd replication in photomixotrophic callus and suspension cultures was analyzed by RNA gel blot hybridization. Fig. 2 shows that as in heterotrophically grown cultures (Mühlbach et al. 1983) PSTVd replicates continously in photosynthetically active cells and is accumulated to levels comparable to green leaves.

3.2. DIFFERENTIAL GENE EXPRESSION IN TOMATO CELLS IN RESPONSE TO PSTVd-INFECTION

From cDNA libraries of PSTVd infected tomato plants several cDNAs could be isolated, which indicate a remarkable influence of PSTVd infection on the accumulation of the corresponding mRNA. RNA gel blot hybridization analyses revealed that, as an example, a ß-glucosidase mRNA was differentially expressed in tomato leaves upon PSTVd infection (Fig. 3A). In very young leaves of the shoot tip, mRNA accumulation was higher in infected (YI) than in uninfected plants (YC), whereas in expanding leaves ß-glucosidase mRNA accumulation was reduced upon viroid infection (EI). Interestingly, in photosynthetically active cell cultures, ß-glucosidase mRNA was only found in uninfected cells, but not in PSTVd infected cells (Fig. 3B). Thus, ß-glucosidase mRNA accumulation in tomato cell cultures was reduced upon PSTVd infection, resembling the situation in expanding leaves. 25S ribosomal RNA accumulation was not affected by the viroid. The mRNA encoding a cathepsin D inhibitor protein (Werner et al. 1992) was accumulated upon PSTVd infection in both, tomato leaves and cell cultures (data not shown).

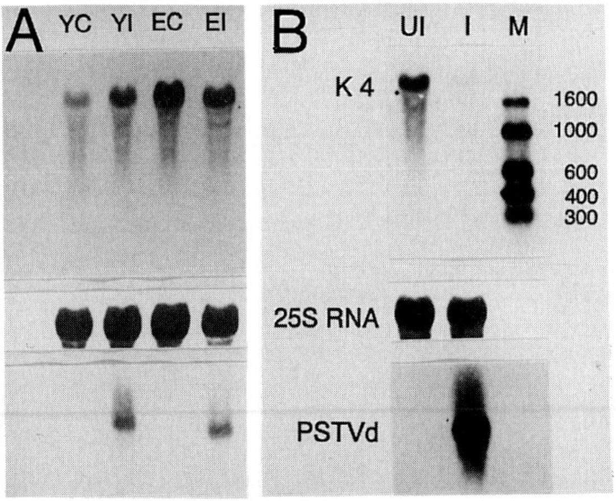

Fig. 3: Influence of PSTVd infection on the accumulation of a ß-glucosidase mRNA in tomato leaves (**A**) and tomato cell suspension cultures (**B**). **YC**: Young leaves from uninfected plants; **YI**: young leaves from PSTVd infected plants; **EC**: expanding leaves from uninfected plants; **EI**: expanding leaves from PSTVd infected plants; **UI**: uninfected cell cultures; **I**: PSTVd infected cell cultures; **K4**: transcripts detected with the ß-glucosidase specific probe K4; **25S RNA**: blots hybridized with a 25S ribosomal RNA specific probe; **PSTVd**: hybridization with a PSTVd specific probe; **M**: RNA size marker, fragment length is indicated by nucleotide numbers.

3.3. ISOLATION OF VIROID RESPONSIVE cDNAs FROM TOMATO CELLS BY IN VITRO SUBTRACTION

From the cDNA clones isolated by in vitro subtraction, 16 individual clones were sequenced. As shown in table 2, five of them were specific for uninfected cell cultures, 7 were specific for PSTVd infected cells, and 4 showed no detectable differences in the screening of enriched cDNA species from uninfected and from PSTVd infected cell cultures. The majority of the sequenced clones had no homologous counterparts in the EMBL and GenBank databases. Two

Table 2: Characteristics of cDNA fragments isolated by in vitro subtraction from uninfected and from PSTVd infected tomato cell cultures

Number of clones	Length of Inserts (bp)	Specificity	Database homologies
5	181 - 320	uninfected cells	No homology
7	101 - 318	PSTVd-infected cells	2 homologies
4	184 - 344	non-specific	No homology

of the PSTVd-infection specific cDNA clones, however, were homologous to two different transposon cDNAs. The clone 7-9 had 71% identity with the deduced amino acid sequence of the *del* transposon of *Lilium henryi* (Smyth et al. 1989), clone 4-10 had 29.3% identity with the retrotransposon Tf1 from *Schizosaccharomyces pombe* (Levin et al. 1990). Hybridization with clone 4-10 as probe revealed that a transcript of 1700 nucleotides length is highly accumulated in PSTVd infected cell cultures, but hardly detectable in uninfected cells (Fig. 4A). In plants, accumulation of mRNA could only be detected in expanding leaves from uninfected tomato, whereas neither young leaves from uninfected, nor young or expanding leaves from PSTVd infected plants showed any detectable hybridization signal (Fig. 4B). Interestingly, the main transcript found in leaves was approx. 4600 nucleotides long.

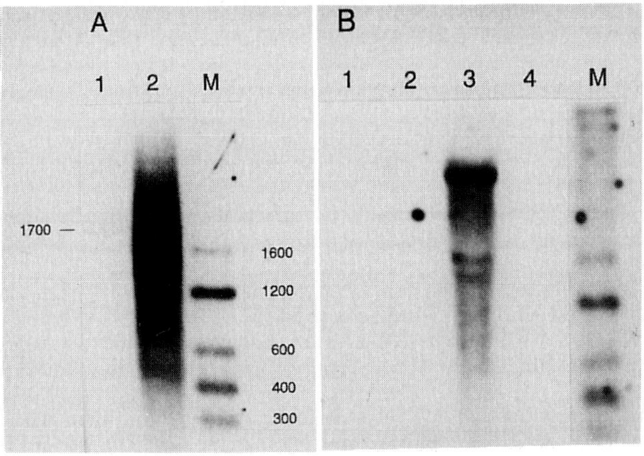

Fig. 4: Influence of PSTVd infection on the accumulation of transcripts detected by the Tf1 transposon homologous cDNA clone 4-10 in tomato cell cultures (**A**) and tomato leaves (**B**).
A) **1**: Uninfected cell suspensions; **2**: PSTVd infected cell suspensions
B) **1**: Young leaves from uninfected tomato plants; **2**: young leaves from PSTVd infected plants; **3**: expanding leaves from uninfected plants; **4**: expanding leaves from PSTVd infected plants; **M**: RNA size marker, fragment length is indicated by nucleotide numbers

4. Discussion

Our study has shown that PSTVd is able to replicate in photosynthetically active suspension cultures of tomato cells, which show a series of changes in the expression of particular genes as compared to uninfected cells. Hybridization analyses of mRNA accumulation revealed that the cathepsin D inhibitor and a ß-glucosidase mRNA were affected in a similar way by PSTVd in leaves and cell cultures. In contrast, the Tf1 transposon homologous clone 4-10 showed completely different responses to PSTVd infection in cell cultures and in leaves with respect to mRNA accumulation and transcript length. Interestingly, some of the clones that respond to PSTVd infection, e.g. the ß-glucosidase and the Tf1 transposon homologue, represent developmentally regulated genes, because their viroid response and overall expression varied in leaves of differing age.

Evidence is now accumulating that viroid infection influences gene expression in plants (Tornero et al. 1994), although the question of viroid pathogenesis is still unanswered. Suspension cultures of photosynthetically active tomato cells, which grow continously under standardized conditions and resemble in several aspects mesophyll cells, will help to elucidate the complex interaction between viroids and host cell metabolism.

Acknowledgements We thank Dr. H.L. Sänger, Martinsried, for a gift of PSTVd, Dr. W. Barz and Dr. W. Hüsemann, Münster, for helpful advice in cell culture. The work of G. Nissen and M. Cerwinski in the greenhouse, the help of C. Adami in photography and of J. Olschewski in cell culture, and the technical assistance of G. Monnier are gratefully acknowledged. This work was supported by grants Mu 559/3-2 and 4-1 from the Deutsche Forschungsgemeinschaft and by grant 0318988C from the German Federal Ministry of Research and Technology. This article is based on doctoral studies by U. G., F. H. and R. W. in the Faculty of Biology, University of Hamburg.

5. References

Herzbeck, H. (1985) 'Untersuchungen zum Kohlenstoffmetabolismus in photoautotrophen Zellsuspensionskulturen von Chenopodium rubrum', Ph.D. Thesis, University of Münster.

Levin, H.L., Weaver, D.C., Boeke, J.D. (1990) 'Two related families of retrotransposons from Schizosaccharomyces pombe', Molecular and Cellular Biology **10**, 6791-6798.

MacDonald, R.J., Swift, G.H., Przybyla, A.E., Chirgwin, J.M. (1987) 'Isolation of RNA using guanidinium salts', in S.L. Berger and A.R. Kimmel (eds.), Methods in Enzymology **152**, Academic Press, San Diego, pp. 219-227.

Mühlbach, H.-P. (1982) 'Plant cell cultures and protoplasts in plant virus research', Current Topics in Microbiology and Immunology **99**, 81-129.

Mühlbach, H.-P., Faustmann, O., Sänger, H.L. (1983) 'Conditions for optimal growth of a PSTV infected potato cell suspension and detection of viroid complementary longer-than-unit-length RNA in these cells', Plant Mol. Biol. **2**, 239-247.

Mühlbach, H.-P., Barth, A., Tank, C. (1992) 'Efficient transfer of infectious potato spindle tuber viroid cDNA and ß-glucuronidase gene to tomato protoplasts by polyethylen glycol', Molec. Biology (Life Sci. Adv.) **11**, 79-90.

Schindler, I., Mühlbach, H.-P. (1992) 'Involvement of nuclear DNA dependent RNA polymerases in potato spindle tuber viroid replication: a reevaluation', Plant Science **84**, 221-229.

Schreiber, U., Schliwa, U., Bilger, W. (1986) 'Continous recording of photochemical and nonphotochemical chlorophyll fluorescence quenching with a new type of modulation fluorometer', Photosynth. Research **10**, 51-62.

Smyth, D.R., Kalitsis, P., Joseph, J.L., Sentry, J.W. (1989) 'Plant retrotransposon from Lilium henryi is related to Ty3 of yeast and the gypsy group of Drosophila', Proc. Natl. Acad. Sci. USA **86**, 5015-5019.

Spiesmacher, E., Mühlbach, H.-P., Tabler, M., Sänger, H.L. (1985) 'Synthesis of (+) and (-) RNA molecules of potato spindle tuber viroid (PSTV) in isolated nuclei and its impairment by transcription inhibitors', Bioscience Reports **5**, 251-265.

Stöcker, S., Guitton, M.-C., Barth, A., Mühlbach, H.-P. (1993) 'Photosynthetically active suspension cultures of potato spindle tuber viroid infected tomato cells as tools for studying viroid - host cell interaction', Plant Cell Reports **12**, 597-602.

Tornero, P., Conejero, V., Vera, P. (1994) 'A gene encoding a novel isoform of the PR-1 protein family from tomato is induced upon viroid infection', Mol. Gen. Genet. **243**, 47-53.

Wang, Z., Brown, D.D. (1991) 'A gene expression screen', Proc. Natl. Acad. Sci. USA **88**, 11505-11509.

Werner, R., Guitton, M.-C., Mühlbach, H.-P. (1993) 'Nucleotide sequence of a cathepsin D inhibitor protein from tomato', Plant Physiol. **103**, 1473.

MOLECULAR ANALYSIS OF THE NITRATE ASSIMILATORY PATHWAY IN SOLANACEOUS SPECIES

M. CABOCHE, P. CRETE, J. D. FAURE, C. GODON, T. HOFF,
A. QUESADA, C. MEYER, T. MOUREAUX, L. NUSSAUME,
H. N. TRUONG, F. VEDELE
Laboratoire de Biologie Cellulaire, INRA
Route de Saint Cyr, 78026 Versailles France

ABSTRACT. The nitrate assimilatory pathway has been the matter of intensive molecular and cellular analysis in solanaceous species. In this report we will present data on the function of the N-terminal part of the enzyme nitrate reductase, on the cloning of genes of the molybdenum cofactor biosythesis, and on the characterization of a transcription factor presumed to be involved in the expression of genes of the nitrate assimilatory pathway.

INTRODUCTION

Nitrate can be used as the sole nitrogen source to sustain growth, in both microorganisms and higher plants. In plants, two successive enzymatic steps reduce nitrate to ammonium, generally in the leaves. First, nitrate is converted into nitrite in a two electron transfer reaction catalysed by nitrate reductase (NR, EC1. 6. 6. 1), a cytoplasmic enzyme. Nitrite is then translocated to the chloroplast, where it is reduced to ammonium by nitrite reductase (NiR, EC1. 7. 7. 1). Ammonium is subsequently incorporated into the amino acid pool through glutamine and glutamate biosynthesis.

NADH:NR is the most common form of nitrate reductase in higher plants, and appears to be a homodimer (monomer size between 100 and 120 kDa), each monomer containing three redox prosthetic groups: FAD, heme and a molybdenum cofactor (MoCo). Many mutants defective in nitrate assimilation have been isolated in different species. These mutants have been found to be defective in nitrate reduction due to disruptions in either the nitrate reductase structural gene (*nia* mutants) or one of the six or seven genes required for synthesis and assembly of the MoCo (*cnx* or MoCo mutants).

Nitrate assimilation is a highly regulated process. In higher plants, factors such as light, cytokinins and circadian rhythm affect NR expression (Caboche and Rouzé 1990). In addition, nitrate induces, and ammonium and/or glutamine down-regulate NR expression (Deng, Faure and Caboche,1993) . This paper will review recent progresses made in our laboratory to characterize nitrate reductase expression and regulation.

M. Terzi et al. (eds.), Current Issues in Plant Molecular and Cellular Biology, 563-570.

RESULTS

1. Relationships between structure and activity of nitrate reductase

A number of mutants defective for nitrate reductase activity were isolated. Among mutants specifically defective for the apoenzyme belonging to the *nia* complementation group, four classes were identified on a biochemical basis. We have recently characterized at the molecular level mutants defective for the so called "terminal transferase" activities allowing the wild type enzyme to reduce nitrate into nitrite in the presence of reduced electron donors (methylviologen, bromophenol blue, etc...). All the five mutations characterized by sequencing affect aminoacid residues from the molybdenum cofactor-containing domain, which are identical in all known homodimeric eukaryotic nitrate reductases, confirming their crucial role in catalysis.

The N-terminal part of plant nitrate reductase is highly divergent in size and sequence to that of fungal nitrate reductases. We have deleted 56 aminoacids of this sequence, including a cluster of acidic aminoacids, and transformed *Nicotiana plumbaginifolia* plants defective for the residant *Nia* gene with a construct allowing the constitutive expression of the N-terminal-truncated cDNA nitrate reductase coding sequence. When transgenic plants expressing a normal or truncated NR were compared for the level of expression of the corresponding NR protein it was found that the truncated NR protein was functional and unaffected by the light regime, as opposed to the normal NR protein which was degraded or inactivated in the dark. The truncated enzyme was also unaffected by agonists or antagonists of NR phosphorylation as opposed to the wild type. Work is in progress to identify by site directed mutagenesis the putative serine residue(s) involved in NR inactivation by phosphorylation.

2. Molybdenum cofactor biosynthesis

Up to six complementation groups were identified among nitrate reductase deficient mutants corresponding to pleotropic mutations affecting XDH as well as NR. The target of these mutations are steps of the biosynthesis of the molybdenum cofactor, and the corresponding loci have been named *cnx A,B,C,D,E,F in Nicotiana plumbaginifolia.* Functional complementation of molybdenum cofactor-deficient mutants of *E. Coli* was used to clone plant homologs to bacterial genes involved in molybdenum cofactor biosynthesis. Two genes strikingly homologous to Moa A and Moa C were cloned and are presently being characterized. Their structure would suggest that the MoCo pathway might be localized in organelles.

3. Transcription of Nia and Nii genes

3.1.Light and carbohydrates

As for light-regulated genes involved in photosynthesis such as ribulose biphosphate carboxylase, NR and NiR transcript levels in leaves of mature green plants decrease in darkness and increase again in the light (Deng et al 1990). Because nitrate assimilation into glutamine and glutamate is largely dependent on the availability of carbons skeletons, we asked to what extent regulation of NR and NiR gene expression by light could be related to carbohydrate supply, which would be limiting after an extended period of darkness. Therefore, we analysed the consequences of providing carbohydrate to detached leaves of dark-adapted *Nicotiana plumbaginifolia* plants. Our results (Vincentz et al 1993) suggest that light regulation of *nia* gene expression is, at least in part, mediated by carbohydrates but that light is most likely the signal missing for a full induction.

3.2 Nitrate inducibility and N-metabolite repression

NR activity is substrate-inducible and NR mRNA accumulates in leaves of plants grown on nitrate-supplemented medium. We studied the expression of NiR mRNA in four *nia* mutants of *N.plumbaginifolia* which differ in the expression level of the NR mRNA (Faure et al 1991). Our results indicate that NR and NiR are co-regulated at the mRNA level in wild-type plants and are still nitrate inducible in NR-deficient mutants. In addition , the lack of NR activity leads to an overproduction of NiR mRNAs. This observation leads us to propose that N-metabolites derived from nitrate assimilation represses the expression of NiR and NR genes. We analysed the effect of expressing an antisense NiR gene in transgenic tobacco, inhibiting nitrate assimilation and N-metabolite production without modifying the NR gene product (Vaucheret et al 1992). The overexpression of the NR mRNA in transgenic plants demonstrate that the NR protein is not involved in its own metabolic repression, in contrast to what has been postulated in the autocatalytic model of NR expression in fungi.

3.3 *Circadian rhythm and N-metabolite repression*

When tobacco or tomato plants are grown under a dark-light regime, the concentrations of NR transcripts are high at the beginning of the day period but almost undetectable at the end of this day period (Galangau et al 1988). Amino acids contents, specially glutamine, show also marked 24h rhythmic variations, inversely correlated to NRmRNA fluctuations (Deng et al 1991). This fluctuation of NR mRNA expression appears to be under the control of a circadian rhythm that can be abolished by biochemical or genetic impairement of NR catalytic activity. Mutations in *nia* or *cnx* genes (Pouteau et al 1989), as well as treatments by tungstate (Deng et al 1989), an inhibitor of NR activity, lead to high and stable expression of NR mRNA. Using detached leaves from *N.plumbaginifolia* plants grown under low light conditions (limited photosynthesis), we tested the effect of glutamine, glutamate and asparagine on NR and NiR gene expression. Our results (Vincentz et al 1993) indicate that glutamine and glutamate down-regulate NR expression in green leaves whereas asparagine has a weaker inhibitory effect. Nitrate and light are strong inducers of *nia* gene transcription. The physiological significance of up-regulation by sugars and down-regulation by N-metabolites may be to regulate the reduction of nitrate in response to changes in the cellular ratio of N to C metabolites.

4. Transcriptional regulation of the expression of *Nia* and *Nii* genes.

Three *nia* genes have been isolated from tobacco and tomato (Vaucheret et al 1989, Daniel-Vedele et al 1989). The general organization of the genes is very similar and an overall homology of 81% is found in the exon sequences. Comparison of the three promoters reveals only one conserved region, containing the TATA and CAAT boxes. Two complementary experiments were performed in order to localize the cis-acting sequences involved in *nia* gene regulations. Four *Nii* genes have also been identified in the genome of tobacco.(Kronenberger et al, 1993). Their promoters do not share aprreciable homologies with *Nia* genes althoug they undergo a similar control of expression (Unpublished data).

4.1 Complementation of a NR-deficient mutant with the tomato nia gene

The E23 *N.plumbaginifolia* mutant is deficient in NR activity and cannot reduce nitrate into ammonium. As a consequence, it cannot grow on its roots in the green house. We have shown (Dorbe et al 1992) that this mutant can be restored for NR activity by transformation with the cloned tomato *nia* gene, surrounded by 3kb upstream and downstream non-coding regulatory sequences. The transgenic plants expressed from undetectable to 22% of the wild type activity and a good correlation was observed between growth rates and levels of NR activities in transgenic plants. The analysis of the transgene expression showed that the tomato *nia* gene transcription was regulated by light, nitrate and circadian rhythm as in tomato plants. These results suggest that all the cis-acting sequences involved in these regulations are contained in this tomato gene.

4.2 Complementation of a NR-deficient mutant with a chimaeric NR gene

The same E23 *N.plumbaginifolia* mutant was restored for NR activity by transformation with a full-length tobacco NR cDNA, fused to the CAMV 35S promoter and to termination signals from the tobacco NR gene (Vincentz and Caboche 1991). The transgenic plants expressed from one-fifth to three times the wild-type NR activity in their leaves. The analysis of chimeric NR gene expression showed that the regulation of NR expression by light, nitrate or circadian rhythm was no longer maintained in transgenic plants. In the same manner, expression of chimeric NR gene was not influenced by darkness and carbohydrate supply (Vincentz et al 1993). These results demonstrate that the *nia* gene expression is mainly mediated by transcriptional regulations. These transgenic plants have normal growth characteristics. However as a consequence of a deregulated expression of the Nia gene transcript, these plants are hypesensitive to chlorate, even when grown on ammonium (Nussaume, Vincentz, and Caboche,1991). Interestingly, a decreased storage of nitrate was reproducibly detected in their leaves (Foyer et al, 1994; Quilleré et al, 1994)

4.3 Use of reporter gene techniques to analyze transcriptional control

A number of attemps to study the promoter of the Nia gene have been limited by the almost systematic extinction of the constructs inserted into the genome of recipient plants. In the few instances where the inserted gene was still expressed regulation of the corresponding construct by nitrate, light, N-metabolites and sucrose was detectable. However none of these regulations was detected in trasient expression assays performed on protoplasts, or in leaf tissues, by particule gun bombardment (Godon, unpublished). Similar observations were made with the Nii promoter, althoug a higher proportion of obtained transformants was still expressing adequately the reporter gene (Truong and Vaucheret, unpublished)

5. Isolation of a tobacco gene homologous to *Are A*, *Nit 2*, and *Gln3*, regulators of nitrogen metabolisms in fungi.

Very little is known about the trans-acting factors that are required for the regulation of NR in higher plants. There are significant similarities between the NR regulation found in plants and fungi. In *Neurospora crassa* and *Aspergillus nidulans*, the NR expression is highly regulated at the mRNA level and requires nitrogen de-repression and simultaneous induction by nitrate. The regulation of the NR and NiR genes is governed on one hand by the pathway-specific positive regulatory genes, respectively *nit-4* and *nirA*, involved in the nitrate inducibility. On the other hand, *nit-2* and *areA*, the major positive regulatory genes are involved in N-metabolites repression.

The nucleotide sequence of the *nit-2* and *areA* genes revealed that they encode transcription factors of the GATA-1 family, characterized by a DNA-binding element consisting of a single Cys2/Cys2 "zinc-finger" and an adjacent basic region . In vitro gel-band mobility-shift and DNA footprinting studies showed that the NIT2 protein recognizes specifically sequences in the 5' region of N-repressible genes of *N.crassa* and of the *A. nidulans* nitrate and nitrite reductase genes. A core consensus sequence, TATCT (or on the complementary strand AGATA), was identified in all of these binding sites.

Recent studies have demonstrated that two fragments of the tomato *nia* gene promoter region were specifically recognized and bound by the NIT2 protein (Jarai et al. 1992). Both fragments contain the putative GATA recognition sequence. These data suggest that a homolog of NIT2, belonging to the GATA-binding protein family, may exist in tomato and other higher plants and play a central role in controlling NR expression.

Using degenerated oligonucleotides encompassing conserved regions in the NIT2 and AREA zinc fingers, we were able to isolate in *N.plumbaginifolia* a partial cDNA coding for a NIT2-like protein. A full length cDNA was then isolated from N. tabacum (Daniel-Vedele and Caboche, 1993). Analysis of the predicted protein sequence reveals the presence of different regions . First, a single putative zinc finger DNA-binding domain, covering 26 residues is localised in the C-terminus of the protein. This domain is composed of two pairs of cysteine residues, separated by a loop of 18 amino acids (instead of 17 amino acids as in the NIT2 protein). Two highly basic regions occur immediately upstream and downstream of this putative zinc finger: respectively, from amino acid 132 to 201 with a net charge of +17 and from amino acid 230 to 266 with a net charge of +9 (basic amino acids are underlined in Fig. 4). This basic region, on the carboxy side of the zinc finger, is also found in the NIT 2 protein and was shown to represent part of the DNA-binding domain. The other basic region, upsteam of the zinc finger, contains the amino acid sequence KEKKRK (169 to 174). This sequence is very similar to the well characterized SV40 nuclear localisation signal, KKKRK. The presence of a putative nuclear localisation signal in the NRG1-NT7 protein suggest the import of the protein from the cytoplasm into the nucleus. An acidic region is localised in the N-terminus of the protein, from amino acid 12 to 48, with a net charge of -13. Such acidic regions have been shown to be responsible for transcriptional activation by the yeast GAL4 and GCN4 proteins. A comparison of the predicted primary amino acid sequence of the large open reading frame of the Ntl1-Nt7 cDNA with the GenEMBL data base (GCG; TFASTA seach) revealed that the zinc finger region is related to previously reported protein sequences. Best scores are obtained with *Are A* and other Zinc finger proteins of the GATA type.

Constructs containing the whole or the DNA-binding domain of NTL1-NT7 driven in anti-sens orientation by the 35S promoter have been introduced via *Agrobacterium tumefaciens* in *N.plumbaginifolia* wild type plants. Primary transformants have been obtained and we are analysing the expression of genes involved in the nitrate assimilation pathway in these transfromants. As a preliminary result yet to be confirmed, we observe, in 3 out of 18 plants, an overexpression of glutamine synthetase. This result may reflect a modification in the regulation of glutamine biosynthesis, the end product of the pathway.

6. Integration of the nitrate assimilatory pathway in plant development

A screen for mutants unable to grow on 1mM nitrate led to the identification of a mutant which is pleiotropically affected in different aspects of early development. The ability of the mutant to grow on 1mM nitrate is prevented by the simultaneous presence of 50mM sucrose in the culture medium (Faure, Jullien and Caboche,1994). A 20 fold increase in sucrose content was observed in the cotyledons of the mutant grown under these non permissive conditions. Nitrate reductase transcript level was accordingly strongly reduced in these conditions. This suggests an impairement of nitrate reductase expression, mediated by sucrose over-accumulation in mutant plants. The pleotropic phenotype of the mutant was reminiscent of that of a cytokinin mutant identified previously in an independant screen. Tests of allelism revealed that these mutants are affected in the very same gene (Faure, Jullien and Caboche,1994). These cytokinins mutants are resistant to a high cytokinin concentrations. Work is underway to identify the target of the mutation and to elucidate the links between cytokinin tolerance and sucrose accumulation.

CONCLUSIONS

The regulation of the nitrate assimilatory pathway appears fairly complex, as one might have expected. From the molecular analysis of this regulation it is clear that a strong metabolic control by N and C metabolites takes place. As a consequence of this regulation, the flow of the pathway is kept under the control of the needs for reduced nitrogen of the developping plant. We have tried to identify some of the genes controling the expression of the pathway. A zinc finger protein appears as a likely candidate for the regulation of the pathway, it is also clear that several elements must contribute to the regulation at the postranscription level.

ACKNOWLEDGEMENTS

This program is partially funded by two EEC grants (BRIDGE contract:BIOT-CT90-0164-C and AIR contract AIR3-CT92-0250)

REFERENCES (work done on nitrate assimilation in Versailles)

Caboche C , Rouzé P (1990) Nitrate reductase: a target for molecular and cellular studies in higher plants. Trends in Genetics 6: 187-192
Daniel-Vedele F, Dorbe M-F, Caboche M, Rouzé P (1989) Cloning and analysis of the tomato nitrate reductase-encoding gene: protein structure and amino acid homologies in higher plants. Gene 85: 371-380
Daniel-Vedele F. and Caboche M (1993): A tobacco cDNA clone encoding a GATA-1 zinc finger protein homologous to regulators of nitrogen metabolism in fungi. Mol Gen Genet 240: 365-373.

Deng MD, Moureaux T and Caboche M (1989) Tungstate, a molybdate analog inactivating nitrate reductase, deregulates the expression of the nitrate reductase structural gene. Plant Physiol. 91: 304-309

Deng MD, Moureaux T, Leydecker MT, Caboche M (1990) Nitrate reductase expression is under the control of a circadian rhythm and is light inducible in *Nicotiana tabacum* leaves. Planta 180: 257-261

Deng MD, Moureaux T, Lamaze T (1991) Diurnal and circadian fluctuation of malate levels and its close relationship to nitrate reductase activity in tobacco leaves. Plant Science 65: 191-197

Deng MD, Faure JD, Caboche M (1993) The molecular aspects of nitrate and nitrite reductase expression in higher plants. In Control of Plant Gene Expression. Chapter 26. D.P. Verma Ed, CRC Press. pp 425-442

Dorbe MF, Caboche M, Daniel-Vedele F (1992) The tomato *nia* gene complements a *Nicotiana plumbaginifolia* nitrate reductase-deficient mutant and is properly regulated. Plant Molec Biol 18: 363-375

Faure J-D, Vincentz M, Kronenberger J, Caboche M (1991) Co-regulated expression of nitrate and nitrite reductases. Plant J 1: 107-113

Faure JD, Jullien M, Caboche M (1994) *Zea* 3 : a pleiotropic mutation affecting cotyledon development, cytokinin resistance and carbon-nitrogen metabolism. Plant Journal 481-491

Foyer C, Lescure JC, Lefebvre C, Vincentz M, Vaucheret H (1994) Adaptations of photosynthetic electron transport, carbon assimilation and carbon partitioning in transgenic *Nicotiana plumbaginifolia* plants to changes in nitrate reductase activity. Plant Physiology In Press

Galangau F, Daniel-Vedele F, Moureaux T, Dorbe MF, Leydecker MT, Caboche M (1988) Expression of leaf nitrate reductase gene from tomato and tobacco in relation to light dark regimes and nitrate supply. Plant Physiol 88: 383-388

Jarai G, Truong HN, Daniel-Vedele F, Marzluf G (1992) NIT2, the nitrogen regulatory protein of *Neurospora crassa*, binds upstream of *nia*, the tomato nitrate reductase gene, in vitro. Curr Genet 21: 37-41

Kronenberger J, Lepingle A, Caboche M, Vaucheret H (1993) Cloning and expression of distinct nitrite reductases in tobacco leaves and roots. Mol Gen Genet 236: 203-208

Nussaume L, Vincentz M, Caboche M (1991) Constitutive nitrate reductase: a dominant conditional marker for plant genetics. Plant Journal 1 (2): 267-274

Pouteau S, Chérel I, Vaucheret H, Caboche M (1989) Nitrate reductase mRNA regulation in *Nicotiana plumbaginifolia* nitrate reductase deficient mutants. Plant Cell 1: 1111-1120

Quilleré I, Dufossé C, Roux Y, Foyer CH, Caboche M, Morot-Gaudry JF (1994) The effects of the deregulation of the NR gene expression on growth and nitrogen metabolism of winter-grown *Nicotiana plumbaginifolia*. Journal of Experimental Botany In press

Vaucheret H, Vincentz M, Kronenberger J, Caboche M, Rouzé P (1989) Molecular cloning and characterization of the two homeologous genes coding for nitrate reductase in tobacco. Mol Gen Genet 216: 10-15

Vaucheret H, Kroneneberger J, Lepingle A, Vilaine F, Boutin JP, Caboche M (1992) Inhibition of tobacco nitrite reductase activity by expression of antisense RNA. Plant J 2: 559-569

Vincentz M, Caboche M (1991) Constitutive expression of nitrate reductase allows normal growth and developement of *Nicotiana plumbaginifolia* plants. EMBO J 10: 1027-1035

Vincentz M, Moureaux T, Leydecker M-T, Vaucheret H, Caboche M (1993) Regulation of nitrate and nitrite reductases expression in *Nicotiana plumbaginifolia* leaves by nitrogen and carbon metabolites. Plant J 3: 315-324

INHERITANCE OF 5-METHYLTRYPTOPHAN-RESISTANCE, SELECTED IN VITRO, IN TOBACCO AND *DATURA INNOXIA* PLANTS

J. M. WIDHOLM AND J. E. BROTHERTON
University of Illinois
Department of Agronomy
PABL, 1201 W. Gregory
Urbana, Il 61801, USA

ABSTRACT. Selection for resistance to the tryptophan analog, 5-methyltryptophan (5MT), has produced cell lines which overproduce free tryptophan (Trp) due to the presence of a feedback altered anthranilate synthase (AS), less sensitive to feedback inhibition. Since AS is the first enzyme in the Trp branch of the shikimic acid pathway and is the feedback control point, the feedback alteration leads to the increased levels of free Trp observed. We have selected 5MT-resistant *Nicotiana tabacum* and *Datura innoxia* cultures and have regenerated plants. The tobacco plants did not have high free Trp or feedback altered AS while the *D. innoxia* plants did. Cultures initiated from these plants, in both cases, were resistant, however. Apparently the feedback resistant AS form was only expressed in tobacco cultured cells, but was expressed in both cells and plants of *D. innoxia.* The resistance trait is inherited by progeny as a single, nuclear, dominant gene in *D. innoxia.* In the case of tobacco the trait appears to be under the control of a nuclear dominant gene or genes but continued segregation upon successive self-pollinations makes the exact mode of inheritance unclear. The trait is also retained for more than 10 years in *D. innoxia* plants maintained through cuttings in the greenhouse.

INTRODUCTION

The selection of plant cell cultures resistant to various amino acid analogs has produced, in many cases, cell lines with elevated levels of the corresponding, natural, free amino acid. Resistance to the tryptophan (Trp) analog, 5-methyltryptophan (5MT), usually results in cell lines with greatly elevated free Trp levels due to the presence of a more feedback-insensitive form of the Trp biosynthetic control enzyme anthranilate synthase (AS) (first reported by Widholm 1972a,b). When such a selection was carried out with *Nicotiana tabacum* (tobacco) and *Datura innoxia*, the leaves of the *D. innoxia* (Ranch et al. 1983) but not the tobacco (Widholm 1980, Brotherton et al. 1986) plants had the feedback altered AS. Both plants did produce 5MT-resistant tissue cultures which also had the altered AS. Both the *D. innoxia* plants and cultures had elevated free Trp levels. Chromatographic studies of the AS enzyme activity showed that the resistance in the tobacco tissue cultures was due to the presence of elevated levels of an AS isozyme which was not expressed in the plant tissues examined, shoot tips, leaves, stems and roots. This isozyme, which was less feedback sensitive, was expressed at very low levels in the wild type tissue cultures, but again not in wild type plants. This change in the level of a naturally occurring feedback insensitive form was also noted in a 5MT-resistant *Solanum tuberosum* suspension culture (Carlson and Widholm 1978).

The 5MT-resistance trait can be of importance since Trp is an essential amino acid for nonruminant animals and is deficient in the seeds of cereals. The 5MT-resistance trait has

M. Terzi et al. (eds.), Current Issues in Plant Molecular and Cellular Biology, 571-576.

also been used as a marker in genetic manipulation studies such as protoplast fusion (Kothari et al. 1986) and in AS compartmentation studies (Brotherton et al. 1986).

The original reports describing the tobacco (Widholm 1980, Brotherton et al. 1986) and *D. innoxia* (Ranch et al. 1983) plants which carry the 5MT-resistance did not show, as we do in this report, that the trait was inherited or that it was stable over long periods in vegetatively propagated plants. The *D. innoxia* plants used in this study are those selected previously by Ranch et al. (1983) but new lines were selected in the case of tobacco to obtain fertile plants and to determine if the characteristics of the 5MT-resistant cultures and plants are the same as those found previously (Widholm 1980, Brotherton et al. 1986).

MATERIALS AND METHODS

The culture initiation, 5MT selection and plant regeneration procedures were as described previously for tobacco (Widholm 1980, Brotherton et al. 1986) and *D. innoxia* (Ranch et al. 1983). The plants were maintained in a greenhouse. Leaf pieces from greenhouse grown plants were surface sterilized or taken directly from sterilely germinated seedlings and placed on MS medium (Murashige and Skoog 1962) with 1.81μM 2,4-D to obtain callus which was then placed into the same liquid medium to obtain suspension cultures. The suspensions were incubated on a gyratory shaker at 130 rpm and all cultures were grown under low continuous fluorescent light at 27-28°C.

The 5MT-resistance of the tobacco lines was determined by placing 0.5g fresh weight cells into 50ml liquid medium containing 46 and 137μM 5MT and 0.5 or 1.0g fresh weight of *D. innoxia* cells were incubated with 100 and 300μM 5MT. The flasks were incubated for 8-14d before the cells were filtered and weighed.

The AS activity was measured as described by Brotherton et al.(1986) except the chorismate concentration of the substrate solution was 8 instead of 80 mM and free Trp was measured using a Waters Pico-Tag system after extraction of fresh tissue with 0.1 N HCl containing norleucine as an internal standard and ultrafiltration.

RESULTS

1. 5MT-Resistant *Nicotiana tabacum*. The selection for 5MT-resistance was carried out by inoculating 36 flasks containing liquid medium of both 46 and 137μM 5MT with 1g fresh weight of *N. tabacum* cells. After about 60d, cells grew in four of the 46μM 5MT flasks and plants were regenerated. One of the lines produced no plants which carried resistance as determined by growth of suspension cultures initiated from leaves in 46 and 137μM 5MT. The other three lines produced a total of 13 plants carrying resistance from a total of 24 tested.

These four original selected cultures and those which showed 5MT-resistance, that had been initiated from leaves, contained higher levels of free trp (usually 5 to 10-fold) than the wild type or the cultures initiated from leaves which did not carry 5MT resistance. This overproduction of Trp also correlated with the presence of less feedback sensitive AS activity. However, when the leaves of the plants that produced the 5MT-resistant cultures were analyzed there was not elevated levels of free Trp nor was there any change in the feedback inhibition pattern of the AS enzyme activity when compared with the wild type. These results are similar to those of Widholm (1980) and Brotherton et al. (1986) with another tobacco selected line and regenerated plant where the feedback altered AS form was not expressed in the plant.

All of the plants regenerated from the 5MT-resistant cultures were male sterile but one plant which carried 5MT-resistance was pollinated with wild type pollen and three of nine progeny tested carried resistance. This plant was self-pollinated and the progeny carrying resistance were also selfed for several generations but homozygosity was not attained.

One of nine plants regenerated from the unselected wild type culture was also found to carry 5MT-resistance apparently as a result of somaclonal variation. This plant was cross-pollinated due to male sterility and one of nine progeny carried resistance. Successive self-pollinations of resistant plants for three generations again did not attain homozygosity as indicated by continued segregation of 5MT-resistant and sensitive progeny. In no cases has 5MT-resistance been observed in cultures initiated from leaves of plants grown from wild type seed.

Reciprocal crosses between one of these resistant plants and the wild type did produce some F1 progeny carrying 5MT-resistance when tested in tissue culture. This result also supports the conclusion that the tobacco 5MT-resistance is under the control of a nuclear, dominant gene or genes.

2. 5MT-Resistant *Datura innoxia*. Suspension cultures were initiated from seven of the original plants which had been regenerated from 5MT-resistant suspension cultures by Ranch et al. (1983). The plants had been maintained from cuttings in the greenhouse until the present time. The newly initiated suspension cultures exhibited 5MT-resistance when grown in 100μM 5MT where the wild type was completely inhibited as shown in Figure 1. Thus the 5MT-resistance was quite stable over a 10 year period.

The resistance was also heritable in reciprocal crosses between the wild type and one of the regenerated 5MT-resistant plants, STR-3/1. When one resistant plant from each of the crosses was self-pollinated there was a 3:1 segregation of resistant to sensitive progeny as six resistant and two sensitive plants were found in both cases. Self-pollination of three of the STR family of plants produced 14 plants which carried resistance and one that was sensitive. The growth data for STR-3/3 and one of its progeny carrying resistance is shown in Figure 1. These results indicate that the 5MT-resistance in *D. innoxia* is controlled by a single, nuclear, dominant gene.

In all cases tested, the 5MT-resistant suspension cultures and leaves of the plants from which the resistant cultures were initiated contained elevated free Trp levels (up to 10 to 70-fold increases). In the case of the cell cultures shown here, STR-3/3 and STR-3/3-4, the free Trp levels ranged from 177-464 nmol per g fresh weight in different analyses while the wild type contained about 18 nmol per g fresh weight. The 5MT-resistant suspension cultures also had AS activity which was less sensitive to feedback inhibition by Trp while the plants were not examined in this study. Figure 2 shows that the AS activity from the wild type culture is more sensitive to feedback inhibition by Trp than is the activity from the 5MT-resistant cultures from the original regenerated plant STR-3/3 and one of its' progeny, STR-3/3-4, which was produced by self-pollination. In the original report by Ranch et al. (1983) the 5MT-resistance was also always accompanied by elevated free Trp and a feedback altered AS in both tissue cultures and regenerated plants.

Inheritance of the 5MT-resistance trait has previously been reported for rice (Wakasa and Widholm 1987; Lee and Kameya 1991), *Zea mays* (Hibberd et al. 1987) and *Arabidopsis* (Kreps and Town 1992). The *D. innoxia* inheritance appears to be normal and the elevated

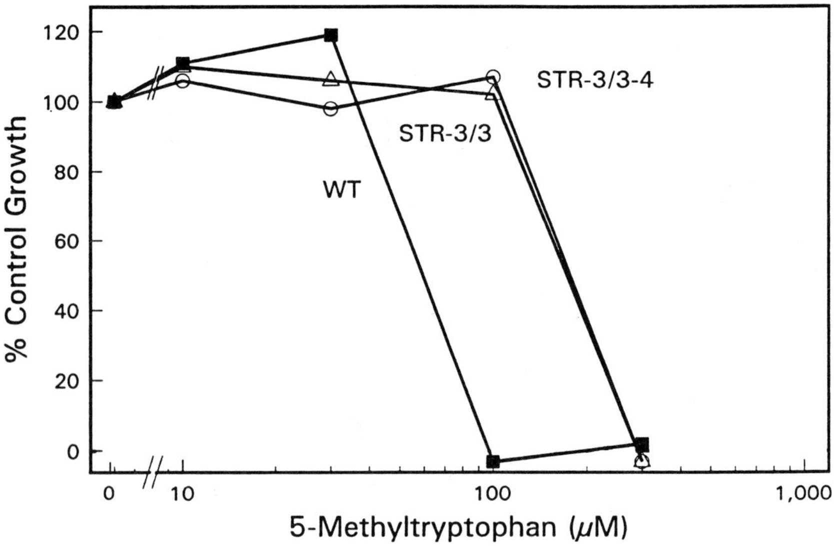

Fig. 1. Growth of suspension cultures in presence of 5-methyltryptophan. Cells (1.0 g fresh weight) were incubated in 100 ml medium for 12 d, collected and weighed. Wild type (■), STR-3/3 (△) and STR-3/3-4 (○).

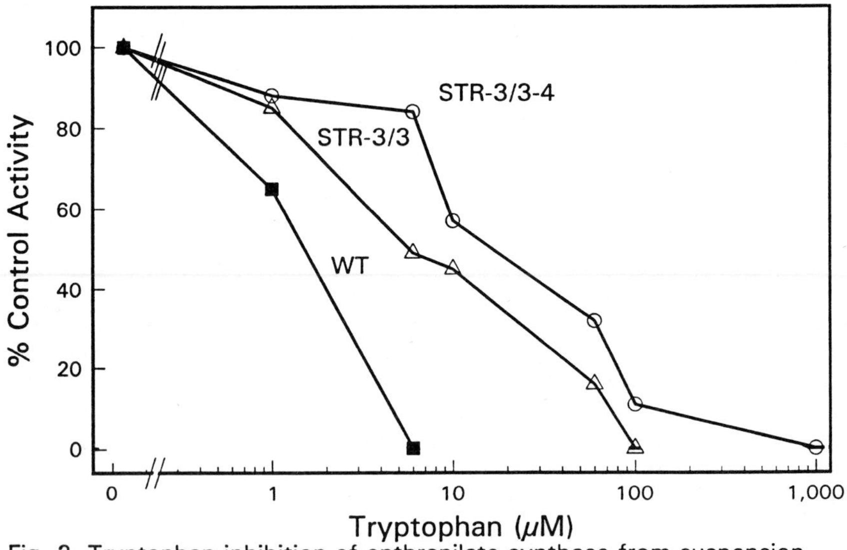

Fig. 2. Tryptophan inhibition of anthranilate synthase from suspension cultures. Wild type (■), STR-3/3 (△) and STR-3/3-4 (○). Enzyme activities per gram fresh weight were 1.7 nkat, 1.2nkat and 1.5nkat, respectively.

free Trp levels do not have a detrimental effect on the plants since they seem to grow normally. The 5MT-resistant rice mutant with elevated free Trp in the leaves always continues to segregate for resistance upon self-pollination so has never become homozygous for the trait (Wakasa and Widholm 1987). The 5MT-resistant rice mutant, obtained by selection for 5MT resistance at the seedling stage by Lee and Kameya (1991), shows normal segregation as if the resistance is controlled by a single, nuclear, dominant gene and homozygotes have been recovered. In this case, however, the elevated free Trp is localized in the seed and is not found in leaves or callus initiated from seeds.

The inheritance of the 5MT-resistance trait in the tobacco described here also does not fit the normal pattern. The gene controlling this does seem to be nuclear and dominant since it is transmitted in reciprocal crosses and in crosses with the wild type. Since several self-pollinations do not readily yield homozygous lines the number of genes involved might be more than one or possibly some sort of lethality is involved, i.e. the homozygous seeds do not develop.

These studies show that Trp overproduction can be accomplished in the plants of *D. innoxia* without detrimental effects or abnormal inheritance patterns. In the case of the tobacco the trait is not expressed in the plants, at least in the tissues examined, but is reexpressed in tissue cultures initiated from these plants. The trait appears to have an effect on the plant, however, since homozygosity has not been attained as one would normally expect following three self-pollinations.

Thus in vitro selection can produce widely different results depending upon the species.

ACKNOWLEDGEMENTS

This work was carried out with the assistance of J.P. Ranch, S. Schechter, R. Finke and W.Q. Zhong with funds from the Illinois Agricultural Experiment Station, the Science and Education Administration of the USDA grant 59-2171-1-1-736-0 from the Competitive Research Grants Office and National Science Foundation grant PCM 80-10927.

REFERENCES

Brotherton, J.E., Hauptmann, R.M. and Widholm, J.M. (1986) Anthranilate synthase forms in plants and cultured cells of *Nicotiana tabacum* L., Planta 168,214-221.

Carlson, J.E. and Widholm, J.M. (1978) Separation of two forms of anthranilate synthetase from 5-methyltryptophan-susceptible and -resistant cultures *Solanum tuberosum* cells, Physiol. Plant. 44,251-255.

Hibberd, K.A., Anderson, P.C., Barker, M. (1987) Tryptophan overproducer mutants of cereal crops, US Patent 4,642,411.

Kothari, S.L., Monte, D.C. and Widholm, J.M. (1986) Selection of *Daucus carota* somatic hybrids using drug resistance markers and characterization of their mitochondrial genomes, Theor. Appl. Genet. 72, 494-502.

Kreps, J.A. and Town, C.D. (1992) Isolation and characterization of a mutant of *Arabidopsis thaliana* resistant to a-methyltyptophan, Plant Physiol. 99,269-275

Lee, H.Y. and Kameya, T. (1991) Selection and characterization of a rice mutant resistant to 5-methyltryptophan, Theor. Appl. Genet. 82,405-408.

Murashige, T. and Skoog, F. (1962) A revised medium for rapid growth and bioassays with tobacco tissue cultures, Physiol. Plant. 15,473-497.

Ranch, J.P., Rick, S., Brotherton, J.E. and Widholm, J.M. (1983) Expression of 5-methyltryptophan resistance in plants regenerated from resistant cell lines of *Datura innoxia,* Plant Physiol. 71,136-140.

Wakasa, K. and Widholm, J.M. (1987) A 5-methyltyptophan resistant rice mutant MTRI, selected in tissue culture, Theor. Appl. Genet. 74,49-54.

Widholm, J. (1980) Differential expression of amino acid biosynthetic control isozymes in plants and cultured cells, in F. Sala, B. Parisi, R. Cella and O. Ciferri (eds.), Plant Cell Cultures: Results and Perspectives, Elsevier/North-Holland Biomedical Press, Amsterdam, pp.157-159.

Widholm, J.M. (1972a) Cultured *Nicotiana tabacum* cells with an altered anthranilate synthetase which is less sensitive to feedback inhibition, Biochim. Biophys. Acta 261,52-58.

Widholm, J.M. (1972b) Anthranilate synthetase from 5-methyltryptophan-susceptible and resistant cultured *Daucus carota* cells, Biochim. Biophys. Acta 279,48-57.

MOLYBDENUM COFACTOR (NITRATE REDUCTASE) BIOSYNTHESIS IN PLANTS: FIRST MOLECULAR ANALYSIS

R. R. MENDEL and B. STALLMEYER
Institute of Botany, Technical University of Braunschweig,
38106 Braunschweig, Germany

ABSTRACT. The aim of this research is the molecular and biochemical description of a new metabolic pathway in plants leading to the moybdenum cofactor (Moco). The Moco is a low-m.w. molybdopterin common to all molybdoenzymes with the exception of nitrogenase. It exhibits no catalytic activity on its own but becomes biologically active on association with an appropriate apoprotein. Nitrate reductase is the most important plant molybdoenzyme. - We applied the approach of functional complementation of Moco mutants of *E.coli* by a plant cDNA expression library and obtained a complete cDNA (2.3 kb) for a Moco gene of *Arabidopsis thaliana* complementing *E.coli mogA*. On the protein level, sequence comparisons of the *Arabidopsis* gene revealed homologies of > 30 % to three *E.coli* molybdenum cofactor proteins and homologies of > 40 % to proteins of an insect and of mammals. This cDNA was recloned into a plant expression vector and used for transient and stable transformations of the six tobacco Moco mutants *cnxA* to *cnxF* in order to complement the homologous locus. To our knowledge this cDNA is the first Moco gene isolated in plants.

1. Introduction

Nitrate reductase is the most important plant molybdoenzyme. It has been studied for many years (Warner and Kleinhofs (1992)): at first physiologically and biochemically, later, after nitrate reductase-defective mutants had become available, the genetics of this enzyme was established, and later the isolation of the nitrate reductase structural gene opened up vast possibilities for plant molecular biology (Caboche and Rouze (1990)). - The analysis of nitrate reductase-deficient mutants in *Nicotiana plumbaginifolia* (Gabard et al. (1987)) has shown that seven genes are involved in the biosynthesis of nitrate reductase: one structural gene (*nia*) encoding the apoprotein, and six genes being responsible for the synthesis of a molybdenum cofactor (*cnxA* to *cnxF*). A mutation in any of these 6 *cnx* loci will lead to the pleiotropic loss of all molybdoenzyme activities in the cell (Müller and Mendel (1989)).

Due to the pleiotropic loss of all molybdoenzymes, a defective Moco has lethal or sublethal consequences for the organism: higher and lower plants and fungi are no longer N-autotrophic, and animals show strong physiological and also morphological abberations leading to severe brain damages during embryogenesis and causing death after birth in most of the cases (Johnson et al. (1993)).

M. Terzi et al. (eds.), Current Issues in Plant Molecular and Cellular Biology, 577-582.

1.1. WHAT IS THE MOLYBDENUM COFACTOR?

The moybdenum cofactor (Moco) is a component common to all molybdoenzymes with the exception of bacterial nitrogenase. It was shown to be a low-m.w. molybdopterin exhibiting no catalytic activity on its own but becoming biologically active on association with an appropriate apoprotein (Rajagopalan und Johnson 1992). Nitrate reductase is the most important plant molybdoenzyme followed by xanthine dehydrogenase. The molybdoenzymes sulfite oxidase, xanthine oxidase and aldehyde oxidase occur mainly in animals, pyridoxal oxidase and xanthine oxidase are important for insects, and several different molybdoenzymes are known in bacteria (Fig. 1).

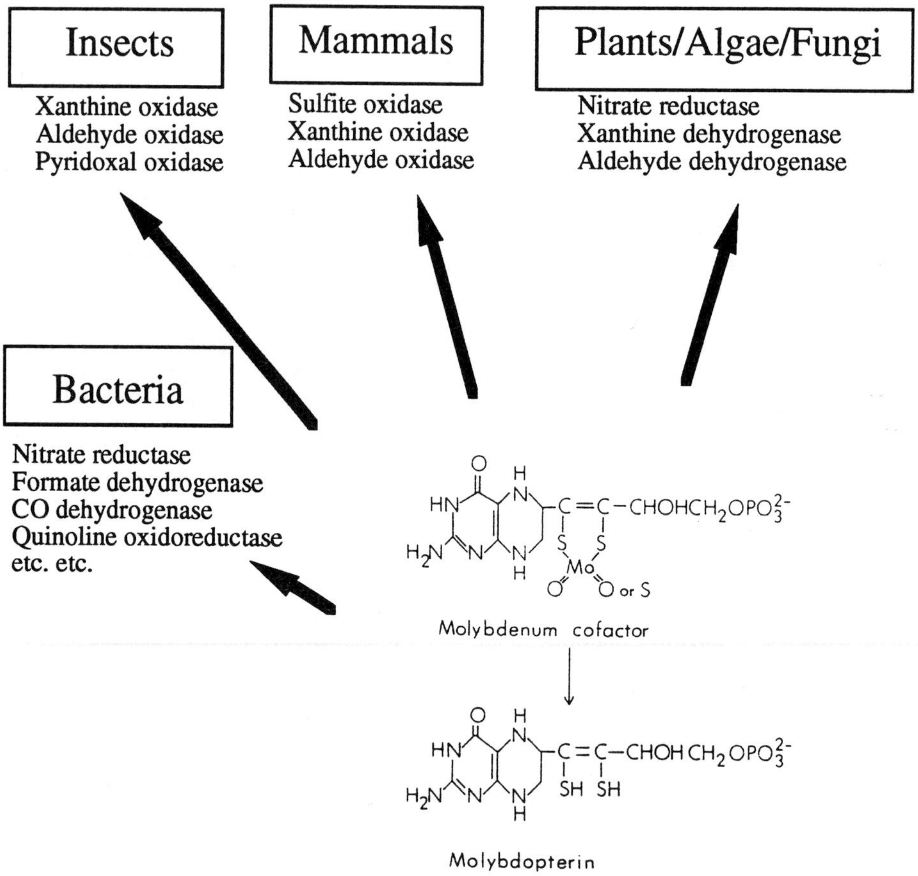

Figure 1. Structure of eukaryotic molybdenum cofactor (Moco) and its occurence in molybdoenzymes. Eubacterial Moco contains an additional nucleotide group attached to the molybdopterin (Rajagopalan und Johnson (1992)), archaebacterial Moco has the same structure as the eukaryotic one (Johnson et al. (1993)).

1.2. WHY IS THE MOLECULAR ELUCIDATION OF MOLYBDENUM COFACTOR-BIOSYNTHESIS OF INTEREST?

The aim of the research presented here is the molecular and biochemical description of the biosynthetic pathway leading to the molybdenum cofactor. The analysis of this new metabolic pathway in plants is of relevance to perhaps all eukaryotes because Moco has the following unique characteristics:

(1) Moco is unique in structure which is similar in all organisms so far examined (eubacterial Moco contains an additional nucleotide group attached to the molybdopterin (Krüger and Meyer (1986), Johnson et al. (1990)), archaebacterial Moco has the same structure as the eukaryotic one (Johnson et al. 1993);

(2) Moco is universal in function in that it forms part of the catalytically active center of very diverse enzymes thus catalyzing reactions on C-, S- and N-atoms:

(3) Moco is ubiquitous in occurrence in that it occurs in all living organisms ranging from bateria to humans and plants.

Moco structure has been first established in animals (see review by Rajagopalan and Johnson (1992)) however Moco biosynthesis could not be studied because of the lack of an appropriate genetic system in animals. These genetic systems, however, have been established in bacteria, fungi, algae, plants and insects. Moco mutants biochemically and genetically best characterized in higher eukaryotes, were generated in the tobacco species *Nicotiana plumbaginifolia*: Moco-specific loci *cnxA* to *cnxF* (Gabard et al. (1987); Müller and Mendel (1989)). The detailed biochemical and genetic analysis of these mutants lead us to a first working model for the pathway of Moco biosynthesis (Mendel 1992). The gene products of the Moco *cnx* genes are unknown, however it can be assumed that they are involved in converting a guanosine derivative into molybdopterin, in transferring sulfur into the molybdopterin and in inserting molybdenum into the cofactor. Compared to eukaryotes, in *E.coli* the situation is more advanced. Using the many Moco-specific mutants and combined with biochemical approaches, a first working model für Moco biosynthesis in *E.coli* was published (Wuebbens and Rajagopalan (1993), Pitterle and Rajagopalan (1993)).

In higher plants (*N. plumbaginifolia* and barley) and in the filamentous fungus *Aspergillus nidulans* 6 Moco gene loci are known, in the green alga *Chlamydomonas reinhardtii* 5 loci and in *E.coli* at least 14 reading frames in 6 operons. It can be concluded, therefore, that the biosynthesis of Moco is a rather complicated, multistep-process that is essential for the organism.

The Moco is an evolutionary highly conserved structure. Thus it is tempting to ask whether also the pathway leading to Moco is conserved, and further, whether also the Moco genes whose gene products are invloved in this pathway are conserved. This requires the isolation and characterization of the Moco genes from organisms of diverse phylogenetic origin. Furthermore, the isolation of plant Moco genes is a prerequisite for completely deciphering the biochemistry of Moco sythesis and opens up possibilities for studying the evolution of this pathway by using the plant genes as probe for isolating Moco genes from other organisms.

2. Results and Discussion

2.1. MOLECULAR ANALYSIS

The isolation of eukaryotic Moco genes is rather difficult because neither the function of Moco genes nor their gene products are known. Mutants defective in Moco-specific loci are the only tool available. First attempts to use these mutants as recipients for shot-gun cloning experiments with a plant genomic library were not successful (Schiemann et al. (1990), Mendel (1992)). Therefore we applied the approach of functional complementation of Moco mutants of *E.coli* by a cDNA expression library of *Arabidopsis thaliana* in order to obtain a cDNA clone for a plant *cnx* gene.

2.2. COMPLEMENTATION OF *E.COLI* MUTANTS BY A PLANT cDNA LIBRARY

Moco-defective *E.coli* strains were used as recipients for the *Arabidopsis* cDNA library (shuttle vector λYES (Elledge et al. (1991), kindly provided by Dr. Ron Davis (Stanford, USA)). These *E.coli* mutants are representative for reading frames within the five Moco-specific *E.coli* loci *moa, mob, mod, moe* and *mog*. These *E.coli* strains are Mu-phage insertion mutants showing in most cases no background activity of NR and no reversion (Stewart and MacGregor 1982).

Bacterial cells possessing a complementing plant sequence should be able to synthesize active Moco and hence should regain nitrate reductase activity which can be utilized as strong selectable marker in a modified liquid selection system during repetitive rounds of enrichment. This selection scheme permits the detection of 1 NR-positive cell among 10^6 NR-defective mutant cells. In the case of *E.coli mogA* it was possible to enrich NR-positive cells that were actively growing in the selection medium. In repetitive experiments it was reproducibly possible to select NR-positive cells after transfecting independent mutant strains for *mogA* with the *Arabidopsis* library in λYES. Plasmid DNA was isolated from the selected NR-positive cells and was used for successfully re-transforming of different *mogA* mutant strains.

2.3. THE SELECTED *ARABIDOPSIS* cDNA ENCODES A PROTEIN THAT SHOWS HOMOLOGIES TO PROTEINs OF *E.COLI, DROSOPHILA* AND MAMMALS

The selected cDNA (2.3 kb) was sequenced. It encodes a protein of 670 amino acids and a molecular weight of 71,283 Dalton. A data bases search showed that there were no homologies to other genes on the DNA level. On the amino acid level, however, there were homologies to the following proteins (listed in the order of decreasing homologies):

- rat neuroprotein gephyrin, linking the postsynaptic glycin receptor to the cellular cytoskeleton (Kirsch and Betz (1993), Kirsch et al. (1993)),
- gene product of *Drosophila* locus *cinnamon* (Moco-specific locus; function unknown) (Kamdar et al. (1994)),
- gene product of *E.coli mogA (*Moco-specific locus; function unknown, perhaps involved in molybdenum metabolism),
- gene product of *E.coli moaB* (Moco-specific locus; function unknown),
- gene product of *E.coli moeA* (Moco-specific locus; function unknown).

After the multiple alignment of amino acid sequences of the above proteins and of the *Arabidopsis* protein encoded by the selected cDNA it turned out that the amino-terminal part of the *Arabidopsis* protein is homologous to the gene product of *E.coli moeA*, whereas the carboxy-terminal part of this protein is homologous to the gene products of *E.coli mogA* and *E.coli moaB* which are also homologues. Thus the functions of three proteins encoded by three separate genes in *E.coli* are combined to a multifunctional, two domain-protein in eukaryotes linked by a short "stuffer" sequence showing no homology. Fig. 2 shows a schematic drawing of these homologies. It is noteworthy that *Drosophila* cinnamon and rat gephyrin also show homologies to the three listed *E.coli* Moco-proteins but that the order of the two domains is reversed as compared to *Arabidopsis*.

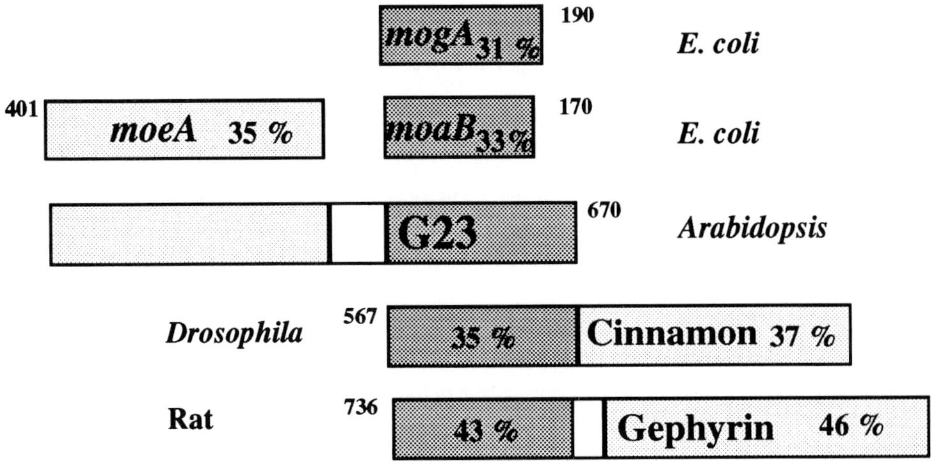

Figure 2. Schematic alignment on the basis of the amino acid sequences of the proteins encoded by the *E.coli* genes *moeA, mogA* and *moaB*, and encoded by the eukaryotic genes for *Arabidopsis* G23, *Drosophila cinnamon* and rat gephyrin. Homologies are given in % identity as compared to the *Arabidopsis* protein. The number of amino acid residues is given in the upper corner of each protein box. Note that in *Drosophila* cinnamon and in rat gephyrin the order of the two domains is reversed as compared to the *Arabidopsis* protein.

A similar approach had been used by us previously with a cDNA expression library of tobacco (*N. tabacum*) where a 800 bp plant sequence was sufficient to functionally complement *E.coli mogA* (Schiemann et al. (1990), Mendel (1992)). In the case of *Arabidopsis* we found that also cDNA clones shorter than the 2.3kb full-length clone were reproducibly able to complement *E.coli mogA*. After determining the DNA sequence of these clones it turned out that they were truncated forms of the full-length 2.3kb clone. This demonstrates that the MogA-homologous domain of the *Arabidopsis* protein alone is able to functionally complement the corresponding*E.coli* defect.

2.4. COMPLEMENTATION OF TOBACCO Moco MUTANTS (*cnx*)

The selected *Arabidopsis* cDNA was recloned into a plant expression vector (with 35S as promoter) and used for transformations of the six *N. plumbaginifolia* Moco mutants *cnxA* to *cnxF* in order to complement the homologous locus (*N. plumbaginifolia*-mutants were chosen because *Arabidopsis*-mutants show a considerably high background activity of nitrate reductase (Crawford (1992)). In transient gene expression assays (particle gun-bombardment of callus cultures), the locus *cnxE* could be complemented up to 10 % of nitrate reductase wild type-activity. However only stable transformations will be the final proof that the selected *Arabidopsis* cDNA is homologous to *N. plumbaginifolia cnxE.*

3. References

Caboche, M., and P. Rouzé (1990) 'Nitrate reductase: a target for molecular and cellular studies in higher plants'. Trends in Genetics 6, 187 - 192.

Crawford, N. (1992) 'Study of chlorate-resistant mutants of Arabidopsis: insights into nitrate assimilation and ion metabolism in plants'. In: Genetic Engineering (Ed.: J.K. Setlow) 14, 89 - 98, Plenum Press, New York.

Gabard, J., F. Pelsy, A. Marion-Poll, M. Caboche, I. DSaalbach, R. Grafe and A. Müller (1988) 'Genetic analysis of nitrate reductase deficient mutants of *Nicotiana plumbaginifolia*: Evidence for six complementation groups among 70 classified molybdenum cofactor deficient mutants. Mol. Gen. Genet. 213, 206 - 213.

Johnson, J.L., K.V. Rajagopalan and S.K. Wadman (1993) 'Human molybdenum cofactor deficiency'. In: Chemistry and Biology of Folates (Ed.: J. E. Ayling), 373 - 378, Plenum Press, New York.

Johnson, J.L., K.V. Rajagopalan, S. Mukund and M.W.W. Adams (1993) 'Identification of molybdopterin as the organic component of the tungsten cofactor in four enzymes from hyperthermophilic Archaea'. J. Biol. Chem. 268, 44848-4852.

Johnson, J.L., K.V. Rajagopalan, and O. Meyer (1990) 'Isolation and characterization of a second molybdopterin dinucleotide: molybdopterin cytosine dinucleotide. Arch. Biochem. Biophys. 283, 542-545.

Kamdar, K.P., M.E. Shelton and V. Finnerty (1994) 'The Drosophila molybdenum cofactor gene *cinnamon* is homologous to three *E.coli* cofactor proteins and to the rat protein gephyrin'. Genetics, in press.

Kirsch, J., I. Wolters, A. Prillar, H. Betz (1993) 'Gephyrin antisense oligonucleotide prevents glycine receptor clustering in spinal neurons'. Nature 366, 745 - 748.

Krüger, B., and O. Meyer (1986) 'The pterin (bactopterin) of carbon monoxide dehydrogenase from *Pseudomonas caboxydoflava*'. Eur. J. Biochem. 157, 121-128.

Müller, A.J. and R.R. Mendel (1989) 'Biochemical and somatic cell genetics of nitrate reduction in Nicotiana', In: Molecular and genetic aspects of nitrate assimilation (Eds.: J.L.Wray and J.R. Kinghorn). Oxford University Press.

Mendel, R.R.{1992) 'The plant molybdenum cofactor (MoCo) - its biochemical and molecular genetics'. In: Plant Biotechnology and Development - Current Topics in Plant Molecular Biology; (Ed.: P.M. Gresshoff), Vol. 1. CRC Press, Boca Raton, London.

Mendel, R.R.(1994) 'Molybdenum cofactor biosynthesis in plants: Molecular analysis'. Pteridines, in press.

Pitterle, D.M., and K.V. Rajagopalan (1993) 'The biosynthesis of molybdopterin in *Escherichia coli*. J. Biol. Chem. 268, 13499-13505.

Rajagopalan, K.V., and J. L. Johnson (1992) 'The pterin molybdenum cofactors'. J. Biol. Chem. 267, 10199-10202.

Schiemann, J., D. Inze and R.R. Mendel (1990): 'Cloning of Moco genes (*cnx*) from tobacco'. 3rd International Symposium on Nitrate Assimilation - molecular and genetic aspects. Bombannes, France.

Stallmeyer, B., A. Nerlich and R.R. Mendel (1994) 'Isolation of a novel Arabidopsis gene involved in molybdenum cofactor biosynthesis' (in preparation).

Stewart, V., and C.H. Macgregor (1982): 'Nitrate reductase in Escherichia coli K-12: Involvement of chlC, chlE, and chlG loci'. J. Bacteriol. 151, 788-799.

Warner, R.L., and A. Kleinhofs (1992) 'Genetics and molecular biology of nitrate metabolism in higher plants'. Physiol. Plant. 85, 245 - 252.

Wuebbens, M.W., and K.V. Rajagopalan (1993) 'Structural characterization of a molybdopterin precursor'. J. Biol. Chem. 268, 13493-13498.

ANALYSIS OF THE STRUCTURE OF THE 5' END OF THE GENE CODING FOR CARROT DIHYDROFOLATE REDUCTASE-THYMIDYLATE SYNTHASE

M.Z. LUO and R. CELLA
Dipartimento di Genetica e Microbiologia
Università di Pavia
Via Abbiategrasso 207
27100 Pavia
Italy

Abstract. Carrot cells contain two transcript types for the *dhfr-ts* gene and evidence was obtained for a plastidial localization of the product of the longer transcripts. This indication contrasted with previous circumstantial evidence which suggested a cytosolic localization. The conflicting evidence was found to be due to the presence of a large intron in the region corresponding to the 5' end of the mRNA. In this paper we report the structure of the genomic region corresponding to the 5' end of the carrot DHFR-TS-encoding gene and the cloning by inverse polymerase chain reaction of the promoter region which has been partially sequenced showing the presence of putative TATA boxes and AT-rich regions. It is also observed that the two *dhfr-ts* gene transcripts contain TC-repeats and a TC-rich stretch, respectively, whose possible role in the regulation of gene expression is discussed.

1. Introduction

Dihydrofolate reductase catalyzes the reduction of dihydrofolate (DHF) to tetrahydrofolate (THF) while thymidylate synthase, the enzyme responsible for the *de novo* synthesis of deoxythymidylate, catalyzes the synthesis of deoxythymidine-monophosphate from deoxyuridine-monophosphate and 5,10-methylenetetrahydrofolate. In this reaction 5,10-methylenetetrahydrofolate acts both as donor of the methyl group and as a reducing agent, giving rise to DHF. Thus TS is dependent on DHFR for the regeneration of THF which is then transformed in 5,10-CH_2-THF.

In carrot, dihydrofolate reductase is present as a bifunctional polypetide of about 58 kDa along with thymidylate synthase (Cella et al., 1988). A cDNA for the carrot protein was obtained through the immunoscreening of a carrot expression library; the sequence analysis of this clone (1978 bp) revealed the presence of an open reading frame of 1584 bp. The deduced 528 amino acids yield a polypeptide of 59,156 Da, a value which is close to 58,400±1,000 obtained by SDS-PAGE of the pure protein. The sequence of the cDNA upstream of the translation start point and of a genomic clone (overlapping the 5' end of the cDNA), obtained by single sided PCR (Luo and Cella, 1994) did not reveal the presence of a larger ORF. These data agreed with previous immunofluorescence localization studies (at the whole cell level) which had shown a cytoplasmic localization for the carrot DHFR-TS. Since in mammals, both cytosolic and mitochondrial isoforms

M. Terzi et al. (eds.), Current Issues in Plant Molecular and Cellular Biology, 583-588.

of DHFR have been demonstrated (Appling 1991), this was regarded as an indication for a cytosolic localization of carrot DHFR-TS.
No indication for the presence of a presequence corresponding to a transit peptide was obtained also in the case of two *Arabidopsis thaliana dhfr-ts* genes (Lazar et al., 1993).
Contrarily to the available evidence, immunogold localization studies on suspension cell thin sections showed a plastidial localization of carrot DHFR-TS; this indication found support in the analysis of cDNAs obtained by rapid amplification of the 5' end of the cDNA (RACE) for carrot DHFR-TS which revealed the presence of two transcript types. The longer transcripts were found to contain an additional transcription start point upstream and in frame with the previously described one, giving rise to an ORF which corresponds to a pre-protein 6300 Da larger than native DHFR-TS (Luo, Orsi and Cella, 1994, manuscript submitted).
An explanation for this conflicting evidence came from the comparison of the sequence of the cDNA for DHFR-TS (Luo et al., 1993) with that of a partially overlapping genomic region, obtained by single-sided PCR (Luo and Cella, 1994): divergence for the last 6 nt of the 5' end of the cDNA and the presence of an AG 3' splicing consensus at the divergence point indicated the occurrence of an intron.
In this paper we report the structure of the genomic region corresponding to the 5' end of the mRNA for carrot DHFR-TS and the cloning of the relative promoter region.

2. Materials and Methods

2.1. PLANT CELL CULTURE

Carrot cell suspensions of *Daucus carota*, cv. Lunga di Amsterdam (E4 line) were cultured as previously described (Luo et al.,1993).

2.2. DNA EXTRACTION

Carrot genomic DNA was extracted as described (Della Porta et al., 1983) and purified by cesium chloride gradient centrifugation (Sambrook et al., 1989).

2.3. POLIMERASE CHAIN REACTION

PCR was carried out with primer P3 (RACE 1) (5'-CCTTGGCGGCACTCGCGTCAG-3', corresponding to the region from -2302 to -2282 of the sequence reported in figure 2) and primer P2 (PROM 2) (5'-CCAGCAATCTCACTAAGT-3', complementary to the region from -1123 to -1140) in a 50 μl volume containing: 10 ng of carrot genomic DNA, 1x PCR buffer (Perkin Elmer Cetus), 0.2mM of each dNTP, 500 ng of each primer, 3 units of *Taq* polymerase (Perkin Elmer Cetus). Cycling conditions were: 2 min at 94°C followed by 35 cycles of 50 sec at 94°C, 50 sec at 50°C, 2 min at 72°C and a 10-min extension step at 72°C.
Inverse PCR was performed as described by Triglia et al (1988) and Ochman et al (1988). Carrot genomic DNA was digested with several 4-base-recognition restriction enzymes (Biolabs), autoligated and precipitated in the presence of 70 ng/ml glycogen (Boehringer). The following two primers were used: P4 (RACE 5), (5'-GCACGCCTGACGCGAGTGCCG-3', complementary to the region from -2276 to-2296) and P5 (RACE 4) (5'-CTCACTTTCACAAACTTTGGC-3', corresponding to the region from -2074 to -2054). Aliquots of 100 ng of the autoligated DNA were added to

the PCR cocktail and amplified as described above except that the annealing temperature was 60°C.
PCR products were separated by 1% agarose gel electrophoresis; DNA was recovered from the cut agarose slice using the Qiaex kit (Diagen), filled-in using the Klenow fragment (Promega) and cloned in the *Sma* I site of pGEM-4Z vector (Promega).

2.4. NUCLEOTIDE SEQUENCE ANALYSIS

DNA sequencing was performed by the standard dideoxynucleotide technique (Sambrook et al., 1989) using T7 sequencing™ kit (Pharmacia) with ^{35}S-dATP. When needed oligonucleotides (17-, 23-mers) were synthesized and used as sequencing primers. Sequence data were analyzed with MacDNASIS v3 program (Hitachi).

3. Results and Discussion

In order to gain information on the structure and expression of the carrot *dhfr-ts* gene we proceeded to the isolation and sequencing of the genomic region corresponding to the 5' end. The strategy used for cloning, which has made used of single-sided PCR, PCR and inverse PCR, is outlined in figure 1. Clones were completely sequenced on both strands except for the one corresponding to the promoter for which only part of the sequence corresponding to the 3' end is presently available (figure 2).

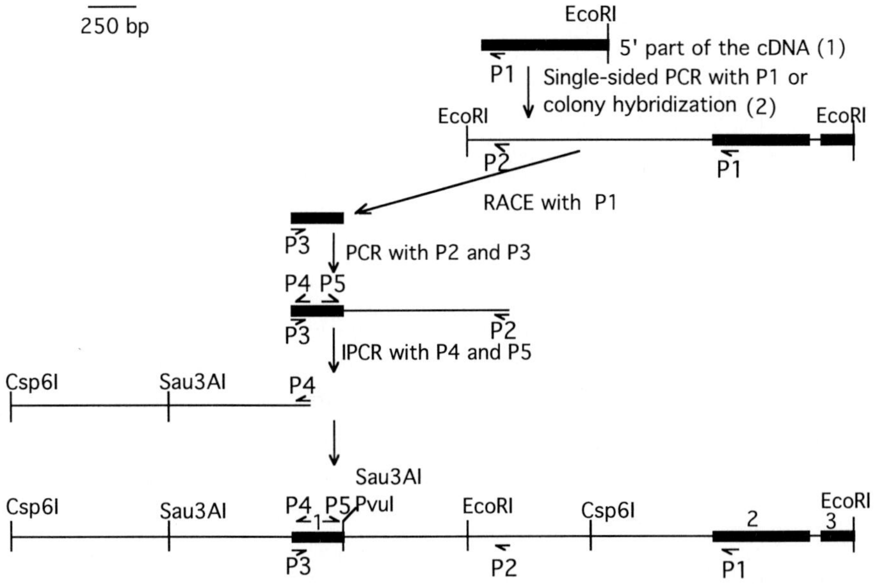

Figure 1. Scheme utilized for cloning the 5' end of the carrot *dhfr-ts* gene. Arabic numbers indicate exons. References: (1), (Luo et al., 1993); (2), (Luo and Cella, 1994).

```
TTTAATGGCAAATTTATTAAGTTGTATTATTATAAAAATAATTTTACATAGATACAATTT-2429

ATTTAATTTTAAATATATATAATTATTTCTAATATTGTAGTCCGTAAATTATTTATTTAA-2369
                                    AT-rich
TTCTATATAAAAGTAGGTTGCTGATGAGAGAAAGGAGAACAAGAGGAGACACAATTCACT-2309
   TATA Box 2                                   |>   |>
GCACTGCCTTGGCGGCACTCGCGTCAGGCGTGCGAATGTGCGATTATCTCTCTCGAAAAT-2249
  |>
CTCTCTCTCTCTCTCTCGGAAATCAATCTCTCTCTCATAGTTTCAAAAACCGCTGCTTGA-2189

GCCACTAGTTTTCCTCACCCGCACACCTTTTCTGCCGTCAATTTCACTACATTTCGAAAT-2129

ACATACATTTTGAATGTATAAAGCCCTAATTTCATCCACATTTCCTCTTCTAAACTCACT-2069
                TATA Box 1                                  |>
TTCACAAACTTTGGCTCTCCGATCGgtaacttccttttgcatatgaatgtttgtttgctt-2009
ttcacaatttacttgtaattgatggtgtgattattaataagaatgagattattgtaatgt-1949
tttcgcctgcattagttaaagcatcaccattacttgcgatgttatgttatagatgcaaag-1889
catgagctcatagatgttgatatatgcaatatttgtttcgcaacttgtttactcaaatta-1829
ctgaacttataacattctactcgtggatttacaaaaatttgtgatattcttagtgtaagg-1769
ttctgattcaactttggttagattctttcaattaagttgtatttgttctttcggagagtt-1709
ttcagactattatttcgtattattgtctataactatagatagttggagttttgcttcttt-1649
tgatggttgaaatttgttgttgctttctgttcttttgatgttttcaccgcattattccga-1589
atttcataagtataagggacataataagttgctgaaactcgtggcttttcatgtgtggat-1529
tgttcagtgttctgtaatagtgacttataaacaagtataagatgcatgtgtgaaagttat-1469
atacttatgtttcacactgtagctatgatctttcttgagataacttgcgtcaacttggga-1409
actttctagttttgaagagaataataagcctgcctgccagcagaattctgctataatgt-1349
tgattgtgtttttgcgcgttttctacttcccttatccacagattttggaaatttgtct-1289
attcatttataaatttggaatataacatattgttctccatctgtccatcactaagaaat-1229
gttattatcttatttgattttacgccttatccatatcttattttccagtttatagact-1169
ttggattttgtatttatgacaggggaatacttagtgagattgctggtttttaattcccaa-1109
ctaaatttgaaaaaaggtcattttacatatctttatctagtttatagacttctcattttg-1049
tatttttgagaagggagagaaagggtgcaaaattgtcggtttttgatttactaaattgaa -989
acaaagtcttcgtgacgtgtcatgaaacgtgttcgatgtgataatcattatttttcttct -929
taaattcatcctctcctatctggacattaatgaggaagttgactttatgcttagtataag -869
aatccatccatcttatactttgccctgctgttcatgttttagtagtgtgttttacaatag -809
tgaattacattattaatactttttggtcaattaccctattggagagtttgtgttatctat -749
catcctcttgtgagtgttgtcggtattcactttctctgcttacgtacttgagtccgtaga -689
gacagagagtagctttccgattttcaccccgaaacacaaataaattgttattgttgacct -629
aaaagcctttcttgatgtttgaatcccaaatttaattagttaaattcaactgtaattttt -569
aaaggtcacttgaaatcttgcacaggcttatatttgacttcttatcttctgtgttctcaa -509
aatctatgtgtaattagtttgatgtgcacaatttttaatgtatagcatttcagtttttat -449
gttctctgcgtccttgtactctaaaagatgttcagttacgcttttggtgaagatttgtgc -389
acattttgttaattaagtggtcaaggtctattaattatcttatcaagtcattctctagtg -329
tttcaatccttttagtcacgttgaatgtgcaaactggcaatacagacttctttctgtatc -269
tatctttattgggaaatattaagtttcttcatccatctacgtttccttgtaatagtaatt -209
gttcttccctgttttggtctgatcttctggttaaattatatctctatttatgaatgcagt -149
tatagattgattaatctttcttgaatttacaattaataatttaagtcatcaatccatcta  -89
ttttgcttttcagGCTTCTTTGTACTTGCTTACCAATTGCAACTTATCATCATCAAACTT  -29

CTCTCATTGCCATCGGTTTACTACATATATGGCTAGCGAACGTCTCGCAAATCCTACTAA   32

TGGAAGTGGCATTACACGCCCTGATCCACAAAGAACTTATCAGGTTGTGGTTGCTGCAAC   92

CCAAAATATGGGTATCGGTAAGGATGGTAAATTGCCCTGGAGGTTACCTTCTGACATGAA  152

ATTTTTCAAGGATGTCACCATGACTACATCAGATCCTTTGAAAAGGAATGCAGTCATAAT  212

GGGTAGGAAAACTTGGGAAAGTATTCCCATTCAACATCGCCCTCTGCCAGGACGTCTTAA  272

TGTTGTTTTGACTCGTTCTGGGAGTTTTGATATTGCAACTGTCGAAAATGTCGTAATATG  332

TGGTAGCATGATTTCTGCTTTGGAATTATTAGCAGGATCTCCTTATTGTGTTTCAGTTGA  392

GAAGGTTTTTGTCATTGGGGGTGGCCAGATATATAGgtgagcaactgagcaacatttggt  452
tgttaaatcctttaccaaggagatcctctcccacgttttaacaatctgttttactaatgc  512
agGGAAGCTCTCAATGCTCCTGGATGTGATGCAGTCCACATCACTGAAATTGAAGAACAC  572

ATAGAATGTGATACCTTCATTCCTCTTCTCGATGAATCAGTTTTTCAGCCATGGTACTCA  632

TCATTCCCATTGGTGGAGAACAAAATTCGTTATTGTTTCACGACTTATGTTCGTGTGAGG  692

AATTC 697
```

Figure 2. Nucleotide sequence of the region corresponding to the 5' end of the *dhfr-ts* gene from carrot. Translation start codons are indicated by bold characters while positions of transcription start sites and TATA boxes are indicated by arrows and written indications. Nucleotides of intron sequences are in lower case lettters. The nucleotide sequence is numbered from the first translational start site. The sequence corresponding to the previously cloned cDNA (Luo et al., 1993) begins at the position -2049.

The sequence reported in figure 2 shows the presence of two introns: the larger one (1968 bp) is situated in the middle of the untraslated region of short *dhfr-ts* transcripts while the smaller intron (86bp) is present at the same position in the DHFR domain as that of *Arabidopsis thaliana dhfr-ts* genes (Lazar et al., 1993).
The available promoter sequence shows the presence of putative TATA boxes, AT-rich regions and a relatively low presence of CpG and CpXpG sites. This is in contrast with the situation observed in mammals where *dhfr* genes are characterized by the presence of CpG islands. Therefore it will be interesting to see the mode of expression of this carrot gene.
An interesting feature of the 5' end of the carrot *dhfr-ts* transcripts is the presence of TC tandem repeats which behave as a polymorphic microsatellite (Luo, Orsi and Cella, 1994, manuscript submitted); TC repeats are homopyrimidine-homopurine tracts which, by binding a homopyrimidine oligodeoxyribonucleotide are thought to form triple helix also under physiological conditions (Moser and Dervan, 1987). For this reason it was proposed that they might play a role in the regulation of gene expression.

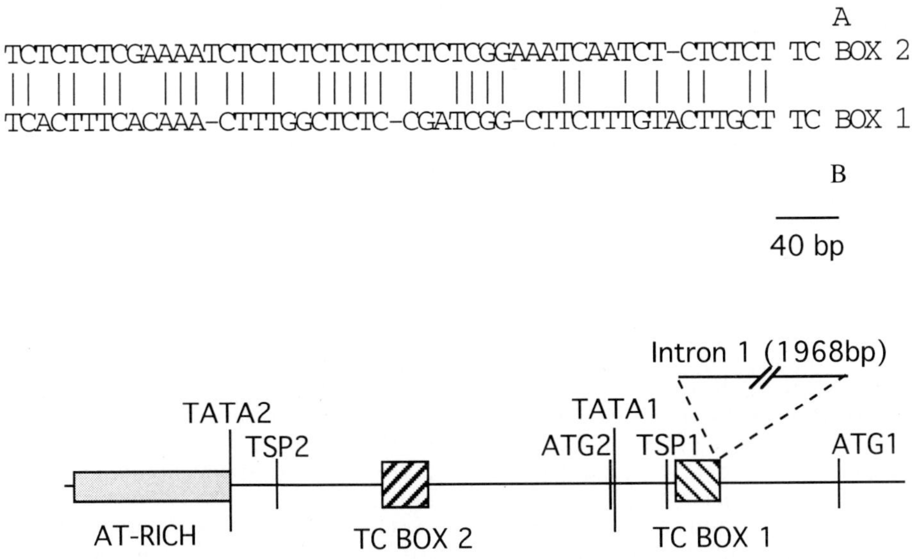

Figure 3. The two sequences reported in panel A (which show a degree of conservation higher than 60%) are located, respectively, from -2262 to -2213 (box 2) and from -2073 to -2044 and from -76 to -59 (box 1) of the genomic sequence reported in figure 2. Their positions on the mRNA are, respectively, from -235 to -186 and -106 to -59. Positions of these boxes relative to transcription and translation start sites are reported in B.

Interestingly, a region which bears a high level of conservation (more than 60%) with the TC-rich region was also found in the sequence corresponding to the untranslated region of transcript 1 (Fig. 3A). These boxes are located between the trascription start site and the translation start site (Fig. 3B).
The occurrence of TC-rich box at the 5' mRNA end, has been observed also in the case of the 3-hydroxy-3-methylglutaryl coenzyme A reductase gene of *A. thaliana* (Enjuto et al., 1994 , L. Balcells, personal communication); thus the question arises whether this might be a feature of other plant genes and which functional role(s) these sequences might play.

4. Acknowledgements

One of us (L. M-Z.) was funded by Ministero degli Affari Esteri (Progr. di Cooperazione allo Sviluppo). The work was supported by grants of MAF, Piano Nazionale di Ricerca, Sviluppo di Tecnologie Avanzate Applicate Alle Piante, CNR and MURST.

5. References

Appling, D. R. (1991) 'Compartmentation of folate-mediated one-carbon metabolism in eukaryotes' FASEB J. 5, 2645-2651.

Cella, R., Nielsen, E. and Parisi, B. (1988) '*Daucus carota* cells contain a dihydrofolate reductase-thymidylate synthase bifunctional polypeptide' Plant Mol. Biol. 10, 331-338.

Della Porta, S., Woo, J. E. and Hicks, J. B. (1983) 'A plant DNA minipreparation: version II' Plant Mol. Biol. Rep. 2, 1-19.

Enjuto, M., Balcells, L., Campos, N., Caelles, C., Arro, M. and Boronat, A. (1994) '*Arabidopsis thaliana* contains two differentially expressed 3-hydroxy-3-methylglutaryl-CoA reductase genes, which encode microsomal forms of the enzyme' Proc. Natl. Acad. Sci. USA 91, 927-931.

Lazar, G., Zhang, H. and Goodman, H. M. (1993) 'The origin of the bifunctional dihydrofolate reductase-thymidylate synthase isogenes of *Arabidopsis thaliana*' Plant J. 3, 657-668.

Luo. M.-Z., Piffanelli, P., Rastelli, L. and Cella, R. (1993) 'Molecular cloning and analysis of a cDNA coding for the bifunctional dihydrofolate reductase-thymidylate synthase of *Daucus carota*', Plant Mol. Biol. 22, 427-435.

Luo. M.-Z. and Cella, R. (1994) 'A reliable amplification technique with single-sided specificity for the isolation of 5' gene-regulating regions', Gene 140, 59-62.

Moser, H. E. and Dervan, P. B. (1987) 'Sequence-specific cleavage of double helical DNA by triple helix formation', Science 238, 645-650.

Ochman, H., Gerber, A. S. and Hartl, D. L. (1988) 'Genetic application of an inverse polymerase chain reaction', Genetics 120, 621-623.

Triglia, T., Peterson, M. G. and Kemp, D. J. (1988) 'A procedure for the in vitro amplification of DNA segments that lie outside the boundary of known sequences', Nucleic Acids Res. 16, 8186.

Sambrook, J., Fritsh, E. F. and Maniatis, T. (1989) 'Molecular cloning. A laboratory manual' II edition, Cold Spring Harbor Laboratory Press.

EMBL Data Library accession number Z33996.

GLUTAMINE SYNTHETASE IN CELLS FROM CARROT (*Daucus carota* L.): INTERACTION BETWEEN PHOSPHINOTHRICIN AND GLUTAMATE

A. FUGGI, M.R. ABENAVOLI, A. MUSCOLO, M.R. PANUCCIO
Dipartimento di Agrochimica ed Agrobiologia
Università di Reggio Calabria
Piazza S. Francesco 1
89061 Gallina di Reggio Calabria
Italy

ABSTRACT. Growth of cells from carrot (*Daucus carota* L., cv. Saint Valery) was strongly inhibited by phosphinothricin, an herbicide inhibiting nitrogen metabolism in plants. The addition of glutamate and the herbicide to the culture medium prevented the inhibition and restored cell growth. Analysis of cells treated with phosphinothricin evidenced the inactivation of glutamine synthetase which was prevented by simultaneous addition of glutamate. To test the effect of phosphinothricin, glutamine synthetase from carrot cells grown in liquid culture has been partially purified and characterized. In ionic exchange chromatography and not denaturing PAGE electrophoresis only one isoform of glutamine synthetase was detected. It was similar to the isoform detected in carrot root. The ATP-dependent irreversible inhibition of the enzyme was also prevented by glutamate. The kinetic analysis of this reaction suggested that glutamate inhibits the inactivation of glutamine synthetase inhibiting, allosterically, the phosphorilation of phosphinothricin besides the competitive interaction for the same site. The inhibition constant of glutamate for the reaction has a value similar to the Km of glutamate determined in the biosynthetic assay.

Introduction

Glutamine synthetase is a key enzyme of nitrogen metabolism in plants. Through the glutamate synthase cycle it is responsible for the assimilation of ammonium derived from the inorganic nitrogen nutrition, photorespiration, catabolism of amino acids and other nitrogen compounds (Lea et al., 1990). This enzyme, serving various physiological processes, occurs in different isoforms that display specific properties and are differently expressed during ontogenesis. Most of the higher plants contain in their leaves two isoforms of the enzyme separable by ionic exchange chromatography: GS_1 and GS_2, located in cytoplasm and in the chloroplast, respectively. Roots, generally, contain only one isoform that, in some case, is similar to the GS_1 of the leaves. Root nodules, however, evidenced also isoforms that were not found in roots and leaves (Cullimore et al., 1990).

Phosphinothricin (PPT; DL-homoalanine-4-yl-(methyl)-phosphinic acid), also known as gluphosinate, is a broad spectrum contact herbicide that selectively inhibits nitrogen metabolism in plants. The L- phosphinothricin, the active enantiomer, analogue of glutamate, acts as potent inhibitor of glutamine synthetase, bringing to glutamine deficiency and ammonia accumulation (Lea and Ridley, 1989 and references therein). In the last years transgenic plants resistant to phosphinothricin have been engineered by introducing the phosphinothricin-N-acetyltransferase gene isolated from the soil bacterium *Streptomyces viridochromogenes,* a natural producer of this

M. Terzi et al. (eds.), Current Issues in Plant Molecular and Cellular Biology, 589-594.

antibiotic as part of a tripeptide (Droge et al., 1992). Genes encoding this enzyme have been also used as markers for positive selection in plant transformation (Hineeke et al., 1994).

The engineered tobacco and carrot plants were resistant to the herbicide. They, however, accumulated the N- acetylated product within the cells (Droge et al. 1992).

Many kinetic studies on the effect of phosphinothricin on glutamine synthetase have analyzed the inhibition under initial rate conditions. Only a few of them have considered the irreversible inhibition of the enzyme that causes the herbicidal action (Lea and Ridley, 1989). Inhibition of carrot cell growth by phosphinothricin is prevented by glutamate and glutamine (Abenavoli et al, 1993). In this paper the irreversible inhibition by phosphinothricin of glutamine synthetase of carrot cells and the effect of glutamate have been studied .

Materials and Methods

Liquid culture or callus from carrot (*Daucus carota* L., cv. Saint Valery) were maintained on Gamborg B5 medium supplemented with 2,4 D (0.5 mg/L), Benzyl amino purine (BAP) (0.25 mg/L) and sucrose (2%) in 250 mL Erlenmeyer flasks. Cells were maintained at 25 °C in the light with a photoperiod of 16 h. In the light period the PAR radiation produced by halogen lamps was 150 µmol m^{-2} s^{-1}.

Enzyme extraction and assays Cells were collected by filtration on filter paper, washed, frozen and subsequently extracted in a mortar with the following buffer: 0,1 M Tris HCl pH 8,2 , EDTA (1 mM), mercaptoethanol (100 mM), $MgCl_2$ (10 mM), glycerol (10%). The extraction ratio was of 3 ml of buffer per g of wet tissue. After centrifugation at 20000 g for 20 min the supernatant was used for the enzyme and protein assays. The enzyme was also partially purified by ionic exchange chromatography through a DE 52 column (6 x 1.5 cm) previously equilibrated with a 5 fold diluted extraction buffer. A linear gradient of 0 - 0.5 M NaCl in the equilibration buffer was used to elute protein. The fractions containing enzyme activity were pooled made 1.5 M NaCl and applied at a Phenyl sepharose column (5 x 1.5 cm).Subsequently they were eluted in the elution buffer supplemented with glycerol up to 50%.

Glutamine synthetase activity was determined at 30 °C by the transferase assay according to Rodhes et al (1975) or by the semibiosynthetic assay according to Robinson et al. (1991). with some modification.

Protein was estimated according to Bradford (1976).

Polyacrilamide gel electrophoresis Non denaturing polyacrilamide (7%) gel on slab gel was conducted according to Davis (1964). GS activity was detected by the transferase assay according to Barrat (1980).

Results

Characteristics of glutamine synthetase from carrot cells grown in liquid medium.

Ionic exchange chromatography on DE 52 of extract of carrot cells grown in liquid suspension revealed a single peak of GS activity at 0.12 M NaCl (not shown). Similar elution pattern was observed for the carrot root enzyme activity. A single band of activity was also shown in not denaturing PAGE electrophoresis. The enzyme was very unstable in crude extract as well as when it was partially purified by ionic exchange chromatography. However, in the extraction buffer supplemented with 50% glycerol it suffered no significant loss of activity even at 55 °C for half an hour, as evidenced for the root isoform of carrot In similar condition the leaf GS suffered a 90% loss of activity (not shown). The enzyme evidenced an optimum pH of 6,9 and 5.9 for biosynthetic and transferase activity, respectively.

Loss of glutamine synthetase activity in carrot cells treated with phosphinothricin and effect of glutamate.

Phosphinothricin added to the colture medium strongly inhibited the growth of carrot cells even at concentration 10 µM (Abenavoli et al.,1993). Simultaneous addition of glutamate or glutamine and inhibitor to the medium prevented inhibition.

A cell suspension culture in the exponential growth phase, in which the glutamine synthetase specific activity was maximal, was treated with PPT (100 µM) and leaved in the same conditions as for growth. At given time intervals aliquot of suspension were collected, washed and used to prepare the enzyme extract. Glutamine synthetase specific activity was determined by the transferase assay. In order to reduce the inhibition by residual PPT present in the cell extract glutamate 20 mM was added to the transferase assay mixture.It did not interfere with the assay. Simultaneous analysis were done in an untreated suspension. The same procedure was used also when the suspension cultures were treated with glutamate and PPT. Fig. 1 shows that application of the inhibitor to a growing culture brought to a time dependent loss of glutamine synthetase activity. A 50% loss of the enzyme activity occurred in half an hour when 100 µM PPT was added. The initial rate of GS inactivation was linearly related to the concentration of PPT (not shown). No significant changes of specific activity of the enzyme occurred in a suspension culture that was not treated with the inhibitor . The loss of glutamine synthetase activity was also accompained by ammonium accumulation in the medium. Simultaneous addition of glutamate and inhibitor reduced the rate of enzyme inactivation (Fig. 1).

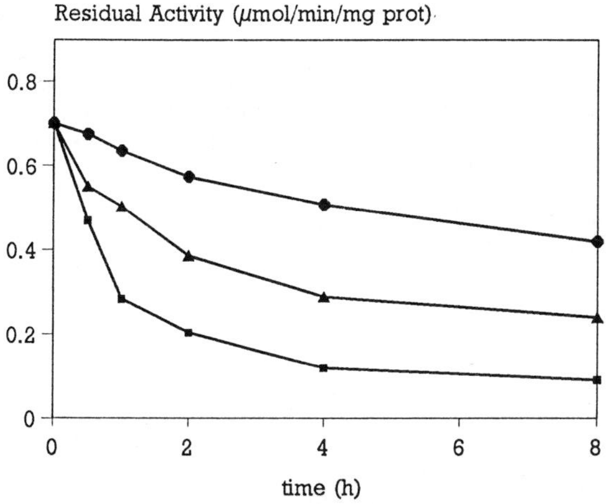

Figure 1. Time course of inhibition of glutamine synthetase activity in cells treated with phosphinothricin 100 µM in the absence (-■-) and in the presence of glutamate 3 mM (-▲-) and 10 mM (-●-). The specific activity of the enzyme was determined in the cell extract using the transferase assay.

Irreversible inhibition of glutamine synthetase from carrot cells by phosphinothricin and effect of glutamate.

The inactivation of glutamine synthetase by phosphinothricin, as by methionine sulphoximine, another glutamate analogue, is due to the formation of an ATP-dependent phosphorylated intermediate of the inhibitor that remains strongly bound to the enzyme. In physiological conditions it cannot be replaced and the enzyme cannot work furtherly (Lea and Ridley, 1989). The reaction takes place also *in vitro*. Glutamine synthetase from carrot cells partially purified as reported in Materials and Methods was incubated at 30 °C with PPT and ATP (5 mM) in a imidazole buffer (0,1 M; pH = 7.2) supplemented with Mg^{2+} (10 mM) and mercaptoethanol (10 mM). Aliquot of the incubation mixture at given time intervals were used to determine the residual transferase activity using the above mentioned modified assay.

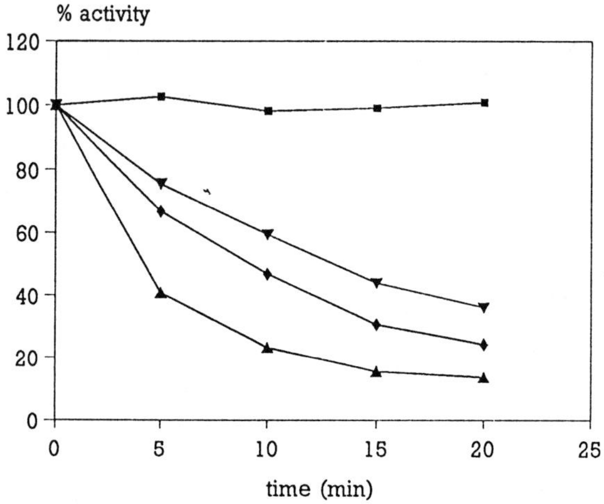

Figure 2 Time dependent loss of glutamine synthetase incubated with phosphinothricin (10 μM) with (▲) and without ATP (5 mM) (■). The effect of glutamate, at 3 mM) (◆) and at 10 mM (▼) in the presence of PPT and ATP as reported above is also shown. The activity is indicated as percentage of the control without ATP.

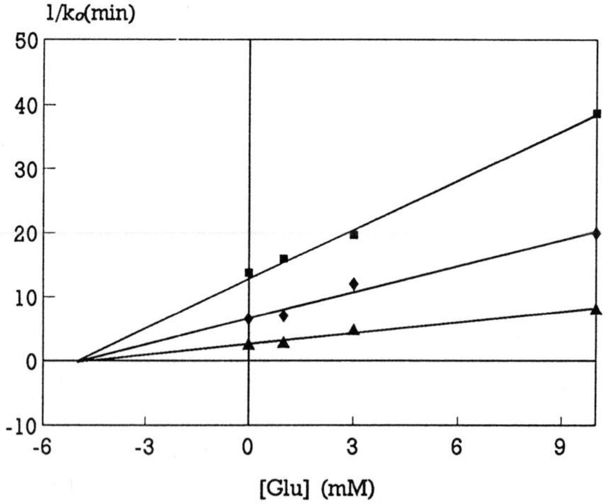

Figure 3. Dixon plot of initial rate of glutamine synthetase inactivation by PPT versus glutamate concentration at [PPT] = 5 μM (■), [PPT] = 10 μM (◆), [PPT] = 20 μM (▲).

Fig. 2 shows that a time dependent loss of enzyme activity occurred during the incubation period. As can be seen a 50% inactivation occurred in about 4 min at 10 μM PPT in presence of

ATP. No significant inactivation occurred in the absence of ATP. Magnesium ion was essential for the reaction (not shown). In addition, glutamate, decreased the inactivation of the glutamine synthetase also *in vitro* (Fig. 2).

The progress curves of inactivation of glutamine synthetase (Fig. 2) were neither linear, nor simply exponential, as reported for the inactivation by phosphinothricin of the GSs from *Triticum aestivum* L.(Manderscheid and Wild, 1986). Therefore, the initial inactivation rate was extrapolated using a polynomial interpolation of the logarithm of the percentage of the initial activity data.

Fig. 3 shows that the Dixon plots of the initial inactivation rate (ko) versus glutamate concentration at different concentrations of PPT were linear. The extrapolated interception point among the straight lines at a value of 1/ko near to zero indicates a pure non competitive type of inhibition by glutamate. The extrapolated inhibition constant for the reaction was Kglu = 4.9±1 mM.

Discussion

Glutamine synthetase in carrot cell grown in standard Gamborg B5 medium appeared as one isoform by ionic exchange chromatography and non denaturing PAGE electrophoresis. No variation could be observed in a culture period in which changes of the enzyme activity between a maximum during the exponential growth and a minimum in the stationary phase were evidenced. Variation of GS activity in a culture period have been also reported in cells grown in high ammonium medium (Robinson et al., 1992). The characteristics of the enzyme in growing cells were similar to those found in root of carrot not only for the electrophoretic mobility, but also for heat stability and other characteristics.

Phosphinothricin strongly inhibits the growth of carrot cells (Abenavoli et al., 1993). Glutamate and glutamine added to the basal medium, however, were able to restore growth. Fig. 1 shows that the treatment of carrot cells with the herbicide brings about to the inactivation of glutamine synthetase the enzyme target for this compound (Lea and Ridley, 1989). Similar results were obtained when carrot cells were treated with methionine-D,L-sulphoximine (not shown). Fig. 1 shows also that glutamate, that restored cell growth was able to reduce the inactivation rate of GS.

Fig. 2 shows that the glutamine synthetase inactivation by PPT occurred also *in vitro* in the presence of ATP and magnesium ions, suggesting that the inactivation is dependent on the formation of the phosphoriled PPT tightly bound to the enzyme that cannot work further on (Manderscheid and Wild, 1986), as for the inactivation of GS by methionine sulphoximine (Ronzio et al., 1969). Glutamate was able to prevent the enzyme inactivation also in vitro (Fig. 2). In particular the Dixon plot (Fig. 3) evidenced an almost pure non competitive inhibition. The value of the inhibiting constant (Kglu = 4.9 mM) was similar to the Km value of the enzyme for glutamate. Knowing that glutamine synthetase is a multimeric enzyme (Lea et al., 1990), these data suggest that allosteric interactions occur among the enzyme subunits even if the enzyme did not show a cooperative kinetics versus glutamate (not shown). The inactivation of GS isoform from carrot cell, in the absence of glutamate, support the mechanism proposed for the root isoform of *Triticum aestivum L.* (Manderscheid and Wild 1986), the rate-limiting step of the process being the formation of the enzyme-inhibitor complex (not shown). The occurrence of a non competitive type of inhibition by glutamate, however, suggests that glutamate, through allosteric interaction inhibits also the step of phosphorilation of PPT, besides the inhibiton for competition to the same site of the enzyme, as evidenced by the kinetic analysis of the inhibition by PPT of the biosynthetic activity of enzyme (not shown). The time dependent inactivation of glutamine synthetase by PPT in the presence of ATP, occurred at high rate even at low concentration of inhibitor. Therefore the inhibition constant of PPT for the enzyme calculated using data derived by the biosynthetic assay method has to be corrected for the effect of the incubation time.

Acknowledgments

This work was supported by "Ministero della Università e della Ricerca Scientifica e Tecnologica" and by "Consiglio Nazionale delle Ricerche" of Italy

References

Abenavoli M.R., Muscolo A., Panuccio M.R. and Fuggi A. (1993) Inhibition by glufosinate of growth of cells cultured in liquid medium from carrot (*Daucus carota* L.) Giornale Botanico Italiano 127: 895-896.

Bradford M.M. (1977) - A rapid sensitive and versatile assay for protein using Comassie brilliant blue G 250. Analytical Biochemistry 79: 544-552.

Barrat D.H.P. (1980) Methods for the detection of glutamine synthetase activity on starch gels. Plant Science Letters 18: 249-254.

Davis B.J. (1964) Disc electrophoresis II. Methods and applicationto human serum proteins. Ann. N.Y. Academy Science 121: 404-436.

Cullimore J.V., Cock J.M., Robbins M.P. and Bennett M.J. (1990) Glutamine synthetase of french bean: from genes to isoenzymes. In: Ullrich W.R., Rigano C., Fuggi A., and Aparicio P.J. eds. *Inorganic Nitrogen in Plants and Microorganisms.* Springer-Verlag, Berlin pp. 273-280.

Droge W, Broer I. and Puhler A. (1992) Transgenic plants containing the phosphinothricin-N-acetyltransferase gene metabolize the herbicide L-phosphinothricin (glufosinate) differently from untransformed plants. Planta 187: 142- 151.

Hineeke M.A.W., Corbin D.K., Armstrong C.L., Fry J.E., Sato S.S., DeBoer D.L., Petersen W.L., Armstrong T.A., ConnorWard D.V., Layton J.G. and Horsch R.B. (1994) Plant transformation In: *Plant Cell and Tissue Culture*, Vasil I.K. and Thorpe T.A. eds. Kluwer Academic Publishers, Dordrecht, pp. 231- 270.

Lea P.J. and Ridley S.M. (1989) - Glutamine synthetase and its inhibition In: Dodge A.D. ed. *Herbicides and Plant Metabolism* Cambridge University Press Seminar Series 38: 137 - 170.

Lea P.J. Robinson S.A. and Stewart G.R. (1990) The enzymology and metabolism of glutamine, glutamate and asparagine. In: *The Biochemistry of Plants*, Miflin B.J. ed., Academic Press, New York, Vol. 16 pp. 121 -169.

Manderscheid R. and Wild A. (1986) Studies on the mechanism of inhibition by phosphynothricin of glutamine ynthetase from *Triticum aestivum* L. J Plant Physiology 123: 135-142.

Rhodes D., Rendon G.A. and Stewart G.R. (1975) The control of glutamine synthetase level in *Lemna minor*. Planta 125: 201-211.

Robinson S.A., Stewart G.R. and Philips R. (1992) Regulation of glutamate dehydrogenase activity in relation to carbon limitation and protein catabolism in carrot cell suspension culture. Plant Physiology 98: 1190-1195.

Robinson S.A., Slade A.P. Fox G.G., Phillips R., Ratcliffe R.G. and Stewart G.R. (1991) - The role of glutamate dehydrogenase in plant nitrogen metabolism. Plant Physiology 95: 509-516.

Ronzio,A.R., Rowe W.B. and Meister A. (1969) Studies on the mechanism of inhibition of glutamine synthetase by methionine-sulphoximine. Biochemistry 8: 1066-1075.

CHANGES OF SECONDARY METABOLISM BY ELICITOR TREATMENT IN *PUERARIA LOBATA* CELL CULTURES

Ushio SANKAWA,* Takashi HAKAMATSUKA, Kenji SHINKAI, Makoto YOSHIDA, Hyung-Hwan PARK and Yutaka EBIZUKA
Faculty of Pharmaceutical Sciences, The University of Tokyo
7-3-1, Hongo, Bunkyo-ku, Tokyo 113, Japan

ABSTRACT. The cell cultures of *Pueraria lobata* contain isoflavone O- and C-malonylglucosides (IMG) as the main constituents. Upon treatment of *P. lobata* cell cultures with an elicitor yeast extract (YE) induced the production of three dimeric isoflavones, kudzuisoflavone A, B and C, which were probably formed by non-specific oxidation of daizein with peroxidase. In contrast a biotic elicitor $CuCl_2$ induced hypersensitive response in the cultured cells and nine isoflavonoids including a phytoalexin tuberosin and the three dimeric daizeins were produced. Treatment of the cell cultures with YE caused rapid and transient decrease of IMG within 4 h. IMG then reaccumulated and its level reached to three times higher than that of control after 100 h. $CuCl_2$ treatment caused rapid disappearance of IMG and no reaccumulation was observed, however enzymes and mRNAs relating to the biosynthesis of isoflavonoids in $CuCl_2$ treated cells are higher or equal to the levels of YE treated cells. ^{14}C-Labelled IMG experiment proved that rapid and transient decrease of IMG resulted in the deposition of isoflavones to insoluble lignocellulose fraction in cell wall, which may be a rapid defense mechanism of plant resistance to outer stress.

1. Introduction

Phenolic compounds of shikimate pathway origin are widely distributed among higher plants and play important physiological roles in plants. The derivatives of cinnamic acids are not only the precursors of phenylpropanoids and lignin, but also they served as a precursor of flavonoids, which are widely distributed among higher plants and possess a variety of physiological roles in plants.[1] In the defense response of leguminous plants isoflavonoids, a class of flavonoid having modified C_6-C_3-C_6 skeleton, are produced as phytoalexins upon physical, chemical and biological stress. [2] In the course of our enzymatic studies on the biosynthesis of isoflavonoids in cell suspension cultures of *Pueraria lobata* (Japanese name kudzu), several different kinds of elicitors were tested to obtain suitable material for enzymatic studies. [3, 4, 5, 6, 7, 8, 9] When *P. lobata* cultured cells were treated with a glycoprotein elicitor prepared from a phytopathogenic fungus *Phytophthora megasperma f. sp. glycinea* (Pmg elicitor) [10], a pterocarpan tuberosin was detected as a phytoalexin. Similar response were observed with $CuCl_2$ treatment.

M. Terzi et al. (eds.), Current Issues in Plant Molecular and Cellular Biology, 595-604.

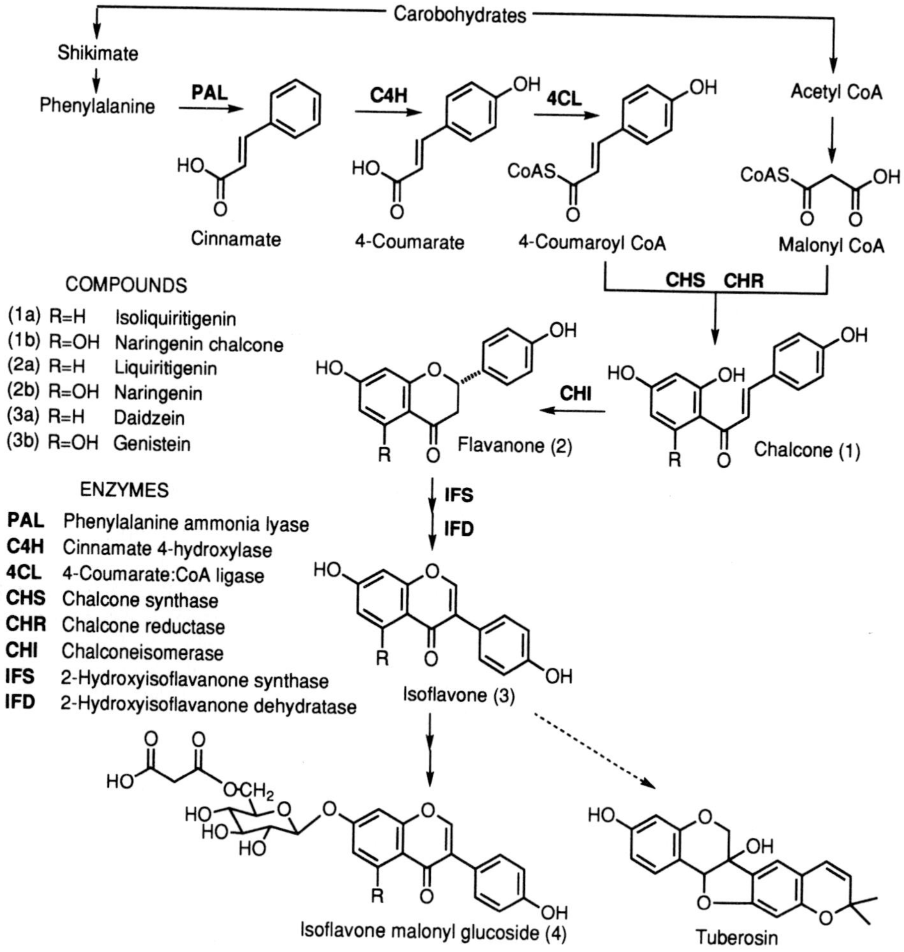

Fig. 1 Biosynthetic Pathway of Isoflavonoids in *Pueraria lobata*

However, cells treated with Pmg elicitor or $CuCl_2$ were not suitable for enzymatic studies, since these condition led to hypersensitive response of cultured cells and caused cell death resulting in poor yield of target enzyme activities. To obtain better yield of the enzymes we attempted the endogenous elicitor which had been prepared from cell wall fraction of cultured *P. lobata* cells by enzymic hydrolysis. [3, 4] Deoxychalcone synthase and isoflavone synthase, key enzymes in the biosythesis of 5-deoxyisoflavonoids (2a, 3a in Fig. 1) were successfully obtained by using endogenous elicitor. [3, 4] Later, a commercially available yeast extract (YE from Difco) was found to have the same effect with the endogenous elicitor and cell response was milder than Pmg and $CuCl_2$. [6] The treatment of *P. lobata* cells with YE elicitors resulted in the activation of enzymes involved in the biosynthesis of isoflavones, such as chalcone synthase (CHS), reductase for deoxychalcone synthase (CHR), chalcone-flavanone isomerase (CHI) and isoflavone synthase (2-hydroxyisofalvanone synthase (IF) and dehydratase(IFD)). HPLC analysis of methanolic extract of the cultured cells treated with YE revealed the presence of

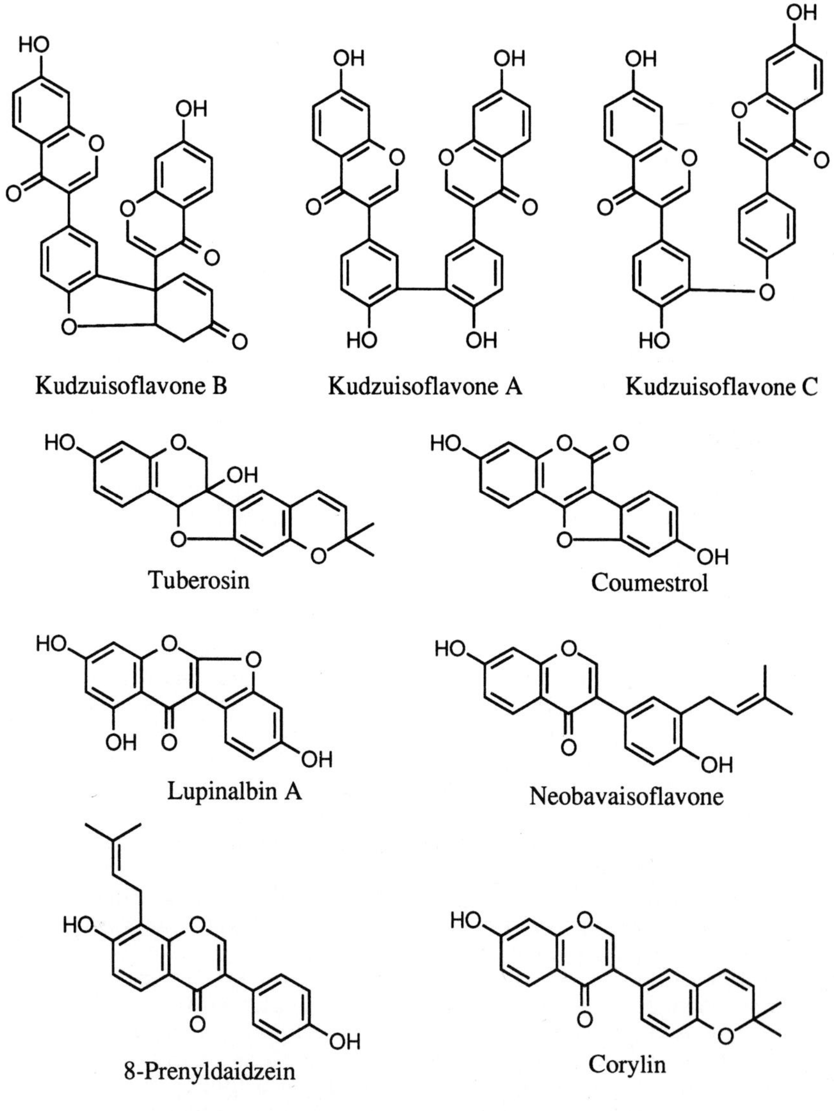

Fig. 2 Elicitor Induced Isoflavonoids in *Pueraria lobata*

several new peaks corresponding to hitherto unknown compounds.[6] The present study was initiated by these observation and extensive investigation has been carried out on the production and metabolism of isoflavonoids in the cultured cells of *Pueraria lobata.*

2. Induction of Isoflavonoid Production by Elicitor Treatment

Repeated cell cultures and elicitation with YE resulted in the accumulation of elicited *P. lobata* cells (ca. 4 kg). Repeated chromatographic separation of acetone extracts of YE-treated *P. lobata* cells afforded three dimers of daizein, kudzuisoflavone A, B and C whose structures were elucidated by NMR interpretation. [6] (Fig. 2) The structures of the daizein dimers indicate that they were formed by the reaction of peroxidase and hydrogen peroxide by so called phenol oxidative coupling reaction. *In vitro* synthesis of the dimeric isoflavones from corresponding monomer daizein has been accomplished with horse radish peroxidase and hydrogen peroxide.[6] The production of isoflavonoids in response to $CuCl_2$ treatment in the cultured cells of *P. lobata* was different from that of YE. HPLC profile of the extracts of *P. lobata* cultured cells treated with $CuCl_2$ instead of YE was more complex. In addition to the dimeric daizeins, kudzuisoflavone A, B and C, other 6 compounds, coumesterol, tuberosin, lupinalbin A, neobavaisoflavone, 8-prenyldaidzein and corylin were isolated and identified. [8] (Fig. 2) The results so far obtained indicate that in addition to the enzyme catalyzing the dimerization of daizein, probably peroxidase, enzymes leading to the formation of pterocarpans and prenylated isoflavones were induced by $CuCl_2$ treatment. The daizein dimers are thought to be formed by the reaction of non-specific peroxidase, since kudzuisoflabone B is optically inactive. Peroxidase induced by YE and $CuCl_2$ treatment may be a responsible enzyme that catalyzes the polymerization of phenylpropanoids forming lignin or its analogues to fortify cell wall and the daizein dimers are thought to be the by-products of this reaction.

3. Isoflavonoid Metabolism by the Treatment of Elicitors

The formation of isoflavonoids induced by elicitor treatment is the results of enzyme inductions leading to the changes of enzyme activities. The time course changes of isoflavonoid contents, enzyme activities and mRNA levels would give useful information on the metabolic alteration caused by elicitor treatment. Extraction of fresh cultured cells of *P. lobata* with mild extraction condition such as sonication in methanol revealed the presence of several new compounds on TLC. It becomes clear that the main glucosides contained in the cultured cells of *P. lobata* are daizein-6"-malonylglucoside, genistein-6"-malonylglucoside and 6"-malonyl-puerarin. [9] Isoflavone-6"-malonylglucosides (IMG) has been know as the stored forms of isoflavones in leguminous plants and they serve as precursors of phytoalexins when the reaction of phenylalanine ammonia lyase was inhibited. [11] In order to clarify the fate of these isoflavone malonylglucosides (IMG) in elicitor treatment, time course changes of malonyl isoflavone conjugates (IMG) were followed by HPLC analysis. A rapid and trasient decrease of IMG is characteristic in YE treatment and the contents of IMG rapidly decreased to 10-20 % of initial level after 4 h. Then IMG started to reaccumulate and reached to a higher level than initial or control values. In some cases IMG accumulated 3 times higher level of the control. (Fig. 3) In contrast $CuCl_2$ treatment caused rapid metabolism of IMG and no reaccumulation was observed as in the case of YE. Therefore YE and $CuCl_2$ are not only different in their production of isoflavonoids such as phytoalexins but also the effect on the metabolism and synthesis of isoflavone malonylglucosides (IMG) is

Fig. 3 Accumulation of Isoflavonoid Malonylglucosides in Elicitor-treated Cell Cultures of Pueraria lobata

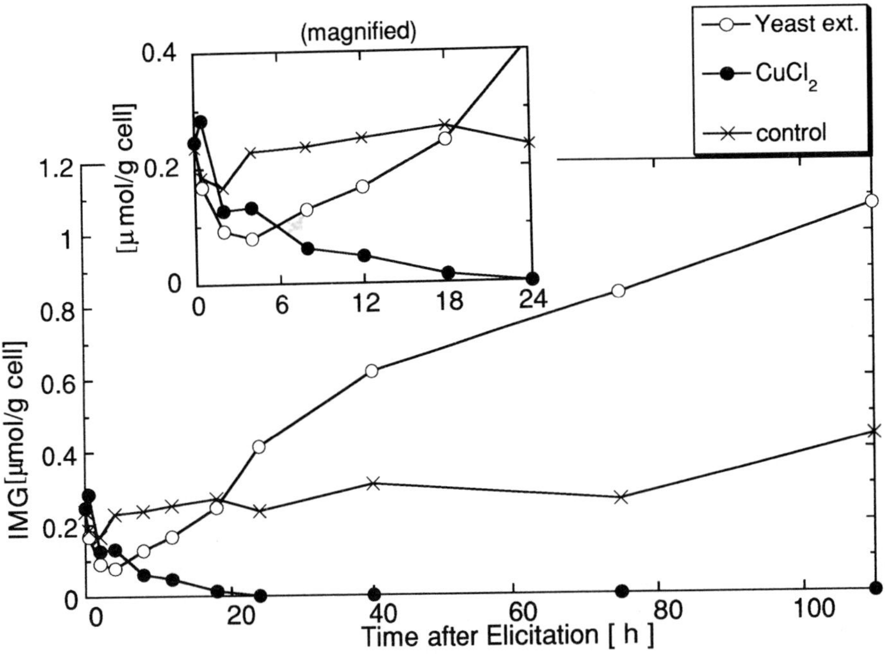

quite different. The difference at compounds level should also reflect in enzyme activities and mRNA levels, and next we investigated the time course changes of enzyme activities and mRNA levels which relate to the biosythesis of isoflavonoids. [12] The elevation of mRNA level upon elicitation with YE was quite rapid and prominent in chalcone synthase (CHS), reductase in deoxychalcone synthase (CHR) and chalcone flavanone isomerase (CHI). They reached to their maxima at about 6 h after elicitation and fell down gradually to the levels different in each enzyme. (Fig. 4) In contrast increase of mRNA level by $CuCl_2$ was slow process compared to YE and formed broad maxima at 20 h after elicitation.[12] The enzyme activities of phenylalanine ammonia lyase (PAL), CHS, CHR, CHI and 2-hydroxyisoflavanone dehydratase(IFD) showed different time course changes which reflect the stability of each enzyme, since the changes of mRNA levels are similar in CHS, CHR and CHI. (Fig. 5) $CuCl_2$ induced rather long lasting enzyme induction in all the enzymes tested. [13] The results so far obtained clearly demonstrated that YE and $CuCl_2$ are intrinsically different elicitors and their regulation mechanisms at the gene level should be different. As it clear from the difference in the production of isoflavonoids, the both elicitor activate enzyme genes relating to isoflavone biosythesis in different fashion and a limited genes are induced by YE while $CuCl_2$ induces wider range of genes leading to the production of phytoalexins.

Recently several compounds have been claimed to be second messenger in elicitation cascade leading to the formation of phytoalexins in various plants. Salicilic acid, abscicic acid, hydrogen peroxide and jasmonic acid were tested to clarify their

Fig. 4 Time Courses of Elicitor-induced CHS, CHR, CHI mRNA in *Pueraria lobata* Cells

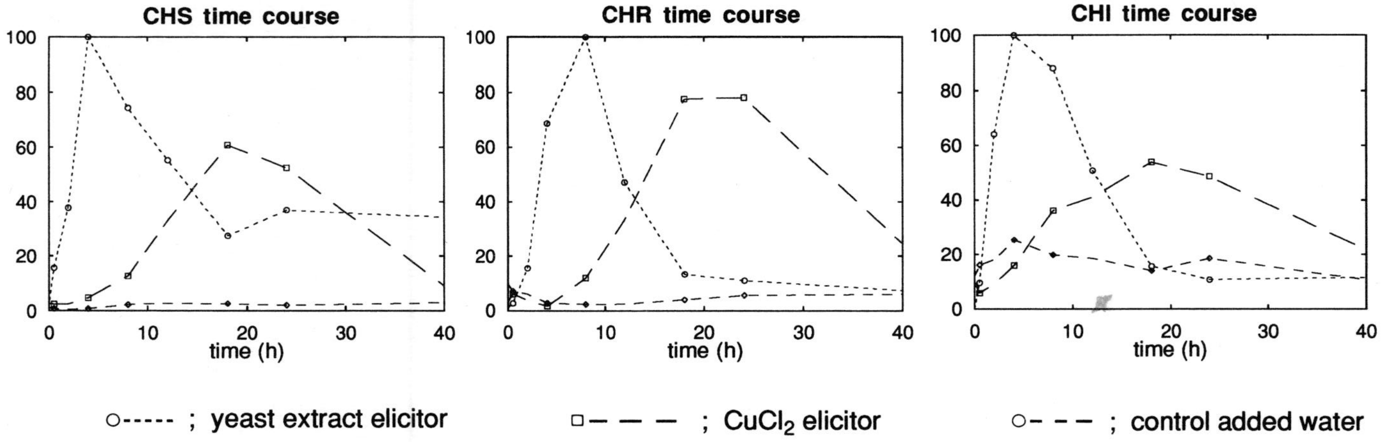

Fig. 5 Elicitor-induced Changes of Enzyme Activities in Cell Suspension Cultures of *Pueraria lobata*

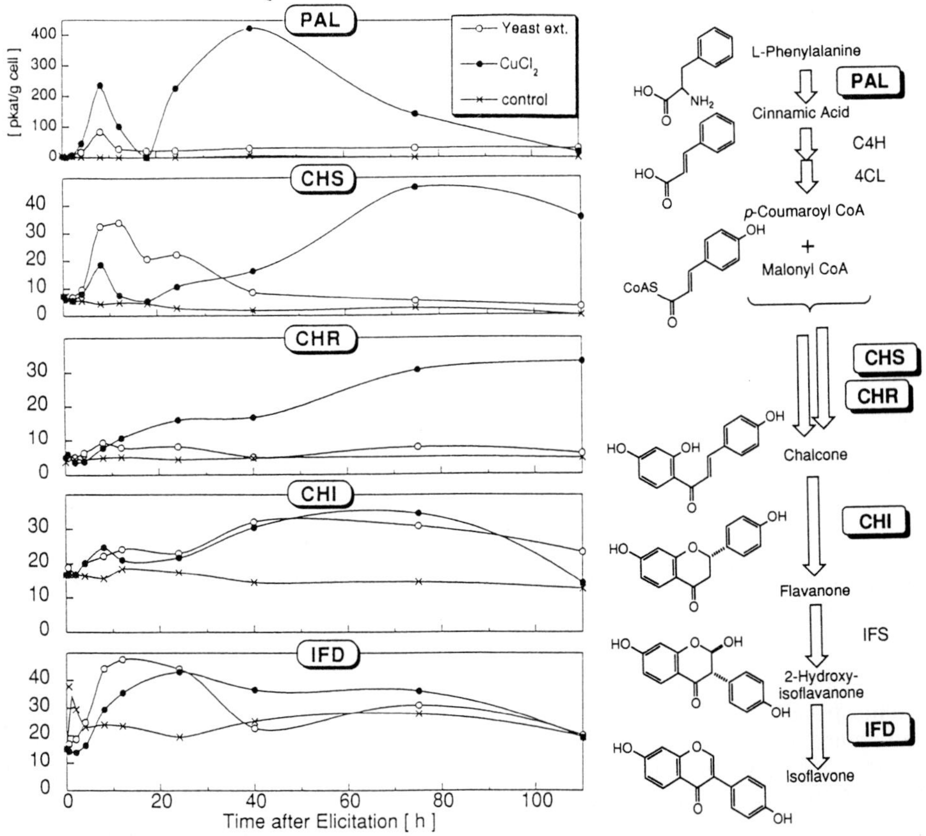

effects on the production of isoflavonoid malonylglucoside (IMG). Of the compounds tested hydrogen peroxide showed almost similar response with YE to *P. lobata* cultured cells and trasient decrease of IMG and reaccumulation was exactly the same with that of YE. [14] The effect of jasmonic acid was quite distinct.[15] Accumulation of IMG was quite prominent and no temporary decrease of IMG was seen contrary to the case of YE and hydrogen peroxide. Jasmonic acid is very effective activator for the biosynthesis of isoflavonoids as well as other secondary metabolites. [15]

4. Metabolic fate of Isoflavone Malonylglucosides by Elicitation with Yeast Extract (YE)

Rapid and transient decrease of IMG upon elicitation with YE occurred within 4 h, however the production of daizein dimers was observed after 18 h. The rapid decrease of IMG is more or less the same in $CuCl_2$ elicitation. In addition to non-synchronous formation of isoflavonoid dimers and phytoalexins the amount of decreased IMG is much lager than that of newly formed isoflavonoids. The medium of 4h-elicited suspension culture was analyzed by HPLC, but no significant difference was observed between elicited and control cells. To clarify whether the elicitor induced rapid IMG metabolism is regulated by *de novo* synthesis of enzymes, effect of cycloheximide was tested.[16] The rapid decrease of IMG induced by YE was not affected by the treatment of cycloheximide (200 μM). In contrast the elicitor induced reaccumulation of IMG was completely suppressed by cycloheximide. The results support that elicitation leads to *de novo* synthesis of enzymes required for isoflavonoid biosynthesis, while the elicitor-triggered decrease of IMG is catalyzed by preexisted enzymes. First enzymes required for the degradation of IMG would be malonylesterase and β-glucosidase to give isoflavones and IMG is converted into free isoflavone which undergoes further modification. β-Glucosidase specific for isoflavone has been purified and characterized in *Cicer arietinum* . [17] The malonylesterase was monitored for 48 h after the addition of YE and no specific increase of the activity was observed compared to the control without elicitor treatment. The enzyme activity of β-glucosidase was found in insoluble preparation of cultured cells and also no inducible effect was shown by YE elicitation. This is quite well in accord to the experiment with cycloheximide, indicating that malonylesterase and β-glucosidase are constantly active in cultured cells, while the enzymes relating to isoflavonoid biosynthesis are synthesized *de novo* by elicitor triggered process.

5. Metabolism of ^{14}C-Isoflavone Malonylglucosides (IMG)

In order to test the metabolic activities of isoflavones, ^{14}C-daizein and ^{14}C-genistein, endogenous isoflavones which had been prepared biosynthetically by feeding ^{14}C-labelled phenylalanine, were fed to 5 day old cultures of *P. lobata* and methanolic extract was analyzed with HPLC after 12 h. More than 70% of applied radioactivity was found in daizein and genistein malonylglucosides (IMG). When the cultured cells containing ^{14}C labelled IMG was treated with YE, radioactivity in methanol soluble fraction decreased rapidly, indicating radioactivity of IMG was converted into methanolic insoluble form. Non-extractable fraction was further treated to obtain hemicellulose and lignocellulose fractions. From the calculation of radioactivity in all the soluble fractions nearly 90% of insoluble portion of radioactivity was shown to be present in lignocellulose fraction. The results suggest that the rapid response by YE elicitation is deeply associated with the defense mechanism of plant-pathogen interaction to fortify cell wall by the deposition of isoflavones. Accumulation of phenolic polymer in response to plant-pathogen interaction was also observed in soybean cotyledons. [18]

6. Conclusion

Treatment of *P. lobata* cell cultures with yeast extract (YE) accumulated three dimers of daizein, which would be formed by the action of non-specific peroxidase reaction. In contrast $CuCl_2$ treatment caused hypersensitive reaction to the cultured cells and resulted in the accumulation of nine isoflavonoids including phytoalexin tuberosin. Time course changes of enzyme and mRNA levels relating to isoflavonoid biosynthesis are quite different in YE and $CuCl_2$ treatment, which may reflect intrinsically different induction at the gene level. Treatment of cell culture with YE and hydrogen peroxide resulted in rapid and temporal decrease of IMG level and IMG level recovered to normal level after 48 h. Hydrogen peroxide may be a responsible compound to induce all the responses observed in the experiments with YE. Jasmonic acid act to increase the content of IMG without temporal decrease of IMG, indicating jasmonic acid represent positive role in the control of isoflavonoid biosynthesis. Reaccumulation of IMG was not observed in $CuCl_2$ treated cells, however enzyme activities relating to isoflavonoid biosynthesis are higher or equal level with those of YE treated cells. The role of enzymes synthesized *de novo* for isoflavone synthesis should be investigated in future. The rapid and temporal decrease of IMG resulted in the deposition of isoflavones to lignocellulose fraction would be catalyzed by peroxidase and dimerization of daizein may be a side reaction of a rapid defense mechanism of plant resistant to fortify cell wall in response to outer stress.

References

[1] Harborn, J.B. and Mabry, T.J., eds., (1982) "The Flavonoids, Advances in Research", Chapman and Hall, London.
[2] Smith, D.A. and Banks, S.W., (1986) Biosynthesis, elicitation and biological activity of isoflavonoid phytoalexin, *Phytochemistry*, **25**, 979-995.
[3] Hakamatsuka, T., Noguchi, H., Ebizuka, Y. and Sankawa, U., (1988), Deoxychalcone synthase from cell suspension cultures of Pueraria lobata, *Chem.Pharm.Bull.*, **36**, 4225-4228.
[4] Hakamastsuka, T., Noguchi, H., Ebizuka, Y. and Sankawa, U., (1989), Isoflavone syntase from cell suspension culutures of *Pueraria lolbata*, *Chem.Pharm.Bull.*, **37**, 249-252.
[5] Hakamatsuka, T., Hashim, M. F., Ebizuka, Y. and Sankawa, U., (1991), P-450-dependent oxidative rearangement in isoflavone biosynthesis: Reconstitutin of P-450 and NADPH:P-450 reductase, *Tetrahedron*, **47**, 5969-5978.
[6] Hakamatsuka, T., Shinkai, K., Noguchi, H., Ebizuka, Y. and Sankawa, U., (1992), Isoflavone dimers from yeast extract-treated cell suspension cultures of *Pueraria lobata*, *Z. Naturforsch.*, **47c**, 177-182.
[7] Hashim, M.F., Hakamatsuka, T., Ebizka, Y., and Sankawa, U., (1990), Reaction mechanism of oxidative rearrangement of flavanone in isoflavone biosynthesis, *FEBS Lett.*, **271**, 219-222.
[8] Hakamatsuka, T., Shinkai, K., Noguchi, H., Ebizuka, Y. and Sankawa, U., Unpublished.
[9] Park, H.-HL., Hkamatsuka, T., Noguchi, H., Sankawa, U. and Ebizuka, Y., (1992), Isoflavone glucosides exist as their 6"-O-malonyl esters in *Pueraria lobata* and its cell suspension cultures, *Chem.Pharm.Bull.*, **40**, 1978-1980.
[10] Kenn, N.T., (1975) Specific elicitors of plant phytoalexin production: Determinants of race specificity in pahtogens?, *Science*, **187**, 74-75.
[11] Mackenbrock, U. and Barz, W., (1991) Elicitor-induced formation of pterocarpan phytoalexins in chidkpea (*Cicer arietinum* L.) cell suspension culures fromn sintitutive isoflavone conjugates upon inhibitionof phenylalanine ammonia lyase, *Z.Naturforsch.*, **46c**, 43.-46
[12] Nakajima, O., Akiyama, T., Hakamatsuka, T., Shibuya, M., Noguchi, H., Ebizuka, Y. and Sankawa, U., (1991) Isolation, sequencing and bacterial expression of cDNA for chalcone synhtase from the cultured cells of Pueraria lobata, *Chem.Pharm.Bull.*, **39**, 191-193 ; Nakajima, O., Shibuya, M., Yoshida, M., Hakamatsuka, T, Ebizuka, Y. and Sankawa, U., unpublished observation.
[13] Yoshida, M., Hakamatsuka, T, Ebizuka, Y. and Sankawa, U., unpublished observation.
[14] Vianello, A. and Macri, F., (1991) Generation of speroxide anion and hydrogen peroside at the surface of plant cells, *J. Bioenergetics and Biomembranes*, **23**, 409-423; Devlin, W.S. and Gusdtine, D.L., (1992) Involvement of the oxidative burst in phytoalexin accumulation and hypersensitive reaction, *Plant Physiol.*,

100, 1189-1195.

[15] Gundlach, H., Müller, M.J., Kutchan, T.M. and Zenk, M.H., (1992) Jasmonic acid is a signal transducer in elicitor-induced plant cell cultures, *Proc.Natl.Acad. Sci.USA*, **89**, 2389-2393.

[16] Park, H.-H., Hakamatsuka, T., Sankawa, U. and Ebizuka, Y., (1994) Rapid metabolism of isofalvonenoids in elicitor treated cell suspension cultures of *Pueraria loata*, submitted for publication.

[17] Hösel, W. and Barz, W., (1975) β-Glucodisase from *Cicer arietinam* L. *Eur.J.Biochem.*, **57**, 607-616.

[18] Graham, M.Y. and Graham, T.L., (1991) Rapid accumulation of anionic peroxidases and phenolic polymers in soybean cotyledon tissues following tratment with *Phytophthora megasperma f. sp. Glyciea* wall glucan, *Plant Physiol.*, **97**, 1445-1455.

EXPRESSION OF HYOSCYAMINE 6ß-HYDROXYLASE IN PLANTS AND IN *E. COLI*

T. HASHIMOTO,* T. KANEGAE, H. KAJIYA, J. MATSUDA & Y. YAMADA
Department of Agricultural Chemistry
Faculty of Agriculture
Kyoto University, Kyoto 606-01, Japan

*Present Address:
Nara Institute of Science and Technology
8916-5 Takayama, Ikoma
Nara 630-01, Japan

ABSTRACT. The tropane alkaloid scopolamine is synthesized at the pericycle of branch roots in certain species of the Solanaceae. The enzyme responsible for the synthesis of scopolamine from hyoscyamine is hyoscyamine 6ß-hydroxylase (H6H). Expression of H6H cDNA in *E. coli* as a fusion protein with maltose-binding protein demonstrated that H6H is a bifunctional oxygenase endowed with strong hydroxylase activity and comparatively weak epoxidase activity. Site-directed mutagenesis and chemical modification further revealed that His-217, Asp-219, and His-274 in H6H that are invariant among structurally related none-heme oxygenases are important for the hydroxylase function, and may be the ligands to the active-site iron.

The 0.8-kbp 5'-flanking region of the *Hyoscyamus niger* H6H gene was fused to the GUS reporter gene, and transferred to three solanaceous species by *Agrobacterium*-mediated transformation systems. Histochemical analysis showed that GUS expression occurred at the pericycle and root meristem of transgenic *H. niger* hairy roots, but only at the root meristem of hairy roots and plants of transgenic tobacco. In transgenic hairy roots and regenerated plants of *Atropa belladonna,* the root meristem was stained with GUS activity, except for a few transformants in which the vascular cylinder was also stained. These studies indicate that the cell-specific expression of the H6H gene is controlled by some genetic regulation specific to scopolamine-producing plants.

1. INTRODUCTION

Hyoscyamine and its epoxide scopolamine are typical tropane alkaloids found in several solanaceous plants. Previous feeding experiments with alkaloid precursors [1, 2] have suggested that the 6,7-epoxide bridge of scopolamine is formed from hyoscyamine by way of 6ß-hydroxyhyoscyamine (Fig. 1). We initially discovered that a 2-oxoglutarate-dependent dioxygenase that hydroxylates hyoscyamine at the 6ß-position in alkaloid-producing root cultures, and named it hyoscyamine 6ß-hydroxylase (H6H; EC 1.14.11.11) [3, 4]. Later, we also found that a similar 2-oxoglutarate-dependent dioxygenase converts 6ß-hydroxyhyoscyamine to scopolamine by dehydrogenation of the

M. Terzi et al. (eds.), Current Issues in Plant Molecular and Cellular Biology, 605-609.

7ß-hydrogen [5]. This epoxidase activity was relatively weak and represented only 1-10% of the hydroxylase activity in partially purified enzyme preparations. Our subsequent experiments using transgenic plants expressing the cloned H6H cDNA from *Hyoscyamus niger* indicated that H6H may be a bifunctional enzyme endowed with strong hydroxylase activity and comparatively weak epoxidase activity [6, 7].

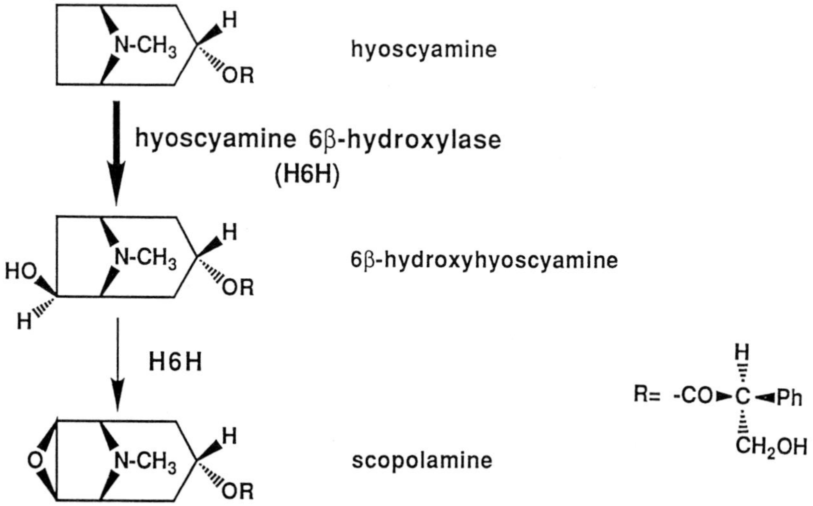

Figure 1. Hyoscyamine 6ß-hydroxylase converts hyoscyamine to scopolamine by way of 6ß-hydroxyhyoscyamine.

For catalysis, H6H requires, in addition to the alkaloid substrate, Fe^{2+}, 2-oxoglutarate, O_2, and ascorbate. The primary structure of *H. niger* H6H shares significant homology to several other 2-oxoglutarate-dependent dioxygenases, such as flavanone 3ß-hydroxylase, deacetoxycephalosporin C synthase, and deacetylcephalosporin C synthase [8]. Similar levels of homology in the amino acid sequences have also been noted between H6H and several oxygenases that require Fe^{2+} and ascorbate, but not 2-oxoglutarate, such as ethylene-forming enzyme and isopenicillin N synthase [8]. Alignment of these amino acid sequences has revealed that two histidine residues and one aspartic acid residue are strictly conserved among them. Although vertebrate 2-oxoglutarate-dependent dioxygenases, such as prolyl 4-hydroxylase, lysyl hydroxylase, and aspartyl ß-hydroxylase, are scarcely homologous in amino acid sequences to these enzymes, two histidines and one aspartic acid can be aligned at similar spacings in their sequences. Thus, these three amino acid residues may play important roles in the structure and function of this broad class of non-heme oxygenases.

H6H mRNA is abundant in cultured roots, present in plant roots, but absent in stems, leaves and cultured cells [8]. Immunohistochemical analysis has further localized the H6H protein to the pericycle in the root, and the pericycle-specific expression of H6H has been proposed to be important for the root-to-shoot translocation of tropane alkaloids through xylem [9]. To understand the molecular mechanism behind this highly cell-specific biosynthesis of scopolamine, the regulatory components that control the expression of the H6H gene in scopolamine-producing plants must first be identified.

2. H6H Is a Bifunctional Enzyme

H. niger H6H cDNA was expressed in *Escherichia coli* as a fusion protein with maltose-binding protein (MBP; Fig. 2). The *E. coli* harboring the expression plasmid pMH1 was cultured at 10 C for 2 days after the addition of 0.4 mM isopropylthio-ß-D-galactoside. Considerable amount of the MBP-H6H chimeric protein was recovered in the soluble fraction, which showed both hydroxylase and epoxidase activities. The epoxidase activity was about 2.4% of the hydroxylase activity. Thus, the H6H cDNA of *H. niger* encodes a bifunctional 2-oxoglutarate-dependent dioxygenase with a strong hydroxylase activity and a relatively weak epoxidase activity [10].

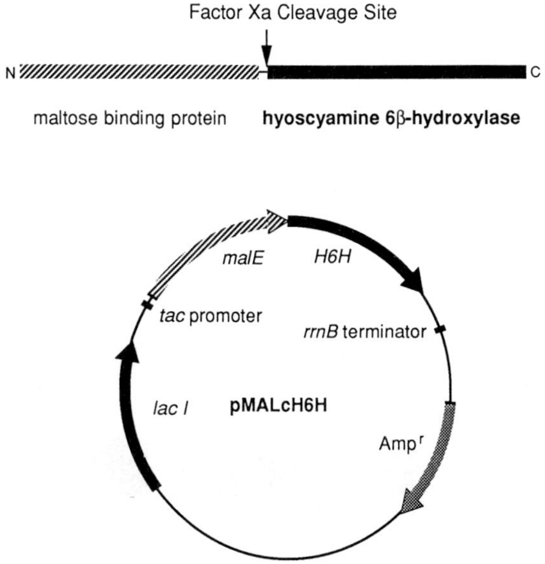

Figure 2. Structure of pMALcH6H that expresses a fusion protein between hyoscyamine 6ß-hydroxylase and maltose-binding protein.

Modification of histidine residues by diethyl pyrocarbonate inactivated the recombinant chimeric H6H. Inactivation was prevented most efficiently by the presence of 2-oxoglutarate. Site-directed mutagenesis of any three strictly conserved amino acid residues (histidine-217 to glutamine, histidine-274 to glutamine, or aspartic acid-219 to either histidine or asparagine) inactivated the H6H, whereas substitution of loosely conserved histidine-66 with glutamine did not decrease the catalytic activity of the enzyme. These results indicate that the three conserved amino acid residues in H6H and related non-heme oxygenases play important roles in catalysis, and may be the ligands to the active-site iron.

3. Species-Dependent Expression of the *H. niger* H6H Promoter

Libraries constructed from *H. niger* genomic DNA was screened with an *H. niger* cDNA probe, and a genomic clone that contained the 5'-portion of the gene was obtained (Fig. 3). The 5'-flanking region from -827 to +108 was translationally fused to the ß-glucuronidase (GUS) reporter gene in pBI101, and introduced to three solanaceous species by Agrobacterium-mediated transformation systems : i.e., *H. niger* and belladonna, which have high and low levels, respectively, of H6H mRNA in the root, and tobacco, which has no endogenous H6H gene [11]. Histochemical analysis showed that GUS expression occurred at the pericycle and root meristem of transgenic *H. niger* hairy roots. Clear GUS staining was observed in one or a few pericycle cells in each transverse section. Notably, many of the stained cells were located opposite primary xylem poles. This indicates physiological heterogeneity in the pericycle cell population, and may correlate two types of pericycle cells that can be distinguished anatomically [12].

GUS activity was observed only at the root meristem of transgenic tobacco plants. In transgenic hairy roots and regenerated plants of belladonna, the root meristem was the only tissue stained with GUS activity, except for a rare transformant in which the vascular cylinder was also stained. 5'-Deletion promoter analysis indicated that the cis-acting element(s) responsible for expression at the pericycle and root meristem are both located within 423-bp upstream of the transcription initiation site, which includes an extremely A/T-rich region. These studies indicate that the cell-specific expression of the H6H gene is controlled by some genetic regulation specific to scopolamine-producing plants.

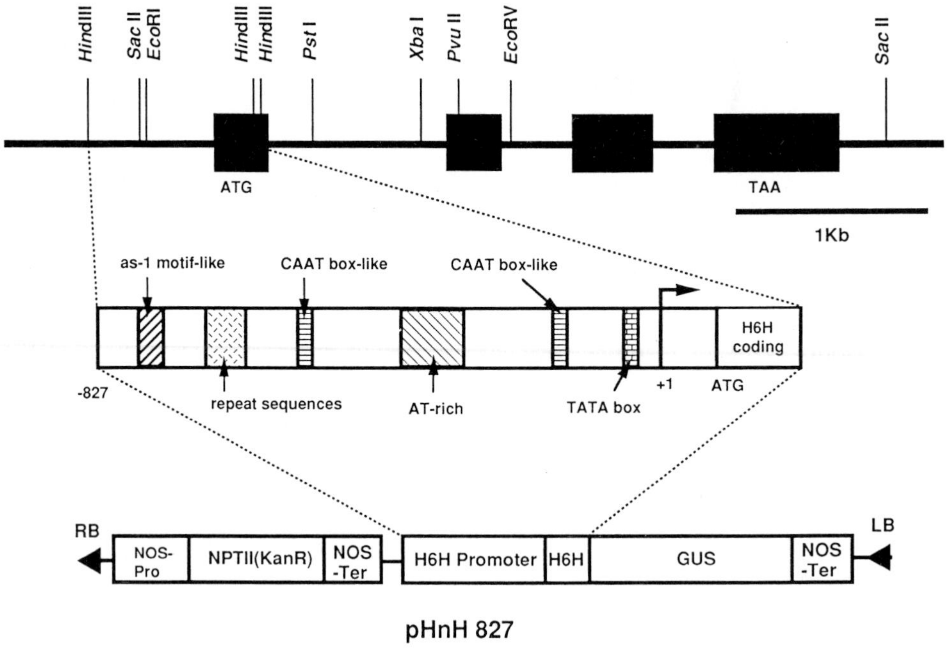

Figure 3. The 5'-flanking region of the *H. niger* H6H gene is fused to ß-glucuronidase reporter gene in a binary vector pHnH 827. Structural features of the 5'-flanking region are also shown. Black boxes in the genome structure (uppermost figure) indicate exons.

4. References

1. Romeike, A. (1960) 'Über ein Zwischenprodukt bei der Epoxydierung des Hyoscyamins in der lebenden Pflanze', Naturwissenschaften **47**, 64-65.
2. Hashimoto, T., Kohno, J. and Yamada, Y. (1987) 'Epoxidation *in vivo* of hyoscyamine to scopolamine does not involve a dehydration step', Plant Physiol. **84**, 144-147.
3. Hashimoto, T. and Yamada, Y. (1986) 'Hyoscyamine 6ß-hydroxylase, a 2-oxoglutarate-dependent dioxygenase, in alkaloid-producing root cultures', Plant Physiol. **81**, 619-625.
4. Hashimoto, T. and Yamada, Y. (1987) 'Purification and characterization of hyoscyamine 6ß-hydroxylase from root cultures of *Hyoscyamus niger* L.', Eur. J. Biochem. **164**, 277-285.
5. Hashimoto, T., Kohno, J. and Yamada, Y. (1989) '6ß-Hydroxyhyoscyamine epoxidase from cultured roots of *Hyoscyamus niger*', Phytochemistry **28**, 1077-1082.
6. Yun, D.-J., Hashimoto, T. and Yamada, Y. (1993) 'Transgenic tobacco plants with two consecutive oxidation reactions catalyzed by hyoscyamine 6ß-hydroxylase', Biosci. Biotech. Biochem. **57**, 502-503.
7. Yun, D.-J., Hashimoto, T. and Yamada, Y. (1992) 'Metabolic engineering of medicinal plants: Transgenic *Atropa belladonna* with an improved alkaloid composition', Proc. Natl. Acad. Sci. USA **89**, 11799-11803.
8. Matsuda, J., Okabe, S., Hashimoto, T. and Yamada, Y. (1991) 'Molecular cloning of hyoscyamine 6ß-hydroxylase, a 2-oxoglutarate-dependent dioxygenase, from cultured roots of *Hyoscyamus niger*', J. Biol. Chem. **266**, 9460-9464.
9. Hashimoto, T., Hayashi, A., Amano, Y., Kohno, J., Iwanari, H., Usuda, S. and Yamada, Y. (1991) 'Hyoscyamine 6ß-hydroxylase, an enzyme involved in tropane alkaloid biosynthesis, is localized at the pericycle of the root', J. Biol. Chem. **266**, 4648-4653.
10. Hashimoto, T., Matsuda, J. and Yamada, Y. (1993) 'Two-step epoxidation of hyoscyamine to scopolamine is catalyzed by bifunctional hyoscyamine 6ß-hydroxylase', FEBS Lett. **329**, 35-39.
11. Kanegae, T., Kajiya, H., Amano, Y., Hashimoto, T. and Yamada, Y. (1994) 'Species-dependent expression of the hyoscyamine 6ß-hydroxylase gene in the pericycle', Plant Physiol. (in press).

PLANT CELL CULTURE COMBINED WITH CHEMISTRY - POTENTIALLY POWERFUL ROUTES TO CLINICALLY IMPORTANT COMPOUNDS

J. P. KUTNEY
Department of Chemistry
University of British Columbia
2036 Main Mall
Vancouver, B.C., V6T 1Z1
Canada

ABSTRACT. The development of plant cell culture methods, in combination with chemistry, affords an attractive and often a powerful route to complex natural products. Several examples of such an interdisciplinary program are cited to illustrate the various types of research strategies which have been pursued. Studies with cell cultures of Catharanthus roseus provide biosynthetic information and subsequently an entry into the efficient synthesis of the clinical anti-cancer drugs vinblastine and vincristine. Experiments with enzymes derived from C. roseus and Podophyllum peltatum cell lines and dibenzylbutanolides as precursors, reveal an attractive route to podophyllotoxin analogues required for synthesis of the anti-cancer drug etoposide. Still other studies with a cell line of Tripterygium wilfordii, an important Chinese herbal plant, allow the production of novel terpenoid systems of pharmacological interest.

1. Introduction

The plant kingdom has, for many years, provided an important source of natural products many of which have formed the basis for development of medicinally important drugs. Unfortunately, Nature often provides such compounds in low yields and the difficulties associated with their isolation from other less interesting co-occurring constituents can present problems particularly when large quantities of the biologically active compound are required. It is possible to alleviate such difficulties by the use of plant cell culture methodology and when these techniques are coupled with chemistry, a powerful route to such natural products and/or their biologically important analogues, is achieved. This lecture will present results to illustrate how the interplay of plant cell culture methodology, in combination with chemistry, can afford interesting routes to clinically important plant derived medicinal agents. A discussion of the various avenues of research will include: i) studies of biosynthesis and application of biosynthetic information toward development of highly efficient syntheses of clinical drugs; ii) use of plant cell cultures or enzymes derived therefrom , as "reagents" in organic synthesis; iii) use of plant cell cultures to produce higher levels of plant derived natural products and novel compounds for pharmacological screening; iv) an illustration to show how well developed cell lines can afford the opportunity to separate pharmacological activities exhibited by complex mixtures generally employed in herbal medicine practices.

The studies involve the clinical anti-cancer drugs vinblastine, vincristine, etoposide, and natural products of a Chinese herbal plant which possess immunosuppressive activity.

M. Terzi et al. (eds.), Current Issues in Plant Molecular and Cellular Biology, 611-616.

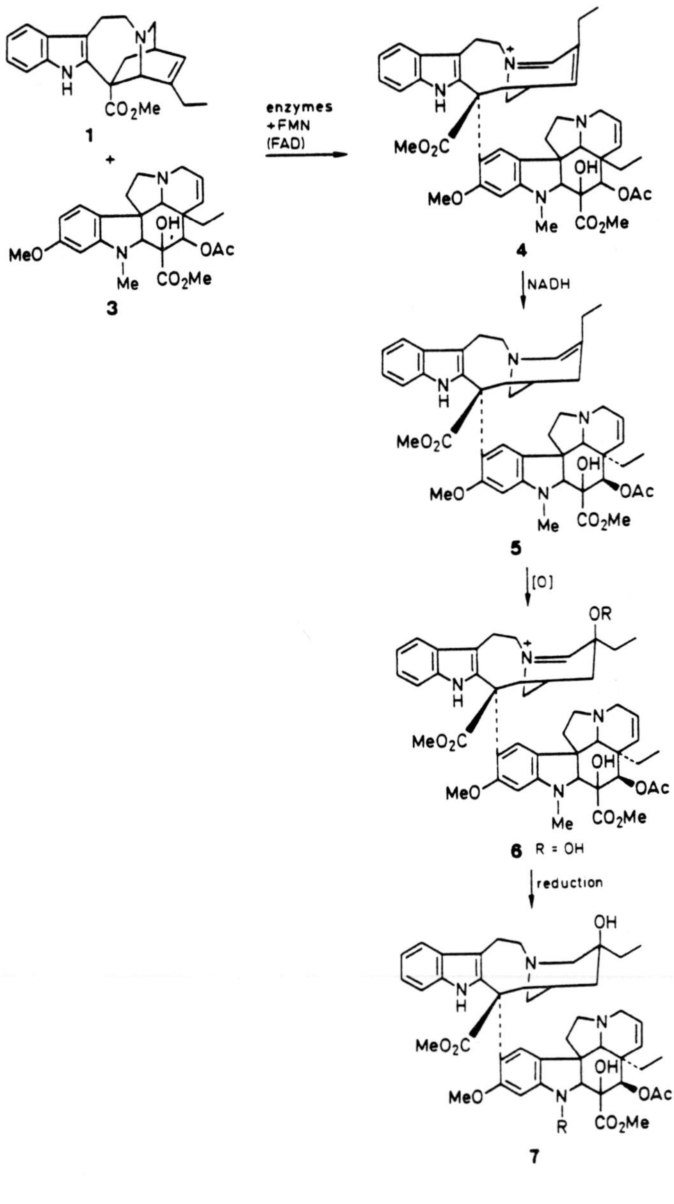

Figure 1. Overall summary of the biosynthetic pathway of vinblastine (**7**) from catharanthine (**1**) and vindoline (3).

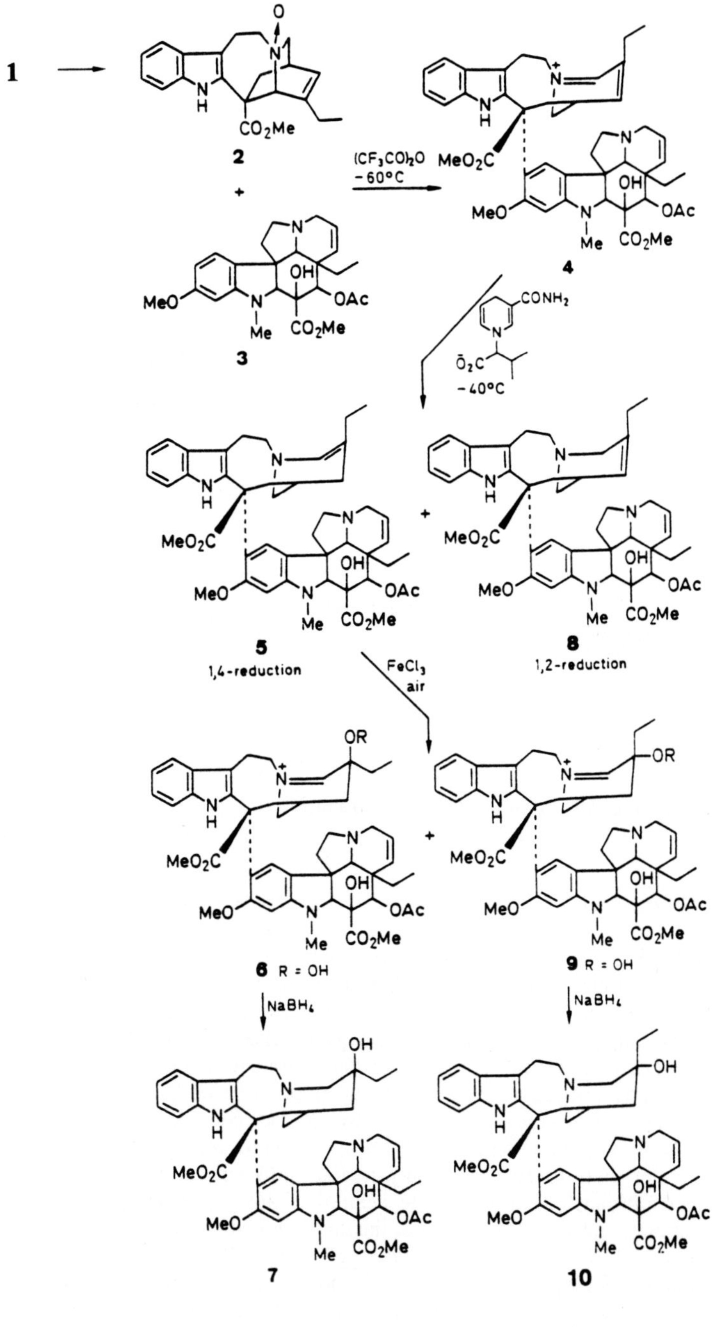

Figure 2. A highly efficient "one-pot" process for the synthesis of vinblastine (7) and leurosidine (**10**) from catharanthine (**1**) and vindoline (**3**).

2. Biosynthetic Information Forms a Basis for Efficient Synthesis of the Vinblastine-Vincristine Family

From a large number of investigations involving enzymes obtained from a stable cell line of Catharanthus roseus and the alkaloids catharanthine (**1**) and vindoline (**3**), we have unravelled the structures of the important late stage intermediates in the biosynthetic pathway of the clinical anti-cancer drugs vinblastine (7, R=CH_3) and vincristine (7, R=CHO). Fig. 1 summarizes the overall sequence involved. From these data, a highly efficient and commercially important "one-pot" process for the synthesis of the clinical drugs was developed (Fig. 2). The overall process, involving five separate chemical reactions and providing a 40% overall yield of vinblastine, demands that each reaction must proceed with yields in excess of 80%. A recent review (1) summarizes these extensive studies and provides citations to pertinent earlier references.

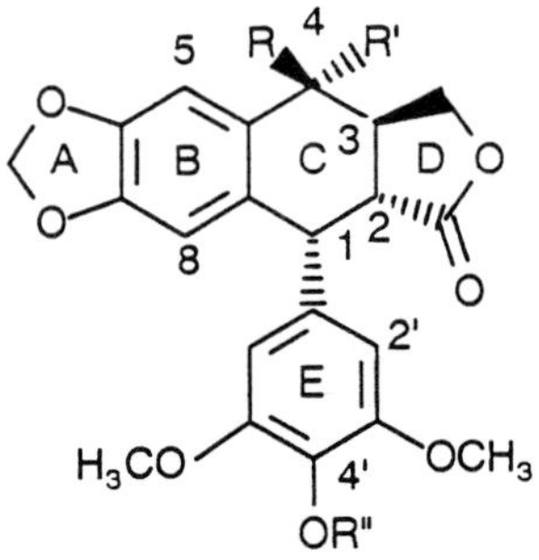

Podophyllotoxin: R=H; R'=OH; R"=CH_3
Epipodophyllotoxin: R=OH; R'=H; R"=CH_3
Deoxypodophyllotoxin: R=R'=H; R"=CH_3
4'-Demethylpodophyllotoxin: R=H; R'=OH; R"=H
4'-Demethylepipodophyllotoxin: R=OH; R'=H; R"=H (12)
Etoposide: R= 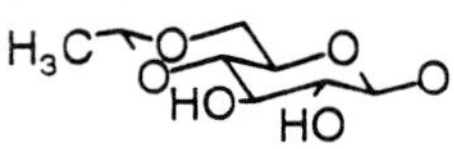 ; R'=H; R"=H (13)

11

Figure 3. The podophyllotoxin family of compounds.

3. Plant Cell Cultures as "Reagents" in Studies Related to Etoposide Synthesis

The podophyllotoxin family (**11**, Fig. 3) of natural products has been extensively studied over the years. One of the important analogues within this family is the clinical anti-cancer drug etoposide (**13**) and studies in our laboratory directed toward the development of an efficient synthesis of **13** are underway. The overall strategy involves the use of appropriate enzymes derived from plant cell cultures as "reagents" in the biotransformation of suitable substrates to end products presently utilized in the commercial production of **13**. Based on biosynthetic information provided from the studies of Dewick and coworkers (2), it appeared that dibenzylbutanolides of general structure **19** (Fig. 4) were appropriate substrates for such studies. A versatile route to **19** and closely related analogues was developed (Fig. 4) and enzyme-catalyzed biotransformations to the desired cyclic podophyllotoxin analogues were undertaken. Utilizing enzymes isolated from the above-noted C. roseus cell line and/or whole cell fermentations with a stable cell line of Podophyllum peltatum, the plant from which the podophyllotoxins are normally isolated (Fig. 3), highly successful biotransformations were achieved (Fig. 5). With P. peltatum cultures, the biotransformation of **20** (R=OH; R'=H; R"=isopropyl, Fig. 5) to **21**, was achieved via batch and semi-continuous processes. Details of these studies are published (1, 3-5).

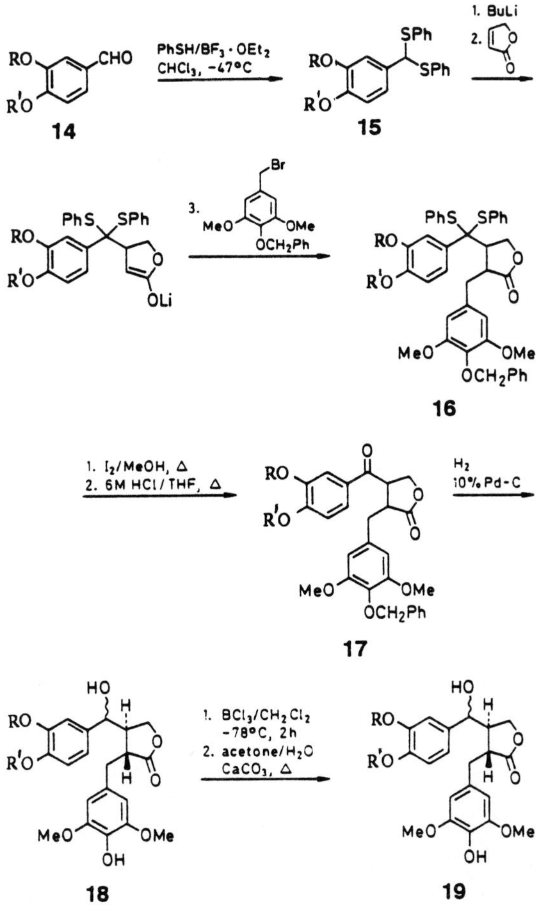

Figure 4. Synthesis of a 4'-dimethylepipodophyllotoxin precursor.

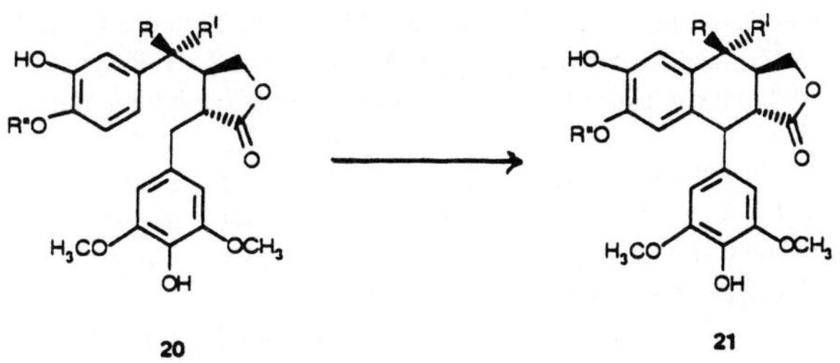

Figure 5. Biotransformation of dibenzylbutanolide **20** with cell free extract of C. roseus cell cultures and whole cells of P. peltatum.

4. Use of Plant Cell Cultures to Produce Higher Levels of Pharmaceutically Interesting Compounds and Separation of Pharmacological Activities in Complex Herbal Medicine Extracts.

Plant cell culture methodology in combination with chemistry, can provide significantly higher levels of plant derived medicinal agents, and their isolation from the culture is much less complicated than from a typical plant extract. For example, in our studies with a well developed cell line of the important Chinese herbal plant *Tripterygium wilfordii* (1, 6), we have obtained the highly interesting diterpene triepoxide tripdiolide (**22**, Fig. 6) in yields 36 times greater than in the living plant. Furthermore, by the isolation of various metabolites from the cultures with the di- and triterpene structural types summarized in Fig. 6, and subsequent biological screening, we have been able to separate their respective pharmacological properties. The diterpenes **22** and **23** are highly active as immunosuppressive agents and likely related to their possible use for rheumatoid arthritis treatment while the triterpenes (**24**, **25**) exhibit anti-inflammatory activity with application for dermatological disorders.

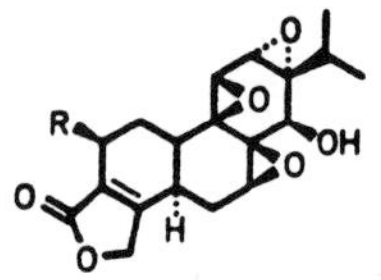

22 R = OH

23 R = H

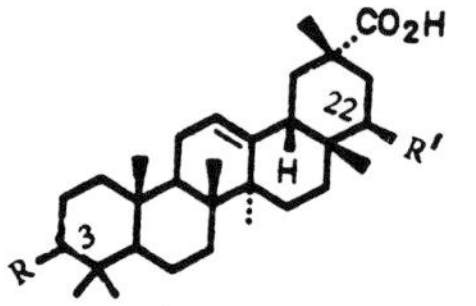

24

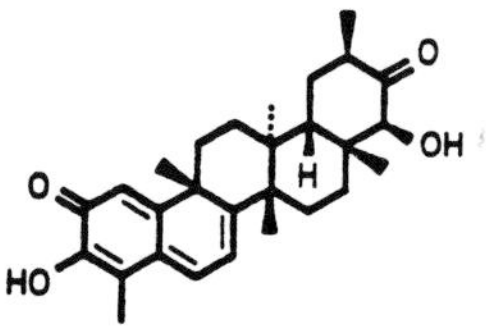

25

Figure 6. A summary of structural types of secondary metabolites isolated from *T. wilfordii* cell cultures.

5. References

1. Kutney, James P. (1993) 'Plant cell culture combined with chemistry - a powerful route to complex natural products'. Acc. Chem. Res. 26, 559-566, and references cited therein.
2. Broomhead, A. J., Rahaman, A. M. M., Dewick, P. M., Jackson, D. E., and Lucas, J. A. (1991) 'Matairesinol as precursor of *Podophyllum* lignans'. Phytochem. 30, 1489-1492.
3. Kutney, James P., Hewitt, Gary M., Jarvis, Terrence J., Palaty, J., and Rettig, S. (1992) 'Studies with plant cell cultures of *Catharanthus roseus*. Oxidative coupling of dibenzylbutanolides catalyzed by plant cell culture extracts'. Can. J. Chem. 70, 2115-2133.
4. Kutney, James P., Arimoto, M., Hewitt, Gary M., Jarvis, Terrence C., and Sakata, K. (1991) 'Studies with plant cell cultures of *Podophyllum peltatum* L. I. Production of podophyllotoxin, deoxypodophyllotoxin, podophyllotoxone, and 4'-demethylpodophyllotoxin'. Heterocycles, 35, 2305-2309.
5. Kutney, James P., Chen, Yung P., Gao, S., Hewitt, Gary M., Kuri-Brena, F., Milanova, Radka K., and Stoynov, N. (1993) 'Studies with plant cell cultures of *Podophyllum peltatum* L. II. Biotransformation of dibenzylbutanolides to lignans. Development of a "biological factory" for ligan synthesis'. Heterocycles, 36, 13-20.
6. Kutney, James P., Hewitt, Gary M., Lee, G., Piotrowska, K., Roberts, M., and Rettig, S. (1992) 'Studies with tissue cultures of the Chinese herbal plant, *Tripterygium wilfordii*. Isolation of metabolites of interest in rheumatoid arthritis, immunosuppression, and male contraceptive activity'. Can. J. Chem. 70, 1455-1480.

MOLECULAR AND METABOLIC CONTROL OF SECONDARY METABOLISM IN *TAGETES*

A.F. CROES, J.J.M.R. JACOBS, R.R.J. ARROO, G.J. WULLEMS
Department of Experimental Botany
NOVAPLANT Cell Biotechnology Group
University, Toernooiveld
6525ED Nijmegen, the Netherlands

ABSTRACT. Secondary metabolism is confined to specialized cell types at specific places in the plant. The concentrations of precursors and end products are determining factors in the metabolic control of synthesis and breakdown of the compounds involved. Molecular control operates at the level of enzyme amount and gene expression. If the secondary product contains an element in its molecule which is derived from a mineral nutrient in the environment, the operation of the control mechanisms can be studied by varying the concentration of that mineral. This is exemplified by thiophene metabolism in transformed root cultures of *Tagetes*. The characteristic groups in the molecule are two five-membered rings with a sulfur atom in them. Using an EMS-mutant we isolated an intermediate with one ring. The rate of thiophene biosynthesis was manipulated by varying the sulfate concentration in the medium. Sulfur limitation led to preferential channeling of sulfur into primary metabolism and a concomitant drop in thiophene biosynthesis. The major part of the reduction was caused by a drop in enzyme activity. Substrate availability played a minor role. The results indicate that sulfur is involved in the molecular control of secondary metabolism in *Tagetes*.

1. Introduction

All secondary metabolic routes derive their precursors from primary metabolism. At the onset of these routes, part of these precursors is channeled in the direction of secondary product formation whereas the rest continues to be converted in the primary pathway. A controlled partitioning is vital to the cell because a too heavy drain of compounds from primary metabolism would endanger cellular functioning and growth. How much product will be formed is ultimately determined by the flux at the rate-limiting step (usually the first reaction [3]), and this in turn depends on enzyme amount and actual activity. The metabolic control of the flux is exerted by factors that influence this activity, such as substrate availability and end product inhibition. The amount of enzyme can be modulated by variation in the rates of transcription and translation. That secondary pathways are only present in some cells of the organism is an extreme manifestation of this type of control. One way to study the operation of the control mechanisms is by reducing the concentration of a substrate in the environment. It is reasonable to expect that such a deficiency leads to preferential channeling of the limiting compound into primary metabolism at the expense of the secondary routes involved. Seconda-

M. Terzi et al. (eds.), Current Issues in Plant Molecular and Cellular Biology, 617-622.

ry metabolism may be directly affected by the availability of less precursor but also indirectly in case the synthesis of one or more enzymes is reduced.

The biosynthesis of thiophenes in *Tagetes* is a suitable model system to study the regulation of secondary metabolism. The compounds are preferentially accumulated in roots and hypocotyls but the level in the leaves is low [4]. Thiophenes are characterized by five-membered rings with a sulfur atom in each ring. The ring is formed from a sulfhydryl group donated by cystein and two adjacent groups in a polyacetylenic chain [1]. The reaction occurs at a branching point in sulfur metabolism: cystein is used for the synthesis of primary compounds such as proteins, and its sulfhydryl group is incorporated in thiophenes. Thiophene synthesis is particularly high in isolated roots [2]. There is no breakdown nor transport in these roots which means that accumulation is a measure of biosynthesis. Incorporation of radiolabeled sulfur from [^{35}S]sulfate provides direct data on the flux of precursors. Substrate availability can be manipulated by varying the sulfate concentration in the medium.

In this paper we describe the limiting factors in thiophene biosynthesis in whole plants and in sulfur-deficient isolated roots. The biosynthetic capacity varied among the organs and in dependence on substrate availability. A molecular mechanism was found to operate in governing synthesis and accumulation.

2. Results

2.1. RELATION BETWEEN SYNTHESIS AND ACCUMULATION

The thiophenes in *Tagetes* are mainly accumulated in the roots and the hypocotyl whereas the concentration in the leaves is low (Table 1). The involvement of biosynthesis in the

TABLE 1. Thiophene content of seedlings of *T. patula*

Organ	[Thiophene] (μmol.g FW^{-1})
Cotelydon	0.16 ± 0.035
Hypocotyl	0.39 ± 0.075
Root	0.77 ± 0.20

TABLE 2. Uptake and incorporation in thiophenes of [^{35}S]sulfate by three-weeks-old plants of *T. patula*

Organ	Uptake (kBq.g^{-1})	Incorporation (kBq.g^{-1})	Incorporation (% of uptake)
Hypocotyl	588	29.6	5.1
Leaf	219	2.6	1.2
Apex	503	6.2	1.8

establishment of thiophene distribution was assayed by incubating young plants of *T. patula* with [^{35}S]sulfate and monitoring the distribution of label and its incorporation in thiophenes in the aerial parts (Table 2). Much sulfate was trapped in the hypocotyl and relatively little was recovered in the leaves. This might be an explanation for the lower rate of thiophene synthesis in the leaves as compared to that in other organs. However, when sulfate incorporation in thiophenes was expressed as a percentage of the total amount of [^{35}S]sulfur in the organ, the lowest activity was again found in the leaves (Table 2). Thus, the concentration of thiophene in these organs correlates with the capability to synthesize the metabolite as well as with the availability of the sulfur source.

2.2. THIOPHENE BIOSYNTHESIS

The bithienyl BBT (Fig. 1) and its derivatives are the main thiophenes in *Tagetes*. Neither the presumptive precursor PYE nor the hypothetical one-ring intermediate accumulate in the tissue. For this reason, we raised mutants with an aberrant thiophene spectrum by mutagenizing seed with ethyl methane sulfonate. The plants grown from this seed were selfed and root

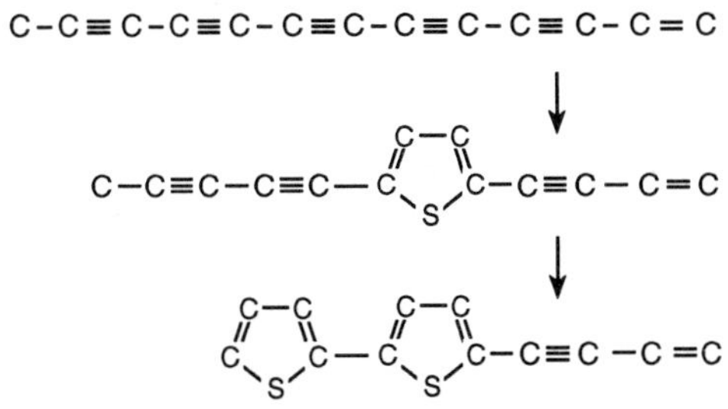

Figure 1. Outline of thiophene biosynthesis. PYE, trideca-3,5,7,9,11-pentaynene; BPT, 2-(but-3-en-1-ynyl)-5-(penta-1,3-diynyl)-thiophene; BBT, 5-(but-3-en-1-ynyl)-2,2'-bithienyl

extracts of M2 plants were analyzed by HPLC (Fig. 2). An unknown compound abundantly present in a mutant and eluting at 13 min was identified as the monothiophene BPT (Fig. 1) by GC-MS and ^{1}H-NMR. When fed to a normal root culture, the compound was readily converted into BBT and its derivatives. The results indicate that the two sulfur atoms in BBT are incorporated in separate steps and that BPT is an intermediate in the chain of reactions.

2.3. SULFUR CHANNELING IN METABOLISM

The partitioning of sulfur between primary and secondary metabolism was assessed by incubating root cultures in media with standard or low sulfate concentration. A 20 to 40-fold reduction in the sulfate concentration had no effect on biomass production, elongation growth or branching but severely affected thiophene accumulation (Table 3). The capacity of the roots to incorporate sulfur into thiophene was assayed after 8 days of growth at 0.05 mM sulfate by exposing tissues to [^{35}S]sulfate at a total sulfate concentration of 30 mM. Thiophene synthesis appeared to be 5 times lower than in the control roots (Table 4). The capacity to synthesize thiophenes was only gradually restored when the roots grown at low sulfate were shifted to

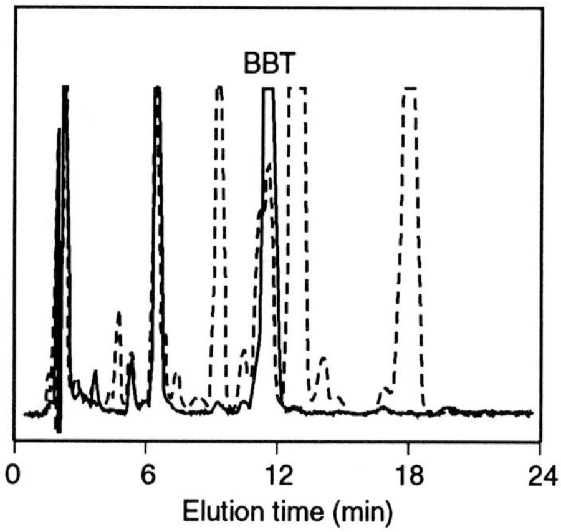

Figure 2. HPLC chromatogram of wild-type (solid line) and mutant (broken line) *T. erecta* root extract. The peak eluting at 13 min is BPT.

TABLE 3. Effect of sulfate on elongation, branching, and thiophene accumulation in *T. patula* roots

[Sulfate] (mM)	Root length (cm)	Laterals (#)	[Thiophene] (μmol.g^{-1})
0.1	6.5 ± 2	32 ± 11	0.6 ± 0.1
0.5	6.3 ± 1	34 ± 10	1.1 ± 0.2
2.0	6.7 ± 1	24 ± 8	2.3 ± 0.4

standard medium. The recovery was completely blocked by the transcription inhibitor cordycepine which indicates that the recovery was the result of an active regulation of transcription. The question whether thiophene synthesis was also limited by low substrate availability, was addressed by measuring the pool of thiol compounds in the cell. The sulfate concentration in the medium had no effect on the size of this pool (data not shown). In principle, sulfate deficiency could reduce the formation of either the first or the second ring only, or the closure of both rings. If only the second step were affected, the monothiophene would accumulate which was not found. However, roots grown at 0.05 mM were less efficient in converting BPT to bithienyls than tissues under standard conditions (Table 5). The combined results suggest that the formation of each of the two rings is under control of the regulating mechanism.

3. Discussion

The distribution of thiophenes in Tagetes plants is determined by the rates of biosynthesis of these compounds in the various tissues. An important element in the regulation of biosynthesis

TABLE 4. Restoration of thiophene synthesis in *T. patula* roots precultured in a low-sulfate (0.05 mM) medium. After preculture, the roots were shifted to standard (2.0 mM) medium with or without cordycepine, 5μg.ml^{-1}.
The incorporation of ^{35}S-sulfur in thiophenes was measured at the time of the medium shift and 24 h later.

Time (h)	[Sulfate] during preculture (mM)	Cordycepine added	S incorporation (nmol.g^{-1}.h^{-1})
0	0.05	-	1.7 ± 0.4
	2.0	-	8.8 ± 0.9
24	0.05	no	5.3 ± 0.6
		yes	1.9 ± 0.3
24	2.0	no	10.1 ± 0.3
	2.0	yes	5.3 ± 0.6

TABLE 5. Effect of sulfate on the conversion of BPT (Fig.1) to bithienyls. Roots precultured on low-sulfate (0.05 mM) medium were fed [^{35}S]BPT.

[Sulfate] (mM)	Bithienyl formed (% of BPT added)
0.05	37.7 ± 0.4
2.0	86.5 ± 3.0

is the partitioning of sulfur over primary and secondary metabolism. The mechanism controlling this channeling operates by differential transcription of genes involved in the formation of the two heterocyclic rings in the thiophene molecule.

The high concentrations of thiophenes in hypocotyl and roots and the low level in leaves are mirrored by the rates of synthesis in these organs. Thiophenes thus appear to be accumulated where they are synthesized. This is also supported by the observation that thiophenes administered to roots of seedlings are rapidly taken up but barely transported to the aerial parts (Jacobs et al., unpublished). The relation between accumulation and synthesis, in combination with the unequal distribution of thiophenes over the plant, indicates that the expression of the biosynthetic pathway is tightly linked to cellular differentiation.

The repeated incorporation of sulfur is the key reaction in the formation of the bithienyl molecule. The reaction rate is governed by the abundance of sulfate in the environment of the root. At first sight, this would appear to be an example of metabolic regulation. However, two lines of evidence argue for a molecular mechanism. First, the concentration of the actual sulfur donor is presumably independent of the sulfate concentration because the thiol content of the cell was not affected by a fourty-fold reduction in the sulfate level. Second, restoration of the low biosynthetic capacity in sulfate-deficient roots to normal levels requires active gene transcription and, as a consequence, protein synthesis. The constancy of the thiol level in the cells even at extreme sulfate limitation also shows the effectivity of the channeling mechanism in maintaining a reduced sulfur pool for protein synthesis.

In the light of the above conclusion, it is reasonable to assume that the molecular control of thiophene synthesis also operates in the leaves. The sulfur level in the leaves is lower than in stem and roots because much sulfate is trapped in the latter organs during xylem transport, and redistribution via the phloem causes sulfur export from the leaves to other organs. It thus may be envisaged that the low remaining sulfur content in the leaves is the effector of the molecular mechanism leading to reduced thiophene synthesis. The rationale behind such a mechanism would be that due to its operation the requirements of primary metabolism are first met. The absence of any effect of severe sulfate deficiency on root growth and development favors this interpretation.

4. References

1. Bohlmann, F., Zdero, C. (1985) 'Naturally occurring thiophenes', in S. Gronowitz (ed.), Thiophene and its Derivatives, John Wiley and Sons, New York, pp. 261--323.
2. Croes, A.F., Van den Berg, A.J.R., Bosveld, M., Breteler, H., Wullems, G.J. (1989) 'Thiophene accumulation in relation to morphology in roots of Tagetes patula', Planta 179, 43-50.
3. Galneder, E., Zenk, M.H. (1990) 'Enzymology of alkaloid production in plant cell cultures', in H.J.J. Nijkamp, L.H.W. van der Plas, J. van Aartrijk (eds.), Progress in Plant Cellular and Molecular Biology, Kluwer Academic Publishers, Dordrecht, pp. 754-762.
4. Tosi, B., Lodi, G., Dondi, F., Bruni, A. (1988) 'Thiophene distribution during ontogenesis of Tagetes patula', in G. Lam, H. Breteler, T. Arnason, L. Hansen (eds.), Chemistry and Biology of Naturally Occurring Acetylenes and Related Compounds, Elsevier, Amsterdam, pp. 209-216.

AQUAPORINS: WATER CHANNEL PROTEINS IN THE PLASMA MEMBRANE AND THE TONOPLAST[1]

M. J. CHRISPEELS, M. J. DANIELS AND C. MAUREL[2]
Department of Biology 0116
University of California, San Diego
9500 Gilman Drive
La Jolla, CA 92093-0116

ABSTRACT. The plasma membrane and the vacuolar membrane (tonoplast) contain water-selective channel proteins called aquaporins. These 27 kDa proteins function as monomers and facilitate osmotic pressure-driven water transport through the membranes. The presence of aquaporins in a membrane increases its water permeability 5 to 10-fold.

Osmotic effects observed with living cells indicate that their plasma membranes are freely permeable to water while essentially creating a barrier to other molecules. The hydraulic conductivity of biological membranes is sometimes still ascribed to the simple diffusion of water molecules through the lipid bilayer. However, more than 30 years ago the idea was advanced that hydraulic water movement through living cells occurs by bulk flow of water through pores in the membrane (Sidel and Solomon, 1957) Certain membranes of animal cells are unusually permeable to water and there is now a substantial body of evidence for the existence of water transport channels in such membranes (reviewed by Finkelstein, 1987, Verkman, 1992). For example, membranes from red blood cells and renal proximal tubules are exceptionally permeable to water; the membranes of the convoluted distal tubules of the kidney are also highly water permeable, and water permeability can be modulated by hormones. The hydraulic conductivity of plant cells has been thoroughly investigated and numerous reviews on this subject have been published (for a recent review, see Steudle, 1992). Many observations on plants and animals support the recent discovery of proteins that form water channels (Fushimi et al., 1993, Maurel et al., 1993, Preston et al., 1992). We proposed to call such proteins "aquaporins" (Agre et al., 1993). These water channel proteins belong to the MIP (MIP = major intrinsic protein) family, an ancient family of membrane proteins (see Reizer et al., 1993 for review). Aquaporins form water-selective channels, allowing water to pass freely, while excluding ions and metabolites. In plants, aquaporins have been demonstrated in the tonoplast (vacuolar membrane) (Höfte et al., 1992) and more recently in the plasma membrane (Daniels and Chrispeels, unpublished results). In animal cells, aquaporins are found in the plasma membranes of specific cell types. It is important to note

[1]Large portions of this paper have been reprinted from Chrispeels, M.J. and C. Maurel (1994), Plant Physiol. 105:9-15, with permission of the copyright holder, the American Society of Plant Physiologists; [2]Present address: Institut des Sciences Végétales, CNRS, 91198 Gif-Sur-Yvette, France.

M. Terzi et al. (eds.), Current Issues in Plant Molecular and Cellular Biology, 623-631.

that such channels permit or facilitate the movement of water through membranes and do not act as pumps. The driving forces behind water movement are hydraulic or osmotic in nature.

Aquaporins are Members of an Ancient Family of Channel Proteins

In earlier reports, we and others have described in plants a family of integral membrane proteins that has cognates in mammals, yeasts and bacteria and is part of the larger MIP family (MIP = major intrinsic protein) (see Reizer et al., 1993 for a recent review). The polypeptide chains of all the MIP proteins span the membrane six times and have amino- and carboxy-termini that face the cytoplasm. In plants, this family is represented by several tonoplast intrinsic proteins, such as α-TIP and γ-TIP, the soybean nodule protein NOD26, proteins such as Rtob7 that are specifically expressed in roots, and proteins that are induced by specific stresses such as irradiation or dehydration, or by darkness.

The functions of many MIP proteins are still unknown, although most of the proteins have been postulated to be involved in transport processes. Several MIP proteins, such as plant γ-TIP and CHIP28 have been shown to be aquaporins, whereas GlpF and MIP itself transport small polyols and ions, respectively. Other members of the MIP family have properties or cellular locations that are suggestive of a transport function: NOD26 is in the peribacteroid membrane of soybean nodules and may allow an exchange of metabolites between the bacteroids and the cytoplasm; α-TIP is found in the protein body membranes of seeds and may play a role in transport processes into or out of the protein bodies during seed development or germination. BIB, the big brain protein of *Drosophila melanogaster* is required for normal brain development and may play a role in cell — cell communication.

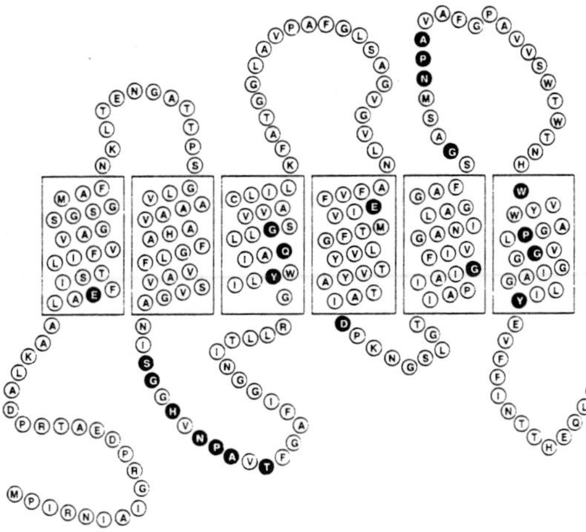

Figure 1. Schematic representation of aquaporin γ-TIP in the tonoplast showing the six putative transmembrane domains. The orientation of the protein is based primarily on extensive studies with the bovine lens protein MIP; the orientation of the sixth transmembrane domain has been confirmed for the seed protein α-TIP. The residues that are conserved in all members of the MIP family are shown as bold circles.

Phylogenetic trees depict the relatedness of the various MIP family proteins as well as the clustering patterns. On such a tree the eight plant proteins and the five animal (mammalian and *Drosophila*) proteins form respectively two clusters, whereas the yeast and bacterial proteins appear less related (Reizer et al., 1993). A phylogenetic tree of the plant proteins shows that proteins with similar patterns of expression such as the seed-specific α-TIPs or the desiccation-induced proteins cluster together (Höfte et al., 1992). Eukaryotic species contain several MIP genes — eight full-length TIP sequences have already been obtained from *Arabidopsis thaliana*, with more probably to come — whereas only a single MIP gene (GLP) has ever been found in bacteria. This led Reizer *et al.* (1993) to speculate that a single MIP gene was vertically transmitted from the prokaryotes to the different eukaryotic kingdoms, and that these genes then duplicated and diverged to yield the different subfamilies.

The Existence of Water Channel Proteins has been Known for Many Years

Bulk flow of water across a membrane occurs in response to an osmotic or hydrostatic gradient. Osmotic water permeability is readily measured in small vesicles or cells by the stopped-flow light scattering technique, a method that relies on the dependence of light scattering on vesicle or cell volume, and is used to quantitate the time course of net water flow which occurs in response to transmembrane osmotic gradients. The osmotic gradients are established by adding an impermeant solute, to the external solution. With the help of other chemical and physical methods (tracer fluxes, ^{1}H-NMR) to measure diffusional and osmotic water transport across biological membranes, biophysicists and cell physiologists have obtained evidence for the existence of facilitated or channel-mediated water transport in several membranes (Finkelstein, 1987, Macey, 1984). Membranes with facilitated water transport share a number of properties (Macey, 1984, Verkman, 1992) that generally support, but do not prove the presence of water channels. For example, water transport across these membranes is inhibited by mercurial sulfhydryl reagents, demonstrating the existence of proteinaceous components in water channels, and the functional unit of the water channel in kidney tubules and red blood cells is 30 kDa, as determined by radiation inactivation (van Hoek et al., 1991).

In spite of this accumulated knowledge about the channels that facilitate water movement, until recently no one succeeded in identifying or isolating any candidate proteins. Expression of mRNAs in *Xenopus* oocytes, led to the identification of water channel proteins in red blood cells (Preston et al., 1992) kidney tubules (Fushimi et al., 1993, Zhang et al., 1993b) and plants (Maurel et al., 1993).

The Activity of Aquaporins can Readily be Demonstrated by Swelling Assays with *Xenopus* Oocytes

Xenopus oocytes are very large cells, measuring 600 to 1300 μm in diameter, that can be micromanipulated with relative ease. This and their high RNA translation capacity make it possible to inject oocytes with exogenous mRNA to study the expression of heterologous proteins. Because the plasma membrane of the oocyte appears to have a low water permeability, these cells are suitable to study the activity of water channel proteins. The common procedure to assay for aquaporin activity is to shift oocytes two to three days after injecting them with aquaporin mRNA, to a hypotonic culture medium by diluting the medium three-fold and to measure the influx of water by determining the changes in cell volume after the shift. The presence of aquaporin in the plasma membrane increases the water permeability Pf from 0.1 to $1.0 - 2.0 \times 10^{-2}$ cm/s (Figure 2). This causes the oocytes to increase in volume by 50% and rupture within 2 to 6 min of hypotonic exposure, whereas control oocytes burst injecting them with aquaporin mRNA, to a hypotonic culture medium by diluting the medium three-fold and to measure the influx of water by determining the changes in cell volume after the shift. The presence of aquaporin in the plasma membrane increases the water permeability Pf from 0.1 to 1.0 - 2.0 × 10^{-2} cm/s (Figure 2). This causes the oocytes to increase in volume by 50% and rupture within 2 to 6 min of hypotonic exposure, whereas control oocytes burst after 45-60 minutes. It is the large size of the oocytes that makes these observations easy. The diameter of an oocyte is 50 times larger than that of a vacuole and 200 times larger than that of a red blood cell. Since the surface to volume ratio decreases with increased cell size, and assuming similar rates of water uptake per unit area of membrane, then a vacuole would burst in 5 to 6 seconds and a red cell in about one second under conditions where oocytes take several minutes to swell and burst.

The mRNAs that have been identified with the oocyte swelling assay as encoding high activity water channels all encode proteins that are members of the MIP family. Significantly, mercurial sulfhydryl reagents inhibit the increase in Pf caused by the presence of the aquaporins, just as they inhibit the permeability of red cell and kidney cell membranes. In addition, all aquaporins have a MW of approximately 27 kDa close to the size of the functional unit of the water channel as determined by radiation inactivation. Together these observations strongly support the conclusion that the aquaporins are the proteins that permit the rapid flow of water through biological membranes.

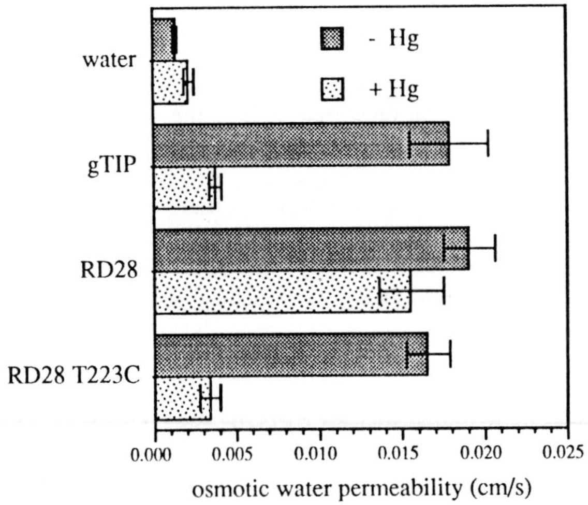

Figure 2. Osmotic water permeability of *Xenopus* oocytes measured in the absence or the presence of mercuric chloride (Hg) after injection with mRNAs for γ-TIP (g-TIP), RD28 or a mutant of RD28 in which threonine 223 is changed to a cysteine residue. γ-TIP is mercury sensitive, RD28 is mercury insensitive and the T223C mutant of RD28 is mercury sensitive.

Aquaporins Form Water-Selective Channels

Swelling of oocytes in the manner described above has been obtained with mRNAs encoding the animal proteins CHIP28 (Preston et al., 1992, Zhang et al., 1993a) and WChP (Fushimi et al., 1993), as well as two plant proteins, γ-TIP (Maurel et al., 1993) and RD28 (Chrispeels et al., 1993). The water transport property of CHIP28 has been confirmed with liposomes containing purified CHIP28 had been incorporated, but such reconstitution experiments have not yet been done for any of the plant proteins.

Efforts to demonstrate the existence of transport activities associated with aquaporins other than water transport have generally been fruitless. In some of these experiments, oocytes that express aquaporins were used for voltage clamp studies or to measure uptake of radioactive metabolites. No ions or metabolites have been found to be co-transported with water (Maurel et al., 1993). Conversely, homologs such as GlpF that transport glycerol, do not transport water. Expression of GlpF in oocytes allows them to take up glycerol in accordance with the known function of GlpF, but does not induce swelling when the oocytes are shifted to a hypo-osmotic medium (Maurel et al., 1994). It appears therefore that aquaporins are channels highly selective for water. However, other channel proteins have been shown to transport water and ions together. Electroosmotic couplings of up to 200 moles of water per mole of cation have been reported for *Nitella* (see Lüttge and Higinbotham (1979) pp. 101-102). Red beet vacuoles have a stretch-activated channel that is sensitive to osmotic pressure and may mediate cation and osmotically-driven water fluxes (Alexandre and Lassalles, 1992).

What is the Molecular Mechanism of Water Transport?

Amino acid sequence comparisons of all members of the MIP family show that there are 22 amino acids that are conserved in all sequences and these are shown in dark circles in Figure 1. These amino acids include a membrane-embedded glutamate (E) in transmembrane segments 1 and 4, a sequence of asparagine-proline-alanine (NPA) in the loops between transmembrane segments 2 and 3 as well as transmembrane segments 5 and 6, (these loops are on opposite sides of the membrane) and a glycine (G) in the middle of transmembrane segments 3 and 6 (Reizer et al., 1993 and Figure 1). A detailed computer analysis of the first and second halves of many MIP proteins shows that the full sequence probably arose by an intragenic duplication event (Reizer et al., 1993). Most interesting is that the two similar halves have opposite orientations in the membrane: NPA is inside in the first half of the molecule, but outside in the second half. This generates a symmetry of the channel with respect to both sides of the membrane.

The low activation energy for water transport through membranes that contain aquaporins (E_a = 4-6 Kcal/mol) is similar to the activation energy for the self diffusion of water or for the viscous transport of water. This implies that as water molecules traverse the water channel, they encounter polar groups similar to the polar environment of the bulk solution external to the membrane (Macey, 1984). Macey (1984), postulated that the water channel should have a radial dimension that lies between 1.5 Å (the radius of a water molecule) and 2.0 Å (the radius of a urea molecule) and that water is constrained to move as a single file of molecules. The ratio of the osmotic to the diffusional permeability indicates that five to nine water molecules should be in the rate-limiting portion of the channel (Finkelstein, 1987).

Water channels are characteristically inhibited by mercury derivatives and this inhibition is reversed by reducing reagents (Macey, 1984). This is also the case for the swelling of oocytes that express plant or animal aquaporins (γ-TIP or CHIP28). Such a result suggests the functional importance of cysteine or methionine residue(s). CHIP has several cysteine or methionine residues, and the mercury-sensitive amino acid residue of CHIP28 has recently been identified as Cys189 (Preston et al., 1992, Zhang et al., 1993b). This residue is located adjacent to the conserved NPA motif in the loop between membrane spanning domains 5 and 6. Substitution of Cys189 in CHIP28 by a serine residue creates an active water channel that cannot be inhibited by mercury (Preston et al., 1992). Replacement of Cys189 by large amino acids such as valine or tryptophane abrogates water transport. Of the four aquaporins that have been tested (CHIP28, WChP, γ-TIP and RD28), only RD28 is insensitive to mercury, and it has no cysteine or methionine residue close to the second conserved NPA motif (Yamaguchi-Shinozaki et al., 1992). Introduction of a cys in this position in RD28 creates a mercury sensitive water channel (Daniels and Chrispeels, unpublished results).

Both MIP26 and CHIP28 (Smith and Agre, 1991) form tetramers, but the monomer appears to be the functional unit of water transport. This is supported by radiation inactivation data (van Hoek et al., 1991) and by the coexpression of CHIP28 and a non-functional mutant, such as the Cys189→Tyr189 mutation. Mutant inactive protein is not negatively dominant and the rate of water transport is proportional to the amount of active protein (Preston et al., 1992, Zhang et al., 1993b).

Expression and Subcellular Location of Plant Aquaporins

The expression pattern of a gene and the subcellular location of the protein it encodes can often give additional clues about the physiological role of the protein. The aquaporin γ-TIP is located in the tonoplast and the gene is most highly expressed either during or immediately after cell elongation (Ludevid et al., 1992). Expression is also higher in the vascular bundles than in the leaf parenchyma or cortex (Ludevid et al., 1992). There is no expression in the root and stem meristems, although these cells have numerous small vacuoles. A detailed study is needed to describe exactly when γ-TIP accumulates: during the vacuolation process or immediately thereafter. The cells of growing tissues require a considerable supply of water and have a high hydraulic conductivity (Cosgrove and Steudle, 1981, Steudle and Boyer, 1985), and this finding is in agreement with the high expression of γ-TIP in such cells.

Yamaguchi-Shinozaki et al. (1992) identified a MIP homolog (RD28) in *A. thaliana* whose mRNA is induced by desiccation. A similar sequence was identified earlier in peas by Guerrero (1990). We demonstrated that the *A. thaliana* sequence encodes a mercury insensitive aquaporin (Figure 2). Antibodies to the C-terminal peptide of this protein react with a 27 kD protein found in roots, leaves, stems and siliques of plants grown under normal conditions, suggesting that the presence of the protein is not dependent on desiccation of the plants (Daniels and Chrispeels, unpublished).

Subcellular fractionation using the two-phase polymer system to obtain pure plasma membranes indicate that RD28 is a plasma membrane protein. Thus, aquaporins are constitutively expressed in the tonoplast and in the plasma membrane.

Do Aquaporins Play a Role in Water Movement within the Plant?

In most plants, a transpiration stream, that starts with the uptake of water from the soil and ends with the loss of water vapor from the leaves, moves through the plant during daylight hours. Movement through the xylem vessels obviously does not involve aquaporins since the vessel elements have no membranes, but how does water get into the xylem and how does it flow from the xylem to and through other cells? There are two possible pathways for osmotic flow of water between tissues: an apoplastic route that encompasses only the cell walls, and a symplastic and transcellular path that involves the cytoplasm as well as the vacuoles (Figure 3). This distinction is somewhat artificial because it is likely that water moves via both paths at all times. Measurements obtained with pressure probes indicate that the preferred route may depend on the species, the plant organ, the physiological condition of the plant, as well as the driving force (hydrostatic or osmotic pressure). In maize and cotton roots, apoplastic transport dominates but in barley and bean, transport is mostly cell-to-cell (see Steudle, 1992) for review and references therein). Plasma membranes are thought to be the primary impediments to water flow, with the role of plasmodesmata still poorly understood. The presence of aquaporins in the tonoplast increases the effective cellular cross-section through which water flows freely once it has passed through the plasma membrane.

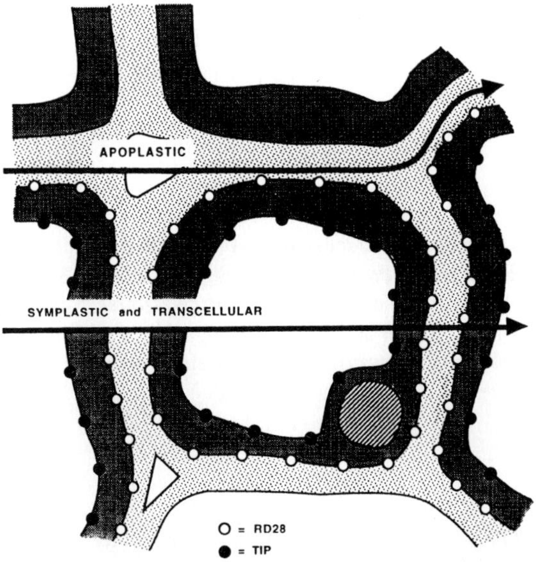

Figure 3. Possible routes for hydraulic water flow through a living tissue: apoplastic and symplastic/transcellular. TIP, shown in the tonoplast, is constitutively expressed, whereas RD28, shown in the plasma membrane is only expressed as a result of desiccation.

Water flow through the cells could be regulated in two different ways: first, by altering the activity of the individual water channel proteins, and second, by changing the abundance of these proteins in the membranes. Steudle and Boyer (1985) suggested that the rate of water movement out of the xylem may be determined by the small cells that surround the vessel elements. Thus, the control of water flow may operate only in those cells, and the expression of γ-TIP as well as the induction of RD28 may be higher in the vascular bundles than in other cells. Results obtained with promoter-GUS fusions with the γ-TIP promoter (Ludevid et al., 1992) are consistent with this interpretation. The presence of aquaporins in the cortex and mesophyll could also significantly reduce the resistance to water flow through the plant.

Water channels may be involved in other processes related to water absorption and flow within the plant tissues. Processes, such as pollen or seed imbibition and germination, stomatal opening and closing will have to be revisited in the light of recent evidence for water channels in plants. The cells of growing tissues require a considerable supply of water and have high Lp values. Expression of aquaporins may be highest in these cells. The use of the existing aquaporin molecular probes, coupled with the *Xenopus* oocyte swelling assay will aid the discovery of new plant aquaporins.

Where Do we Go from Here?

The identification of aquaporins has opened an exciting new chapter in plant-water relations. We can envision how water flow within the plant could be modified by altering the regulation, expression or subcellular location of specific aquaporins. We now need to establish that aquaporins are the molecules that permit and modulate cell-to-cell water flow and the hydraulic conductivity of the plant. This can be done with transgenic plants that express different forms of these proteins (e.g., mercury sensitive and mercury insensitive) in different membranes. By combining molecular biology with plant physiology, it should be possible to determine the role that aquaporins play in water transport in the plant. It will be of major interest to determine what is the limiting factor: the generation of water potential gradients or the hydraulic conductivity of the cellular membranes.

Acknowledgements

We thank Drs. H. Höfte and J. Reizer for stimulating discussions about TIP proteins and aquaporins and Dr. Ernst Steudle and Ted Hsiao for their critical reading of the manuscript. Research in our laboratory has been consistently supported by the National Science Foundation, the United States Department of Agriculture and the U.S. Department of Energy. C. Maurel was supported by a long term EMBO fellowship. M. Daniels is a predoctoral trainee of the National Institutes of Health.

References

Agre, P., Sasaki, S. and Chrispeels, M. J. (1993) 'Aquaporins — A family of water channel proteins', Am. J. Physiol., 265, F461.

Alexandre, J. and Lassalles, J.-P. (1992) 'Hydrostatic and osmotic pressure activated channel in plant vacuole', Biophys. J., 60, 1326-1336.

Chrispeels, M. J., Maurel, C., Daniels, M. J. and Mirkov, E. T. (1993) 'The vacuolar membrane of plant cells has an aquaporin or water channel protein', Mol. Biol., 4, 425a.

Cosgrove, D. J. and Steudle, E. (1981) 'Water relations of growing pea epicotyl segments', Planta, 153, 343-350.

Finkelstein, A. (1987) 'Water movement through lipid bilayers, pores, and plasma membranes. Theory and reality', Distinguished Lecture Series of the Society of General Physiologists, John Wiley & Sons, Inc., pp. 1-228.

Fushimi, K., Uchida, S., Hara, Y., Hirata, Y., Marumo, F. and Sasaki, S. (1993) 'Cloning and expression of apical membrane water channel of rat kidney collecting tubule', Nature, 361, 549-552.

Guerrero, F. D., Jones, J. T. and Mullet, J. E. (1990) 'Turgor-responsive gene transcription and RNA levels increase rapidly when pea shoots are wilted. Sequence and expression of three inducible genes', Plant Mol. Biol., 15, 11-26.

Höfte, H., Hubbard, L., Reizer, J., Ludevid, D., Herman, E. M. and Chrispeels, M. J. (1992) 'Vegetative and seed-specific isoforms of a putative solute transporter in the tonoplast of *Arabidopsis thaliana*', Plant Physiol., 99, 561-570.

Ludevid, D., Höfte, H., Himelblau, E. and Chrispeels, M. J. (1992) 'The expression pattern of the tonoplast intrinsic protein γ–TIP in *Arabidopsis thaliana* is correlated with cell enlargement', Plant Physiol., 100, 1633-1639.

Macey, R. I. (1984) 'Transport of water and urea in red blood cells', Amer. J. Physiol., 246, C195-C203.

Maurel, C., Reizer, J., Schroeder, J. I. and Chrispeels, M. J. (1993) 'The vacuolar membrane protein γ-TIP creates water specific channels in *Xenopus* oocytes', EMBO J., 12, 2241-2247.

Maurel, C., Reizer, J., Schroeder, J. I., Chrispeels, M. J. and Saier, M. H., Jr. (1994) 'Functional characterization of the *Escherichia coli* glycerol facilitator GlpF in *Xenopus* oocytes', J. Biol. Chem., in press.

Preston, G. M., Carroll, T. P., Guggino, W. B. and Agre, P. (1992) 'Appearance of water channels in *Xenopus* oöcytes expressing red cell CHIP28 protein', Science, 256, 385-387.

Reizer, J., Reizer, A. and Saier, M. H., Jr. (1993) 'The MIP family of integral membrane channel proteins: Sequence comparisons, evolutionary relationships, reconstructed pathway of evolution, and proposed functional differentiation of the two repeated halves of the proteins', Crit. Rev. Biochem. Mol. Biol., 28, 235-257.

Sidel, V. W. and Solomon, A. K. (1957) 'Entrance of water into human red cells under an osmotic pressure gradient', J. Gen. Physiol., 41, 243-257.

Smith, B. L. and Agre, P. (1991) 'Erythrocyte Mr 28,000 transmembrane protein exists as multisubunit oligomer similar to channel proteins', J. Biol. Chem., 266, 6407-6415.

Steudle, E. (1992) 'The biophysics of plant water; compartmentation, coupling with metabolic processes, and water flow in plant roots', in Somero, G. N., Osmond, C. B. and Bolis, C. L. (eds.), Water and Life: A Comparative Analysis of Water Relationships at the Organismic, Cellular and Molecular Levels, Springer-Verlag, Berlin, pp. 173-204.

Steudle, E. and Boyer, J. S. (1985) 'Hydraulic resistance to radial water flow in growing hypocotyl of soybean measured by a new pressure-perfusion technique', Planta, 164, 189-200.

van Hoek, A. N., Hom, M. L., Luthjens, L. H., de Jong, M. D., Dempster, J. A. and van Os, C. H. (1991) 'Functional unit of 30 kDa for proximal tubule water channels as revealed by radiation inactivation', J. Biol. Chem., 266, 16633-16635.

Verkman, A. S. (1992) 'Water channels in cell membranes', Annu. Rev. Physiol., 54, 97-108.

Yamaguchi-Shinozaki, K., Koizumi, M., Urao, S. and Shinozaki, K. (1992) 'Molecular cloning and characterization of 9 cDNAs for genes that are responsive to desiccation in *Arabidopsis thaliana*: Sequence analysis of one cDNA clone that encodes a putative transmembrane channel protein', Plant Cell Physiol., 33, 217-224.

Zhang, R., Skach, W., Hasegawa, H., van Hoek, A. N. and Verkman, A. S. (1993a) 'Cloning, functional analysis and cell localization of a kidney proximal tubule water transporter homologous to CHIP28', J. Cell Biol., 120, 359-369.

Zhang, R., van Hoek, A. N., Biwersi, J. and Verkman, A. S. (1993b) 'A point mutation at cysteine 189 blocks the water permeability of rat kidney water channel CHIP28k', Biochemistry, 32, 2938-2941.

THE EFFECT OF BREFELDIN A ON THE GOLGI APPARATUS IN NORWAY SPRUCE CELLS

O. GORBATENKO & I. HAKMAN
Department of Botany
Stockholm University
S-106 91 Stockholm
Sweden

ABSTRACT. The effect of the fungal antibiotic Brefeldin A (BFA), was studied in two different cell systems of Norway spruce (Picea abies): embryogenic tissue cultures and root meristems. We analyzed the Golgi apparatus (GA) ultrastructurally with TEM in tissues after exposure to 50, 100 and 200 µg/mL of BFA for 1 and 3 hr. Striking morphological modifications of the GA occurred in both cell systems. Often the GAs were seen to have: reduced number of cisternae; curled cisternae, which also became very electron-dense, narrow and elongated; abnormal-looking cisternae closely positioned between two ER strands; and cisternae fragmentation. These abnormalities were more pronounced in cells treated with the higher concentrations of BFA, and in the tissue culture system. Such cells, after treatment with 200 µg/mL BFA, had almost completely distorted GAs, and many clusters with a large number of vesicles present. After a recovery period of 1 hr, the observed effects of BFA were reversed. When comparing the extent of changes of the GA in the two tissue systems, we conclude that the cultured cells are more sensitive to BFA. Other organelles seemed not to be affected by BFA.

1. Introduction

There are still many problems to be solved before the biogenesis of the Golgi apparatus (GA) will be established. The GA plays a central role in the endomembrane system of plant and animal cells with a major function to biochemically process molecules, received from ER, and then dispatch them to their final distinations [10].

Recently, the relationship between the control of membrane traffic and the maintenance of organelle structure has been investigated with the use of Brefeldin A (BFA) [7], an antibiotic synthesised by a variety of fungi.

It has been found that BFA has profound and dramatic effects on the secretory pathway in mammalian cells [6, 8], by inhibiting secretion. BFA also causes massive morphological changes of endomembrane structures in mammalian cells, where typically the GA disintegrates by the formation of tubules, and many Golgi enzymes are redistributed to the ER [9]. In plant cells the dissociation of the Golgi apparatus induced by BFA has been shown to be associated with a redistribution of Golgi glycoproteins into discrete and vesiculated cytoplasmic compartments [11]. All these processes, both in mammalian and plant cells, are reversible. The morphological transformation of GAs in response to BFA suggests that Golgi bodies themselves exist at a fine balance maintained by the assembly of cytosolic proteins. BFA, therefore, provides a powerful tool for studying Golgi dynamic and function.

The aim of the present work was to investigate the effect of BFA on the GA in two different cell systems in Norway spruce (Picea abies): embryogenic tissue cultures and root meristems.

M. Terzi et al. (eds.), Current Issues in Plant Molecular and Cellular Biology, 633-638.

2. Materials and methods

2.1. PLANT MATERIAL

2.1.1. *Cultured Cells.* Embryogenic tissue cultures of Norway spruce (Picea abies (L.) Karst.) were initiated in September 1991 and maintained on half-strength LP medium (LP 1/2 + N) containing 1% sucrose, 2 mg/L NAA, 1 mg/L BA, 0.3% Phytagel, as described earlier [4]. Cell-line 91-1v was used in this study. This cell-line was chosen among other cell-lines after a series of experiments where the cell-lines´ embryo reproducibility was tested.

2.1.2. *Root-tip Cells.* Norway spruce seeds (Emmaboda) were grown for 10 days in a day-light regime on moist vermiculite at room temperature, before root tips were excised.

2.2. BREFELDIN A TREATMENT

Brefeldin A (BFA) (Epicentre Technologies, Madison, WI, USA) was dissolved in ethanol and, as a 10 mg/mL stock solution, stored at -20°C.

In order to test the sensitivity of the different cell-systems for BFA, different concentrations of BFA, 50, 100 and 200 μg/mL, in liquid DCR-medium [3] was applied to the tissues.

Root tips cut from 10-day-old seedlings and pieces of embryogenic calli ($1mm^3$) were immersed in BFA-solution for 1 or 3 hr. Controls consisted of tissues subjected to DCR-medium.

In some experiments the reversibility of BFA-effects was examined. For these experiments, root tips and calli were carefully washed after the BFA-treatment and left to recover for 1 hr in DCR-media before fixation for microscopy.

2.3. ELECTRON MICROSCOPY

For general ultrastructural observations untreated and BFA-treated root tips and cultured cells were EPON-embedded essentially as described in [6].

Ultra-thin sections were cut with a diamond knife JUMDI (JUM, Stockholm, Sweden) on an ultramicrotom 2088 Ultratome RV (LKB, Bromma, Sweden), collected on formvar-coated copper mesh grids and stained with uranyl acetate for 15 min and lead citrate for 3-5 min. Specimens were examined at 60 kV in a ZEISS EM10 transmission electron microscope.

3. Results

3.1. BFA-TREATMENT

To examine the effects of BFA on the general architecture of the plant cells, we have analysed the ultrastructural changes in EPON-embedded embryogenic tissue cultured cells and root tip cells of Picea abies, that were exposed to 50, 100 and 200 μg/mL of BFA for 1 or 3 hr. These concentrations were chosen based on the effects of BFA on plant cells as described in [5, 11].

Typical morphology of the Golgi stacks of the untreated embryogenic tissue cultured cells is illustrated in Figure 1. In these cells the Golgi stacks consisted of 4 to 7 flattened cisternae. The discrete morphological differences in staining pattern of *cis, medial,* and *trans* Golgi cisternae and the TGN is evident, also the general morphology of the ER is clearly distinguished. The dispersed organisation of Golgi stacks, typical for plant cells, was observed (not shown), although they were unequally distributed within the cells.

A few striking morphological modifications were clearly observed in BFA-treated embryogenic tissue cultured cells and root tip cells.

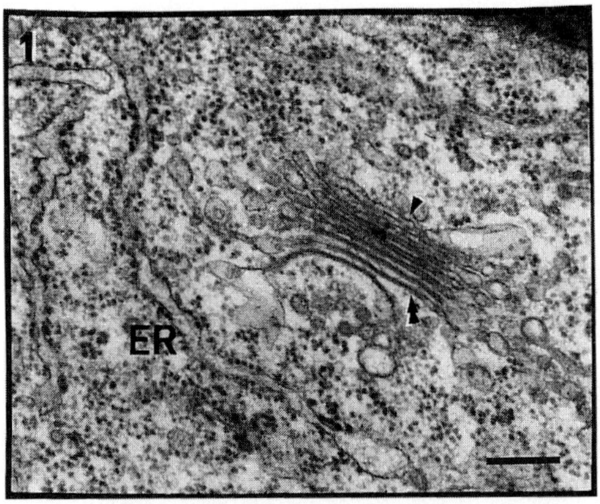

Fig. 1. Typical morphology of the Golgi stacks in embryogenic tissue cultured cells; arrowhead - *cis*-cisternae; double-arrowhead - *trans*-cisternae; ER - endoplasmic reticulum. Bar = 0.2 µm.

3.1.1. *Embryogenic tissue culture.* After treatment with 50 µg/mL of BFA the number of the Golgi stacks in the cells had not changed much in comparison to the control, although almost all of the stacks exhibited an abnormal arrangement. Typical changes were: curled cisternae, a reduced number of cisternae that often also had become longer (up to 4-5 times) than those seen in the control cells (c. f. Figures 1 and 2 a,b). These elongated cisternae had a similar staining feature as *trans* cisternae and showed fragmentation in the form of electron-dense vesicles. Clusters of vesicles also occurred. Often close associations of prolonged cisternae with one or two ER strands were observed (Figs. 2 a,b).

Cells treated with 100 µg/mL BFA exhibited more extensive Golgi modifications. They had fewer Golgi stacks, many clusters of vesicles that were both homogene (Fig. 2 c) and heterogene (Fig. 2 d) in vesicular composition. The morphological features of Golgi stacks in this experiment were otherwise similar to cells treated with 50 µg/mL BFA.

Even more pronounced effects on Golgi morphology was found in cells treated with 200 µg/mL BFA. In these cells Golgi stacks were rarely observed and the occurrence of the clusters of vesicles was enhanced.

3.2. RECOVERY FOLLOWING BFA TREATMENT

After a recovery treatment from BFA for 1 hr the typical Golgi structure was restored. However, the degree of restoration correlated with the BFA concentration used in the previous treatment experiment.

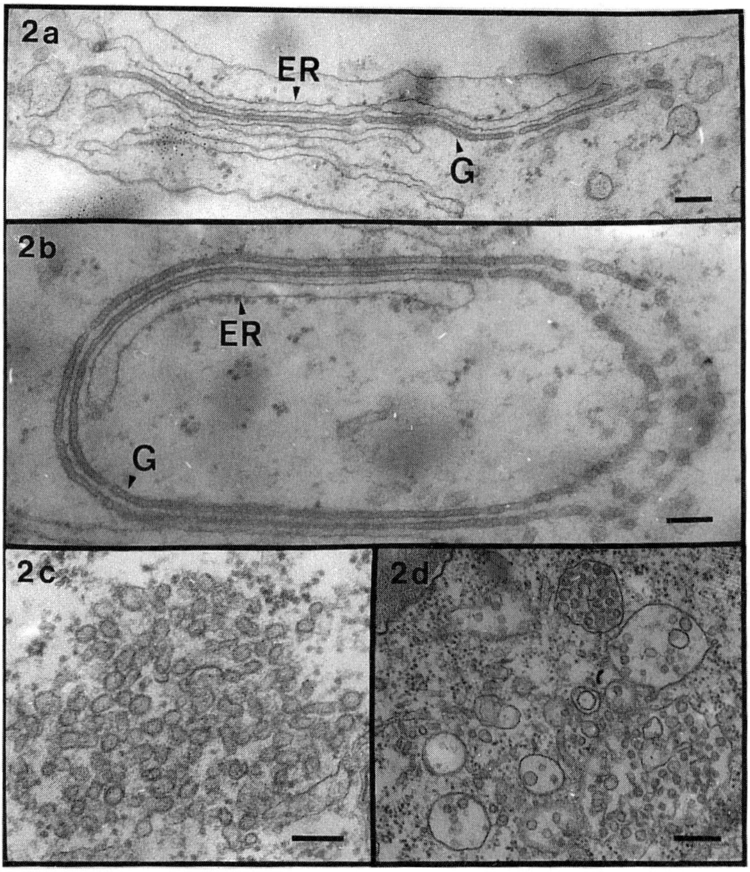

Figs. 2. Tissue cultured cells after BFA-treatment.

2 a. Close association of an elongated Golgi cisternae (G) with strands of the endoplasmic reticulum (ER), (BFA 50 μg/mL). Bar = 0.1 μm.

2 b. Curved and partially vesiculated Golgi body (G), (BFA 50 μg/mL). Bar = 0.1 μm.

2 c. Homogene vesicle cluster, (BFA 100 μg/mL). Bar = 0.1 μm.

2 d. Heterogene vesicle cluster, (BFA 100 μg/mL). Bar = 0.2 μm.

Cells treated with 50 μg/mL BFA and then recovered showed many Golgi stacks with a morphology that was typical for control cells. Correlation between ER and Golgi distribution in these cells was also similar to control cells. Some cells were seen to have ribosomes regularly associated with the ER, and many multivesicular bodies (MVB), often close to the plasma membrane (not shown). At higher BFA concentrations the number of restored Golgi stacks was much lower. In 200 μg/mL BFA-treated and recovered cells, still many curled Golgi cisternae and clusters of vesicles could be observed.

In some experiments embryogenic tissue cultured cells were treated with BFA (50, 100 and 200 μg/mL) for 1 hr. Nearly the same extent of morphological changes of the Golgi stacks occurred after 1 and 3 hr.

3.2.1. *Root tip cells.* BFA-treated root tip cells showed similar Golgi structure modifications as the BFA-treated embryogenic tissue cultured cells, although the extent of these modifications were less in the root tip cells. For example, similar Golgi arrangements as observed in cultured cells after treatment with 50 μg/mL BFA, were only observed in root tip cells after treatment with 100 or 200 μg/mL BFA. BFA-recovery experiments with root tip cells also revealed much fewer Golgi. Instead many Golgi, typical for BFA-treated cells, were still present.

4. Discussion

The present work demonstrates profound effects of BFA on Golgi in Picea abies, both in embryogenic tissue cultured cells and in root tip cells. The results agree with those previously reported in other plant cell systems [1, 11].

Brefeldin A generally induces morphological changes followed by complete disintegration of the Golgi bodies. However, the fate of Golgi membrane components or secretory products usually sorted by the Golgi remains unknown. Perhaps, the destiny of Golgi membrane components can be detected by using antibodies against plant Golgi specific structural proteins.

In plant cells, BFA has so far not been observed to induce tubular connections between Golgi and ER followed by retrograde transport of membrane from Golgi to ER (agrees with our result), as in animal cell system [8]. Driouich and colleagues [1] postulated that plant Golgi stacks are less likely to be resorbed into the ER both because their association with the ER is of a more transient nature and that they have an inherently more stable structure than their animal counterparts. There is only one natural occurrence of plant Golgi stack dissociation reported in the literature, which is during seed desiccation [2]. Therefore, since desiccation is one of the developmental processes during conifer embryogeny that we are interested in, these question will be further explored.

In our studies we could not see any close spatial relationship between ER and Golgi in control cells. However, close association of Golgi with one or often two ER strands has been revealed in BFA-treated cells. It seems not to be an occasional event since many such arrangements were seen in the BFA-treated cells. The nature of this relationship between ER and Golgi stacks is unknown.

The mechanism responsible for keeping the Golgi cisternae together, leading to Golgi structural stability, is also not known, although it has been suggested that intercisternal elements, as found in slime secreting root tip cells, may play such a role [12].

From our investigation, BFA-induced modifications in Golgi structure appeared similar in cultured cells and root tip cells. However, the intensity of the response to BFA-treatment differed in these two cell systems. The degree of morphological modifications also varied slightly between cells in the same sample. Generally, root tip cells showed less sensitivity to BFA. A possible explanation for this phenomenon can be the more complex morphology of the root tips compared to the embryogenic tissue cultures and/or that they differed in uptake. Higher concentrations of BFA and/or longer exposure time were required to induce the same extent of Golgi morphological changes in root tip cells as in cultured cells.
After treatment with BFA, no significant morphological modifications in other cell organelles were detected in the present study whereas other researchers have postulate swelling of ER strands [1].

5. References

1. Driouich, A., Zhang, G.F., Staehelin, L.A. (1993) "Effect of brefeldin A on the structure of the Golgi apparatus and on the synthesis and secretion of proteins and polysaccharides in sycamore maple suspension cultured cells" Plant Physiol. 101, 1363-1373.
2. Griffing, L.R. (1991) "Comparisons of Golgi structure and dynamics in plant and animal cells" J. Elect. Microsc. Techn. 17, 179-199.
3. Gupta, P.K. and Durzan, D.J. (1985) "Shoot multiplication from mature trees Douglas-fir (Pseudotsuga menziesii) and sugar pine (Pinus lambertiana)" Plant Cell Rep. 4, 177.
4. Hakman, I., (1993) "Embryology in Norway spruce (Picea abies). An analysis of the composition of seed storage proteins and deposition of storage reserves during seed development and somatic embryogenesis" Physiol. Plant. 87, 148-159.
5. Horsley, D., Coleman, J., Evans, D., Crooks, K., Peart, J., Satiat-Jeunemaitre, B., and Hawes C., (1993) "A monoclonal antibody, JIM 84, recognizes the Golgi Apparatus and plasma membrane in plant cells" J. Exper. Bot. 44, 223-229.
6. Hunziker, W., Whitney, J.A. and Mellman, I. (1991) "Selective inhibition of transcytosis by Brefeldin A in MDCK cells" Cell 67, 617-628.
7. Klausner, R.D., Donaldson, J.G., Lippincott-Schwartz, J. (1992) "Brefeldin A: Insights into the control of membrane traffic and organelle structure" J. Cell Biol. 5, 1071-80.
8. Lippincott-Schwartz, J., Yuan, L., Tipper, C., Amherdt, M., Orci, L. and Klausner, R.D. (1991) "Brefeldin A's effects on endosomes, lysosomes, and the TGN suggest a general mechanism for regulating organelle structure and membrane traffic" Cell 67, 601-616.
9. Pelham, H.R.B. (1991) "Multiple targets for brefeldin A" Cell 67, 449-451.
10. Pfeffer, S.R. and Rothman, J.E. (1987) "Biosynthetic protein transport and sorting by the endoplasmic reticulum and Golgi" Annu. Rev. Biochem. 56, 829-852.
11. Satiat-Jeunemaitre, B. and Hawes, C. (1992) "Reversible dissociation of the plant Golgi apparatus by Brefeldin A" Biol. Cell 74, 325-328.
12. Staehelin, L.A., Giddings, T.H., Kiss, J.Z., Sack, F.D. (1990) "Macromolecular differentiation of Golgi stacks in root tips of Arabidopsis and Nicotiana seedlings as visualised in high pressure frozen and freeze-substituted samples" Protoplasma 57, 75-91.

Brefeldin A differentially affects protein secretion from suspension cultured tobacco cells (*Nicotiana tabacum* L.)

I. Kunze, C. Horstmann, R. Manteuffel, G. Kunze, and K. Müntz

Institute of Plant Genetics
and Crop Plant Research
D-06466 Gatersleben
Germany

ABSTRACT. Proteins secreted by suspension cultured tobacco cells were analysed. Nineteen clearly separated polypeptide bands (designated SN1 to SN19) between 6 and 100 kDa were reproducibly found after SDS-PAGE. Brefeldin A (BFA) strongly inhibits protein secretion including secretion of apoplastic peroxidase which was used as an indicator for BFA effects on glycosylated proteins passing through the endomembrane system. The labelling of secreted proteins is strongly decreased by short-term (20 μM BFA, up to 4 h) as well as by long-term (2 μM BFA, up to 3 d) incubation with the drug whereas neither the uptake into cells nor the incorporation of ^{35}S-methionine into cellular proteins was affected by the drug. In contrast to the majority of secretory proteins the polypeptides SN1, SN8, SN9 and SN11 were not significantly affected by the drug. The differential BFA effects can not be related to proteolysis in the culture medium. N-terminal sequences of SN9 and SN11 are similar to a region within chitin binding domains of class I chitinases of tobacco.

1. Introduction

In plant cells BFA inhibits protein secretion (Driouich et al. 1993). So far no detailled analysis of the effect of BFA on the secretion of a single specific plant protein has been reported, and it has not been investigated as to whether the transport of all secreted proteins is decreased, or whether the transport of some is strongly inhibited while that of others is not affected. We investigated how on way to the cell surface an apoplastic peroxidase known to be glycosylated and to pass the endomembrane system is affected by BFA, and demonstrated that in contrast to the majority of the proteins secreted into the medium of suspension cultured tobacco cells a small number of proteins is barely affected by BFA.

2. Materials and Methods

2.1. Preparation of secretory proteins

Culture medium was taken from tobacco suspensions (*Nicotiana tabacum* L. cv. Havana) which had doubled their fresh weight after approximately 5 d of culture. About 200 ml medium were passed through a 0.22 μm Millipore filter and lyophilized. Subsequently, the dry material was dissolved in about 7 ml of 10 mM sodium phosphate buffer, pH 7.0. Insoluble lipids and low molecular compounds were removed by three extractions with hexane and subsequent gel filtration over a Sephadex G25 column. The eluate was dialysed against distilled water under sterile conditions and finally lyophilized. The cell mass corresponding to 200 ml medium amounted from 80 to 120 g fresh weight. The dye binding method of Bradford (1976) was

M. Terzi et al. (eds.), Current Issues in Plant Molecular and Cellular Biology, 639-645.

used for quantitative analysis of proteins in the cell culture medium with bovine serum albumin (BSA) as a standard protein.

2.2.Brefeldin A treatment

In order to analyse the time and concentration dependence of BFA effects upon peroxidase secretion, in long-term experiments, the cells were embedded into alginate. Analysis of secreted proteins in both long-term and short-term experiments were carried out without alginate embedding.Cells were carefully washed with RM medium, free of growth regulators and taken either for embedding or for analysis of secretory proteins.

2.3. Radioactive protein labelling

Cells were sampled by sieving and washing. Subsequently, they were resuspended and cultivated for 1 to 2 d in growth regulator-free RM medium containing 0.37 Mbq (10 µCi) ^{35}S-methionine. Proteins from 500 µl aliquotes of the culture medium were precipitated by 10 % trichloroacetic acid (TCA, final concentration), in the presence of 1 mg/ml of sodium deoxycholate (DOC). After sedimentation of the precipitate, it was washed twice with a mixture of ether and ethanol (4:1 v/v) and dissolved in sample buffer according to Laemmli.

2.4. Enzyme activity determination

Peroxidase: Activity of peroxidase was used as a marker for secretory proteins, and assayed according to Hendriks et al. (1990). α-Mannosidase, was used as a marker for vacuolar enzymes, and assayed according to Gerhardt and Heldt (1984).

2.5. Amino acid sequence analysis

Secretory proteins were separated by SDS-PAGE as described and electro-transferred to PVDF-membrane. The membrane was stained with amidoblack and stained bands corresponding to SN9 and SN11, respectively, were excised with razor blade and loaded in a sequencer (Modell LF 3400 D/ Beckman) using special program for PVDF-type blot membranes.

3. Results

3.1. Polypeptide pattern of secretory proteins from the medium of suspension cultured tobacco cells

Proteins secreted into the suspension culture medium may be contaminated by proteins from broken cells. A vacuolar enzyme, α-mannosidase, which only appears in the medium when cells are damaged was used as an indicator for contamination (Tague and Chrispeels 1987). Medium was only used for the analysis of secretory proteins if the α-mannosidase activity did not exceed 0.1 % of the activity found in the corresponding cells.

Nineteen clearly separated polypeptide bands (designated SN1 to SN19 in Fig.1) were reproducibly found after electrophoretic fractionation of the proteins under denaturing and reducing conditions on 12.5 % polyacrylamide gels.

Autoradiography indicated that the majority of the polypeptide bands became labelled by ^{35}S-methionine during the 4 h cultivation period (Fig.2 a, lane 1). Most of the secreted proteins are glycosylated (not shown).

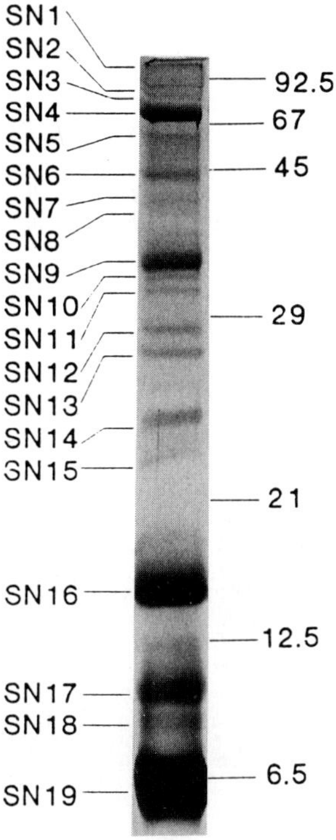

Fig. 1:Pattern of polypeptide bands obtained from secreted proteins from the medium of suspension cultured tobacco cells. SDS electrophoresis of 15 µg protein on 12.5 % polyacrylamide gel slabs under denaturing and reducing conditions. The gels were first stained with Coomassie Brillant Blue R 250 and subsequently with silver nitrate.

3.2. BFA strongly and differentially inhibits protein secretion into suspension culture medium

The labelling of secreted proteins is strongly decreased by short-term as well as by long-term incubation with the drug whereas neither the uptake into the cells nor the incorporation of ^{35}S-methionine into cellular proteins was affected by the drug (not shown). The degree of labelling depression increases in a time- and concentration-dependent manner during short-term inhibition with BFA.

Electrophoretic analysis of ^{35}S-methionine-labelled secreted proteins was performed with long-term incubation using 2 µM BFA since the effect of the drug on peroxidase secretion turned out to be completely reversible and cell damage was found to be neglegible at this concentration (not shown). After 2 d of inhibitor treatment the cells were transferred into fresh medium containing BFA and the ^{35}S-methionine where they were cultivated for an additional 1 and 2 d period, respectively. Proteins precipitated from 500 µl aliquotes of the corresponding medium were loaded into the respective gel slots.

The autoradiographs in Fig.2 indicated a differential long-term effect of the drug on secreted polypeptides as visualized by SDS-PAGE. Whereas no decrease of methionine-incorporation for bands SN8, SN9 and SN11 was seen (Fig.2b, lane 2) as compared with the BFA-free control (lane 1) low level methionine labelling indicated a strong effect on the majority of the polypeptides. Quantitative densitometric analyses revealed that under BFA influence the relative

intensity of these „non-affected" bands increased on autoradiographs from gels with secreted proteins.
After 2 h of 20 µM BFA treatment with parallel ^{35}S-methionine labelling only SN9 gave a signal on the autoradiographs of the control (not shown). Prolonging the incubation up to 4 h led to the labelling of practically all the secretory protein bands (Fig.2a, lane 1) that also appeared on the stained control gels shown in Fig.1. If the 20 µM BFA was included into the cell cultivation medium during the 4 h of incubation the majority of labelled polypeptides from secretory proteins did not appear on the autoradiographs, except the bands SN1, SN8, SN9 and SN11.

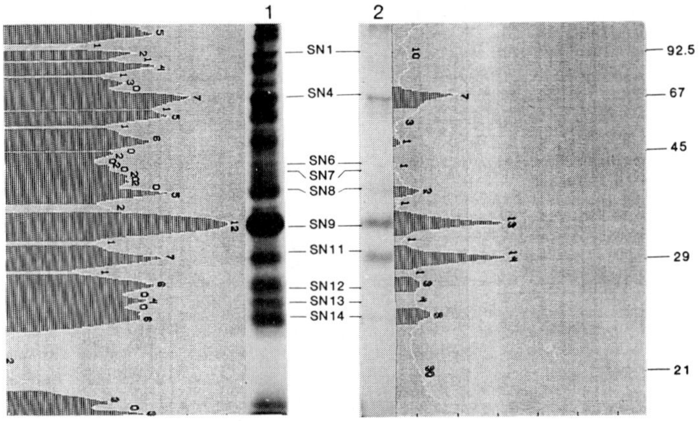

a

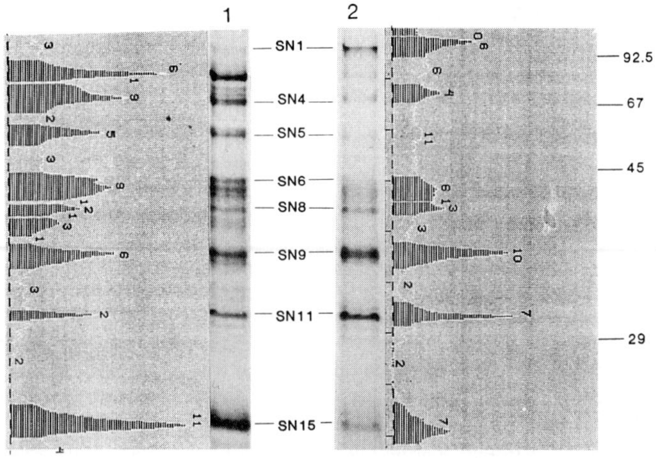

b

Fig. 2: Differential ^{35}S-methionine labelling of polypeptides from secreted proteins after BFA treatment of suspension cultured tobacco cells. Autoradiography of polypeptides that were electrophoretically fractionated under denaturing and reducing conditions. a) Comparison of banding patterns and corresponding densitometric scans, respectively, after 4 h cultivation without (lane a1) and with 20 µM BFA (lane a2). b) Comparison of banding patterns and corresponding densitometric scans after 48 h incubation without (lane b1) and with 2 μM BFA (lane b2).

3.3. BFA reversibly affects secretion of apoplastic peroxidase

Apoplastic peroxidase was choosen as an indicator for the BFA-induced inhibition of a protein known to be a glycoprotein which passes through the endomembrane system on its way to the cell surface (van Huystee 1987, Mäder 1976).

Cells were cultivated in the presence of 2 to 8 μM BFA. During about 30 h the activity of peroxidase measured in the medium decreased in a concentration dependent manner of BFA. Afterwards the cells were subcultivated without BFA over different time periods with transfer into fresh medium. No recovery of the cells from experiments with 6 and 8 μM was possible, whereas complete reversibility could be achieved at BFA concentrations of 2 and 4 μM (not shown).

3.4. Microsequencing of SN9 and SN11

The identified sequences of N-terminal amino acids of SN9 and SN11 are completely similar to a region within chitin binding domains of CHI1 and CHI2 of tobacco (Fig. 3), both belonging to class I chitinases. Class I chitinases include basic endochitinases normally located in vacuoles.

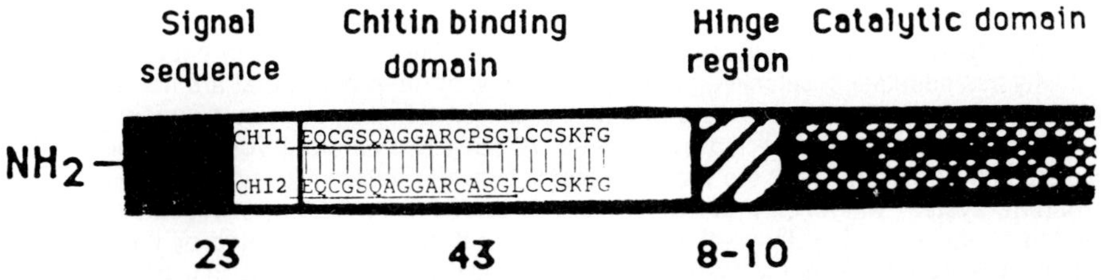

Fig. 3: Model for primary preprotein structure of class I tobacco chitinases and alignment of a part of chitin binding domains of CHI1 and CHI2. The underlined sequences are homologeous to the N-terminal sequences of SN9 and SN11.

4. Discussion

The fungal metabolite Brefeldin A (BFA) interferes with the protein transfer through the endomembrane system, also known as the secretory pathway of intracellular protein transfer, and mainly induces dysfunction of the Golgi-apparatus, which should similarly affect all secretory proteins on their way from the rER to the cell surface. This hypothesis has not been verified experimentally. Only Driouich et al. (1993) investigated the effect of BFA on the protein secretion from plant cells. The drug strongly decreased the amount of ^{35}S-methionine-labelled proteins that were secreted into the medium of suspension cultured sycamore maple cells. In these experiments the secretory proteins were not fractionated further. We investigated for the first time the effect of BFA on the secretion of peroxidase, an enzyme which is known to pass the secretory pathway of plant cells, and show that several polypeptides appear in the medium of suspension cultured tobacco which are barely affected by the drug`s action. The BFA-induced changes in the pattern of secreted polypeptides can not be interpreted as the result of an increased and selective protein breakdown as well as results of a differential uptake of the radioactive precursor into the cell or incorporation into cellular proteins (not shown).

Four of the secreted proteins seem to be practically unaffected by the BFA-dependent inhibition of protein transfer through the secretory pathway. Whereas in short-term experiments with high BFA concentrations, some decrease in ^{35}S-methionine-labelling of these polypeptides was observed, no reduction of labelling after long-term incubation at low BFA concentrations could be detected. SN9 and SN11 show strong homology to a region within the chitin binding domains of tobacco genes CHI1 and CHI2 encoding endochitinases. These class I chitinases are in general located in vacuoles. Other known chitinases of tobacco are class II and class III chitinases. Both types appear to be localized in the intracellular space. Class I and II chitinases are expressed in an organ-specific and age-dependent manner in uninfected plants (see for reviews Sahai and Manocha, 1993).In contrast to class I, classII and III chitinases lack the chitin binding domain. Therefore SN9 and SN11 should belong to class I. However it has to be noticed, that the chitin binding domain of class I chitinases is flanked by imperfect direct repeats, which are known to cause transpositions into other genes. Chitinases are known to pass through the endomembrane system on their way to vacuoles or out of the cells.

How can we explain the differential BFA inhibition of protein secretion from tobacco suspension-cultured cells? Basically, two possibilities have to be taken into consideration. 1)To a certain degree cells become deficient after BFA treatment and some proteins leak out into the medium. Although Driouich et al. (1993) did not detect an indication for the fusion of Golgi-derived vesicles with the plasmalemma under the influence of relatively high BFA concentrations in suspension cultured sycamore maple cells, it can not be excluded that polypeptides which normally pass through the endomembrane system are also getting into the medium if the ER to Golgi passage is blocked by BFA, The high degree of ^{35}S-methionine labelling of polypeptides SN1, SN8, SN9 and SN11 could be taken to reflect a secondary adaptation of the cells to BFA inhibition during long-term incubation at low levels of the drug. But it is difficult to explain by this argument why this adaptation selectively for only very few of the proteins that are normally secreted from control cells. 2) It can be assumed that cells normally secrete some proteins on a pathway which is not or barely affected by the drug. At least two different vesicle-dependent mechanisms could be assumed for the protein transfer between different compartments of the endomembrane system where only the major one is affected by BFA. Consequently, proteins transported by the second system will be secreted even if the first is inhibited by the drug. This interpretation is supported by the finding of Gomez and Chrispeels (1993) that proteins of the vacuolar lumen and membrane are transported to this compartment on pathways including the endomembrane system which are differentially affected by BFA. Further experiments should clarify the differential effect of BFA upon protein secretion.

5. Acknowledgements

We gratefully acknowledge the skilfull technical assistance of H. Bohlmann and S. Förster. The experimental work was supported by the Ministry of Science and Research of The Land Sachsen-Anhalt, Magdeburg grant no. 236A0731 and by the Deutsche Forschungsgemeinschaft grant no.20632 (Germany).

6. References

Bradford, M.M. (1976) `A rapid and sensitive method for the quantitation of microgram quantities of protein utilizing the principle of protein-dye binding` Anal. Biochem. 72, 248-254.

Driouich, A., Zhang, G.F., Staehelin, L.A. (1993) `Effect of Brefeldin A on the structure of the Golgi apparatus and on the synthesis and secretion of proteins and polysaccharides in sycamore maple (*Acer pseudoplatanus*) suspension-cultured cells` Plant Physiol. 101, 1363-1373.

Gerhardt, R., Heldt, H.W. (1984) `Measurement of subcellular metabolic levels in leaves by fractionation of freeze-stopped material in monoaqueous media` Plant Physiol. 75, 542-547.

Gomes, L., Chrispeels, M.J. (1993) `Tonoplast and soluble vacuolar proteins are targeted by different mechanisms` Plant Cell 5, 1113-1124.

Hendriks, T., de Jong, A. Wijsman, H.J.W., van Loon, L.C. (1990) ´Antigenic relationships between petunia peroxidase a and peroxidase isoenzymes in other *Solanaceae*` Theor. Appl. Genet. 80, 113-120.

van Huystee, R.B. (1987) `Plant peroxidases`, in Isoenzymes. Current topics in biological and medical research 16, 241-245. Alan R. Liss, Inc., New York.

Mäder, M. (1976) `Die Lokalisation der Peroxidase-Isoenzymgruppe $G_1$ in der Zellwand von Tabak-Geweben` Planta 131, 11-15.

Sahai, A.S., Manocha, M.S. (1993) `Chitinases of fungi and plants: their involvemant in morphogenesis and host-parasite interaction` FEMS Microbiology Reviews 11, 317-338.

Tague, B.W., Chrispeels, M.J. (1987) `The plant vacuolar protein, phytohemagglutinin, is transported to the vacuole of transgenic yeast` J. Cell. Biol. 105, 1971-1980.

A TONOPLAST INTRINSIC PROTEIN, α-TIP, IS PRESENT IN SEEDS AND SOMATIC EMBRYOS OF NORWAY SPRUCE (PICEA ABIES)

OLIVIUSSON P. and HAKMAN I.
Dept. of Botany, Stockholm University
S-106 91 Stockholm
Sweden

ABSTRACT. The major tonoplast intrinsic protein, TIP, belongs to a family of channel-forming proteins with homology in a variety of organisms. Different isoforms are present in seeds and in vegetative tissues. By using antibodies against a seed-specific TIP, α-TIP, from seeds of Phaseolus vulgaris (a gift from Dr. Chrispeels), we have investigated the occurrance and accumulation of α-TIP in developing megagametophytes and in zygotic embryos of Picea abies, and also in young seedlings. α-TIP accumulation in the seed material seemed to occur concomitantly with the deposition of storage proteins in the tissues. During development of the megagametophytes, two proteins that differed slightly in their molecular weight were detected by Western blot analysis. In the seedlings (13-day-old), we found α-TIP to be present in both the roots and the hypocotyls but not in the cotyledons. Somatic embryogenesis is used in our laboratory as a model system to study embryo development in conifers. We found α-TIP also to be present in mature somatic embryos, but not in the younger stages of the embryos.

1. Introduction

Reserve food like protein, fat and starch are deposited in the seed tissues in large quantities during seed development. The storage proteins are deposited in protein storage vacuoles (PSVs), membrane-bounded organelles, and will, together with the other reserve materials, be mobilized and degraded during germination and early seedling growth. Protein accumulation in Picea abies megagametophytes (female gametophytes) coincides with early embryo development, and a few weeks later also the embryos start accumulate storage material [2, 3].

We use somatic embryogenesis as a model system to study conifer embryology. Embryogenic tissue cultures can relatively easy be initiated from immature or mature zygotic embryos. Such cultures, when maintained on proliferation medium only produce the earlier stages of somatic embryos. By transferring embryogenic cultures to a medium containing abscisic acid (ABA) and with an increased osmotic strength, immature somatic embryos undergo a maturation process thus giving us the possibility to investigate embryo development independently of the season and eventual fluctuation of the flowering. By manipulation of the culture conditions somatic embryos can synthesize and accumulate storage proteins in PSVs similar to zygotic embryos [1].

An abundant, highly conserved vacuolar membrane protein, called TIP (tonoplast intrinsic protein), has been found in seed tissues from a variety of species including monocotyledons,

M. Terzi et al. (eds.), Current Issues in Plant Molecular and Cellular Biology, 647-652.

dicotyledons and also a conifer, Pinus halpensis [5]. The seed-specific TIP (α-TIP) from Phaseolus vulgaris has been revealed to be an abundant channel-forming protein in the tonoplast [9, 5, 6]. The protein has a molecular weight of 27 kD, and shares regions of significant homology to other membrane proteins of the MIP (major intrinsic protein) superfamily which includes membrane proteins from diverse organisms [10].

In the present study we have investigated the accumulation of α-TIP in developing seed tissues and somatic embryos. Also seedling materials were analysed for the protein. By immunolocalization, antibodies against α-TIP were found to bind to the tonoplast of PSVs.

2. Materials and methods

2.1. PLANT MATERIAL

Seed cones from one Norway spruce [Picea abies (L.) Karst.] tree growing in a seed orchard 10 km north-west of Uppsala, Sweden were collected at about weekly intervals between May 27 and August 3, 1993. Collected cones were used immediately or stored for one to three days in paper bags at 4°C before dissection. Developing seeds were removed and megagametophytes isolated. From July 13, the zygotic embryos were isolated from the surrounding gametophytic tissue and treated separately. The tissues were frozen in liquid nitrogen and stored at -80°C. The seed developmental stages at the different collection times were essentially as described in [2]. The exact date of fertilization was not determined but probably occured around June 23. However, some further development of the seeds also could have occured during storage, between collection and the actual isolation of the tissues.

Mature seeds were germinated on moist vermiculite at room temperature and seedlings were grown at a 16/8 h day-night light regime for 13 days before tissues were harvested and frozen in liquid nitrogen. The seedlings were cut into cotyledons, hypocotyls and roots that were treated separately.

2.2. SOMATIC EMBRYO DEVELOPMENT

Embryogenic tissue cultures were initiated and maintained as described previously [2]. For embryo maturation somatic embryos were transferred to full-strength LP-medium containing 7.6 μM ABA and 90 mM sucrose. Somatic embryos were treated as the zygotic embryos.

2.3. IMMUNOBLOT ANALYSIS

Total protein extracts from developing and mature megagametophytes and zygotic embryos, 13-day-old seedling material and developing somatic embryos were separated by SDS-PAGE (15% gels) and electrophoretically transferred to Immobilon membranes by semi-dry electroblotting (Sartoblot, Sartorius) according to the manufacturer. After completion of the transfer procedure the blots were treated essentially as described earlier [3]. An antisera against α-TIP, kindly provided by Dr Chrispeels (Univ. of California, San Diego, La Jolla, USA), described in [5] was used at a 1:1000 dilution in TBST with 1% BSA. A goat anti-rabbit IgG coupled to horseradish peroxidase (Amersham) was used at a 1:300 dilution in TBST with 1% BSA as secondary antibody. Peroxidase was stained using HRP colour development solution (Bio-Rad).

2.4. IMMUNO-GOLD LOCALIZATION

Mature megagametophytes were embedded in LR White and processed for microscopy as described earlier [2]. Ultrathin sections for electron microscopy were incubated with α-TIP antisera diluted 1:50 in TBST with 1% BSA. After primary immunolabeling, the grids were rinsed in TBST, and goat anti-rabbit IgG-gold (AuroProbe EM GAR G10) diluted 1:1 in TBST with 1% BSA was used as secondary antibody. After staining with 5 % uranyl acetate for 1 h, the sections were examined in a Zeiss EM10 transmission electron microscope.

3. Results

3.1. ACCUMULATION OF α-TIP IN MEGAGAMETOPHYTES DURING SEED DEVELOPMENT

At the first two collection dates the megagametophytes were at a prefertilization stage and probably still in a free nuclear stage. Later during seed development the megagametophytes started to accumulate storage proteins (Fig. 1a). The immunostained blot in Figure 1b indicates the presence of a protein, around 26 kD, already at the first collection date. However, this band was stained more intensively at later collection dates. From June 21 the protein started to accumulate in the megagametophyte. Between June 21 and July 8 two proteins were seen to stain with the antibodies. The higher molecular weight protein (27 kD) became more dominant in the tissue as seed development progressed. The time for the appearence of both bands coincides with the fertilization which, in Uppsala, usually occurs around mid-summer. The exact time for the fertilization was, however, not determined.

3.2. ACCUMULATION OF α-TIP IN ZYGOTIC EMBRYOS DURING SEED DEVELOPMENT

Zygotic embryos were isolated from the surrounding gametophytic tissue as they reached the club-shaped stage, being 1.5-2 mm in length (including the suspensor). For embryo developmental stages see also [2]. Figure 2a shows total proteins after separation with SDS-PAGE from embryos collected at different dates. At later stages of seed development, about three weeks after the megagametophytes had started to accumulate storage proteins, proteins of similar molecular weight also became prevalent in the embryos (Fig. 2a, lane 4), which agrees with earlier findings [1, 2]. The α-TIP also became more pronounced in the embryo at this later developmental stage (Fig. 2b, lane 4), although it was detectable on immunoblots also at the earlier stages (Fig. 2b, lanes 1-3).

3.3. α-TIP IS PRESENT IN ROOTS AND HYPOCOTYLS OF SEEDLINGS

Contrary to the results with developing Phaseolus vulgaris seedlings [5] and non-seed tissues of Arabidopsis thaliana [4] both roots and hypocotyls of young Norway spruce seedlings contained proteins immunoreactive with the α-TIP antibodies (Fig. 3b, lanes 1 and 2). The immunstaining signal of α-TIP was very strong in the root extract and somewhat weaker in the hypocotyl. Both proteins have a molecular weight around 27 kD. However, no protein could be detected in extracts from the cotyledons (Fig. 3b, lane 3).

3.4. α-TIP IS PRESENT IN MATURE SOMATIC EMBRYOS

Proliferating embryogenic tissue cultures were transferred to medium containing 7.6 μM ABA and 90 mM sucrose for maturation of the somatic embryos [1]. Embryogenic tissues growing on proliferation medium (Fig. 3, lane 1) did not show any immunostaining with α-TIP antibodies. On the other, hand isolated somatic embryos grown for 56 days on maturation medium, showed a cross-reactive band of 27 kD.

3.5. IMMUNO-GOLD LOCALIZATION OF α-TIP

The antibodies specifically labeled the tonoplast surrounding the PSVs (Fig. 5), in the mature megagametophytes. A background labeling was also observed in the cell walls and sometimes in some areas within the PSVs.

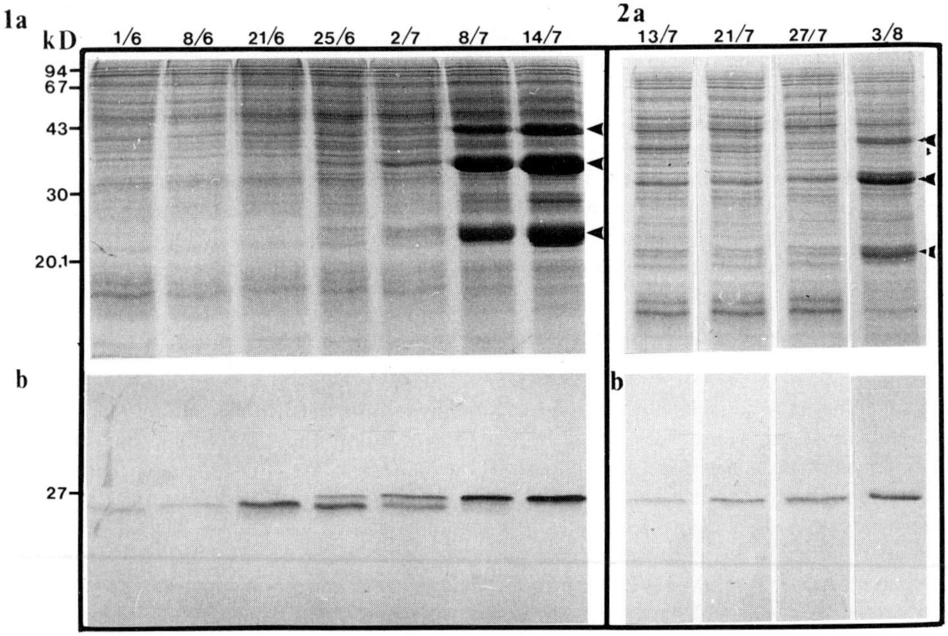

Figs 1-2. SDS-PAGE (15% gel) of total protein extracted from megagametophytes (Fig. 1a) and zygotic embryos (Fig. 2a) during seed development 1993. Arrowheads indicate storage proteins of 42, 33 and 22 kD. Immunodetection of α-TIP in megagametophytes (1b) and zygotic embryos (2b). Note the change in molecular weight of the proteins in the megagametophytes between 21/6-8/7.

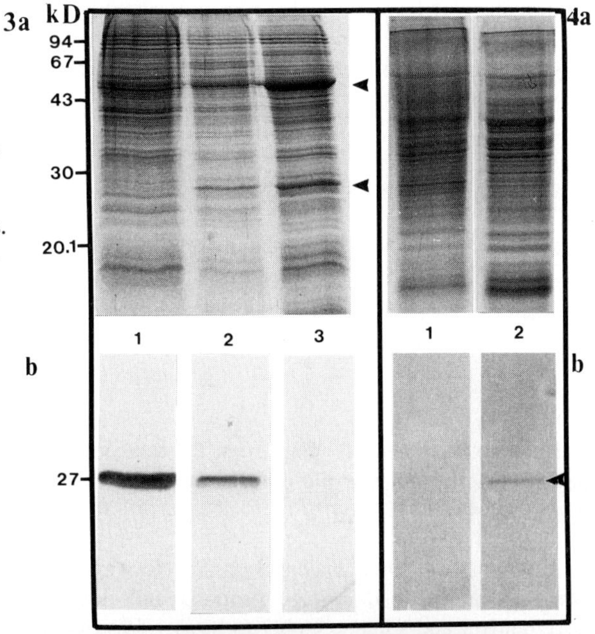

Fig. 3a. SDS-PAGE (15% gel) of total protein extracted from different parts of 13-day-old seedlings. Lane 1: roots, lane 2: hypocotyls, lane 3: cotyledons. The large subunit of Rubisco (55 kD) and the chlorophyll *a/b* binding protein (28 kD) are indicated to the right (arrowheads).
b, Immunodetection of TIP in seedlings. No immunostaining could be detected in cotyledons.

Fig. 4a. SDS-PAGE (15% gel) of embryogenic callus on proliferation medium (lane 1) and mature somatic embryos after 56 days on maturation medium (lane 2).
b, Immunodetection of TIP. Immunostaining was only observed in mature somatic embryos (lane 2, arrowhead).

5

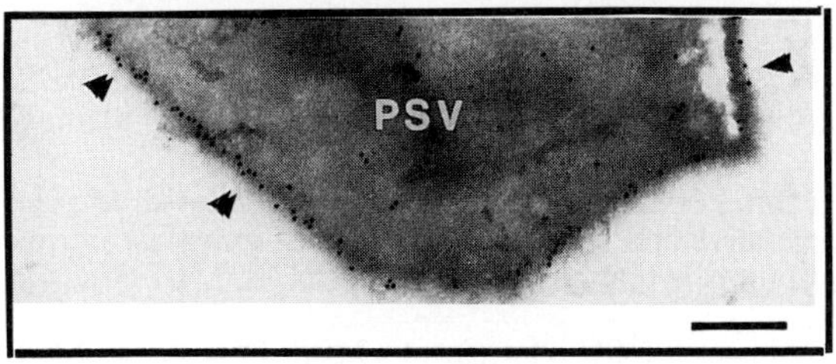

Fig. 5. Electron micrograph of a mature megagametophyte immuno-gold labeled for TIP. Note labeling of the tonoplast (arrowheads). Bar = 0.2 μm.

4. Discussion

The present investigation has shown that the megagametophyte and the zygotic embryo of Picea abies contain a protein recognized by Phaseolus vulgaris α-TIP antibodies. This result agrees with previous findings that TIP is a highly conserved protein found in a variety of plant species [5]. During seed development two proteins, differing slightly in their molecular weight, stained with the antibodies. A slower moving protein appeared around the time of fertilization. As the seeds matured only the slower moving band could be detected. Dimerization and/or aggregation of membrane proteins is not uncommon, and could result in the double bands seen on the blots. However, the immunostaining pattern (Fig. 1b) also raises the hypothesis that the protein could change during seed development, or that two different proteins are present at different time during the development of the tissue. In the megagametophyte accumulation of α-TIP occured at a stage of vacuolar fragmentation into PSVs, which thus increase the membrane area (see Fig. 1b, lane 3). This result agrees with those in [8]. A possible function of TIP as a solute transporter or in stabilizing the tonoplast has been suggested [5]. Recently TIP specific for vegetative tissues, has been shown to act as a water channel [7].

Interestingly, also somatic embryos capable of accumulating storage proteins and withstanding severe dehydration contained TIP. A better understanding of the maturation process during conifer embryogeny is vital and research in our laboratory is concerned with these aspects.

5. Acknowledgements

Research support was provided by grants from the Swedish Agricultural and Forestry Research Council, Carl Trygger Research Foundation, and by the Commission of European Communities, Contract MA2B-CT91-0093, to I. H.

6. References

1. Hakman, I., Stabel, P., Engström, P., & Eriksson, T. 1990. Storage protein accumulation during zygotic and somatic embryo development in Picea abies (Norway spruce). - Phys. plant. 80: 441-445.

2. Hakman, I. 1993a. Embryology in Norway spruce (Picea abies). An analysis of the composition of seed storage proteins and deposition of storage reserves during seed development and somatic embryogenesis. - Phys plant. 87: 148-159.

3. Hakman, I. 1993b. Embryology in Norway spruce (Picea abies). Immunochemical studies on transport of a seed storage protein. - Phys. plant. 88: 427-433.

4. Höfte, H., Hubbard, L., Reizer, J., Ludevid, D., Herman, E. M. & Chrispeels, M. J. 1992. Vegetative and seed-specific forms of tonoplast intrinsic protein in the vacuolar membrane of Arabidopsis thaliana. - Plant Physiol. 99: 561-570.

5. Johnson, K. D., Herman, E. M. & Chrispeels, M. J. 1989. An abundant, highly conserved tonoplast protein in seeds. - Plant Physiol. 91: 1006-1013.

6. Johnson, K. D., Höfte, H. & Chrispeels, M. J. 1990. An intrinsic tonoplast protein of protein storage vacuols in seeds is structurally related to a bacterial solute transporter (GlpF). - Plant Cell 2: 525-532.

7. Maurel, C., Reizer, J., Schroeder, J. I. & Chrispeels, M. J. 1993. The vacuolar membrane protein γ-TIP creates water specific channels in Xenopus oocytes. - EMBO J. 12: 2241-2247.

8. Melroy, D. L. & Herman, E. M. 1991. TIP, an integral membrane protein of the protein-storage vacuols of the soybean cotyledons undergoes developmentally regulated membrane accumulation and removal. - Planta 184: 113-122.

9. Mäder, M. & Chrispeels, M. J. 1984. Synthesis of an integral protein of the protein-body membrane in Phaseolus vulgaris cotyledons. - Planta 160: 330-340.

10. Pao, G. M., Wu, L.-F., Johnson, K. D., Höfte, H., Chrispeels, M. J., Sweet, G., Sandal, N. N. & Saier, M. H. Jr. 1991. Evolution of the MIP family of integral membrane transport proteins. - Mol. Microbiol. 5: 33-37.

THE CHARACTERISTICS OF ATROPINE METABOLISM IN CULTURED TISSUES AND SEEDLINGS OF *DUBOISIA*

Y. KITAMURA, R. YAMASHITA, C. YABIKU, H. MIURA & M. WATANABE
School of Pharmaceutical Sciences, Nagasaki University,
Bunkyo-machi 1-14, Nagasaki 852, Japan

1. Introduction

The intact plant has communication between organs with respect to transport and metabolism of secondary products, but isolated organs do not. In fact, it is well established that tropane alkaloid hyoscyamine is biosynthesized in the root of tropane alkaloid-producing plants and then transported to and accumulated in the aerial parts (1,2). Study with cultured tissues of *Duboisia* clarified that conversion of hyoscyamine into scopolamine occurs in the shoots as well as in the roots (3). Previously, we showed that transport and metabolism of atropine (*dl*-hyoscyamine) were much more dynamic in the intact *Duboisia* plants than in the cultured tissues. Atropine was transported not only from root to leaf but also from leaf to root, through the xylem and phloem, respectively (4,5). Furthermore, atropine was hydrolyzed to tropine and tropic acid by atropine esterase in the roots of intact plants but not in the cultured roots which are able to produce atropine, suggesting that the intact plant roots decompose atropine which is transported back to the roots from the aerial parts (6).

These findings show that studies using both cultured tissues and intact plant give us more precise information on the dynamics of plant metabolites. Here we compared atropine metabolism in various cultured tissues such as cell suspension, root, shoot and leaf cultures with that in seedlings of *D. myoporoides*, using ^{14}C-labeled atropine as a tracer.

2. Materials and Methods

2.1. TRACER FEEDING

Tissue cultures. Cell suspensions induced from callus (7), roots (8), shoots (7) and leaves isolated from the shoots (each tissue 0.3g fr. wt. /flask) were inoculated into the fresh liquid MS media (25ml/flask) containing suitable plant hormones as follows: cells (2,4-D 1 mg/L), roots (IBA 2 mg/L + gibbelleric acid 1 mg/L) and both shoots and leaves (BA 0.2 mg/L). After 7 days in preculture, 0.5 ml of [carbonyl-^{14}C]atropine sulfate solution (1 mg/ml, 29.6 kBq/ mg) was added to each flask and cultured for an additional 3 to 28 days.

M. Terzi et al. (eds.), Current Issues in Plant Molecular and Cellular Biology, 653-658.

Seedlings. Two-to 3 month-old seedlings after germination were used for tracer feeding experiments. The same amount of the tracer described above was supplied to the seedling via roots as reported elsewhere (4), and then cultured in water for 1 to 11 weeks.

2.2. EXTRACTION AND COUNTING OF RADIOACTIVITY

After harvest, all cultures were separated into the tissue and the medium. Lyophilized tissues were extracted four times with 80% MeOH, the extract was evaporated to dryness, and the residue was dissolved in distilled water. This aqueous solution as well as culture medium was fractionated into the three organic acid, alkaloidal and the aqueous residue components as previously reported (9). Radioactivity in the 80% MeOH extract, tissue residue, filtered medium and each fraction separated was measured by liquid scintillation counting.

2.3. ANALYSIS OF RADIOACTIVE METABOLITE

The organic acid, alkaloidal and residual aqueous fractions were applied to silica gel-TLC plates, which were developed with MeOH-benzene (1:1), $CHCl_3$-EtOH-28%NH_4OH (85:14:1) and BuOH-AcOH-H_2O (4:1:1), respectively. Labeled compounds were located by autoradiography. Radioactive spots corresponding to atropine, scopolamine and aposcopolamine were scraped off, the alkaloids were eluted from the gel with MeOH and their radioactivity was determined by liquid scintillation counting.

Four non-alkaloidal radioactive metabolites found in aqueous solution were isolated and then treated with 10% KOH to analyze the alkali hydrolyzed products by TLC-autoradiography.

3. Results

3.1. TISSUE CULTURES

Uptake and Secretion. All cultured tissues grew well in the liquid medium containing ^{14}C-labeled atropine, which they took up from the medium at the early stage of growth. After 7 days in culture, both shoot and leaf cultures stored all radioactivity in their tissues, but not in the media. On the other hand, around 10% of the total radioactivity was left in the media of the cell suspension and root cultures after that, although the tissues propagated. The suspension cells and cultured roots seem to transport the alkaloid not only from the medium to the tissues but also in the opposite direction.

Secretion of radioactivity from the cells was determined by repetition of medium exchange, showing radioactivity was found in the tracer-free medium freshly added and cultured for a while (Fig. 1). Such a secretion was also detected from the cultured roots, judging from the finding of the metabolites in the medium.

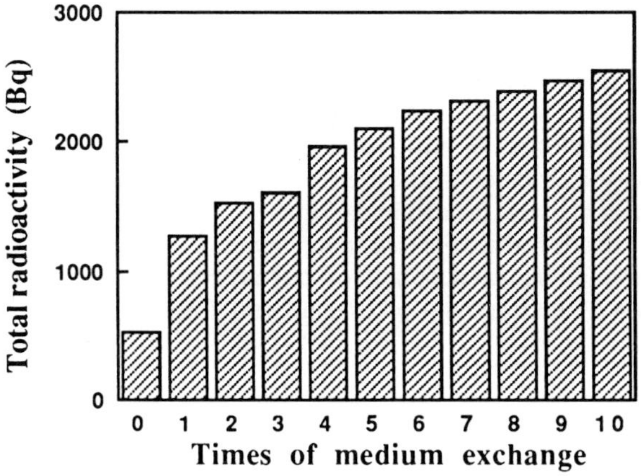

Fig. 1. Radioactivity recovered from the medium of suspension cell cultures fed with ^{14}C-atropine

Metabolism. Since no radioactivity was detected in any tissue residue or in non-alkaloidal fractions, only the alkaloidal fraction was analyzed. TLC-autoradiograms of alkaloid fractions showed that all cultures except suspension cells produced scopolamine as the main metabolite, and root and shoot cultures also formed aposcopolamine. Aposcopolamine was detected only in the medium of root cultures. Scopolamine contents increased in all organized tissues during culture, although aposcopolamine contents did not. The roots had the highest ability to metabolize atropine into scopolamine, followed by the shoots and the leaves (Fig. 2).

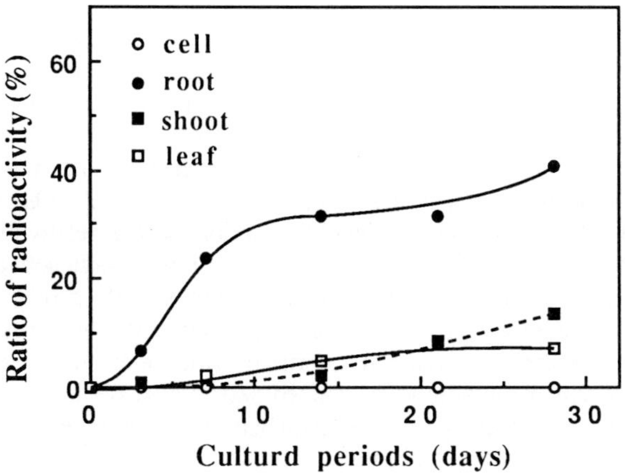

Fig. 2. Scopolamine formation by various tissue cultures fed with ^{14}C-atropine

3.2. INTACT PLANT

Distribution of radioactivity. ^{14}C-labeled atropine, fed though the roots where atropine is biosynthesized, distributed to all parts of the plant. All radioactivity was extracted with 80% MeOH and no activity was detected in the tissue residue. The extract was fractionated to show that radioactivity distributed to all three organic acid, alkaloidal and aqueous residue fractions; the radioactivity in alkaloidal fraction was the highest, followed by that in the aqueous residue fraction. Radioactivity in aqueous residue fraction increased gradually corresponding to decrease of radioactivity in alkaloidal fraction during culture. Distribution of radioactivity in organic acid fraction was trace throughout the culture (Fig. 3).

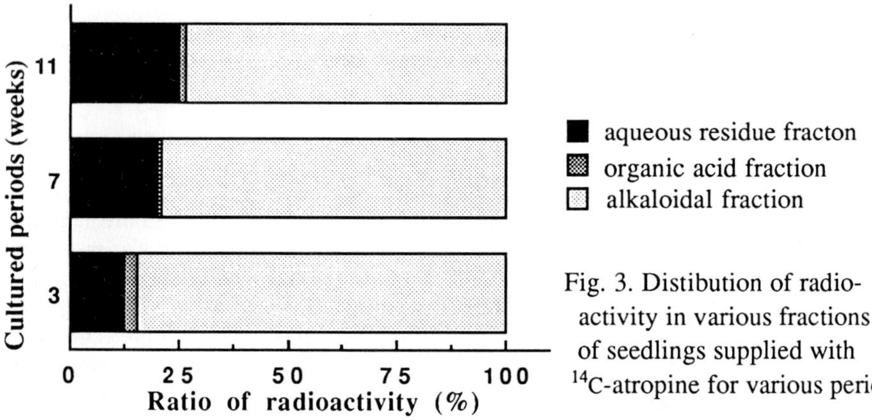

Fig. 3. Distibution of radioactivity in various fractions of seedlings supplied with ^{14}C-atropine for various periods.

Metabolism. Scopolamine and aposcopolamine were found as major alkaloidal metabolites similar to organ cultures. Formation of labeled scopolamine reached about 50% of total radioactivity in the alkaloidal fraction after 5 weeks in culture, although aposcopolamine did not increase so much.

When the non-alkaloidal metabolite in the organic acid fraction where radioactivity was trace was examined, a radioactive spot corresponded to tropic acid. Other non-alkaloidal metabolites contained in the aqueous residue fraction where relatively high radioactivity existed were surveyed, showing the presence of four radioactive bands on the autoradiogram. These were not identical with the water soluble primary metabolites such as sugars, amino acids and proteins. Then supposing that the tropic acid produced by atropine esterase in the roots may be esterified with some unknown sugar or amino acid, each radioactive metabolite was isolated and treated with alkali, from which the organic acid substance was re-extracted and supplied to TLC-autoradiography. The result showed all metabolites had tropic acid moiety in their ester-derivatives.

Fig. 4 shows our proposed metabolic pathway of atropine in seedlings.

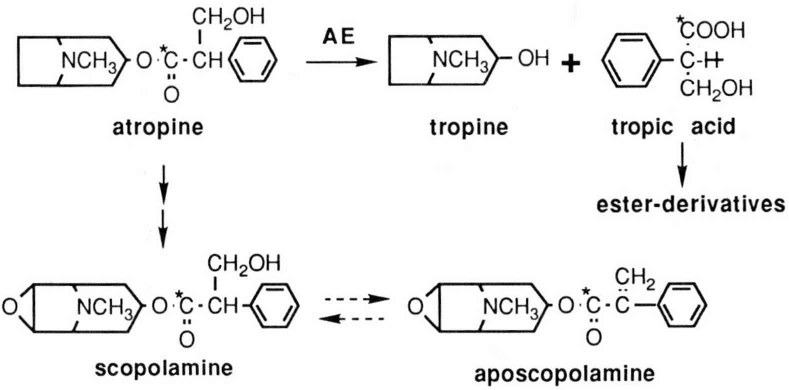

Fig.4. Possible metabolic pathway in seedlings

4. Discussion

Atropine uptake, secretion and metabolism by the various cultured tissues cell suspension, root, shoot and leaf cultures were compared with those by seedlings of *D. myoporoides* .

Every plant tissue of *Duboisia*, either differentiated or dedifferentiated, or with or without organ-organ communication, incorporated atropine into the tissue. However, excretion of alkaloid was detected only from cultured cells and roots. Cultured shoots and leaves must have a system against alkaloid loosing from their tissues. Shoots and leaves of intact plants must have the same abilities as their cultures, though they can communicate with and transport alkaloids to the roots. There was a characteristic differences between cultured roots and intact plant roots with respect to atropine transport. Cultured roots excreted the alkaloids from the tissues, while roots of intact plants did not seem to excrete the alkaloids, but transported them to the aerial parts.

All cultures except suspension cells produced scopolamine as the main metabolite, and root and shoot cultures also formed aposcopolamine. On the other hand, seedlings formed not only the alkaloidal metabolites scopolamine and aposcopolamine, but also non-alkaloidal metabolites, one of which corresponded to tropic acid and another metabolite was determined as an esterified products between tropic acid and unknown substances. Since our previous works showed that roots of intact plant but not cultured roots decompose atropine which must be transported back from aerial parts (4-6), it is reasonable that tropic acid and its derivatives were found only in the seedlings. A trace of tropic acid was detected, indicating that the acid produced by atropine esterase must be converted to stable ester-derivatives at once.

The characteristic differences between isolated roots and intact plant roots with respect to atropine secretion and metabolism indicate that organ-organ communication is important and that the root plays the most active role for atropine dynamics in *Duboisia.*

5. References

1) James, J. O. (1950) 'Alkaloids in plants' in Manske, R.H.F. (ed), The alkaloids, Academic press, New York, Vol.1, pp.15-90.

2) Wink, M. (1987) 'Physiology of the accumulation of secondary metabolites with special reference to alkaloids' in Constabel, F. and Vasil, I. K. (eds), Cell culture and somatic cell genetics of plants, Academic Press, Vol.4, pp. 17-42.

3) Yamada, Y. and Endo, T. (1984) 'Tropane alkaloid production in cultured cells of *Duboisia leichhardtii*' Plant Cell Rep. 3, 186-188.

4) Kitamura, Y., Matuo, N., Takashi, T. and Miura, H. (1991) 'The characteristics of atropine transport in seedlings and regenerated plantlets of *Duboisia myoporoides* ' J. Plant Physiol. 137, 613-618.

5) Kitamura, Y., Yamashita, R., Miura, H. and Watanabe, M. (1993) 'Phloem transport of tropane and pyridine alkaloids in *Duboisia myoporoides* ' J. Plant Physiol. 142, 635-637.

6) Kitamura, Y., Sato, M. and Miura, H. (1992) 'Differences of atropine esterase activity between intact roots and cultured roots of various tropane alkaloid-producing plants' Phytochemistry 31, 1191-1194.

7) Kitamura, Y. (1988) ' *Duboisia* spp.: In vitro regeneration, and the production of tropane and pyridine alkaloids ' in Y.P.S. Bajaj (ed), Biotechnology in agriculture and forestry vol.4. Medicinal and aromatic plants I, Springer-Verlag, pp.419-436.

8) Kitamura, Y., Sugimoto, Y., Samejima, T., Hayashida, K. and Miura, H. (1991) 'Growth and alkaloid production in *Duboisia myoporoides* and *D. leichhardtii* root cultures ' Chem. Pharm. Bull. 39, 1263-1266.

9) Kitamura, Y., Taura, A., Kajiya, Y. and Miura, H. (1992) 'Conversion of phenylalanine and tropic acid into tropane alkaloids by *Duboisia leichhardtii* root cultures' J. Plant Physiol. 140, 141-146.

ENVIRONMENTAL CONTROL FOR LARGE SCALE PRODUCTION OF IN VITRO PLANTLETS

T. KOZAI, Y. KITAYA, K. FUJIWARA & J. ADELBERG
Faculty of Horticulture, Chiba University, Matsudo,
Chiba 271 Japan

ABSTRACT Chlorophyllous cultures have high photosynthetic ability. In many cases, they grow faster under photoautotrophic (sugar–free medium) conditions than under heterotrophic and photomixotrophic culture conditions when provided proper in vitro physical/chemical environment for promoting the photosynthesis of chlorophyllous cultures. It has also been shown that problems of physiological and morphological disorders of in vitro plantlets, such as hyperhydricity and abnormal stem elongation, can be solved in many cases by controlling the in vitro environment properly. Acclimatization ex vitro is unnecessary if the in vitro environment is controlled to complete the acclimatization in vitro. Control of the in vitro environment is essential to realize full potential for large scale production of quality plantlets in vitro at minimal costs. Proper understanding of eco–physiological relationships among environmental factors and physiological characteristics of cultures in the vessel is a key to develop a large scale production system of micropropagated plantlets.

1. INTRODUCTION

Development of automated systems for a large–scale micro–propagation is essential to drastically reduce production costs and to improve quality of micropropagated propagules. Recently, research has been conducted on automation of a large–scale micropropagation system for reducing manual operation including handling of somatic embryos, microcutting and transplanting (Kurata and Kozai, 1992; Hayashi et al., 1992; Aitken–Christie et al., 1994). Research on automation in bioreactor systems for measurement and control of the liquid medium environment has also been conducted to produce quality microtubers, bulblets, meristematic nodules or somatic embryos at low costs (Aitken–Christie at al., 1994). In bioreactor systems, cultures contain little chlorophyll and are basically grown in the liquid medium heterotrophically. Thus, the control of aerial environment in the bioreactor for promoting photosynthesis of cultures is not essential.

On the other hand, it has recently been shown that the in vitro environment, especially aerial environment in vitro, affects the photosynthesis and thus growth and development of chlorophyllous cultures (shoots, leafy explants, plantlets, etc.). The control of in vitro environment is critically important to improve the growth, development and morphology of chlorophyllous plantlets in vitro (Kozai, 1991a). These findings have led to a concept of photoautotrophic (sugar–free) micropropagation of chlorophyllous plantlets (Kozai, 1991b; 1991c).

M. Terzi et al. (eds.), Current Issues in Plant Molecular and Cellular Biology, 659-667.

Control of in vitro environment is also important to improve quality of plantlets in vitro. However, not much research has been done on automation in measurement and control of the in vitro environment during micropropagation of chlorophyllous plantlets. In this article, features and effects of in vitro environment on growth, development and quality of chlorophyllous plantlets is summarized and the importance of automatic control of the in vitro environment in a large-scale micropropagation is discussed.

2. FACTORS AFFECTING THE GROWTH, DEVELOPMENT AND MORPHOLOGY OF IN VITRO PLANTLETS

Major environmental factors in micropropagation are classified in Fig. 1. The initial value of each environmental factor may be known at the beginning of a culture period but its variation in time and space affects the growth, development and morphology of plantlets in vitro. In turn, the variation of in vitro environment is affected by the presence of plantlets, vessel type and the environment outside the vessel. For example, time courses of concentrations of medium components are affected by the growth of cultures and medium volume, and vice versa. These relationships among environmental factors inside and outside the vessel and other related variables and parameters are shown as a relational diagram in Fig. 2 (Aitken-Christie et al., 1994). Detailed description of the relationships and their effects are given in Aitken-Christie et al. (1994).

General features of the in vitro environment in conventional micropropagation are listed in Table 1. Shoots and plantlets in vitro respond to such environmental conditions typically as shown in Table 2. Problems requiring environmental control solutions in a large scale production of in vitro plantlets are summarized in Table 3.

3. ADVANTAGES OF PHOTOAUTOTROPHIC MICROPROPAGATION

Recent research revealed that chlorophyllous shoots had remarkable photosynthetic ability (Kozai, 1991a) and sometimes grew better under photoautotrophic (with sugar-free medium) than hetero- or photomixotrophic condition when the in vitro environment was controlled properly for photosynthesis.

CO_2 concentrations in an air-tight vessel containing chlorophyllous shoots or plantlets generally decrease sharply with time at the onset of photoperiod and reach the CO_2 compensation point within a few hours. Thus, shoots in vitro cannot photosynthesize mainly due to the low CO_2 concentration, not due to their little photosynthetic ability. Under such low CO_2 concentration conditions, photosynthesis is not increased with increase in photosynthetic photon flux density. Besides, the net photosynthetic rate of shoots, not the photosynthetic ability, decreases with the presence of sugar in the medium. Thus, CO_2 enrichment under relatively high photosynthetic photon flux densities (150–250 $\mu mol\ m^{-2}\ s^{-1}$) greatly promotes the photosynthesis and thus growth of chlorophyllous shoots in vitro when cultured photoautotrophically.

A slight decrease in relative humidity in the vessel by moderate vessel ventilation under enriched CO_2 and elevated photosynthetic photon flux density promotes the transpiration from shoots. Thus, water and nutrient uptake from the medium, and the shoot growth. Lowering the relative humidity also improves epicuticular wax formation and stomatal functioning etc. After all, most problems indicated in Tables 1, 2 and 3 are solved or partly solved by the above environmental control.

In most cases, the above environmental control can be more easily implemented and the benefits of environmental control are generally greater under photoautotrophic than under hetero- or photomixotrophic condition. In addition to the growth promotion, photoautotrophic micropropagation gives many other advantages over hetero- or photomixotrophic micropropagation (Table 4).

4. ADVANTAGES OF ENVIRONMENTAL CONTROL IN VITRO

Photoautotrophic micropropagation is advantageous over heterotrophic micropropagation only when the in vitro environment is controlled properly for photosynthesis of chlorophyllous cultures. It should also be noted that the control of in vitro environment is advantageous for promoting not only photoautotrophic growth but also heterotrophic and photomixotrophic growth of cultures. Furthermore, the control of in vitro environment is beneficial to complete the in vitro acclimatization of micropropagated plantlets, thus, eliminating the procedures for ex vitro acclimatization. The control of in vitro environment is beneficial to preserve quality of micropropagated plantlets in vitro during storage at relatively low temperatures (0–10°C) (Aitken–Christie et al., 1994). Environmental control in vitro is also advantageous for controlling embryogenesis, morphogenesis and differentiation or development of cultures, and for controlling physiological, anatomical and biochemical features of cultures (Aitken–Christie et al., 1994).

When micropropagated plantlets are transferred for acclimatization ex vitro, the morphology of plantlets is as important as their physiological characteristics. Required morphological characteristics of transplants are, in general, (1) a thick main stem with reduced shoot length or short plantlet height, and (2) low ratio of aerial to root weight ratio with relatively large leaf area. In reality, however, micropropagated plantlets do not meet these requirements (Kurata and Kozai, 1992). The reduction in shoot length has often been achieved by application of plant growth regulators (growth retardants). However, its widespread application is not recommended from the viewpoint of reduction in possibilities of environmental pollution. Thus, the control of shoot length by environmental control is becoming an essential technique. Plantlets with reduced shoot length are obtained by controlling temperature, relative humidity, and light environment. Shoot length is decreased by keeping dark period temperature higher than photoperiod temperature, and with decreasing relative humidity. Shoot length can also be decreased by increasing light intensity and/or the ratio of red to far–red photon flux density, and by giving the light from the sides (sideward lighting system) (Kurata and Kozai, 1992).

5. ENVIRONMENTAL MEASUREMENT AND CONTROL FOR LARGE SCALE PRODUCTION

Environmental measurement and control systems currently used in micropropagation have been mostly introduced, with some modifications, from those used in areas of environmental plant physiology and fermentation technology. For the measurement and control of the in vitro environment, however, sensors, actuators, control units, and amounts of sampling gases/liquid should be as small as possible not to disturb the in vitro environment and to control the in vitro environment accurately. These technologies are well documented in Aitken–Christie et al. (1994). For large scale production of plantlets in the near future, application of micro–computerized measurement and control systems should be dispensable.

In areas of greenhouse crop research, many mathematical models have been developed for simulating interactions between plant growth/development and the environment, and they are found to be useful for developing practical environmental control strategies in greenhouse horticulture. However, in areas of micropropagation research, limited numbers of mathematical models have been developed and much more research needs to be conducted to develop mathematical models for environmental control purposes.

6. MEMBRANE INTERFACE BETWEEN MEDIA AND PLANTLET

Microporous polypropylene membranes, as an alternative support material to agar, allows exchanges between the tissue and liquid media. Hydraulic conductance of media across the micropores regulates availability of dissolved nutrients and water to the tissue. Inert, autoclavable vessel inserts are commercially available in either rigid support or buoyant forms.

Membrane-based culture allows simultaneous advantages of both solid (e.g. agar-gel) and liquid (e.g. shake flask or bioreactor) systems. (1) Media composition can be monitored and modified during the culture regime. (2) Tissue grows in aerial environment which can promote normal shoot physiology (e.g. photosynthesis and transpiration). Water lost from vessel by gas exchange with culture room can be directly compensated by addition of liquid water to the media. Media amendments could deliberately reduce sugar levels for gradual induction of photoautotrophism, or similarly, plant growth regulators or nutrients can be altered as plantlets mature. (3) Membrane vessels can order plant material to obtain growth response data during culture (e.g. fresh weight or image analysis). Specimens are in fixed position so sampling error and rotational affects are not a concern. Decisions with regard to harvest can be implemented with respect to individual specimens. (4) Unlike gelled media, tissue growth on a fixed position of the support does not create gradients in available nutrients if media is changed or circulated periodically. Generally, availability of nutrients should be greater than gelled media.

A membrane based bioreactor has been designed and tested to demonstrate some potentials in shoot propagation systems. With a variety of herbaceous plants, periodic infusions of fresh media increased yield several-fold (Hale et al., 1993). With micropropagation of *Cattleya* orchid, vegetative growth was promoted and time required in greenhouse for flowering was reduced (Adelberg et al., 1992; 1993). With watermelon, shoot bud proliferation and shoot elongation were shown to be heavy users of nitrogen salts, with preferential uptake of ammonium during shoot proliferation (Desamero et al., 1993). Further applications include (1) the use of flow systems to remove deleterious exudates from the plant tissue, (2) somatic embryo systems for transplant production with data collection and mechanical harvest facilitated by fixed position of plant material on membrane surface, and (3) structural fabrications to control spread of microbial infestation within reactor vessels.

7. GENERAL ENVIRONMENTAL CONTROL PROCEDURES FOR A LARGE-SCALE PRODUCTION SYSTEM OF MICROPROPAGATED PLANTLETS

The followings are an example of general environmental control procedures to be employed for a large scale production system of micropropagated plantlets:

(1) Environmental control for production of non-chlorophyllous cultures including somatic embryos and non-chlorophyllous buds/meristematic nodules on the sugar-containing medium under aseptic conditions.

(2) Environmental control for making the non-chlorophyllous cultures chlorophyllous on the sugar-containing medium under high CO_2 concentration, high relative humidity, medium light intensity conditions. Chlorophyllous cultures include germinated somatic embryos, leafy stem cuttings, etc.

(3) Transplanting of chlorophyllous cultures on either a plug tray (with several hundred cells containing fibrous medium) or on a supporting sheet including a microporous membrane (for fixing propagules) with nutrient solution, under pathogen-free conditions.

(4) Environmental control for growth of chlorophyllous cultures to plantlets on the sugar-free medium under CO_2 enriched and high light intensity.

(5) Environmental control for acclimatization of the plantlets at relatively low relative humidities and high light intensities under pathogen free conditions.

(6) Environmental control for storage of the micropropagated plantlets, if needed, at low temperatures and low light intensities to preserve the dry weight and quality of the plantlets.

(7) Environmental control during transportation for shipping to the consumers.

(8) Transplanting of the acclimatized plantlets to the greenhouse or field conditions, using an automatic transplanting machine.

- In vitro Environment
  - Aerial
    - Physical
      - Temperature
      - Light
        - Photosynthetic
        - Photomorphogenic
      - Gaseous composition
        - Carbon dioxide
        - Oxygen
      - Water potential (Relative humidity)
      - Gas diffusivity (Air movement)
      - Air pressure
      - Thermal radiation
    - Chemical
      - Ethylene
      - Other gases
    - Biological
      - Microorganisms
      - Insects
  - Root zone
    - Physical
      - Temperature
      - Water potential
        - Osmotic
        - Matric
      - Pressure
      - Diffusivity of medium components
      - Hardiness of the medium
      - Hydraulic conductivity
      - Shearing stress in moving liquid medium
    - Chemical
      - Inorganic elements
        - Macroelements
        - Microelements
        - Dissolved gases
      - Organic elements
        - Sugar
        - Growth regulators
        - Gelling agent
      - pH
    - Biological
      - Pathogens
      - Non pathogens
        - Symbiotic
        - Competitive association

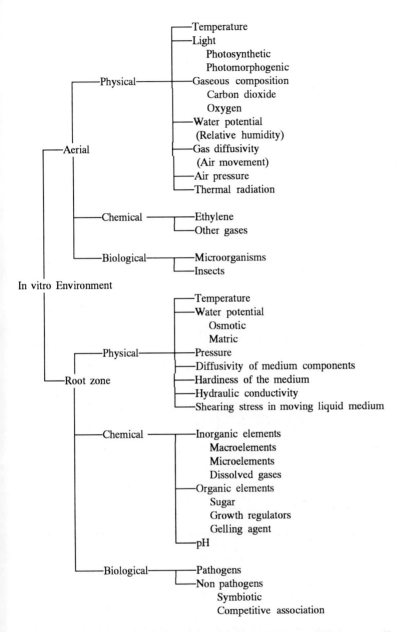

FIGURE 1. Classification of the in vitro environmental factors which have an effect on their growth, development and morphogenesis (Aitken–Christie et al., 1994).

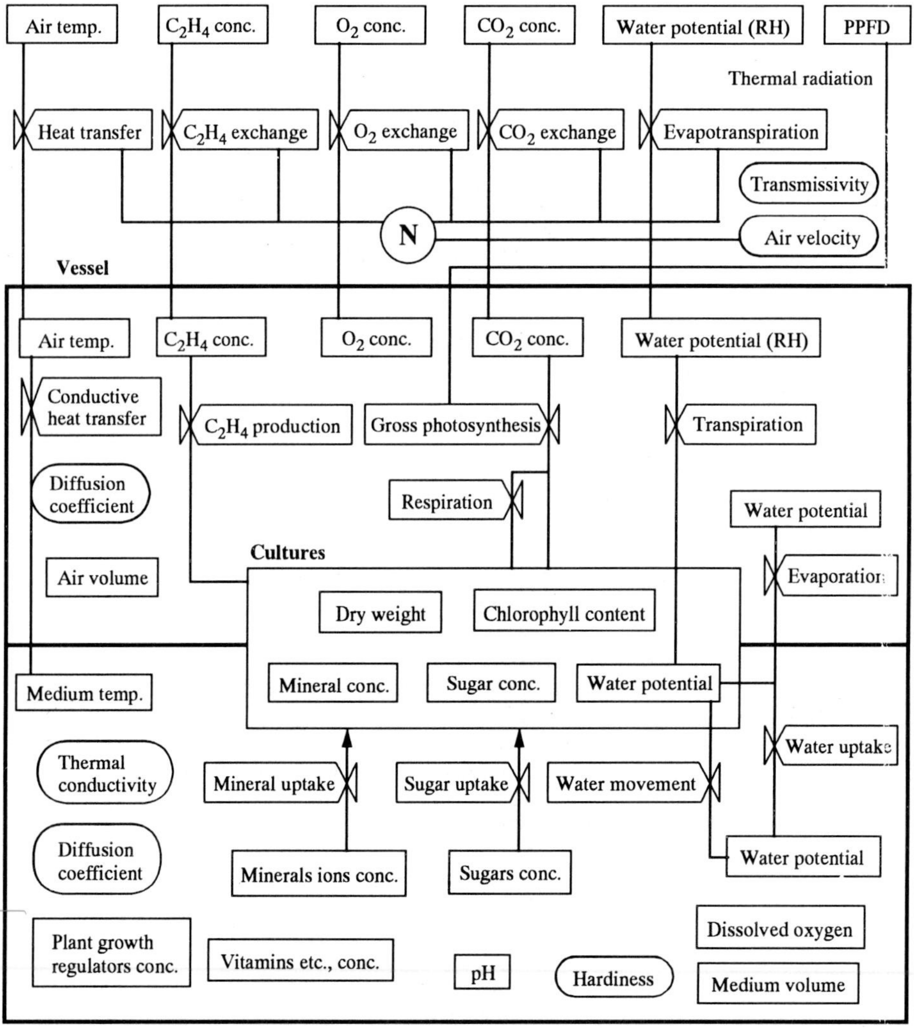

PPFD: Photosynthetic photon flux density

N: Number of air changes per hour of the vessel

RH: Relative humidity. Water potential of the air is determined as a function of RH and air temperature. Water potential of culture medium is a sum of osmotic, matric and pressure potentials.

FIGURE 2. Relational diagram showing flows of mass and energy in and around a culture vessel ecosystem. Rectangle symbols denote state variables and valve symbols denote rate variables. Ellipse symbols denote coefficients and parameters (Aitken–Christie et al., 1994).

TABLE 1. General features of the in vitro environment in conventional micropropagation

State variables	Rate variables
AERIAL ENVIRONMENT	FLOW OF MATERIAL AND ENERGY
1) high relative humidity (high water potential)	1) Low transpiration rate
2) constant air temperature	2) Low net thermal radiation flux
3) Low CO_2 concentration in light and high CO_2 conc. in dark	3) Low net photosynthetic rate
	4) Low photosynthetic photon flux
4) High C_2H_4 concentration	5) High dark respiration rate
ROOT ZONE (MEDIUM) ENVIRONMENT	------------------------------
1) High sugar concentration	1) Low sugar uptake rate
2) High mineral ion concentration	2) Low mineral ion uptake rate
3) Low dissolved oxygen conc.	3) Low water uptake rate
4) High conc. of phenolic or other toxic substances	4) Low transport rates of other medium components
5) Low density of microorganisms	
6) High concentrations of plant growth regulators	

Notes) State variables quantify conserved properties of the system. Rate variables quantify the rate of change with time of the state variables.

TABLE 2. Typical responses of shoots/plantlets in vitro in conventional micropropagation

Tissue level	Whole plantlet level
1) Little epicuticular formation	1) Low growth rate
2) Stomatal malfunction	2) Succulent growth (thin, tall and succulent shoot)
3) Low chlorophyll concentration	3) Physiological and morphological disorders such as hyperhydricity
4) Low percent dry matter	
5) Restricted leaf area expansion	4) Incomplete rooting/few secondary roots
6) Less number of stomata on leaves than normal	5) High variation in size, shape and developmental stage
7) Poorly-structured spongy and palisade tissues	6) Occurrence of mutation
	7) Death of plants due to contamination

TABLE 3. Problems requiring environmental control solutions in a large scale production of in vitro plantlets

1) A low growth rate of cultures at each culture stage
2) A significant percentage of dead cultures due to bacterial or fungal contamination and/or physiological/morphological disorders such as hyperhydricity in vitro
3) A significant percentage of dead plantlets due to environmental stress at the acclimatization stage ex vitro
4) A large variation in size, shape and quality of cultures and significant labor costs for selecting, sorting and transplanting cultures at each culture stage
5) Significant energy costs for lighting, air conditioning, autoclaving, washing, drying, transportation, etc.
6) Significant costs for sugar, mineral nutrients, plant growth regulators, agar, vessels, vessel caps, etc.
7) An overproduction or shortage of plantlets at various times due to unpredictable circumstances
8) Significant space and its related costs required for multiplication, rooting and acclimatization

TABLE 4. Advantages of photoautotrophic micropropagation over conventional, heterotrophic or photomixotrophic micropropagation

1) Growth and development of chlorophyllous cultures in vitro are promoted.
2) Physiological and/or morphological disorders are reduced.
3) Relatively uniform growth and development are expected.
4) Leafy single node stem cuttings can be used as explants.
5) Procedures for rooting and acclimatization are simplified.
6) Application of growth regulators and its related mutation can be minimized.
7) Loss of cultures in vitro due to microbial contamination is minimized.
8) A large vessel can be used with minimum risk of contamination.
9) Environment control of the vessel becomes easier due to minimized contamination and use of a large vessel.
10) The control of growth and development by means of environmental control becomes easier.
11) Asepsis in the vessel is not required as long as pathogens are excluded from the vessel.
12) Automaton, robotization and computerization become easier with use of a large vessel.

8. REFERENCES

Adelberg, J. and J. Darling (1993) *In vitro* Membrane Treatment Accelerates Flowering of *Laeliocattleya* (El Cerrito × Spring Fires). Amer. Orc. Soc. Bul. 62:920–923.

Adelberg, J., N. Desamero, A. Hale, and R. Young (1992) Orchid micropropagation on polypropylene membranes. Amer. Orc. Soc. Bul. 61(7):688–695.

Aitken–Christie, J., T. Kozai and M. A. L. Smith (eds.) (1994) Automation and Environmental Control in Plant Tissue Culture, Kluwer Academic Publishers, Dordrecht, The Netherlands.

Desamero, N., J. Adelberg, A. Hale, R. Young and B. Rhodes (1993) Nutrient Utilization in Liquid/membrane System for Watermelon Micropropagation. Plant Cell Tissue and Organ Culture. 33:265–271.

Hale, A., R. Young, J. Adelberg, R. Keese and D. Camper (1992) Bioreactor Development for Continual–flow, Liquid Plant Tissue Culture. Acta Hort. 319:107–112.

Hayashi, M., A. Kano and E. Goto (eds.) (1992) Acta Horticulturae 319 (Proc. of International Symposium on Transplant Production Systems – Biological, Engineering and Socioeconomic Aspects –, Yokohama, Japan), ISHS, Wageningen, The Netherlands, 694pp.

Kozai, T. (1991a) Autotrophic micropropagation (In: Bajaj (ed.) Biotechnology in agriculture and forestry, Vol. 17), Springer–Verlag, New York, p313–343.

Kozai, T. (1991b) Controlled environments in conventional and automated micropropagation (In: Cell culture and somatic cell genetics of plants, Vol. 8), Academic Press, Inc., New York, p213–230.

Kozai, T. (1991c) Micropropagation under photoautotrophic conditions (In: Debergh and Zimmerman (eds.), Micropropagation: Technology and application, Kluwer Academic Publishers, Dordrecht, The Netherlands. p447–469.

Kurata, K. and T. Kozai (eds.) (1992) Transplant Production Systems, Kluwer Academic Publishers, Dordrecht, The Netherlands. 335pp.

LARGE SCALE PRODUCTION OF SECONDARY METABOLITES

M.A.L. Smith
University of Illinois
Urbana, IL 61801 USA

ABSTRACT. Several new in vitro plant products have now reached the stage of semi-commercial or pilot scale production, patents for production have escalated, and biotransformation is a premier approach for promising plant derivatives such as chemoprotective alkaloids. The research community still has only an incomplete picture of actual commercial progress, since most current industrial investigations fall under the realm of proprietary research. Strong collaborations between plant scientists and engineers have resulted in: 1) a new appreciation for the physical and chemical microenvironmental factors required to promote secondary metabolite production, and 2) more sophisticated technologies for controlling those factors. These events presage the imminent introduction of a wider range of new in vitro-produced plant products in the global marketplace.

1. Introduction

Researchers to date have described the in vitro production and recovery of valuable products that range from potent chemoprotective agents to flavors, fragrances, pigments, sweeteners, aromatic complexes, enzymes, and biocontrol agents. When compared to technologies for harvest and recovery of product from whole plants in nature, in vitro systems promised to circumvent problems of variable product quality (due to vagaries of climate, pests, and disease), availability (of wild or rare target plants), seasonal constraints, political and geographic restrictions, and unacceptable inherent odors or flavors in an extract. Yet with few exceptions, the large scale gains anticipated for plant cell culture production systems have not been realized, even after years of systematic research aimed at system optimization.

This paper notes the few recognized, full-scale commercial systems for plant secondary metabolite production, and highlights some emerging commercial and semi-commercial ventures in this arena. The traditional in vitro approaches to product recovery are now changing: 1) advanced systems are using biotransformation of plant cells to achieve product recovery, and 2) new projects are using plant cells, as microbial-style factories, to produce non-plant products which are in high demand. Technological advances that promise to break through current limitations, shift the balance of factors in favor of natural product recovery in vitro, and permit future scale-up efficiency for in vitro plant products are outlined.

2. Overview of In Vitro Secondary Metabolite Systems

Only three cases of large scale, commercial production of secondary compounds from plants in vitro are recognized: berberine, saponin, and shikonin (Table 1).

M. Terzi et al. (eds.), Current Issues in Plant Molecular and Cellular Biology, 669-674.

Berberine is produced in vitro by two members of the Ranunculaceae. Cell cultures of *Coptis japonica* produce and store the berberine within the cells; *Thalictrum minus* cells are capable of secreting berberine extracellularly, where it interacts with the medium components to form nitrate or chloride salts. The latter affords an enormous advantage towards commercial development of cell culture/product recovery systems, and in this case berberine can be produced in a continuous flow 4,000 liter bioreactor (continuous product recovery). High-producing cells are easily marked visually, then isolated and cloned, because berberine can be unambiguously detected under a fluorescent microscope.

Shikonin, a red naphthoquinone pigment from *Lithospermum erythrorhizon*, is produced in ER vesicles, so although high-producing cells occur infrequently in a total population, they are at least very easy to identify. Repeated selection and single cell cloning has resulted in stabilization of quite high yielding culture lines (1400 g shikonin/two weeks in a 750 ml bioreactor; Payne et al., 1991). The obvious color identification advantage has facilitated rapid progress at identification of factors which elicit or increase shikonin production. Saponin & other ginseng (*Panax ginseng*) secondary products are located in root systems in vivo. Intensively-managed root cultures (20,000 L bioreactors) and also cell cultures are cultivated to produce the product (Payne et al., 1991).

These three commercial operations feature some unique production advantages, that have contributed to greater ease in commercialization. For both berberine and shikonin, cell cultures synthesize a visually-identifiable product, which greatly facilitates selection for high yielding cell lines, and allows headway towards greater production efficiency. In the case of berberine, the ability to excrete product extracellularly facilitates economical recovery. Technology for using highly productive transformed root cultures has been used for both ginseng metabolites and shikonin.

3. Emerging Commercial/Semi-Commercial Ventures

Several new in vitro plant products have just reached commercial or semi-commercial production levels. New commercial patents are being filed at a brisk pace for natural plant secondary metabolites (primarily for medicinal compounds such as indole alkaloids or morphine).

ESCAgenetics Corporation (San Carlos, CA) has proven the technical feasibility of producing natural vanilla flavor complexes and vanillin in mechanical impeller-driven 72 L bioreactors with further scale up in progress, and has formed a scale-up agreement with Quest International Division of Unilever. The new products PhytoVanillin ™ and PhytoVanilla ™ are geared to compete with synthetic vanillin and expensive natural vanilla extract. Callus, originally derived from vegetative explants of *Vanilla planifolia*, accumulates the vanilla flavor components derived in vivo from the vanilla bean (Sahi 1994). This is a very interesting venture, because quite frequently, callus from vegetative explants cannot synthesize the complex phytochemical from a fruit, flower or seed, or expresses a much simpler profile than that found in vivo.

Taxol production in vitro, and the development and isolation of proprietary taxanes, is the pursuit of numerous university laboratories, and commercial firms now report that they are at semi-commercial status of product development and delivery (Table 1). On the laboratory scale, taxol has been produced from cell cultures in airlift bioreactors, but currently mechanical impeller-driven bioreactors are used in pilot scale manufacturing. ESCAgenetics reports production in 2500 L stir tank reactors; the largest in North America, and has formed a cooperative agreement in Korea for mass scale production. Similarly, Phyton Catalytic has announced operation of a 75000 L facility in Germany (Diversa), which is the largest known operation in the world for taxol production.

Both sanguinarine and anthocyanin production are smaller-scale operations which nonetheless have commercial implications. Commercial-scale in vitro production of sanguinarine, an effective component in toothpastes and oral rinses for treatment of gingivitis and dental plaque, has been developed by Vipont Research Laboratories, using cell cultures of *Papaver somniferum*. Although in general the industrial use of natural plant anthocyanins as a colorant still can not be justified in terms of cost, researchers at Nippon Paint Company reported use of extracted anthocyanins from high-producing *Euphorbia millii* cell cultures to dye textiles specifically for an expensive "kusakizome" method that absolutely requires natural pigments. For this specialty application, use of the cell culture system could be appraised as a commercial application (Herman 1993). The emerging applications nearing commercial utility go well beyond these examples, with production of essential oils (geranium), vincristine, drug components, and several flavors coming close to commercial introduction. The research community still has only an incomplete picture of actual commercial progress, since current industrial investigations fall under the realm of proprietary research.

4. Plant Cells to Produce Valuable Non-plant Products

While lower market values are a major impediment to mass plant secondary metabolite production, the market for valuable non-plant proteins (for medical applications) is far more lucrative. The major advantage of plant systems for production of these products (as opposed to production in mammals or mammalian cell cultures) is that the product does not risk contamination by other mammalian pathogens (e.g. HIV or Hepatitis B).

Recombinant DNA strategies were used both to produce human serum albumin in *Solanum* tissues, and vertebrate antibodies in plants. A strategy used for model mouse immunoglobulin genes transformed *Nicotiana* leaf disks (with cDNAs from a catalytic antibody cloned to a plant expression vector, via co-cultivation with *Agrobacterium tumefaciens*). The best producing plants are subsequently cloned as suspension cultures (Hiatt & Mostov 1993). Plant systems attach entirely different sugar residues on an antibody protein than mammalian cells; the result in terms of compatibility remains to be determined. Antibodies against *Streptococcus* (causal agent of dental caries) are produced in transformed plant cell cultures with high levels of product in the supernatant which can be removed and used in crude form.

A very new, emerging application is the use of plant cell cultures to produce mammalian candidate vaccines. In vivo, whole plants infected with viral pathogens can produce single protein antigenes, but production is slowed since the virus fails to spread well through the complex, different tissue types. Alternatively, simple tobacco suspension cultures can be infected with virus, and uniform spread throughout all cells in suspension is achieved (Fitchen 1994). A hybrid virus is produced, which combines animal and plant components, with the coat protein over the mammalian virus switched, which confers antigenic value. Currently, model vaccines effective against mouse diseases are produced in vitro routinely in high yields.

5. Technological Innovations for Commercial Feasibility

Investigators cite the lack of adequate technology and instrumentation as major hindrances to scale-up. Plant systems can not capitalize on technological advances from animal and microbial fronts unless plant systems are regeared to retrofit existing bioreactor technology. A new plethora of innovations, and priority status of efforts by pharmaceutical concerns, presages the progress to come in the next several years.

Biotransformation is a major approach to factory-scale production of promising plant derivatives such as chemoprotective alkaloids. This requires only that plant cells perform one or more steps in final product synthesis; enzymes present in the plant cells can convert the precursor into the final product. Biotransformation has the capacity to synthesize novel compounds in vitro via the feeding of synthetic or precursor analogues from other organisms to plant cell cultures. Although this is a viable option, there is no commercial application reported to date, *and* any production of novel compounds will mandate regulatory testing and approval before medicinal or food additive use could be envisioned. Elicitation via biotic agents induces secondary compounds in cultured cells which may have protective roles in nature, especially the biocontrol phytochemicals or stress-related metabolites (e.g. pigments).

Genetic engineering/transformation is also a strategy (noted in the previous section) used in current research schemes for plant metabolites in vitro. The required gene encoding a limiting enzyme should be present in all cells in a culture. Genetic engineering could be a means to develop cell lines where the enzyme-encoding gene(s) is expressed even in undifferentiated tissues, which uncouples product synthesis from differentiation. Hairy root cultures (transformed by *Agrobacterium rhizogenes*) grow at a rate comparable to suspension cultures, and can continuously grow and accumulate valuable secondary products.

Until fairly recently, airlift, bubble column, or modifications were conjectured to be essential for plant cells due to shear stress. Currently, stir tank impellers are successfully used for a wide range of plants; increasing blade size has been found to reduce shear. A helical configuration has been cited as particularly efficient for mass transfer of medium & gasses, yet is less damaging to cells than other impellers used routinely for microbes. Geometry of the helical ribbon impeller results in upward liquid pumping at the blades upon rotation, and circumvents typical losses in productivity associated with scale up from shake flasks to large volume (Jolicoeur et al. 1992).

Objective, non-intrusive characterization of the bioprocess requires autosampling and on-line monitoring of process parameters (O_2, gas mixing, etc.) and cell status (yield, size, etc.), coupled to knowledge based expert systems. An autosampling mechanism used in tandem with color machine vision characterization of bioreactor cultures of both microbial cells and pigment-producing plant cells is illustrated in Fig. 1A. Cells and liquid media flow continuously across a viewing stage into a flow cell with a 300 μm depth; flow cell dimensions are increased for aggregated cell cultures. A schematic of an image from pigmented cell cultures (Fig. 1B) illustrates categorization of entities in culture for characterization of culture status. When coupled to instruments for autoejection and recovery based on visual analysis (Cantliffe et al. 1993), the technology is capable of accomplishing cell line selection. A similar strategy was crucial for isolating high yielding berberine lines.

Figure 1. A. On-line vision control system for sampling from a bioreactor non-intrusively. B. Schematic of an image from a pigmented cell culture. Pigmentation, degree of pigmentation, and entity size are automatically categorized by component from the images. Reprinted from Smith et al., 1993.

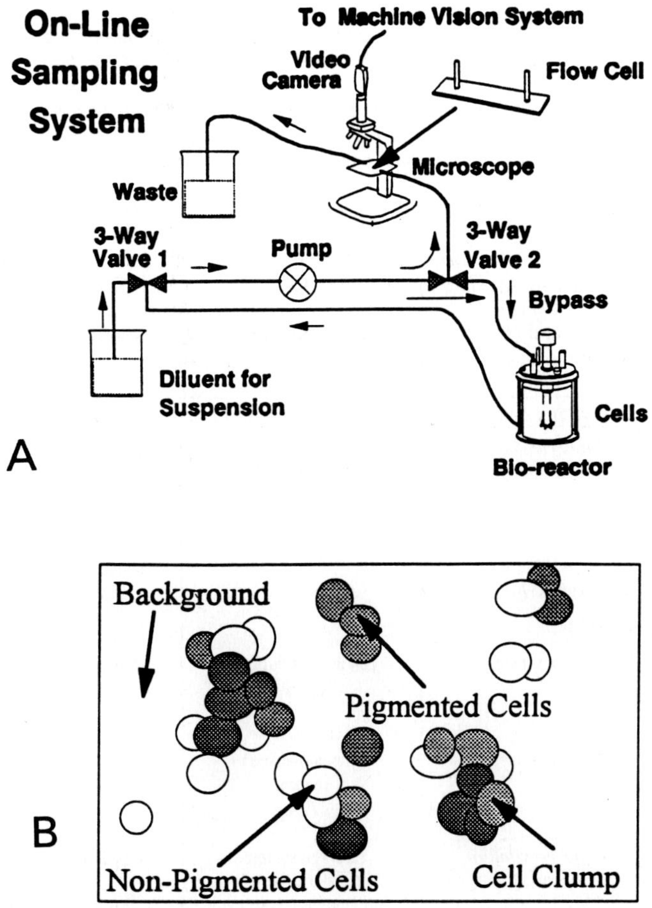

References

Cantliffe DJ, Bieniek ME & Harrell RC (1993) A systems approach to developing an automated synthetic seed production model. p. 160-196. In: Suh WY, Liu JR and Komamine A. (eds.) Advances in developmental biology and biotechnology of higher plants. Korean Society of Plant Tissue Culture.

Fitchen, J (1994) Production of monoclonal antibodies and candidate vaccines in plants and plant cell cultures. In Vitro Cellular & Developmental Biology: Program Issue 30A:36.

Herman, EB (1993) Recent Advances in Plant Tissue Culture II. Secondary Metabolite Production 1988-1993. Agritech Consultants, Inc., Shrub Oak.

Hiatt, A & Mostov K (1993) Assembly of multimeric proteins in plant cells:characteristics and uses of plant-derived antibodies. p. 221-237. In: (Hiatt, A., ed.) Transgenic Plants. Fundamentals and Applications. Marcel Dekker, Inc. New York.

Jolicoeur, M, Chavarie C, Carreau P & Archambault J (1992) Development of a helical-ribbon impeller bioreactor for high-density plant cell suspension culture. Biotech.& Bioengineering 39:511-521.

Payne GF, Bringi V, Prince C & Shuler ML (1991) Plant cell and tissue culture in liquid systems. Hanser Pub.,N.Y. 346 p.

Sahi O (1994) Plant tissue culture. p. 239-267. In: (Gabelman A, ed.). Bioprocess Production of Flavor Ingredients. John Wiley & Sons, Inc.

Smith, MAL, Reid JF, Hansen AC, Li Z & Madhavi DL (1993) Vision characteristics for analysis of pigment-producing cell cultures. ASAE Meeting Paper #933513.

Table 1. Commercial and emerging/semi-commercial large-scale secondary metabolite production schemes for plant cells in vitro.

Product	cell culture source	bioreactor /[1] maximum volume	producer(s)
berberine	*Thalictrum minus, Coptis japonica*	batch & continuous flow; impeller-driven 4000 L	Mitsui Petrochemical Industries, Japan
shikonin	*Lithospermum erythrorhizon*	batch 750 mL	Mitsui Petrochemical Industries, Japan
ginseng saponins, biomass	*Panax ginseng*	cell & root cultures 20,000 L	Nitto Denko, Japan
vanillin	*Vanilla planifolia*	impeller-driven reactor 72 L	ESCAgenetics, USA
taxol	*Taxus brevifolia; Taxus sp.*	2-stage system impeller driven & airlift reactor systems 2500 L/75000 L	ESCAgenetics, USA Phyton Catalytic, USA Nippon Oil Company, Japan
anthocyanins	*Euphorbia millii*	rotary culture system	Nippon Paint Company, Japan
sanguinarine	*Papaver somniferum*	airlift system 300 L	Vipont Research Labs,USA

[1]Proprietary restrictions from industrial concerns preclude a detailed listing of bioreactor configurations currently employed in production.

SOMES2: IMAGE-ANALYSIS-BASED SOMATIC EMBRYO SORTER

Y. IBARAKI, M. FUKAKUSA, K. KURATA
University of Tokyo,
Department of Agricultural Engineering
Yayoi 1-1-1, Bunkyo-ku, Tokyo 113
Japan

ABSTRACT. A system to mechanically separate normally shaped torpedo stage embryos from immature and malformed embryos based on image analysis was developed and named SOMES. In SOMES, an image of a somatic embryo was acquired when flowing in a thin glass tube. To improve the accuracy of the embryo evaluation, SOMES was improved and named SOMES2. SOMES2 can acquire images of an embryo from two angles simultaneously. A thinning-based algorithm to evaluate an embryo from two angle images was developed. In an experiment using carrot, SOMES2 judged 56 % of normal embryos as normal and was supposed to raise the percentage of normal embryos from 30 % to 80 %.

1. Introduction

Somatic embryogenesis is a promising tool for micropropagation. It is, however, a fact that asynchrony of embryo development and lack of uniformity in quality of produced embryos restrict its commercial application. A suspension obtained as a product of somatic embryo induction may contain immature embryos, malformed embryos and even nonembryo objects. Therefore, to make somatic embryogenesis a viable transplant production method, a technique to select normal embryos out of the suspension has been desired. A prototype system to mechanically separate normally shaped torpedo stage embryos from immature and malformed embryos based on image analysis was developed and named SOMES (SOMatic Embryo Sorter, Kurata and Ibaraki (1993)). In SOMES, an image of a somatic embryo was acquired when flowing in a thin glass tube. In a previous experiment using carrot somatic embryos, SOMES raised the percentage of normal embryos from 18 % to 56 %. This result showed the effectiveness of SOMES, which, however, was not satisfactory for practical application to transplants or artificial seed production. SOMES could select only 40 % of normal embryos. This is in a striking contrast to 90 % coincide with human judgment obtained using images of embryos at rest (Kurata et al. (1993)). The discrepancy shows the difficulty of applying image analysis technique to moving objects. The main reason of misjudgments in SOMES was that SOMES's judgment was based on one image of an moving embryo. A somatic embryo showed different appearances when seen from different angles. Figure 1 shows images of the same carrot torpedo stage embryo seen from different angles. A normal embryo's cotyledonary part is separated into two lobes (leaf primordium). However, when seen from a

M. Terzi et al. (eds.), Current Issues in Plant Molecular and Cellular Biology, 675-680.

certain angle, the two lobes overlap, leading to a misjudgment. Moreover, an observation from only one angle restricted occasions to find secondary embryos on the embryo surface. A somatic embryo image from one angle was not sufficient for the embryo evaluation. Therefore, use of two images of an embryo acquired from different angles was expected to enhance the accuracy of SOMES. The objectives of this study were to develop a new criterion for evaluating embryos using two images and to analyze how SOMES judgment would be improved. An improved SOMES was named SOMES2.

2. Materials and Methods

2.1. SOMATIC EMBRYO PREPARATION

Carrot (*Daucus carota* L. cv. Kinkoh–yonsun) was used as a model plant. Hypocotyls of 10 days old seedlings from sterilized seeds were sliced into 5 mm segments and transplanted into callus induction medium (Murashige and Skoog (1962) medium (hereafter MS medium) with 1 μM 2,4–dichlorophenoxyacetic acid, 3% sucrose, and 0.8% agar). They were cultured under total darkness and 27 °C air temperature for about twelve weeks. Callus were subcultured at an interval of about two weeks in liquid medium with the same components except for agar. For somatic embryo induction, embryogenetic cells were separated from a callus suspension by sieving and centrifuging. Callus was sieved using 125, 63 and 38 μm stainless steel meshes. Callus remaining on the 38 μm mesh was transferred into somatic embryo induction medium (MS medium with 3% sucrose) and centrifuged three times at 1000 g for 5 minutes. Selected embryogenic cells were inoculated at the rate of one unit of packed cell volume to 1000 unit of medium, and cultured between 2 and 3.5 weeks. They were sieved with 435 and 800 μm polyester meshes and those remaining on the 435 μm mesh were used for SOMES2 experiment.

2.2. SOMES IMPROVEMENT

2.2.1. *Improvement of A Imaging Cell.* A mirror was fixed along the imaging cell at an angle of 45 degrees to the direction from which images was taken (Figure 2). This enabled us to acquire simultaneously two images of an embryo from two directions making an angle of 90 degrees. A thinner glass tube (inside diameter 1.5 mm) than that used in SOMES (square cross–section 2.8 mm x 2.8 mm) was used for the imaging cell to prevent embryo tumbling and to acquire more sharply focused images. Embryos' tumbling in the imaging cell caused blurring in an acquired image, leading to a misjudgment for the embryo evaluation in SOMES. Use of the thinner imaging cell was expected to reduce embryo tumbling.

Figure 3 schematically shows the system configuration of SOMES2. SOMES2 consisted of some reservoirs for embryos, tubes connecting them, the image acquisition part, and the image processing and flow controlling part. The image acquisition part consisted of a imaging cell, a microscope, a lighting device, and a CCD camera. For image analysis, a work station (Sun Microsystems, SPARK station 2 GX) with an image acquisition card (Sun Microsystems, Video Pix) was used. To simplify embryo's image acquisition and separation of a target embryo, embryos were introduced singularly into the imaging cell. This was achieved by the following manner. Liquid medium containing embryos was stirred in the embryo reservoir and pushed by the peristaltic pump (Carrier flow 1) to a passage confirmation sensor (hereafter

PC sensor), which optically detected objects in the flow. If an embryo passed the PC sensor, the sensor sent a signal to a solid–state timer and stopped the Carrier flow 1. The embryo was then transferred slowly to the imaging cell by Carrier flow 2 controlled by the timer. A color image (8 bits for R, G, and B) of the imaging cell was acquired by the CCD camera (Elmo, ES302) mounted on the microscope (magnification x10). If the embryo was judged as normal by the computer, Carrier flow 3 transferred the embryo to the harvest reservoir. A high speed electrical shutter of the CCD camera (1/1000 s) was used to prevent blurring images. The field of view was about 1.5 cm (parallel to the imaging cell) x 0.56 cm (perpendicular to the cell) area, which corresponded to the 640 x 240 pixels (spatial resolution; about 23μm /pixel). Half strength MS medium with 1.5 % sucrose was used as a carrier liquid. The flow rates of Carrier Flows 1, 2 and 3 were adjusted to be 90 mL/min., 0.5 mL/min. and 120 mL/min., respectively.

2.2.2. *Image Analysis Process.* A thinning–based algorithm to evaluate embryos developed by Kurata et al. (1993) was applied to SOMES2 with a slight modification. Thinning of the image was a transformation of the original image to its skeleton, which was composed of lines with a width of one pixel (Figure 4). This algorithm by Kurata et al. (1993) was based on the fact that normal torpedo stage embryos have Y–shaped skeletons.

Images of an embryo from two angles, a direct image and a reflected image by the mirror (hereafter, reflected image), were acquired at an interval of 0.6 seconds. A decision whether an embryo existed in the field of view or not was made by counting the pixel number whose green intensity had changed more than a predetermined value during the image acquisition interval. If an embryo existed in the imaging cell, its image was smoothed (see Kurata and Ibaraki (1993)). Then, scanning area for the thinning procedures was restricted to enhance computation efficiency. Each skeleton induced by the thinning procedure was classified as follows: 1) I–shaped skeleton which consisted of one line with no branch, 2) Y–shaped skeleton which had three branches and satisfied two criteria, branch length related criterion and symmetry related criterion (see Figure 5), and 3) others. Then, the combination patterns of two skeletons for an embryo were classified as follows: 1) "YI" : one skeleton was Y–shaped and another was I–shaped, 2) "YY" : both skeletons were Y–shaped, 3) "II": both skeletons were I–shaped, and 4) others. Number of normal embryos in each combination pattern was examined and the criterion of normal embryos based on the skeleton pattern combination was investigated.

2.2.3. *SOMES2 experiment.* An experiment using a video image was conducted. Output of the CCD camera was recorded on a video tape. The video image was sent to the work station and processed in the above mentioned way. Two skeletons were obtained for each embryo. Three performance indices were introduced for evaluating the accuracy of the judgment: the successful ratio for normal embryos α (number of embryos judged as normal out of normal embryos / total number of normal embryos), the successful ratio for malformed embryos β (number of embryos judged as malformed out of malformed embryos / total number of malformed embryos), and the error ratio δ ($(1-\beta)/\alpha$). The error ratio δ was introduced because, using this index, the ratio of normal embryos in the selected embryos, N_s, can be easily estimated from the ratio normal embryos in the original suspension, N_o, by the following relation: $N_s = N_o / (N_o + \delta(1-N_o))$. Thus, the lower error ratio means the higher homogeneity of the selected embryos.

3. Results and Discussion

Figure 6 shows examples of embryo images from direct view and reflected view by the mirror, their thresholded images and the induced skeletons. Table 1 shows the skeleton combination patterns of direct images and reflected images. A total of 359 embryos were tested and compared with human evaluation which resulted in 107 normal and 252 malformed embryos. Table 1 shows that more than half of normal embryos were in YI group and that about half of malformed embryos belonged to II group. This suggested that the criterion of normal embryos would be either YY pattern or (YI+YY) pattern. In the former case embryos belonging to YI group would be judged as normal and in the latter case YY-belonging embryos would be also judged as normal. Table 2 shows the performance indices of YI pattern and (YY+YI) pattern. For comparison, results of Y pattern (based on one image) are also shown. Including YY in the criterion in addition to YI slightly increased α and decreased β. The error ratio δ, the index of inhomogeneity of the selected embryos, increased by inclusion of YY. These results suggested that YI pattern should be the criterion of normal embryos. Table 2 also shows that use of two images remarkably improved the accuracy in the judgment. As a result, SOMES2 is expected to raise the percentage of normal embryos from initial 30 % to 80% .

In the above experiment, a thinner imaging cell was used to prevent tumbling of embryos, compared to that used in SOMES. However, embryos were still imaged at various orientations to the cell. Use of a much thinner imaging cell is expected to make embryos flow with their axes nearly parallel to the cell and raise the successful ratios. In this case, however, refraction of light at curved wall of the imaging cell may cause distortion of the images. Another anticipated problem is the blockage of the cell by over-sized embryos. These problems are left for future studies.

4. Conclusion

Use of two images of an embryo for embryo evaluation improved the accuracy of the judgement. A new criterion for normal embryos was that one of the two skeletons was Y-shaped and the other was I-shaped. To further improve the accuracy of the embryo evaluation, more precise regulation of embryos' orientation in the imaging cell is needed.

References

Kurata, K., Komine, M., Liyanage, H., and Ibaraki, Y. (1993) 'A thinning-based algorithm for evaluating somatic embryos', Transaction of ASAE, 36, 1485-1489.

Kurata, K., and Ibaraki, Y. (1993) 'Automatic selection and dielectrophoretic handing of somatic embryos', submitted to Plant Cell, Tissue and Organ Culture.

Murashige, T. and Skoog, F. (1962) 'A revised medium for rapid growth and bioassays with tobacco tissue culture', Physiol. Plant, 15, 473-497.

TABLE 1. Results of classification of embryos based on combination patterns of the skeletons.

Embryo type	Embryo numbers				
	YI	YY	II	Other	Total
Normal embryos	60	10	29	8	107
Malformed embryos	15	7	124	106	252

TABLE 2. Performance indices.

	α	β	δ
YI	0.56	0.94	0.11
YI+YY	0.65	0.91	0.14
Y (one image based)	0.41	0.91	0.22

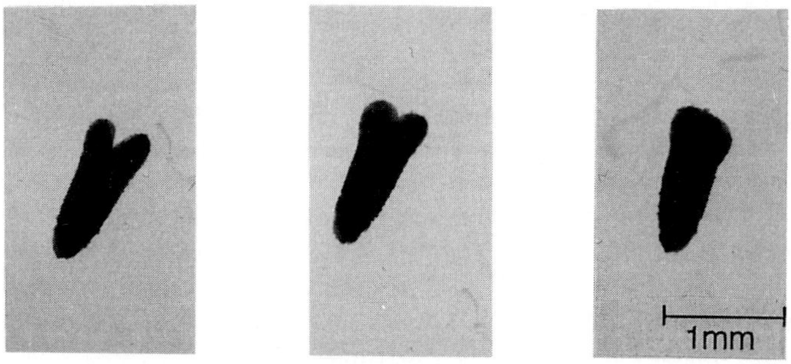

Figure 1. Images of the same carrot embryo seen from different angles.

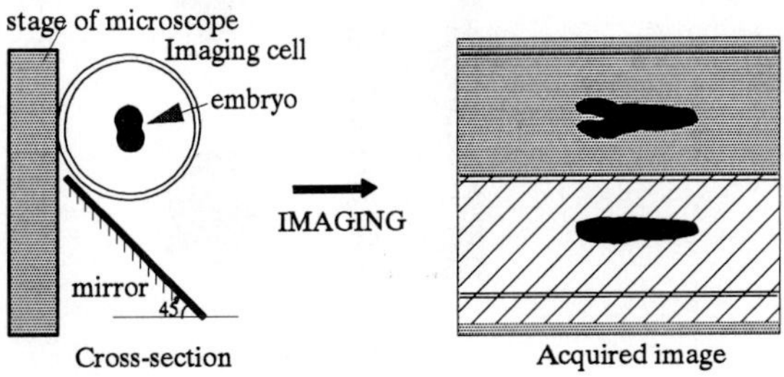

Figure 2. Imaging cell with a mirror.

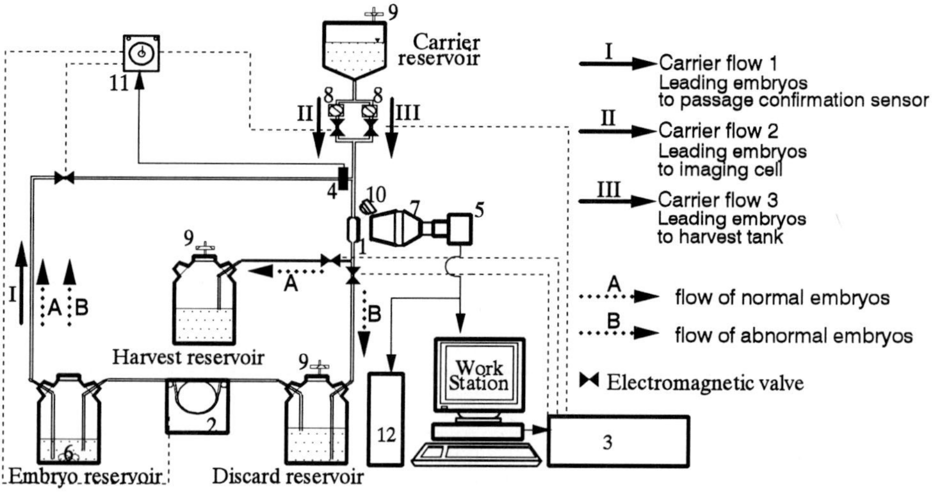

Figure 3. Schematic illustration of SOMES2. 1: Imaging cell with a mirror, 2: Peristaltic pump, 3: I/O device, 4: Passage confirmation sensor, 5: CCD camera, 6: Stirrer, 7: Microscope, 8: Flow controller, 9: Air filter, 10: Light, 11: Solid-state timer, 12: Video recorder.

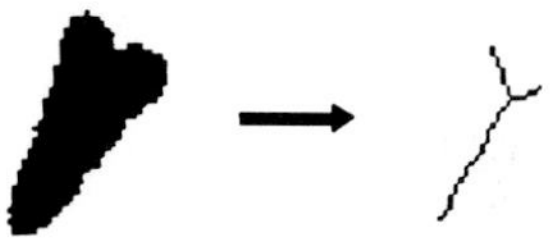

Figure 4. A skeleton induced by the thinning procedure from a thresholded embryo image.

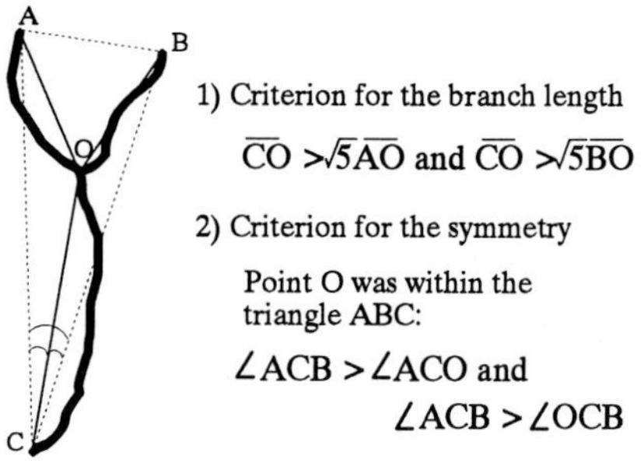
A
B
O
C
1) Criterion for the branch length
$\overline{CO} > \sqrt{5}\,\overline{AO}$ and $\overline{CO} > \sqrt{5}\,\overline{BO}$
2) Criterion for the symmetry
Point O was within the triangle ABC:
$\angle ACB > \angle ACO$ and
$\angle ACB > \angle OCB$

Figure 5. Criteria for jugding Y shpaed skeleton (Kurata and Ibaraki (1993))

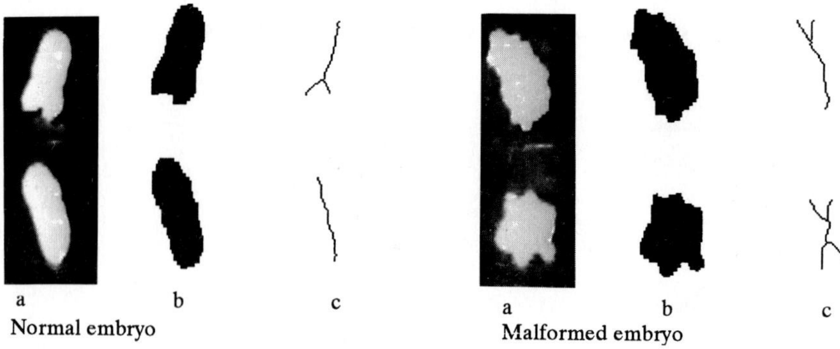

Figure 6. Examples of direct images and reflected images (below the direct images).
a) embryo images reproduced from video record, b) thresholded images, c)induced skeletons.

NUTRITIONAL ASPECTS OF DAUCUS CAROTA SOMATIC EMBRYO CULTURES PERFORMED IN BIOREACTORS

J. ARCHAMBAULT[1], L. LAVOIE, R.D. WILLIAMS, & C. CHAVARIE
École Polytechnique de Montréal
BIOPRO R & D Laboratories
P.O. Box 6079, Station Centre-Ville
Montréal, Québec, Canada, H3C 3A7
[1]*University du Québec à Trois-Rivières*
Engineering Department
P.O. Box 500
Trois-Rivières, Québec, Canada, G9A 5H7

ABSTRACT. Daucus carota somatic embryos (SE) were produced in 2-L helical-ribbon-impeller bioreactors operated under low shear mixing (60 RPM) and low dissolved oxygen concentration (20% of air saturation) by surface oxygenation only. These conditions have been found previously to yield better SE cultures. These cultures were performed using Gamborg's B5 medium supplemented with 30 g•L^{-1} sucrose and increasing concentrations of ammonium ion. Results showed increasing true SE production with respect to increasing NH_4 concentration as well as better production of bioreactor as compared to flask cultures. Furthermore, chromatographic analysis of their medium allowed determination of their nutritional behavior with respect to all macronutrients. Thus limiting (carbohydrates, NH_4, PO_4 and Ca) as well as less used (Na and Cl) extracellular nutrients could be identified. This approach offers tremendous opportunities for the development of improved media as well as for the optimization of plant cell culture bioprocesses.

1. INTRODUCTION

The eventual commercialization of artificial seeds will require the automated large scale production of normal, torpedo shaped, quiescent or dormant somatic embryos (SE) capable of uniform and synchronized regeneration into normal plants in the field. Production of high quality SE is said to be one of the major obstacles to the successful exploitation of this technology [16]. Embryogenic cultures display numerous problems, including lack of synchrony, high heterogeneity, abnormal embryo and plant development, precocious germination and low conversion into normal plants (typically ≤ 30-70%) [8,15]. Most of these difficulties have been ascribed to the inherent high developmental plasticity of these delicate plant organs, which makes SE highly sensitive to their culture protocols [1,4,19].

Consequently, there is a critical need to study these cultures under closely monitored physico-chemical conditions in order to understand the effect of these operational parameters on this process for eventual optimization for quality and productivity. In this respect, it appears that physical culture conditions may play an important and subtle, but underestimated role in this sensitive plant development process [3,5]. Thus, we have shown recently [3] that embryogenic Eschscholtzia californica cell cultures carried out in a helical-ribbon-impeller (HRI) bioreactor [12] displayed markedly poorer quality upon increasing the mixing speed from 60 to 100 RPM. This result illustrates the high sensitivity of this type of culture to mixing conditions especially when considering that this impeller and bioreactor configuration is characterized by significantly lower mixing shear than most

M. Terzi et al. (eds.), Current Issues in Plant Molecular and Cellular Biology, 681-687.

conventional bioreactors [3,13]. Similarly, low rate sparging (0.05 VVM, $k_La \sim 6\ h^{-1}$) resulted in a low quality embryogenic culture. The negative effects of these operating conditions on this production were ascribed mainly to the low, but still excessive shear experienced by the embryogenic cells and/or embryogenic aggregates which partly inhibited the development of somatic embryos.

In the same study, we have also found that the main effect of the dissolved oxygen (DO) concentration on this culture process seems to be nutritional. High DO conditions (≥ 60% of air saturation) of flask and bioreactor cultures favored higher undifferentiated biomass production and associated faster nutrients uptake, than low DO (~ 10-20%) cultures, at the expense of slowly differentiating embryogenic cell clusters. Controlled low DO bioreactor cultures, on the other hand, resulted in limited undifferentiated biomass formation (< 5%) and higher and more normal embryo production with lower precocious germination.

Obviously, the next target of our project is the study of the chemical environment of embryogenic cultures performed under suitable physical and gas transfer conditions. This was the objective of this communication using the known embryogenic biomodel D. carota.

2. MATERIALS AND METHODS

2.1 Plant cell cultures

Daucus carota callus cultures were generated from surface sterilized explant tissues placed on Gamborg's B5 [9] solidified medium supplemented with 0.2 $mg{\bullet}L^{-1}$ 2,4 dichlorophenoxyacetic acid, 8 $g{\bullet}L^{-1}$ agar and 30 $g{\bullet}L^{-1}$ sucrose. Suspension cultures were obtained using the same medium. These were maintained in 500-mL flasks containing 200 mL suspension, subcultured (10% (V/V)) every 14 days and agitated at 120 RPM. Embryogenic liquid cultures were achieved by inoculating fresh medium of the same basic formulation as above, but for ammonium and potassium nitrate concentrations without growth regulator, with 5% (V/V) of a 14-day old stock suspension culture. The NH_4 concentration was increased above that of the basic medium formulation by addition of NH_4NO_3 and corresponding reduction of the concentration of KNO_3 to maintain nitrate concentration at the normal B5 level of 25 mM. For the experiment with no ammonium, the sulfate content of B5 in the form of $(NH_4)_2SO_4$ (1mM) was supplied as K_2SO_4.

Bioreactor cultures were performed using a 2-L version of the original 10-L HRI bioreactor developed previously [12]. They were inoculated, together with four 500-mL flask control cultures, as indicated above and carried out at 60 RPM (Tip speed = 36 $cm{\bullet}s^{-1}$, $k_La = 0.9\ h^{-1}$) and at a controlled DO concentration of 20 ± 1% of air saturation by surface oxygenation only, with automatic regulation of the gas phase composition. Conductivity and pH were continuously monitored on-line using appropriate probes. The two series of duplicate flask control cultures were performed using the same medium as the bioreactor culture and basic B5 medium, respectively. All cultures were carried out at 26°C without light. All media were steam sterilized (1 h).

2.2 Analytical

All cultures were sampled every three days. One-to-three mL volumes were examined visually to count well formed, torpedo shaped somatic embryo. Twenty harvested somatic embryos were placed on solidified B5 medium without growth regulator for plant regeneration. The residual uniform samples were assayed for pH, conductivity, packed cell volume (PCV) (by centrifugation at 200 g for 5 min.) and filtered wet and dry (24 h, 80°C) overall biomass concentrations. Filtered medium samples were analyzed by conventional methods [2] for carbohydrates, phosphate, nitrate and ammonium ion concentration. These samples were also analyzed by ion chromatography [3] for all other macronutrients (Na, NH_4, K, Mg and Ca; Cl, NO_3 and SO_4). All extracellular nutrient concentrations reported were not corrected for the absorption of water by the growing biomass.

3. RESULTS

3.1 Bioreactor vs Flask Control Cultures

A series of six bioreactor experiments were performed using the same cell line at subcultures 7 to 13 and B5 medium with increasing ammonium concentrations from 0 to 20 mM. As shown in Figure 1 for Experiment J (NH_4 = 15 mM), all bioreactor cultures outperformed significantly flask control cultures in embryo production, homogeneity of the resulting suspension and lower undifferentiated biomass production. In all cases, SE production was limited, at least, by extracellular carbohydrates availability (Figure 4). Consequently, low DO supply conditions of bioreactor cultures yielded better embryo production, as shown previously [3,10,14], as compared to higher DO (unpublished) flask cultures.

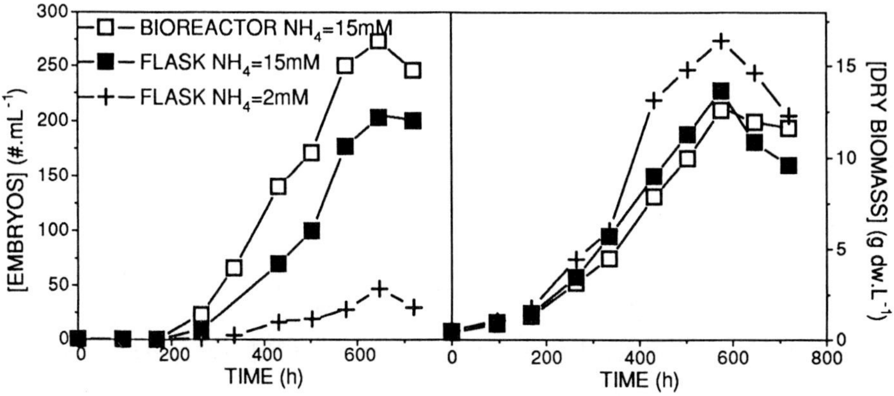

Figure 1. Production and growth curves of bioreactor and flask cultures of Experiment J.

3.2 Effect of Ammonium Ion Concentration

Figure 2 illustrates the effect of increasing the NH_4 concentration on embryo production. In all cases, a lag phase of ~ 200 h was observed which likely resulted from culture adaptation and early cation, mostly ammonium, uptake and corresponding H^+ exchange [18] and acidification of the culture (Figure 5) at low initial biomass concentration (< 1-3 $gdw.L^{-1}$). Thereafter, embryo production rate increased up to a maximum at NH_4 concentration of 5 mM and was constant above.

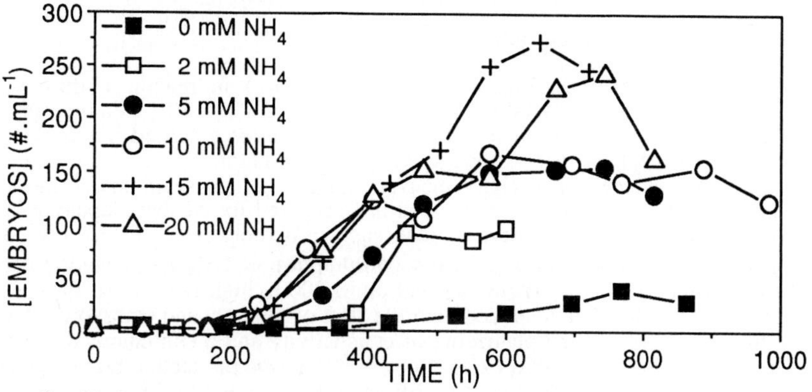

Figure 2. Embryo production curves of bioreactor cultures.

Maximum SE production occurred apparently at a NH_4 concentration of 15 mM (Figure 3). Similar results have been reported by others for small scale solid and liquid cultures, who found varying levels of NH_4 concentration (1 to 12.5 mM) for maximum production [6,11,17,20,21 and others]. Since all bioreactor cultures displayed similar specific growth rates (~ 0.12-0.15 d^{-1}) and maximum biomass concentrations (~ 11-13 gdw . L^{-1}), it appears that increasing NH_4 concentration favored the differentiation of a larger portion of the biomass into somatic embryos. This is illustrated by the increasing average embryo yield and specific production rate curves of Figure 3. Consequently, ammonium seems to be the first nutrient limiting SE production.

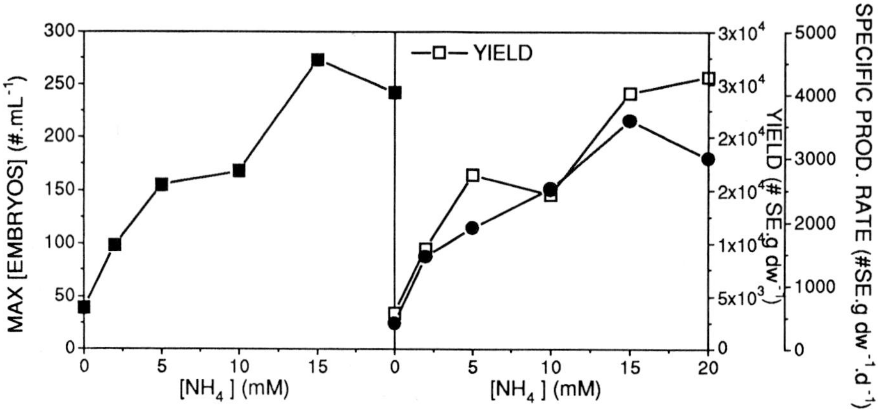

Figure 3. Somatic embryo production with respect to initial NH_4 concentration.

Figure 4 details the biomass and embryo production patterns of Experiment J bioreactor culture. Embryo production paralleled biomass formation, both of which were limited by the availability of extracellular carbohydrates. Consequently, higher levels of the latter may have favored higher SE production, especially at higher NH_4 concentration. Furthermore, the total biomass produced occupied up to 60-70% of the culture volume, 80% of which was made of somatic embryos. Remarkably, the efficiency of regeneration into plants of the embryos produced was ~ 80-100% up to extracellular carbohydrates limitation of the culture at 600 h. Thereafter, the quality of this production, in terms of plant regeneration, dropped significantly. Similar patterns were observed for all other bioreactor and flask cultures, except for those carried out using a medium devoid of ammonium which displayed the poorest production and quality.

3.3. Macronutrients Uptake of an Embryogenic Culture

These patterns are illustrated in Figure 5 for Experiment J bioreactor culture which displayed the highest, but still not totally differentiated (Figure 4 : ~ 80%), embryo production. Similar patterns were observed for other bioreactor cultures carried out at different NH_4 concentrations, except for that performed without ammonium.

According to on-line conductivity measurement, a lag phase in major inorganic nutrients uptake occurred up to 140 h, which can be accounted for by the adaptation of the cells to their new physico-chemical environment, sucrose hydrolysis (~ 100 h), low initial biomass concentration (~ 1 gdw•L^{-1}) and early consumption of NH_4 and resulting culture medium acidification (pH ~ 5.0). Thereafter and up to 300 h, high NO_3, SO_4, PO_4, K and NH_4 uptake was observed which coincided with the start of biomass and embryo production as well as a first rapid decline and plateauing of conductivity and a continuous decrease of pH to its lowest value (~ 4.8). At PO_4 exhaustion (~ 300 h), the production rate of embryos increased while most ions, including Na, Mg and Ca, were highly consumed with accompanying decrease in conductivity and increase in pH. As indicated above, biomass

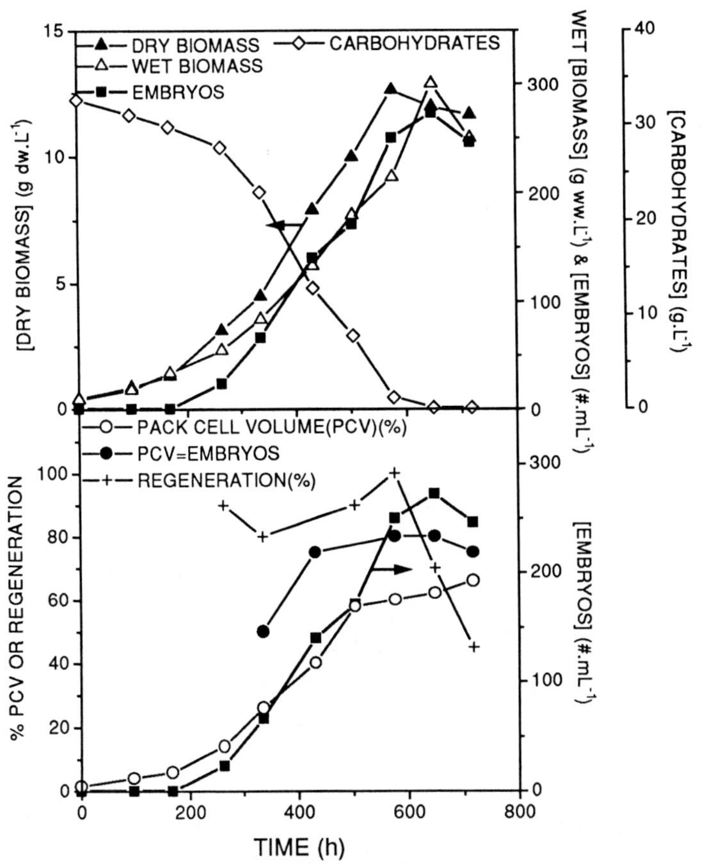

Figure 4. Growth and embryo production curves of Experiment J bioreactor culture.

formation and embryo production ceased upon extracellular depletion of NH_4 and carbohydrates at 600 h. At that stage, NO_3, SO_4 and Na uptake ended while Mg and Ca consumption appeared to continue to depletion. Surprisingly, chloride uptake seemed to start only after most anions (~ 80%) and NH_4 were consumed at ~ 500 h and was maximum at carbohydrates depletion. How this exhaustion of extracellular NH_4 and carbohydrates and late high chloride uptake may be linked to the decrease in plant regeneration efficiency of embryos produced (Figure 4) remains an interesting question.

4. DISCUSSION

The results presented above confirm our previous findings that better embryogenic cultures can be produced in bioreactors, under low shear mixing, without gas bubbling and at controlled low DO concentration, than in flasks [3]. This was achieved using cell lines not specifically selected for their embryogenic potential, and inocula neither sieved nor washed to remove residual growth regulators. This was observed for all bioreactor cultures. Obviously, better cell lines and inocula treatment will yield improved cultures.

On the other hand, this study into the nutritional aspects of embryogenic bioreactor cultures performed under improved physical and gas transfer conditions illustrates the importance of their chemistry on this delicate differentiation process. Thus, restricted growth through limited DO supply appears as valid method to produce richer embryogenic

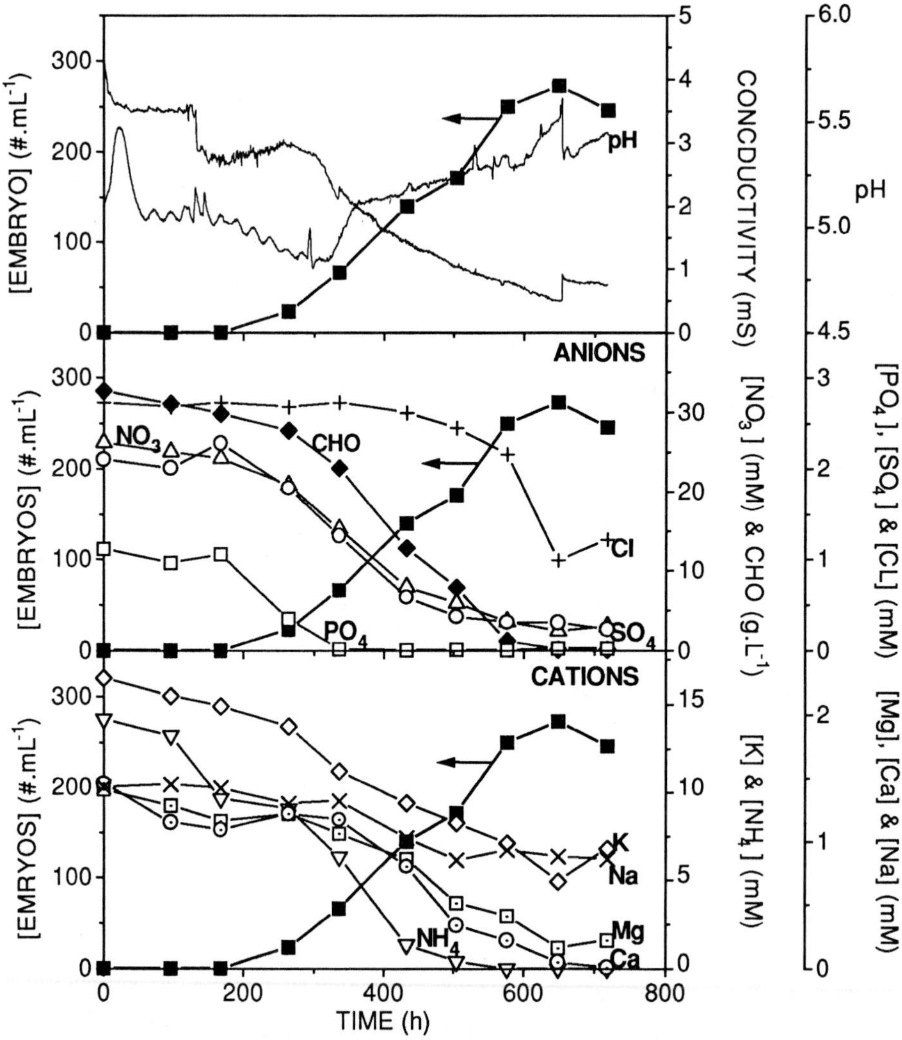

Figure 5. Conductivity, pH and nutrients uptake curves of Experiment J.

cultures. Furthermore, improved embryo production and high plant regeneration efficiency seem to require balanced feeding of embryogenic cultures with most major nutrients, especially NH_4 and carbohydrates. The requirement for chloride remains questionable. However, this approach needs a better understanding and proper modelling of this complex biological process for best production. In addition, other culture and product quality objectives, such as improved homogeneity, limitation of germination and induction of quiescence or dormancy and dessication tolerance using, for example, differential feeding of growth regulators such as abscisic acid [7], will need to be factored in for optimal production of the resulting bioprocess.

REFERENCES

1. Ammirato P.V. (1987), "Organizational events during somatic embryogenesis". In : C.E. Green, D.A. Somers, W.P. Hackett and D.D. Biesboer (eds), Plant Tissue and Cell Culture, A.R. Liss Inc., New York, pp. 57-81.
2. Archambault J. (1991). "Large scale (20L) culture of surface immobilized Catharanthus roseus cells". Enz. Microbiol. Technol. 13, 882-892.
3. Archambault J., Williams R.D., Lavoie L., Pépin M.F. and Chavarie C. (1994). "Production of somatic embryos in a helical ribbon impeller bioreactor". Biotech. Bioeng. (in press).
4. Carman J.G. (1990). "Embryogenic cells in plant tissue culture : occurrence and behavior". In vitro cell. Devel. Biol. 26, 746-753.
5. Chen T.H.H., Thompson B.G. and Gerson D.F. (1987). "In vitro production of alfalfa somatic embryos in fermentation systems". J. Ferment. Technol. 65, 353-357.
6. Dougall D.K. and Verma D.C. (1978). "Growth and embryo formation in wild carrot suspension cultures with ammonium as a sole nitrogen source" In vitro 14, 180-182.
7. Finkelstein R. and Cranch M.L. (1987). "Hormonal and osmotic effects on developmental potential of maturing rapeseed". Hortscience 22, 797-800.
8. Fujii J.A.A., Slade D.T., Redenbaugh K. and Walker K.A. (1985). "Artificial seeds for plant propagation". Trends in Biotechnol. 5, 335-339.
9. Gamborg O.L., Miller R.A. and Ajima K. (1968). "Nutrient requirements of suspension cultures of soybean root cells". Exp. Cell. Res. 50, 151-158.
10. Greidziak U., Diettrich B. and Luckner M. (1990). "Batch cultures of somatic embryos of Digitalis lanata in gaslift fermenters. Development and cardenolide accumulation". Planta Med. 56, 175-178.
11. Halperin W. and Wetherell D.F. (1965). "Ammonium requirements for embryogenesis in vitro". Nature 205, 519-520.
12. Jolicoeur M., Chavarie C., Carreau P.J. and Archambault J. (1992). "Development of a helical-ribbon-impeller bioreactor for high density plant cell suspension culture". Biotech. Bioeng. 39, 511-521.
13. Kamen A.A., Chavarie C., André G. and Archambault J. (1992). "Design parameters and performance of a surface baffled helical ribbon impeller bioreactor for the culture of shear sensitive cells". Chem. Eng. Sci. 47, 9-11.
14. Kessel R.H.J. and Carr A.H. (1972). "The effect of dissolved oxygen concentration on growth and differentiation of carrot (Daucus carota) tissue". J. Exp. Bot. 23, 996-1007.
15. Nouaille C. and Pétiard V. (1988). "Semences artificielles : rêves et réalités". Biofutur 67, 33-38.
16. Redenbaugh K. (1993). "Introduction". In : K. Redenbaugh (ed) Synseeds. Applications of Synthetic Seeds to Crop Improvement. CRC Press, Boca Raton, pp. 3-7.
17. Smith D.L. and Krikorian A.D. (1989). "Release of somatic embryogenic potential from excized zygotic embryos of carrot and maintenance of proembryogenic cultures in hormone-free medium". Amer. J. Bot. 76, 1832-1843.
18. Smith F.A. and Raven J.A. (1976). "H^+ transport and regulation of cell pH". In : U. Lüttge and M.G. Pittman (eds), Transport in Plants II. Part A : Cells. Springer-Verlag, New York, pp. 317-346.
19. Thorpe T.A. (1988). "In vitro somatic embryogenesis". Animal and Plant Sciences, 3, 81-88.
20. Walker K.A. and Sato S.J. (1981). "Morphogenesis in callus tissue of Medicago sativa: the role of ammonium ion in somatic embryogenesis". Plant Tissue and Organ Culture 1, 109-121.
21. Wetherell D.F. and Dougall D.K. (1976). "Sources of nitrogen supporting growth and embryogenesis in cultured wild carrot tissue". Physiol. Plant. 37, 97-103.

AUTHOR INDEX

KEYWORDS INDEX

Current Plant Science and Biotechnology in Agriculture

1. H.J. Evans, P.J. Bottomley and W.E. Newton (eds.): *Nitrogen Fixation Research Progress.* Proceedings of the 6th International Symposium on Nitrogen Fixation (Corvallis, Oregon, 1985). 1985 ISBN 90-247-3255-7
2. R.H. Zimmerman, R.J. Griesbach, F.A. Hammerschlag and R.H. Lawson (eds.): *Tissue Culture as a Plant Production System for Horticultural Crops.* Proceedings of a Conference (Beltsville, Maryland, 1985). 1986 ISBN 90-247-3378-2
3. D.P.S. Verma and N. Brisson (eds.): *Molecular Genetics of Plant-microbe Interactions.* Proceedings of the 3rd International Symposium on this subject (Montréal, Québec, 1986). 1987 ISBN 90-247-3426-6
4. E.L. Civerolo, A. Collmer, R.E. Davis and A.G. Gillaspie (eds.): *Plant Pathogenic Bacteria.* Proceedings of the 6th International Conference on this subject (College Park, Maryland, 1985). 1987 ISBN 90-247-3476-2
5. R.J. Summerfield (ed.): *World Crops: Cool Season Food Legumes.* A Global Perspective of the Problems and Prospects for Crop Improvement in Pea, Lentil, Faba Bean and Chickpea. Proceedings of the International Food Legume Research Conference (Spokane, Washington, 1986). 1988 ISBN 90-247-3641-2
6. P. Gepts (ed.): *Genetic Resources of* Phaseolus *Beans.* Their Maintenance, Domestication, Evolution, and Utilization. 1988 ISBN 90-247-3685-4
7. K.J. Puite, J.J.M. Dons, H.J. Huizing, A.J. Kool, M. Koorneef and F.A. Krens (eds.): *Progress in Plant Protoplast Research.* Proceedings of the 7th International Protoplast Symposium (Wageningen, The Netherlands, 1987). 1988 ISBN 90-247-3688-9
8. R.S. Sangwan and B.S. Sangwan-Norreel (eds.): *The Impact of Biotechnology in Agriculture.* Proceedings of the International Conference The Meeting Point between Fundamental and Applied in vitro Culture Research (Amiens, France, 1989). 1990. ISBN 0-7923-0741-0
9. H.J.J. Nijkamp, L.H.W. van der Plas and J. van Aartrijk (eds.): *Progress in Plant Cellular and Molecular Biology.* Proceedings of the 8th International Congress on Plant Tissue and Cell Culture (Amsterdam, The Netherlands, 1990). 1990 ISBN 0-7923-0873-5
10. H. Hennecke and D.P.S. Verma (eds.): *Advances in Molecular Genetics of Plant–Microbe Interactions.* Volume 1. 1991 ISBN 0-7923-1082-9
11. J. Harding, F. Singh and J.N.M. Mol (eds.): *Genetics and Breeding of Ornamental Species.* 1991 ISBN 0-7923-1094-2
12. J. Prakash and R.L.M. Pierik (eds.): *Horticulture – New Technologies and Applications.* Proceedings of the International Seminar on New Frontiers in Horticulture (Bangalore, India, 1990). 1991 ISBN 0-7923-1279-1
13. C.M. Karssen, L.C. van Loon and D. Vreugdenhil (eds.): *Progress in Plant Growth Regulation.* Proceedings of the 14th International Conference on Plant Growth Substances (Amsterdam, The Netherlands, 1991). 1992 ISBN 0-7923-1617-7
14. E.W. Nester and D.P.S. Verma (eds.): *Advances in Molecular Genetics of Plant–Microbe Interactions.* Volume 2. 1993 ISBN 0-7923-2045-X
15. C.B. You, Z.L. Chen and Y. Ding (eds.): *Biotechnology in Agriculture.* Proceedings of the First Asia-Pacific Conference on Agricultural Biotechnology (Beijing, China, 1992). 1993 ISBN 0-7923-2168-5

Current Plant Science and Biotechnology in Agriculture

16. J.C. Pech, A. Latché and C. Balagué (eds.): *Cellular and Molecular Aspects of the Plant Hormone Ethylene.* 1993 ISBN 0-7923-2169-3
17. R. Palacios, J. Mora and W.E. Newton (eds.): *New Horizons in Nitrogen Fixation.* Proceedings of the 9th International Congress on Nitrogen Fixation (Cancún, Mexico, 1992). 1993 ISBN 0-7923-2207-X
18. Th. Jacobs and J.E. Parlevliet (eds.): *Durability of Disease Resistance.* 1993 ISBN 0-7923-2314-9
19. F.J. Muehlbauer and W.J. Kaiser (eds.): *Expanding the Production and Use of Cool Season Food Legumes.* A Global Perspective of Peristent Constraints and of Opportunities and Strategies for Further Increasing the Productivity and Use of Pea, Lentil, Faba Bean, Chickpea, and Grasspea in Different Farming Systems. Proceedings of the Second International Food Legume Research Conference (Cairo, Egypt, 1992). 1994 ISBN 0-7923-2535-4
20. T.A. Thorpe (ed.), *In Vitro Embryogenesis in Plants*, 1995. (forthcoming) ISBN 0-7923-3149-4
21. M.J. Daniels, J.A. Downie and A.E. Osbourn (eds.), *Advances in Molecular Genetics of Plant-Microbe Interactions.* Volume 3. 1994 ISBN 0-7923-3207-5
22. M. Terzi, R. Cella and A. Falavigna (eds.), *Current Issues in Plant Molecular and Cellular Biology.* Proceedings of the VIIIth International Congress on Plant Tissue and Cell Culture (Florence, Italy, 1994). 1995 ISBN 0-7923-3322-5

KLUWER ACADEMIC PUBLISHERS – DORDRECHT / BOSTON / LONDON